INSECTS AND WILDLIFE

INSECTS AND WILDLIFE

ARTHROPODS AND THEIR RELATIONSHIPS WITH WILD VERTEBRATE ANIMALS

John L. Capinera

Professor and Chairman
Entomology and Nematology Department
University of Florida
Gainesville, Florida, USA

⊛WILEY-BLACKWELL

A John Wiley & Sons, Ltd., Publication

Library of Congress Cataloguing-in-Publication Data

Capinera, John L.
   Insects and wildlife : arthropods and their relationships with wild vertebrate animals / John L. Capinera.
      p. cm.
   Includes bibliographical references and index.
   ISBN 978-1-4443-3299-5 (hardcover : alk. paper) – ISBN 978-1-4443-3300-8
(pbk. : alk. paper)
   1. Insects–Ecology.   2. Vertebrates–Ecology.   3. Vertebrates–Food.   4. Insects as carriers of disease.   5. Insecticides–Toxicology.   6. Wildlife conservation.   I. Title.   II. Title: Arthropods and their relationships with wild vertebrate animals.
   QL496.4.C36 2010
   595.7'178–dc22

                                                                    2009041246

A catalogue record for this book is available from the British Library.

Set in 9 on 11 pt Photina by Toppan Best-set Premedia Limited
Printed and bound in Singapore by Ho Printing Singapore Pte Ltd

1—2010

CONTENTS

Preface, ix

Acknowledgments, xi

SECTION 1: INTRODUCTION TO THE ARTHROPODS, 1

CHAPTER 1 INSECTS AND THEIR RELATIVES, 3
Naming of Taxa, 3
Arthopoda, 6
Arachnida, 6
Crustacea, 8
Diplopoda, 9
Chilopoda, 11
Entognatha, 11
Insecta, 14
Classification of Insects, 14
Characteristics of the Major Groups of Insects, 15
Evolution of Insects, 27
Insect Biogeography, 32
Summary, 33
References and Additional Reading, 33

CHAPTER 2 STRUCTURE AND FUNCTION OF INSECTS, 34
Integument, 34
Molting, 35
Body Regions, 35
The Head, 37
The Thorax, 41
The Abdomen, 43
Internal Anatomy, 45
Muscular System, 45
Fat Body, 47

Digestive System, 47
Circulatory System, 50
Ventilatory System, 51
Nervous System, 52
Vision, 54
Glandular Systems, 55
Polyphenism or Polymorphism, 57
Communication, 58
Sociality, 61
    Ants, 62
    Social Bees and Wasps, 62
    Termites, 63
Metamorphosis, 63
Reproductive System, 65
Eggs of Insects, 66
Excretory System, 69
Thermal Biology, 69
Feeding Ecology, 71
    Scavenging, 73
    Feeding Belowground, 73
    Feeding in Aquatic Habitats, 74
    Feeding on Living Plants, 75
    Feeding on Blood, 78
    Predation and Parasitism, 79
Summary, 79
References and Additional Reading, 81

SECTION 2: FOOD RELATIONSHIPS, 83

CHAPTER 3 FOOD RESOURCES FOR WILDLIFE, 85
Assessment of Insectivory, 85
    Methods for Determining the Abundance of Insects, 86
    Methods for Determining Wildlife Diets, 91

Nutritional Value of Insects, 97
Importance of Insects in the Diets of
Wildlife, 98
Summary, 103
References and Additional Reading, 103

CHAPTER 4   WILDLIFE DIETS, 105
Analysis of Amphibian and Reptile Diets, 105
Analysis of Mammal Diets, 107
Analysis of Bird Diets, 126
Analysis of Fish Diets, 152
The Benefits of Insects for Wildlife Survival and
Reproduction, 156
How Insects Avoid Becoming Food for
Wildlife, 158
    Crypsis, 158
    Aposematism, 159
    Mimicry, 159
    Flight and Startle Behavior, 160
    Physical and Chemical Defenses, 162
    Group Actions, 163
    Nocturnal Activity, 165
Summary, 165
References and Additional Reading, 166

**CHAPTER 5   INSECTS IMPORTANT AS
FOOD FOR WILDLIFE, 171**
Aquatic Insects, 171
    Mayflies (Order Ephemeroptera), 171
    Stoneflies (Order Plecoptera), 172
    Dragonflies and Damselflies (Order
    Odonata), 173
    Bugs (Order Hemiptera), 175
    Alderflies, Dobsonflies, and Fishflies (Order
    Megaloptera), 176
    Beetles (Order Coleoptera), 176
    Flies (Order Diptera), 177
    Caddisflies (Order Trichoptera), 177
Terrestrial Insects, 179
    Termites (Order Isoptera), 179
    Cockroaches (Order Blattodea), 180
    Grasshoppers, Katydids, and Crickets (Order
    Orthoptera), 181
    Earwigs (Order Dermaptera), 184
    Barklice or Psocids (Order Psocoptera), 186
    Bugs (Order Hemiptera), 186
    Lacewings, Antlions and Mantidflies (Order
    Neuroptera), 188
    Beetles (Order Coleoptera), 189

    Moths and Butterflies (Order Lepidoptera), 190
    Flies (Order Diptera), 192
    Wasps, Ants, Bees, and Sawflies (Order
    Hymenoptera), 192
Summary, 194
References and Additional Readings, 197

**CHAPTER 6   INSECTS AND
ECOSYSTEMS, 198**
Insects and Decomposition, 201
    Decomposition of Plant Remains, 201
    Decomposition of Excrement (Dung), 202
    Decomposition of Carrion, 204
Nutrient Cycling, 206
Herbivory by Insects, 210
    The Importance of Herbivory, 210
    Plant Compensation, 213
    Insect Outbreaks, 214
Plant Diseases and Insects, 215
Pollination and Seed Dispersal by Insects, 217
Invasiveness of Insects, 218
    Pathways of Invasion, 219
    Ecological and Taxonomic Patterns
    of Invasion, 221
    Establishment and Spread, 222
    Latency Among Invaders, 223
    Why Invasive Species become so Abundant,
    225
    Impacts of Invaders, 226
Summary, 227
References and Additional Reading, 227

**SECTION 3: ARTHROPODS AS DISEASE
VECTORS AND PESTS, 231**

**CHAPTER 7   TRANSMISSION OF
DISEASE AGENTS TO WILDLIFE BY
ARTHROPODS, 233**
Arthropod Feeding Behavior, 233
Disease in Wildlife, 235
Virulence, 236
Disease Hosts, 239
Disease Transmission, 240
Causes of Disease, 241
The Nature of Parasitism, 241
Parasite-Induced Changes in Host Behavior, 242
Summary, 243
References and Additional Reading, 244

CHAPTER 8 INFECTIOUS DISEASE AGENTS TRANSMITTED TO WILDLIFE BY ARTHROPODS, 245

Viruses, 245
  Myxomatosis, 248
  Avian Pox, 248
  West Nile Virus, 249
  Yellow Fever, 250
  St. Louis Encephalitis, 250
  Hemorrhagic Disease, 251
Bacteria, 252
  Tularemia, 252
  Anaplasmosis, 254
  Lyme Disease, 254
  Plague, 255
  Avian Botulism, 258
Fungi, 259
  Aflatoxin Poisoning, 259
Summary, 261
References and Additional Reading, 261

CHAPTER 9 PARASITIC DISEASE AGENTS TRANSMITTED TO WILDLIFE BY ARTHROPODS, 263

Protozoa, 263
  American Trypanosomiasis, 263
  African Trypanosomiasis, 266
  Avian Malaria, 269
  Toxoplasmosis, 271
Helminths, 274
  Spirocercosis, 276
  Dirofilariasis, 276
  Elaeophorosis, 277
  Lancet Fluke, 279
  Dog Tapeworm, 279
  Giant Thorny-headed Worm, 279
Summary, 283
References and Additional Reading, 283

CHAPTER 10 ARTHROPODS AS PARASITES OF WILDLIFE, 285

Mites and Ticks (Arachnida: Acari or Acarina: Several Orders), 289
  Mites, 289
    Mange Mites, 289
    Respiratory Mites, 291
    Ear Mites, 291
    Bird Mites, 292
    Sarcoptic Mange Mite, 292

Ticks, 293
  Taiga Tick, 296
  Wood Tick, 297
  Blacklegged Tick, 297
Insects (Insecta), 298
  Lice (Phthiraptera), 298
  Bugs (Hemiptera: Reduviidae, Cimicidae, and Polyctenidae), 302
    Assassin Bugs, Subfamily Triatominae – Kissing or Blood-Sucking Conenose Bugs (Hemiptera: Reduviidae), 302
    Bed Bugs, Swallow Bugs, and Bat Bugs (Hemiptera: Cimicidae and Polyctenidae), 303
  Flies (Diptera: Several Families), 304
    Mosquitoes (Diptera: Culicidae), 305
    Black Flies (Diptera: Simuliidae), 308
    Biting Midges (Diptera: Ceratopogonidae), 310
    Phlebotomine Sand Flies (Diptera: Psychodidae: Phlebotominae), 312
    Horse Flies and Deer Flies (Diptera: Tabanidae), 314
    Tsetse Flies (Diptera: Glossinidae), 316
    Muscid Flies (Diptera: Muscidae), 318
    Stable Fly, 319
    House Fly, 320
    Blow Flies (Diptera: Calliphoridae), 321
    New World Screwworm Fly, 322
    Flesh Flies (Diptera: Sarcophagidae), 324
    Bot and Warble Flies (Diptera: Oestridae), 325
    Louse Flies (Diptera: Hippoboscidae), 329
  Fleas (Siphonaptera), 330
  Other Taxa of Occasional Importance, 332
    Eye Gnats (Diptera: Chloropidae), 333
    Snipe Flies (Diptera: Rhagionidae), 333
    Bees and Wasps (Hymenoptera: Various Families), 333
    Ants (Hymenoptera: Formicidae), 333
    Dermestids (Coleoptera: Dermestidae), 334
Summary, 334
References and Additional Reading, 335

SECTION 4: PEST MANAGEMENT AND ITS EFFECTS ON WILDLIFE, 339

CHAPTER 11 PESTICIDES AND THEIR EFFECTS ON WILDLIFE, 341

Pesticides, 343
Insecticide Mode of Action, 345

Persistence of Insecticides, 349
Acute Effects of Insecticides, 350
Sublethal Effects of Insecticides, 354
Other Pesticides, 355
Indirect Effects of Pesticides on Wildlife, 356
Insecticides in The Food Chain, 357
Risks of Insecticides, 359
Resistance to Insecticides, 361
Summary, 362
References and Additional Reading, 363

CHAPTER 12   ALTERNATIVES TO INSECTICIDES, 366
Environmental Management or Cultural Control, 366
Physical and Mechanical Control, 370
Host Resistance, 371
Semiochemicals, 373
Biological Control, 375
Area-Wide Insect Management, 379
Integrated Pest Management (IPM), 381
Preventing versus Correcting Problems, 382
Summary, 383
References and Additional Reading, 383

SECTION 5: CONSERVATION ISSUES, 385

CHAPTER 13   INSECT–WILDLIFE RELATIONSHIPS, 387
How Wildlife Affect Insect Survival, 387
  Naturally Occurring Predation by Wildlife on Insects, 387
    Western Pine Beetle and Woodpeckers, 392
    Spruce Budworm, Birds, and Mammals, 392
    Gypsy Moth, Birds, Mammals, and Beneficial Insects, 393
    Rangeland Grasshoppers and Birds, 393
    Crop-Feeding Aphids and Birds, 393
    Crop-Feeding Caterpillars, Spiders, and Birds, 394
    Tropical Forest Floor-Dwelling Insects, Lizards, and Birds, 394

    Tropical Forest Insects, Bats, and Birds, 394
    Aquatic Insects, Ducks, and Fish, 395
    Predation of Animal Ectoparasites by Birds, 395
  Introduction of Vertebrates for Biological Suppression of Insects, 396
How Insects Affect Wildlife Survival, 397
  Predation by Insects on Wildlife, 397
    Effects on Terrestrial Wildlife, 397
    Effects on Aquatic Wildlife, 400
  Symbiotic Relationships Between Insects and Wildlife, 400
  The Benefits of Insects for Habitat Conservation, 402
  The Benefits of Insects for Wildlife-Based Recreation, 406
Summary, 407
References and Additional Reading, 408

CHAPTER 14   INSECT AND WILDLIFE CONSERVATION, 410
Other Economic Benefits of Insects, 410
  Pollination, 410
  Honey, 412
  Silk Production (Sericulture), 413
  Shellac and Lacquer, 414
  Dyes, 414
  Food for Humans and Domestic Animals, 414
  Medical Treatment, 416
Conservation of Insects, the 'Smallest Wildlife', 416
  Conservation Status, 418
  Advancing the Conservation of Insects, 419
  Conservation of Bumble Bees, 422
  Conservation of Butterflies, 423
  Conservation of Beetles, 424
Managing Insect Resources for the Benefit of Wildlife, 425
  Principles, 426
  Practices, 426
Summary, 434
References and Additional Reading, 435

Glossary, 437

Index, 457

PREFACE

David Grimaldi and Michael Engel wrote in *Evolution of the insects* (Cambridge, 2005) that 'if ants, bees and termites alone were removed from the earth, terrestrial life would probably collapse'. This is a powerful statement, reflecting a sentiment that often is not appreciated by those who dislike 'bugs'. Nevertheless, it is the intent of this book to show how insects are an integral and important part of most ecosystems. In particular, the wild animals we cherish, and sometimes upon which we depend, would not exist without insects. This is not to say that insects are always beneficial, but they are more beneficial to wild animals than imagined by many. When Grimaldi and Engel penned the aforementioned statement about the importance of insects, they were mostly concerned about decomposition, nutrient cycling, and plant growth and reproduction. In the ensuing chapters, I will reinforce their statement with supporting information on the importance of insects to these ecological processes, but will also discuss additional functions such as conversion of plant biomass to more nutritious animal matter, feeding by vertebrate wildlife on insects, disease transmission by insects from and to wildlife, and the roles of insects in environmental conservation. Indeed, without insects the world would be a far different place. As you read through this book, I am confident that you will gain a new perspective on insects, particularly as they relate to wildlife (wild vertebrate animals).

I prepared this book with the intent to increase awareness of the importance of arthropods to vertebrate wildlife. There is a rich literature surrounding this subject, but curiously there has never been an attempt to bring it all together. Similarly, although many introductory entomology texts have been published, they have all centered on insects, or on protection of plants, people and livestock from insect damage. Here I discuss insects from the perspective of wildlife. My hope is that wildlife managers and vertebrate-oriented biologists will find it more informative and relevant to read about entomology within the context of a broader ecology: the interrelationships of a plant resource base, herbivores and predators of all types, and the need to conserve and preserve natural resources – including insects.

ACKNOWLEDGMENTS

This project benefited greatly from the reviews provided by several University of Florida colleagues: Don Forrester, Department of Infectious Diseases and Pathology; Holly Ober, Department of Wildlife Ecology and Conservation; Russ Mizell, Jim Nation, Paul Choate, Howard Frank, Don Hall, and Mike Scharf of the Department of Entomology and Nematology. Pam Howell proof-read the document, and Hope Johnson and Kay Weibel prepared the graphic material.

INTRODUCTION TO THE ARTHROPODS

Chapter 1

INSECTS AND THEIR RELATIVES

Insects are invertebrate animals within the class Insecta and the phylum Arthropoda of the kingdom Animalia. Phylum Arthropoda consist of more than insects, of course, but to most people all arthropods are 'bugs,' those small but annoying and damaging organisms that lack fur and feathers, and therefore seem rather alien. Arthropods have jointed legs and generally are hard to the touch because they wear their skeleton externally. Think about shrimp, crabs, and lobsters if you are having trouble envisioning this, as they are arthropods, too. However, most people don't think of shrimp, crabs, and lobsters as bugs. To most people bugs are insects, ticks, spiders, centipedes, millipedes and similar terrestrial organisms. Here we will learn about the true insects and their near relatives, and you will learn what distinguishes bugs from similar organisms.

NAMING OF TAXA

But first we need to discuss the naming of organisms. The purpose of naming always has been largely practical. We need to have a way of identifying and describing organisms accurately but concisely. Secondly, we want to describe the relationship of organisms to other organisms, usually by giving their biological position from an evolutionary perspective. Initially, organisms typically were grouped according to their appearance, but as we learned more about them and their inter-

Insects and Wildlife: Arthropods and Their Relationships with Wild Vertebrate Animals, 1st edition. By J.L. Capinera. Published 2010 by Blackwell Publishing.

relationships, their positions relative to others often changed. For example, organisms that dwell within ant nests often look quite like ants. Apparently, if you look and smell like an ant, you can gain easy access to ant nests, and the food riches contained there. However, if they are only ant mimics but actually beetles, we want to call them beetles, not ants. So in many cases physical appearance is not adequate, and we need to know about relatedness.

Do you find scientific terminology and the classification of organisms somewhat overwhelming? If you do, you may find it comforting to know that this is normal. All these hard-to-pronounce names are difficult for everyone. Nevertheless, we need to have a way to organize the diversity of life into manageable groups of related organisms or **taxa** (taxon is the singular form). And insects are truly diverse! Figure 1.1 shows the number of species of common groups of animals. There are far more insects species than other animals. I have used a conservative estimate for the number of species of insects, about one million. In all likelihood, the actual number is probably 3–4 million, as most tropical insect species have yet to be named. As you can see, the number of many other animal species such as reptiles, birds, fish, and mammals is almost insignificant compared to insects.

The insects and their relatives are arranged into taxa that organize similar organisms into categories. The principal taxa, in descending (most inclusive to least) order, are:

Kingdom
  Phylum
    Class
      Subclass
        Infraclass
          Series

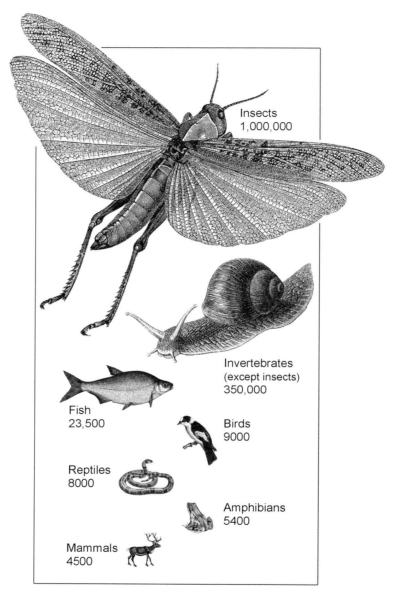

Fig. 1.1. The number of species in some important animal taxa. Note that the size of each image is proportional to the number of species in each group.

Superorder
Order
Suborder
Infraorder
Family
Subfamily
Tribe
Genus
Species

Not all of these categories are always used; the most commonly used taxa are class, order, family, genus and species.

Every organism has a **scientific name**, consisting of a genus and species designation. This two-part name follows a system called **binomial nomenclature**. The rules of binomial nomenclature are standardized to reduce potential confusion, and to confer information about the relatedness of the organism to other organisms. Additional information may be included along with the scientific name. So, you might see something like 'Schistocerca americana (Drury) (Orthoptera: Acrididae).' The italics signify the genus (which is capitalized) and species (not capitalized). The name following the genus and species is the person who originally described the species; this person is sometimes called the 'author' of the name. If the scientific name (genus and species) is unchanged since the original description, the describer's name lacks parentheses. If the describer's name has been modified (usually this indicates transfer to a different genus) then this change is reflected by the parentheses surrounding the author's name. Therefore, in our example, '(Drury)' signifies that the scientific name has been changed since the original description. In fact, it was originally described as *Libellula americana* by D. Drury in 1773. After studying the genus, however, in 1899 S.H. Scudder placed this species in the genus *Schistocerca*, where it now remains. The names within the next set of parentheses are the order and family designations: order Orthoptera, and family Acrididae. This latter information is not always given, but it is beneficial to do so because it tells you about the relationship of this insect relative to other insects. In this case, *S. americana* belongs to a family that contains many 'short-horned grasshoppers' in an order that contains grasshoppers, crickets and katydids. Sometimes a third italicized name is included: the subspecies.

The **species** is the most fundamental taxon, but for something so fundamental it is surprisingly difficult to define. Most biologists subscribe to the 'biological species concept,' which states that a species is an interbreeding group of organisms that is reproductively isolated from other such groups. Species most often develop through geographic isolation (**allopatric speciation**) from other parts of their interbreeding population. After sufficient isolation and incremental change they can no longer interbreed so they become a new species. Often accompanying this isolation are changes in appearance or behavior, which helps us to identify that they are different species. Sometimes the changes are so subtle that we have trouble identifying groups as being different. Conversely, sometimes we are fooled by the different appearance of organisms into thinking that they are different species when they are only environmentally induced differential expressions of the same species (polyphenisms, see Chapter 2). Geography is not the only means of isolation leading to formation of new species, of course. Another important means occurs when species adopt different host plants, leading to isolation based on feeding behavior, as this can result in host specific associative mating. Speciation that occurs without geographic isolation occurring is called **sympatric speciation**. Sympatric speciation also commonly results when insects become separated in time (developing at different times of the year) or from the use of different chemical attractants (pheromones).

Species that have populations differing in appearance are sometimes divided into subspecies. A **subspecies** is usually little more than a color variant, and can interbreed successfully with other subspecies in the same species. Subspecies may be indicative of speciation in progress, however. Butterflies commonly display regional color variations, so subspecies designations are especially frequent in this taxon.

Other categories also exist, and are used when it is necessary or convenient. For example, the class Insecta, along with the class Entognatha (the collembolans, proturans, diplurans) are often placed together into the superclass Hexapoda. In fact, the entognathans are sometimes considered to be insects. The subclass Pterygota is sometimes divided into two divisions, consisting of the hemimetabolous orders and the holometabolous orders. Also, related families are often grouped into superfamilies. Possibly the only level that can be assessed objectively is the species, and even that can be argued. Species are grouped into genera, genera into families, and so forth, but taxonomists differ in the importance of characters used to cluster the taxa, so different arrangements are possible. The names of most orders end in -ptera; of families in -idae; of subfamilies in -inae, and of tribes in -ini.

Remember, all this naming protocol is simply to create order and to show relatedness. Although it may seem confusing, it is really intended to inform you. Admittedly, it takes a while to catch on.

ARTHOPODA

The Arthopoda are a large (about two to three times all other animal species combined) and diverse phylum. All arthropods have some things in common. Among the characteristics that arthropods share, but which help to separate them from other invertebrates, are:
• bilateral symmetry (the left and right halves are mirror images);
• the integument (external covering) contains a great deal of chitin, a structural polysaccharide, and functions as an external skeleton;
• segmented bodies;
• jointed appendages that assist with walking and feeding; the name 'Arthropoda' is derived from Greek words meaning 'jointed feet';
• a dorsal brain but a ventral nerve cord.

They also have some differences that allow us to separate them into groups (taxa). The principal arthropod taxa of relevance to insects are:

Phylum Arthropoda
   Subphylum Trilobita – trilobites (these are extinct)
   Subphylum Chelicerata
      Class Merostomata – horseshoe crabs
      Class Arachnida – arachnids (scorpions, spiders, ticks, mites, etc.)
      Class Pycnogonida – sea spiders
   Subphylum Crustacea – crustaceans (amphipods, isopods, shrimp, crabs, etc.)
   Subphylum Atelocerata
      Class Diplopoda – millipedes
      Class Chilopoda – centipedes
      Class Pauropoda – pauropods
      Class Symphyla – symphylans
      Class Entognatha – collembolans, proturans, diplurans
      Class Insecta – insects

Trilobites (subphylum Trilobita) have been extinct for 250 million years, but they once were very common and over 17,000 species are known. Probably because they had a tough integument (body covering), they

preserved quite well, and are second only to the dinosaurs as well-known fossils.

The chelicerate arthropods (subphylum Chelicerata) have two principal body segments, the **cephalothorax** (fused head and thorax) and the **abdomen**. They have feeding structures called **chelicerae**, which are often fang-like and good for grasping and piercing, but not for chewing. They also possess **pedipalps**, additional segmented appendages that assist in feeding and which may possess claws. The chelicerate arthropods usually have four pairs of legs, but lack **antennae** (elongate sensory appendages located on the head). The pedipalps, however, often perform sensory functions.

The crustaceans also have a cephalothorax but possess feeding structures called **mandibles**. Mandibles are jaw-like structures that are good for grasping, holding, and masticating food. The number of legs present is variable, but they have two pairs of antennae.

The classes in the subphylum Atelocerata also have mandibles, and variable numbers of legs, but possess only one pair of antennae. The more important arthropods are briefly discussed below prior to a more detailed discussion of insects.

ARACHNIDA

This class is the most important and familiar of the chelicerate classes, containing many familiar forms (Fig. 1.2). They tend to be carnivorous, and often the digestion occurs largely outside the body (predigestion) because enzymes are secreted into the prey, with the resulting fluid taken up. Four pairs of legs are normal, although the occurrence of enlarged pedipalps in some may give the appearance of five pairs of legs. The most important arachnids are:

Order Scorpiones, the scorpions
Order Pseudoscorpiones, the false scorpions
Order Solifugae, the sun spiders or wind scorpions
Order Uropygi, the whip scorpions
Order Aranaea, the spiders
Order Opiliones, the harvestmen or daddy longlegs
Order Acarina, the mites and ticks

The **scorpions** (order Scorpiones) seem to be the oldest terrestrial arthropods, and may have been the first to conquer land. They are nocturnal (night-active) and secretive, hiding by day in burrows and beneath stones and logs. They are found widely in both arid and

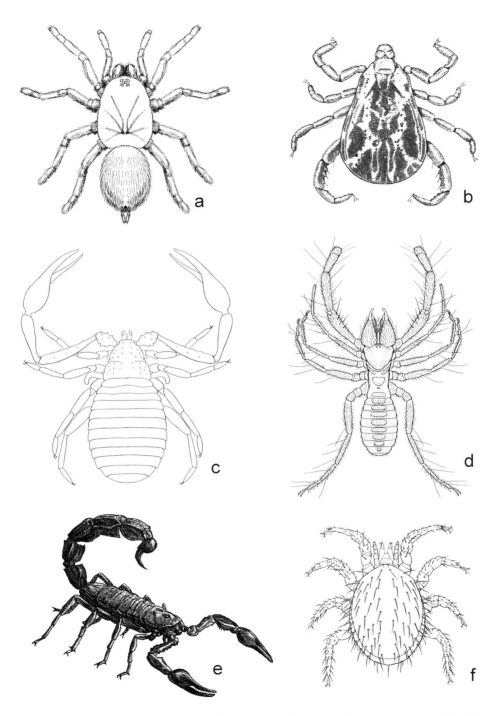

Fig. 1.2. Representative arachnids (phylum Arthropoda, class Arachnida): (a) a spider (order Aranaea); (b) a tick (order Acarina); (c) a pseudoscorpion (order Pseudoscorpiones); (d) a sun spider (order Solifugae); (e) a scorpion (order Scorpiones); (f) a mite (order Acarina).

moist environment. Scorpions are well known for their long abdomen that terminates in a sting apparatus. The venom of scorpions is sufficient to kill many invertebrates, but not usually dangerous to humans, producing pain equivalent to a yellow jacket. A few species, however, are quite dangerous, and can kill a human. In the southwestern USA, species of *Centruroides* fall into this 'quite dangerous' category. The front appendages, which appear to be legs, are really enlarged mouth structures (pedipalps) bearing claws for capturing prey. They feed mostly on arthropods, but occasionally on small vertebrates. Scorpions can be up to 18 cm in length. There are about 2000 species of scorpions.

The **pseudoscorpions** (order Pseudoscorpiones) are smaller than scorpions, barely exceeding 8 mm in length. About 3400 species are known. Superficially they resemble scorpions because they bear enlarged pincer-like pedipalps, but their abdomen is not elongate and they lack a sting apparatus. They are commonly found in leaf debris, but due to their small size often are overlooked. They feed on small arthropods.

The **solifugids** (order Solifugae) live in warm, arid environments, and shelter belowground. They are large, up to 7 cm long. They are best known for their large, conspicuous chelicerae. These are in the form of two pairs of pincers that articulate vertically. Like scorpions, they will feed on both arthropods and vertebrates. Unlike scorpions, they are principally diurnal (day-active). About 1000 species have been described.

The **whip scorpions** (order Uropygi) resemble scorpions except that the abdomen terminates in a long flagellum (the 'whip'). Some are known as 'vinegaroons' because they spray acetic acid as a defensive measure. They vary from small to large in size. Only about 100 species are known.

The **spiders** (order Araneae) are among the most widely known arthopods. This order is quite species-rich, with over 40,000 named species. They range in size from 0.5 mm to perhaps 9 cm. They feed mostly on insects, but sometimes capture small vertebrates, including birds. Many, but not all, depend on production of silk to ensnare prey. Many hunting spiders such as wolf, crab and jumping spiders stalk their prey, and while they may rest within a web or even tie their captured prey with silk, they do not rely on a web for prey capture. The hunting spiders tend to be heavy bodied. In contrast, the web-producing spiders usually have long, slender legs. Spider webs vary in complex-

ity, but most contain both dry and adhesive strands. The chelicerae of all spiders possess poison glands. The venom of most spiders is not toxic to humans, but a few are quite poisonous. The eyes of spiders are more developed than most arachnids, though they are unable to form an image due to insufficient number of receptors. They usually occur in two rows of four eyes. Courtship of spiders is often a complex process. Eggs are often deposited within an egg sac, and when the young **spiderlings** (young spiders) hatch they disperse. A common method of dispersal involves ballooning. When **ballooning**, the spiderling climbs to an elevated location and releases a strand of silk. If the breezes are sufficiently strong to tug on the strand of silk, the spider releases its hold on the substrate and is carried in the wind.

The **harvestmen** or daddy longlegs (order Opiliones) are a small (about 6400 species) but well-known group of long-legged, spider-like arachnids. They vary in size but some are quite large. However, the size is deceptive; the body is always small and the legs are always long. More than most arachnids, harvestmen are omnivorous. Although usually feeding on insects, they also eat dead animal matter and plant material. Thus, unlike spiders, which suck in predigested prey, harvestmen can ingest particulate matter.

The **mites** and **ticks** (order Acarina) are unquestionably the most important arachnids. They occur in all habitats, and often are very numerous. There are about 50,000 described species, but most mites have yet to be described. As might be expected with such a large group, they are quite varied in appearance. Mites feed on economically important plants and animals, and although ticks do not attack plants, ticks are very important ectoparasites of vertebrates. Ticks are second only mosquitoes in importance as vectors of animal disease. Mites are usually only 1 mm in length, whereas ticks can reach up to 3 cm in length. Perhaps the most distinctive feature of ticks and mites is the lack of body segmentation. Abdominal segmentation is not visible, and even the two major body regions are joined imperceptibly. Ticks and mites are discussed more fully in Chapter 10.

CRUSTACEA

The **crustaceans** (subphylum Crustacea) are a diverse group numbering about 52,000 described species. The Crustacea are the only large group of arthropods that

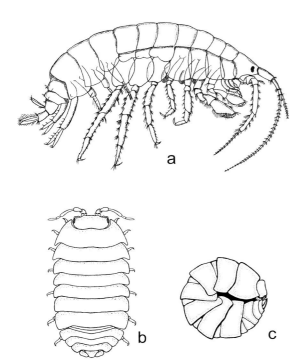

Fig. 1.3. Representative crustaceans (phylum Arthropoda, subphylum Crustacea): (a) a common aquatic crustacean, the amphipod *Gammarus* sp. (class Malacostraca: order Amphipoda); (b, c) common terrestrial crustaceans, woodlice or pillbugs, *Armadillidium* sp. (class Malacostraca: order Isopoda).

are primarily aquatic. Most occupy the marine environment but they are also numerous in fresh-water aquatic habitats. Some are terrestrial, and woodlice (class Malacostraca, order Isopoda), which are sometimes called pillbugs and sowbugs, are examples of common terrestrial crustaceans (Fig. 1.3). Woodlice can be quite important in decomposition of organic materials, and often are quite abundant on the forest floor. However, whereas crustaceans are very abundant in the marine environment, insects dominate terrestrial environments. Small crustaceans can obtain adequate oxygen through their body covering. Larger crustaceans, however, use gills associated with their legs to move water and take up oxygen. Crustaceans may have two body regions (cephalothorax and abdomen), or three (head, thorax, abdomen). Some of the important groups of crustaceans are:

Class Branchiopoda, the tadpole shrimps, water fleas, others
Class Ostracoda, the seed shrimps
Class Copepoda, the copepods
Class Cirripedia, the barnacles
Class Malacostraca, the mantis shrimps, amphipods, isopods, krill, crabs, lobsters, shrimps

Crustaceans are quite variable in appearance, but have two pairs of antennae. The thorax is often covered by a dorsal carapace that may extend over the sides of the animal. The cuticle of the larger crustaceans is calcified, a characteristic lacking in most other arthropods. Some crustaceans swim, but most simply crawl. The adults of most have a pair of compound eyes. Crustacea are significant mostly as a food resource for animals dwelling in aquatic and marine habitats. Some, such as lobsters, crabs, shrimp, and crayfish are economically important food for humans. Many of the terrestrial forms may help in decomposition of detritus, however, and a few consume plants.

DIPLOPODA

Millipedes (class Diplopoda) have numerous legs, but not the thousand suggested by their common name. The occurrence of numerous legs is a characteristic of this group, but none have more than 375 pairs, and most have considerably fewer. In distinguishing this group from the similar-appearing centipedes, the presence of two pairs of legs per body segment is the key character used to identify millipedes (Fig. 1.4). There are about 10,000 species known around the world.

Millipedes are quite diverse morphologically, though they all consist of a long chain of rather uniform body segments, and lack wings. Some are rather short, and may be covered with feather- or scale-like adornment. Others look greatly like woodlice (Isopoda, the pillbugs), and even roll into a ball in the manner of pillbugs. Most, however, are elongate and thin in general body form. There are three basic body regions: the head, which bears a pair of moderately long antennae; the body, consisting of numerous leg-bearing segments, and which normally are rather cylindrical but sometimes bears prominent lateral projections; and the telson, or posterior body segments bearing the anus. The integument is very hard.

The life cycle of millipedes is often long. Many live for a year, but some persist for 2 to 4 years before attaining maturity. In a *Julus* sp. studied in a temperate

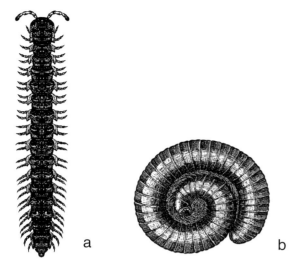

a   b

Fig. 1.4. Representative millipedes (phylum Arthropoda, class Diplopoda): (a) normal posture; and (b) coiled defensive posture of *Oxidus* sp.

environment, oviposition took place in April, with instar 6 to 7 attained by winter and instars 8 to 9 by the second winter. They overwintered as instars 9–11, then mated and oviposited the following spring, their third year of life, before they perished.

Adult millipedes vary considerably in size, often measuring from 10–30 mm in length, but in some species exceeding 100 mm. Their color ranges from whitish to brown and black. The sexes are separate. The external genitalia of adult millipedes are located between the second and third pairs of legs. Some adult millipedes have the ability to molt from a sexually active adult to an intermediate stage that is not functional sexually. Parthenogenesis (production of eggs or young without fertilization by males) occurs in some species and some populations, but this is not usual. Millipedes lack a waxy cuticle and are susceptible to desiccation. They have glands, with openings usually located laterally, which secrete chemicals that are toxic and may immobilize predatory arthropods such as spiders and ants.

Millipedes are common animals, and are found wherever there is adequate food. They are detritivores, normally feeding on dead plant material in the form of leaf litter. However, they occasionally graze on roots and shoots of seedlings, algae, and dead arthropods and molluscs. They move rather slowly. They are selective in their consumption of leaf litter, preferring some leaves over others. They also tend to wait until leaves have aged, and are partially degraded by bacteria and fungi. Thus, they function principally as decomposers, hastening the break-up of leaf material into smaller pieces, and incorporating the organic matter into the soil. Whether they derive most of their nutritional requirements from the organic substrate or the micro-organisms developing on the substrate is uncertain. Millipedes also tend to consume their own feces, and many species fare poorly if deprived of this food source.

Millipedes sometimes are viewed as a severe nuisance as a result of exceptional abundance in an inappropriate location such as in yards, homes, or commercial or food processing facilities. Millipedes can exist in tremendous quantities in the soil and become a problem only when they come to the surface and disperse as a group. This often occurs following abnormally large rainfall events, though hot and dry conditions also are sometimes suspected to be a stimulus for dispersal. Several species of millipedes are reported to be injurious to plants. Probably the most important is garden millipede, *Oxidus gracilis* Koch (Diplopoda: Paradoxosomatidae). It apparently was accidentally introduced to most temperate areas of the world from the tropics, probably via transport of specimen plants. In cool areas it is principally a greenhouse pest.

Millipedes produce various foul-smelling fluids (sometimes including hydrogen cyanide) from openings along the sides of their body. Despite the formidable chemical defenses of millipedes, several natural enemies are known. Small vertebrates such as shrews, frogs, and lizards eat millipedes. Invertebrate predators such as scorpions (Arachnida), ground beetles (Coleoptera: Carabidae), and rove beetles (Coleoptera: Staphylinidae) also consume millipedes, though ants are usually deterred. Some diseases and parasitic flies also are known to affect millipedes.

The toxic secretions of millipedes do not go unnoticed by wildlife such as the New-World capuchin monkeys, *Cebus* spp. The monkeys anoint themselves with *Orthoporus dorsovittatus* millipede secretions (benzoquinones) by rubbing crushed millipedes over their body. This repels mosquitoes, which are a serious nuisance, but perhaps more importantly this repellent is thought to reduce the probability that monkeys will be infested by bot flies. Some bot flies (Diptera: Oestridae) capture mosquitoes and lay an egg on the mosquito. When the mosquito feeds, the bot fly egg hatches, and the young bot fly enters the wound created by the mosquito's feeding. Thus, reducing the feeding by mosquitoes also reduces the infestation by bot flies. Millipedes, therefore, are natural sources of insect repellency that are used by monkeys to their advantage. Often monkeys will pass a crushed millipede to others in the group, creating a significant social interaction. Malagasy lemurs and some birds also take advantage of millipedes by self-anointing with insect-repellent or insecticidal secretions.

CHILOPODA

Centipedes (class Chilopoda) possess but one pair of legs per segment (Fig. 1.5), a feature that allows them to be easily distinguished from the superficially similar millipedes (Diplopoda). Like many other arthropods, but unlike entognathans and insects, the centipedes bear a head and a long trunk with many leg-bearing segments. The head bears a pair of antennae, and sometimes ocelli, but not compound eyes. The mouthparts are ventral, and positioned to move forward.

Centipedes usually are 1 to 10 cm in length, but may be larger in the tropics, where they can attain a length of up to 26 cm. Covering the mouthparts of centipedes is a pair of poison claws. They are derived from the first pair of trunk appendages, but are involved in feeding. A poison gland is found within the base of the claw. Centipedes are well known for their poison claws, but they have other defenses as well. In some, the posterior-most legs may be used for pinching, and repugnatorial glands on the last four legs are common. As with millipedes, defensive secretions may include hydrocyanic acid. About 3000 species are known.

Life histories of centipedes are poorly known, but longevity is often 3 to 6 years, and it commonly takes more than a year to attain maturity. Some centipedes produce a cavity in soil or decayed wood in which to brood their egg clutch. The female guards the eggs until the young hatch. In the remaining taxa, the eggs are deposited singly in the soil.

Centipedes are predaceous. As might be expected of predators, they move very quickly. Most feed on arthropods, snails, earthworms and nematodes, but even toads and snakes are consumed by some. The antennae and legs are used to detect prey. The poison claws are used to stun or kill the prey. Though painful, the bite of centipedes is normally not lethal to humans, resembling the pain associated with a wasp sting.

Centipedes require a humid environment. Their integument is not waxy, and their spiracles do not close. Hence, they are found belowground, in sheltered

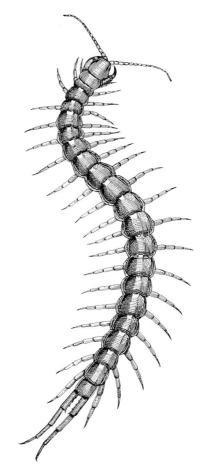

Fig. 1.5. A representative centipede (phylum Arthropoda, class Chilopoda): *Scolopendra* sp.

environments, or are active aboveground principally at night. Some centipedes have adapted to a marine existence, living among algae, stones and shells in the intertidal zone. Apparently, they can retain sufficient air during high tides, or capture a sufficiently large bubble of air to allow submersion.

ENTOGNATHA

The entognathans (class Entognatha) consist of three orders of six-legged arthropods that are closely related to insects: Collembola, Protura, and Diplura. Indeed, sometimes they are included with the insects in the class Insecta. However, it is better to consider the enognathans and insects part of the same superclass, the **Hexapoda**, thereby recognizing that while they are closely related, they are not truly insects. The Entognatha have their mouthparts sunk into the head (**entognathous**), unlike the insects, which have their mouthparts extruded (**ectognathous**). Also, unlike most insects, they lack wings, and the immature forms greatly resemble the adult forms.

Springtails and proturans exhibit **anamorphosis**, a type of development that adds a body segment after a molt. In some cases, the adult stage continues to molt throughout the remainder of its life. Sperm transfer in entognathans is indirect, with sperm deposited in stalked droplets (spermatophores) on the ground. In most cases, sperm uptake by the female is a passive process, but in some of the more advanced species, the male guides the female to his sperm.

Springtails (order Collembola) are a primitive entognathous order of hexapods that are among the most widespread and abundant terrestrial arthropods. They are even found in Antarctica, where arthropods are scarce. Generally they are considered to be useful, as they assist in the decomposition of organic materials, and few are pests. There are about 6000 species of springtails.

Springtails are small animals, measuring only 0.25 to 10 mm in length, and usually less than 5 mm long. They are either elongate or globular in body form (Fig. 1.6). Their color varies greatly, and though often obscure, some are brightly colored. The antennae are short to medium in length, and consist of four to six segments. The compound eyes are small, with only a few facets per eye. The mouthparts are basically the biting (chewing) type, but sometime extensively modified, and somewhat enclosed by the head. The legs are small, and lack extensive modifications. Springtails often are equipped with a jumping apparatus that serves as the basis for the common name. This apparatus consists of a spring-like **furcula** originating near the tip of the abdomen that is flexed ventrally and held by a catch, the **tenaculum**. When the tenaculum releases, the furcula springs with a snap that propels the insect forward. In addition, springtails possess a small ventral tube-like structure on the first abdominal segment, called a **collophore** or ventral tube. The collophore has various functions, including water absorption and excretion, and possibly adhesion to smooth surfaces. The immatures usually resemble the adults in

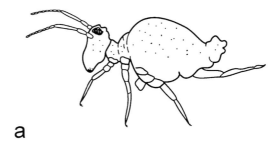

a

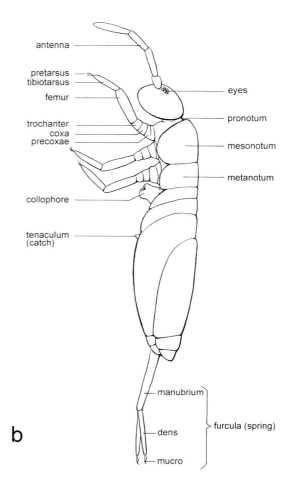

b

Fig. 1.6. Typical springtails (class Entognatha, order Collembola): (a) lateral view of a globular springtail; (b) diagram of elongate springtail showing key morphological features.

external morphology. There are three major body types found among springtails: globular, elongate, and grub-like.

Springtails occur in numerous habitats, including the water surface of fresh water, the tidal region of salt water, in soil, leaf debris, bird nests, beneath bark, and occasionally on foliage. They feed primarily on lichens, pollen, fungi, bacteria, and carrion, though a few feed on seedling plants or are carnivorous. Springtails often aggregate in large groups, and this can be observed on the surface of water, snow, or on organic material. The purpose or cause of aggregations is unknown.

Proturans (order Protura) are minute soil-inhabiting hexapods that lack eyes and antennae, and possess a 12-segmented abdomen. The first three abdominal segments have small leg-like appendages that are capable of movement. The first pair of legs have enlarged foretarsi that are covered with many types of setae and sensilla and function as antennae. Protura have a worldwide distribution with over 500 described species. Soil animals are not well known, and undoubtedly there are many hundreds of species yet to be found in the tropics as well as temperate areas.

Proturan life history is poorly understood, including their diet. Many species can be found in leaf litter, soil that is rich in organic matter, and dead wood. Most species appear to have modifications for feeding on fungi; however, some species have piercing or grinding structures. Like most soil arthropods, proturans most likely feed on a variety of materials including plants and fungi as well as scavenging on dead arthropods.

Diplurans (order Diplura) number about 660 species from throughout the world, though this group is not well known. They resemble proturans and silverfish (Insecta: Thysanura) but the presence of only two cerci rather than three caudal filaments distinguishes diplurans from silverfish (Fig. 1.7). They are distinguished from proturans by the presence of antennae.

Diplurans are usually less than 10mm in length, though their size range is 2 to 50mm. They are elongate, soft-bodied, and brownish. The integument is thin, and few scales are found on these animals. The body regions, including the three thoracic segments, are distinct. The head is distinct, and the antennae long, many-segmented, and slender. Compound eyes and ocelli are absent. The three pairs of legs are similar in appearance and moderately short. The tip of the abdomen bears a pair of cerci. The cerci vary greatly among the taxa, from long, many-segmented antenna-like appendages to stout, rigid forceps used for prey

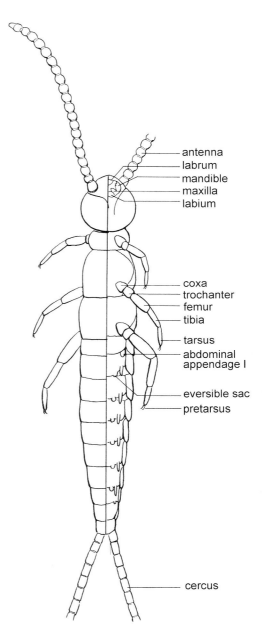

Fig. 1.7. Typical dipluran (class Entognatha, order Diplura) with dorsal view on left and ventral view on right.

capture. Most of the abdominal segments also bear small appendages. The immatures greatly resemble the adult stage, differing principally in the number of antennal segments.

These small animals are found among fallen leaves and in decaying vegetation, under logs and stones, and in soil and caves. They vary in dietary habits; some feed on vegetation whereas others eat other small soil-inhabiting arthropods. Eggs are deposited in the soil. They can be long-lived, with the life span requiring 2–3 years in some species.

INSECTA

The insects (class Insecta) are characterized by having three body regions (head, thorax, abdomen), three pairs of legs attached to the thorax, one pair of antennae, and external mouthparts that include mandibles. Most insects are terrestrial and possess wings, at least in the adult stage. The immature stages often differ considerably from the adult stage. However, as you might expect from a group with over a million species, there is tremendous variability among insects in their structure, physiology and development, and ecology. The following synopsis will attempt to describe some of the common patterns found in their structure, physiology and development, and ecology. Where relevant, some of the important differences among taxa will be noted, but this overview is designed mostly to show the fundamental characteristics of insects.

CLASSIFICATION OF INSECTS

The classification of insects is often debated, but a common arrangement and the derivation (from Greek or Latin) for the names of the taxa in the class Insecta follow. Also, the vernacular (common) name is given for each order.

Class Insecta

Subclass Apterygota: Greek *a* (without) + *pteron* (wings)

Order Archeognatha: Greek *archaios* (primitive) + *gnathos* (jaw) – the bristletails

Order Zygentoma: Greek *zyg* (bridge) + *entoma* (insect) – the silverfish

Subclass Pterygota: Greek *pteron* (wing)

Infraclass Paleoptera: Greek *palaios* (ancient) + *pteron* (wing)

Order Ephemeroptera: Greek *ephermeros* (short-lived) + *pteron* (wing) – the mayflies

Order Odonata: Greek *odon* (tooth) (referring to the mandibles) – the dragonflies and damselflies

Infraclass Neoptera: Greek *neos* (new) + *pteron* (wing)

Series Exopterygota: Greek *exo* (outside) + *pteron* (wing)

Superorder Plecopteroidea

Order Plecoptera: Greek *plecos* (plaited) + *pteron* (wing) – the stoneflies

Order Embiidina: Latin *embios* (lively) – the webspinners

Superorder Orthopteroidea

Order Phasmida: Latin *phasma* (apparition or specter) – the stick and leaf insects

Order Mantodea: Greek *mantos* (soothsayer) – the mantids

Order Mantophasmatodea: from Mantodea + Phasmatodea – the gladiators

Order Blattodea: Latin *blatta* (cockroach) – the cockroaches

Order Isoptera: Greek *iso* (equal) + *pteron* (wing) – the termites

Order Grylloblattodea: Latin *gryllus* (cricket) + *blatta* (cockroach) – the rock crawlers

Order Orthoptera: Greek *orthos* (straight) + *pteron* (wing) – the grasshoppers, katydids, and crickets

Order Dermaptera: Greek *derma* (skin) + *pteron* (wing) – the earwigs

Order Zoraptera: Greek *zoros* (pure) + *a* (without) + *pteron* (wing) – the angel insects

Superorder Hemipteroidea

Order Psocoptera: Latin *psocos* (book louse) + Greek *pteron* (wing) – the barklice, booklice, or psocids

Order Thysanoptera: Greek *thysanos* (fringed) + *pteron* (wing) – the thrips

Order Hemiptera: Greek *hemi* (half) + *pteron* (wing) – the bugs

Order Phthiraptera: Greek *phtheir* (louse) + *a* (without) + *pteron* (wing) – the chewing and sucking lice

Series Endopterygota: Greek *endo* (inside) + *pteron* (wing)

Superorder Neuropteroidea

Order Megaloptera: Greek *megalo* (large) + *pteron* (wing) – the alderflies and dobsonflies

Order Raphidioptera: Greek *raphio* (a needle; referring to the ovipositor) + *pteron* (wing) – the snakeflies

Order Neuroptera: Greek *neuron* (nerve) + *pteron* (wing) – the lacewings, antlions, and mantidflies

Superorder Coleopteroidea

Order Coleoptera: Greek *coleos* (sheath) + *pteron* (wing) – the beetles

Order Strepsiptera: Greek *strepti* (twisted) + *pteron* (wing) – the stylopids

Superorder Panorpoidea

Order Mecoptera: Greek *mecos* (length) + *pteron* (wing) – the scorpionflies

Order Trichoptera: Greek *trichos* (hair) + *pteron* (wing) – the caddisflies

Order Lepidoptera: Greek *lepido* (scale) + *pteron* (wing) – the butterflies and moths

Order Diptera: Greek *di* (two) + *pteron* (wing) – the flies

Order Siphonaptera: Greek *siphon* (tube) + *a* (without) + *pteron* (wing) – the fleas

Superorder Hymenopteroidea

Order Hymenoptera: Greek *hymen* (membrane) + *pteron* (wing) – the wasps, ants, bees, and sawflies

CHARACTERISTICS OF THE MAJOR GROUPS OF INSECTS

The major groups of insects are characterized by a number of fundamental differences. Some of the major differences, and the relationships of taxa, are shown diagrammatically in Fig. 1.8. Characteristics of the insect orders most important to wildlife are shown in Table 1.1. Additional information on the important insect families is provided in Chapters 5 (Insects important as food for wildlife) and 10 (Arthropods as parasites of wildlife).

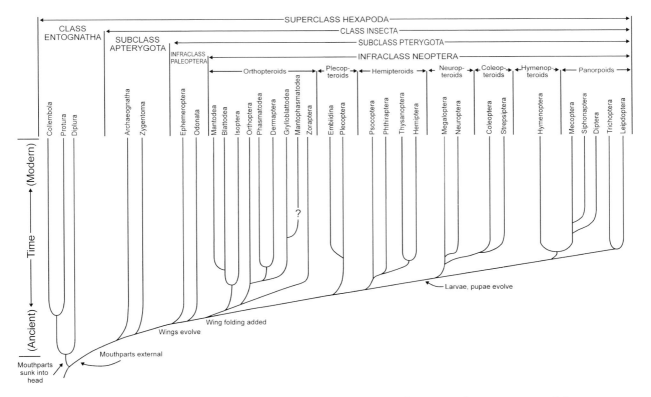

Fig. 1.8. Diagram of the possible phylogeny of the hexapods, showing relationships among the major groups and the temporal occurrence of major evolutionary steps in insect structure and development.

Table 1.1. The orders of insects most important to wildlife, and the key characteristics of the adult stage.

Order	Common name	Example	Metamorphosis
Ephemeroptera	Mayflies		Hemimetabolous
Odonata	Dragonflies, damselflies		Hemimetabolous
Plecoptera	Stoneflies		Hemimetabolous
Blattodea	Cockroaches		Hemimetabolous

.

Forewings	Hindwings	Antennae	Mouthparts	Other features
Triangular, membranous, many veins and cross-veins	Smaller, rounded, sometimes absent	Very short, bristle-like	Chewing, but not functional in adults	Two or thee long slender cerci
Long, slender, membranous, many veins and cross-veins	Similar to forewings	Very short, bristle-like	Chewing	Two or more cerci, but very short
Slender, membranous, with many veins	Usually wider than forewings	Long, thread-like	Chewing	Two cerci, length varies, no ovipositor
Usually long, thickened, with few veins	Usually wider than forewings	Long, thread-like	Chewing	Two short cerci, no external ovipositor

Table 1.1. *Continued*

Order	Common name	Example	Metamorphosis
Isoptera	Termites		Hemimetabolous
Mantodea	Mantids or praying mantids		Hemimetabolous
Phasmida	Stick and leaf insects		Hemimetabolous
Orthoptera	Grasshoppers, katydids, crickets		Hemimetabolous

Forewings	Hindwings	Antennae	Mouthparts	Other features
Long, narrow	Long, narrow	Medium length, thread-like	Chewing	Short cerci, no external ovipositor
Long, slender, membranous	Wider than forewings	Medium to long, threadlike	Chewing	Prothorac legs strongly developed and bearing spines
Variable, often lacking	Variable, often lacking	Long, thread-like	Chewing	Cryptic, resembling twigs or leaves
Usually long and slender, thickened, with veins if present	Usually wider than forewings, membranous if present	Medium to long, thread-like	Chewing	Two short cerci, ovipositor present and sometimes long

Table 1.1. *Continued*

Order	Common name	Example	Metamorphosis
Dermaptera	Earwigs		Hemimetabolous
Psocoptera	Barklice, booklice, or psocids		Hemimetabolous
Hemiptera	Bugs, cicadas, leafhoppers, treehoppers, whiteflies, scales, aphids, etc.		Hemimetabolous
Phthiraptera	Chewing and sucking lice		Hemimetabolous

Forewings	Hindwings	Antennae	Mouthparts	Other features
Short, thickened if present	Long, folding, membranous if present	Medium, thread-like	Chewing	Two large, pincer-like cerci, no ovipositor
Membranous, with few veins if present	Smaller, but similar to forewings if present	Long, thread-like	Chewing	No cerci, ovipositor small or absent
Thickened or membranous if present, but often absent	Membranous and shorter than forewings if present, but often absent	Variable length, thread-like	Piercing-sucking	No cerci, ovipositor may be present
None	None	Short	Chewing or piercing-sucking	No cerci, ovipositor present or absent

Table 1.1. *Continued*

Order	Common name	Example	Metamorphosis
Thysanoptera	Thrips		Generally regarded as hemimetabolous, but somewhat intermediate
Megaloptera	Alderflies, dobsonflies, fishflies		Holometabolous
Neuroptera	Lacewings, ant lions		Holometabolous
Coleoptera	Beetles, weevils		Holometabolous
Trichoptera	Caddisflies		Holometabolous

Forewings	Hindwings	Antennae	Mouthparts	Other features
Very slender with wide fringe of hairs	Very slender with wide fringe of hairs	Short with 6-10 segments	Piercing-sucking, but asymmetrical	Very small, generally less than 5 mm long
Elongate, membranous, with many veins	Similar to forewings	Long, thread-like	Biting	Larvae with paired piercing-sucking mouthparts
Long, membranous, many cross-veins	Similar to forewings	Usually long	Biting	Larvae with paired piercing-sucking mouthparts
Thickened, hardened, veins usually absent	Medium length, membranous, folding if present	Variable in length and shape, often elaborate	Chewing	Larvae may lack legs, often called grubs
Elongate, few cross-veins, many hairs present	Similar to forewings but shorter	Long, thread-like	Chewing	Larvae aquatic, usually build and live within cases

Table 1.1. *Continued*

Order	Common name	Example	Metamorphosis
Lepidoptera	Butterflies, moths		Holometabolous
Diptera	Flies, midges		Holometabolous
Siphonaptera	Fleas		Holometabolous
Hymenoptera	Wasps, bees, ants, sawflies		Holometabolous

Forewings	Hindwings	Antennae	Mouthparts	Other features
Slender to broad, usually covered with scales	Similar to forewings but usually shorter, broader	Long, may be thread-like, plumose, or clubbed	Tubular, siphoning	Larvae usually with prolegs, called caterpillars
Membranous, relatively few veins	Functional wings absent	Often long and thread-like, some short with a bristle	Variable	Larvae legless, often with reduced head and called maggots
None	None	Short	Piercing-sucking	Legless larvae have well-developed head
Membranous, usually long	Membranous, smaller than forewings	Medium length, usually thead-like	Usually chewing, sometimes sucking	Head of larvae well developed, legs variable, may be called grubs

The members of subclass Apterygota lack wings, possess rudimentary abdominal appendages, practice indirect insemination, and molt throughout their life. In contrast, the subclass Pterygota possess wings (or did at one time), generally lack abdominal appendages, practice direct insemination via copulation, and molt only until sexual maturity is attained. So the apterygotes are really quite different from the other insects. They have little or no importance to wildlife. They are rarely seen except in homes where cardboard and other paper products accumulate, especially if accompanied by moisture.

Infraclass Paleoptera consist of primitive insects (evolved early in the lineage of insects), and are comprised of the orders Ephemeroptera and Odonata. The wings often cannot be flexed over the back when the insect is at rest. The immature stages of existing paleopterans live in aquatic habitats. These are important components in aquatic systems, and the adults of Odonata are also important predators.

The infraclass Neoptera, on the other hand, are a diverse groups of relatively modern insects. They can flex their wings over the body. They display development in which the immatures are similar to the mature form (Exopterygota), or the immatures differ markedly in appearance from the adults (Endopterygota).

Superorder Plecopteroidea, consisting of Plecoptera and Embiidina, are closely related to superorder Orthopteroidea. Both suborders have chewing mouthparts; complex wing venation, with the hindwings larger than the forewings; and cerci. They differ, however, in that in the Plecopteroidea the forewings are not thickened, and external male genitalia are lacking. The Plecoptera are important in streams and rivers as a food resource for fish. Embiidina, though interesting, are of little consequence to wildlife.

The members of superorder Orthopteroidea (Polyneoptera) have chewing mouthparts, long antennae, complex wing venation, large hindwings, thickened forewings, large cerci, and nymphs with ocelli. Several orders are considered to be orthopteroids: Phasmida, Mantodea, Mantophasmatodea, Blattodea, Isoptera, Grylloblattodea, Orthoptera, Dermaptera, and Zoraptera. In many older classification systems, they were all considered to be part of the order Orthoptera. This is a relatively primitive group. Many of these taxa are important herbivores, predators, or detritivores, so they are important in ecosystem function. Orthoptera and Isoptera are among the most important sources of wildlife food.

Superorder Hemipteroidea (Paraneoptera) are also a large group, consisting of the orders Psocoptera, Hemiptera, Thysanoptera, and Phthiraptera. Unlike the Orthopteroidea, they lack cerci, and have styletlike structures associated with their mouthparts (though in Psocoptera, chewing mouthparts are preserved). Some groups are well designed for piercing the host and sucking liquids. Though this and the preceding superorders are considered to be hemimetabolous, many Thysanoptera and a few Hemiptera are physiologically holometabolous. The Psocoptera, Hemiptera, and Thysanoptera are important plant feeders. A few Hemiptera and all of the Phthiraptera feed on wildlife.

The superorder Neuropteroidea are holometabolous, and consist of the orders Megaloptera, Raphidioptera, and Neuroptera. Both aquatic and terrestrial forms occur in this superorder. The wings tend to bear numerous cells. Megaloptera and Raphidioptera are small taxa, and not very important. Neuroptera are best known for the predatory members of the order that feed on terrestrial insects, especially aphids.

Superorder Coleopteroidea consist of the orders Coleoptera and Strepsiptera. They are similar in that the metathorax is developed for flight, and bears functional wings. The forewings of the Coleoptera are reduced to hard wing coverings, and even more reduced in the Strepsiptera, to small club-like appendages. They contrast strongly in that the Coleoptera are the largest order, and Strepsiptera one of the smaller orders. The Coleoptera are very diverse and important ecologically. Only a very few can be viewed as injurious to wildlife, but many are important in ecosystem function and as wildlife food. The Strepsiptera are relatively unimportant parasites of insects.

Panorpoidea is another large superorder, consisting of the orders Mecoptera, Trichoptera, Lepidoptera, Diptera, and Siphonaptera. They share few common characters, however, such as a tendency for a reduced meso- and metasternum, some similar wing vein elements, and the terminal abdominal segments tend to function as an ovipositor. Trichoptera and Diptera are important food resources for fish, and Lepidoptera for birds. Diptera are the most important pests and disease vectors of humans and wildlife.

Lastly, the superorder Hymenopteroidea consist only of the order Hymenoptera, though this is a very divergent, complex group. They bear numerous Malpighian tubules, unlike all other endopterygotes, which have only four to six tubules. Also, their wing venation is often greatly reduced, and wings may be entirely absent in members of some groups. Hymenoptera are important sources of food for birds, mammals, reptiles, and amphibians. Some of the ants and wasps can be troublesome to wildlife, however.

EVOLUTION OF INSECTS

The time line that describes the history of the earth has been divided into large blocks of time, but each large block is normally subdivided, and subdivided again, for convenience (Fig. 1.9). The generally accepted divisions are eon, era, period, epoch, and age. The names given to the block of time often have historical significance, and may be associated with occurrence of different fossils. For example, the Phanerozoic eon also consists of three major divisions: the Cenozoic, the Mesozoic, and the Paleozoic eras. The 'zoic' part of the word comes from the root 'zoo,' meaning animal. 'Cen' means recent, 'Meso' means middle, and 'Paleo' means ancient. These divisions

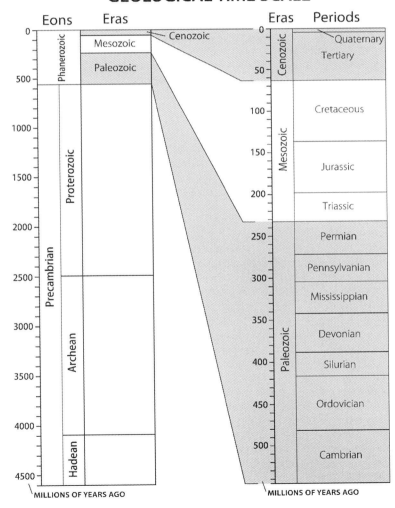

Fig. 1.9. The geological time scale.

reflect major changes in the composition of ancient faunas, with each era associated with domination by a particular group of animals. The Cenozoic has sometimes been called the 'Age of Mammals,' the Mesozoic the 'Age of Dinosaurs,' and the Paleozoic the 'Age of Fishes.' This is not entirely accurate, though there is some basis for these designations. Different spans of time on the geological time scale are usually delimited by major geological or paleontological events, such as mass extinctions. For example, the end of the Cretaceous period of the Mesozoic era is marked by the demise of the dinosaurs and of many marine species. Also, unlike most time lines, the time is expressed not in years from a past event, but from the present. Thus, periods or events are commonly described in millions of years ago (mya).

The oldest known meteorites and lunar rocks are about 4.5 billion years old, but the oldest portions of Earth currently known are 3.8 billion years old. Some time during the first 800 million or so years of its history, the surface of the Earth changed from liquid to solid. Once solid rock formed on the Earth, its geological history began. This most likely happened about 3.8–4 billion years ago, but firm evidence is lacking. The oldest time period, the Hadean eon, is not a geological period *per se*. No rocks on the Earth are this old – except for meteorites. During the Hadean time, the Solar System was forming, probably within a large cloud of gas and dust around the sun. The Archean eon was marked by formation of land masses as the earth's crust cooled and plates began to form. The atmosphere was hostile to life as we know it today, consisting mostly of methane, ammonia, and other toxic gases. The only life known from this early period are bacteria and bacteria-like archaea, commencing about 3.5 billion years ago. Things got interesting only in the Proterozoic eon, when life became more plentiful and the first more advanced life (eukaryotic) forms began to appear and oxygen began to accumulate. Eukaryotic life forms, including some animals, began to appear perhaps as long ago as one billion years ago, but certainly by 500 mya.

The Paleozoic era was interesting because well-preserved fossils document this period. The seas were dominated by trilobites, brachiopods, corals, echinoderms, mollusks, and others, and toward the end of this period life appeared on land. On land, the cycads (primitive conifers) and ferns were abundant. The Mesozoic saw the radiation and disappearance of dinosaurs, mammals appeared, while more advanced land plants such as ginkos, ferns, more modern conifers, and eventually the angiosperms began to appear.

The Cenozoic, the most recent era, is divided into two main sub-divisions: the quaternary and the tertiary periods. Most of the Cenozoic is the Tertiary, from 65 million years ago to 1.8 million years ago. The Quaternary includes only the last 1.8 million years. The Cenozoic is particularly interesting to biologists because most of the life forms we see today developed in this period. It has been called the **'age of insects'** due to the development of great diversity, but could also be known as the age of flowering plants, birds, etc., because most of the flora and fauna we see today evolved during this period. The **phylogeny**, or history, origin, and evolution of insects, is shown diagrammatically in Fig. 1.8.

The last 10,000 years (the Holocene) is sometimes known as the 'age of man' and is also the time period since the last major ice age. The time period before the Holocene, the Pleistocene, is noteworthy because though much of the recent flora and fauna is the same as today, some interesting and now extinct megafauna were present, including mastodons, mammoths, sabretoothed cats, and giant ground sloths. The human species, *Homo sapiens*, also expanded during this time period, and as mentioned previously, there was a significant ice age period.

From an entomological perspective, the Phanerozoic eon (Table 1.2) was an exciting time. Arthropods ventured onto land during the Paleozoic, perhaps 400 mya, though the Silurian entomofauna consisted of primitive myriapods and arachnids. Fossil hexapods have been recovered from the Devonian, most notably springtails from a type of stone called 'chert'. Insects proliferated rapidly during the remainder of the Paleozoic and thereafter. Interestingly, during the Mississippian (also called the Early Carboniferous) we have no fossil evidence of insects, whereas in the Pennsylvanian (also called late Carboniferous) we have numerous records of early (mostly now extinct) insect groups (e.g., protodonata and protorthopterans from deposits in France). At the close of the Paleozoic, the Permian period, the environment of earth was undergoing significant change, most notably a less tropical climate. Numerous insects from many deposits around the world document over a dozen orders of insects, including the occurrence of 'giant' insects.

The Triassic period of the Mesozoic era saw a warming of the earth, and fossil deposits document the occurrence of early insects such as Blattodea and some Orthoptera, Coleoptera, Odonata, Plecop-

Table 1.2. Important time periods of the Phanerozoic eon (543 million years ago to present).

Cenozoic Era (65 mya to today)	Quaternary Period (1.8 mya to today)
	Holocene Epoch (10,000 years to today)
	Pleistocene Epoch (1.8 mya to 10,000 yrs)
	Tertiary Period (65 to 1.8 mya)
	Pliocene Epoch (5.3 to 1.8 mya)
	Miocene Epoch (23.8 to 5.3 mya)
	Oligocene Epoch (33.7 to 23.8 mya)
	Eocene Epoch (54.8 to 33.7 mya)
	Paleocene Epoch (65 to 54.8 mya)
Mesozoic Era (248 to 65 mya)	Cretaceous Period (144 to 65 mya)
	Jurassic Period (206 to 144 mya)
	Triassic Period (248 to 206 mya)
Paleozoic Era (543 to 248 mya)	Permian Period (290 to 248 mya)
	Carboniferous Period (354 to 290 mya)
	Pennsylvanian Epoch (323 to 290 mya)
	Mississippian Epoch (354 to 323 mya)
	Devonian Period (417 to 354 mya)
	Silurian Period (443 to 417 mya)
	Ordovician Period (490 to 443 mya)
	Cambrian Period (543 to 490 mya)
	Tommotian Epoch (530 to 527 mya)

tera, Neuroptera and Grylloblattodea. Transition into the Jurassic was not abrupt for insects, and the fossil record documents few marked changes, but increased radiation.

The Cretaceous period is notable for the radiation of angiosperms that took place. Because many insects are intimately associated with plants through plant feeding and pollination, they were profoundly affected by the availability of these new resources. Many of the modern taxa became established during this period, though more modern taxa such as some Diptera and Lepidoptera radiated later in the Cenozoic. One very noteworthy feature of the Cretaceous is the great availability of amber. The spread of resin-producing trees through this period and into the Tertiary provided an excellent preservation medium for insects. Thousands of species and perhaps 30 orders have been recovered from amber deposits around the world. As insects transitioned from the Cretaceous to the Cenozoic era, the earth witnessed the appearance of 'modern' insect groups such as termites (Isoptera), scale insects and bat bugs (Hemiptera), fleas (Siphonaptera), lice (Phthiraptera), flies (Diptera), and bees and ants (Hymenoptera).

Insects, or at least arthropods, have existed for perhaps 400 million years, making them among the oldest land animals. Most modern orders of insects had appeared by 250 mya, and some modern families can be traced back at least 120 mya. This is quite old compared to mammals, which did not appear until about 120 mya, with modern orders of mammals appearing only about 60 mya. However, unlike some ancient animals (e.g., coelocanth, *Latimeria chalumnae*) or plants (e.g., ginkgo, *Ginkgo biloba*) they are not all relics of earlier times. Many evolved relatively recently, and continue to evolve.

The evolution of insects is complex, and aspects of it continue to be debated. As an example of how these strange-looking beasts came to be, consider the development of the general insect body form: the three tagma we now call the head, thorax and abdomen. The theoretical evolution of the insect body form as we know it today, from an annelid-like ancestor, is shown in Fig. 1.10. There are five major steps:
• Initially the organism was worm-like, long, and segmented. This insect ancestor had a ventral mouth and a simple head-like structure called a prostomium that contained the principal sensory organs.

• In the second step, the organism acquired append-ages on nearly all the body segments. Antennal organs developed on the prostomium.

• In the third step, the union of the prostomium with other anterior body segments commenced. This was the beginning of the complex, composite head structure found in modern insects.

• In the fourth step, the trunk segments differentiated to form the thorax, which became the body segment responsible for locomotion. The body segments anterior to the thoracic segments continued to differentiate into structures that aided feeding, while the body segments posterior to the thorax differentiated into the abdomen, with greatly diminished appendages.

• Finally, the composite head appeared, consisting of the consolidation of the anterior segment into a single unit but bearing mouthpart appendages derived from several body segments. The thorax bore the legs, and in some lineages wings also evolved here. The abdominal appendages had mostly disappeared.

We know something about the evolution of insects because in some cases they are well preserved in mineral deposits or in amber. When insects perish in lakes or marine environments, they may settle to the bottom and be covered with very fine sediments. If they do not decompose and are undisturbed for millions of years, they become fossilized. If uncovered later, we may find a cast or mold of the insect showing

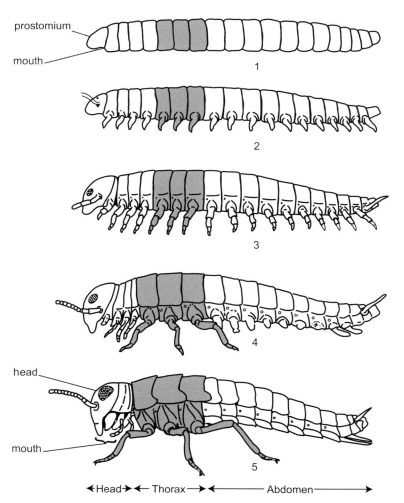

Fig. 1.10. The possible evolution of body form in insects, beginning with an annelid-like ancestor. See text for details. (adapted from Snodgrass 1993).

its form and some relief (such as the venation of the wings), but no color; these are called impressions. However, sometimes the remains of the cuticle are preserved, so some aspects of color are apparent; these are called compressions. The part of insects that preserves well is the hard exoskeleton. The sections of exoskeleton are connected by soft, membranous tissue, which provides the insect with the opportunity for mobility. These soft tissues do not preserve well, so they disappear, leaving sections of insects behind to be preserved. Therefore, often we find just pieces or sections of insects rather than entire, intact bodies. However, sometimes the entire body is replaced by minerals, providing three-dimensional artifacts; this is called petrification. The most celebrated insect fossils are likely amber inclusions. Amber is ancient tree resin. When the resin was fresh, it was sticky, much like present-day exudates of conifers. Insects sometimes became entangled in the surface of the sticky tree exudates, and eventually were engulfed by the flowing resin. This resulted in preservation of insects from up to 250 mya. Amber inclusions often provide nearly perfect detail, including color, so they are exceptionally valuable for studying ancient life.

Why have insects been so successful at persisting, speciating, and exploiting resources relative to other forms of life? The inordinate abundance and diversity of insects is spectacular, and several factors seem to be of paramount importance:

• *Size.* The small size of insects allows them to exploit nearly all food resources. For example, a single seed may support the complete dietary needs of an insect.

• *Longevity.* Insects often complete develop (mature) quickly. This allows insects to take advantage of transient resources. For example, flies may complete their larval development is as little as 3–4 days, before fruit, dung, or standing water disappears.

• *Exoskeleton.* The body plan of insects includes an integument (body covering) that provides a support structure (equivalent to the skeleton of vertebrate animals), protection (essentially a layer of armor that deters predation), and resistance to water loss (the integument has an external layer of wax). This is a very efficient combination that makes insects tougher and hardier than most other animals.

• *Wings.* The occurrence of wings in insects provides the unusually good opportunity to disperse; this allows them to take advantage of transient resources, to escape predation, and to find mates. Food, predation,

and reproduction are the major factors governing the abundance of animals. No other invertebrates have wings, and few other animals, only birds and bats. Many have speculated that the evolution of wings in birds and bats is directly related to their insectivorous habits.

• *Ectothermy.* Insects do not have a constant body temperature, unlike mammals and birds. This feature reduces their energetic (food) requirements because they do not have to burn calories to maintain body temperatures, and allows them to survive long periods without food during inclement weather.

• *Diapause.* Insects can undergo arrested development, which means that their progression toward developmental maturity and reproduction ceases temporarily. During diapause, their metabolism falls to a very low level, allowing them to survive long periods of inclement weather and the absence of food.

• *Chemoreception.* Insects have highly developed chemical senses. Everyone knows that bloodhounds and many other animals can smell odors that humans cannot perceive. Insects are even more sensitive to odors. Unlike humans, who are vision-dependent, the world of insects is based more on odors. For example, although we might not be able to see into dense vegetation, insects can perceive hosts that are hidden from view because they have such an acute sense of smell. Thus, they are unusually well adapted to find hosts and mates, and to communicate to one another by releasing and sensing chemicals.

• *Evolutionary precedence.* Organisms can monopolize resources by evolutionary precedence. It is difficult for organisms to displace other organisms that are already well adapted for a certain niche. For example, insects have made virtually no progress in exploiting marine resources because crustaceans and other marine organisms have monopolized this environment. Similarly, because insects were among the first terrestrial organisms to exploit the terrestrial environment, it is very difficult for other organisms to displace insects in terrestrial environments.

• *Flowering plants*. The flowering plants, or angiosperms, radiated (evolved into new species) at the same time insects were radiating. It is difficult to know which was more important in this radiation of plant and insects, but clearly there often is an interdependency of insects and terrestrial plants. Certainly angiosperms provided niches that did not exist previously, allowing insects to speciate.

INSECT BIOGEOGRAPHY

The study of spatial patterns of animal and plant occurrence is called **biogeography**. **Ecological biogeography** is concerned with the occurrence of organisms in relation to their environment. For example, the changing distribution of organisms in relation to global warming is an increasingly important aspect of ecological biogeography. The other major form of biogeography is historical biogeography. **Historical biogeography** is the distribution of taxa in relation to the geological history of the earth. Here we are mostly concerned about historical biogeography as background to the zoogeographic realms.

The tenets of historical biogeography have long been postulated, but it was not until the 1960s that geologists gained full understanding of the movement of continents, and provided a geological mechanism for patterns of biodiversity that had long puzzled biologists. Basically, the earth's crust has been in constant motion since its formation, resulting in the movement of the land masses. This movement is called continental drift, and resulted in the formation of mountain ranges where crustal elements collided, and troughs where crustal elements diverged.

A single large land mass existed on earth about 250 million years ago, called Pangaea. Pangaea broke up into two supercontinents, Laurasia in the northern hemisphere and Gondwana in the southern hemisphere. These, in turn, broke into the modern continents, which continue their slow movement. The significance of continental drift is that there are biological similarities in flora and fauna derived from earlier times when continents were physically joined, and they do not necessarily correspond to the modern continents. Thus, for example, the fauna of Africa and South America share some common ancestry despite their current geographic distance.

Entomologists commonly make reference to **zoogeographic realms** (or regions or provinces) that relate to historical biogeography. The fauna within a realm share common phylogeny, and developed in relative isolation from other realms. Traditionally, the zoogeographic realms are:
• *Australian realm*. Australia and nearby islands;
• *Oriental realm*. India and Southeast Asia through Indonesia;
• *Ethiopian realm*. Central and southern Africa;
• *Palearctic realm*. Europe and Asia except for Southeast Asia, and including the Arabian Peninsula and Northern Africa;
• *Nearctic realm*. North America except for southern Mexico and Central America;
• *Neotropical realm*. South and most of Central America, including the Caribbean region.

The major realms are shown in Fig. 1.11. The Nearctic and Palearctic realms are often combined into the **Holarctic realm**, as the fauna are really quite similar, despite the present distances between the continents. Also, sometimes the realms are divided further, reflecting smaller areas with unusual fauna. For example, Madagascar is often given a separate designation, as is the southern tip of Africa, and both of these areas have unique flora and fauna.

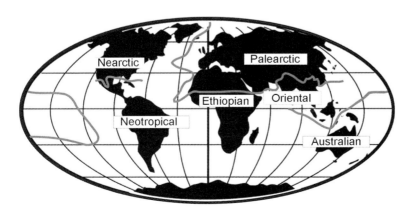

Fig. 1.11. The major zoogeographic realms on earth.

SUMMARY

• Insects are named and categorized into like groups to facilitate information access. The scientific name is based on binomial nomenclature, and all insects have genus and species designations that comprise their scientific name.

• The phylum Arthropoda is composed of organisms with jointed legs and an external skeleton. In addition to insects, it consists of several important classes including the arachnids (spiders, ticks, mites, scorpions), crustaceans (isopods, shrimps, crabs, etc.), millipedes, centipedes, and entognathans.

• The entognathans (class Entognatha) consist of three orders of six-legged arthropods that are closely related to insects: Collembola, Protura, and Diplura. Indeed, sometimes they are considered to be insects. They differ from insects in having their mouthparts sunk into the head (entognathous), unlike the insects, which have their mouthparts extruded (ectognathous). Also, unlike most insects, they lack wings, and the immature forms greatly resemble the adult forms.

• The class Insecta displays three body regions, three pairs of legs, one pair of antennae, and mouthparts containing mandibles. They are extremely diverse, but many possess wings. Presently, 30 orders of insects are recognized, and nearly one million species have been described.

• Insects seem to have evolved about 400 million years ago, during the Paleozoic Era. The modern orders were present 250 million years ago. Insects are much older than mammals, which first appeared about 120 million years ago. Modern orders of mammals appeared about 60 million years ago.

• We know something about the evolution of insects because they are well preserved in mineral deposits and in amber. Insects underwent spectacular radiation in the Cretaceous Period, 144 to 65 million years ago. Some of the features contributing to the success of insects were their small size, short life span, tough exoskeleton, wings, ectothermic temperature relations, ability to undergo diapause, and well developed chemical senses. Other important factors were evolutionary precedence (they were among the first animals to inhabit the terrestrial environment), and the radiation of angiosperms (which provided many new niches).

REFERENCES AND ADDITIONAL READING

Daly, H.V., Doyen, J.T., & Purcell III, A.H. (1998). *Introduction to Insect Biology and Diversity*. Oxford University Press, Oxford, UK.

Elzinga, R.J. (1994). *Fundamentals of Entomology*, 6th edn. Pearson-Prentice Hall, Upper Saddle River, New Jersey, USA.

Gillot, C. (2005). *Entomology*, 3rd edn. Springer, Dordrecht, The Netherlands.

Grimaldi, D. & Engel, M.S. (2005). *Evolution of the Insects*. Cambridge University Press, Cambridge, UK.

Hopkin, S.P. (1997). *The Biology of the Springtails*. Oxford University Press, Oxford, UK.

Hopkin, S.P. & Read, H.J. (1992). *The Biology of Millipedes*. Oxford University Press. Oxford, UK.

Lewis, J.G.E. (1981). *The Biology of Centipedes*. Cambridge University Press. Cambridge, UK.

Pennak, R.W. (1978). *Fresh-Water Invertebrates of the United States*, 3rd edn. John Wiley & Sons Ltd, New York.

Rossi, A. (2008). Speciation processes among insects. In Capinera, J.L. (ed.) *Encyclopedia of Entomology*, 2nd edn., pp. 3478–3480. Springer Science & Business Media B.V., Dordrecht, The Netherlands.

Snodgrass, R.E. (1993). *The Principles of Insect Morphology (reprint)*. Cornell University Press, Ithaca, New York, USA.

Chapter 2

STRUCTURE AND FUNCTION OF INSECTS

Insects often seem to be unusual, even unique creatures. In other respects, they are quite like other terrestrial animals, concerned with the need to acquire appropriate levels of nutrition and water, the need to survive, and the need to propagate. Here we will review the fundamental aspects of insect structure and function that allow them to survive and compete successfully with other animals.

INTEGUMENT

One of the unusual features of insects, and of arthropods in general, is the rigid integument (Fig. 2.1). The body wall, or **integument**, is hardened over most of the body to form a series of plates, called **sclerites**. The plates on the upper (dorsal) surface are called terga or tergites). Plates on the lower (ventral) surface are called sterna or sternites. Plates found on the sides (laterally) are called pleura or pleurites. The plates are useful because they provide a degree of protection, but such rigid structures can work against insects by limiting movement.

The sclerites are connected by areas of integument where a layer called the exocuticle is absent, decreasing the rigidity of the integument and allowing the integument to remain somewhat flexible. These soft, flexible areas are usually called **intersegmental membranes**. The outer area of the integument is secreted by the **epidermal cells**, the innermost living

Insects and Wildlife: Arthropods and Their Relationships with Wild Vertebrate Animals, 1st edition. By J.L. Capinera. Published 2010 by Blackwell Publishing.

portion of the integument. The nonliving external area of the integument is called the **cuticle**. The principal regions of the cuticle are the thin waxy **epicuticle** externally, and a thick, rigid interior region that initially is called the **procuticle**. The procuticle differentiates into two layers: the outer region of the procuticle is called the **exocuticle** and the inner region is called the **endocuticle**. These two layers look slightly different, but their chemical composition is about the same, consisting mostly of various proteins and a polysaccharide called **chitin**. One of the most important proteins is called **resilin**, a rubberlike material that provides elasticity. The cuticle is covered by a thin waxy layer that provides waterproofing, and sometimes is topped by a cement layer. The integument is separated from the internal organs of the insect by a membrane called the **basement membrane**.

The functions of the integument are several, but perhaps most important is that it functions as a skeleton for these animals, providing support for the muscles and organs. Hence, the integument is often referred to as an **exoskeleton**. Admittedly, the skeleton is unusual because it is external, unlike the internal skeleton of mammals, but it is not very different from the external skeleton of lobsters, crabs, and shrimps. The integument sometimes has infoldings called **apodemes**, and these internal ridges serve to strengthen the integument, and also may serve as points of anchor for muscles. These infoldings usually are marked externally by narrow linear depressions called **sutures**. Sutures also sometimes mark the boundaries of plates, thus delimiting areas of the integument. In addition to supporting the body, the integument provides protection from injury, and reduces water loss to a very low level. The integument is not uniform, as some areas are differentiated. Not only are some areas thickened, but

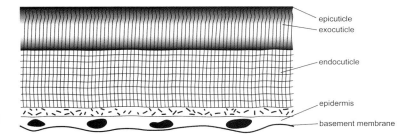

epicuticle
exocuticle
endocuticle
epidermis
basement membrane

Fig. 2.1. Diagram of the integument of insects.

there are thin areas where sensory structures occur or where secretions are released from the epidermal cells via pore canals. The integument also imparts color to the animal. The integument typically has various rigid, pointed outgrowths called **spines**, and sometimes has movable ones called **spurs**.

MOLTING

The problem with being encased in a fairly rigid integument is that growth is severely limited. A certain amount of growth can occur because of the elastic nature of the intersegmental membranes. However, to allow significant increase in size, the insect must shed its old cuticle (the nonliving part of the integument) and produce a new, larger body covering. Insects accomplish this by producing a new, larger but soft integument beneath their old rigid body covering, then shedding the old one and expanding to accommodate the new larger integument. This process is regulated by hormones, particularly ecdysone (this is discussed further under glandular systems, below).

The epithelial cells produce the cuticle, so this area of the integument is central to the entire molting process. The first important step in molting is called **apolysis**, which is the separation of the epidermis from the old cuticle. The space that is created between the epidermis and the old cuticle during apolysis is called the **exuvial space**, a region where **molting fluid** is secreted by the epidermal cells. After apolysis, the epidermal cells begin to secrete the new cuticle. First deposited is the outer layer of the epicuticle, then the inner epicuticle is formed. This is followed by secretion of the procuticle. Thus, the new cuticle is produced starting with the outer layers and working inward. Now the molting fluid is activated, which digests the

old endocuticle. Up to 90% of the cuticle is digested by protease and chitinase enzymes, and recycled to help construct a new procuticle. After all the layers are in place, the insect produces a layer of wax that is secreted through pore canals onto the surface of the new epicuticle. This protects the insect from desiccation. Finally, the insect sheds its old cuticle, called the **exuviae**, in a process called **ecdysis**. Ecdysis is a tricky process, as the insect must escape from its old covering. Often it anchors the old integument and crawls out, and may use gravity to aid in its escape by hanging from a branch (Fig. 2.2). However, it must split the old cuticle somewhere that will allow escape, often in the head region. After it first escapes the old integument it is white or pale in color, and soft-bodied; such insects are said to be **teneral**. Finally, the molting insect must expand its body size while the new cuticle dries and hardens, because once it hardens not much more growth is possible until the next molt. So insects swallow air or water, and expand their body to swell to its maximum size while their body cover hardens. Slowly the insect cuticle hardens and darkens during a chemical process called **sclerotization**, which cross-links the proteins to create a new rigid exoskeleton. After this physical expansion and sclerotization, insects have some opportunity to add body tissue before they need to molt again.

BODY REGIONS

The principal body regions of insects are the head, thorax and abdomen (Fig. 2.3). Each region, or functional unit, is called a tagma (plural, **tagmata**), and the process of the individual segments functioning as a unit is called **tagmosis**. The presence of these three body regions is not always apparent, however. The

Fig. 2.2. A katydid (Orthoptera: Tettigoniidae) undergoing ecdysis. Note how it uses gravity to escape from the old integument after the cuticle split near the head. Typically, it remains hanging until the new cuticle dries and partially hardens.

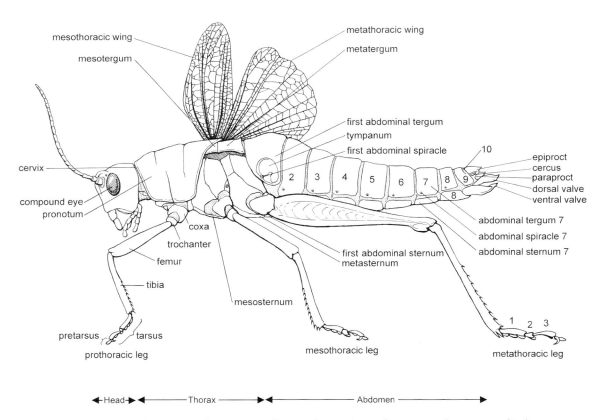

Fig. 2.3. The principal body regions and components of a typical insect (a grasshopper, *Romalea microptera* [Orthoptera: Romaleidae]). The abdominal segments are numbered.

head may be small or hidden from view when examined from above, or the front wings may cover both the thorax and abdomen, giving the impression that there is only a single large segment. Alternatively, when viewed from below it is evident that insects consist of quite a large number of segments, some of which are fused. Even after fusion, it usually is possible to recognize three thoracic segments and about 11 abdominal segments. The six segments that fused to form the head are mostly unrecognizable. Legs, wings, and antennae are the appendages that are most evident on the tagmata, but often mouthparts and cerci can be seen.

THE HEAD

The **head** is the most anterior (pertaining to the front) of the principal body regions. It contains the feeding appendages, the most important sensory organs, and the brain. The eyes and antennae are perhaps the most obvious sensory organs (Fig. 2.4), but other organs such as the mandibular and maxillary palps also are important in 'smelling' and 'tasting'. They are conveniently situated to be in close proximity, and connected to the brain, of course.

Insects have various types of eyes. In adults, a large pair of **compound eyes** is usually quite evident, with one eye on each side of the head. However, they may also have less complex structures, the ocelli and stemmata. **Ocelli** are very small eyes that are found only on the top of the head. **Stemmata** are also small eyes, though only found on the side of the head of insect larvae and often occurring in clusters. The visual acuity of the eyes varies, with the compound eyes providing the best resolution.

The **antennae** are highly modified, paired, segmented sensory appendages. The components of the antennae are usually divided into the **scape**, or the basal (point of attachment) segment, the **pedicel**, the second segment, and the **flagellum**, the remainder of the antenna. The different shapes of the antennae often are useful in distinguishing different taxa. Some common types of antennae are shown in Fig. 2.5. More importantly, of course, the shapes reflect the ecology, behavior and sensory needs of the different insects. The antennae are greatly reduced on some immature insects.

The **mouthparts** are at least as diverse as the antennae, and are formed from numerous segments. The most primitive type of mouthpart is the chewing type

(Fig. 2.4), and although the components often are quite modified and specialized in later evolving mouthparts, often the original structures are recognizable. With **chewing mouthparts**, the most important structures are the **mandibles**, which are heavily sclerotized jaws, and **maxillae**, a secondary pair of jaws. The **labrum** is a flap that closes the mouth cavity from above, and the **labium** similarly closes the mouth cavity from below. **Palps** associated with the maxillae and labium are small antennae-like sensory structures.

Insects have **ectognathous** mouthparts, which means that the major components are normally visible (Fig. 2.6), despite variation in form. In addition to the chewing type of mouthparts, **sucking mouthparts** are common. In this case, one or more parts of the feeding structures are modified into a tubular structure that allows ingestion of liquid. One important variation on sucking mouthparts is called **piercing-sucking mouthparts**. These mouthparts are pointed at the apex and can pierce plant or animal tissue to obtain liquid food (Figs. 2.7, 2.8). Insects with piercing-sucking mouthparts inject saliva and ingest either plant sap, blood, or liquified body contents from their hosts. Sometimes that sap comes from individual cells, which is the case with thrips (Thysanoptera), but other times it comes from the plant vascular system (either xylem or phloem), which is the case with bugs (Hemiptera). Some beetles called weevils have chewing mouthparts that are found on the tip of an elongate, beak-like structure. Although superficially resembling piercing-sucking mouthparts, they serve only to allow the beetle to chew deeply into its host plant. Another major type of sucking mouthparts consists of **siphoning mouthparts** or a **tubular proboscis**. The proboscis is composed of two maxillary galea that when held together form a hollow food tube. This siphoning structure is best developed in the moths and butterflies (Lepidoptera) (Fig. 2.9), which can uncoil a long tube to collect nectar from deep within flowers. This is also evident, to a lesser degree, in bees (Hymenoptera, various families). Some flies have yet another type of feeding apparatus called **sponging mouthparts** (Fig. 2.10). In these flies, liquified materials are sucked up without penetration. Finally, it is worth mentioning **mouth hooks**, which are rather unique feeding structures found in the larvae (maggots) of some higher flies. Mouth hooks tear the tissues of the host, allowing ingestion of semi-liquid material.

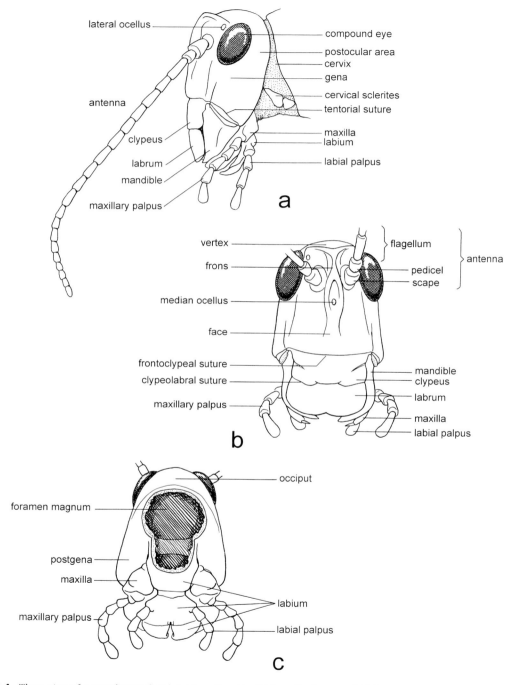

Fig. 2.4. Three view of a grasshopper head: (a) from the side; (b) from the front; and (c) from the back (the shaded region is the point of connection to the thorax).

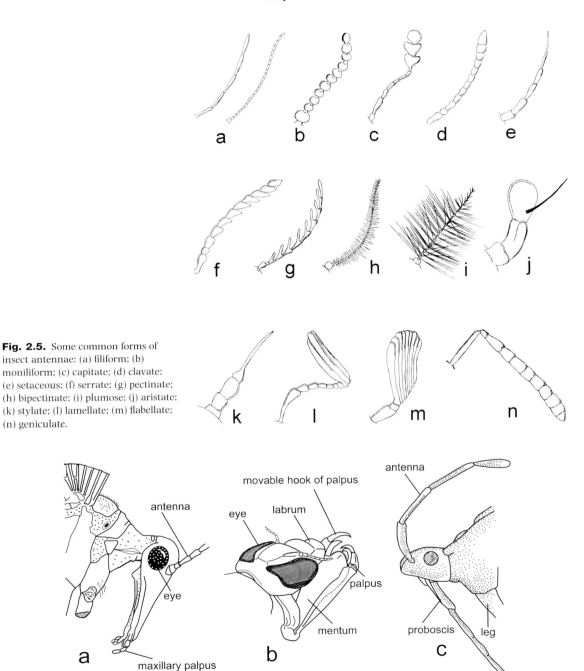

Fig. 2.5. Some common forms of insect antennae: (a) filiform; (b) moniliform; (c) capitate; (d) clavate; (e) setaceous; (f) serrate; (g) pectinate; (h) bipectinate; (i) plumose; (j) aristate; (k) stylate; (l) lamellate; (m) flabellate; (n) geniculate.

Fig. 2.6. The orientation of insect mouthparts is variable. Shown here are typical (a) hypognathous, or downward-oriented mouthparts of a scorpionfly (Mecoptera); (b) prognathous, or anterior-oriented mouthparts of a dragonfly larva (Odonata); and (c) opisthognathous, or posterior-oriented mouthparts of a bug (Hemiptera).

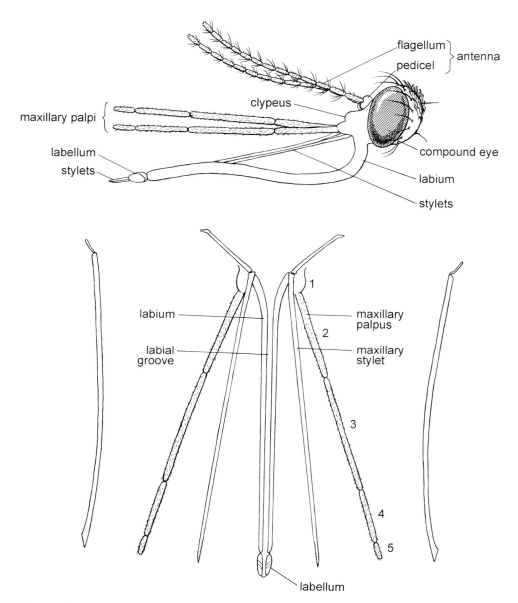

Fig. 2.7. Diagram of the piercing-sucking mouthparts found in mosquitoes (Diptera: Culicidae). In the lower part of this figure the mouthpart components are teased apart to show the mouthpart elements. Note that, although they are derived from same elements found in chewing mouthparts, they are very different in form.

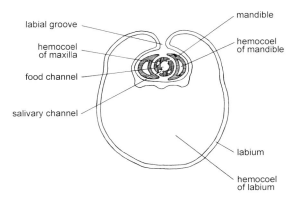

labial groove

hemocoel
of maxilla

food channel

salivary channel

mandible

hemocoel
of mandible

labium

hemocoel
of labium

Fig. 2.8. Diagram of a cross-section of the piercing-sucking mouthparts found in cicadas (Hemiptera: Cicadidae). As is the case with mosquitoes, they are highly modified from the mouthpart components found in chewing insects, and they form hollow tubes through which salivary liquids can be secreted and food liquids can be withdrawn.

THE THORAX

The **thorax** is found immediately behind the head, and is the body region that bears the jointed legs and (if present) the wings. The thorax is divided into three segments, called (from front to back) the **prothorax**, **mesothorax**, and **metathorax**. At least in adults, each segment bears a set of legs. Wings normally are attached to the mesothorax and metathorax (two pairs of wings), or just to the mesothorax (one pair of wings) (Fig. 2.11). The most common modifications of the thorax are associated with the prothorax, which often is greatly enlarged, sometimes expanded forward to cover part or all of the head, or expanded back over the mesothorax and metathorax.

The **legs** have six basic components called (beginning at the base) the **coxa, trochanter, femur, tibia, tarsus**, and **pretarsus** (Figs. 2.3, 2.13). The largest segments are the femur (plural, femora) and the tibia (plural, tibiae), but the tarsus (plural, tarsi; essentially the 'foot') is important, as is the apex of the leg, the pretarsus, which often takes the form of hooks, which we call claws. The legs are highly modified for running, digging, leaping, swimming, capturing prey, and other functions. The immature stages of some insects lack legs; fly larvae (maggots) are a good example of legless

forms. Other immatures have leg-like structures usually called **prolegs**; caterpillars are a good example of forms bearing prolegs. Prolegs assist with walking, but are simple extensions of the abdomen and are not jointed like the true legs.

The most unique structures in insects are the flight structures, the **wings**, although they are not found in all taxa and are undeveloped or incompletely developed in the immature stages. Wings are actually expansions of the lateral body walls, thickened outgrowths of the integument on the mesothorax and metathorax. Wings (Fig. 2.12) usually are membranous and usually transparent, but may be pigmented. The wings also bear veins. The **wing veins** act as tubular braces to stiffen the wing and support the thin interveinal areas, which are called **cells**. The veins allow blood to be pumped through them, which is critically important during expansion of the wings when the insect transforms into the adult stage. The veins also contain trachea and nerves. There also are folds in the wings: flexion lines, where bending occurs during flight, and folding lines, where the wing folds during periods of rest. Most insects, when at rest, close their wings over their body. Longitudinal folding is most common, but in beetles (Coleoptera) and earwigs (Dermaptera) the wings are folded transversely so they can fit beneath the elytra. The arrangement of wing veins and cells varies among taxa, and are useful characters for identification of insects. Sometimes the wings are modified. The most common modifications are basal thickening of the front wings (e.g., the **hemelytra** of some Hemiptera) or entirely (e.g., the **elytra** of Coleoptera (Fig. 2.13)). Often, the hind wings are smaller than the front wings. In flies, the hind wings are reduced to small vibrating structures called **halteres** that help to provide stability when these insects take flight.

Expression of the wings varies: length, width, shape, and distribution of veins are highly variable among species. Sometimes there is considerable variability even within species. Aphids, for example, may have winged and wingless generations depending on the season of the year, aphid density, and the condition of the host plant. Winged individuals are called **alatae** (adjective, alate), and wingless forms are called **apterae** (adjective, apterous). In some species, populations vary, and wings may be shortened and nonfunctional in some areas and not in others. When both pairs of wings are reduced, they are said to be **brachypterous** or micropterous.

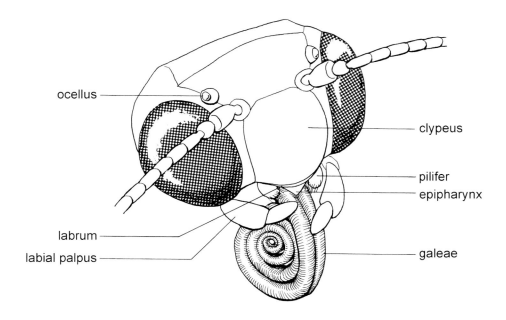

ocellus

clypeus

pilifer
epipharynx

labrum

galeae

labial palpus

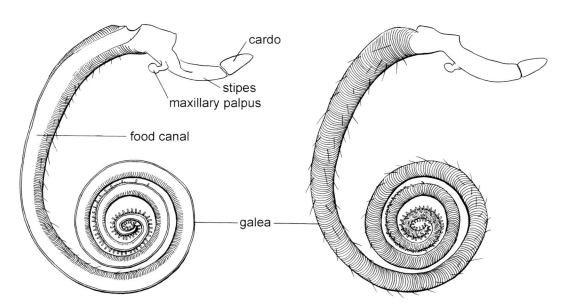

cardo

stipes
maxillary palpus

food canal

galea

Fig. 2.9. Diagram of the coiled, siphoning mouthparts of a butterfly (Lepidoptera), which are often called a proboscis. The lower portion of the figure shows the component parts and removal of one galea from the proboscis to show the food canal.

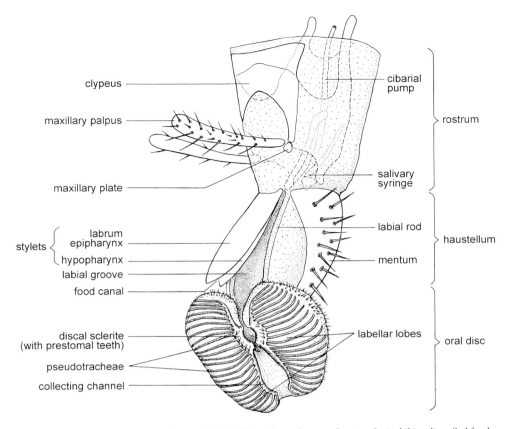

clypeus

maxillary palpus

maxillary plate

stylets {
labrum
epipharynx
hypopharynx

labial groove

food canal

discal sclerite
(with prestomal teeth)

pseudotracheae

collecting channel

cibarial
pump

rostrum

salivary
syringe

labial rod

haustellum

mentum

labellar lobes

oral disc

Fig. 2.10. Diagram of the sponging mouthparts of a fly (Diptera), another mechanism for imbibing liquefied food.

THE ABDOMEN

The **abdomen** is the third and largest major body region, and consists of several to many segments (Fig. 2.13). It contains most of the digestive and reproductive organs. Many of the abdominal segments bear small lateral openings called **spiracles**. The spiracles are the entry point for air that enters into the insect's ventilatory (breathing) system. Carbon dioxide is also expelled from the insect through the spiracles. Nearly all insects ventilate the tracheae by active body and muscular movements, although in very small insects passive diffusion of air into trachea is adequate to ventilate the body. Among the Hymenoptera that sting or oviposit within other insects, the abdomen is greatly constricted where it joins the thorax. This basal con-

striction of the abdomen is called the **petiole**, and apparently allows the insect to better flex its abdomen both for parasitizing its prey and for defense.

Some insects such as grasshoppers and moths possess a large acoustical organ called the **tympanum** (plural, tympana) on each side of the first abdominal segment (Fig. 2.3). These acoustical organs can be found elsewhere, however, including on the front legs of insects closely related to grasshoppers, the katydids and crickets. Among the Orthoptera (grasshoppers, katydids and crickets), this organ is used for mate location. However, among moths it is used to detect echo-locating bats, and to avoid predation.

Additional structures called **caudal appendages** are sometimes found terminally. A pair of lateral appendages called **cerci** sometimes occurs near the

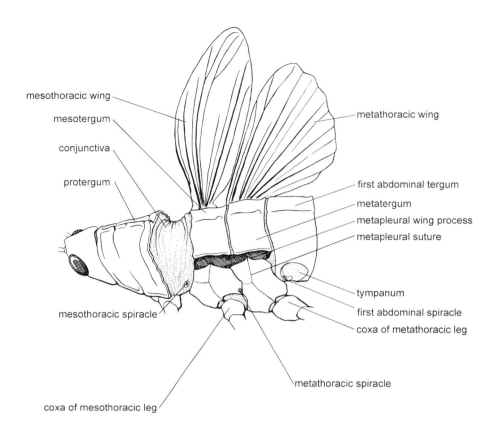

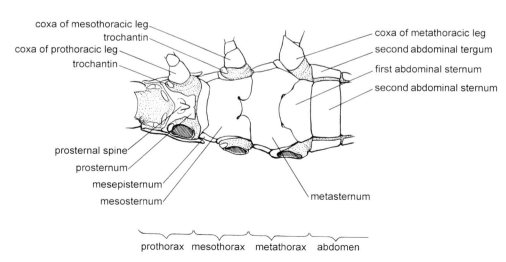

prothorax   mesothorax   metathorax   abdomen

Fig. 2.11. The thorax of a grasshopper viewed (a) from above, and (b) from below. Note that each of the three thoracic segment bears a pair of legs, but that the wings are attached to the mesothoracic and metathoracic segments. This is the typical arrangement in insects.

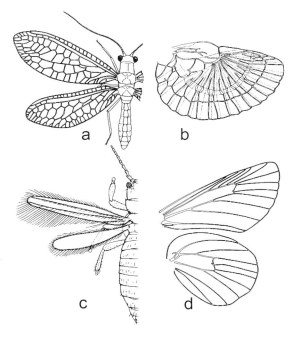

Fig. 2.12. Representative wings found in insects: (a) lacewings (Neuroptera: Chrysopidae), a group bearing numerous cells in their wings; (b) earwigs (Dermaptera), a group with folding hindwings that collapse to fit under the small forewings; (c) thrips (Thysanoptera), a group with few veins but considerable fringe; (d) butterflies, a group with relatively 'typical' wing venation.

end of the abdomen, which may be quite small or quite large, depending on the taxon. Also found terminally on the abdomen are the **genitalia**, structures associated with mating. Many insects also have a well-developed structure located terminally for depositing their eggs, or ova, called the **ovipositor**. For insects that deposit their eggs within a substrate the ovipositor may be quite elongate. The ovipositor of some wasps, bees, and ants is modified into a defensive structure called a **sting**. Only females possess ovipositors, so only females can sting. Stinging insects have associated with the sting a gland that produces **venom**. Venoms are biologically active substances that are injected. The venom can be quite toxic, in some cases equaling or exceeding that from the most poisonous snakes. Insects are capable of injecting much less venom than snakes, of course.

INTERNAL ANATOMY

The insect body generally is an elongate, cylindrical structure (Fig. 2.14). Its shape is determined by the integument and associated muscles. As is the case with the wide variation in external appearance among taxa, there is considerable variation in internal structure. The same basic systems are normally present in all insects, of course, but their expression varies with their behavior and ecology. Similarly, the systems found in the immature and adult stages may be quite similar in some insects or quite different in others, depending not only on the development of reproductive structures in adults but also on the feeding ecology on the insects. There are several important internal systems that together provide many of the unusual characteristics that make an insect different from most other animals.

MUSCULAR SYSTEM

Muscles are found throughout the insect body, though they are best developed in the head (for ingestion) and the thorax (for locomotion) (Fig. 2.15). Muscles are pale in color and striated, forming flat or strap-like bundles. The naming system of muscles is confusing. They tend to be named in some cases for the function they perform, but in other cases based on their point of attachment to the integument. Skeletal muscles tend to be segmented, corresponding to the body segments. The most important skeletal muscles are the cephalic, thoracic (including flight) and abdominal muscles. They are anchored to various points of the integument, including the infoldings (apodemes) of the integument. Cephalic muscles control the activities of the head, such as moving the head, moving the mouthparts, and moving the antennae. Thoracic muscles control the legs and wings.

The wing muscles may act directly on the wing (so-called **direct flight muscles**) or indirectly (so-called **indirect flight muscles**). The indirect flight muscles are of two principal types: the dorsoventral and the longitudinal (Fig. 2.16). They alternately deform the shape of the integument, which in turn provides energy to help power the wings. In most insects, when the dorsoventral muscles contract, the upper portion of the thoracic segment (notum) is pulled downward, driving the wings upward (Fig. 2.16a). This occurs because the pleural (side) region of the insect acts as a fulcrum, and the wings respond like a

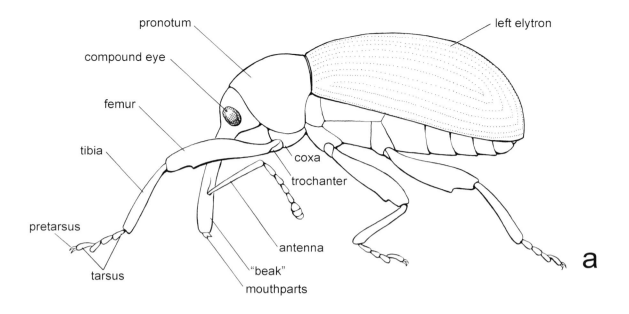

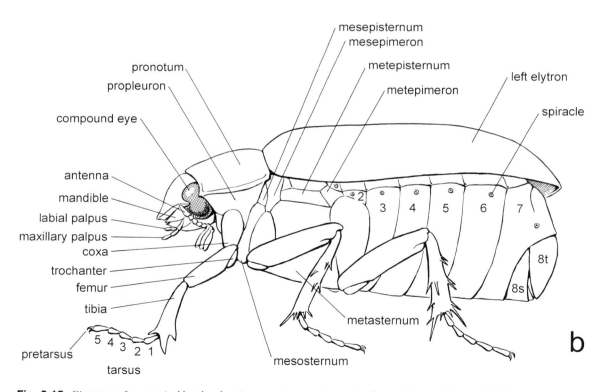

Fig. 2.13. Diagrams of two typical beetles showing some important aspects of morphology: (a) a weevil (Coleoptera: Curculionidae), with its chewing mouthparts at the end of an elongate 'beak' and thickened front wings (elytra) covering some of the thorax and all of the abdomen from above; and (b) a June beetle (Coleoptera: Scarabaeidae), with more typical mouthpart arrangements and the spiracles exposed along the sides of the abdomen.

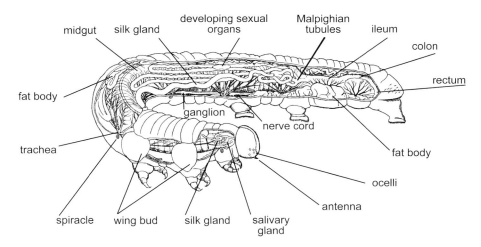

Fig. 2.14. A typical caterpillar (Lepidoptera) with integument partially removed to expose the internal anatomy.

child's 'seesaw' or 'teeter-totter'; a downward push on one end elevates the other end. Likewise, when the longitudinal muscles contract, the upper portion of the notum is arched upward, driving the wings downward (Fig. 2.16b). The indirect muscles usually are more important for flight than the direct muscles because some of their energy is captured by the elasticity of the cuticle. The direct flight muscles pull directly on the wings. Although they provide the muscle for lift in some insects, usually they act mostly to control twisting of the wings, allowing for propulsion.

The **abdominal muscles** connect abdominal segments and serve to bend and contract the abdomen. In addition, there are **visceral muscles** associated with such organ systems as the gut, heart, and gonads. They may encircle the organs, or run longitudinally, but they usually are small muscles and not well ventilated.

FAT BODY

The fatty tissues of insects are found widely in the body, but often are collected into tissues called **adipose tissue** or **fat body** (Figs. 2.14, 2.15b). These tissues consist principally of lipids, glycogen and protein. These nutrient stores are used when a great deal of energy or nutrients are needed, such as during metamorphosis or egg production. They are not simply a storage area for energy, however, and sometimes assume other functions, such as storage of insect waste products (uric acid) or as a site for harboring beneficial microorganisms (in special cells called bacteriocytes or mycetocytes). Usually, fat bodies are whitish or yellowish in color, and form irregular masses or lobes. They are especially pronounced in immature forms of insects.

DIGESTIVE SYSTEM

The **digestive system** (also called alimentary system or alimentary canal) extends the length of the body from the mouth to the anus. In general, is consists of a straight tube, though there are various points where it may be dilated, convoluted, or branched (Fig. 2.17).

The digestive system begins at the mouth or buccal cavity, and salivary glands often provide enzymes and lubricants at this point. Generally, the alimentary system is said to consist of three parts: the **foregut**, **midgut**, and **hindgut**. Sometimes the buccal cavity is considered to be part of the foregut. The foregut (also known as the stomodeum) also consists of the pharynx, esophagus, crop, proventriculus, and esophageal valve. This section serves mostly to transport, store, and sometimes filter the ingested food, which then enters the midgut (ventriculus). The midgut is the 'stomach' of insects, and is the principal site of digestion. It may possess gastric caeca, bladder-like pouches that increase the surface area of the midgut. The posterior region, or hindgut, consists of an anterior portion of intestine, or ileum, followed by the posterior intes-

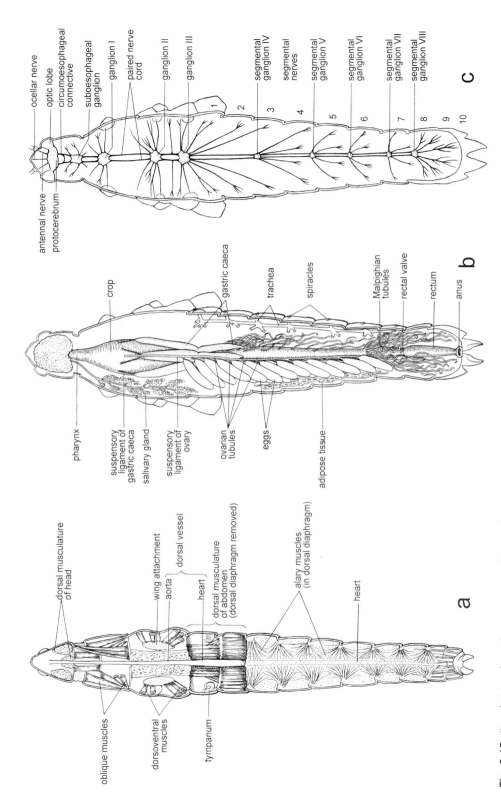

Fig. 2.15. Dorsal views of grasshopper anatomy showing the principal elements of the (a) muscular; (b) digestive and reproductive; and (c) nervous systems.

c

- ocellar nerve
- optic lobe
- circumoesophageal connective
- suboesophageal ganglion
- ganglion I
- paired nerve cord
- ganglion II
- ganglion III
- segmental ganglion IV
- segmental nerves
- segmental ganglion V
- segmental ganglion VI
- segmental ganglion VII
- segmental ganglion VIII

- antennal nerve
- protocerebrum

1
2
3
4
5
6
7
8
9
10

b

- crop
- gastric caeca
- trachea
- spiracles
- Malpighian tubules
- rectal valve
- rectum
- anus

- pharynx
- suspensory ligament of gastric caeca
- salivary gland
- suspensory ligament of ovary
- ovarian tubules
- eggs
- adipose tissue

a

- dorsal musculature of head
- wing attachment
- aorta
- dorsal vessel
- heart
- dorsal musculature of abdomen (dorsal diaphragm removed)
- alary muscles (in dorsal diaphragm)
- heart

- oblique muscles
- dorsoventral muscles
- tympanum

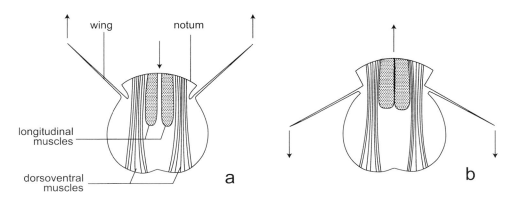

Fig. 2.16. Diagram of the function of indirect flight muscles: (a) contraction of the dorsoventral muscles pulls the notum down, driving the wings up; (b) contraction of the longitudinal muscles reverses the process, driving the notum up and the wings down.

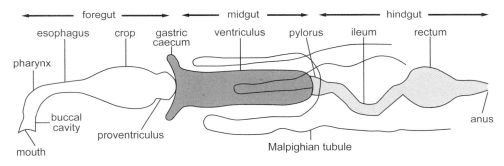

Fig. 2.17. Diagram of the insect digestive (alimentary) system.

tine, or colon. The anterior and posterior regions of the intestine are not always differentiated. The beginning of the hindgut is marked by the attachment of Malpighian tubules. The hindgut terminates with a chamber called a rectum, which regulates movement of waste materials to the anus.

Digestion occurs in the crop portion of the foregut, or in the midgut. The midgut of many insects is lined with a semipermeable membrane called the **peritrophic membrane**. The peritrophic membrane protects the midgut lining from abrasion by ingested food, but allows enzymes secreted by the midgut to move through this protective layer and to attack the food and digest it. Some insects, and many other arthropods, also display **extraoral digestion**, or digestion that occurs before consumption. Saliva containing enzymes is secreted from the salivary glands or regurgitated

from the midgut, and can begin the digestion process even before the food is ingested. Digestive enzymes are injected into the prey by predatory insects such as assassin bugs (Hemiptera: Reduviidae) or secreted onto the surface of the food by some flies. Thus, the insects suck up both the partially digested (predigested) food and the digestive enzymes, with the latter available to be recycled for additional digestion on the inside of the insect. The digestive processes of insects are adapted for their diets, of course, so the enzymes that are present reflect whether they are feeding on plant or animal material, and on such specialized but hard-to-digest food items as hair, feathers, wax, and wood. In some cases, most notably termites, symbiotic microorganisms are present in the insects to assist in digestion.

Absorption of digested food occurs principally in the midgut, but also in the hindgut. Most food is broken down into simple sugars and amino acids before absorption, although some simple lipids are absorbed without digestion. Water is also absorbed from the feces in the final region of the alimentary canal, the rectum, before the water materials are excreted. Water is often a limiting factor for insects due to their high surface to volume ratio, which favors water loss. Both their ventilatory system and the coating of their integument favor water conservation, but extraction of water from the food is also vital.

It is difficult to generalize about insects because they are so diverse, and water conservation is an excellent example of their diversity. The piercing-sucking, sap-feeding insects tap into the vascular system (usually xylem but sometimes phloem) of plants, where water is transporting sugars and other nutrients around the plant. The nutrients in the vascular tissues of the plant are often diluted by the aqueous component of the sap, so the problem that the sap feeders face is too much water, and they must deal with quickly eliminating the water but conserving the nutrients. Many sap-sucking insects have a **filter chamber**, a bypass of part of the midgut to allow excess fluid to be quickly passed through the system. Some nutrients are absorbed, but not all, so insects with a filter chamber tend to excrete copious amounts of fluid still containing a good deal of sugars and amino acids, but especially sugars. The small sticky droplets of excreta that commonly rain down gently beneath trees in the summer result from this process. This sugar-laden sticky secretion is called **honeydew**. Honeydew is a nuisance when it drips onto automobiles, but also is a food resource. Fungi growing on the honeydew deposited on plants are called sooty mold, and develop into a black layer that impede photosynthesis of the foliage. On the other hand, some insects feed on the honeydew, so perhaps it benefits insect biodiversity.

CIRCULATORY SYSTEM

The **circulatory system** of insects usually consists mostly of a single **dorsal vessel** to move the blood through the body. It extends from the posterior region forward to the head, and consists of two principal components, a pumping organ or heart posteriorly, and a large vessel for transport, or aorta, anteriorly (Fig. 2.18). The blood, or **hemolymph**, circulates freely in the spaces between the organs once it leaves the aorta, rather than being pumped though a complex system of arteries and veins, as is found in mammals and birds. After circulating, it is again taken up by the heart from the posterior of the insect through openings called ostia, and propelled anteriorly. However, diaphragms in various parts of the body also serve to direct the flow of blood, and blood is also pumped through the veins of the wings. In some insects, accessory pulsatile organs occur to increase movement of the hemolymph. These sac-like structures function independently of the heart, and assure that blood flows into the appendages.

The hemolymph contains blood cells, called **hemocytes**. Insect blood cells act to plug up wounds, and also can function like the lymph system of vertebrates, engulfing or encapsulating foreign particles, thereby protecting the insect from invaders. The insect hemolymph also serves to transport nutrients and hormones though the body. However, the blood is not too involved

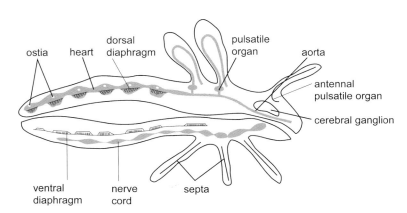

Fig. 2.18. Diagram of the insect circulatory system.

in the transport of oxygen, which is not very soluble in hemolymph. Its role in transport of carbon dioxide is greater because it is more soluble.

The hemolymph contains large quantities of a disaccharide, **trehalose**. Ingested food is broken down to monosaccharide sugars, primarily glucose, in the midgut. The monosaccharides will pass readily through the midgut epithelium into the blood if the concentration of monosaccharide level is lower in the blood. To maintain this passive flow of nutrients (no energy is expended moving the sugars across a membrane from a high concentration to a low concentration) insects maintain a low concentration of monosaccharides in the blood by converting the glucose to trehalose. Because many insects use trehalose to facilitate nutrient uptake, and trehalose is circulated through the body to provide nutrition to cells elsewhere, trehalose is viewed as the insect's 'blood sugar.' Nutrients similarly are moved into the hemolymph, and then circulated through the body to where they can be used, using this passive transport system.

VENTILATORY SYSTEM

The **ventilatory** (respiratory) **system** serves to exchange gases, though insects lack anything as sophisticated and centralized as the lungs found in mammals and birds. In insects, gas exchange is accomplished through a system of **trachea** (large vessels) and **tracheoles** (small vessels). The ventilatory system is analogous to a system of arteries, veins, and capillaries, with air (rather than blood) provided to the organs, tissues and cells of the body. The trachea take on a silvery appearance when filled with air. The trachea open to the outside of the body through openings called **spiracles**. The spiracles have closure mechanisms that allow the insect to have control over gas exchange, but this is primarily a water conservation device. Rather than having the internal regions of the insect continuously exposed to the potentially desiccating outside air via the tracheal system, the insect releases carbon dioxide and takes in air in pulses, which minimizes water loss. The spiracles are arrayed laterally along the thorax and abdomen. Although some trachea penetrate deeply into the insect to provide ventilation to internal organs, most insects have interconnected longitudinal tracheal trunks along the length of the body. These longitudinal trunks are often the largest trachea, with smaller branches originating from them, and providing gas exchange throughout the body. A cross-section of an insect's abdomen, illustrating these relationships, is shown in Fig. 2.19. Air sacs occur in many insects. **Air sacs** are dilations of the trachea, and constitute a significant air reservoir. Pressure on the air sacs provided by muscles and integ-

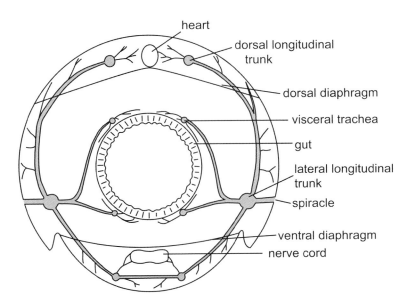

Fig. 2.19. Diagram showing cross-section of an insect's abdomen showing the principal elements of the ventilatory system.

ument helps to move air through the ventilatory system.

Air exchange is modified in aquatic insects. In aquatic insects, the exchange of gases can occur in tracheal gills; these may be found in several locations including the normal positions of the spiracles, or from the sides of the body, or associated with the rectum. Sometimes gas exchange occurs directly through the cuticle. Some aquatic fly larvae, particularly mosquitoes, have their functional spiracles located posteriorly in a structure called the **respiratory siphon**. The respiratory siphon makes contact with the air, allowing gas exchange, whereas the rest of the insect's body is submerged in water. Some water-dwelling insects have a film of water surrounding them when they are submerged, and exchange gases with this bubble.

NERVOUS SYSTEM

The **central nervous system** of insects is primarily located ventrally along the length of the body, so insects are said to have a ventral nerve cord (Fig. 2.15c). This is quite different from vertebrates, which have a dorsal nerve cord. The ventral nerve cord of insects consists of a series of ganglia, usually one per body segment, connected by a double nerve cord. A **ganglion** (plural, ganglia) is a mass of nervous tissue.

The insect nerve cell consists of a cell body, called the **perikaryon**, and projections called **axons** (Fig. 2.20). The axon often has branches, and ends in a branched or tree-like structure called the **terminal arborization**. Nervous impulses are transmitted along the axon after being received by one end, called the **dendrite**. When the signal reaches the other end, the terminal arborization, a neurotransmitter chemical is released to bridge the minute gap between it and the next nerve cell (dendrite). The gap or space across which the neurotransmitter diffuses is called a **synapse**. The signal, once received by the next nerve in the series, travels along the length of the axon until it reaches the other end of the nerve and stimulates production of more neurotransmitter, stimulating another nerve cell. Thus, signals move sequentially from nerve to nerve until they are received at the appropriate reception site.

One of the most important neurotransmitters is **acetylcholine**. Once a stimulus travels the length of an axon, acetylcholine is produced at the end of the nerve and diffuses across the synapse to stimulate another nerve, thereby continuing the signal. Normally the nerve releases an enzyme, **acetylcholine esterase**, after it releases acetylcholine. This enzyme stops the impulse and restores the nerve to its original condition so it will be receptive to another signal. Meanwhile, of course, the signal is moving along from nerve to nerve, transmitting the signal, followed by subsequent releases of acetylcholine esterase and restoration of nerves to a resting state. Some of the modern insecticides are acetylcholine esterase inhibitors, so they act on the nervous system by preventing release of the enzyme that restores nerves to their resting state. Other commonly used insecticides act by disrupting the function of the nerve axon. Hence, in the popular literature many insecticides are called 'nerve poisons,' and unfortunately they sometimes affect the nerves of vertebrates as well as the nerves of insects.

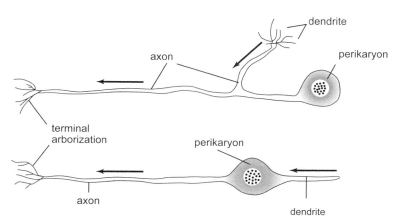

Fig. 2.20. Diagram of some insect nerve cells. Arrows indicate the direction of the nerve signal.

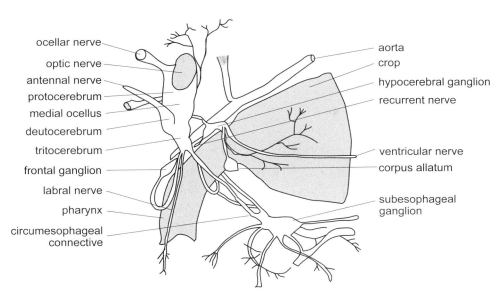

ocellar nerve
optic nerve
antennal nerve
protocerebrum
medial ocellus
deutocerebrum
tritocerebrum
frontal ganglion
labral nerve
pharynx
circumesophageal
connective

aorta
crop
hypocerebral ganglion
recurrent nerve
ventricular nerve
corpus allatum
subesophageal
ganglion

Fig. 2.21. Diagram of the insect brain, viewed from the side, showing the principal nerves and connection to the ventral nerve cord.

Nerves also radiate to all regions of the body, though each ganglion tends to serve principally the functions of the segment in which it is found. In the head, enlarged ganglia form the brain and subesophageal ganglion. The brain is located dorsally, and consists of three fused ganglionic masses called the protocerebrum, the deutocerebrum and the tritocerebrum (Fig. 2.21). The **protocerebrum** is the largest section of the brain, and receives the nerves from the compound eyes laterally and from the ocelli dorsally. Within the protocerebrum are the corpora pendunculata, sections of nervous tissue that are critical for the integration and coordination of behavior. Also in the protocerebrum are the pars intercerebralis, nervous tissue that contain neurosecretory cells used to communicate to endocrine glands. The **deutocerebrum** connects to the antennae of insects. The antennae of insects are the main sensory structures. Although insects are capable of vision, insects especially excel at chemical perception, and the antennae play a large (but not exclusive) role in this ability. Finally, the **tritocerebrum** connects to portions of the mouthparts. The tritocerebrum also connects to the subesophageal ganglion, the first ventral segment of the ventral nerve cord. Even more than the tritocerebrum, this ventral ganglion controls the mouthparts. Thus, the brain controls the impor-

tant functions of the head, including vision, smell, and feeding (in part). However, the brain is connected via nerves to the ventral nerve cord, and via neurosecretory cells to endocrine glands, so it serves as a coordination center for the entire body.

As mentioned previously, the nerves connect to the sense organs, and though there are many types of sense organs in insects, the compound eyes and antennae are most important. The eyes of insects are quite unlike the eyes of humans and wildlife. Although we do not fully understand insect vision, it is apparent that the **compound eye** actually sees multiple images. Each compound eye consists of several to many ommatidia. Each **ommatidium** is a separate visual unit, clustered together to form a compound eye. Insects can detect form, color and movement with their compound eyes, but their ability to see form is compromised by the design of the eye, which tends to portray objects as broken into small pieces. This type of vision is called mosaic vision, and how well the brain integrates the multiple images is not really known. Insects generally see a different portion of the spectrum than humans, instead being very receptive to short-wave (ultraviolet) and less sensitive to long-wave (red) wavelengths. Insects also can detect the polarization of light, something many animals cannot detect. Polarized light is parallel wavelengths moving in a single plane, normally parallel to the sun's rays. The ability to detect the location of the sun by detecting polarized light undoubtedly aids in navigation.

Interestingly, insects have other eyes besides the large **compound eyes** that generally are obvious dorsally and laterally on the head. Two types of simple eyes are also found on many insects: ocelli and stemmata. **Ocelli** are very small eyes that are sensitive to changes in light intensity, but insensitive to form. Generally found only on the top of the head, they are thought to keep the insect oriented properly, especially when flying. **Stemmata** are also small eyes, though only found on the side of the head of insect larvae and often occurring in clusters. They function like compound eyes, but are thought to have less visual acuity because they are few in number.

Antennae are rich in nerves. This, combined with their shape, makes them sensitive to certain stimuli. For example, the feathery antennae of some insects allow them to have a great deal of surface area and to be receptive to chemicals (many beetles and moths) or sounds (male mosquitoes in search of females). Other senses provided by antennae include touch, temperature, and humidity perception. The second antennal segment, the pedicel, contains a mass of nerve cells called **Johnston's organ**. This organ detects movement of the flagellum, and is thought to measure speed while flying, to detect ripples in the water for swimming insects, and several other functions.

Other sense organs are scattered over the insect's body. They are especially numerous on the insect's appendages. Insects lack a nose or tongue, so they smell and taste with various sense organs located not only on their antennae, but on their feet, and sometimes elsewhere. Thus, although insects have many of the same sensory abilities as humans and wildlife, they are not as visually oriented as vertebrate animals, and more chemically oriented.

VISION

The most important visual organs of insects are the compound eyes. Externally, the compound eye appears to be composed of a number of hexagons, called **facets**, but this is just the surface (lens) of a number of individual units called **ommatidia**. The number of ommatidia in an eye varies greatly among species. Each ommatidium (Fig. 2.22a) consists of several

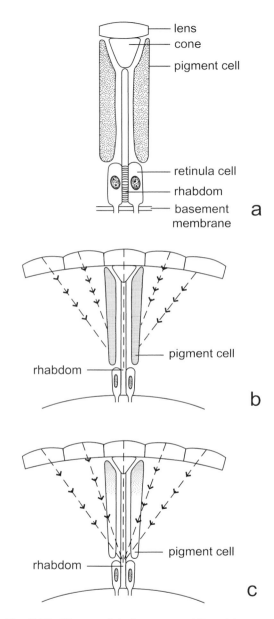

Fig. 2.22. Diagram of the insect ommatidium: (a) its component parts; (b) blocking by pigment of light entering from the lenses of adjacent ommatidia in a compound eye; and (c) migration of pigment, allowing light from adjacent ommatidia to stimulate the rhabdom.

structures, including the cornea or lens, the cone or other structures that bend or shape the light, receptor cells, and usually pigment-containing cells. Light is received by a light-sensitive portion of the retinula cells called the rhabdom or rhabdomere. The retinula cells connect directly to the insect brain without involving synapses, so their response time is short. Light stimulates a light-sensitive pigment called rhodopsin when the retinula cells are illuminated, and this stimulus is translated into a nervous impulse. The pigment cells around the retinula cells shield the photoreceptors from light (Fig. 2.22b), thereby reducing the ability of light from nearby ommatidia to stimulate the photoreceptor. This shielding effect works well during periods of adequate light, but insects sometimes need to be active in low light or darkness, so the shielding pigment moves (migrates) and allows more light to enter from adjacent ommatidia (Fig. 2.22c), thereby allowing the insect to perceive things better. There is a cost to this 'night vision', however, as visual acuity diminishes under such circumstances.

The compound eyes are best at detecting motion. With several to many individual ommatidia, stimulation of different ommatidia surely translates into motion detection. Visual acuity is lacking, and most insects probably lack the ability to discern shape very well, if at all. On the other hand, insects can perceive color, ultraviolet (UV) light, distance, and polarized light. Light polarization is related to the location of the sun, and is used as a navigational aid.

Ocelli and stemmata assist in vision, but are not very effective visual organs. Ocelli, found on the top of the head, are sensitive to changes in light intensity, but insensitive to form. Stemmata, found on the side of the head of insect larvae, function like compound eyes, but are thought to have less visual acuity because they are few in number.

GLANDULAR SYSTEMS

There are many important secretions from various glands in the insect body. **Exocrine glands** release chemical to the outside, whereas **endocrine glands** release chemicals internally. Among the important exocrine glands are the salivary glands, silk glands, poison glands, odor glands, wax glands and pheromone glands. **Salivary glands** are located in the head and thorax and empty into the region of the mouth. Sometimes the salivary glands are very large. **Silk glands** similarly can be quite large, depending on the silk production of the insects. Silk glands produce fibrous protein material, which is often used to construct cocoons. Normally, silk is released from the labium, but in some insects the tarsi or anus is the source. **Poison glands** sometimes produce secretions associated with setae, such as the venom found in caterpillars with stinging hairs or spines. Some secretions are sticky and aid in insect adhesion. Poison glands also may be found in association with the ovipositor or sting, and are modified accessory reproductive glands. **Odor glands** are quite variable. Those associated with glandular scales on wings of butterflies are quite small, whereas those associated with the osmeterium of swallowtail caterpillars, and the stink glands of stink bugs, are larger. Many glands have a defensive function. **Wax glands** are found commonly in Hemiptera and bees. Wax secretions range from a powdery covering to large and complex body coverings. Insects often produce **semiochemicals**, chemical messengers that are released from the body and perceived by another organism. Semiochemicals that function intraspecifically (within the same species) are called **pheromones**, and are produced in **pheromone glands**. The most important pheromones are sex pheromones, which bring together the opposite sexes for mating; alarm pheromones, which incite alarm among members of the same species, especially colonial insects; and aggregation pheromones, which bring together both sexes, usually to a location where there is ample food. Semiochemicals that function interspecifically (between different species) are called **allelochemicals**.

Among the endocrine glands are the corpora allata, corpora cardiaca, and the prothoracic glands (Fig. 2.23). **Corpora allata** are small exocrine glands associated with the brain that produce juvenile hormones. **Corpora cardiaca** are small organs associated with the brain that store and release prothoracicotropic hormones (PTTH) which activate the prothoracic glands. The **prothoracic glands** are found in the thorax of insects and secrete molting hormone (ecdysone) or a closely related ecdysteroid. Products from neurosecretory cells imbedded in the brain and other nervous tissue affect the function of these glands.

Hormones are chemicals that are released from endocrine glands and which affect another part of the body. The hormones cause differential expression of genes, with the hormones released periodically in pulses, and the timing of hormone release critical in

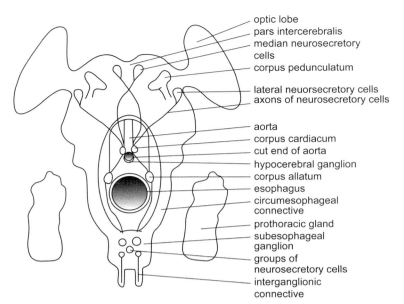

optic lobe
pars intercerebralis
median neurosecretory
cells
corpus pedunculatum
lateral neurosecretory cells
axons of neurosecretory cells

aorta
corpus cardiacum
cut end of aorta
hypocerebral ganglion
corpus allatum
esophagus
circumesophageal
connective
prothoracic gland
subesophageal
ganglion
groups of
neurosecretory cells
interganglionic
connective

Fig. 2.23. Diagram showing a cross-section of the insect brain with the most important endocrine glands and neurosecretory cells.

affecting development. There are periods of time (so-called 'critical periods') when various genes seem to be sensitive to the hormones, and can be activated to express various larval, pupal or adult characteristics.

Although insects produce many hormones that regulate many physiological and biochemical processes of their body, three of the most most important hormones in insects are juvenile hormones, PTTH, and ecdysone. **Juvenile hormones** from the corpora allata preserve larval characters during the molting cycle, and inhibit metamorphosis (change in body form) including the premature expression of imaginal discs. **Imaginal discs** are groups of undifferentiated cells in holometabolous insects that are expressed, usually resulting in formation of adult characters such as wings, when hormonal conditions no longer suppress their expression. When juvenile hormone concentrations are high during the critical periods, the insect tends to retain its juvenile characteristics although it may increase in size. For example, with high concentrations of juvenile hormone in the blood, small larvae molt to larger larvae, but not to pupae. **PTTH** (prothoracicotropic hormone, or brain hormone) stimulates the prothoracic glands to secrete ecdysone or a similar ecdysteroid. When environmental or physiological cues are appropriate (e.g., increasing day length or stretching of the abdomen) the brain may be stimulated to release PTTH, triggering ecdys-

one production by the prothoracic glands. **Ecdysone** (or ecdysone-like compounds called ecdysteroids) initiates molting, and controls differentiation of the tissues. When ecdysone is released and the juvenile hormone concentration is high, the insect molts to another immature stage. However, if the juvenile hormone titer drops, release of ecdysone stimulates molting to a new and different stage or body form because new genes are expressed. This change in developmental stage (form) is called **metamorphosis**. Thus, it is the combination of different hormones and their concentrations at critical periods that control the progression from immature to adult. Ecdysone stimulates molting, but the outcome of the molt is determined by the juvenile hormone concentration. Because insects display different patterns of development (i.e., ametabolism, hemimetabolism, holometabolism; see Metamorphosis, below) the specific patterns of hormone release vary, but the overall relationship of juvenile hormones and ecdysone is consistent among the different insects. Juvenile hormones and ecdysone also have important roles in regulating reproduction, affecting such important features as maturation of the reproductive organs and deposition of yolk in eggs. There are many other hormones found in insects, but these three control some of the most important physiological processes and help define the developmental characteristics of insects.

POLYPHENISM OR POLYMORPHISM

Polyphenism is the condition of having two or more discrete phenotypes, or physical appearances, without intermediates. This is also known as polymorphism. Although some phenotypes show gradual change in response to environmental variation, without producing discretely different subsets, some phenotypes produce discretely different (lacking intermediate forms) intraspecific variation, and it is this latter condition that is called 'polyphenism.' Often these insects are so different in appearance that they have been incorrectly described as different species, and it is only when they are cultured under controlled conditions that we learn that we can turn one 'species' into another 'species.'

Sometimes evolution can channel organisms into a stabilized phenotype, with little variation, that is well developed for a certain function or environment. Alternatively, it can lead to a more flexible phenotype, producing a more variable organism. An example of variable phenotypes is seen in Fig. 2.24, which shows the variable nature of the life cycle of an aphid, *Aphis fabae* (Hemiptera: Aphididae). Both winged and wing-

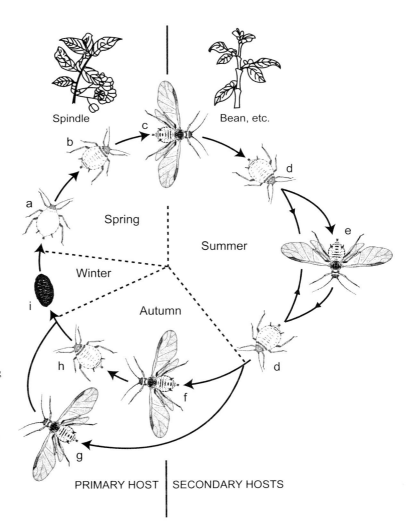

Fig. 2.24. The life cycle of a host plant-alternating aphid, *Aphis fabae* (Hemiptera: Aphididae): (a) a female emerging from an overwintering egg, producing (b) wingless female offspring without mating; (c) eventually winged females are produced that migrate to summer host plants; (d) wingless females continue to be produced for several generations until (e) host plant deterioration or (f) autumn, when winged females are again produced; back on the overwintering host (h) females mate with (g) males, the only time sexual reproduction occurs, and (i) overwintering eggs are produced.

Table 2.1. Some examples of polyphenisms, the environmental cue that triggers the polyphenic developmental switch, and the selective force that induces the polyphenic response (adapted from Nijout, H.F. 2003).

Polyphenism	Environmental stimulus	Selective agent (to which the polyphenism is an adaptation)
Seasonal	Photoperiod, nonlethal temperature	Lethal temperature, food scarcity
Phase in aphids	Crowding, temperature, photoperiod	Food quantity or quality
Phase in locusts	Crowding, food quality	Food quantity or quality
Phase in caterpillars	Food quality	Predation
Horn length	Food quantity or quality	Mating success
Wing length	Crowding, photoperiod	Food quantity or quality
Caste in ants (soldiers)	Food quantity or quality	Food quality, pheromones
Caste in ants (reproductives)	Pheromones, overwintering	Reproduction
Caste in bees	Nutrition, pheromones	Reproduction
Diapause	Photoperiod, nonlethal temperature	Lethal temperature

less forms, as well as sexual and asexual reproduction, occur seasonally. As is the case with aphids, often the species displaying polyphenism have more than one generation per year, thereby occurring during periods of the year with different weather conditions (e.g., spring and autumn), so it is exposed to different environmental stimuli. Polyphenisms are adaptations to reliable and somewhat predictable variations in the environment. Some examples are given in Table 2.1. Note that the inducing factor (the environmental cue) is not the same as the selective environment (the selective force that has resulted in the occurrence of the alternate phenotype).

Many insects have evolved a critical period in development when they are sensitive to inducing stimuli, and this critical period occurs well in advance of the occurrence of the alternative phenotype. Thus, token stimuli such as day length signal the onset of lethal temperatures, the selective agent.

Although environmental cues determine the expression of the polyphenic trait (e.g., presence or absence of wings, caste, diapause), the developmental switch that leads to alternative phenotypes is regulated by hormones. Insects express both larval and adult polyphenisms. The stimulus for adult polyphenism occurs in an immature stage, and at a hormone-sensitive period an alternative development pathway is initiated. The hormones controlling polyphenic development generally are the same as those controlling molting and metamorphosis: ecdysone and juvenile hormone. The hormones trigger different patterns of gene expression, leading to alternative phenotypes.

Most people see polyphenism expressed in the different forms, or castes, of social insects (Fig. 2.25). It can be advantageous to an organism to have the ability to develop multiple phenotypes. In social insects, there may or may not be marked differences in appearance of the subsets, but they definitely display differences in behavior and biology. Regulation of caste is determined by hormones, which trigger different patterns of gene expression, leading to alternative phenotypes. The labor of reproduction, foraging, nest maintenance, and defense is often distributed based on caste formation. Though not as apparent in the non-social insects, polyphenism can be equally important. For example, the change in wing presence or wing length in aphids and other bugs (Hemiptera), or the swarming behavior of locusts (Orthoptera), are vital aspects of their biology that affect their continued existence in their varying habitats.

COMMUNICATION

Communication occurs when insects signal one another using visual, auditory or chemical stimuli. Signals may be intra- or interspecific, and short- or long-range. Signals may be relatively simple or

Fig. 2.25. Some termite (Isoptera: Rhinotermitidae) castes: (a) king (dark individual) and soldier (with large mandibles) of *Reticulitermes hageni*; (b) winged adult of *Reticulitermes flavipes*; (c) queen and larva of *Reticulitermes flavipes*; (d) king, queen, and larvae of *Reticulitermes hageni*; (e) workers and soldier of *Reticultermes flavipes*; (f) nymph of *Reticultermes hageni* (photos of *R. flavipes* by J. Castner, of *R. hageni* by L. Buss).

complex, and may involve learning. **Learning** is change in behavior as a result of prior experience. Physiological signals such as hormones are sometimes considered to be communication, but for our purposes we will confine the discussion of communication to signals between individuals, not within an individual. See discussion on glandular systems (above) for information on hormonal stimuli.

Visual communication is signaling that depends on vision, and usually occurs between insects that are diurnal. Displays of flight by butterflies, flies, and grasshoppers that attract mates are examples of visual communication. Such visual signals may also be accompanied by sounds, as when bandwinged grasshoppers produce sound while hovering and displaying their wings. One of the most interesting forms of visual communication involves fireflies (Coleoptera: Lampyridae), which are nocturnal. Both sexes use bioluminescent organs to signal, and to locate one another for mating. Different species have different flash patterns. Some fireflies are predators, and use deceptive flash patterns to lure other firefly species to their doom. Visual communication also is commonly used to ward off predation, but of course this is interspecific communication, and is discussed in Chapter 3 under "How insects avoid becoming food for wildlife".

Acoustical communication is based on the ability to produce vibrations in air, water, or solid substrates, and to sense (hear) the vibrations. Only arthropods and vertebrate animals are able to communicate using sound. About ten orders of insects display acoustic communication. Sound dissipates, and is distorted and reflected as it travels. The loss of sound energy is called attenuation. Sound typically is used more for short distance communication than for long distance communication, though there are exceptions to this statement.

To create sound, a vibrating structure is needed. Usually this involves a **stridulatory apparatus**. The chitinous exoskeleton of insects is suitably rigid to serve as a stridulatory apparatus, and in most insects the major stridulatory components are a file and a scraper. The **file** is a series of small teeth and the **scraper** is a ridge that is rubbed along the file. The teeth of the file vibrate, sending out sound waves that are perceived as 'sound.' Among the many insects that stridulate, sound is created using structures throughout the body, including the antennae, mouthparts, legs, and wings. A variation on this is used by various bugs, particularly cicadas (Hemiptera: Cicadidae), which possess tymbals.

A **tymbal organ** is a ribbed, chitinous membrane that vibrates when activated by muscles.

Some insects use percussion for sound production. Some grasshoppers, for example, snap their wings, producing a clicking sound when their wing membranes are popped taut. Other insects bang their wings, appendages or head, or drum their abdomen on a substrate to create percussion. Still others expel air from their trachea to produce sound.

Sound production is only useful if it is detectable. Detection is accomplished by mechanoreceptors associated with the insect's nervous system. The receptors commonly are associated with a **tympanum**, a membrane that is stretched over a cavity and moves in response to sound waves. It functions much like the eardrum of vertebrates. Other receptors are located on the feet, and sense vibration of the substrate. Among flies such as mosquitoes, the plumose antennae perceive vibration and allow hearing.

Humans cannot hear many of the sounds produced by insects. Sounds produced by cicadas and crickets are notable exceptions, as they are readily perceivable. Other perceivable sounds, usually emitted at a high frequency, are often described as clicks, squeaks, or a scraping sound. A surprisingly large number of insects communicate via the substrate vibration, usually plant tissue, and produce low frequency sounds that are less audible to humans. This probably works well for local intraspecific communication among gregarious insects, and does not advertise the presence of the sound emitter to potential predators.

Sound production is often used to facilitate mate finding. Not only can sound be used to bring prospective mates together, but to segregate different species and function as a species isolating mechanism. Thus, calling males (usually the noisier sex) can display their presence, species, sexual readiness, and even quality as a mate. Some Orthoptera (crickets, katydids and grasshoppers) can only be distinguished by their calling behavior. The other principal use for sound production is predator avoidance. The clicking by large caterpillars (Lepidoptera: Saturniidae, Sphingidae, and others) or click beetles (Coleoptera: Elateridae) is thought to be a startle defense, allowing these insects to escape from the temporarily surprised predator.

Chemical communication is universally important in insects, and is based on the emission and perception of odors. Chemicals released by one organism and perceived by another are called **semiochemicals**. Semiochemicals that act intraspecifically (among members of the same species) are called **pherom-ones**. **Primer pheromones** stimulate long-term physiological effects such as maturation or caste determination. These typically are mediated by hormones and are not reversible. **Releaser pheromones** stimulate immediate responses in behavior, and are reversible. Releaser pheromones are very common among insects, and communicate many types of information. Some cause aggregation, as when both sexes of bark beetles (Coleoptera: Curculionidae: Scolytinae) simultaneously attack a tree and overwhelm its defenses. Social insects such as ants (Hymenoptera: Formicidae), and subsocial insects such as tent caterpillars (Lepidoptera: Lasiocampidae) produce trail pheromones that communicate to their relatives how to find good sources of food. Other pheromones stimulate alarm behavior, as when aphids (Hemiptera: Aphididae) suddenly move away or drop from a plant, or when bees and wasps (Hymenoptera) swarm out of a hive aggressively. Some pheromones are used to mark where eggs have been laid inside fruit or insects; this discourages other insects from depositing more progeny than can survive within that food resource.

The most common releaser pheromones are sex pheromones. **Sex pheromones** bring together the opposite sexes for mating. Sex pheromones usually are released by a receptive female, and release is terminated once she has mated. Sex pheromones are volatile, low molecular weight lipids, species specific, and released at a certain time of day. Usually, sex pheromones are blends of different chemicals. By using chemical blends and different but consistent release times, many possible combinations of pheromone release can be developed, which allows species to maintain species segregation. In fact, sex pheromones are the primary means by which closely related species achieve reproductive isolation. Specialized sensilla on the antennae detect the pheromone chemicals. Males that sense the pheromone travel upwind to its source to find the female. The act of mating may be stimulated by different, close-range sex pheromones, or by other stimuli including visual, tactile and auditory stimuli. Many sex pheromones have been identified and synthesized, and are used to attract and capture insects, primarily for population monitoring but also in 'attract-and-kill' systems. At other times, the crop environment is saturated with sex pheromone, causing disorientation of the insects and a disruption of mating. Over 2000 sex pheromones have been identified from insects.

Some semiochemicals act interspecifically (between different species); these are called **allelochemicals**. There are two principal types of allelochemicals: kairomones and allomones. **Kairomones** are allelochemicals that benefit the perceiving organism. Host volatiles produced by plants that allow insects to find them for food are examples of kairomones. Perhaps not surprisingly, chemicals can be used in more than one manner, so pheromones produced by insects for aggregation, trail following, or oviposition marking are sometimes used as kairomones by natural enemies to find and feed upon the pheromone-producing insects. **Allomones** are allelochemicals that benefit the producing organism. Chemicals that deter feeding by insects on plants are examples of allomones. The odor of onion, for example, signals to many insects that the onion plant is not edible. As previously noted, however, chemicals sometimes are used in more than one manner, so insects that specialize on onions as a food resource use onion volatiles as a feeding stimulant, a kairomone.

SOCIALITY

Cooperation among different individuals of a species is called **sociality**. The level of sociality varies greatly, however, among insects. Most insects are solitary, with cooperation among individuals limited to little more than mating. In contrast, some insects protect their offspring (brood), or provide food to them for a period of time, but then leave before the offspring attain maturity. This limited type of care is called **subsocial behavior**. Subsocial behavior is fairly common, occurring among some beetles, bees, wasps, bugs, earwigs, crickets, cockroaches, and others. A more advanced form of sociality is called **parasocial behavior**, which basically involves formation of aggregations or colonies of insects in addition to brood care. Parasocial behavior can, in turn, be subdivided into degrees of social behavior (i.e., communal, quasi-social, semisocial) but fundamentally they are similar and simply differ in the degree of development of caste differentiation, with semisocial behavior having the most advanced level of caste development. This type of sociality is found in some bees and wasps. The most advanced form of social development in insects is called **eusocial behavior**. Euscocial insects display cooperative care of the young, overlap of generations (the parents and their offspring work cooperatively), and reproductive activities that involve the labor of queens, males/kings, and nonreproducing workers. As is the case with parasocial behavior, eusocial behavior can be subdivided into different degrees or levels of sociality (i.e., primitively eusocial and highly eusocial), based primarily on the longevity of the colony. Highly eusocial insects have colonies that persist for more than a year, and sometimes several years.

Sociality is often marked by occurrence of different castes within colonies of insects. Caste formation is regulated by insect hormones, and is manifested by differences in behavior, and sometimes by differences in appearance. Sociality is best developed in ants, social bees and wasps, and termites. The evolution of sociality is not completely understood, but is certainly based on a shift from individual organisms to the unit (colony) as the unit of natural selection. The altruistic behavior (e.g., workers sacrificing their reproductive potential to support the colony) is perhaps understandable based on the high degree of relatedness of colony members (nearly all are sisters) and the higher survival potential of organisms working together.

Ants

Among the ants, female castes are expressed as worker, soldier and queen castes. Males do not display different castes. Soldiers are often referred to as major workers, with the coexisting smaller members called minor workers. Not all castes occur in all species. Castes are determined by a number of factors, including larval nutrition, winter chilling, post-hibernation temperature, egg size, queen age, and queen influence. The function of castes or 'caste polyethism' is well described by the aforementioned designations. Queens are mostly concerned with production of eggs, though early in the life of the colony the queen may perform various tasks, and some grooming of workers may occur indefinitely. Soldiers are specialized for colony defense, and often bear a large head and over-sized mandibles to aid in this task. Workers repair the nest, gather food, and tend larvae, pupae, and the other colony members. Workers may also be involved in colony defense, especially the larger workers. Males exist only to fertilize queens. Temporal changes also occur over the course of the ant's lifespan; this is called 'age polyethism.' For example, young workers tend to work inside the nest, whereas older worker

tend to forage outside. Over the years, some authors have recognized variants, phases, or anomalies within castes in an attempt to recognize small differences in behavior and morphology, but these are not generally used.

Social Bees and Wasps

Among social bees and wasps, the more primitively eusocial groups lack morphological differences but display different behaviors. More interesting is the general lack of the worker subcastes such as those that are found in ants and termites. This is despite the fact that bees and wasps, at least the species with very large colonies, display a sophisticated division of labor (polyethism). In bees and wasps, however, the division of labor is based less on production of castes, and more on temporal polyethism. In temporal polyethism the same individual passes through different stages of specialization as it grows older. Some differences exist based on size, however, with larger individuals tending to forage more and smaller individuals tending to conduct nest work and brood care.

Among social bees and wasps, there is a correlation between caste evolution and colony size. Noted sociobiologist and ant specialist E.O. Wilson divided this into four steps:
- Colony size of 2 to 50 adults – species with females that are semisocial or begin life as workers and later become egg layers.
- Colony size of 10 to 400 adults – externally, the queen is still identical to the worker caste, but there is functional differentiation of the worker caste from the queen. The egg-laying females maintain the workers in a subordinate position by aggressive dominance behavior. This can be expressed by the stealing and eating of eggs laid by rivals. Temporal polyethism is weakly developed among workers.
- Colony size of 100–5000 adults – some external differentiation of queens and workers is evident, and this is under the control of nurse workers that feed larvae differently. Queens do not display dominance behavior. Temporal polyethism is weak among workers.
- Colony size of 300–80,000 adults – queen and worker dimorphism is strong. Queen dominance is absent and queens maintain control with pheromones. Temporal polyethism is strongly developed among workers.

Termites

Termites and ants have similar caste systems, despite being phylogenetically distant from each other. Both have evolved a soldier caste with specialized head structure and behavior, and both are populated primarily by similar-looking but behaviorally versatile workers. Their systems of temporal polyethism also are similar. Termites differ, however, in that males do not exist solely for fertilization.

Termite castes have some unique features. In the lower termites, the reproductives secrete sex-specific pheromones that inhibit metamorphosis of the immatures into additional reproductives. Termites also are capable of producing 'supplementary reproductives.' If the primary reproductives are removed, fertile but wingless individuals of both sexes develop in the colony. Thus, termite colonies display 'immortality;' they may never completely perish because reproductives can be generated as necessary.

The classification of termite castes follows. The larva (a wingless nymph, as these are hemimetabolous insects) lacks evidence of wings and of the features that characterize soldiers. The nymph (brachypterous nymphs) develops from the larval stage but possesses wing buds initially, and wing pads after some molts. Eye differentiation also occurs at this stage. A worker stage occurs in the higher termites, but not the lower. Workers lack wings, and eyes are reduced or lacking. The head and mandibles are well developed. Lacking the worker stage, the lower termites have instead a stage called pseudergate. Pseudergates develop from nymphal stages or larvae. Soldiers have morphological features that are specialized for defense. This includes large mandibles, large heads, and glands capable of discharging defensive secretion. Primary reproductives are derived from colony-founding queens and males. If the primary reproductives are removed from the colony, supplementary reproductives can appear. The supplementary reproductives take three forms: (1) the adultoid reproductive, in the higher termites only, appears identical to the primary reproductive but changes behavior in the absence of the primary reproductive; (2) the nymphoid reproductive is a supplementary male or female derived from a nymph and retains wing buds; (3) the ergatoid reproductive is also a supplementary male or female, but is larval in form and lacks wing buds. Some forms of termites are shown in Fig. 2.25.

The primary reproductives construct an initial cell and rear the first brood, providing them not only with food but the protozoans necessary for independent feeding on cellulose. The first brood workers (or worker-like pseudergates or nymphs) soon take over responsibilities for foraging, nest construction and nursing. The queen and male become specialized reproductive organisms. Interestingly, the worker caste is morphologically uniform but behaviorally diverse when species are compared, whereas the soldier caste is morphologically diverse but behaviorally uniform. Soldiers can use their mandibles effectively in defense against insects their own size, or in the case of those practicing chemical defense, their glands can secrete or spray a number of bioactive substances to deter intruders.

METAMORPHOSIS

As noted previously, gene expression is controlled by hormone levels. Insects transition from the immature to the adult stage through a series of molts, with the outcome (body form and function) determined by hormones. The immature forms are called larvae in some insects, but nymphs in others. **Larvae** and **nymphs** are active feeding stages but are sexually immature. Larvae transition through a pupal stage before molting into the adult, but nymphs do not. The pupal stage is a nonfeeding, inactive stage. The stage of the insect between molts is called the **instar**. The interval (period of time) between molts is called the **stadium**.

The outcome of molting varies, with three basic forms of metamorphosis commonly found in insects: ametabolous, hemimetabolous, and holometabolous insects (Fig. 2.26). **Ametabolous** insects show little or no change in body form as the insect transitions through the immature stages to the adult. The adult is basically a larger immature form. This is not so unusual, as humans and wildlife develop similarly. The typical ametabolous insect hatches from an egg, progresses through several larval (immature) instars, and then attains the adult stage, all without (other than hatching from an egg) significant change in body form. The adults are wingless.

Hemimetabolous insects undergo a partial change in form as they transition to the adult. The immature stages, or nymphs, after hatching from the egg are quite similar to the adults, mostly lacking the fully formed wings and external genitalia that occur on the adult. They also tend to be found in the same habitats and feeding on the same hosts as the adults. These insects are also said to have 'incomplete metamorpho-

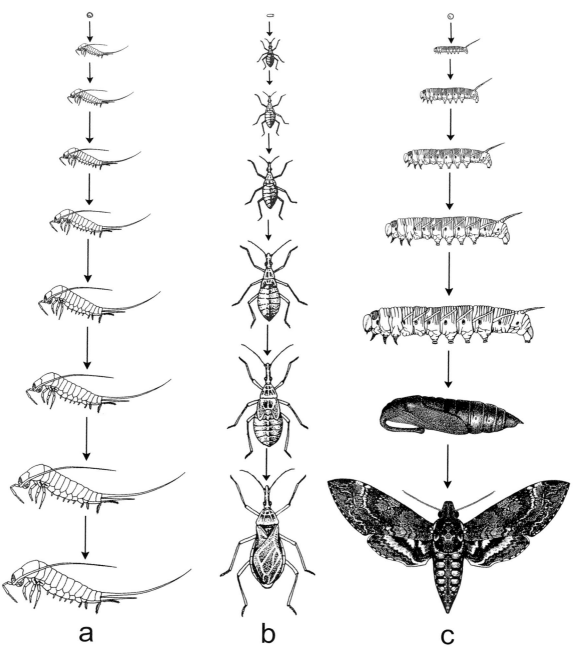

Fig. 2.26. Metamorphosis in: (a) ametabolous; (b) hemimetabolous; and (c) holometabolous insects.

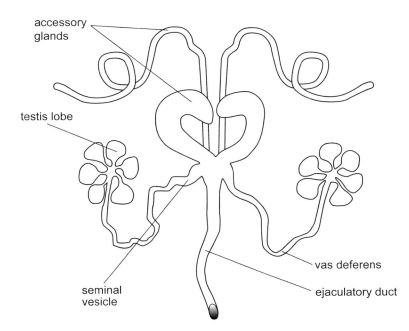

accessory glands

testis lobe

vas deferens

ejaculatory duct

seminal vesicle

Fig. 2.27. Diagram of the male reproductive system in *Tenebrio* sp. (Coleoptera: Tenebrionidae).

sis.' The immature stages of the aquatic orders Ephemeroptera, Plecoptera, and Odonata are sometimes called 'naiads,' but these also are hemimetabolous insects.

Holometabolous insects undergo a radical change in body form, and are said to have 'complete metamorphosis.' The immature or larval stages are quite different from the adults in form and function. They often are found in different habitats, or at least feed on different food. A major reorganization of the body is needed to transition from the larva to the adult stage, so an inactive, nonfeeding stage called the **pupal stage** occurs during the transition. Interestingly, the pupal stage often looks very much like the adult if it is examined closely, although it lacks adult coloration and mobility. The pupa also tends to be enclosed in a cell in the soil or debris (e.g., the pupal cell of some beetles), within the epidermis of the last larval instar (e.g., the puparium of higher flies) or within a cocoon (e.g., some moths, fleas, and other insects).

REPRODUCTIVE SYSTEM

The reproductive system of insects is quite variable among species, but unlike most other systems found in insects, also differs markedly between the sexes. The **male reproductive system** consists of paired **testes** composed of testicular follicles, paired vasa deferentia, a seminal vesicle, an ejaculatory duct, and accessory glands that open into the genitalia (Fig. 2.27). Sperm are produced in the testicular follicles, travel down the vas deferens to the seminal vesicle where they are stored, and then released, often in association with a secretion of the accessory glands. In most insects, sperm are transferred internally to the female's genital tract during copulation. The sperm may be stored in the spermatheca of the female for later use, and sometimes a long period such as an entire winter is passed before the female releases the sperm from the spermatheca to fertilize her eggs. In the more primitive insects such as silverfish, the sperm are not transferred directly to the female. Instead, the sperm are deposited in a 'packet' or spermatophore, often on a substrate or suspended on threads, to be picked up by females. One of the most remarkable methods of insemination occurs in bedbugs, wherein the female's abdomen is penetrated by the male's genitalia and sperm are injected. This unusual means of fertilization is called 'hemocoelic insemination'.

The **female reproductive system** consists of paired **ovaries** composed of ovarioles, paired oviducts, a

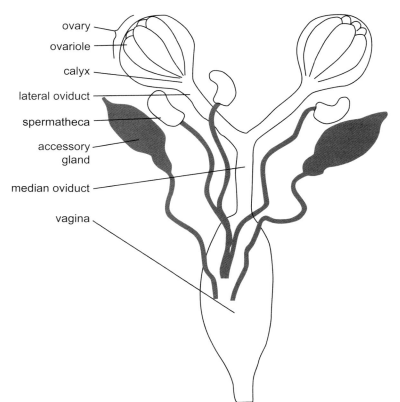

ovary
ovariole
calyx
lateral oviduct
spermatheca
accessory gland
median oviduct
vagina

Fig. 2.28. Diagram of the female reproductive system in *Rhagoletis* sp. (Diptera: Tephritidae).

common oviduct and vagina, accessory glands, spermatheca, sometimes a bursa copulatrix, and sometimes an ovipositor (Fig. 2.28). The eggs are produced in the ovarioles, travel down a paired oviduct to the common oviduct, where they may be fertilized by sperm that have been stored in the **spermatheca**. Females have different types of ovarioles, and methods of egg deposition. Often the eggs receive an adhesive from accessory glands, or are deposited in a protective case (the **ootheca**) that is derived from secretions of the accessory glands. Oothecae are especially common among cockroaches (Blattodea), mantids (Mantodea), and grasshoppers (Orthoptera). Reproduction by deposition of eggs is called **oviparity**.

Eggs are not the only means of reproduction, however. In some cases, the eggs are retained within the female's genital tract until they hatch, so the female deposits a larva or nymph instead of an egg (this process is called **ovoviviparity**). Some females are able

to provide nourishment to the developing embryo that originates from outside the egg (this process is called **viviparity**). One of the most important means of reproduction in insects is **parthenogenesis**, which is development without fertilization. There are many variations of parthenogenesis. Less common is **polyembryony**, in which more than one individual develops from a single egg.

EGGS OF INSECTS

Nearly all insects produce eggs during the adult stage, though some seemingly can produce living offspring indefinitely by retaining their eggs until they hatch. A few species retain not only their eggs internally until after they hatch, but the larval stage as well, depositing partly grown progeny that are nearly ready to pupate (e.g., sheep keds and tsetse flies). Eggs are a common

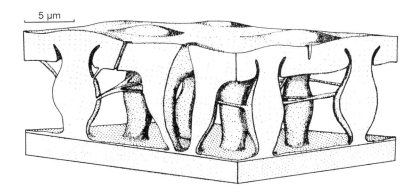

5 μm

Fig. 2.29. Diagram showing cross-section of an egg chorion with air-containing meshwork.

means of passing through unfavorable (dry season or winter) periods when food is unavailable, though because other stages of insects can also enter diapause, egg-overwintering is by no means universal.

The egg shell, or **chorion**, of insects is proteinaceous. It apparently lacks chitin. The insect egg shell suffers from the same problem faced by other stages, and other organisms: the gas exchange needed for respiration also results in dehydration (water loss). Thus, insects have evolved means to facilitate gas exchange while minimizing water loss. Their challenge is appreciably more difficult than for birds and reptiles because the insect eggs are much smaller and therefore have a much larger surface to volume ratio, leading to greater potential for water loss. Insects tend to have more complex egg shells than birds and reptiles to accommodate this challenge.

Most terrestrial eggs, but not most eggs laid in water, have air-containing meshworks within the chorion (Fig. 2.29). The chorionic meshwork contains a layer of gas, and has holes (aeropyles) that connect it to the outside. The holes, which measure less than a micron to several microns in size, provide continuity with the ambient atmosphere, allowing gas exchange. Their small size and small number help in water conservation. Eggs that are found in water or likely to be submerged often are adapted to function with the aid of a plastron. An egg **plastron** is the surface characteristics of the egg that cause retention of a film (bubble) of gas having extensive water-air interface, thereby allowing extraction of oxygen from water. The plastron usually consists of hairs or meshworks, and the volume is not depleted during use. It functions as a **physical gill**; as the insect withdraws oxygen from

the plastron the relative concentration of nitrogen increases, stimulating more oxygen to flow into the plastron from the surrounding water, and nitrogen to flow out of the plastron into the water.

The chorion must, in some cases, also allow uptake of water or liquid nutrients from the environment. **Hydropyles**, structures that allow the uptake of water, are found on some eggs. Water is absorbed when the embryo is rapidly growing, and the egg enlarges in size. The eggs of most aquatic insects absorb water, but terrestrial insects that deposit their eggs where moisture may occur (e.g., grasshopper eggs in soil) also may have eggs that enlarge due to uptake of water.

The chorion also must allow the sperm to pass into the egg. The egg is fertilized after the chorion is deposited. Thus, insects have a small hole (or holes) called the **micropyle** located at the anterior end of the egg that allows sperm to pass into the egg to accomplish fertilization.

On the outside of the chorion, the female may also secrete glue that attaches the egg to a substrate. Also secreted in some cases are jelly-like materials that harden to form oothecae, pods, or egg cases containing the individual eggs. The glands that secrete these are known by various names including accessory, mucous, cement, and colleterial glands. Some different types of eggs are shown in Fig. 2.30.

The number of eggs produced by a female insect (fecundity) varies considerably. Social insects are the most fecund. Although data on fecundity of social insects is often lacking, data on honey bees are reliable, so the estimate of up to 2000 eggs per day, or 220,000 per year, or about 600,000 eggs in a lifetime seems reasonable. Termites, on the other hand, can produce

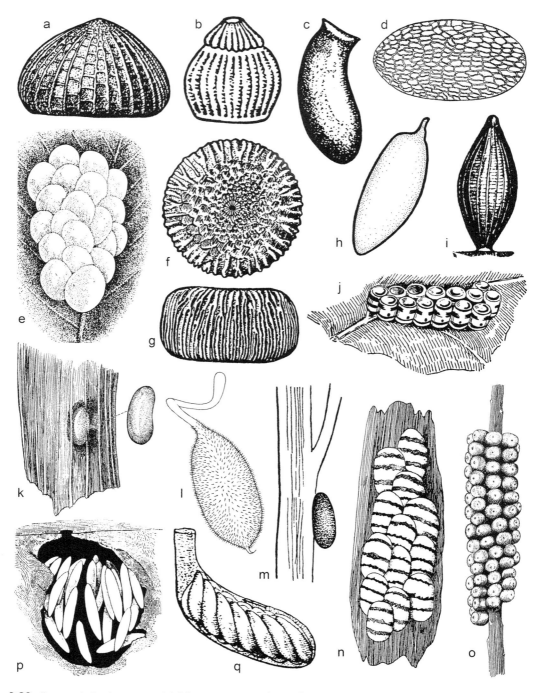

Fig. 2.30. Representative insect eggs: (a) fall armyworm, *Spodoptera frugiperda* (Lepidoptera: Noctuidae); (b) beet armyworm, *Spodoptera exigua* (Lepidoptera: Noctuidae); (c) garden fleahopper, *Halictus brachtatus* (Hemiptera: Miridae); (d) beet leafminer, *Pegomya betae* (Diptera: Anthomyiidae); (e) garden webworm, *Achyra rantalis* (Lepidoptera: Crambidae); (f) hop vine borer, *Hydraecia immanis* (Lepidoptera: Noctuidae) dorsal view; (g) hop vine borer, *Hydraecia immanis* (Lepidoptera: Noctuidae) lateral view; (h) greenhouse whitefly, *Trialeurodes vaporariorum* (Hemiptera: Aleyrodidae); (i) alfalfa caterpillar, *Colias eurytheme* (Lepidoptera: Pieridae); (j) harlequin bug, *Murgantia histrionica* (Hemiptera: Pentatomidae); (k) southern corn billbug, *Sphenophorus callosus* (Coleoptera: Curculionidae); (l) *Eurytoma* sp. (Hymenoptera: Eurytomidae); (m) spotted asparagus beetle, *Crioceris duodecimpunctata* (Coleoptera: Chrysomelidae); (n) southwestern corn borer, *Diatraea grandiosella* (Lepidoptera: Crambidae); (o) range caterpillar, *Hemileuca oliviae* (Lepidoptera: Saturniidae); (p) Mediterranean fruit fly, *Ceratitis capitata* (Diptera: Tephritidae); (q) migratory grasshopper, *Melanoplus sanguinipes* (Orthoptera: Acrididae).

30,000 eggs per day and live 15 to 20 years, so their reproductive output is likely unsurpassed. Ants have a more modest reproductive output, perhaps several hundred eggs per year and 5000–6000 over the course of a lifetime. Non social insects are much less fecund, often producing 100 eggs or less per female.

The site of egg deposition is incredibly varied, reflecting the vast diversity of life styles displayed by insects. Deposition ranges from apparently random, as when stick insects (Phasmida) drop eggs to the ground while feeding in tree tops, to highly selective, as when wasp (Hymenoptera) hyperparasites (parasites of parasites) select specific physical locations on specific ages of parasitoids within certain hosts, and also check to determine that the prospective host is not already parasitized before depositing the egg. Some of the interesting locations for egg deposition include:

• Under water, glued to a substrate such as plant tissue. This is common among insects where both the adult and immature stages dwell in the water, such as predaceous diving beetles (Coleoptera: Dytiscidae) and backswimmers (Hemiptera: Notonectidae).

• Dropped into water, or near the water and in a location likely to be flooded. This occurs in mosquitoes (Diptera: Culicidae), horse flies and deer flies (Diptera: Tabanidae), and caddisflies (Trichoptera).

• Deposited on or in a host insect. This is common among parasitic Hymenoptera and Diptera, and they may deposit their egg free within the host insect's hemocoel, within a specific organ, attached to the body wall internally or externally, or only within a certain stage (e.g., the host egg).

• The parasitic insect oviposits on the host while the latter is flying. This requires that the female be equipped with claspers to hold the host while egg deposition occurs, as in thick-headed flies (Diptera: Conopidae).

• The insect is phoretic rather than parasitic, and oviposits on an insect to take advantage of the phoretic insect's behavior. This occurs with human bot flies (*Dermatobia hominis* (Diptera: Oestridae)) in South America, which oviposit on mosquitoes that then seek out humans and other mammals, the eventual host of the bot fly. The truly interesting aspect of this is that the bot fly egg hatches only when the mosquito is blood-feeding, and falls onto the host where it burrows in and feeds.

• The female deposits her eggs on the back of the male, as in some giant water bugs (Hemiptera: Belostomatidae), which then provide aeration and protection for the eggs.

EXCRETORY SYSTEM

Excretory systems serve to eliminate metabolic wastes so they do not harm the excreting organisms. Also, they work to maintain water balance and physiological homeostasis, or maintenance of a constant physiological environment. Excretion refers not only to defecation, or the production of solid, undigested food (feces). It also includes the movement of salts, water, and nucleic acids across membranes and out of the body. The important organs involved in excretion are the Malpighian tubules and the rectum (hindgut).

Nitrogenous waste products are potentially toxic to the producing organisms. When water is not limiting, as is the case with aquatic insects and plant sucking insects, ammonia is the common water product. However, for most terrestrial insects, uric acid is excreted. Birds also excrete uric acid.

Malpighian tubules play a very important role in excretion. They are 'blind' (one end is closed) tubules that attach to the digestive system at the juncture of the midgut and hindgut. They serve to extract water, salts, sugars, amino acids, and nitrogenous wastes from the hemolymph and channel them into the alimentary canal and into the excreta. Sometimes the Malpighian tubule is associated with other tissues. The **hindgut** contains rectal pads that also serve to absorb water from the feces before they are excreted. In insects that live in dry environments, the tip of the Malpighian tubule may be embedded in the wall of the hindgut. This offers the additional opportunity for the insect to reabsorb water that might be excreted. On the other hand, for insects that ingest large volumes of water but dilute nutrients (e.g., sap-sucking insects), the basal areas of the Malpighian tubules and parts of the hindgut come into contact with the midgut. This offers the opportunity for the insect to shunt water out of the body quickly, leaving more concentrated nutrients in the midgut. This water-shunting region of the digestive tract is called the **filter chamber**.

As noted previously, the rectum is the final region of the alimentary canal. Here the waste is dehydrated even further and compressed by muscles into a pellet before it is excreted.

THERMAL BIOLOGY

Small organisms like insects have large surface to volume ratios. Thus, they are poor energy storers, and so are exceptionally susceptible to environmental conditions. It is not surprising, then, that insects generally are **ectothermic**, which means that the environment determines their body temperature. Small uninsulated flies, for example, never have their body temperature more than 1 °C higher than ambient air temperature. Reptiles and amphibians also are ectothermic. This is quite different from **endothermic** animals such as mammals and birds, which maintain a relatively constant body temperature. There are advantages and disadvantages to each of these approaches to thermoregulation. Ectotherms have the advantage of not needing to expend energy to maintain body temperature, and this is vitally important in the winter or dry months when food might not be available. Thus, insects and other ectotherms can survive a nonfeeding period of dormancy without massive stores of food reserves. In order for this evolutionary strategy to work, though, you must be able to withstand freezing conditions or escape to a site where you will not freeze. On the other hand, endotherms can be active under all environmental conditions, which allows them to hunt, feed, or escape predation even if the weather is cold. For this alternative strategy to be viable, the organism must be able to access food even in the winter.

The development rate of ectothermic organisms is determined largely by temperature. Linking temperature to development provides predictability to insect life processes, and this linkage is called **insect phenology**. The term 'phenology' is not restricted to insects, of course, as other organisms such as plants develop similarly, giving rise to the parallel term 'plant phenology.' Growth is largely a function of how temperature affects enzymatic function. At some low temperature (the **lower developmental threshold**) the enzymes cannot function (often about 7–10 °C in temperate climates). Slightly above this threshold temperature some enzymatic action occurs, so there is slow development. For many insects, as temperatures increase between about 12–30 °C there is a linear relationship between increasing temperature and increased development rate. Then, at some higher temperature the developmental rate no longer increases in a linear manner and a maximum development rate is achieved (the **upper developmental threshold**). In temperate climatic regions, the upper developmental threshold is often about 35 °C. If time (the x-axis) is plotted against development rate (the y-axis) a sigmoid curve is derived. There are three

important facts here: (1) the linear relationship between temperature and development rate provides a great deal of predictability over much of the normal temperature range; (2) there are species-specific differences in the minimum and maximum developmental thresholds; and (3) the amount of heat needed to progress through any development stage or life cycle is a constant for any species.

The amount of heat needed to complete development in any life stage, or the entire life cycle, can be expressed as a function of time and temperature. The number of heat units (also called degree days or day degrees) above the developmental threshold when multiplied by the number of days of exposure is called the thermal summation or **thermal constant**. Basically, a life stage can be completed in a few days at a high temperature, or many days at a cool temperature, but a specific total heat requirement is required to move from any stage to the next stage. The significance of all this is that if the threshold values and thermal constants have been derived it is possible, using thermal time, to predict when insects will emerge, oviposit, commence feeding, and complete their life cycle. The relationship between temperature and growth rate is not completely linear, of course, so predictions based on thermal summations are not exact. Nevertheless, if you are attempting to monitor or manage pests, this is very useful information, and the process can be made more exact by using nonlinear models.

Not all insects are completely ectothermic, or at least they display behaviors and physiological or morphological modifications that ameliorate their ectothermy. Insects commonly will bask in the sun, including spreading their wings or angling their body to intercept as much incident sunlight as possible. This helps them warm up, of course, so they can be active. The bodies of night-flying flying insects such as moths (Lepidoptera) often are heavily insulated by hairs or scales. Insulation is important because muscles generate heat when they are used, so flying insects are able to maintain body temperatures appropriate for flight well into the cool night due to their insulation. Body temperatures 20–30 °C higher than ambient air temperature can be found in some insects after flight. This self-generated heat derived from muscle activity is called **metabolic heat**. Sometimes insects will move their appendages slowly or seemingly 'shiver' as they work their muscles to increase their metabolic heat. Overheating can also be an issue in the thorax where flight muscles are active, but they transfer heat to other parts of the body using their hemolymph, helping to dissipate the excessively high temperatures. Insects will also display behavior that minimizes heat exposure if it is too hot. Stilting, or standing as far as possible from a hot substrate is one means of minimizing body temperature. Burrowing deep into the soil is another means of escaping radiant heat. They also climb vegetation to get away from hot soil, and where the cooling effects of wind can be beneficial.

Surviving periods of adverse weather is one of the greatest challenges for insects, especially surviving freezing temperature. Cold hardiness varies among insects. Those that perish when exposed to freezing temperatures are called freezing-susceptible, whereas those that survive freezing are called freezing-tolerant. There are several physiological and ecological factors associated with surviving cold weather. Insects often evacuate their guts as they enter into cold weather. The presence of food is thought to serve as a focal point for ice formation, which can physically disrupt the insect when ice crystals form internally. Overwintering insects also tend to have higher concentrations of sugars and lipids in the hemolymph, which retards freezing. Most important among the freeze inhibitors is a sugar alcohol, **glycerol**. Glycerol retards the rate of freezing, lowers the temperature necessary for freezing to occur, and reduces damaging intracellular freezing. Insects that survive freezing have reduced water content and elevated glycerol content. Overwintering insects also tend to seek shelter from rapid changes in temperature, so they often enter logs or soil, or burrow beneath leaf litter. Snow cover is important as it acts as a good insulator, protecting such insects from extremely cold air. These behaviors often reduce the need for physiological protection from freezing. Nevertheless, winter mortality is often quite high.

Insects undergo periods of arrested or suspended development when inclement weather prevails, or when environmental cues indicate that inclement weather will occur. Photoperiod, temperature, and change in host nutritional quality are the most common cues that stimulate arrested development. Arrested development often occurs in the winter, of course, and is genetically programmed to last a certain amount of time or until the insect is exposed to a certain amount of cold. Such relatively irreversible arrested development is called **obligatory diapause**. This is a nice adaptation to keep the insect from emerging prematurely, as might occur during an unseasonably mild period in early spring. A more reversible form of arrested development is called **facultative diapause**, as this is terminated whenever favorable conditions reappear. Facultative diapause occurs in insects that experience more transient periods of inclement, but less life-threatening weather. Obligatory diapause normally is long-lasting, and occurs in only one stage, though the stage varies among species. Facultative diapause is more likely to be short-lasting, and may involve multiple stages within the same species. Facultative diapause is also called dormancy, and if it occurs in the summer it is called estivation. Among mammals, dormancy or facultative diapause is usually called hibernation.

The number of generations (life cycles) that an insect species can complete in a year is often enforced by diapause, and is called **voltinism**. In temperate climates, many species are **univoltine**, having one generation per year. With species that develop rapidly, or occur in warmer climates, many generations per year may occur; these insects are called **multivoltine**. Some species require more than 1 year to complete their life cycle, and are called **perennial**. Terrestrial insects are not usually perennial, but this form of development is common in aquatic insects, especially those living in cold water.

FEEDING ECOLOGY

Feeding ecology is central to the relationship of insects and wildlife. Insects are fed upon by wildlife, feed on wildlife, and have profound influences on the availability of plant resources that provide food and cover for wildlife. Thus, the focus not only of this section, but of the remainder of this book, is on feeding and its myriad interactions.

One of the principal reasons that insects have become exceptionally species-rich is that they have evolved diverse forms of behavior and ecology, complementing their diverse morphologies, and allowing them to take advantage of nearly every terrestrial and fresh-water resource. Among the important niches that insects have exploited successfully are scavenging and feeding on detritus, feeding belowground, feeding in aquatic habitats, feeding on living plants, feeding on blood, and predation and parasitism. Most people tend to think of insects as plant pests, and surely a great number of insects are associated with plants. However,

Table 2.2. Feeding habits of insects and their close relatives (** indicates major food resource; * indicates minor food resource) (adapted from Southwood 1973).

	Animal	Blood	Detritus	Fungi	Algae/lichens	Mosses/ferns	Seed plants
Chilopoda	**						
Diplopoda			**				
Protura			**				
Diplura	**		**				
Collembola	*		**	**	**		*
Archaeognatha			**				
Zygentoma			**				
Ephemeroptera	*		**				
Odonata	**						
Plecoptera	**		**		**		
Grylloblattodea			**				
Mantodea	**						
Blattodea			**				
Orthoptera	**		**			*	**
Mantophas-matodea	**						
Isoptera			**				
Dermaptera	**		**				*
Phasmida							**
Embiidina	**		**				
Psocoptera			*	**	**		
Zoroptera				**			
Hemiptera	**	*		*	*	*	**
Thysanoptera	**		*	**			**
Phthiraptera		**	**				
Mecoptera	**		**				
Megaloptera	**						
Neuroptera	**						
Trichoptera	**		**		*		*
Lepidoptera	*		*	*	*	*	**
Coleoptera	**		**	**	*	*	**
Strepsiptera	**						
Diptera	**	**	**	**	*	*	**
Siphonaptera		**	**				
Hymenoptera	**		*	*		*	**
Total major	19	3	20	6	3	0	8
Total minor	3	1	4	3	5	6	3

it is interesting to look more closely at insect (arthropod) food associations (see Table 2.2).

Here you can see that *from a taxonomic perspective* feeding on plants is not as commonplace as you might expect. The distribution of the major feeding habits of common arthropod groups is heavily weighted to feeding on detritus (scavenging) and animals (carnivory). It appears that the more primitive insects largely were unable to take advantage of the radiation of seed plants (Spermatophyta), focusing instead on scavenging and carnivory. However, what is not so apparent from this table is that the taxa that were able to adapt to the evolution of plants were able to thrive and to radiate, so the largest number of *species* are found in the more modern orders such as Coleoptera, Hemiptera, Diptera, Lepidoptera, Hymenoptera (see

Feeding on plants, below). Following is discussion of some of the adaptations of insects that allow these diverse forms of existence.

Scavenging

A large number of taxa can be classified as scavengers, organisms that feed on detritus or debris. This includes dead animals, whether they are small (other insects) or large (mammals). Insects may possess unusual digestive enzymes that aid in the conversion of unusual foods (e.g., bird feathers, wood) into body tissues, but more often they are able to eat these strange foods because they harbor symbiotic microorganisms that assist in the digestion. These microorganisms are so important that insects have evolved means of transferring the microorganisms to their offspring. A common means is by feeding their fecal material, which usually contains the microbes, to their young. Examples of insects that acquire part of their gut flora in this manner are cockroaches (Blattodea) and termites (Isoptera). This assures that the digestive system of the young insects will be inoculated. Insects also may have special morphological adaptations that allow them to harbor the microorganisms when they disperse. Pouches containing fungi (called mycangia) are commonly found in bark beetles (Coleoptera: Curculionidae: Scolytinae), for example, so they can inoculate newly invaded trees with fungi. The bark beetles feed on the fungi growing on dying and dead trees. Insects and decomposition is discussed in more detail in Chapter 6, 'Insects and ecosystems.'

One of the most overlooked features of scavenging by insects is that in many cases they are harvesting microorganisms when they scavenge on detritus. The leaf litter, dung, or decomposing mouse that they feed on serves as a convenient matrix or substrate for fungi, bacteria, and yeasts. Microorganisms can produce a nutritious 'brew' or 'soup' for insect scavengers that meet the nutritional requirements of the more primitive arthropods, so from a taxonomic perspective, detritus feeding is the most common way to make a living. Not coincidentally, it allows these less advanced arthropod taxa to avoid dealing with the chemical defenses evolved by plants to fend off herbivory. Detritus feeding allows the insects to avoid anti-herbivore chemistry, with the microorganisms overcoming the chemical defenses and converting plant nutrients into microbe tissues that are easily assimilated into arthropod tissues.

Feeding Belowground

Subterranean feeding is quite common among insects. Two substantial advantages of feeding in the soil are that moisture conditions are higher, so desiccation is not so great an issue, and it is harder for predators to locate insects belowground. Many insects are adapted for burrowing beneath the soil to feed on bulbs, tubers and roots of plants, or to ingest organic matter such as leaf detritus that occurs at the soil surface, or to feed on other soil organisms including other arthropods. Larvae of scarab beetles called white grubs (Coleoptera: Scarabaeidae), larvae of click beetles called wireworms (Coleoptera: Elateridae), and larvae of flies called root maggots (Diptera: Anthomyiidae) are examples of belowground-feeding insects. When we look at plants we tend to forget how extensive the belowground component of plants can be, and how large a food resource exists out of our sight. The extreme situation occurs with grasslands, where the belowground plant biomass usually exceeds the aboveground biomass. Even in forests, which have most of their plant biomass aboveground, considerable energy is stored belowground in roots.

However, there is a third important reason to feed belowground, at least for root feeders. The roots of plants are not as well defended against herbivory. Roots and other belowground tissues usually lack the physical and chemical defenses found in foliar tissues (see Feeding on plants, below).

In addition to ingesting root tissue, belowground we can also find insects that tap into the nutrients that flow in the vascular tissue of roots. Aphids (Hemiptera: Aphididae), ground pearls (Hemiptera: Margarodidae), and cicadas (Hemiptera: Cicadidae) are examples of piercing-sucking insects that feed on roots. Cicadas are well-known for the periodic mass emergence of adults, after feeding belowground for up to 13 years. Root-feeding aphids, however, may never come to the soil surface, instead depending on ants (Hymenoptera: Formicidae) to tend them and move them from plant to plant through their subterranean tunnels. The aphid-tending ants harvest sugar-rich honeydew from the aphids.

Many insects use soil as a benign, secure place to hide, pupate, or deposit eggs. After feeding aboveground at night, many moth larvae called cutworms burrow belowground during the daylight hours. Sometimes they even bury their host plant material belowground. This behavior serves to protect them from the sharp eyes of birds, or from parasitic insects that are active during the day. Pupation is a particularly difficult time for insects as they transition from immature to mature forms because they are helpless and unable to flee or defend themselves during this stage. Thus, if you are a predatory insect, the belowground environment offers considerable food resources. The larvae of ground beetles (Coleoptera: Carabidae), blister beetles (Coleoptera: Meloidae), rove beetles (Coleoptera: Staphylinidae), and bee flies (Diptera: Bomyliidae) are examples of insects that tend to feed belowground on other insects. Subterranean termites normally live belowground because of the favorable high humidity, but may venture aboveground to seek food. Often they construct tunnels out of soil particles when they move aboveground, as this allows them to maintain high and favorable humidity while they feed.

The ants, as well as the ground nesting bees and wasps, have a similar relationship with the soil, though unlike termites, the adults of these insects are desiccation resistant. They tend to use the soil as a nesting matrix or substrate, with the immature stages living belowground but the adults feeding aboveground. Food, whether in the form of seeds or insect prey, is moved belowground to supply their larvae with nutrition. Some ants also store food belowground to eat during periods of inclement weather.

Feeding in Aquatic Habitats

Insects are basically terrestrial animals, and only secondarily aquatic. Their relatively impermeable cuticle and the interior location of their gas exchange apparatus (tracheal system), though necessary for terrestrial life, are not ideal for aquatic life. Thus, we see evolution of various adaptations to overcome these handicaps. Some insects have evolved systems that allow them to live in the water by taking excursions to the surface and collecting oxygen at the water–air interface. This is commonly seen in mosquito (Diptera: Culicidae) larvae and pupae, where they have respiratory siphons or tubes that reach the atmosphere. Some insects have extremely long respiratory tubes, such as are found in some syrphid larvae (Diptera: Syrphidae), the so-called rat-tailed maggots. A few mosquitoes, beetles and caterpillars actually penetrate the underwater portions of aquatic plants to obtain oxygen stored in tissues there. Others capture a bubble of water from the surface. They may use this as a source of oxygen until it is

depleted, or use it as a **physical gill**, a bubble that has its oxygen replenished by oxygen that is dissolved in the water as the insect continues to remove oxygen from the bubble. Thus, they can remain submerged much longer than one would expect if they were simply dependent on the initial oxygen content of the bubble. A variation on the physical gill concept is called **plastron respiration**. With a plastron, the hairs covering the insect body serve to hold air and to repel water. These insects tend to live in fast-flowing, oxygen rich streams, so the thin layer of air readily allows the oxygen in the water to diffuse into the plastron as the insect uses oxygen. This allows the insect to remain submerged, not having to come to the surface. Insects also may have spiracular or tracheal gills. Spiracular gills are extension or invaginations of the spiracles. This increases the area of water-air interface, allowing greater ventilatory exchange. Some flies and beetles display this adaptation. Tracheal gills are extensions of the body wall, forming specialized ventilatory organs. For this to work efficiently, the cuticle must be thinner than normal, and a large number of trachea must be concentrated in the gill area. Many of the most important aquatic insects have tracheal gills, including the immatures of mayflies (Ephemeroptera), dragonflies and damselflies (Odonata), alderflies and relatives (Megaloptera). Also, some spongillaflies (some Neuroptera), stoneflies (Plecoptera), caterpillars (Lepidoptera) and beetles (Coleoptera) have tracheal gills. Tracheal gills are rare in flies (Diptera) and absent from the aquatic bugs (Hemiptera). The two most common types of tracheal gills are plate-like lateral extensions of the abdomen (found in mayfly nymphs) and caudal projections (found in dragonfly and damselfly nymphs). Finally, some insects have cutaneous ventilation. This occurs in insects that remain in the water and do not need the wax and cement layers of the cuticle that is found on other insects, so their epicuticle is much more permeable to gas exchange. Such diffusion through the body surface is most common in very small species, or those inhabiting cold, highly oxygenated water.

Aquatic insects occur frequently in both **lotic** (running water) and **lentic** (standing water) freshwater habitats. Other than the ventilatory adaptations discussed previously, the only other significant adaptations are related to their need to be able to swim efficiently. Aquatic insects sometimes display a drop-shaped body, or a slight variation on this, which provides minimal resistance when swimming. The wings are not used in the water, other than in the case of a few parasitic wasps. Instead, the legs are used like oars. Efficient 'rowing' by aquatic beetles and bugs occurs when the swimming legs have greatest surface area during the power stroke and diminished surface area during the recovery stroke. This usually is accomplished with hair fans, clusters of hairs that flare out during the power stroke. In contrast, flies move by coiling and uncoiling their bodies, and mayflies and damselflies by swimming with undulating movements of their abdominal segments. Dragonflies are unusual in that they propel themselves by forcing water from the anus. Oftentimes insects simply walk along the bottom of the pond or stream. Some insects occurring in streams do not move during their immature stages, instead remaining attached to the substrate and harvesting food as it floats by.

Insects develop in nearly all types of aquatic habitats, ranging from temporary, shallow bodies of water to deep lakes and, of course, to running water. Their occurrence is variable, however. Bugs and beetles typically occur only at shallow depths, rarely exceeding a few meters, and usually must surface periodically to acquire oxygen. On the other extreme are the flies, particularly the chironomid (Chironomidae) and chaoborid midges (Chaoboridae), which can occur at great depths. The other aquatic insects occur in shallow or intermediate depths. Their immature stages obtain food in several different ways, but are usually grouped by function rather than by taxon into shredders, collectors, grazers, and predators. **Shredders** feed on coarse debris, usually leaf litter and associated microbes. They tend to be found in the headwaters of streams, where the debris is larger. **Collectors** feed on smaller debris. Collectors can be found anywhere, but are the dominant nonpredatory form in the lower reaches of watercourses. **Grazers** scrape algae from the surface of rocks, wood, and plants. **Predators** feed on the other forms of life and are found everywhere. Thus, aquatic insects clearly display the aforementioned taxonomic trend of feeding mostly on detritus and associated microbes, or in being carnivorous. Those that feed on plants mostly avoid higher plants and eat algae. The adults of many aquatic insects live only briefly and do not feed.

Feeding on Living Plants

Why do more taxa of arthropods not feed on living plants? Clearly, plants can't run away from insect her-

bivores, so they should be easy prey. Living plants are apparent resources to insects, even allowing for their somewhat limited visual acuity. Certainly, plants can be perceived accurately and from a long distance by the exquisite chemical detectors possessed by insects. The answer is that there are some important evolutionary hurdles that insects must overcome in order to feed on plants.

Among the most important problems insects must overcome in feeding on plants are the nutritional levels of the plant tissue, the physical defenses of the plants, and the chemical defenses of the plants. Plant tissues are less satisfactory nutritionally than are animal tissues or microbes. The levels of protein found in plant tissue, relative to microbes and other animals, is quite low. Fat levels also are notably low in plants as compared to animal tissues. Therefore, it is difficult for insects that feed on foliar tissues of plants to extract enough nutrition to grow properly, and to reproduce.

Also, the epidermis of foliage is thick and tough, and often equipped with a slippery, waxy layer or trichomes (hairs) that physically or chemically deter insects from attaching or feeding successfully. Hooked and pointed hairs will catch and impede insects; glandular hairs often exude sticky exudates that entrap insects.

But most important are the so-called **secondary plant substances**, which are allelochemical compounds (non-nutritive compounds produced by plants that affect other organisms, in this case insect preference and plant suitability) such as alkaloids, phenolics, resins, tannins, essential oils, and glucosinolates. Secondary plant substances play no role in plant primary metabolism, hence the curious 'secondary' designation. Nevertheless, they are quite important to plants as they affect the activity of plant feeding insects, plant disease organisms, and other plant parasites. They often act as feeding deterrents or as toxins, but also may impede digestion (digestibility-reducing substances). Some of the more toxic secondary plant compounds have been extracted from plants and used commercially as potent insecticides, including nicotine, rotenone, ryania, sabadilla, and pyrethrins.

Allelochemical compounds often determine what insects eat. Some insects deal with the presence of allelochemicals by eating only small quantities of a particular plant, then move to feed on something else. This limits the amount of a particular toxin that is ingested. Insects that feed on a number of plants, particularly those that feed on plants from more than one plant family, are said to be **polyphagous** or **generalists**.

Other insects adapt to the presence of allochemicals by producing chemical detoxification systems that neutralize feeding deterrents and toxins produced by plants. This likely comes at an energetic cost, but allows the insects to specialize on a plant where other insects cannot readily feed. Thus, competition for food is reduced, giving the specialists an advantage. Sometimes the toxins present in the bodies of these herbivores, although not affecting the feeding insects, deter feeding by predators on the herbivores, providing the plant-feeding insects with yet an additional advantage for specializing its feeding behavior. The different plants in a plant family often share common chemistry, so it is not unusual for insects that feed regularly on one plant species to be able to feed on related species. Insects that limit their feeding to related plants, normally a single plant family, are said to be **oligophagous**. In a sense, oligophagous insects are specialists, at least relative to generalists. However, there are also insects that feed only on a single species of plant. These are said to be **monophagous**, and they are true **specialists**.

Clearly, higher plants are not simply passive prey for herbivores. Most coevolved with herbivores, particularly insects but more recently mammals, so they have many adaptations that enhance their survival. As noted previously, those insects that are able to detoxify the secondary plant substances, or otherwise deal with the antiherbivore defenses, also are able to extract sufficient nutrition for growth and reproduction from plant tissues, are free to take advantage of the abundance of plants, and can multiply. Thus, we have some taxa of insects that evolved the ability to exploit plants, and they have prospered. Among these are some groups of Hemiptera, Diptera, Hymenoptera and Coleoptera, plus nearly all of the Lepidoptera. Consequently, this relatively small number of higher taxa (orders) contains a very large number of species (Table 2.3) because they evolved at a time when land plants were also rapidly radiating.

Insects exploit plants in many ways, and they can be sorted into feeding guilds based on their behavior. Among the important feeding guilds are:

• **Leaf mining insects**. These insects feed between the upper and lower epidermis of leaves. Not surprisingly, they usually are either quite small or quite flattened. The mine starts quite small and gets larger as the insect grows. Different species produce quite different mining patterns, but each species creates a consistent and characteristic pattern. Leaf miners generally are larvae of flies (Diptera), beetles (Coleoptera), and moths (Lepidoptera).

• **Skeletonizing insects**. These insects chew off the surface of leaves but the leaf veins remain uneaten. Sometimes they feed only on one surface (usually the lower surface), but the remaining leaf tissue on the opposite side of the leaf perishes. Remnants of the foliage remain, particularly the veins or 'skeleton' of the leaf, providing the basis for the name of this type of leaf feeding. Skeltonizers generally are beetles (Coleoptera) and sawflies (Hymenoptera) but sometimes moths (Lepidoptera), especially in the early instars.

• **Leaf chewing insects**. Although many different types of feeding involve feeding on leaves, only the insects that totally remove sections of tissue are called leaf chewers. Such insects typically consume entire leaves or large sections of leaves. Leaf chewers typically are beetles (Coleoptera), grasshoppers (Orthoptera), sawflies (Hymenoptera), and larvae of moths and butterflies (Lepidoptera).

• **Boring insects**. Insect that tunnel into stem or trunk tissues are usually called borers. Sometimes they feed on the cambium, or inner bark, but sometimes they burrow deeply. When they tunnel into stems they sometimes disrupt the conducting properties of the branch, causing death of the distal portions. Similarly, when enough borers feed on the cambium, they disrupt the translocation of water and food for entire trees, causing tree death. Borers usually are beetles (Coleoptera) or larvae of moths (Lepidoptera).

• **Piercing-suckering insects**. Insects with needle-like piercing-sucking mouthparts tap into the xylem or phloem of plants, removing sap. Often they stunt or cause other deformities of growth without killing the tissue. They also are efficient transmitters of plant viruses, which can be much more injurious to plants than the insect feeding alone. The common piercing-sucking insects are bugs (Hemiptera) and thrips (Thysanoptera). The order Hemiptera is very diverse, and the important plant feeding insects are found in many different-looking groups including whiteflies (Aleyrodidae), adelgids (Adelgidae), aphids (Aphididae), soft scales (Coccidae), armored scales (Diaspididae), mealybugs (Pseudococcidae), psyllids (Psyllidae), planthoppers (Fulgoroidea), cicadas (Cicadidae), spittlebugs (Cercopidae), treehoppers (Membracidae), leafhoppers (Cicadellidae), leaf-footed bugs, (Coreidae), and stink bugs (Pentatomidae).

• **Gall-forming insects**. Insects and other arthropods can induce abnormal, cancer-like growth of plants

Table 2.3. The insect orders and approximate numbers of described species.

Order	Common name	Number worldwide	% of total
Apterygota (ametabolous insects)			
Zygentoma	silverfish	320	<0.1
Archaeognatha	bristletails	250	<0.1
Exopterygota (hemimetabolous insects)			
Ephemeroptera	mayflies	2,100	0.2
Odonata	dragonflies, damselflies	5,500	0.5
Plecoptera	stoneflies	2,000	0.2
Embiidina	webspinners	300	<0.1
Phasmatodea	stick and leaf insects	3,000	0.3
Mantodea	mantids	1,500	0.2
Mantophasmatodea	gladiators	13	<0.1
Blattodea	cockroaches	4,000	0.4
Isoptera	termites	2,900	0.3
Grylloblattodea	rock crawlers	26	<0.1
Orthoptera	grasshoppers, katydids, crickets	25,000	2.7
Dermaptera	earwigs	1,200	0.1
Zoraptera	angel insects	30	<0.1
Psocoptera	book lice, bark lice	2,000	0.2
Hemiptera	bugs	100,000	10.9
Thysanoptera	thrips	4,000	0.4
Phthiraptera	chewing lice, sucking lice	3,500	0.4
Endopterygota (holometabolous insects)			
Megaloptera	alderflies, dobsonflies, fishflies	190	<0.1
Raphidioptera	snakeflies	188	<0.1
Neuroptera	lacewings, ant lions, mantidflies	5,000	0.5
Coleoptera	beetles	370,000	40.0
Strepsiptera	stylospid	600	<0.1
Mecoptera	scorpionflies	480	<0.1
Trichoptera	caddisflies	7,000	0.7
Lepidoptera	moths, butterflies	150,000	16.4
Diptera	flies	100,000	10.9
Siphonaptera	fleas	2,300	0.2
Hymenoptera	wasps, ants, bees, sawflies	120,000	13.1

cells. Protracted secretion of hormone-like substances is responsible for these abnormalities. Galls are mostly aesthetic issues unless they become extremely numerous. The common gall-forming insects are gall wasps (Hymenoptera: Cynipidae) and gall midges (Diptera: Cecidomyiidae), but some psyllids (Hemiptera: Psyllidae), adelgids (Hemiptera: Adelgidae), aphids (Hemiptera: Aphidae), flies (Diptera), sawflies (Hymenoptera) and especially mites (Acari: Eriophyiidae) cause galling on plants.

• ***Pollen feeders***. Although we commonly associate bees (Hymenoptera: Apidae) with pollen collection, they are only one of a very large number of arthropods that eat pollen. Beetles (Coleoptera), flies (Diptera), thrips (Thysanoptera), and even mites (Acari) are among the arthropods that frequently consume pollen. Plant pollen can be a good source of protein, with many plants producing pollen with at least 20% crude protein, and some with over 30% crude protein. Fat and amino acids similarly can be high, but variable among plant species. Thus, although pollen can be an important food source, there is considerable variation in the nutritional value among plants, so not all pollen is very nutritious.

• *Nectar feeders*. Nectar from floral nectaries serves as an attractant to pollinators. Insects with tubular mouthparts, such as moths and butterflies (Lepidoptera), can extract such liquids efficiently even from deep within elongate flowers. However, bees (Hymenoptera: Apidae) and flies (including mosquitoes) (Diptera) can also imbibe nectar from floral nectaries, though usually not from deep-throated blossoms. The extrafloral nectaries found on the stems and foliage of plants are even more accessible, often producing small droplets of sweet liquid that are readily available to ants (Hymenoptera: Formicidae), beetles (Coleoptera) and other insects. Nectar is sugar-rich, and also contains important amino acids and other nutrients. Bees can obtain amino acids from pollen, but moths and butterflies, due to their tubular mouthparts, normally ingest only nectar. Thus, butterfly-pollinated plants provide nectar richer in amino acids than do bee-pollinated plants.

• *Fruit or seed feeders*. The reproductive tissues of plants, both fruit and seed, are often fed upon by insects. You might expect these tissues to be especially well defended by plants, and the seeds normally are, but the primary function of the fruit is to induce animals (usually birds and mammals) to eat the fruit and seeds so as to transport the seeds to a new location where they will be excreted with their solid wastes. Although the seeds are often well defended chemically, some insects are always able to circumvent the defensive chemistry. Some species feed exclusively on the seeds whereas others consume only the fruit or both the fruit and seed.

• *Root-feeding insects*. The root feeders were discussed previously (see Feeding belowground). The common root-feeding insects tend to be beetles (Coleoptera) and flies (Diptera), but some caterpillars (Lepidoptera), bugs (Hemiptera), mole crickets (Orthoptera: Gryllotalpidae) and others also feed on plant roots.

Feeding on Blood

Blood feeding is not widespread among arthropods. Among insects it occurs mostly among flies (Diptera), particularly in mosquitoes (Culicidae), sand flies (Psychodidae), deer- and horse flies (Tabanidae), black flies (Simulidae), biting midges (Ceratopogonidae), tsetse flies (Glossinidae), stable and horn flies (Muscidae), and louse flies (Hippoboscidae). In most cases, it is only the females that blood feed, but in the tstetse flies and in the muscid flies, both sexes feed on blood. In addition to flies, blood feeding commonly occurs in fleas (Siphonaptera), sucking lice (Phthiraptera), some bugs (Hemiptera: particularly Cimicidae and Reduviidae), but occasionally elsewhere (e.g., blood feeding moths, *Calyptra* spp. [Lepidoptera: Noctuidae]). Mites and ticks (Acari) also can be important blood feeders. The significance of blood feeding is not in the number of taxa or species involved. The significance is based on the ability of blood-feeding insects to transmit diseases; this ability greatly enhances the impact of blood feeders on animal populations. Blood-feeding insects affect not only mammals and birds, but also reptiles and amphibians, and even fish on occasion. These insects are not limited to feeding on blood. They commonly feed on sugar, which can be obtained from floral nectaries, extrafloral nectaries, aphid honeydew, juices from fruits, and other plant exudates.

Blood feeding, though essential for many insects, is variable in occurrence. For example, in many ticks a blood meal is necessary between each stage of growth. The larva, pupa and adult each must feed, and so are called three-host ticks. For two-host ticks, the larva does not leave the host, so the larvae and pupa both feed on the same host, but then the pupa departs and molts to an adult that seeks the second host. Some species are one-host ticks. However often ticks leave their host between blood meals, it seems to be a seemingly risky strategy considering that they do not fly and therefore are not very mobile and able to locate another host. They compensate for this by being long-lived and hardy, waiting patiently for prey to come within range of their grasp so they can attach and feed. Likewise, blood feeding is commonly needed to complete production of eggs by blood-feeding flies; there may even be a direct correlation between the size of the blood meal and the number of eggs produced. However, many species of Diptera can produce their first batch of eggs without feeding on blood. This capacity is called *autogeny*. If blood is a prerequisite to egg production, this is called *anautogeny*.

Perhaps because mosquitoes are the most commonly encountered blood feeders, we tend to think of blood feeding insects as piercing-sucking insects that puncture the blood vessel to remove blood. This type of feeding is found in mosquitoes (Diptera: Culicidae), sucking lice (Phthiraptera), fleas (Siphonaptera), and bugs (Hemiptera). However, many insects feed from pools of blood that develop when the animal's epidermis is cut or abraded by tearing or cutting mouth structures. Such insects then tend to lap up or suck up the blood from the pool forming on the surface of the epidermis. This pool-feeding behavior is found in horse- and deer flies (Diptera: Tabanidae), black flies (Diptera: Simulidae), and some other flies.

Predation and Parasitism

Over one-fourth of all insects are **entomophagous** (insect-feeding). Although they are greatly outnumbered by **detritophagous** (detritus-feeding), **mycetophagous** (fungus-feeding), and **phytophagous** (plant-feeding) species, it is not likely that competition *per se* is the basis for the large number of predatory and parasitic insects. Rather, the nutritional advantages of feeding on animal tissue likely account for entomophagous behavior, despite the ability of potential prey to flee and to fight back. **Predatory insects** are those that regularly feed on other insects, and must consume several to complete their development. **Parasitic insects** are usually called parasitoids to differentiate them from true parasites, which do not kill their hosts. **Parasitoids** develop on or in a host insect and kill it, but complete their development on a single host. Most parasitoids of insects develop inside their host, so they are called **endoparasitoids**. Some parasitoids reside externally, though also feeding on the host, and so they are called **ectoparasitoids**. Feeding by insects on conspecifics (organisms of the same species) is called **cannibalism**, or intraspecific predation.

Predation is well developed in many orders, and in virtually all orders if cannibalism is included. Parasitism is found mostly among wasps (Hymenoptera), stylopids (Strepsiptera), some flies (Diptera), though occasionally elsewhere. However, as with phytophagy, specialized predation and parasitism is best developed in the more advanced (later evolving) groups. Cannibalism is a normal phenomenon for many arthropods, not an anomaly. Cannibalism has been documented in many insect orders, including Odonata, Orthoptera, Thysanoptera, Hemiptera, Trichoptera, Lepidoptera, Diptera, Neuroptera, Coleoptera, and Hymenoptera. As noted previously, insects have difficulty in extracting adequate nutrition from leaf tissue, and often benefit from being a carnivore or a cannibal. This often is manifested in more rapid growth, larger size, and increased reproduction when insect tissues are consumed.

SUMMARY

• The integument, or body covering, consists of hardened plates connected with flexible intersegmental membranes. The integument consists of a living portion called the epidermal cells, but consists mostly of a nonliving portion called the cuticle.
• One problem with having a rigid body covering is that it impedes growth. Insects must molt (shed their old integument, and produce a new one) in order to grow appreciably. The molting process is regulated by hormones, particularly ecdysone, and about 90% of the old integument is digested and recycled into creation of a new integument.
• The three principal body regions (tagma) of insects are the head, thorax, and abdomen. The head contains most of the sensory organs, particularly the eyes and antennae, and the mouthparts. The thorax bears the structures that provide locomotion: three pairs of legs, and (if present) one or two pairs of wings. The abdomen contains the organs of digestion, excretion, and reproduction. Some structures found on the abdomen are associated with mating or oviposition, and some have a defensive structure called a sting. Stinging insects can produce potent venom.
• Internally, insects have the same life systems found in all animals, but often they are modified to suit the unique ecology of insects. The most important life systems are the muscular, digestive, circulatory, ventilatory, nervous, glandular, reproductive, and excretory systems.
• The muscles may be directly responsible for locomotion, or act indirectly in conjunction with the elasticity of the cuticle, to power flight. The muscles also assist in feeding, circulating blood, and maintaining posture, among other functions.
• The digestive system is sometimes highly modified to reflect the diet of the insect. Digestion occurs principally in the central portion of the alimentary canal, the midgut. The midgut is protected from abrasion by a semipermeable membrane called the peritrophic membrane. Sometimes insects display extraoral digestion, secreting enzymes and then sucking up digested host liquids.
• Blood (hemolymph) is circulated freely in insects rather than in veins. A heart muscle provides propulsion of the blood, however. The blood cells also act like the lymph system of vertebrates, attacking foreign substances such as micro-organisms.

• Ventilation is accomplished through a series of small tubes called trachea, and smaller tubes called tracheoles. These tubes deliver oxygen to tissues and cells thoughout the body, and remove carbon dioxide. Oxygen enters the tubular system though openings called spiracles; carbon dioxide is expelled through the same openings.

• The nervous system of insects functions similar to that of higher animals, with impulse transmission both electrical and chemical in nature. The center of perception and coordination is the brain. It is located in the head, where it connects directly to the eyes and antennae. However, the ventral nerve cord also consists of considerable nervous tissue, and innervates much of the body.

• Insects have several types of eyes, but the compound eyes are the largest type, and the most important in adults. The functional unit of the eye is called the ommatidium, and compound eyes consist of many ommatidia. Insect compound eyes perceive motion, color, depth, and light polarization.

• Insects possess exocrine glands, releasing secretion to the outside of their body, and endocrine glands, that release secretion internally. Among the important exocrine secretions are saliva, silk, poison, odor, wax, and pheromones. Among the important endocrine secretions are juvenile hormones, prothoracicotropic hormone, and ecdysteroids. The hormones regulate the developmental progression of insects through the immature stages to the adult form, a process called metamorphosis.

• Insects sometimes display two or more discrete different phenotypes, or physical appearances. This is called polyphenism or polymorphism. Polyphenisms are adaptations to reliable variations in the environment, but are induced by hormones that trigger different patterns of gene expression.

• Communication or signaling by insects takes many forms, but generally falls into the categories of visual, acoustical, or chemical communication. Many insects possess a stridulatory apparatus for sound production and a tympanum for hearing. Many more insects produce intraspecific chemicals called pheromones, usually to communicate the availability of mates, food, or danger.

• Cooperation among insects, or sociality, varies considerably. Although most insects are solitary, some display limited brood care (subsocial behavior), provide brood care and form colonies (parasocial behavior), or have brood care, colonies and caste systems (eusocial behavior). Sociality is best developed in ants, bees and wasps, and in termites.

• The process of metamorphosis is regulated by hormonal control of genes, and varies among insects. The most primitive insects show little change in form as they grow (ametabolous development), but some show moderate change in form (hemimetabolous) or radical change in form (holometabolous development). The immatures of hemimetabolous insects are called nymphs, whereas holometabolous insects have larval and pupal stages preceding the adult stage.

• The male reproductive system consists principally of testes, and produces sperm. The female reproductive system consists principally of ovaries and produces eggs, although it also collects and stores sperm from the male. The mode of reproduction is quite variable in insects, and females of different species may deposit eggs, larvae or nymphs. Some species do not require fertilization for reproduction, and others alternate between sexual and asexual reproduction..

• Reproduction in insects usually includes production of eggs. The egg is protected by a covering (shell) called the chorion. The chorion is adapted to allow gas exchange but to limit water loss, and bears a penetration (micropyle) that allows entry of sperm for fertilization of the egg.

• The most important components of the insect excretory system are the Malpighian tubules. They serve to extract water, salts, sugars, amino acids, and nitrogenous wastes from the hemolymph and channel them into the alimentary canal and then into the excreta.

• Insects generally are ectothermic; their body temperature largely reflects the conditions of their environment. Ectothermic animals typically have a temperature below which they do not develop and a temperature above which they will not develop. In between these temperature extremes, the relationship between insect development rate and environmental temperature is largely linear, providing predictability to insect forecasting.

• Insects may escape adverse environmental conditions by entering diapause, a form of arrested development. Many insects survive freezing weather during diapause by elevating the concentration of glycerol in their blood. The occurrence of diapause affects the number of complete life cycles per year (voltinism).

• Insects are varied in their food habits. Scavenging (feeding on detritus), carnivory (feeding on animals, especially other insects), and phytophagy (feeding on plants) all are common feeding behaviors.

REFERENCES AND ADDITIONAL READING

Blum, M.S. (2008). Allelochemicals. In Capinera, J.L. (ed.) *Encyclopedia of Entomology*, 2nd edn., pp. 121–133. Springer Science & Business Media B.V., Dordrecht, The Netherlands.

Blum, M.S. (2008). Multifunctional semiochemicals. In Capinera, J.L. (ed.) *Encyclopedia of Entomology*, 2nd edn., pp. 2506–2513. Springer Science & Business Media B.V., Dordrecht, The Netherlands.

Chapman, R.F. (1998). *The Insects: Structure and Function*, 4th edn. Pergamon Press, New York.

Daly, H.V., Doyen, J.T., & Purcell III, A.H. (1998). *Introduction to Insect Biology and Diversity*. Oxford University Press, Oxford, UK.

Elzinga, R.J. (1994). *Fundamentals of Entomology*, 6th edn. Pearson–Prentice Hall, Upper Saddle River, New Jersey, USA.

Gillot, C. (2005). *Entomology*, 3rd edn. Springer, Dordrecht, The Netherlands.

Goula, M. (2008). Acoustical communication in Heteroptera (Hemiptera: Heteroptera). In Capinera, J.L. (ed.) *Encyclopedia of Entomology*, 2nd edn., pp. 23–33. Springer Science & Business Media B.V., Dordrecht, The Netherlands.

Harbourne, J.B. (1993). *Introduction to Ecological Biochemistry*, 4th edn. Academic Press, London, UK.

Hinton, H.E. (1981). *The Biology of Insect Eggs*. Pergamon Press, Oxford, UK.

Juniper, B. & Southwood, T.R.E. (1986). *Insects and the Plant Surface*. Edward Arnold, London, UK.

Kime, R.D. & Golovatch, S.I. (2000). Trends in the ecological strategies and evolution of millipedes (Diplopoda). *Biological Journal of the Linnean Society* **69**, 333–349.

Legg, D. (2008). Phenology models for pest managment. In Capinera, J.L. (ed.) *Encyclopedia of Entomology*, 2nd edn., pp. 2834–2841. Springer Science & Business Media B.V., Dordrecht, The Netherlands.

Lehane, M. (2005). *The Biology of Blood-Sucking in Insects*, 2nd edn. Cambridge University Press, Cambridge, UK.

McAuslane, H.J. (2008). Pheromones. In Capinera, J.L. (ed.) *Encyclopedia of Entomology*, 2nd edn., pp. 2842–2846. Springer Science & Business Media B.V., Dordrecht, The Netherlands.

Merritt, R.W., Cummins, K.W., & Berg, M.B. (2008). *An Introduction to the Aquatic Insects of North America*, 4th edn. Kendall/Hunt Publishing Company, Dubuque, Iowa, USA.

Nation, J.L. (2008). *Insect Physiology and Biochemistry*, 2nd edn. CRC Press, Boca Raton, Florida, USA.

Nijout, H.F. (2003). Development and evolution of adaptive polyphenisms. *Evolution and Development* **5**, 9–18.

Sanborn, A. (2008). Thermoregulation in insects. In Capinera, J.L. (ed.) *Encyclopedia of Entomology*, 2nd edn., pp. 3757–3760. Springer Science & Business Media B.V., Dordrecht, The Netherlands.

Sanborn, A. (2008). Acoustic communication in insects. In Capinera, J.L. (ed.) *Encyclopedia of Entomology*, 2nd edn., pp. 33–38. Springer Science & Business Media B.V., Dordrecht, The Netherlands.

Scharf, M.E. (2008). Neurological effects of insecticides and the insect nervous system. In Capinera, J.L. (ed.) *Encyclopedia of Entomology*, 2nd edn., pp. 2596–2605. Springer Science & Business Media B.V., Dordrecht, The Netherlands.

Slansky Jr., F. & Rodriguez, J.G. (1987). *Nutritional Ecology of Insects, Mites, Spiders, and Related Invertebrates*. John Wiley & Sons Ltd, New York, USA.

Southwood, T.R.E. (1973). The insect/plant relationship – an evolutionary perspective. In H.F. van Emden (ed.) *Insect/Plant Relationships*. Symposia of the Royal Entomological Society, No. **6**, pp. 3–30. Blackwell Scientific Publications, Oxford, UK.

Teal, P.E.A. (2008). Sex attractant pheromones. In Capinera, J.L. (ed.) *Encyclopedia of Entomology*, 2nd edn., pp. 3349–3354. Springer Science & Business Media B.V., Dordrecht, The Netherlands.

Truman, J.W. & Riddiford, L.M. (1999). The origins of insect metamorphosis. *Nature* **401**, 447–452.

Valderrama, X., Robinson, J.G., Attygalle, A. B., & Eisner, T. (2004). Seasonal appointment with millipedes in a wild primate: A chemical defense against insects? *Journal of Chemical Ecology* **26**, 2781–2760.

Ward, J.V. (1992). *Aquatic Insect Ecology. 1. Biology and Habitat*. John Wiley & Sons Ltd, New York, USA.

Weldon, P.J., Aldrich, J.R, Klun, J.A., Oliver, J.E., & Debboun, M. (2003). Benzoquinones from millipedes deter mosquitoes and elicit self-anointing in capuchin monkeys (*Cebus* spp.). *Naturwissenschaften* **90**, 301–304.

Williams, D.D. & Feltmate, B.W. (1992). *Aquatic Insects*. CAB International, Wallingford, UK.

Willis, J.H. & J.S. Willis, J.S. (2008). Metamorphosis. In Capinera, J.L. (ed.) *Encyclopedia of Entomology*, 2nd edn., pp. 2350–2354. Springer Science & Business Media B.V., Dordrecht, The Netherlands.

FOOD RELATIONSHIPS

Chapter 3

FOOD RESOURCES FOR WILDLIFE

Wildlife often seem to be limited by availability of suitable 'habitat,' which consists of shelter, food, and water. However, food limitations seem to be of paramount importance. What animals eat is of fundamental importance to most aspects of their behavior and ecology. The availability and choice of food influences their habitat associations, niche and population structures, foraging and dispersal behaviors, and population dynamics. The nature of wildlife feeding behavior ranges from strict plant feeding, **herbivory** (or **phytophagy**), to strict animal feeding, or **carnivory**, with a myriad of variations, including consumption of both animals and plants, or **omnivory**. Wildlife can be quite selective in their food selection (**specialists**), or feed broadly on many foods (**generalists**). This variability in feeding behavior is common within large taxa such as orders and families but also occurs within some species. The occurrence of various feeding behaviors of birds and mammals is shown in Table 3.1. This table presents the proportion of families that display tendencies to eat diet classes such as fruit, insects, etc. It is apparent that over 35% of bird families eat insects, a much greater proportion than any other diet, and of course they eat quite a lot of fruit (berries). Mammals, on the other hand, seem to be about equally divided between eating insects and vegetation. In both cases, however, insects clearly are important diet elements.

Feeding on insects is called **insectivory** and insect-feeding animals are called **insectivores**. As we will see, wildlife often select a mixed diet of insects, other arthropods such as spiders and millipedes, and other invertebrates such as molluscs and annelids (worms). Thus, a more correct description might be to call them **invertivores**. However, this distinction is rarely made, and generally it is satisfactory to talk about insectivory because insects tend to be the principal invertebrate diet items for terrestrial wildlife. Some aquatic wildlife, especially shorebirds, depart from this generalization.

Although not addressed in Table 3.1, the abundance of amphibians such as frogs, salamanders and newts is commonly known to be associated with insect abundance, and even smaller reptiles such as skinks and lizards, and some smaller snakes and turtles, are commonly acknowledged to feed regularly on insects. Fresh-water fish are strongly influenced by the availability of aquatic insects. Surprising to many, however, is the importance of insects as a dietary component for larger forms of wildlife. In particular, the role of insects as food for mammals (Class Mammalia) and birds (traditionally Class Aves, but now regarded by some as part of the class Reptilia) is often underappreciated.

ASSESSMENT OF INSECTIVORY

What wildlife eat is a function of both preference and availability. Food **preference** is the selection of dietary items irrespective of availability. Some wildlife are strongly adapted, either morphologically or behaviorally, to feed on certain insects. Thus, they may feed preferentially on certain prey, such as wood-boring beetles or ants. Others feed more widely, perhaps influenced more by habitat than by taxa. For example, some wildlife prefer open grasslands, whereas others prefer aquatic habitats, dense vegetation, or even only

Insects and Wildlife: Arthropods and Their Relationships with Wild Vertebrate Animals, 1st edition. By J.L. Capinera. Published 2010 by Blackwell Publishing.

Table 3.1. Diet preferences of birds and mammals expressed as the number of families feeding on each diet class (adapted from Fleming 1991).

Taxon (Number of families)	% of families consuming					
	Mostly fruit	Fruit & other	Insects	Vegetation	Vertebrates	Other
Mammals (107)	0.9	18.7	33.6	32.7	3.7	10.3
Birds (135)	16.3	19.3	35.6	5.9	5.9	17.0

the upper strata of trees. They cannot display their diet preference, of course, without the insects being adequately available. **Availability** is the abundance of potential prey when and where the predator is searching. Availability of a prey item in the appropriate habitat is an important factor in determining whether or not wildlife will eat that prey. If the prey are abundant, but not found in the appropriate habitat, they may not be located.

Often it is desirable to have information on availability of insect prey to determine whether wildlife are displaying preference for a particular prey item, or simply taking advantage of a particularly plentiful food resource. This requires qualitative and quantitative assessment of prey populations.

Methods for Determining the Abundance of Insects

From the perspective of the wildlife biologist or the entomologist, abundance is usually determined by various relative and absolute sampling methods. **Relative sampling measures** provide indices of abundance such as the numbers of insects per surface area of a sticky trap, numbers of insects per pitfall trap-night, or numbers per ten sweeps with a sweep net. These efforts tell us only that the insects are more, or less, abundant than at another location or another time. Some of the common methods of obtaining relative measures of abundance are:

• *Sticky traps.* An adhesive is applied to a surface, commonly flat, spherical, or cylindrical, and suspended 1–2 m above the soil surface where flying insects that alight are captured. The color of the trap influences attractiveness, so this technique is better for some insects than others. Sticky traps are messy to work

with and the insects are often damaged if they are removed. Small insects may not adhere, and very large ones may escape. Ground dwelling and non-flying insects are under-estimated. However, these traps are simple to construct.

• *Water-pan traps.* A bowl containing water, and usually a small amount of soap to reduce surface tension and antifreeze as an evaporation retardant, is placed on the surface of the soil to capture insects that are alighting. Like the sticky traps, capture rates are influenced by color. The advantage is that the insects are well preserved. The disadvantage is that there may be overflow in a rainstorm.

• *Beating and shake-cloth.* For beating, cloth stretched over a frame is held beneath a branch that is then beat with a rod. For shaking, the same idea applies but the cloth is spread on the soil and the branch is simply shaken over the cloth. When a branch is struck or shaken, many insects instinctively drop, so they can be collected on the beating cloth. This is a good way to capture beetles and many small wingless insects, but winged individuals soon take flight and can be lost.

• *Sweep-net.* A net is commonly used to capture insects that are sheltered in vegetation because it is inexpensive and easy to use. In this case, the vegetation is beat with the net so it must be well constructed and not prone to tear or rip. This works well with reasonably short vegetation, but samples mostly insects in the upper strata of vegetation.

• *Pitfall traps.* These are steep-sided containers that are sunk in the soil with the opening at the soil surface. They capture surface-dwelling arthropods that stumble in and cannot escape. A killing or preserving liquid may be used to keep captured insects from killing one another or taking flight. Holes may be included to allow rainfall to escape rather than overflowing the container. Diverters may be used with the containers to steer the insects to the trap, thereby increasing the number of captures.

• *Emergence traps.* When insects pupate or over-winter below-ground, in water or in plant materials, it is sometimes useful to use emergence traps to assess abundance and emergence times. Commonly, cone-shaped traps topped by insect capturing devices are placed over the item of interest to assess the number and time of emergence of insects.

• *Interception traps.* Netting or clear plastic may be suspended in such a manner as to intercept flying insects. Malaise traps are large net structures that not only interrupt flight but divert any insects that alight on the net to a capture device, usually a jar with a small entrance hole and from which it is difficult for insects to escape. Panel traps contain a liquid-containing trough at the base so when the flying insect hits the transparent panel it drops into the liquid and is captured. Insects that drop readily, such as beetles, are easily captured with panel traps. In contrast, those that alight and climb when they encounter a barrier, such as flies and wasps, are more readily captured with Malaise traps.

• *Baited traps.* Various devices can be baited with food such as fruit, sugar-water, tuna fish, or carrion to attract insects. More than with most traps, baited traps capture only a subset of the total insect fauna.

Absolute sampling measures are designed to iden-tify or capture all the insects present in a defined area. They tend to be more labor-intensive and expensive, but have the advantage of providing higher quality data on insect abundance. This allows direct compari-son among different insect taxa because the insect abundance data are not biased by the capturing or trapping device, as is the case with relative sampling measures. Some examples include:

• *Direct observation.* Careful visual examination of soil, water, or small plants can be effective if the insects are not too small or cryptic. Examination of one square meter sample units is appropriate for terrestrial studies where the vegetation is not too tall or dense. For taller vegetation such as trees, clipping branches with pole pruners followed immediately by visual examination can be effective. Another variation is to enclose vegeta-tion in a bag, then clip it, and return the vegetation to the laboratory for examination. Fumigation or freezing of the bag and contents ensures that there will be no escape.

• *Suction traps.* Stationary suction traps powered by a fan can be used to sample a known volume of air.

This works well for small flying insects such as aphids and small flies. Portable suction traps (often called 'd-vacs') can be used to vacuum a unit of habitat such as a plant or small plots.

• *Night cages and drop cages.* A cage can be posi-tioned over vegetation at night, or carefully positioned or dropped over a section of vegetation, capturing all insects within the cage. Sleeves in the side of the cage allow the investigator to collect the insects captured, including vacuuming the vegetation and soil surfaces. Cage capture techniques work well if the nights are cool and the insects are inactive. Even during the day, if the cage is suspended from long handles or poles, it is usually possible to position the cage without disturb-ing the insect fauna. Sometimes the cage is pre-posi-tioned and dropped at a later date using a remote release mechanism.

• *Pesticide knockdown.* Application of fast-acting, broad-spectrum pesticides such as pyrethroids can be used to kill insects for collection and examination, especially if they are contained by a cage or bag.

• *Extraction.* Insects can be driven from soil or veg-etation samples by heat or chemicals, and in some cases water can be used to separate arthropods from these substrates. The most common technique, called the Berlese or Tullgren funnel, uses a heat source to dry the substrate from the top, encouraging the insects to move away and fall into the tunnel where they are directed into a receptacle, often a jar containing alcohol.

No sampling method is without pitfalls. Relative methods of sampling are most often used because they are easy or inexpensive. However, they invariably are biased for or against certain insect taxa or certain behaviors, so interpretation of data collected in this manner must be used cautiously. Absolute methods of sampling give a much better picture of the insect fauna present, but are much more difficult and expensive to implement in most environments (see Table 3.2 and Figs. 3.1, 3.2). In either case, knowledge of insect behavior and ecology is important if meaningful con-clusions are to be made. Also, identification of insects only to a higher-level taxon (e.g., order) often is not very satisfactory. For interrelationships to be fully understood, species-level resolution, or at least family-level resolution, of insect prey can be important. As is the case with vertebrates (e.g., consider the diversity shown by carnivorous mammals and passerine birds) often there is too much diversity within orders and families of many insect groups to allow adequate

Table 3.2. Summary of some arthropod sampling techniques commonly used in diet studies (adapted from Cooper and Whitmore 1990).

Method	Arthropods sampled	Advantages	Disadvantages
Sticky trap	flying or otherwise active	inexpensive; easily deployed	messy to use; trap color and environmental temperature affect capture rates
Malaise trap	flying	easy to maintain; captures many insects	expensive to purchase and hard to deploy
Shake-cloth	foliage-dwelling	inexpensive; good for sessile species	not reliable for active insects or those occurring high in canopy
Sweep-net	foliage-dwelling	simple, inexpensive	affected by foliage height
Emergence traps	arthropods emerging from soil, water or plant materials	accurate, inexpensive and good for assessing timing of emergence	large number required to assess accurately
Pole pruning	foliage-dwelling, particularly sessile species	inexpensive	large number required, good for sessile species
Branch-clipping	foliage-dwelling	more sensitive than pole-pruning	best for sessile species
Interception traps	good for flying insects	collects great diversity of species	only flying species collected; can be expensive
Baited traps	species attracted to baits	may attract from a distance, increasing capture rate	selective, attracting only certain species
Portable suction	foliage-dwelling	good when used within sampling cage	expensive
Stationary suction	flying	gives volumetric assessment and time data	expensive
Direct observation	surface-dwelling (foliage and soil)	can compare abundance of different taxa	prone to sampling bias
Cage capture	foliage and soil-dwelling	good for direct comparison of taxa; efficient capture	time consuming, limiting number of samples possible
Pesticide knockdown	foliage-dwelling	samples diverse taxa with minimal bias	can underestimate especially active or sessile spp.
Berlese–Tullgren funnel	mobile, nonflying	easy to use, extracts microarthropods efficiently	must apply heat slowly to allow movement rather than death

insight into ecology based on order and family determination; thus, genus or species determinations may be required. In many cases, the most satisfactory studies involve those conducted in relatively simple environments or involving only a few species. This allows better assessment of insect populations because the sampling techniques can be optimized for the fauna of interest. Wildlife often feed on certain functional feeding groups (i.e., classifications based on how or where they feed) rather than their exact taxon. Thus, we might expect to find woodpeckers accessing large numbers of woodboring beetles (Coleoptera: Curculionidae, Buprestidae, Cerambycidae) and woodboring caterpillars (Lepidoptera: Sesiidae, Cossidae) because they occur commonly in the environment where woodpeckers hunt. Thus, in this sense, the woodboring beetles and woodboring moths have more in common than they do with other insects in their respective orders.

Fig. 3.1. Examples of several types of arthropod sampling methods: (a) a beat-bucket in which plant foliage is shaken against the sides of a bucket to dislodge and count or capture arthropods; (b) a beat-cloth in which a piece of white cloth is spread over the soil surface and plants are shaken or beaten to dislodge arthropods; (c) a beat-net for capture of arthropods dislodged by beating foliage; (d) a plant cage for enclosing a unit of habitat so arthropods can be dislodged from the enclosed plant with little likelihood of them escaping; (e) example of a Berlese–Tullgren extraction device in which arthropods are driven from vegetation, litter or soil by light and heat applied above, and captured below; (f) use of a sweep net, probably the most commonly used device for capturing insects; (g) a back-pack mounted 'D-vac' suction sampler; (h) a high-powered suction sampler mounted on wheels and designed for field crop use; (i) removing soil to a predetermined depth to sample for soil-dwelling arthropods; (j) use of a 'quadrat', a device used to delimit an area of sampling; (k) visual inspection of a leaf showing whiteflies (Hemiptera: Aleyrodidae) attached on the underside of the vegetation; (l) visual inspection of an entire plant (adapted from Naranjo 2008).

Fig. 3.2. Examples of several types of traps used for sampling arthropods: (a) a canopy trap for sampling insects flying upwards from a field; (b) an emergence trap for sampling adults as they emerge from the soil following pupation; (c) a window-pane trap for assessing the movement of flying insects; (d) a modified Malaise trap for capturing flying insects; (e) an example of a sticky trap with insects adhering to the sticky surface; (f) a suction trap for aerial sampling of flying insects (adapted from Naranjo 2008).

Sampling efforts must reflect the nature of these functional feeding groups.

Availability of prey is just one indicator of potential diet because wildlife are not indiscriminate feeders. Although wildlife are influenced by availability and opportunity, they often display strong innate preferences. Thus, it is usually necessary to obtain direct measures of consumption of insects by wildlife to assess preference accurately.

Many wildlife studies do an excellent job of characterizing the consumption of insects (arthropods) by vertebrates, but do a poor job at characterizing the background population of invertebrate animals. Often researchers sample with easy-to-use techniques such as sticky boards or sweep nets, which are not appropriate for soil-dwelling arthropods, or perhaps use pitfall traps, which are good for ground-dwelling insects but not effective for foliage-dwelling insects. If we are to draw accurate conclusions about dietary 'preference' we must be certain to characterize the potential invertebrate diet components accurately.

Methods for Determining Wildlife Diets

There are various measures available to accomplish identification of wildlife diet, but like insect sampling, each has its advantages and disadvantages (Table 3.3). Among the common methods of analysis are stomach samples, emetics and gastric lavage (stomach pumping), pellet analysis, direct observation of feeding, ligature, and excrement (fecal) analysis.

The major source of dietary information for vertebrates traditionally has been stomach contents. However, killing vertebrates for scientific study is increasingly less acceptable, and certainly a major problem for threatened and endangered species. Also, stomach analysis, like all methods of diet analysis, has limitations. Studies of food habits in Canadian and Australian ducks, for example, showed that the perception of food preference changed markedly depending on where in the digestive tract the food sample originated. Gizzard analysis is very common, but invertebrate food items are often digested before they reach the gizzard. This leads to underestimation of invertebrate food and overestimation of seed, as the latter is resistant to digestion.

Emetics can be administered to vertebrates; these cause regurgitation of stomach contents and allows the investigator to obtain items that were more recently eaten, overcoming much of the differential digestion issue. Although this may seem preferable to harvesting the animal, emetics also can be damaging, causing stress and sometimes death. Stomach pumps have also been used successfully for some species, but not for others.

Some raptors, and also owls, gulls, and sometimes other birds (e.g., corvids, herons, sandpipers, kingfishers, flycatchers, honeyeaters) commonly swallow their food whole and then regurgitate undigested food afterwards. The regurgitated matter is called a **pellet** or **bolus**. By carefully examining the pellets of regurgitated food remains, the nature of the diet can often be determined.

Table 3.3. Summary of some analysis techniques commonly used for determination of wildlife diets.

Method	Advantages	Disadvantages
Stomach removal	food items in relatively good condition	must kill animal; differences in rates of digestion may skew findings
Emetic or stomach pump	less destructive than stomach removal; food items usually in good condition	requires animal capture and handling; emetics can be dangerous to animals
Food pellet	relatively easy to collect once favored resting/nesting sites are determined	applicable only to some animals; mostly suitable for less digestible diet items
Observation	little risk of harming animal under observation	quantity and quality of data often insufficient; especially difficult for nocturnal animals
Ligature	diet items in very good condition	frequent access to animals necessary
Excrement (fecal) collection	little risk of harming animal under observation; applicable to most animals	may require animal capture; mostly suitable for less digestible diet items

Observing and photographing food that animals are about to consume is another option for diet studies, though this is a challenge due to the small sample sizes possible with this technique, and the inability to examine carefully the diet items. Often this leads to broad generalizations such as order-level identification of dietary material.

Sometimes the young of wildlife study subjects are allowed to accept but not to swallow food provided by adults. For example, a ligature or neck-collar (such as a twist-tie or pipe-cleaner) may be positioned gently around the throat of a nestling, allowing the young bird to breathe and to accept any food items presented by the parents, but not allowing the bird to swallow the offering. This has limited application, of course, due to the need to disturb the nest, to retrieve the food items, and yet to allow the immature to obtain adequate sustenance.

Also, fecal material (excrement) can be collected and analyzed for dietary components. Feces usually are called **droppings** or **guano** when obtained from birds, but **scats** or **dung** when obtained from mammals, though terminology is not always consistent. Fecal materials are often obtained by capturing and holding the animals in bags (birds) or cages (mammals) for a few hours, though some mammal scat is often large enough to persist for some time, so it can be collected from the natural environment.

Some newer techniques under investigation include polymerase chain reaction (PCR) and stable isotope ratios. Molecular techniques such as PCR can be used to distinguish components of the stomach or feces. However, this is not generally used, and while excellent at establishing identities, it is less satisfactory at measuring quantities. Analysis of stable carbon and nitrogen isotope ratios offers a way of determining general patterns of feeding such as animal versus plant sources of food. The advantage is that it assesses assimilated food, not just the food present in the gut. The disadvantage is that it lacks the power of discrimination provided by stomach and fecal analyses.

However diet information is collected, it is important to consider that a single study rarely provides accurate description of dietary habits. Large numbers of animals must be observed because not only is there individual variability in food selection behavior among individual animals, but opportunities for wildlife to select prey may vary among individuals and throughout the year. It is not uncommon, for example, to have the stomach of one animal completely filled with a particular diet item, and yet for that item to be nearly absent from all other animals in a collection. There may be differences in dietary habits between immatures and adults of the same species of wildlife, or between males and females. Even among adult females, diet may vary among non-reproductive periods, during pregnancy, and during lactation (when female mammals produce milk). Animals that have a wide geographic distribution will have greatly different opportunities to feed due to differing availabilities of prey. If a prey item is abundant, animals may sometimes learn to select the abundant prey to the exclusion of other prey until the supply is depleted, but then shift to specialize on another abundant source of food. Diet information must be averaged over individual animals, locations, and times to obtain a good picture of dietary habits. Thus, for within-season sampling at a single location it is probably necessary to make observations on a minimum of 50 animals, though this is best calculated based on the variance of the diet. Though modern studies often limit sampling to a season (usually summer), animals must eat throughout the year! Some birds, for example, feed extensively or exclusively on insects in the summer months, only to switch to fruit in the autumn and seeds in the winter. In tropical areas, the wet and dry seasons often impose the equivalent changes in diet. So except for animals that hibernate (many amphibians and reptiles, plus a few mammals) sampling must be done over the entire year to learn not only what the dietary preferences might be, but what might truly limit food intake and survival of wildlife populations. Hundreds or even thousands of samples may be needed to account for geographic variation and seasonality. This is not to say that broad trends in diet selection cannot be discerned easily. For example, it is readily apparent that snail kite, *Rostrhamus sociabilis*, eats little but snails, whereas pelicans, *Pelecanus* spp., and cormorants, *Phalacrocorax* spp., specialize on fish. Also, fish-eating birds are not likely to eat ants, and large carnivores do not often eat grass seeds. However, to determine whether or not a species such as purple martin, *Progne subis*, prefers mosquitoes may take considerable sampling, including periods when the prey are relatively rare and abundant (contrary to popular lore, they do not eat many mosquitoes).

An example of how extensively diet can vary can be seen in a study of wood duck, *Aix sponsa*, conducted in Missouri, USA. Wood ducks feed mostly by pecking at objects on the water surface. In the case of females (but not males), increased consumption of invertebrates occurred during the pre-laying, laying and post-laying periods. Invertebrate consumption peaked at egg-laying, and comprised about 76% of the food consumed. Most of the invertebrates consumed were insects, especially aquatic species, not terrestrial. Invertebrates were also consumed by both males and females during the non-breeding season, although they were often terrestrial species. In the spring, 51% of the invertebrates were aquatic, whereas in the autumn, 81.4% were terrestrial (averaged over sexes in both cases). Thus, not only was there seasonal variation, but the reproductive state and sex of the ducks affected their diet preference.

Most diet analyses are conducted with stomach or fecal samples. Both types of samples are examined under a dissecting microscope for constitutive elements. Usually, suspension of the stomach or fecal samples in water or alcohol is adequate to separate arthropod remnants from other dietary components. Identification is challenging because the food items are crushed and shredded, and often partly digested (see Fig. 3.3). Usually a reference collection of known food elements is created by feeding the insects to animals and then collecting the feces, or by simulating the digestive process by breaking up the insects with metal balls (bearings) or other crushing techniques. Insects are among the dietary items that are most easily identified, at least to the level of order or family. Foliar material also is fairly identifiable, and seeds are easily determined. Nectar and fruit can be especially difficult to identify. As noted previously, food items differ in their digestibility, so although beetle components often resist digestion, earthworms disappear rapidly. Caterpillar bodies, but not the head and legs, are quite digestible. Small insects are more likely to be underestimated than large insects. Therefore, digestibility is an important consideration, especially when examining fecal samples. Table 3.4 lists some of the structures shown to be useful in arthropod identification.

How useful are analyses of insect fragments found in the diet of wildlife? Clearly, they are the most reliable methods of diet analysis currently available, and some knowledge of insect morphology is necessary to identify the fragments. Experience has shown that with guidance and training, however, extensive knowledge of insect morphology is not needed to identify diet elements. Patience and attention to detail are required!

An example of the insight that can be gained can be seen by examining a study of the foraging ecology of California gnatcatcher, *Polioptila californica californica*.

This study analyzed fecal samples from adults and juveniles collected by mist-netting, or from chicks in nests. Over a two-year period, arthropods from nine orders were identified in the fecal samples. The taxa differed considerably in how often they occurred in the diet of these birds (see Table 3.5). Numerically, the small members of the insect order Hemiptera, especially families Cicadellidae and Issidae, were most 'important' in the diet of gnatcatchers, and at least eight species from five families were identified. However, spiders (Araneae) were also very numerous, and eight families were represented in the fecal samples. The beetles (Coleoptera) were abundant, and were represented mostly by the weevils (Curculionidae) and leaf beetles (Chrysomelidae). Gnats (some of the smaller members of order Diptera) were poorly represented in gnatcatcher diets, but this sort of discrepancy is commonplace with vernacular (so-called 'common') names of animals. Gnatcatchers seem to be surface gleaners, collecting insects from the surface of plants rather than capturing them in flight.

Several methods are used to express consumption of food by wildlife. Commonly we see the amount of each type of food expressed as **volume, bulk,** or **mass** (weight) per consumer. This is excellent because it is an expression of both number and size of food items. Mass or weight is easily measured with a balance or scale; volume or bulk can be determined by adding the diet items to a volumetric flask containing water or alcohol and recording the increased volume. Alternatively, we sometimes see consumption expressed as **frequency of consumption**, which is the average proportion (%) or number of food items per consumer, not their mass or volume. The frequency of consumption data are meaningful, but they do not allow us to understand the importance of each type of food unless we also have some idea of the size of a food item. To accomplish either of these methods, the researcher must be able to sort all items into one diet category or another, and often this is not possible because they are broken into very small pieces or largely digested. Therefore, we also see consumption expressed as **frequency of occurrence** or **prevalence**, which expresses the proportion (%) of a population that contains a particular food item. This latter method is easiest to use because it does not require that all diet items be identified, as each consumer is recorded as simply having a diet item, or not. It provides the least degree of precision, however. Each of these three methods provides some insight into dietary habits of

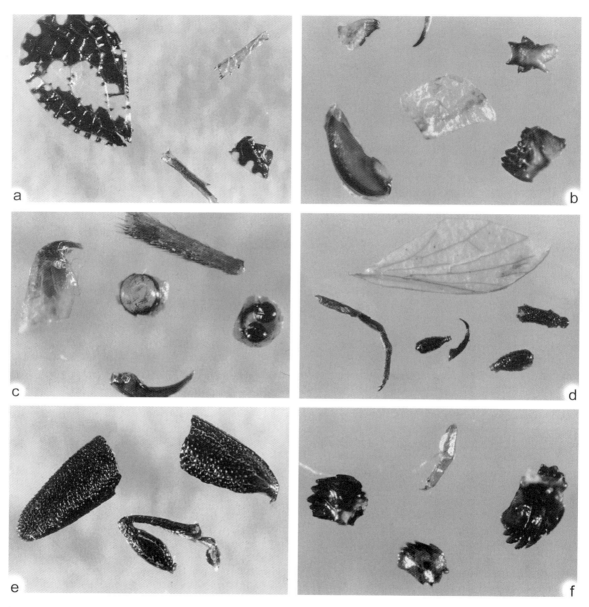

Fig. 3.3. Arthropod fragments (described clockwise beginning with the upper left-hand fragment for each image) found within fecal samples of California gnatcatchers: (a) *Dictyssa obliqua* (Hemiptera: Issidae) entire wing, hind tibia fragment, wing fragment, hind tibia fragment; (b) (Orthoptera: Gryllidae) mandible, tibial spine, tibial joint, mandible, pronotal plate, and (Orthoptera: Acrididae) hind femur; (c) (Araneae) chelicera with fang, eye, leg fragment, female epigynum, fang; (d) (Diptera) wing of suborder Nematocera, fragmented tarsus (tarsomeres and claw likely from Asilidae), intact tarsus; (e) (Coleoptera: Mordellidae) elytral fragment, (Coleoptera: Chrysomelidae) elytral fragment and intact leg; (f) (Coleoptera and Lepidoptera) larval mandible, leg, large mandible, small mandible (adapted from Burger *et al.* 1999).

Table 3.4. Some arthropod structures and features commonly found in bird droppings (adapted from Ralph *et al.*1985).

Group	Structure	Description of contents
Psocoptera	mandible	Small, translucent, but dark on the two points.
Hemiptera	'rib'	The strong, curved apodeme (an internal ridge of the exoskeleton) associated with the hind leg is distinctive for each family (nymphal psyllids lack this).
Adult	clavus, corium	These are easily recognized without the membrane attached.
Nabidae	foreleg	Tibia has two long rows of small, black teeth.
Male	clasper	Part of the male genitalia, visible at tip of abdomen.
Cicadellidae	hindleg	Tibia has rows of prominent spines, marked by dark bumps where the spine has been knocked off.
Delphacidae	hindleg	Tibia has a large, toothed, movable spur at apex. Tibia and tarsal segments have several large apical teeth.
Cixiidae	hindleg	Similar to delphacid's, but tooth pattern distinguishable, and spur lacking.
Psyllidae		
Adult	hindleg	In our species, the tibia has two or four small, dark, apical spurs.
Nymph	wing pad	For our one species with free-living nymphs, this dorsal covering
	abdominal tergite	of the abdomen was the most numerous fragment.
Neuroptera		
Chrysopidae		
Nymph	mandible, maxilla	This smooth, sickle-shaped piece (four per individual) occurs alone, showing the flat surface that matches its dorsal or ventral mate, as well as attached to its mate, forming a rounded sickle.
Coleoptera	tarsus, mandible	Beetle mandibles are so diverse as to defy generalization. They sometimes differ between adult and larva of the same species. They differ from larval Lepidoptera in being usually more elongate and bearing teeth or grinding surfaces somewhere besides the apical edge. They must also be distinguished from Hymenoptera and larval Tipulidae mandibles.
Carabidae		
Adult	foreleg	Tibia of our species is distinctively notched.
	hind trochanter	Found separate, as well as attached to the coxa.
Curculionidae		
Adult	tarsus	In our fauna, penultimate tarsal segment has two large, flat lobes.
Lepidoptera		
Adult	wing scale	Wings are covered with countless small scales, which may be swallowed even if the bird tears off the wings.
Larva	mandible	Most commonly shaped like baseball glove, or broad scoop, with
(=caterpillar)		one or more teeth along the cutting edge and a spherical knob at one of the basal corners.
	front	A triangular sclerite on the front of the head.
	spiracle	A dark, elliptical ring.
	crochet	Many occur on each proleg.
	anal comb	Only some families have this.
Diptera		
Adult		
Cyclorrhapha	antenna	Apical segment is acorn-shaped, is often encountered.
(many sturdy,	wing	Leading edge has small but stout, curved bristles.
large flies)	bristle (seta)	These are numerous, sometimes found still attached to legs. They are strong, black, slightly curved, tapered.
Larva	entire body	Surprisingly, these did occur in droppings.
Tipulidae	mandible	Finger-like teeth.
	head	V-shaped incisions along posterior edge of head capsule.

Table 3.4. *Continued*

Group	Structure	Description of contents
Hymenoptera Adult		
wasps	head	Hard, hypognathous, with a distinct, round foramen where it connects with thorax.
	mandible	Generally longer and slenderer than those of Coleoptera or Lepidoptera, with two apical teeth.
Ants	thorax	The hump or node on the gaster is distinctive.
Araneida	chelicera	Even when fangs are absent, chelicerae are distinguishable by their slightly asymmetric but conical shape and sometimes an arrangement of spines.
	fang	Curved and sharp, this piece sometimes resembles tarsal claws.
	leg	Leg segments tend to be straight-sided, whereas those of insects usually taper at the joints. Spiders' also are usually hairy. Simple tarsus with two claws is diagnostic.
Male	pedipalp	Male genitalia in the spherical or egg-shaped terminal segment occur frequently.

Table 3.5. Diet composition of California gnatcatchers, *Polioptila californica californica*, based on the occurrence of items in their droppings (adapted from Burger *et al.* 1999).

Order	Frequency of consumption (% of items in diet)	Biomass of items (% of total mass)
Hemiptera	28	11
Araneae	18	18
Miscellaneous	13	8
Coleoptera	12	15
Orthoptera	8	37
Various larvae (mostly Lepidoptera)	7	1
Diptera	6	7
Adult Lepidoptera	6	3
Hymenoptera	2	<1

consumers, and some investigators report all three methods of tabulating dietary habits, which is an ideal approach. The use of different methods sometimes makes comparisons of different research studies difficult to interpret, and the problem is only exacerbated by the inconsistent use of terminology; unfortunately this occurs frequently.

Study of primate diets is usually conducted differently. Stomach analysis is not common because researchers are reluctant to kill or handle these animals excessively, many of which are not abundant. Scats are sometimes collected for analysis, though the typical problem of differential digestion affects interpretation of such data. But, because the behavior of these animals often can be observed easily, **time budgets** are sometimes presented. Thus, data on the time primates spend feeding on fruit, foliage, or insects may be presented as an indication of dietary preference or importance. Time budgets may also be developed for other wildlife, but this is not done commonly.

Numbers of insects eaten, or frequency of occurrence of animal stomachs with a particular item present, may be indicative of what wildlife will accept, but are probably not as indicative of the 'importance' of the dietary items as are measures of diet volume, bulk, or mass (weight). In terms of nutritional value and energy gain, large items are often more significant than small items. Sometimes researchers will devise an index that factors in both occurrence and biomass, but this is not much different than the traditional volume or mass assessment. The significance of food item size versus food item number can be seen clearly in the gnatcatcher example, wherein insects in the order Hemiptera (mostly leafhoppers) are most numerous, but insects in the order Orthoptera (grasshoppers and crickets) constitute the largest mass of the food consumed. Needless to say, our perspective of the 'importance' of food items changes with this additional information on mass of the diet items. Significantly, gnatcatchers chose larger prey items for their chicks than for their own consumption. This is a common pattern with birds, and usually is explained based on energy costs or chick safety. Foraging by adults for their offspring would be optimized if the maximum amount of food could be provided to the chicks using the smallest number of trips, as this would maximize calories provided to the young per calories consumed by the parent. Additionally, predators may locate prey nests based on activity of the adults, and also due to the noisy begging activity of chicks, so minimizing trips to the nest may reduce the likelihood of discovery by predators. Digestibility probably should be considered in assessing diet suitability and optimization because in some cases a significant proportion of the diet item (especially the integument) is not very digestible and therefore of little nutritional value. However, there seems to be no popular technique for assessing diet digestibility in insectivorous wildlife studies.

A central concept in ecology is that animals will forage for food in a manner that maximizes the intake of energy per unit of time or energy expended. Finding, capturing, and consuming or transporting food takes time and effort, and it is advantageous to optimize calorie intake. This concept is called **optimal foraging theory**. The nature of the prey influences the selection process, so prey selection is not as simple as capturing the largest prey. Not only does prey size have to be balanced with prey availability, but anti-predator defenses such as cryptic coloration and toxins are important. In some cases 'handling-time', the time taken to subdue a prey organism and perhaps remove its legs or wings prior to ingestion or feeding to young, should be considered. Less well appreciated are nutritional needs beyond calories, but wildlife are constrained by other nutritional needs such as protein, which is one reason why insects are so often consumed.

The problem of differential digestion remains unresolved unless efforts are made to examine the digestion rate of diet components. This can be accomplished by comparing esophagus, stomach, and excrement samples to determine if some items are digested more quickly. Alternatively, the investigator can conduct controlled feeding studies by feeding known diet mixes to captive animals and then comparing stomach contents or excrement to the known contents of the ingested food. Some researchers have developed formulae to correct for differential digestion.

NUTRITIONAL VALUE OF INSECTS

Most animals are quite similar in their dietary requirements. They must consume some materials (called essential nutrients) but can biosynthesize others. Amino acids are the building blocks of proteins, and are usually obtained by eating proteins. Dietary proteins vary in suitability, depending on their digestibility and amino acid content. Thus, not all insects provide equivalent nutrition when consumed by wildlife. Wildlife also ingest carbohydrates, mostly in the form of sugars, for a source of energy, although carbohydrates can be derived from fats and amino acids. Other requirements include sterols, lipids, fatty acids, vitamins, and minerals. To some degree, insects can satisfy most of these needs when they are part of the wildlife diet, and are exceptionally good sources of most nutrients except carbohydrates and calcium. Wildlife often ingest insect dietary items along with other food to provide supplemental protein and fat, as well as vitamins and minerals. Perhaps no better example exists than the feeding behavior of **granivorous** (seed or grain-feeding) birds when they are nesting. The chicks of granivorous birds are often fed mostly insects, as these rapidly growing animals need a diet that is rich in protein and fat. **Nectarivorous** (nectar-feeding), **frugivorous** (fruit-feeding), **fungivorous** (fungus-feeding), and **folivorous** (foliage-feeding) species similarly need protein to feed their young, and often obtain it by feeding insects to their young. Thus, we see marked changes in the foraging pattern of such birds, and sometimes both the nestlings and adults feed heavily on insects during this period. Arthropods are surprisingly digestible to birds. The average metabolizable energy coefficients (a measure of digestibility) according to food type consumed are: nectar, 0.98; arthropods, 0.77; vertebrates, 0.75; cultivated seed, 0.80; wild seeds, 0.62; fruit pulp and skin, 0.64; whole fruits, including seeds, 0.51; and leaf material, 0.35 (higher values indicate greater digestibility).

The levels of nutrients provided to wildlife by some insects are shown in Table 3.6. Insects generally contain 55%–85% moisture. The nutrient with the highest concentration usually is protein, in large measure because the integument is protein-rich, and ranges from about 20%–80% on a dry weight basis. The other major component of most insects is fat, comprising 2%–60% of the insect's dry weight. Female insects often contain more fat than males. Ash is not usually abundant because insects lack the calcified internal skeletons found in vertebrates. Values for fiber are quite variable, with some variance likely due to the nature of the food found in the intestinal tract of the insects. Although insects tend to have low levels of calcium they have high levels of phosphorus and other minerals such as magnesium, sodium, potassium and chloride. They also contain relatively high levels of trace minerals such as iron, zinc, copper, manganese and selenium. Insects are a good source of amino acids and provide good quantities of the essential amino acids. Digestibility of insect protein is less than some diets favored by humans, such as egg, meat and milk, but higher than that of many vegetable-based proteins. Experimental research with animals such as rats, chickens and fish demonstrates that insect-based diets provide for good growth. The vitamin levels in insects are less well known, but vitamin A and carotenoids (which can be converted to vitamin A) are abundant, though vitamins B and E are low. The fatty acid composition of insects is often high, but diet-dependent. Chitin is the second most abundant biopolymer in nature, second only to cellulose. Chitin is found in the integument of arthropods, but also in algae, fungi, protozoa, and molluscs. Insectivorous and omnivorous wildlife produce the enzyme chitinase, which is neces-sary for the digestion of chitin. Animals that are not insectivorous or omnivorous lack chitinase.

As an example of food selection in relation to nutrition, consider the results of a study of co-occurring grassland birds conducted in Nebraska, USA. The birds studied were grasshopper sparrow, *Ammodramus savanarrum*; lark sparrow, *Chondestes grammacus*; and western meadowlark, *Sturnella neglecta*. These birds differed in their propensity to consume insects, with seeds comprising 5% of the adult meadowlark's diet, 33% of the adult grasshopper sparrow's diet, and 61% of the adult lark sparrow's diet. In all cases, however, nestlings were fed exclusively on insects. In each bird species, grasshoppers and adult beetles were the numerically dominant prey items although these insects were not the most abundant in the grasslands. Bugs (Hemiptera) were next in importance for the smallest bird species, the grasshopper sparrow, but the other two species chose caterpillars as their next-favored prey. All the birds avoided the smallest prey items, selecting mostly prey of intermediate size, but the larger birds also favored the largest prey. The small grasshopper sparrows spent considerable time removing chitinous portions of their prey, thereby concentrating the nutrients they ingested. This may also explain why all birds avoided the smaller prey, as there tends to be a higher relative concentration of the poorly digestible chitin in smaller insects. Clearly, these birds did not feed strictly on an opportunistic basis, and the prey selection process is complex and varies among species.

IMPORTANCE OF INSECTS IN THE DIETS OF WILDLIFE

The importance of insects and related invertebrates as food items for wildlife varies considerably. Insects are consumed regularly not only by insectivores, but by omnivorous reptiles, birds, fish, and smaller mammals. For some omnivores, especially larger carnivorous mammals, insects are an alternate source of food when more desirable, larger food items are not available. However, insects are a critical, essential source of nutrition among a sizable fraction of wildlife, and especially for the younger animals. As an example of the nutritional significance of insect-based food, consider the data presented in Table 3.7, showing some dietary requirements for several species of fish and (for comparison purposes) for rat.

Table 3.6. Nutrient content of selected insects (calculated on a dry weight basis except for moisture) (adapted from Finke 2008).

I. Moisture, crude protein, crude fat, fiber, and ash content (%) of selected insect species.

Order and species	Moisture	Protein	Fat	Fiber	Ash
Lepidoptera					
Bombyx mori	76.3	25.8	38.5		2.1
Catasticta teutila		60.0	19.0	7.0	7.0
Coleoptera					
Rhynchophorus palmarum	71.7	25.8	38.5		2.1
Tenebrio molitor	63.7	65.3	14.9	20.4	3.3
Isoptera					
Cortaritermes silvestri	77.8	48.6	6.9		8.5
Nasutitermes corniger	75.3	66.7	2.2	27.1	4.6
Hymenoptera					
Apis melifera	65.7	60.0	10.6		17.4
Atta mexicana		46.0	39.0	11.0	4.0
Diptera					
Drosophila melanogaster	67.1	56.3	17.9		5.2

II. Mineral content (mg/kg) of selected insect species.

Order and species	Ca	P	Mg	K	Na	CL	Fe
Lepidoptera							
Anaphe venata	400	7,300	500	11,500	300		100
Galleria mellonella	590	4,700	760	5,320	400	1,540	50
Nudaurelia oyemensis	1,490	8,710	2,660	11,070	1,400		87
Coleoptera							
Phyllophaga rugosa	430		1,900	11,510	790		170
Rhynchophorus palmarum	1,000	4,800	3,100	6,800	2,600		34
Tenebrio molitor	640	7,630	1,670	9,370	1,740	5,260	60
Orthoptera							
Acheta domestica	1,320	9,580	1,090	11,270	4,350	7,370	63
Gryllotalpa africanus	2,360	8,820		9,300	3,320		1,148
Isoptera							
Macrotermes subhyalinus	400	4,420	4,210	4,810	19,880		76
Nasutitermes corniger	2,000	4,000	1,300	6,100	2,400		394

Table 3.6. *Continued*

III. Amino acid content (%) of some insect species. Note: not all essential amino acids are shown).

Order and species	Argenine	Histidine	Leucine	Methionine	Phenylalanine	Tryptophan	Valine
Lepidoptera							
Galleria mellonella	1.71	0.80	2.99	0.53	1.28	0.29	1.64
Nudaurelia oxemensis	3.88	1.11	5.05	1.44	3.58	0.98	5.86
Coleoptera							
Rhynchophorus palmarum	1.62	1.02	1.62	0.27	0.73	0.25	0.81
Tenebrio molitor	2.81	1.87	5.40	0.83	1.71	0.72	4.13
Orthoptera							
Acheta domesticus	4.06	1.56	6.66	0.97	2.11	0.42	3.47
Sphenarium histro	5.08	1.46	6.70	0.77	9.01	0.46	3.93
Diptera							
Hermetia illucens	2.07	1.74	3.48	0.89	2.44		2.44
Musca autumnalis	1.30	1.46	2.94	1.36	2.19		2.34

By comparing the nutrient requirements of fish (Table 3.7) with the nutrient content of insects (Table 3.6) you can see that insects are usually amino acid-rich, and capable of satisfying the nutritional needs of wildlife. Indeed, the principal reason that wildlife eat larger animals is related to the energy efficiency of capturing larger prey items (relatively more nutrition per energy investment). Especially for endothermic (warm-blooded) animals, the energy requirements are quite high, so larger prey are needed.

There are important **temporal shifts** in the diet of wildlife. Food availability changes throughout the year. As an example, Fig. 3.4 shows the changing dietary habits of antelope ground squirrel, *Ammospermophilus leucurus*, over the course of a season in Nevada, USA. Although seeds and green vegetation are consumed through the entire year, green vegetation is not such a large part of their diet when arthropods are being eaten. Arthropods become a significant part of their diet in late summer and fall. Our perception of the importance of these diet elements could easily be distorted if the study included only a portion of the year. Unfortunately, most studies of mammals are conducted for only a portion of the year.

Food availability can change even over the course of a single day, sometimes in a cyclical pattern. Our perception of diet preference can be influenced by natural activity patterns among insect prey. Thus, studies in Wisconsin, USA showed that bluegill (*Lepomis macrochirus*) fed little at night, so samples collected in the morning contained less food. However, bluegill stomachs contained relatively more zooplankton in morning samples. Larger fish, which tended to remain along the periphery of the lake, took advantage of the migration of zooplankton away from the lake edges beginning at sunrise to capture these food items.

Much like the zooplankton in the bluegill study mentioned above, insect activity levels typically display a rhythm of increase and decrease over the course of the day. Daybreak or sunset often are peak times for adult insect emergence (a period of great susceptibility to predation), and therefore wildlife feeding. This has important implications for diet analysis. If wildlife are sampled shortly after emergence of a prey item they

Table 3.7. Dietary requirements of some essential amino acids (g/100 g dietary protein) and protein (% dry wt.) for fish species and rat (adapted from National Research Council 1993).

Species	Arginine	Histidine	Isoleucine	Leucine	Lycine	Methionine	Phenylalanine	Threonine	Tryptophan	Valine	Protein
carp	3.8	1.4	2.3	4.1	5.3	1.6	2.9	3.3	0.6	2.9	40.0
Japanese eel	4.5	2.1	4.0	5.3	5.3	3.2	5.8	4.0	1.1	4.0	37.7
Chinook salmon	6.0	1.8	2.3	4.0	5.0	4.0	5.2	2.2	0.5	3.3	40.0
Rainbow trout	3.5	1.6	2.4	4.4	5.3	2.7	3.1	3.4	0.5	3.1	40.0
Channel catfish	4.3	1.5	2.6	3.5	5.1	2.3	5.0	2.2	0.5	2.9	24.0
Rat	1.5	3.0	3.8	6.8	7.6	4.6	6.8	3.8	1.5	3.0	13.2

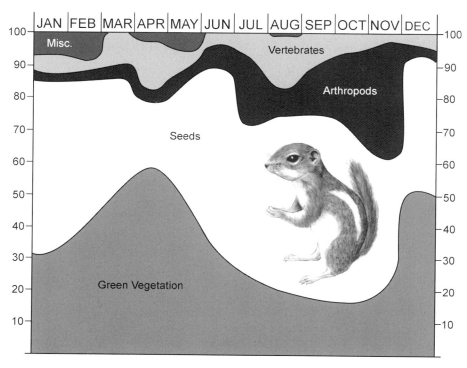

Fig. 3.4. The volume of different food items found in the stomachs of antelope ground squirrel in Nevada, USA (adapted from Bradley 1968).

more likely will have identifiable items in their stomach than if there has been several hours since emergence/ingestion, because digestion will have made the prey item unrecognizable. Therefore, to maximize the likelihood of identification of prey items, it is best to sample wildlife when they are actively feeding.

Studies that are restricted to one geographic area also can present a distorted perception of what wildlife eat because there are **geographic shifts** in diet that are often based on differing availability of prey. While few scientists have the luxury of studying wildlife dietary over a broad geographic range, many scientists addressing the same issue at different locations can certainly provide an accurate collective assessment of animal dietary habits if consistent methods are used for data collection

Finally, it is worth mentioning again that there are important **ontogenetic shifts** (development-related) in diet. Previously we saw that granivorous birds often fed their young mostly insects. This phenomenon is by

no means limited to birds, as it is well documented in reptiles, fishes, and mammals. Young reptiles cannot subdue larger prey and so are often dependent on invertebrates, usually insects, for food. For example, young of the yellow-bellied slider, *Trachemys scripta*, inhabit shallow water where they can feed on aquatic insects and crustaceans (plus the terrestrial insects that find themselves helpless in the water) whereas the older turtles live in deeper water and are principally herbivorous (Fig. 3.5). This carnivorous/insectivorous habit is not based strictly on availability, however, because when growth of young turtles fed only grass shrimp, *Palaemontes paludosus*, or duckweed, *Lemna valdiviana*, was compared, the animal-based diet fostered more rapid growth. Digestive processing was more efficient on the animal diet, with young turtles acquiring three times as much dry matter, organic matter, energy and nitrogen. The shrimp diet provided higher levels of nitrogen and calcium, both vital for growth of juvenile turtles. Most importantly, growth of

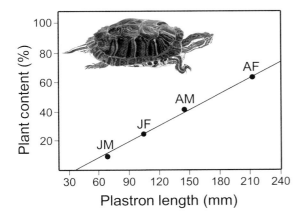

Fig. 3.5. The volume of plant material in the stomach of red-eared turtle, *Pseudemys scripta*, in relation to turtle size (length). Data points are juvenile males (JM), juvenile females (JF), adult males (AM), and adult females (AF) (adapted from Clark and Gibbons 1969).

juveniles was 3.2 times faster on shrimp-based diets. Because juvenile animals have higher mortality rates, rapid growth allows animals to pass through the more vulnerable young stages of growth, increases ultimate size, and improves reproductive output and survivability.

So why not be strictly carnivorous? Well, besides the availability issue (plants are more abundant and *much* easier to catch!), there are nutritional advantages to feeding on plants, too. Going back to our yellow-bellied slider example, other research has shown that the digestive efficiency of turtles is higher on a diet of duckweed plus insects than on insects alone or duckweed alone. Many animals benefit from mixed diets, and omnivorous feeding behavior is not accidental. Adults of yellow-bellied sliders, even though they feed primarily on plants, continue to ingest some animal matter.

SUMMARY

• Food is of fundamental importance to the biology of wildlife. Wildlife display a wide range of feeding behaviors ranging from a great deal of specialization (specialists) to feeding broadly (generalists). They also vary from feeding mostly on plants (herbivory) to animals (carnivory) with varying degrees of selectivity. Some of the common specializations include insectivory (feeding on insects), picivory (fish), frugivory (fruit), fungivory (fungus), and granivory (seed).

• Food selection by wildlife is related to both preference and availability. Assessment of both abundance of potential food and the composition of the diet are necessary to understand dietary habits. To assess the abundance of insects available for consumption by wildlife, a number of sampling techniques usually must be employed to take advantage of the differing biologies and behaviors of insects. Insect sampling methods usually fall into two general categories: relative and absolute sampling methods.

• Wildlife diets are assessed by removal or emptying of the stomach, by examination of excrement or food pellets, and by direct observation of what is being consumed. Examination of stomach/crop contents or of excrement are most common, necessitating reconstruction of diet from fragments. Differing rates of digestion associated with different diet items complicate accurate assessment of diet.

• Common methods of diet analysis include calculation of the volume or weight (mass) consumed, the frequency of consumption of each diet item, and the frequency of occurrence of each diet item among potential consumers. Time budgets sometimes are used to indicate importance of diet components to wildlife.

• Wildlife optimize their food intake based on energy and nutrition derived, balanced against energy expended in obtaining food. Insects provide wildlife with higher levels of nutrition than most plant-based foods. Wildlife commonly make changes in their feeding habits based on changes in availability (temporal shifts or geographic shifts) or of developmental needs (ontogenetic shifts).

REFERENCES AND ADDITIONAL READING

Baumann, P.C. & Kitchell, J.F. (1974). Diel patterns of distribution and feeding of bluegill (*Lepomis macrochirus*) in Lake Wingra, Wisconsin. *Transactions of the American Fisheries Society* **103**, 255–260.

Bjorndal, K.A. (1991). Diet mixing: nonadditive interactions of diet items in an omnivorous freshwater turtle. *Ecology* **72**, 1234–1241.

Bouchard, S.S. & Bjorndal, K.A. (2006). Ontogenetic diet shifts and digestive constraints in the omnivorous freshwater turtle *Trachemys scripta*. *Physiological and Biochemical Zoology* **79**, 150–158.

Bradley, W.G. (1968). Food habits of the antelope ground squirrel in southern Nevada. *Journal of Mammalogy* **49**, 14–21.

Briggs, S.V., Maher, M.T., & Palmer, R.P. (1985). Bias in food habits of Australian waterfowl. *Australian Wildlife Research* **12**, 507–514.

Burger, J.C., Patten, M.A., Rotenberry, J.T., & Redak, R.A. (1999). Foraging ecology of the California gnatcatcher deduced from fecal samples. *Oecologia* **120**, 304–310.

Clark, D.B. & Gibbons, J.W. (1969). Dietary shift in the turtle *Pseudemys scripta* (Schoepff) from youth to maturity. *Copeia* **1969**, 704–706.

Cooper, R.J. & Whitmore, R.C. (1990). Arthropod sampling methods in ornithology. *Studies in Avian Biology* **13**, 29–37.

Dickman, C.R. & Huang, C. (1988). The reliability of fecal analysis as a method for determining the diet of insectivorous mammals. *Journal of Mammalogy* **69**, 108–113.

Finke, M.S. (2008). Nutrient content of insects. In Capinera, J.L. (ed.), *Encyclopedia of Entomology*, 2nd edn., pp. 2623–2646. Springer Science & Business Media B.V., Dordrecht, The Netherlands.

Fleming, T.H. 1991. Fruiting plant–frugivore mutualism: the evolutionary theater and the ecological play. In Price, P.W., Lewinsohn, T.M., Fernandes, G.W., & Benson, W.W. (eds.) *Plant–Animal Interactions: Evolutionary Ecology in Tropical and Temperate Regions*, pp. 119–144. John Wiley & Sons, Ltd., New York, USA.

Kaspari, M. & Joern, A. (1993). Prey choice by three insectivorous grassland birds: reevaluating opportunism. *Oikos* **68**, 414–430.

Montague, T.L. & Cullen, J.M. (1985). Comparison of techniques to recover stomach contents from penguins. *Australian Wildlife Research* **12**, 327–330.

Moreby, S.J. (2004). Birds of lowland arable farmland: the importance and identification of invertebrate diversity in the diet of chicks. In van Emden, H.F. & Rothschild, M. (eds.). *Insect and Bird Interactions*, pp.21–35. Intercept, Andover, Hampshire, UK.

Naranjo, S.E. (2008). Sampling arthropods. In Capinera, J.L. (ed.), *Encyclopedia of Entomology*, 2nd edn., pp. 3231–3246. Springer Science & Business Media B.V., Dordrecht, The Netherlands.

National Research Council (1993). *Nutrient Requirements of Fish*. The National Academies Press, Washington DC, USA. **116** pp.

Ralph, C.P., Nagata, S.E., & Ralph, C.J. (1985). Analysis of droppings to describe diets of small birds. *Journal of Field Ornithology* **56**, 165–174.

Schwenk, K. (2000). *Feeding. Form, Function, and Evolution in Tetrapod Vertebrates*. Academic Press, San Diego, California, USA. **537** pp.

Wolda, H. (1990). Food availability for an insectivore and how to measure it. *Studies in Avian Biology* **13**, 38–43.

Yalden, D.W. & Morris, P.A. (2003). The analysis of owl pellets, 3rd edn. *The Mammal Society Occasional Publications* **13**, 1–28.

Chapter 4

WILDLIFE DIETS

Detailed data on the diets of vertebrate wildlife exists in the scientific literature, but rarely is this information found in the more general compilations of animal biology. This is especially true for invertebrate diet components, which are generally grouped into 'insects' or 'invertebrates.' Here I provide an overview of wildlife diets, emphasizing species that feed entirely or partly on insects. By examining the tables you can easily see the importance of insects, the diversity of prey consumed, and some patterns of feeding.

ANALYSIS OF AMPHIBIAN AND REPTILE DIETS

Amphibians and reptiles are often grouped together for convenience because they are ecologically similar, terrestrial, ectothermic ('cold-blooded') vertebrates. Class Amphibia consists of order Anura, the frogs and toads; order Caudata, the salamanders; and order Gymnophiona, the caecilians. The frogs and salamanders are well-known animals that are found in most moist habitats. Toads typically inhabit drier areas than do frogs and salamanders. The caecilians are not well known. They are rare, legless, worm-like topical animals that generally live belowground. Amphibians generally are insectivorous. The larger species may depart from this behavior by eating other items, but arthropods are central to their survival. Unlike some taxa (e.g., birds) there is not much morphological

Insects and Wildlife: Arthropods and Their Relationships with Wild Vertebrate Animals, 1st edition. By J.L. Capinera. Published 2010 by Blackwell Publishing.

adaptation apparent among amphibians with different feeding behaviors.

Frogs and toads are carnivorous after metamorphosis from the tadpole stage, and invertebrates are the dominant prey. They also are generalists, feeding on almost anything they can capture and ingest. There are a few that specialize on certain prey, however, such as springtails, ants or termites. Adult frogs and toads have sticky tongues and use rapid tongue protraction and retraction to capture prey. They do not masticate their prey; it is simply swallowed. Interestingly, the traditional view that tadpoles feed only on algae and detritus is being challenged. Increasingly we find that tadpoles are protein limited, and benefit from items mixed in with the detritus and algae. Stable isotope techniques, which measure assimilation rather than just ingestion, document that considerable amounts of animal food are consumed. Carnivory, cannibalism and scavenging of animal tissues are behaviors found in many tadpoles, but the importance of carnivory is not fully known.

Nearly all salamanders have a terrestrial stage or are terrestrial throughout their life. Terrestrial salamanders, and most aquatic salamanders, are strict carnivores. They feed mostly on insects, but larger species may eat molluscs and vertebrates. Salamanders have weak jaws, and usually capture their prey with their tongue. They are capable of rapid tongue protraction and retraction. Aquatic salamanders feed on fly larvae, mayflies, caddisflies, crustaceans, and other small animals. They seem to be opportunistic feeders, and all larvae and most adults of aquatic species ingest their prey by suction.

Caecilians are mostly terrestrial, but some are semi-aquatic and a few are known to be entirely aquatic. They generally burrow close to the surface of the soil.

Their diet is not well known, but they feed on earthworms, insects, crustaceans, and even vertebrates. Like so many other carnivores, they seem to be opportunistic.

Reptiles are a more diverse taxon, although they also lack much evidence of morphological adaptions to diet. Class Reptilia consist of order Crocodilia, the caimans, crocodiles, etc.; order Rhynchocephalia, the tuataras; order Squamata, the amphisbaenians, lizards and snakes; and order Testudines, the tortoises and turtles. Again, not all taxa are well-known. Tuataras are lizard-like and known only from New Zealand. Amphisbaenians are legless lizards that are sometimes called worm-lizards, and known mostly from South America and Africa. Reptiles vary considerably in their dietary habits. Reptiles are thought by many to have evolved as arthropod-feeding organisms, but as some evolved into larger and larger organisms their dietary habits expanded. As a general rule, smaller reptiles eat arthropods, and lizards continue to eat mostly insects although many are omnivorous, eating a variety of animal (often including other lizards) and plant food. Among lizards, there is a fairly strong relationship between body size and feeding behavior, with the smaller species favoring carnivory and the larger herbivory. In the case of omnivorous and herbivorous species, fruit as well as vegetation may be consumed. Nevertheless, small invertebrates are the principal food of the orders Squamata and Rhynchocephalia.

The thought of crocodile, alligator, and caiman feeding evokes images of people or poodles being swallowed, but these large carnivores are subject to the same dietary constraints as other animals. Crocodiles widely exceed the mass of all other existing reptiles, attaining a length of up to 10 m. Despite their immense size when mature, crocodilians display the same ontogenetic (development-related) shifts as most other vertebrates. When young, they consume small prey, particularly aquatic and shoreline-dwelling insects. However, as they grow they transition to feeding on crustaceans, frogs and fish before moving to birds and mammals. In brackish water, young crocodilians feed on insects and crustaceans. In fresh water, insect and crustacean food is complemented with tadpoles, snails and frogs. Research conducted in Florida, USA, showed that invertebrates comprised 63.6% of the volume of juvenile American alligator (*Alligator mississippiensis*) stomachs, with insects comprising 33.6%. For the younger hatchling alligators, however, insects accounted for 71.2% of the volume of food ingested.

Amphisbaenians are often found in association with ants, and in some cases numerous ants have been found in their stomachs, but termites also are eaten. In the case of the amphisbaenian *Amphisbaena alba* from Trinidad, this reptile actually feeds on the beetles living in association with ant nests rather than on the leaf-cutter ants, *Atta* spp. Amphisbaenians may feed more broadly on insects than is generally acknowledged, but they have been too poorly studied to know their dietary habits.

Snakes, which evolved from lizards, are strictly carnivorous, with the smaller species often eating arthropods but the larger species eating amphibians, reptiles, birds and mammals. It is noteworthy that even though some snakes specialize on fish, birds or mammals, when they are young and too small to eat these large animals they will eat arthropods or other small prey. This dietary habit is often overlooked. Snakes often eat any animal that they can swallow, and are renowned for their ability to eat impossibly large organisms. They are exceptional in that they may ingest items up to 50% of their own body mass. This is something that lizards never attempt. Lizards favor small prey items. The lower jaw of snakes is not tightly joined, but only loosely connected by a ligament. Their mouth opening thus can be stretched to allow it to swallow abnormally large objects. Despite this flexibility, or maybe because of it, some snakes are specialized in their food habits. Table 4.1 shows an overview of snake diet habits. Note that invertebrates seem to be relatively unimportant in the diets of snakes in several areas of the world. North America seems to be an exception, with 35% of the snakes consuming invertebrates, although this may be because there has been more study of snakes in North America than elsewhere.

Also, snakes are rather exceptional in that they commonly produce venoms that are used to immobilize prey. Some studies have found that venoms are quite effective against insects. Several species from Europe in the genus *Vipera* (*V. renardi, V. lotievi, V. kaznakovi, V. orlovi*) feed on insects, especially orthopterans, and their venom is moderately or highly toxic to crickets. Species in this genus of snake that do not feed on insects do not have venom that affects crickets. Thus, the effectiveness of their venom is adapted to their prey, being most effective against normal prey species. This specialization of venom is also found among lizard, bird, and mammal-feeding snakes.

Turtles vary in their diet specialization. The strictly land-dwelling species are usually called 'tortoises,' and

Table 4.1. Dietary habits of terrestrial snakes in Australia, China, Africa, and North America (from Shine 1977).

Region	No. snakes	% of snake species feeding on each prey type						
		Invertebrates	Fish	Amphibia	Lizards	Snakes	Birds	Mammals
Australia	53	15	2	42	81	23	21	34
China	81	4	15	41	31	15	11	26
Uganda	75	11	5	37	37	15	9	40
Africa	88	14	2	39	63	13	17	43
North America	95	35	14	20	22	9	15	43

are primarily herbivorous. The turtles usually live in association with water, though to varying degrees. Those that dwell entirely in association with water (aquatic species) are usually mostly carnivorous, whereas those that live mostly on land but may enter water occasionally (semi-aquatic species) have more omnivorous habits. The young of most turtles tend to be more insectivorous than the larger, older individuals. Having said this, it is important to note that the vernacular surrounding the term 'turtle' varies, so what is called a turtle in one country may be called a tortoise in another.

Some examples of the dietary habits of amphibians and reptiles are shown in Table 4.2. Except for snakes and turtles, research showed that the amphibians and reptiles were highly insectivorous (invertivorous). The turtles, more so than most other reptiles, were herbivorous. Taken as a whole, a broad range of invertebrates were eaten. However, some trends in insectivory are evident. Toads, being mostly terrestrial, consumed many insects but particularly large amounts of ant and beetle material. Frogs, on the other hand, occurred in both aquatic and terrestrial environments, and had broader dietary habits. Salamanders consumed many different terrestrial arthropods, but generally ate small flightless organisms. Lizards, like toads, fed broadly but mostly on ants and beetles. In the dietary literature, most snakes are not reported to feed on invertebrates to any degree, though some clearly are specialized to take advantage of this resource. Crickets, grasshoppers and their relatives (Orthoptera) seem to be favorites for snakes, although flat-headed snake (*Tantilla gracilis*) consumed a large number of beetles (Coleoptera), rough green snake (*Opheodrys aestivus*) consumed a large number of caterpillars (Lepidoptera), and blind snakes (*Ramphotyphlops*) specialized on ants. As men-

tioned earlier, young alligators (*Alligator mississipiensis*) consumed insects, especially giant water bugs (Hemiptera: Belostomatidae). Lastly, aquatic insects were important in the diet of turtles.

Ontogenetic shifts in diet are commonplace with amphibians and reptiles that grow very large. As discussed in the previous chapter, there is a trend in increased herbivory (thereby lower levels of insectivory) as red-eared turtles (*Pseudemys scripta*) transition from juveniles to adults. Also, females are larger than males, and so eat proportionally less animal material. Earlier we mentioned that insectivory in alligators diminished with age, and that there was a corresponding increase in consumption of vertebrates other than fish. This is shown in Fig. 4.1. The increase in consumption of noninsect invertebrates shown in Fig. 4.1 is due principally to acceptance of snails as the juvenile alligators increase in size.

ANALYSIS OF MAMMAL DIETS

Successful growth and survival among animals require that more energy be obtained through feeding than is expended in the acquisition and digestion of food. Although this is true for all animals, it is of greatest significance in mammals and birds because, as endotherms, they expend more energy in body maintenance and homeostasis. In temperate climates, these animals also tend to store food as fat for periods when food is not readily available. Some mammals also enter **hibernation**, a reversible state of inactivity and metabolic depression. Ectotherms such as invertebrates, amphibians, reptiles, and fishes, in contrast, require much less food to survive non-feeding periods. It is likely no coincidence that anteaters (order Pilosa, sub-

Table 4.2. The stomach contents of some amphibians and reptiles. The stomach content values are based on volume (% of total stomach contents) or mass (% of stomach content total dry weight) except for snakes, where frequency of consumption (the average proportion or number of food items per consumer) is given (adapted from various sources: see references).

	% of stomach contents				
	Total animal	Coleoptera: beetles, weevils	Lepidoptera: caterpillars, etc.	Hymenoptera: bees, wasps, ants	Hemiptera: bugs
Japanese common toad, *Bufo japonicus*	100	62.6 (44.9 ground, 4.2 rove, 4.2 scarab beetles)	1.5	7.4 (4.4 ants)	9.0
Mexican marbled toad, *Bufo marmoreus*	100	42.7	3.4	30.79 ants	3.0
Cane toad, *Bufo marinus*	100	61.2	2.7	33.3 (28.5 ants, 5.3 bees)	1.3
Daruma pond frog, *Rana porosa brevioda*	100	11.8	2.9	7.4 (4.4 ants)	
Vaillant's frog, *Rana vaillanti*	100	8.8		5.9	17.6
Brown's leopard frog, *Rana brownorum*	100	19.6	10.2	10.2	4.4
Southern leopard frog, *Rana sphenocephala*	99.8	22.1	11.8 caterpillars, 1.4 adults	7.4 (4.4 ants)	3.9
Northern leopard frog, *Rana pipiens pipiens*	74.5	22.6	0.6	5.7	4.0
Blanchard's cricket frog, *Acris crepitans blanchardi*	98.6	12.9 (2.6 ground, 1.5 leaf beetles, 2.0 weevils)		7.4 (4.4 ants)	10.5 (5.6 leafhoppers, 4.7 aphids)
Northern spring peeper, *Pseudacris crucifer crucifer*	100	9.9	12.8 (3.5 cutworms, 2.0 loopers)	6.9 (2.6 ants)	12.9 (4.4 leafhoppers, 2.7 damselbugs, 2.0 seed bugs, 1.5 aphids)
Oklahoma salamander, *Eurycea tynerensis*	100	1.2		0.8	
Northern slimy salamander, *Plethodon glutinosus*	100	8.5		3.3 (3.0 ants)	2.5

Orthoptera: grasshoppers, crickets	Diptera: flies	Other insects	Other arthropods	Other animals	Plant
0.7	0.1		5.2 millipedes, 0.1 spiders, 0.2 centipedes	5.2 worms	
0.2	0.1	15.4 termites, 1.9 lacewings	1.2 spiders, 1.0 harvestmen		
		1.3 dragonflies, 5.3 cockroaches, 21.4 earwigs	13.4 crustaceans	21.4 molluscs	
31.6		6.1 dragonfly nymphs, 2.5 earwigs	5.7 isopods, , 3.2 spiders, 2.0 millipedes	7.8 frogs, 4.1 leeches, 2.1 worms	
8.8	8.8		29.5 crustaceans, 8.8 spiders	11.8 molluscs	
10.2	1.5		19/0 millipedes, 4.4 crustaceans, 13.2 spiders		
12.8	13.5	1.5	2.1 crustaceans, 2.1 millipedes, 0.5 centipedes	3.3 frogs, 3.1 molluscs, 0.8 earthworms, 0.8 fish	
2.5	3.5	18.2 "larvae", 5.9 damselflies	4.0 spiders, 0.6 other	1.8 molluscs, 0.8 worms,	25.5
5.5 (5.0 grasshoppers)	21.2 (5.8 flower, 1.5 black, 1.5 spear-winged)	3.1 dragonflies, 3.9 caddisflies)	2.7 springtails		1.0
	4.7	2.5 scorpionflies,	28.8 spiders, 10.6 harvestmen, 3.9 springtails, 3.4 crustaceans		
	19.1 midges, 0.2 others	46.5 mayflies, 3.2 stoneflies, 1.2 caddisflies	17.6 isopods, 6.0 amphipods, 1.0 other crustaceans	2.8 snails	
	2.4	7.8 "larvae"	60.9 millipedes, 5.5 centipedes, 2.7 woodlice, 1.7 spiders	3.8 snails	

Table 4.2. *Continued*

	% of stomach contents				
	Total animal	**Coleoptera: beetles, weevils**	**Lepidoptera: caterpillars, etc.**	**Hymenoptera: bees, wasps, ants**	**Hemiptera: bugs**
Red-cheeked salamander, *Plethodon jordani*	100			5.1 (4.9 ants)	2.6
Broad-headed skink, *Eumeces laticeps*	100	18.0 (4.4 phengodid larvae, 4.2 longhorn, 3.1 click, 1.4 darkling beetles)	12.0	0.9	2.3
Grand ctenotus skink, *Ctenotus grandis*	76.9	5.4		9.2	1.4
Clay-soil ctenotus skink, *Ctenotus helenae*	97.3	3.7		1.9	4.7
Mexican fringe-toed lizard, *Uma exsul*	75.5	9.4	1.2	40.4 (29.2 ants)	8.8
A Brazilian skink, *Mabuya agilis*	100	0.2	0.8	3.8 (mostly ants)	1.2
Greater earless lizard, *Cophosaurus texanus*	100	2.7	60.7	12.2 (0.9 ants)	6.2
Tropical sand lizard, *Liolaemus lutzae*	58.9	12.1		11.5 (8.8 ants)	4.5
Bonaire whiptail lizard, *Cnemidophorus murinus*	16.0		mostly caterpillars		
Mountain spiny lizard, *Sceloporus jarrovi*	86.1	17.4		46.9 (mostly ants)	8.5
Mexican knob-scaled lizard, *Xenosaurus grandis*	100	12.7	11.4	5.8	0.8
Newman's knob-scaled lizard, *Xenosaurus newmanorum*	95.8	29.6	21.8		8.3
Western yellow-bellied racer, *Coluber constrictor mormon*	100				

Orthoptera: grasshoppers, crickets	Diptera: flies	Other insects	Other arthropods	Other animals	Plant
	2.8	18.9 "larvae"	46.1 millipedes, 6.3 centipedes, 0.6 woodlice, 3.1 harvestmen, 1.8 spiders	10.2 worms, 5.7 snails	
25.3 (20.4 crickets)		13.6 cockroaches, 3.9 earwigs	9.4 woodlice, 2.6 harvestmen, 2.0 spiders	7.8 lizard	
8.4		24.9 termites	10.2 centipedes, 3.7 spiders, 1.0 scorpions	10.3 lizard	23.1 seed & berries
38.0	0.5	30.6 termites, 0.8 cockroaches	10.8 centipedes, 3.1 spiders, 2.6 scorpions		2.7 seed & berries
4.2			2.3 spiders	1.2 molluscs	24.5
27.6		14.9 cockroaches	35.2 spiders, 1.8 woodlice	1.2 molluscs	
	13.6	1.2	1.2 spiders		
0.6	0.4	14.2	5.3 crustaceans, 3.6 spiders		41.1
					74.0
1.6	1.2		4.4 spiders, 2.0 crustaceans	2.7	13.9
22.7	14.6		10.6 millipedes or centipedes, 1.1 spiders	19.6 snails	
26.5	2.5			1.5 snails	4.2
35.6 grasshopper, 25.5 crickets, 25.5 Jerusalum crickets, 3.0 other, 1.9 katydids				7.5 rodents, 0.7 frogs, 0.3 snakes	

Table 4.2. *Continued*

	% of stomach contents				
	Total animal	Coleoptera: beetles, weevils	Lepidoptera: caterpillars, etc.	Hymenoptera: bees, wasps, ants	Hemiptera: bugs
Flat-headed snake, *Tantilla gracilis*	100	46.8 darkling, 10.8 click, 21.5 com-clawed beetle larvae			
Yucátan cricket-eating snake, Symphimus mayae	100				
Rough green snake, *Opheodrys aestivus*	100		40.6 caterpillars, 3.6 moths		
Southern blind snake, *Ramphotyphlops australis*	100			98 ants, mostly larvae and pupae	
Northeastern blind snake, *Ramphotyphlops polygrammicus*	100			90.9 ants, mostly larvae and pupae	
Banded rock rattlesnake, *Crotalus lepidus klauberi*	100				
American alligator (juvenile), *Alligator mississipiensis*	100	9.0 (1.8 predaceous water, 1.4 water scavenger beetles)			10.8 water bugs
False map turtle, *Graptemys pseudogeographica*	50.0				
Ouachita map turtle, *Graptemys ouachitensis*	58.5				
Murray turtle, *Emydura macquarii*	34.0	0.9	0.2	0.2	
Krefft's river turtle, *Emydura krefftii*	80.0				2.0 water boatmen

Orthoptera: grasshoppers, crickets	Diptera: flies	Other insects	Other arthropds	Other animals	Plant
			11.4 centipedes	3.8 other	
28.4 katydids, 17.6 crickets, 6.8 Jerusalem crickets, 4.1 grasshoppers, 14.9 other		5.4 mantids, 13.5 walking sticks	4.1 spiders		
15.6 grasshoppers & crickets	108	9.0 dragonflies, 2.7 mantids	24.6 spiders, 2.5 harvestmen,	2.0 snails	
		9.1 termites			
			28.3 centipedes	56.0 reptiles, 13.8 rodents, 1.9 birds	
2.3		3.7 dragonflies, 7.9 other	15.5 crustaceans	16.5 reptiles, 16.1 fish, 14.5 snails, 3.8 amphibians	
		17.5 mayfly larvae, 5.0 caddisfly, 3.0 damselfly larvae		19.0 molluscs, 12.0 fish	42.0 plants, 8.0 other
		41.5 mayfly larvae, 8.0 caddisfly, 0.5 damselfly larvae		3.5 molluscs, 5.0 fish	31.5 plants, 10 other
	0.3	0.02 caddisflies 8.8 crustaceans	0.6 millipedes	26.0 carrion	26.0 filamentous algae, 12.0 small algae, etc., 21 other plants
	11.0 other	18.0 dragonflies and damselflies, 9.0 caddisflies, 2.0 mayflies	25.0 crustaceans		20.0

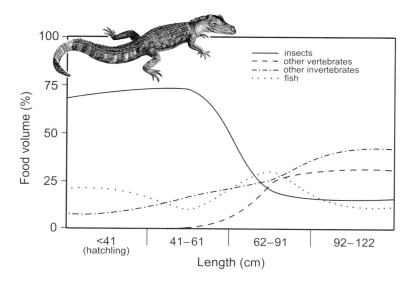

Fig. 4.1. The volume of major food items in juvenile American alligator, *Alligator mississippiensis*. Insects are a major source of food for young alligators, but other invertebrates (crustaceans, snails) and nonfish vertebrates increase in importance as the alligators increase in size (adapted from Delany 1990).

order Vermilingua) have an unusually low body temperature relative to other mammals. It must be quite a challenge to maintain a large body by feeding on such small prey, and lower body temperatures translate into less energy needed for maintenance.

Invertebrates, and insects in particular, can be an important part of the diet of mammals. For some, invertebrates are simply a 'snack,' a quick meal to hold them over until a more desirable and substantial meal can be obtained. For others, though, invertebrates are either the mainstay of their diet or a regular component. A synopsis of insectivory in mammals (Class Mammalia) follows:

• On one extreme, we have several small, insectivorous mammalian orders formerly called Insectivora, a convenient grouping that is now largely abandoned but which contained a diverse assemblage of small insect-eating mammals such as hedgehogs, shrews, tenrecs and moles. Modern classification systems treat them as several orders, including Afroscoricida, the tenrecs and golden moles; Erinaceomorpha, the gymnures and hedgehogs; Macroscelidae, the elephant-shrews; Scandentia, the tree shrews; and Soricomorpha, the shrews and moles. These animals are largely dependent on insects for food, hence the old order name 'Insectivora.' Likewise, the order Chiroptera, the bats, are usually insect-dependent, though some feed on other foods (e.g., fruit, fish). Order Pholidota consists of 8 species of unusual insectivores covered with

plate-like scales; these animals are known as pangolins. Order Tubulidentata contains a single insectivorous species, *Orycteropus afer*, known as aardvark.

• On the other extreme we have herbivorous animals in the order Artiodactyla, the even-toed hoofed animals. These include ruminants such as the families Girafidae, giraffes; Cervidae, the caribou, moose, elk, and deer; Antilocapridae, pronghorn; and Bovidae, including American bison, waterbucks, wildebeest, gazelles, and muskoxen. Also included in this order are families of nonruminating animals such as Suidae, hogs; Tayassuidae, peccaries; Hippopotamidae, hippopotamuses; and Camelidae, camels and llamas. Likewise, in the order Perissodactyla there are the families Equidae, or horses and zebras; Tapiridae, the tapirs; and Rhinocerotidae, the rhinoceroses. Few of these animals (mostly the hogs and peccaries) feed deliberately on insects, instead eating grass and forbs.

• In between these extremes, we have many taxa that display broad dietary habits (omnivory). Included here are the orders Carnivora, the meat-eaters, containing the families Canidae, or wolves, jackals, and foxes; Ursidae, the bears and giant pandas; Primates, the apes, monkeys, lemurs, tarsiers, and others; Procyonidae, the coatis, raccoons and lesser pandas; Mustelidae, the martens, weasels, badgers, and otters; Mephitidae, the skunks; Felidae, the lions, cheetahs, leopards, and other cats; and Hyaenidae, the hyenas. Also omnivorous is the order Rodentia, which contains the families Sciuridae, the chipmunks, squirrels, and marmots; Cricetidae, the New World mice, voles, lemmings, muskrats and gerbils; Muridae, the Old World rats and mice; Castoridae, the beavers; family Geomyidae, the pocket gophers; and many others. Another group with varying food habits is the order Pilosa (Edentata), or toothless mammals, containing Bradypodidae, the sloths; and (although completely insectivorous) Myrmecophagidae, the anteaters. The order Cingulata contains the armadillos. The marsupials, or pouched mammals, are now found in several orders, including Didelphimorphia, American marsupials; Diprotodontia, kangaroos, possums, wombats and koala; and Peramelemorphia, bandicoots. It is this third category of mammals, the omnivores, that is particularly interesting because many use insects as a portion of their diet.

Examples of stomach and excrement analyses for mammals are shown in Tables 4.3 and 4.4, with the former presenting frequency data and the latter presenting volumetric data. Most mammals are herbivores, but no herbivores are shown in the tables. Instead, the dietary habits of some omnivorous and carnivorous species are highlighted

The consumption data provide important information on the dietary habits of omnivorous and carnivorous mammals. Not surprisingly, these frequency data show the importance of insects in the diet of some bats (order Chiroptera). A wide breadth of feeding can be seen in Fig. 4.2, which shows how a number of insect orders are taken by big brown bat, *Eptesicus fuscus*, studied in West Virginia, USA. Beetles, being the most speciose and often the most abundant insects, commonly are the most abundant prey for insectivorous bats. It is not surprising then, that insectivorous bats are sometimes said to be largely opportunistic, feeding on whatever is available. In support of this, research in central Texas, USA on Brazilian free-tailed bats, *Tadarida brasiliensis*, showed that their consumption of corn earworm (*Helicoverpa zea*) and fall armyworm (*Spodoptera frugiperda*) (both Lepidoptera: Noctuidae) moths increased dramatically when flights of moths arrived from Mexico. Their percentage of feces volume increased from 14.8% to 43.0% immediately after the influx of moths. These bats actually fed most often on beetles, but the beetles were smaller than the moths, so the highest volume of feces was comprised of moths.

Bats do not always feed opportunistically, however. Big brown bat research conducted in Canada showed that whereas beetles constituted only 1.9% of the insects flying in the environment during the study and flies constituted 79.3%, over 50% of the volume of insects and 66% of the number of insects in bat droppings were beetles. Thus, these bats demonstrated preference for beetles and consumed flies at much less than their relative abundance. Interestingly, bats differ in jaw thickness, tooth size and number, and skull robustness. Those with thinner jaws, smaller teeth, and weaker skulls avoid insects with a thicker integument such as beetles, taking instead soft-bodied insects such as moths. Other bats (e.g., white-lined bat, *Vampyrops lineatus*) feed exclusively on fruit, and some 'fruit-eating' bats take both insects and fruit. Other bats eat nectar and pollen, blood, or vertebrates such as frogs, lizards and rodents. A few bats catch fish, and some use their feet to pluck insects from the surface of water. Nevertheless, insectivory is the most common feeding habit, with about 70% of bat species considered insectivorous.

Moles are voracious animals, often consuming 70%–100% of their weight daily in various invertebrates, but rarely plant matter. The shallow tunnels commonly encountered near the surface of the soil are feeding tunnels created by moles searching for food; they may retreat to deeper tunnels when not feeding. The diet analyses for the two mole species (order Soricomorpha) in Table 4.5 are interesting, as eastern mole, *Scalopus aquaticus*, appeared to be largely insectivorous in a South Carolina study, whereas Townsend's mole, *Scapanus townsendii*, fed predominantly on earthworms in Oregon (see also Fig. 4.3). However, this likely reflects food availability because other studies have found eastern mole to eat considerable numbers of earthworms. Shrews (order Soricomorpha) are also well known for their voracious appetite, and eat other small vertebrates, invertebrates and seeds. The two species shown in Tables 4.5–4.6 are similar in food preferences, and although largely insectivorous, eat many other types of invertebrates. Plant material was taken infrequently.

Rodents (order Rodentia) traditionally have been considered to be herbivores, but increasingly their omnivorous tendencies are becoming known. This omnivorous behavior can be seen clearly in Tables 4.3–4.4, as insects were commonly consumed by mice, chipmunks, and squirrels. Not surprisingly, there is strong seasonality in their diet (e.g., antelope ground squirrel, Fig. 4.4, with more green vegetation in the spring and arthropods in the autumn). As suggested by their name, grasshopper mice (*Onychomys* spp.) have long been known to be carnivorous, but they are

Table 4.3. Diet of some mammals based on frequency of consumption (% of individual diet items occurring in stomachs or scats) (adapted from various sources; see references).

Mammal	Arthropods	Earthworms	Other animals	Fruit, seed	Acorns, nuts	Other
Brazilian free-tailed bat, *Tadarida brasiliensis*	20.8 moths, 19.6 beetles, 9.2 flies, 18.6 bugs-Heteroptera, 15.5 bugs-other, 7.1 ants and relatives					
Greater spear-nosed bat, *Phyllostomus hastatus*	60.0 beetles, 16.9 ants, 8.4 cockroaches, 3.7 bugs, 3.1 moths					
Pale spear-nosed bat, *Phyllostomus discolor*	74.0 beetles, 11.6 ants, 6.9 termites, 2.3 bugs, 2.3 lacewings, 2.3 moths					
Golden-tipped bat, *Kerivoula papuensis*	96.3 spiders, 16.0 beetles, 8.9 moths, 10.5 other					
Pallas's long-tongued bat, *Glossophaga soricina*	40.0 (16.0 ants, 13.0 termites)			fruit: 29.0 *Vismia*, 28.0 other		
Seba's short-tailed bat, *Carollia perspicillata*	6.0 beetles, 4.0 ants and termites			fruit: 74 *Vismia*, 11 *Solanum*		
White-lined bat, *Vampyrops lineatus*				fruit: 54.0 *Solanum*, 40.0 other		
Townsend's mole, *Scapanus townsendii*	31.0 insects, some centipedes	85.6	2.6 slugs			2.6 (esp. bulbs)
Eastern mole, *Scalopus aquaticus*	70.6 scarab larvae, 27.3 other beetle larvae, 63.1 ants, 25.1 adult scarabs, 22.7 other beetles, 16.3 caterpillars, 8.6 flies, 5.3 ground beetles, 47.9 centipedes	8.3	2.7 snails	4.8 seeds		15.0 fungus
Short-tailed shrew, *Blarina brevicauda*	18.6 beetles, 14.5 caterpillars, 3.6 flies, 5.9 centipedes, 4.5 other	54.3	38.9 slugs and snails, 6.7 sowbugs, 20.8 other			24.4 fungi, 29.4 other plants

Species						
Least shrew, *Cryptotis parva*	29.4 caterpillars, 4.6 crickets, 4.6 ground beetles, 8.2 other adult beetles, 7.3 beetle larvae, 3.7 fly larvae, 2.8 leafhoppers, 6.4 unidentified Hemiptera, 6.4 aphids, 3.7 plant bugs	15.6	3.7 slugs & snails, 7.3 centipedes, 2.8 woodlice, 11.0 spiders	2.8	4.6 acorns	0.9 fungi
Woodland jumping mouse, *Napaeozapus insignus*	40.8 beetles, 30.1 caterpillars, 2 other		8.7 unknown	seed: 33.0 unknown , 13.6 *Impatiens*, 3 other; fruit: 29.1 *Rubus*, 3.9 *Fragaria*		77.7 fungus
Grasshopper mouse, *Onychomys leucogaster* & *O. torridus*	38.7 grasshoppers, 20.7 beetles, 17.0 caterpillars, 2.0 ants, 0.7 flies, 2.5 other		3.1 mammals			11.1 plant, inc. 4.9 grain
White-footed mouse, *Peromyscus leucopus*	41.6 beetles, 19.1 caterpillars, 13.3 centipedes, 2.5 spiders, 1.6 millipedes		8.3 slugs& snails, 19.1 unknown, 1.6 sowbugs	seed: 31.4 various; fruit: 16.6 *Rubus*, 19.1 *Viburnum*		34.1 plants
House mouse, *Mus musculus*	43.7 caterpillars, 7.6 beetles, 1.9 bugs, 1.9 flies, 1.9 centipedes, 1.3 ants, 1.2 other	1.3	0.8 animal flesh, 5.2 unknown	65.0 various seeds, esp. grass		25.1 plants
Prairie deer mouse, *Peromyscus maniculatus bairii*	32.0 caterpillars, 16.0 beetles, 5.9 bugs, 7.9 flies, 3.2 centipedes	5.0	12.7 animal flesh	60.3 seeds		38.4 plants
Mantled squirrel, *Citellus lateralis*	10.2		3.1 animal flesh	35.7 seeds		78.0 fungi, 44.0 plants
Yellow pine chipmunk, *Eutamias amoenus*	47.1			62.7 seeds		43.1 fungus, 17.7 plant
Lodgepole chipmunk, *Eutamias speciosus*	43.7			48.7 seeds		44.7 fungus, 6.2 plant

Table 4.3. *Continued*

Mammal	Arthropods	Earthworms	Other animals	Fruit, seed	Acorns, nuts	Other
Least chipmunk, *Eutamias minimus jacksoni*	30.0 grasshoppers, 25.0 beetles, 24.0 caterpillars, 15.0 ants and bees, 6.0 flies, 10.0 others		10.0 spiders, 6.0 birds	fruit: 44.0 *Vaccinum*, 35.0 *Rubus*, 23.0 *Chiogenes*; 16.0 seeds		
Virginia opossum, *Didelphis virginiana*	87.6 (esp. grasshoppers, ground beetles, longhorn beetles)		29.0 snails, 28.2 mammals, 18.9 reptiles, 8.9 birds	50.6 wild fruit, 12.7 seed		
Nine-banded armadillo, *Dasypus novemcinctus*	100 beetles, 74.7 ants and wasps, 50.6 crickets and grasshoppers, 54.3 flies, 43.2 caterpillars, 14.8 cicadas, 4.9 termites	34.6	60.5 millipedes, 30.9 woodlice, 21.0 vertebrates, 11.1 slugs & snails			13.6 plants
Striped skunk, *Mephitis mephitis*	43.5 (mostly white grubs and grasshoppers)	0.9	35.3 mammals, 11.3 carrion, 2.3 birds, 0.7 reptiles	40.9 wild fruit, 16.0 seeds		11.3 grasses
Northern raccoon, *Procyon lotor*	23.3 caterpillars, moths, other	8.5	10.3 snails, 14.7 crayfish	35.7 grapes, 15.6 corn	25.3 acorns	occasional vertebrates
Collared peccary, *Pecari tajacu*	3.6					4.5 grass, 12.3 broad-leaf plants, 11.1 woody plants, 21.8 cactus

Species	Insects	Other invertebrates	Mammals / birds	Fruit	Nuts / acorns	Grass / carrion / plant
Black bear, *Ursus americanus*	3.5 bee, 2.2 beetle	2.2 crayfish		27.6 wild cherry, 25.2 apple, 13.9 grape, 2.2 dogwood	58.4 acorns, 7.1 beechnuts	occasional vertebrate
American badger, *Taxidea taxus*	9.0		80.5 pocket gopher, 22.0 vole, 8.0 mice, 5.5 squirrel, 2.0 woodchuck			6.5 plant
Swift fox, *Vulpes velox*	80.5 grasshoppers and crickets, 67.8 beetles		42.4 rodents, 38.9 birds, 27.8 rabbits, 7.8 reptiles			34.4 grass & seeds
Red fox, *Vulpes vulpes*	5.3	1.4	40.3 mice, 17.2 rabbit, 4.3 squirrel, 2.5 birds, 1.9 porcupine	8.7 fruit	1.4 nuts	24.7 grass, 8.2 carrion
Gray fox, *Urocyon cinereoargenteus*	23.1		41.4 rabbit, 50.0 mice, 1.2 squirrel, 20.7 birds	29.3 dry fruit, 32.9 fleshy fruit		7.3 carrion
Coyote, *Canis latrans*	8.9		38.9 ungulates, 11.5 rabbits, 12.7 squirrel			8.8
Mexican wolf, *Canis lupus baileyi*	3.2		69.6 ungulates, 5.1 rabbits, 2.8 squirrel			6.9

Table 4.4. Diet of some mammals based on volume (average % volume of individual diet elements found in stomachs or scats) (adapted from various sources; see references).

Mammal	Arthropods	Earthworms	Other animals	Fruit, seeds	Acorns, nuts	Other
Brazilian free-tailed bat, *Tadarida brasiliensis*	28.4 moths, 26.3 beetles, 9.2 flies, 17.8 bugs-Heteroptera, 9.0 bugs-other, 11.6 ants and relatives					
Trawling long-fingered bat, *Myotis capaccinii*	71.0 flies (mostly midges), 7.2 moths, 6.6 spiders, 4.0 caddisflies, 17.1 other					
Eastern mole, *Scalopus aquaticus*	31.1 scarab larvae, 4.5 other beetle larvae, 15.4 ants, 4.2 adult scarabs, 2.6 other beetles, 4.1 caterpillars, 1.4 flies, 0.6 ground beetles, 12.7 centipedes	2.9	1.0 snails	2.3 seeds		3.4 fungus
Short-tailed shrew, *Blarina brevicauda*	5.9 beetles, 4.3 caterpillars, 1.5 flies, 1.1 beetles, 2.0 other	31.4	27.1 slugs and snails, 8.1 other			8.8 fungi, 5.4 other plants
Least shrew, *Cryptotis parva*	17.9 caterpillar larvae, 3.2 crickets, 4.9 beetle larvae, 2.7 ground beetles, 1.2 scarab beetles, 3.0 other beetles, 2.4 fly larvae, 3.8 aphids, 2.8 plant bugs, 1.0 leafhoppers	11.2	3.3 slugs and snails, 3.6 centipedes, 1.9 woodlice, 6.8 spiders	1.2 seeds	3.5 acorns	0.9 fungi
Woodland jumping mouse, *Napaeozapus insignus*	10.3 caterpillar, 7.5 beetles, flies and spiders trace			24.2 seeds, 11.2 fruit		33.3 fungi
White-footed mouse, *Peromyscus leucopus*	15.5 beetles, 5.5 caterpillars, 1.3 spiders, 1.1 millipedes, 7.1 centipedes		2.7 slugs and snails, 7.0 unknown, 1.7 sowbugs	Seed: 26.9 various; fruit: 5.2 *Rubus*, 4.5 *Viburnum*		18.7 plants
House mouse, *Mus musculus*	14.6 caterpillars, 3.7 beetles, 0.9 bugs, .07 centipedes, 0.6 ants, 0.7 other	0.5	0.3 animal flesh, 1.0 unknown	42.5 various seeds		8.4 plants
Prairie deer mouse, *Peromyscus maniculatus bairii*	16.8 caterpillars, 4.8 beetles, 1.7 bugs, 1.0 flies, 1.2 centipedes	1.7	3.5 animal flesh	31.3 various seeds		19.8 plants
Mantled squirrel, *Citellus lateralis*	6.0			4.3 seed		61.3 fungi, 21.7 plant
Yellow pine chipmunk, *Eutamias amoenus*	20.0		2.0 animal flesh	41.3 seed		27.7 fungi, 7.0 plant

Species	Insects		Other animals	Fruit/seeds	Nuts/acorns	Other
Lodgepole chipmunk, Eutamias speciosus	15.7			29.3 seed		31.7 fungi
Virginia opossum, Didelphis virginiana	34.2 (esp. longhorn and ground beetles, grasshoppers, bugs)		32.3 mammals, 10.0 reptiles, 4.9 birds, 3.9 snails	7.3 seeds, 6.8 wild fruit		occasional vertebrate, rarely plants
Nine-banded armadillo, Dasypus novemcinctus	45.0 beetles, 6.9 flies, 2.6 ants and wasps, 1.2 caterpillars,1.0 cicadas, 0.6 crickets & grasshoppers	4.2	1.2 spiders, 1.4 millipedes, 0.3 woodlice			occasional vertebrates
Northern raccoon, Procyon lotor	8.4 caterpillars, moths, others	4.3	2.9 snails, 4.5 crayfish	10.5 corn, 26.0 grapes, 2.1 other berries	18.8 acorns	occasional vertebrate
Black bear, Ursus americanus	0.3 bees, 0.1 beetles		0.1 crayfish	26.8 wild cherry, 6.3 apple, 1.4 grape, 2.2 dogwood	44.5 acorns, 10.2 beechnuts	occasional vertebrate
Red fox, Vulpes vulpes	3.4	0.8	29.3 mice, 27.4 rabbit, 2.9 squirrel, 1.8 porcupine, 1.7 birds	5.3 fruit	0.4 nuts	24.7 grass
Gray fox, Urocyon cinereoargenteus	4.1		32.4 rabbit, 18.2 mice, 0.6 squirrel, 7.4 birds	10.9 dry fruit, 11.1 fleshy fruit		5.4 carrion
Raccoon dog, Nyctereutes procyonoides	7.7		25.5 rodents and shrews, 26.2 waterfowl, 4.9 other birds, 2.2 rabbits, 0.8 frogs and reptiles	23.3 berries		5.5 carrion, 3.1 cereal, 0.7 eggs
Spectral tarsier, Tarsius spectrum	17.1 caterpillars, 14.5 moths, 13.1 termites, 11.3 beetles, 10.7 grasshoppers, 6.1 cockroaches, 4.9 katydids and crickets, 4.8 cicadas, 1.9 walkingsticks, 2.4 spiders					

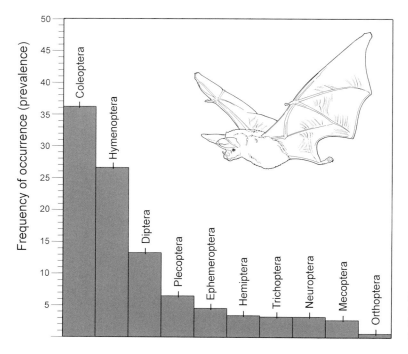

Fig. 4.2. Frequency of occurrence of various items in the stomachs of big brown bat, *Eptesicus fuscus* (adapted from Hamilton 1933).

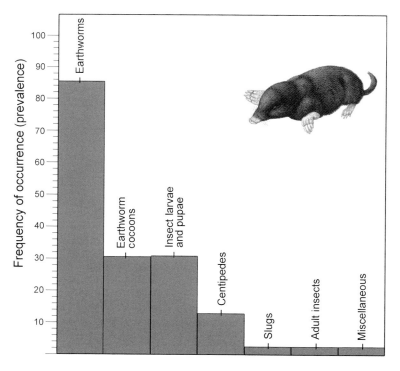

Fig. 4.3. Frequency of occurrence of various items in the stomachs of Townsend's mole, *Scapanus townsendii*, in Oregon (adapted from Wight 1928).

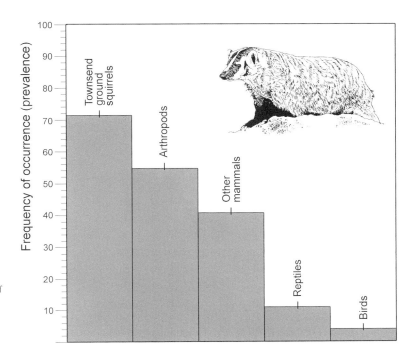

Fig. 4.4. Frequency of occurrence of various items in the stomachs of American badger, *Taxidea taxus*, in Idaho (adapted from Messick and Hornocker 1981).

unusual mostly by virtue of their tendency to eat arthropods almost exclusively.

Some of the most common medium-sized mammals, namely Virginia opossum, *Didelphis virginiana* (order Didelphimorphia), nine-banded armadillo, *Dasypus novemcinctus* (order Cingulata), striped skunk, *Mephitis mephitis* (order Carnivora, family Mephitidae), and northern raccoon, *Procyon lotor* (order Carnivora, family Procyonidae) can be seen to ingest insects readily. These mammals are ground-dwellers that take advantage of the larger insects found in or on the soil, principally ground and wood-boring beetles, grasshoppers, crickets, and sometimes ants. Raccoons sometimes frequent marshy areas, where they prey heavily on eggs of nesting birds, but in these situations they also take advantage of aquatic insects such as immature dragonflies (Odonata), predaceous diving beetles (Coleoptera: Dytiscidae), and giant water bugs (Hemiptera: Belostomatidae). Skunks often are considered to be the most beneficial omnivores because they feed frequently on insects, but opossums and armadillos also eat large quantities of insects. Insects and bird eggs are eaten principally during the summer; vertebrates during the winter.

In contrast, some of the larger mammals such as collared peccary, *Pecari tajacu* (order Ariodactyla, family Tayassuidae), black bear, *Ursus americanus* (order Carnivora, family Ursidae), and American badger, *Taxidea taxus* (order Carnivora, family Mustelidae), consume insects less often. A study of badger feeding behavior in Idaho (Fig. 4.4) found that Townsend ground squirrels, *Spermophilus townsendii*, were the most frequent prey item, exceeding the total number of arthropods. Especially if the size of the much larger ground squirrels is factored in, ground squirrels are a much more important prey item. The foxes (order Carnivora, family Canidae) are quite variable, with swift fox, *Vulpes velox*, eating insects frequently, but red fox, *Vulpes vulpes*, rarely taking insects and gray fox, *Urocyon cinereoargenteus*, intermediate in frequency of insect consumption. Similarly, in a Finnish study insects comprised a small proportion of the scats of raccoon dog, *Nyctereutes procyonoides* (Table 4.4). The other canids listed in Table 4.3 (coyote, *Canis latrans*, and Mexican wolf, *Canis lupus baileyi*) seem relatively unlikely to consume insects. The felids or cats (Carnivora: Felidae) seem to avoid insects and other arthropods.

Though considered to be omnivorous, many primates are entirely herbivorous. A major exception is the tarsiers, *Tarsius* spp. These small primates are found on the islands of Southeast Asia where they feed primarily on insects but also on other small animals. Research on spectral tarsier, *Tarsius spectrum*, revealed that a number of insects were acceptable (Table 4.4), though their diet was affected by availability. They seemed to prefer grasshoppers, caterpillars and moths, or at least they ate these preferentially during the wet season when insects were most abundant. In contrast, during the dry season they ate proportionally more ants, termites and beetles. During the dry season they were also forced to forage on the ground and to eat more but smaller insects.

Clearly, insectivory is not uncommon among primates. Marmosets, for example, are well known to eat insects. In Brazil, buffy tufted-eared marmoset, *Callithrix aurita*, ate mostly gum exudated from *Acacia paniculata* vines, but also fruits and animal prey. It devoted 50.5%, 11%, and 38.5% of its time spent feeding to these foods, respectively. The animal food consisted primarily of insects, and consisted mostly of caterpillars (33%), katydids (5%), and bugs (4%). The common squirrel monkey, *Saimiri sciureus*, devoted 61.4% of its time to foraging for insects and 21.1% eating them, whereas 7.7%, 3.0%, and 5.8% of its time was used for locomotion, rest, and social behavior, respectively. This Brazilian squirrel monkey study also determined (based on observation) that the predominant food item was arthropods (47.3% of its intake) with fruit, flowers and mammals (bat) comprising 39.2%, 12.8%, and 0.7% of the diet, respectively. However, there were seasonal shifts in feeding, with the greatest amount of insectivory in the wet season when insects were abundant, and mostly fruit feeding in the dry season when insects were harder to catch and insect foraging success diminished. The preferred insects were in the orders Orthoptera and Lepidoptera (together these comprised three-fourths of the arthropod diet). Study of this same species of squirrel monkey in Surinam also documented an insectivorous tendency, with 72% of its time devoted to foraging or feeding on insects, and only 28% to seeking and eating plant material. In this same environment, the tufted capuchin, *Cebus apella*, allocated 47.3% of its time to insect feeding, whereas weeping capuchin, *Cebus nigrivittatus*, and red-handed tamarin, *Saguinus midas*, devoted only 33.3% and 32.2% of its time to insects, and the balance to plant feeding.

Brown mouse lemur, *Microcebus rufus*, studied in Madagascar was found to maintain a mixed diet of fruit and insects throughout the year. Fruits from an epiphytic semiparasitic plant (*Bakerella* spp.) were particularly important. Insects alone were found in 14.4% of fecal samples, fruit alone occurred in 21.8% of samples, and both insects and fruit were observed in 61.9% of samples. The identifiable insects most commonly recovered from fecal material were beetles (Coleoptera), crickets and relatives (Orthoptera and allies), ants and relatives (Hymenoptera), and stink bugs (Hemiptera: Pentatomidae). Most lemurs tend to be mostly herbivorous, however. For example, the brown lemur (*Eulemur fulvus*), sifaka (*Propithicus verreauxi*), and ring-tailed lemur (*Lemur catta*) were found to eat negligible amounts of insects when studied in Madagascar. This is not to say that lemurs don't benefit from insects, however. The lesser mouse lemur, *Cicrocebus murinus*, was shown to feed preferentially on plants inhabited by a honeydew-secreting insect, *Flatidia coccinea* (Hemiptera). The secretions coat the foliage and twigs of trees inhabited by this insect, and the lemurs harvest the sweetfood resource selectively. Other lemurs aso partake of this sweet material.

Lorises are a small and poorly known group of primates. Research conducted in India shows that Mysore slender loris, *Loris lydekkerianus*, is almost exclusively faunivorous, favoring insects. About 96% of the feeding events involved animal food, with 62.9% involving ants and temites. Not surprisingly, a high proportion of these were insects occurring in aggregations, as this is an efficient way to collect insects. The researchers noted that some feeding events were followed by sneezing, slobbering, or urine-washing, which they inferred to indicate that these insects were distasteful. (The basis for primates washing their hands and feet in their own urine is poorly understood, but believed to provide them with comfort.) The researchers also were able to examine the stomach contents of one female loris. They found that 63% of the stomach contents were comprised of ants (Hymenoptera Formicidae) and termites (Isoptera), with 12% beetles (Coleoptera) and 9% Orthoptera.

It also is interesting to mention the feeding behavior of chimpanzees, *Pan* spp., in Africa. Remarkably, chimps use tools to break into termite mounds and to retrieve termites for food. Large variation in termite consumption has been documented (there also is large variation in availability) but some adults spend more than 10% of their time retrieving termites. Termites are the fifth most common food for females. Females with dependents were the most frequent chimp visitors to termite mounds. *Macrotermes* spp., the largest African termites (Isoptera: Termitidae), are most favored. The chimps access both belowground and aboveground termite colonies, using a large stick to puncture the colony and inserting a small stick to 'fish' for termites. They consume the termites that cling to the tip of the small stick. Some chimps seem to chew the tip of the 'fishing stick' until it is frayed like the tip of a paintbrush. It has been suggested that the frayed ends are more effective at extracting insects, but this is uncertain. In some populations, chimpanzees also gather ants and bees (Hymenoptera) in the same manner. In the Democratic Republic of Congo, animal material comprised only 9% of the diet in feces, but the researchers acknowledged that animals were likely underestimated in the diet. Indeed, beeswax was found in 55% of the fecal samples. The remains of five types of insects in the feces were documented, as was the use of sticks by chimpanzees to dig up subterranean bee nests. The chimpanzees benefit from eating insects by obtaining several important amino acids that are not present in adequate amounts in vegetation (e.g., histidine, leucine, lysine, and threonine). Also, the honey they consume from the bee nests has very high caloric value. At least in the Congo region, chimpanzees eat more insects during the rainy season, a period when fruit is relatively scarce but insects are abundant.

We might not expect really large primates such as gorillas to eat small insects due to the apparent inefficiency of gathering such small food items. Indeed, the mountain gorilla, *Gorilla beringei beringei*, rarely ingests insects. However, western gorilla, *G. gorilla gorilla* and Grauer's gorilla, *G. beringei graueri*, display regular and deliberate insectivory. For example, examination of the fecal samples from *G. gorilla gorilla* showed that 78% contained insect fragments, mostly from social insects. Ants and termites were prominent (found in 61 and 39% of feces, respectively), but beetles (22%) and orthopteroids (39%, including cockroaches) also were abundant.

Likewise, although it would seem to be inefficient for large animals such as bears to search for and consume insects, there are times when it is profitable. Everyone knows that bears are attracted to beehives, and it is generally assumed that it is the honey that they seek when they tear apart and consume beehive material. After all, for nearly a century children in many countries have grown up with author A.A Milne's 'Winnie-the-Pooh' stories, made even more popular by Walt Disney's adaptions of this lovable character. In these stories, it is honey that Winnie seeks. However, although bears are attracted to beehives, they actually feed preferentially on bee brood, the larvae of the bees. There is a great deal of protein and fat in a beehive, just what a hungry bear needs!

It is not only bee aggregations that attract bears. Both grizzly bear, *Ursus arctos horribilis*, and black bear, *Ursus americanus*, eat the moth *Euxoa auxiliaris* (Lepidoptera: Noctuidae). The larvae of this moth are called army cutworms, and live on the Great Plains of western North America where they feed on grasses and forbs, and sometimes damage wheat crops. When the moths emerge in the early summer they migrate hundreds of kilometers to the Rocky Mountains, where they feed on the nectar of flowering plants through most of the summer, only to migrate back to the grasslands in autumn where they deposit eggs and perish, completing the migratory cycle. The moths are nocturnal, and during they day they aggregate in natural shelters. Often, thousands of moths can be found in a hollow log or in a natural cavity formed by rocks. The bears in the Rocky Mountains are quick to take advantage of this food source, and often are discovered feasting on aggregations of these prey during the daylight hours, when the moths do not fly. Thus, when insects are present in adequate quantity, even carnivores as large as grizzly bears will feed greedily on them.

It is interesting to compare the diet frequency data with analyses based on volumetric assessment of diet. Many of the studies shown in Table 4.3 also presented volumetric data (Table 4.4). In many respects, the diet profile does not change with this different measure of analysis. For example, in the frequency analysis the most 'important' diet items for the Brazilian free-tailed bat, *Tadarida brasiliensis*, were moths, beetles, and bugs in the subfamily Heteroptera. These groups were also assessed as the most-consumed prey items in the volumetric analysis, and in the same order of importance. With most comparisons of diet, however, the frequency values are larger than the volumetric values. Thus, the frequency of scarab beetle larva consumption by eastern mole, *Scalopus aquaticus*, was 70.6% whereas the proportion of the stomach occupied by scarabs was 31.1%. Looking down the list of mammals in Tables 4.3 and 4.4, it is apparent that in nearly all cases the frequency values are larger than the volumetric values. This is due to the small size of most insects. These numerous small meals may well prove vital to keeping wildlife active and healthy while searching for the more infrequent, larger meals. The frequency and volumetric values are highly correlated, though the correspondence of values is not perfect. This can be seen by examining the Virginia opossum (*Didelphis virginiana*) consumption values; the frequency data suggest a higher level of insect and fruit consumption than do the volumetric data relative to consumption of mammals. So both methods of analysis provide useful information which, taken together, help us to understand the importance of various diet components.

Although diet analysis is critically important in understanding wildlife behavior, ecology, and conservation, consider the data presented in Fig. 4.5, which shows dietary information for striped skunk, *Mephitis mephitis*. Here we see the types of issues already mentioned: the disparity between frequency and volumetric data, seasonal differences, and the differences (geographic disparity) between studies. One study was conducted in California (Fig. 4.5a, b) and the other in New York (Fig. 4.5c, d). In both cases, insects occurred more frequently than any other dietary item (Fig. 4.5b, d). However, on a volumetric basis, insects were not the largest component, instead ranking only second most important among the dietary elements in California ('waste' consisted of dirt and parasitic worms, so was disregarded), and sixth in importance in New York (Fig. 4.5a, c). The New York population apparently had ready access to fruit and carrion, which was not the case in California, thus affecting the ranking of 'important' foods. This clearly demonstrates the importance of studying animal diets at several locations. It is also important to note that the data presented in many of these tables are based on individual studies, and may not be indicative of wildlife feeding patterns everywhere.

ANALYSIS OF BIRD DIETS

Like mammals, birds must eat frequently to maintain metabolism and homeostasis. Small birds, in particular, have need for frequent meals. They also store fat in their bodies in the fall, but not to support them while hibernating. Unlike mammals, they often migrate to warmer climates in the winter, allowing them to access better food supplies. Their fat reserves support them during the long flights associated with migration.

Birds are quite diverse in their feeding behavior, and display many adaptations of their feeding structures.

Bill, leg, and feet structures are sometimes strongly modified for food acquisition. Some of the obvious modifications are the large, heavy bill and associated cranial modifications among woodpeckers to support drilling into wood; the large hair-like bristles surrounding the bill of whippoorwill, *Caprimulgus vociferous*, that help to funnel large insects into its mouth during flight; the spatulate bill of spoonbills, *Platalea* spp., that is well equipped with sensory papillae for detecting small objects such as crustaceans and insects in murky water; and the strong, grasping feet of raptors that allow them to capture mobile prey. Less obvious are the benefits of the long narrow bills of shorebirds that allow them not only to pluck small animals from water or among debris, but to apply surface tension to water for transport of small invertebrates into the mouth. Surface-tension feeding works because water is adhesive to the surface of the bird's bill. As the bird slowly opens its bill, the droplet of water containing an invertebrate is stretched. The increase in potential energy that results from the increasing surface area of the stretched droplet drives the prey and droplet into the bird's mouth.

Insectivory by birds is better acknowledged than it is for mammals. For example, it is reported that 80% of the birds breeding in Central Europe feed on insects at least temporarily, and an assessment of North American birds indicated that 61% were primarily insectivorous, 28% were partially insectivorous, and only 11% were not insectivorous. A synopsis of insectivory in birds (class Aves) follows:

• The taxa that are widely acknowledged as commonly feeding on insects include the orders Apodiformes, swifts and hummingbirds; Caprimulgiformes, the nightjars, goatsuckers, or nighthawks; Cuculiformes, cuckoos and relatives; and Piciformes, woodpeckers and relatives.

• Taxa not usually feeding on insects include orders Ciconiiformes, the herons and storks; Coraciiformes, the kingfishers; Columbiformes, the doves and pigeons; Falconiformes, the falcons, eagles, hawks, and vultures; Gaviiformes, the loons; Pelecaniformes, the pelicans and cormorants; Phoenicopteriformes, the flamingos; Psittaciformes, the parrots; Sphenisciformes, the penguins; Procellariiformes, the petrels and albatrosses; and Strigiformes, the owls.

• There are a few taxa that have broad dietary habits that include insects, including orders Anseriformes, the ducks, geese, and swans; Galliformes, the pheasants, turkeys and quail; Charadriiformes, the shore-

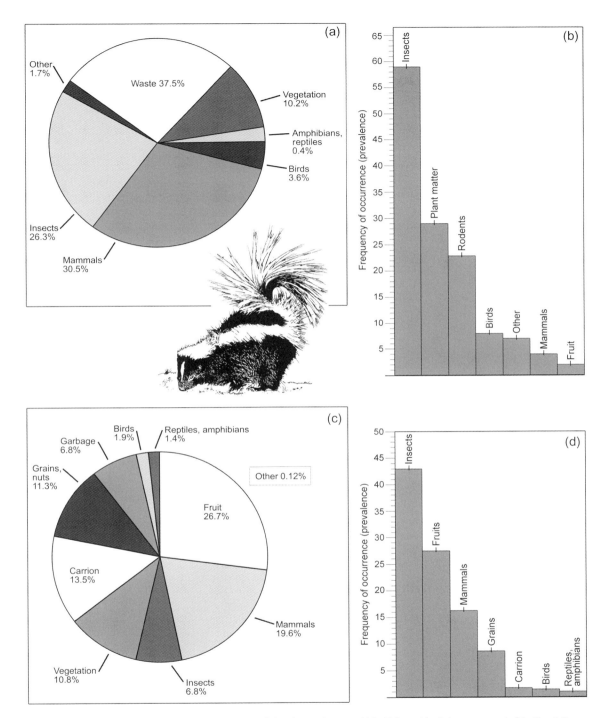

Fig. 4.5. Comparative dietary information for striped skunk. *Mephitis mephitis*. Volumetric data are presented in Fig. 4.5 a, c; frequency of occurrence in Fig. 4.5 b, d. One study was conducted in California (Fig. 4.5 a, b) mostly during the fall and winter, and the other in New York during the fall and winter (Fig. 4.5 c) or spring and summer (Fig. 4.5 d) (adapted from Dixon 1925 and Hamilton 1936).

Table 4.5. The stomach contents (% of total volume) of some bird species. Numbers do not always add up to 100% due to inclusion of sand, or twigs and other organic material eaten inadvertently. Consumption of cultivated fruit is overestimated due to the inability to differentiate source; strawberries, blackberries and raspberries were always tabulated as being cultivated, even though all occur in wild form (adapted from various sources; see References and particularly McAtee (1917) for a listing of most early publications).

| Bird | % of stomach contents | | | | | |
	Total animal	Coleoptera: beetles, weevils	Lepidoptera: caterpillars, etc.	Hymenoptera: bees, wasps, ants	Hemiptera: Bugs	Orthoptera: grasshoppers, crickets
Hairy woodpecker, *Picoides villosus*	77.7	41.4 (31.0 wood boring, 3.0 weevils)	9.9	17.1 ants	2.4	
Downy woodpecker, *Picoides pubescens*	76.1	21.5 (13.9 wood boring)	16.5	13.4 ants	8.6	
Pileated woodpecker, *Dryocopus pileatus*	72.8	22.0		39.9 ants		
Red-headed woodpecker, *Melanerpes erythrocephalus*	33.8	19.0	1.6	6.8 (5.2 ants)	1.9	3.6
American three-toed woodpecker, *Picoides dorsalis*	94.1	71.1 (60.7 wood-boring)	14.5 (wood-boring)	8.3 ants		
California woodpecker, *Melanerpes uropygialis*	22.6	3.0		14.5 (8.1 ants)		
Lewis's woodpecker, *Melanerpes lewis*	37.5	9.1 (6.7 preda-tory)		23.5 (11.9 ants)	1.4	3.2
Red-bellied woodpecker, *Melanerpes carolinus*	30.9	10.3	2.9	8.0 (6.5 ants)	1.9	5.9
Yellow-bellied sapsucker, *Sphyrapicus varius*	49.3	5.4		36.9 (34.3 ants)	0.9	
Red-cockaded woodpecker, *Picoides borealis*	88.1	15.6		51.7 ants	6.9	
Northern flicker, *Colaptes auratus*	60.9	4.1	1.3	49.7 ants	0.8	2.4
Ruby-throated hummingbird, *Archilochus colubris*	94.3			36.3 parasites	8.9 leafhoppers	
Anna's hummingbird, *Calypte anna*	100			35.0	17.3	
Loggerhead shrike, *Lanius ludovicianus*	95	16.0	7.0	11 ants, wasps		43.0

Diptera: Flies	Other insects	Other	Total plant	Weed seed	Fruit	Other
	2.0	3.5 other arthropods	22.3	4.5	5.2 wild fruit	1.4 corn
	4.8	4.1	23.9	5.9	5.8 wild fruit, 8.2 acorns and nuts	1.6 grain
	10.9		27.1		22.6 wild fruit	
	1.0		66.2		16.9 wild fruit, 23.3 acorns and nuts	4.2 grain, 18.3 other
			5.9			
	4.5		77.4		53.3 acorn, 23.0 cultivated fruit	
			62.5		14.6 wild fruit and acorns, 10.9 cultivated fruit	
		2.3 other arthropods	69.1	2.1 poison ivy	30.7 acorn, 27.3 wild fruit	3.9 corn
	5.4		50.7		28.1 wild fruit, 6.1 acorns and seeds	16.5 cambium
	14.6 (grasshoppers, termites, etc.)		13.9		4.0	9.9 conifer seed
		1.5	39.1	6.6	9.2 fruit, mostly wild, 1.8 acorns and nuts	1.1 grain
2.6		43.5 spiders	5.7			nectar not measured
45.3		2.0 spiders				nectar not measured
	5.0	12 mice, shrews, lizards; 2 spiders				

Table 4.5. *Continued*

	% of stomach contents					
Bird	Total animal	Coleoptera: beetles, weevils	Lepidoptera: caterpillars, etc.	Hymenoptera: bees, wasps, ants	Hemiptera: Bugs	Orthoptera: grasshoppers, crickets
Eastern kingbird, *Tyrannus tyrannus*	88.9	25.3	3.2	32.4	3.8	11.8
Western kingbird, *Tyrannus verticalis*	90.6	17.0	7.3	31.4 bees, wasp	5.4	27.7
Scissor-tailed flycatcher, *Tyrannus forficatus*	91.6	13.7	4.6	12.8 (mostly bees, wasps)	10.2	46.1
Ash-throated flycatcher, *Myiarchus cinerascens*	92.3	7.3	17.1	26.9 bees, wasps	20.1	5.1
Western yellow-bellied flycatcher, *Empidonax flaviventris*	99.3	7.3	6.6	38.7 bees, wasps	8.4	
Olive-sided flycatcher, *Contopus cooperi*	99.9	6.2	4.1	82.6	3.2	1.1
Eastern phoebe, *Sayornis phoebe*	89.0	13.3	8.2	35.0 wasps	10.6	2.4 crickets
Say's phoebe, *Sayornis saya*	99.8	15.7	12.1	30.7 (mostly bees and wasps)	4.5	15.4
Black phoebe, *Sayornis nigricans*	99.4	13.3	8.2	30.8	10.6	2.5
Eastern wood pewee, *Contopus virens*	98.9	14.2	12.3	28.2 (7.0 parasitic wasps)	5.9	3.4
Eastern bluebird, *Sialia sialis*	68.0	20.9 (8.0 ground)	10.5	5.0 (3.4 ants)	2.7	22.0
Western bluebird, *Sialia mexicana*	81.9	24.0	20.2	6.7 (5.4 ants)	6.4	21.3
Mountain bluebird, *Sialia currucoides*	91.6	30.1	14.5	13.3 (12.5 ants)	3.9	23.0
American robin, *Turdus migratorius*	42.4	16.7 (5.0 ground, 5.5 May beetles)	9.0	2.7	2.2	4.7
Townsend's solitaire, *Myadestes townsendi*	35.9	10.7 (5.9 ground, 2.0 scarab)	12.9	5.2, mostly ants	3.5	1.0
Wood thrush, *Hylocichla mustelina*	59.6	20.4 (2.2 ground)	9.4	12.8 (8.9 ants)	1.3	2.3
Veery, *Catharus fuscescens*	57.3	14.7 (2.5 weevils, 0.8 ground)	11.9	13.6 (10.3 ants)	1.3	4.9
Swainson's thrush, *Catharus ustulatus*	63.5	16.3 (5.3 weevils, 3.1 ground)	10.3	21.5 (15.2 ants)	3.7	2.4

Diptera: Flies	Other insects	Other	Total plant	Weed seed	Fruit	Other
3.2	7.7	1.5	11.1		10.7	
	1.3		9.4		9.4	
3.8	4.9		3.9			
12.8	2.9		7.6 wild fruit, seeds			
31.3	1.6	4.5 spiders	0.7		0.7	
0.9	1.8		0.1			
18.0			11.0		11.0 wild	
16.7	3.5	1.3	0.2			
28.3	5.4	0.3	0.6	0.3	0.3	
29.9	2.6	2.2	1.0	0.2	0.8	
0.3	9.0	5.7	32.0 wild fruit, weed seed			
0.7	0.1	2.1 (1.9 spiders)	18.1	1.2	14.8	
0.9	2.9		8.4 (mostly fruit)			
3.1	4.0 other insects and invertebr.		57.8		42.0 wild, 8.0 cultivated	
trace	trace	2.9 spiders	64.1	some	58.7 wild	
2.7	1.0	8.5 (esp. spiders, millipedes)	40.4		33.5 wild, 3.7 cultivated	
0.8	1.0	6.3 spiders, 2.7 other invertebr.	42.7	7.2	23.2 wild, 12.1 cultivated	
	0.5	2.2 spiders, 0.3 other invertebr.	36.5	4.0	19.7 wild, 12.6 cultivated	

Table 4.5. *Continued*

Bird	% of stomach contents					
	Total animal	Coleoptera: beetles, weevils	Lepidoptera: caterpillars, etc.	Hymenoptera: bees, wasps, ants	Hemiptera: Bugs	Orthoptera: grasshoppers, crickets
Hermit thrush, *Catharus guttatus*	64.5	15.1 (3.1 weevils, 3.4 scarabs, 3.0 ground)	9.5	17.9 (12.5 ants)	3.6	6.3 mostly crickets
Horned lark, *Eremophila alpestris*	20.6	6.5 (4.5 weevils)	10.0		1.0	2.5
Red-winged blackbird, *Agelaius phoeniceus*	26.6	10.1	5.9			4.7
Bobolink, *Dolichonyx oryzivorus*	57.1	19.0	13.0	7.6		11.5
Brown-headed cowbird, *Molothrus ater*	22.3	5.3				11.0
Yellow-headed blackbird, *Xanthocephalus xanthocephalus*	33.7	7.8	4.6			11.6
Rusty blackbird, *Euphagus carolinus*	53.0	13.8	2.5			12.0
Common grackle, *Quiscalus quiscula*	30.3	13.5	2.3			7.3
Eastern meadowlark, *Sturnella magna*	71.7	18.0 (3.0 weevils, 4.0 May and 7.0 ground beetles)	8.0	3.0 ants, 1.5 bees, wasps	4.0	29.0 inc. crickets
Baltimore oriole, *Icterus galbula*	83.4	18.0 (2.0 weevils, 4.5 click, 3.5 May, 3.0 leaf beetles)	34.0	11.0	6.0	4.0
Blue jay, *Cyanocitta cristata*	22.0	10.5				4.4
American crow, *Corvus brachyrhynchos*	31					4.9
Yellow-billed cuckoo, *Coccyzus americanus*	92.0	3.2	65.6	0.9	12.2	14.3
Purple martin, *Progne subis*	100	12.5 (5.2 scarabs)	9.4 moths	19.5 bees & wasps, 3.5 ants	14.6 (mostly stink, negro, lace bugs)	1.1

Diptera: Flies	Other insects	Other	Total plant	Weed seed	Fruit	Other
3.0	0.3	8.8 spiders and others	35.5	8.1	26.2 wild, 1.2 cultivated	
	1.0		79.4	69.3		
	4.1	1.3 other invertebr.	73.4	54.6		13.9 grain, 4.3 other
	4.6	1.4	42.9	16.2		12.4 grain
	2.3	1.1	77.7	60.0		16.5 grain
			66.3	27.1		38.1 (mostly oats)
	13.7	4.0 arthropods, 7.0 other	47.0	6.0		24.4 grain, 16.6 other
	2.3	1.5 arthropods, 3.1 other	69.7	4.2	2.9 cultivated, 2.1 wild, 14.0 acorns, nuts	46.5 grain (37.2 corn)
	9.7	5.0 spiders, millipedes	26.5	11.0		14.1 grain
4.0		6.0 spiders	16.6 mostly fruit	some	both wild and cultivated	some
		eggs and nestlings of other birds	68.0	7.0 wild	43.0 acorns and other nuts	18.0 grain
	15.1 caterpillars, beetles, others	7.4 crustaceans, plus 4.0 fish, molluscs, amphibians, carrion	69 (wild fruit, corn most common)			
			8.0			
16.1 (mostly crane, muscid, robber flies)	15.1 dragonflies, 8.1 other					

Table 4.5. *Continued*

Bird	% of stomach contents					
	Total animal	Coleoptera: beetles, weevils	Lepidoptera: caterpillars, etc.	Hymenoptera: bees, wasps, ants	Hemiptera: Bugs	Orthoptera: grasshoppers, crickets
Cliff swallow, *Petrochelidon pyrrhonota*	99.3	26.9 (4.9 scarabs)	0.5	20.5 bees & wasps, 8.2 ants	26.3	0.7
Barn swallow, *Hirundo rustica*	99.8	15.6 (6.2 scarabs, 1.9 weevils)	2.4	12.8 bees & wasps, 9.9 ants	15.1 (mostly stinkbugs & leafhoppers)	0.5
Tree swallow, *Tachycineta bicolor*	80.5	14.4 (5.9 scarabs, 1.9 weevils)	1.1	7.6 bees, wasps, 6.4 ants	5.6	0.4
Bank swallow, *Riparia riparia*	99.8	17.9 (5.5 scarabs, 5.8 weevils)	1.2	13.4 ants	7.9	
Northern mockingbird, *Mimus polyglottos*	47.8	12.0	9.5	3.0 bees & wasps, 4.5 ants		14.8
Brown thrasher, *Toxostoma rufum*	62.6	18.1	5.9	2.5 (1.4 ants, 0.9 bees, wasps)	1.5	2.4
Gray catbird, *Dumetella carolinensis*	44.0	14.0	5.0	10.0 ants	2.0	4.0
Northern cardinal, *Cardinalis cardinalis*	29.0	10.5	5.1	0.9	3.7	
Pyrrhuloxia, *Cardinalis sinuatus*	28.8	4.7	10.3		1.5	11.5
Painted bunting, *Passerina ciris*	20.9	2.5 weevils	3.1 caterpillars			
Rose-breasted grosbeak, *Pheucticus ludovicianus*	52.0	35.9	3.8	6.4	2.4 scale insects	
Chipping sparrow, *Spizella passerina*	38.0	11.0	14.2	11.8 mostly ants	7.5	
Dark-eyed junco, *Junco hyemalis*	23.0	6.0	9.4	2.0		
White crowned sparrow, *Zonotrichia leucophrys*	7.4	1.4	3.5	1.9	0.5	
House finch, *Carpodacus mexicanus*	2.4				mostly aphids	
Tufted titmouse, *Baeolophus bicolor*	66.6	7.1 (4.9 weevils)	38.3	12.5 bees, wasps, sawflies	4.0	
Yellow-rumped warbler, *Dendroica coronata*	84.7	6.7	14.2	26.2 ants, wasps	19.6	
Ruby- crowned kinglet, *Regulus calendula*	94.0	13.0	3.0	32.4	25.7	

Diptera: Flies	Other insects	Other	Total plant	Weed seed	Fruit	Other
13.9	2.9		0.7			
39.5 (mostly crane, deer, robber flies)	4.0 dragonflies, 3.7 other		0.2			
40.5 (mostly crane, deer, robber flies)	406		19.5		berries, inc. 16.9 bayberry	
26.6 (mostly blow & crane flies)	2.1 dragonflies, 12.6 other		0.2			
		2.0 spiders	52.2		42.6 wild, 3.3 cultivated	
1.7	2.9	2.4 crustaceans and molluscs	35.0	32 (19.9 wild)		2.6
		9.0	55.0	53.0 (35.0 wild)		2.0
		2.8	71.0	36.4	24.2 wild	8.7
			71.2	69.0 (mostly grasses)		2.0 grain
		various taxa	79.1	67.0 foxtail, 5.9 other		
		3.3	48.0	15.7	19.3 wild	5.1 grain, 7.9 other
3.0			62.0	53.0		4.0 grain
	7.3		77.0	61.8		8.0 grain
			92.6	74.0	4.5	8.6 grain
			97.6	86.0	10.0	0.25 grain
			33.4	4.1	5.2 wild fruit, 23.4 acorns, nuts	
16.4			15.3	9.2	5.0	
			6.0			

Table 4.5. *Continued*

Bird	Total animal	Coleoptera: beetles, weevils	Lepidoptera: caterpillars, etc.	Hymenoptera: bees, wasps, ants	Hemiptera: Bugs	Orthoptera: grasshoppers, crickets
						% of stomach contents
European starling, *Sturnus vulgaris*	57.0	21.0 (incl. 8.5 weevils, 5.7 ground, 2.2 scarabs)	6.0			12.4
House wren, *Troglodytes aedon*	98.0	22.0	16.0	4.0 ants	12.0	25.0
Carolina wren, *Thryothorus ludovicianus*	94.2	13.6	21.2	4.6 ants	18.9	12.6
California quail, *Callipepla californica*	3.0			1.0 ants		
Northern bobwhite, *Colinus virginianus*	15.1	6.8	0.9		2.8	3.7
Horned grebe, *Podiceps auritus*	100	23.3 mostly aquatic				
Wood duck, *Aix sponsa*	51.5	21.7 aquatic	3.7		1.5	8.7
Canvasback duck, *Aythya valisineria*	60.3					
Redhead duck, *Aythya americana*	70.0					
Lesser scaup, *Aythya affinis*	99.0				1.6	
Pink-eared duck, *Malacorhynchus membranaceus*	99.6	2.3 water beetles			10.0 aquatic bugs	
Grey teal, *Anas gibberifrons*	35.6	32.7 flies (mostly mosquitoes)	1.1			
Black tern, *Chlidonias niger*	100	14.5 (6.0 diving, 5.0 scarab, 3.5 leaf)				12.0
Franklin's gull, *Larus pipixcan*	94.5	larvae are important			at some locations important	43.4
Killdeer, *Charadrius vociferus*	97.7	37.1				

Diptera: Flies	Other insects	Other	Total plant	Weed seed	Fruit	Other
	5.9	11.4 millipedes	43.0		23.8 wild, 4.4 cultivated	13.6 other, 1.2 grain
		19.0 esp. spiders	2.0 (prob inadvertent)			
3.0	0.2	2.1 often millipedes	5.8 inc. acorns			
	2.0		97.0	62.5	2.3	37.8 grass, grain, other
	0.7		85.0	50.0	12.5	16.0 grain, 6.5 buds
	12.0 aquatic	27.8 fish, 20.7 crawfish, 13.8 other crustaceans				
7.1 various		4.7 isopods, 2.9 other crustaceans & molluscs	48.5			mostly seeds of silver maple, watershield and elm
4.0 midge larvae	27.0 caddisflies, 4.7 mayfly nymphs	28.0 snails	29.7			mostly pondweed tubers
7.3 midge larvae	53.3 caddisfly nymphs	3.6 snails, 3.0 other	30.3			Mostly pondweed buds
14.3 midge larvae, 6 caddisflies		34.3 aphipods, 12.3 snails, 20.3 leeches,1.0 other	1.0			Mostly nutlets from milfoil
60.7 (mostly midge larvae)		9.5 molluscs, 16.6 crustaceans	0.3			
			63.9			
8.0	13.0 mayflies, 20.0 dragonflies	3.0 crustaceans, 19.0 small fish				
			5.5, incidental			
11.9	17.6	21.1	2.3			

Table 4.5. *Continued*

Bird	% of stomach contents					
	Total animal	Coleoptera: beetles, weevils	Lepidoptera: caterpillars, etc.	Hymenoptera: bees, wasps, ants	Hemiptera: Bugs	Orthoptera: grasshoppers, crickets
Semipalmated sandpiper, *Calidris pusilla*	99.2	27.0 water scavenger, 5.0 ground, 3.3 other			16.6 backswimmers	
Pectoral sandpiper, *Calidris melanotos*	89.5	8.0		2.1	1.3	
Sharp-tailed sandpiper, *Calidris acuminata*	96.1	8.8		1.8		
Stilt sandpiper, *Calidris himantipus*	70.1	15.5 predaceous diving				
Red knot, *Calidris canutus*	84.5	1.2				
Long-billed dowitcher, *Limnodromus scolopaceus*	85.9	10.6				
Short-billed dowitcher, *Limnodromus griseus*	87.9					
Wilson's snipe, *Gallinago delicata*	83.2	15.9			3.3	
American woodcock, *Scolopax minor*	89.5	6.2	3.3			
American coot, *Fulica americana*	10.6					
Cattle egret, *Bubulcus* (*Ardeola*) *ibis*	100.0				11.4 cicada	64.0 grasshoppers & crickets
Red-capped parrot, *Purpureicephalus spurius*	7.2 (mostly psyllids, beetles, caterpillars)					

Diptera: Flies	Other insects	Other	Total plant	Weed seed	Fruit	Other
21.6	12.5	13.0 snails	0.8			
54.5		22.3 amphipods	10.5 (mostly algae)			
39.1	11.8 caddisflies	14.2 molluscs, 18.1 crustaceans	3.9			
72.8 bloodworms (midge larvae), 0.8 mosquito larvae		7.1 snails	29.9			
12.7	0.9	59.0 molluscs, 6.9 crustaceans, 4.1 other	15.2			
58.3	2.0	7.7 crustaceans, 3.9 molluscs, 3.2 marine worms	14.0 aquatic			
18.0	8.9	27.4 marine worms, 20 molluscs, 6.1 crustaceans, 4.3 other	12.1			
20.2	5.2	11.7 crustaceans, 6.7 molluscs, 11.3 earthworms, 2.6 fish	16.8			
6.8	1.9	67.8 earthworms	10.5			
	7.2	2.9 molluscs, 1.1 other	89.4 podweed, sedges & algae			
4.1	1.3	1.7 spiders, 1.0 ticks, 3.6 frogs, 0.3 skinks				
			92.8		68.1 seeds & fruit	24.7

Table 4.6. Diet of some fresh-water fish based on frequency of consumption (% diet elements found in stomachs) (adapted from various sources: see references).

Fish	Plants	Plankton	Fish	Molluscs	Crustaceans	Insects	Other invertebrates	Other
Bluegill, *Lepomis macrochirus*					29.6 cladocerans	3.4 ants, 0.3 midges		32.5
Pumpkinseed, *Lepomis gibbosus*				48.7	4.6 amphipods, 2.1 pelecypods	13.7 ants 9.8 midges, 6.1 beetles, 3.5 caddisflies		8.7
Green sunfish, *Lepomis cyanellus*				28.6		25 midges, 10.7 beetles, 10.7 dragonflies, 3.6 ants, 7.1 other	3.6	
Black crappie, *Pomoxis nigromaculatus*				0.1	80.9 cladocerans, 17.3 copepods, 0.2 ostracods	1.4 midges, 0.1 other		
Yellow perch, *Perca flavescens*					11.7 copepods, 0.2 ostracods	42.3 midges, 44.6 other		
Largemouth bass, *Micropterus salmoides*			0.1		55.1 copepods, 34.7 cladocerans	2.2 midges, 8.8 other		
Rainbow trout, *Salmo gairdneri*			1		4	27 caddisflies, 13 blackflies, 9 midges, 9 stoneflies, 19 other		5
Cutthroat trout, *Salmo clarkii*			21		1	19 caddisflies, 6 blackflies, 13 midges, 15 mayflies, 6 stoneflies, 11 other		4
Brown trout, *Salmo trutta*						38.9 mayflies, 26.4 caddisflies, 15.6 black flies, 7.7 midges, 11.5 other		

Species					
Coho salmon, *Oncorhynchus kisutch*	7.9	19.2 cladocerans	67.2		5.7 fish eggs
Alabama sturgeon, *Scaphirhynchus suttkusi*	2.0	0.1 copepods	61.7 biting midges, 19.6 other midges, 8.6 mayflies		
Ten-spined stickleback, *Pygosteus pungitius*	2	16 cladocerans, 22 copepods, 11 ostracods, 20 other	21 midges, 8 other	2 worms	
Quillback, *Carpoides cyprinus*	1.7	37.4 cladocerans, 36.5 copepods	25.6 midges		1.4 amphibians
Bigmouth buffalo sucker, *Ictiobus cyprinellus*	8.3	81.2	7.6 midge larvae		
Shorthead redhorse sucker, *Moxostoma macrolepidotum*	5.8	28.5	48.6 midges, 17.0 caddisflies, 1.4		
Whitesucker, *Catostomus commersoni*	11.1	11.4	49.7 midges, 1.0 caddisflies		2.4 diatoms
A cave catfish, *Trichomycterus itacarambiensis*	2.5		44.5 (7.5 midges, 7.5 beetles)	30.0 worms, 7.5 spiders	7.5 other
Brown bullhead, *Ameiurus nebulosus*		70.9 cladocerans, 16.1 copepods, 0.2 ostracods	9.6 midges, 0.1 other		

birds; Gruiformes, the coots, cranes and rails; Phoenicopteriformes, flamingos; Struthioniformes, the emus, ostriches and cassowaries; Podicipediformes, the grebes; and Passeriformes, the perching or 'song' birds. The order Passeriformes contains over half the total species of birds, however, so this group is of special significance and there are many passerine birds that feed on insects and their close relatives. Among the important passerine bird families are Furnariidae, the ovenbirds; Thamnophilidae, the antbirds; Tyrannidae, the phoebes, kingbirds, and some flycatchers; Corvidae, the crows and jays; Laniidae, the shrikes; Vireonidae, the vireos; Bombycillidae, the waxwings; Certhiiidae, the creepers; Muscicapidae, the Old World flycatchers; Mimidae, the mockingbirds and thrashers; Reguliidae, the kinglets; Sittidae, the nuthatches; Sturnidae, the starlings, Troglodytidae, the wrens; Cardinalidae, the cardinals and grosbeaks; Emberizidae, the buntings and American sparrows; Estrildidae, the estrildid finches; Fringillidae, the finches; Icteridae, the blackbirds and others; Parulidae, the New World warblers; Polioptididae, the gnatcatchers; Thraupidae, the tanagers; Alaudidae, the larks; Hirundinidae, the swallows; Paridae, the chickadees and titmice; and Turdidae, the thrushes. Insect feeding is common among all these families, and even among species that often are not thought of as insectivorous.

The data available on avian diets are considerably greater than on amphibian, reptile, and mammal diets. In large measure this is due to the efforts of United States government biologists in the late nineteenth and early twentieth centuries. The US Department of Agriculture created a 'Section of Economic Ornithology and Mammalogy' within the Division of Entomology in 1885. The purpose of this section was to investigate and manage both beneficial and harmful activities of wildlife. This evolved into the 'Bureau of Biological Survey' by 1906, then a section for 'Economic Investigations in Ornithology' in 1916, and in 1921 a 'Division of Food Habits Research' was created. Eventually these agencies would be blended into the Department of the Interior and then into the US Fish and Wildlife Service.

Much of the early emphasis of these agencies was devoted to documenting the benefits of avifauna. Today we take for granted the protection accorded wildlife, and the regulation of hunting. But prior to the enabling of the Migratory Bird Treaty Act of 1916, which prohibited killing of select migratory species, all birdlife was considered freely available to be hunted.

This act (initially involving only the USA and Canada but later extended to include other countries) made it unlawful to pursue, hunt, take, capture, kill or sell certain birds. The statute did not discriminate between live or dead birds and also granted full protection to any bird parts, including feathers, eggs and nests. As inconceivable as it seems now, until this Act was implemented it was common to have dead songbirds such as robins, goldfinches, waxwings, and bobolinks freely available for purchase in food markets. Songbirds were bundled like bunches of carrots and sold for inclusion in stews, soups, and pies. At this time, 'market hunters' made their livelihood by shooting and netting vast quantities of birdlife. The coastal marshlands were especially favored hunting sites, and incredible quantities of shorebirds were harvested for market. Other hunters killed simply for the sport, not bothering to harvest their prey. Lastly, but by no means insignificantly, women's fashions of the period usually featured feathered adornments. Many species were severely persecuted by hunters so their feathers could adorn the hats of fashionable women.

The early government biologists sought to document the dietary habits of birds, hoping to discriminate among the pests species that perhaps warranted killing because of their destruction of crops, and the much more numerous species that were beneficial to farmers and to society because they fed upon insects or weed seed. Keep in mind that at this time there were few effective means of killing insects or weeds. Effective insecticides and herbicides were not available in this era, and consumption of crop-feeding insects and weed seeds by birds were important benefits. So this early literature is filled with detailed references to the destruction of pest insects and other dietary details that would help educate the American public about the benefits of wild birds. This information was promulgated widely, and the regional and national Audubon Society chapters were very active in this period in helping to educate Americans about the benefits of birds. Although much of this original literature can be found only in academic libraries, anyone interested in gaining a sense of this issue should consult *Birds of America* (1917 and 1936, edited by T.G. Pearson) which remains generally available in bookstores that sell used books.

The information in Table 4.5 is gleaned largely from the research of these early US government researchers. Its significance is that it often represents collections from throughout the year and from throughout the country. In this respect, it is unsurpassed, often representing analysis of hundreds or even thousands of birds of the same species. For example, government biologist F.E.L. Beale reported on the analysis of 37,825 bird stomachs! Also, because the bird stomachs were processed by specialists in the same manner, and the data treatment was generally uniform, it is easier to make comparisons among the diets of different species.

As we saw with the data on mammals, the trends in consumption of insects vary among higher taxa (families) of birds, but differences among similar birds are also evident. Also, significant temporal differences in diet are commonplace among some taxa, but not others. Figures 4.6–4.9 show year-long diet trends among several birds, with their food intake divided simply into animal and plant materials. For species such as hairy woodpecker, *Picoides villosus*; eastern phoebe, *Sayornis phoebe*; Carolina wren, *Thryothorus ludovicianus*; American woodcock, *Scolopax minor*; and Wilson's snipe, *Gallinago delicata*; the diet remains relatively uniform and rich in animal life. For others such as cedar waxwing, *Bombycilla cedrorum*; ruffed grouse, *Bonasa unbellus*; and blue-winged teal, *Anas discors*; the diet remains relatively uniform, but based principally on vegetation. Perhaps most interesting are species such as red-headed woodpecker, *Melanerpes erythrocepahus*; northern mockingbird, *Minus polyglottos*; eastern meadowlark, *Sturnella magna*; chipping sparrow, *Spizella passerina*; and sora, *Porzana carolina*; in which marked changes in diet occur, corresponding with the increased availability of insects in the spring and summer months and the need to feed offspring nutritious animal-based food. Most of the 'animal' consumption by these species was insect food, but other arthropods, earthworms and even vertebrates sometimes made up a portion of the diet. Also, more detailed diet analyses are given for selected species (blue jay, *Cyanocitta cristata* (Fig. 4.10); red-winged blackbird, *Agelaius phoeniceus* (Fig. 4.11); and bobolink, *Dolichonyx oryzivorus* (Fig. 4.12)) showing the resolving power of the data generated by these early economic ornithologists. Note especially the separation of insect biomass into pest and useful species.

All the woodpeckers shown in Table 4.5 fed on insects, but some to a much greater degree. When they fed on insects, they often selected wood boring beetles and caterpillars. However, the importance of ants to woodpeckers was surprising. Anyone who has observed northern flicker, *Colaptes auratus*, for any length of time would not be surprised that ants were an important part of its diet because these birds often are seen feeding on the soil surface, where ants are abundant. However, ants are also important components of the diet of pileated woodpecker, *Dryocopus pileatus*, and red-cockaded woodpecker, *Picoides borealis*. These birds largely feed on the surface of tree trunks, so they must glean the ants from among those species that forage high in the trees. Acorns and other nuts are important food sources for some woodpeckers.

Data on two species of hummingbirds show that insects are regularly consumed. This might be surprising to many because (like bears!) they often are thought of as preferring sweet liquids (nectar). Whereas a species found commonly in eastern North America, ruby-throated hummingbird (*Archilochus colubris*), chose mostly small parasitic wasps and spiders, a species from western North America known as Anna's hummingbird, *Calypte anna*, chose small flies in addition to parasitic wasps. What both species had in common, of course, was that these small birds fed upon some of the smallest arthropods.

Loggerhead shrike, *Lanius ludovicianus*, is unusually predatory, feeding only on animal life. Its favored food is grasshoppers, though it feeds on small rodents, reptiles and birds in addition to other insects. It is notorious for impaling captured prey on thorns of trees, and the barb structures of barbed wire. The impaling behavior serves mostly as an aid for consumption, but males apparently use such caches to demonstrate their fitness to females. Grasshoppers, crickets (both Orthoptera) and large beetles (Coleoptera) are among the prey items commonly impaled.

The family Tyrannidae is a large family in North America. It contains the phoebes, kingbirds, and some flycatchers. They are nearly completely insectivorous. They feed on a wide assortment of insects, including feeding often on flies. However, they take bees and wasps more than other insects, and also more frequently than most other birds. Thus, they might better be called beecatchers or waspcatchers. They should not be confused with the bee-eaters (Coraciiformes: Meropidae), however, a family of birds from Africa, southern Europe, southern Asia, and Australia. The bee-eaters feed especially on honey bees, a destructive habit not so often found among the American Tyranidae.

The family Turdidae is a small, but important part of American avifauna because some of it members are common or well known. It contains the thrushes, veery (*Catharus fuscescens*), solitaire (*Myadestes*

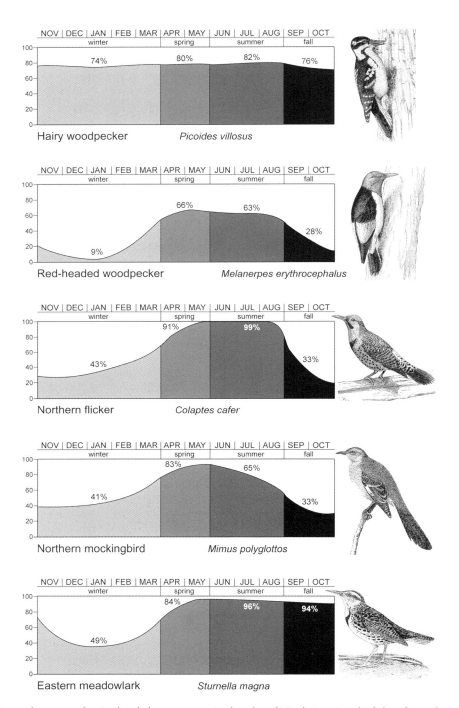

Fig. 4.6. Seasonal patterns of animal and plant consumption by selected North American birds based on volumetric stomach analysis. The shaded portions represent the proportions consisting of animal matter; the unshaded portions are plant material. Also shown are the average proportions of animal (mostly insect) consumption for each season (adapted from Martin *et al.* 1961).

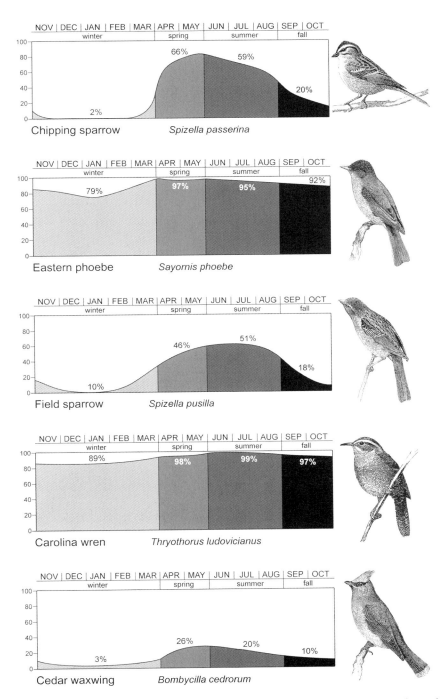

Fig. 4.7. Seasonal patterns of animal and plant consumption by selected North American birds based on volumetric stomach analysis. The shaded portions represent the proportions consisting of animal (mostly insect) matter; the unshaded portions are plant material. Also shown are the average proportions of animal consumption for each season (adapted from Martin *et al.* 1961).

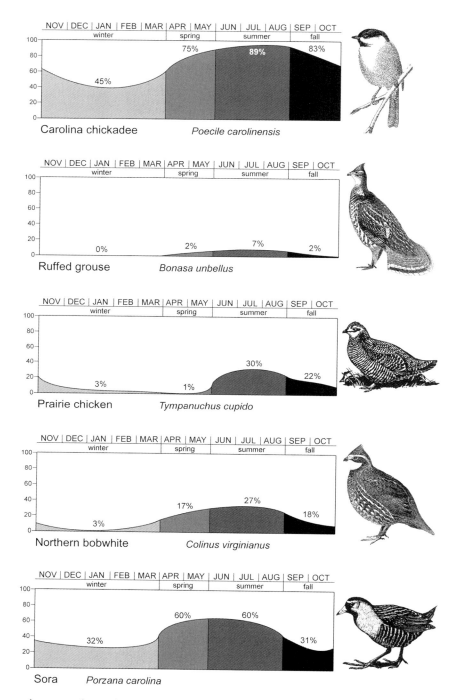

Fig. 4.8. Seasonal patterns of animal and plant consumption by selected North American birds based on volumetric stomach analysis. The shaded portions represent the proportions consisting of animal (mostly insect) matter; the unshaded portions are plant material. Also shown are the average proportions of animal consumption for each season (adapted from Martin *et al.* 1961).

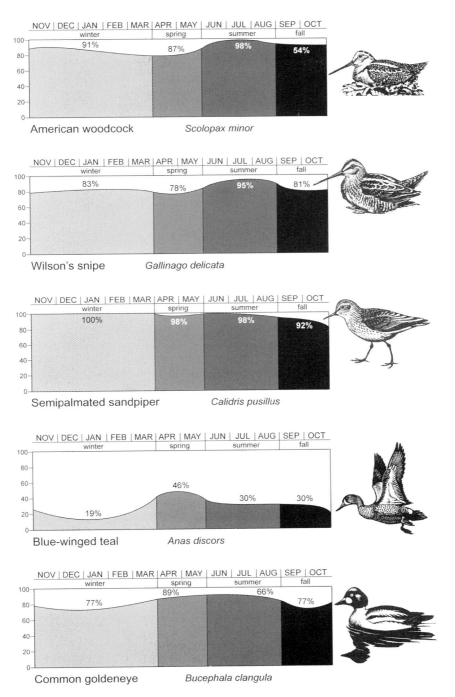

Fig. 4.9. Seasonal patterns of animal and plant consumption by selected North American birds based on volumetric stomach analysis. The shaded portions represent the proportions consisting of animal (mostly insect) matter; the unshaded portions are plant material. Also shown are the average proportions of animal consumption for each season (adapted from Martin *et al.* 1961).

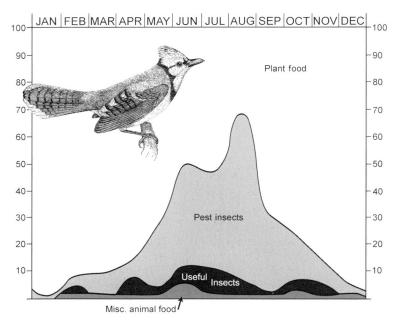

Fig. 4.10. Pattern of food consumption by blue jay, *Cyanocitta cristata*, showing the seasonal shift in the importance of insects and other components of the diet, including the relative importance of pest and beneficial insects, based on volumetric (%) stomach analysis (adapted from Beal 1897).

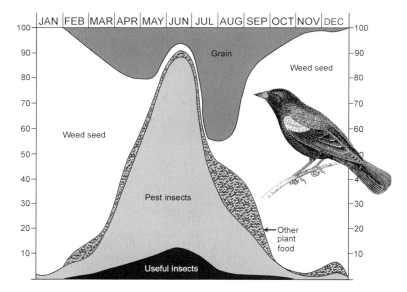

Fig. 4.11. Pattern of food consumption by red-wing blackbird, *Agelaius phoeniceus*, showing the seasonal shift in the importance of insects and other components of the diet, including the relative importance of pest and beneficial insects, based on volumetric (%) stomach analysis (adapted from Beal 1900).

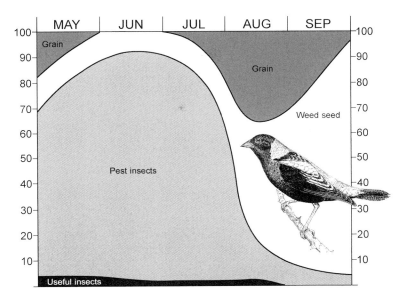

Fig. 4.12. Pattern of food consumption by bobolink, *Dolichonyx oryzivorus*, showing the seasonal shift in the importance of insects and other components of the diet, including the relative importance of pest and beneficial insects, based on volumetric (%) stomach analysis. Bobolinks migrate to South America in the winter, so the data reflect only the portion of the season when they are in North America (adapted from Beal 1900).

townsendi), and the well-known American robin (*Turdus migratorius*) and bluebirds (*Sialia* spp.). More than any other family, these birds feed broadly on all the major orders of insects. This family is less insectivorous than the Tyrannidae, feeding commonly on fruit and seeds.

The horned lark, *Eremophila alpestris* (family Alaudidae) is found on the prairie and steppe regions of North America, where it is quite common. It is principally a seed-eating species, but in the summer its consumption of insects increases, largely because it feeds insects to its young. In the winter this bird forms flocks, which in combination with its open habitat makes it susceptible to hunting. Formerly, this was one of the species most commonly hunted and trapped for food markets.

The family Icteridae is a highly variable taxon. It consists of the blackbirds, grackles, bobolink, meadowlarks, orioles, and cowbirds. The blackbirds, grackles, and cowbirds tend to feed mostly on plant material. They take weed seed, but sometimes incur the wrath of agriculturalists by feeding on grain, including ripening corn (maize) or germinating seeds. About one-third of their diet consists of insects, which is taken in the summer months when they are raising their chicks. They take mostly grasshoppers, beetles, and caterpillars during this period. In contrast, the meadowlarks, orioles, and bobolink are mostly insectivorous. Meadowlarks inhabit open, grassy areas so it is not surprising that they consume a great number of grasshoppers. In contrast, orioles favor trees where they collect mostly beetles and caterpillars.

The family Corvidae consists of jays and crows. These birds are true omnivores, with few items not being consumed at one time or another, including larger insects such as grasshoppers. The members of the family Mimidae, represented here by northern mockingbird (*Mimus polyglottos*), brown thrasher (*Toxostoma rufum*) and gray catbird (*Dumetella carolinensis*) also have fairly broad dietary habits, with few arthropod taxa escaping consumption.

In contrast to the corvids, the yellow-billed cuckoo, *Coccyzus americanus* (family Cuculidae), and the swallows (including purple martin, *Progne subis*) (family Hirundinidae), are strongly insectivorous. The Hirundinidae are colonial, and often found in association with humans. Purple martin, in particular, is rarely found nesting anywhere except in nest boxes provided by humans. These swallows have a special fondness for diurnal prey such as flies, wasps, and bugs. They usually avoid plant and ground-inhabiting insects such as caterpillars and grasshoppers, and nocturnal insects such as moths. It is noteworthy that mosquitoes are conspicuously absent from the lists of insects eaten by swallows, including purple martin. Not only do they not eat many mosquitoes, but they feed readily on dragonflies, which are noted mosquito predators.

This does not detract from the benefits provided by these birds, as they consume great numbers of pest insects. However, the human admirers of purple martins have greatly overstated their benefit relative to biting flies.

Three closely related families of small seed-eating passerine birds are Fingillidae, Emberizidae, and Cardinalidae. Fringillidae are represented here by house finch (*Carpodacus mexicanus*). Emberizidae are represented by the sparrows and the dark-eyed junco (*Junco hyemalis*). The Cardinalidae are represented by northern cardinal (*Cardinalis cardinalis*), pyrrhuloxia (*Cardinalis sinuatus*), rose-breasted grosbeak (*Pheucticus ludovicianus*), and painted bunting (*Passerina ciris*). Though favoring seeds, all these birds also consume insects, especially when raising their young. They show no strong preferences among insect prey, other than to avoid the larger insects such as grasshoppers.

The European starling, *Sturnus vulgaris* (family Sturnidae) is a highly invasive species that has spread through much of North America after a deliberate introduction to New York in 1890. It has a bad reputation, but this is mostly due to displacement of indigenous species (competitive exclusion) rather than its dietary habits. Its diet consists largely of insects and wild fruit, so it causes little direct harm except occasionally when flocking. It especially favors beetles and grasshoppers as food items. Interestingly, it was the introduction of this species that prompted the development of regulations governing animal importations by the US government.

Some of the smaller passerine birds include the yellow-rumped warbler (*Dendroica coronata*) (family Parulidae), ruby-crowned kinglet (*Regulus calendula*) (Reguliidae), house wren (*Troglodytes aedon*) and Carolina wren (*Thryothorus ludovicianus*) (both Troglodytidae), and tufted titmouse (*Baeolophus bicolor*) (Paridae). They are highly insectivorous, but display no particular preference for any insect taxa.

Two representatives of the family Odontophoridae are included in Table 4.5: California quail, *Callipepla californica*, and northern bobwhite, *Colinus virginianus*. Both seem to feed largely on weed seeds, but take a few arthropods. These species are typical of those displaying ontogenetic shifts in their diet. Animal material is crucial during the first few weeks of life but gradually is replaced by plant material as the birds mature, with adults consuming very little animal matter. Without access to invertebrates during their first few weeks, however, they do not survive.

The duck species (order Anseriformes, family Anatidae) considered in Table 4.5 include wood duck, *Aix sponsa*; canvasback duck, *Aythya americana*; redhead duck, *Aythya americana*; lesser scaup, *Aythya affinis*; pink-eared duck, *Malacorhynchus membranaceus*; and grey teal, *Anas gibberifrons*. They all consumed a considerable amount of animal life, often predominantly insects. They consumed mostly aquatic beetles or aquatic flies, sometimes specializing on one or the other of these two aquatic food sources. The horned grebe, *Podiceps auritus* (order Podicipediformes, family Podicipedidae), included a considerable amount of fish it its diet, but otherwise was similar to the ducks in its dietary habits.

A study by US Department of Agriculture biologists involving 7998 ducks representing 18 species was conducted in the USA and Canada. This large study revealed that the total animal portion of duck diets was, on average, about 26.9%, with insects and molluscs each comprising about 10%, crustaceans about 3.4%, and other animals the remainder. The pattern of insect consumption varied in an interesting but perhaps predictable manner: in coastal locations ducks consumed relatively more molluscs and crustaceans, whereas ducks in more interior locations consumed relatively more insects. Thus, ducks from inland areas of the eastern USA consumed about 13.5% insects and those from inland western USA populations consumed about 16.5% insects. Ducks from eastern Canada consumed about 17.7% insects, and from western Canada about 27.2% insects. These levels of insect consumption appear to be considerably less than the amounts reported from the duck species shown in Table 4.5, but this large study was flawed by not including many ducks collected in the summer, when insect consumption is highest.

Several birds often found in association with water are included in Table 4.5, including black tern, *Chlidonias niger* (family Laridae); Franklin's gull, *Larus pipixcan* (Laridae); killdeer, *Charadrius vociferous* (Charadriidae); semipalmated sandpiper, *Calidris pusilla* (Scolopacidae); pectoral sandpiper, *Calidris melanotos* (Scolopacidae); sharp-tailed sandpiper, *Calidris acuminata* (Scolopacidae); stilt sandpiper, *Calidris himantipus* (Scolopacidae); red knot, *Calidris canutus* (Scolopacidae); long-billed dowitcher, *Limnodromus scolopaceus* (Scolopacidae); short-billed dowitcher, *Limnodromus griseus* (Scolopacidae); Wilson's snipe, *Gallinago delicata* (Scolopacidae); American coot, *Fulica americana* (Rallidae); and cattle egret, *Bubulcus ibis*

(Ardeidae). These birds inhabit not only the seacoast, however, as they occupy lake margins and dry pastures as well. American woodcock dwells in wooded areas. Nearly all feed predominately on invertebrates, and although their principal food items sometimes are insects, they also eat molluscs, crustaceans, other animal life, and sometimes a considerable amount of plant material.

Although Table 4.5 principally contains omnivorous species, even in orders and families that are considered to be fruit or meat-eating we do find some species that eat insects on occasion. For example, Table 4.5 also includes an Australian bird, *Purpureicephalus spurius* (order Psittaciformes, family Psittacidae), known as red-capped parrot. This parrot, like most parrots, feeds predominantly on fruit and seeds, though it does ingest some insects.

Though not represented in Table 4.5, it is probably instructive to mention the diurnal (day-active) raptors (order Falconiformes), especially American kestrel (sparrow hawk), *Falco sparverius* (family Falconidae). American kestrel, unlike most falcons, consumes insects regularly, particularly grasshoppers (some have suggested that 'grasshopper hawk' would be a more appropriate name). Often up to 80% of the prey are insects, mostly grasshoppers (often 40% of total prey). Following are some data representing the frequency of occurrence of various prey found in an analysis of 703 kestrels: Orthoptera (principally Acrididae) 69.8%, Coleoptera (mostly ground-dwelling species) 29.4%, spiders 21.7%, Lepidoptera (mostly caterpillars) 21.3%, Odonata (dragonflies) 5.9%, mammals (mostly mice) 27.3%, reptiles 8.5%, and birds 4.3%. Note that these small birds also capture smaller birds (hence the alternate common name, 'sparrow hawk'), as well as rodents, reptiles, and occasionally amphibians. These vertebrate prey, though not numerous, contribute much more biomass (volume) per item. Consequently, vertebrates are usually credited with being more important food. Also, some kites feed largely on insects, including swallow-tailed kite, *Elanoides forficatus*, and Mississippi kite, *Ictinia mississippiensis* (both order Falconiformes, family Accipitridae), though their diet is less well documented. Other hawks that feed on insects include red-shouldered hawk, *Buteo lineatus*; broad-winged hawk, *Buteo platypterus*; northern harrier, *Circus cyaneus*; white-tailed hawk, *Buteo albicaudatus*; and crested caracara, *Caracara cheriway*; though not to the degree of American kestrel and the kites. Most other hawks, falcons, ospreys and eagles are larger, and shun insects or eat them only rarely.

The nocturnal (night-active) ecological counterpart of the Falconiformes is the owls (order Strigiformes). The smaller owl species eat insects, sometimes nearly exclusively. Among those that are most insectivorous are flammulated owl, *Otus flammeolus*, and elf owl, *Micrathene whitneyi*, both from western North American. These small owls eat moths, beetles, crickets, grasshoppers, caterpillars, centipedes, millipedes, spiders, and scorpions. An Asian species, *Otus elegans*, also is documented to feed almost entirely on arthropods. Some owl species that include insects in their diet include burrowing owl, *Athene cunnicularia*, found widely in the western hemisphere; western screech owl, *Megascops kennicotti*, and northern pygmy owl, *Glaucidium californicum*, from western North America; eastern screech owl, *Megascops asio*, from eastern North America; northern saw-whet owl, *Aegolius arcadius*, which occurs widely in North America; tawny owl, *Strix aluco*, from Europe and Asia; African wood owl, *Strix woodfordii*, from southern Africa; tropical screech owl, *Megascops choliba*, from South America; and ferruginous pygmy owl, *Glaucidium brasilainum*, from Central and South America. Insectivorous owls occurring in temperate areas typically migrate to more tropical areas as their food source diminishes in abundance during winter, as do most insectivorous birds.

The overall assessment of insect consumption of birds shows some distinct patterns. Not surprisingly, bird diet strongly reflects the arthropods present in the environment normally occupied by the birds. To a lesser degree, the size of the bird affects the arthropods taken, with smaller birds usually selecting smaller prey. Less obvious, perhaps, is the unusually high frequency of consumption of grasshoppers (Orthoptera: Acrididae) and ants (Hymenoptera: Formicidae). Also, although beetles are consumed by most taxa, the colorful beetles, particularly the abundant leaf beetles (Coleoptera: Chrysomelidae), are not often eaten, whereas the drab ground beetles (Carabidae) and scarab beetles (Scarabaeidae) frequently show up in the diets of birds. This is even more striking when one considers that leaf beetles and most birds are diurnal whereas ground beetles and scarabs are nocturnal. Although perhaps not evident from Table 4.5 but as shown in Figs. 4.10–4.12, the consumption of beneficial insects by birds is minimal. This was a major concern of the early economic ornithologists because there were reports of birds feeding on bees, and of course ground beetles are commonly taken, so their reports commonly address this issue. Almost without exception, they show only minor consumption of insects that could be perceived as being economically important.

ANALYSIS OF FISH DIETS

Fish are ectothermic vertebrates that are adapted to living in water, and that possess gills for their entire lives. They often possess a covering of scales, sometime plates, but sometimes neither. They are quite diverse, and occupy many different habitats. Here we are interested mostly in fresh-water fish, not because salt-water fish are unimportant, but because insects are important mostly for fresh-water species. Fresh water occupies relatively little of the earth's surface, though about 40% of the fish species occur in fresh-water habitats. Salt-water oceans cover about 70% of the earth's surface, and contain about 97% of all water. About 0.5% of the fish species migrate back and forth between fresh- and salt-water habitats.

There is considerable controversy about the higher classification of fishes. Chondrichthyes are one important class of fishes. They have a cartilaginous skeletal system, and are represented by sharks, skates and rays. More important are the Actinopterygii (Osteichthyes), the ray-finned fishes, which have a calcified (bony) skeleton. Some of the important orders of ray-finned fishes are Acipenseriformes, the paddlefish and sturgeons; Elopiformes, the tarpons; Anguilliformes, the eels; Clupeiformes, the herrings and anchovies; Salmoniformes, the salmon and trout; Cypriniformes, the minnows, suckers, carps, and loaches; Siluriformes, the catfish; Gadiformes, the cods and pollacks; Gasterosteiformes, the sticklebacks, pipefishes, and seahorses; Esociformes, the pike; Pleuronectiformes, the flatfish and flounders; and Perciformes, the perches, mackerels, marlins, tunas, barracudas and others. Perciformes are the most important group, containing 6500 of the approximately 23,500 species of fishes.

Fish generally are carnivorous, although many feed on zooplankton and insects rather than higher animals. Most fish that feed on insects have a tubular mouth that allows them to suck in their relatively small prey. This is known as suctorial feeding. However, some **picivorous** (fish-feeding) species also eat insects, especially when young. Picivorous fish tend to have a larger mouth and larger teeth. In contrast, herbivorous fish tend to have smaller, flattened teeth or may even lack teeth. Fish from temperate areas typically display a low level of herbivory, but fish in tropical areas have a high proportion of plants in their diet.

The digestive system of fishes is adapted to their diet. Picivorous species have a relatively short intestinal tract. It is slightly longer for species feeding on insects, crustaceans and worms. It is still longer for omnivores, and longest for herbivores or others eating a relatively indigestible diet. Some fresh-water fishes and their diets are shown in Tables 4.6 and 4.7.

The invertebrates fed upon by fish often are quite different than those eaten by terrestrial wildlife. Fresh-water fish eat large numbers of larval and pupal flies (Diptera), principally midges (Chironomidae), and also the immature stages of caddisflies (Trichoptera), mayflies (Ephemeroptera), stoneflies (Plecoptera), and dragonflies/damselflies (Odonata). They also consume the winged adult forms of these insects as well as terrestrial insects if they have opportunity, but many are principally dependent on the immature, water-dwelling stages. The examples below portray mostly fishes that have fed heavily on aquatic insects, but it is important to note that this is not always the case. For example, a study in Japan of rainbow trout, *Oncorhynchus mykiss*, found that 77% of the invertebrate biomass consumed consisted of terrestrial species rather than aquatic. Consumption of aquatic invertebrates exceeded consumption of terrestrial insects only when the availability of terrestrial insects decreased, with consumption of aquatics peaking about midnight daily. Not surprisingly, there are strong seasonal effects associated with insect consumption, at least in temperate climates. Commonly, terrestrial insect inputs are relatively strong in the spring, summer and autumn, with aquatic insects comprising the major or only source of insect food in the winter.

Also, it is useful to note that fish, like other vertebrates, are not completely opportunistic. Fish display preferences that cause them to eat prey items out of proportion to the abundance of the prey. For example, young Arctic grayling, *Thymallus arcticus*, in northern Canada generally selected the largest prey they could consume. These fish consumed much fewer small crustaceans and nematodes, and much more insect material, than availability might suggest. Among the insects consumed, the young grayling chose midge (Diptera: Chironomidae) and blackfly (Diptera: Simulii-

Table 4.7. Diet of some fresh-water fish based on stomach volume or mass (average % of stomach contents consisting of a diet component) (adapted from various sources; see references).

Fish	Plants	Plankton	Fish	Molluscs	Crustaceans	Insects	Other invertebrates	Other
Bluegill, *Lepomis macrochirus*	16		6	6 snails	19	40		13
Pumpkinseed, *Lepomis gibbosus*	4			58 snails	1	22		16
Black crappie, *Pomoxis nigromaculatus*			23		12	49		16
White crappie, *Pomoxis annularis*		25.6	38.3		20.2	5.2		2.2
Green sunfish, *Lepomis cyanellus*	24.3			6.7 snails		17.3 dragonflies, 10.8 wasps, etc., 15.5 unident.,	6.9	6.9 unident. animals
Yellow perch, *Perca flavescens*			59	1 snails	1	37		3
Largemouth bass, *Micropterus salmoides*	1		96					3 frogs
Alabama sturgeon, *Scaphirhynchus suttkusi*			27.8			7.7 biting midges, 4.1 other midges, 9.9 mayflies, 4.2 dragonflies		
Threespine stickleback, *Gasterosteus aculeatus*	4.7	25.8	1.8	6.1	18.1	24.7 immature, 2.6 adult	12.9 worms	3.3
Black bullhead, *Ameiurus melas*	13		4	16 snails	30	2	8.7	35
Bigmouth buffalo sucker, *Ictiobus cyprinellus*		12.8			71.9	29.4 midges		

Table 4.7. *Continued*

Fish	Plants	Plankton	Fish	Molluscs	Crustaceans	Insects	Other invertebrates	Other
Shorthead redhorse sucker, *Moxostoma macrolepidotum*	3.2				12.7	57.7 midges, 25.2 caddisflies		
Whitesucker, *Catostomus commersoni*	8.1				5.7	68.5 midges, 1.9 caddisflies		1.2 diatoms
Zebrafish, *Danio rerio*	1.7 vascular, 4.0 algae	0.2			14.7 cladocerans, 12.2 ostracods, 6.3 copepods	20.5 miscellaneous, 3.6 flies, 1.4 wasps, etc.,	2.0 invertebrate eggs, 0.5 spiders	
Coho salmon, *Oncorhynchus kisutch*			24.3		4.3 cladocerans	70.7		0.7 fish eggs
Brown trout, *Salmo trutta*						41.0 midges, 31.0 caddisflies, 18.5 stoneflies, 4.5 mayflies, 3.0 beetles, 1.5 black flies		
Common whitefish, *Coregonus lavaretus*		61		11	3	25 (23 larvae, 2 pupae)		
Burbot, *Lota lota*			20	42				
Northern pike, *Esox lucius*			90			38		10 frogs

dae) larvae and pupae, but showed a weak avoidance of mayflies (Ephemeroptera).

The perciform fishes shown in the tables are the bluegill, *Lepomis macrochirus*; pumpkinseed, *Lepomis gibbosus*; green sunfish, *Lepomis cyanellus*; crappies, *Pomoxis* spp.; yellow perch, *Perca flavescens*; and largemouth bass, *Micropterus salmoides*. They consumed large numbers of aquatic insects, but also fed on crustaceans and molluscs. As mentioned earlier, assessment of diet is strongly influenced by how data are tabulated. In Table 4.6 you see little evidence of fish-feeding (picivory) or plant-feeding (herbivory) because few of these items were found in the stomachs of these fish. However, fish and plants can be much larger than insects and crustaceans, so when volume is considered (Table 4.7) the importance of picivory and herbivory are more apparent. Note also that ants were often found in the stomachs of fish; these usually are winged forms that inadvertently alighted on water.

The order Salmoniformes is represented in these tabulations by three species of trout, *Salmo* spp.; common whitefish, *Coregonus lavaretus*; and coho salmon, *Oncorhynchus kisutch*. They consumed large amounts of aquatic insect material, as did the Perciformes, although crustaceans were less plentiful and molluscs were absent. Some salmon and trout spend a large portion of their lives at sea. We have avoided discussion of marine fish species because insects are rare in marine environments. However, these sea-going fresh-water fish (**anadromous** species) also feed in coastal areas. In such environments they continue with their insectivorous habits. A study of brown trout, *Salmo trutta*, from Norway showed that insects were numerically important prey of coast-dwelling trout, but of course these often were winged terrestrial insects such as flies and winged ants, though some were immature aquatic insects that washed down rivers and streams to marine environments. In this study, 23.9 % of the trout contain fish in their stomachs, whereas crustaceans and insects were contained in 20.3% and 19.8%, respectively.

Another Norwegian study of a salmonid, the Arctic charr, *Salvelinus alpinus*, is interesting because the researchers compared the diets and growth rates of fish from the same gene pool that remained in lakes year-round or migrated to the sea for a few weeks (anadromous individuals) during the summer. The lake-dwelling charr ate mostly insects (58% of their diet, mostly chironomids), and little zooplankton (about 13%). The anadromous individuals ate mostly zooplankton (88%) and few insects (4%). Copepods and krill comprised most of the marine diet. Anadromous behavior is believed to have evolved to allow fish to take advantage of the greater availability of marine organisms during a period when fresh water habitats may have limited food availability. Indeed, the growth rate of anadromous fish was higher during the first three weeks of their sea-residency. However, growth nearly ceased prior to their return to fresh water, whereas fish remaining in lakes had a more uniform growth rate. Also, the fish living in lakes had fuller stomachs, relative to their sea-going siblings, throughout the summer months. The reason why the growth rate of anadromous fish dissipates after only a short period at sea is unknown, but may be linked to the physiological costs of living in a hyperosmotic environment.

The only representative of the order Acipenseriformes shown in the tables is Alabama sturgeon, *Scaphirhynchus suttkusi*. This primitive fish seemed largely dependent on insects based on frequency analysis, but when food volume was considered fish became relatively important.

The stickleback (Gasterosteiformes), catfish (Siluriformes), bullhead, suckers and zebrafish (Cypriniformes) were similar to most other fishes, feeding mostly on aquatic insects and crustaceans. However, in most of these fishes plankton was also a significant diet component.

The diets of burbot (*Lota lota*), an example of the order Gadiformes, and northern pike (*Esox lucius*), an example of the order Esociformes, are shown in Table 4.7. Burbot had a broad diet consisting of fish, molluscs, and insects whereas pike fed almost entirely on fish.

Though not apparent from the tables, fish display the same ontogenetic shifts in diet that we have seen with other wildlife. For example, burbot studied in Finland ate mostly insect larvae when small, (<17 cm) but fish when larger (>26 cm) (Fig. 4.13a). Likewise, small whitefish (<19 cm) ate zooplankton but intermediate-sized fish ate crustaceans, and larger fish (> 29 cm) ate insect larvae and molluscs (Fig. 4.13b). Note that in the whitefish example, insect larvae are much larger than zooplankton. Also, in the Norwegian brown trout study mentioned above, the proportion of sea-dwelling trout feeding on insects fell from 40.6% in 1–2-year-old fish, to 17.4, 7.9, and 2.0% in 3-, 4-, and 5-year-old fish, respectively, demonstrating a clear ontogenetic shift in dietary habits.

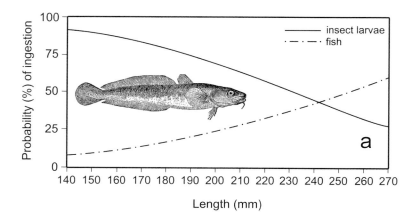

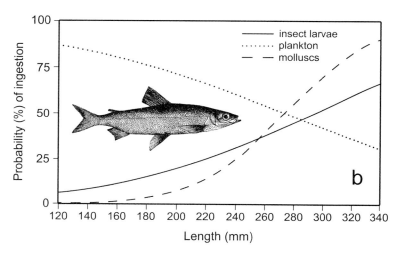

Fig. 4.13. Probability of ingesting different foods as a function of size in burbot, *Lota lota* (A) and common whitefish, *Coregonus lavaretus* (B) in Finnish lakes (adapted from Tolonen *et al.* 1999).

THE BENEFITS OF INSECTS FOR WILDLIFE SURVIVAL AND REPRODUCTION

The availability of insect-based food resources can be important for survival and reproduction of wildlife. Though the importance of insects as food items for many types of vertebrates has been shown previously, the nature of the effect varies and the benefits on wildlife fitness are not often fully documented except for birds, which are well studied. For example, the effect of food supply on the breeding biology of western kingbird, *Tyrannus verticalis*, was studied for 3 years in Arizona, USA. Sites with higher densities of insect biomass had earlier initiation of egg clutches, larger clutch sizes, higher nestling growth rates, and higher survival rates. At higher insect densities, the adult birds spent less time searching and fed the nestlings more often. Similar responses of birds to insect food supply have also been reported for such birds as house martin, *Delichon urbica*; tree swallow, *Tachycineta bicolor*; and northern oriole, *Icterus galbula*, among others.

For a more detailed example of the importance of insects, consider Savannah sparrow, *Passerculus sandwichensis*, breeding in Alaska, USA, which experienced alternating periods of grasshopper abundance and decline. The abundance of migratory grasshopper, *Melanoplus sanguinipes* (Orthoptera: Acrididae), ranged from about $25/m^2$ during periods of great abundance to about $1/m^2$ during periods of low abundance.

During periods of great abundance, mostly grasshoppers were captured and fed to nestlings. During periods of low grasshopper abundance, the sparrows fed their young more caterpillar (Lepidoptera) and sawfly (Hymenoptera) larvae and flies (Diptera) (Table 4.8). In contrast, the proportion of the diet comprised of spiders (Archnida) was equivalent in both periods. Average clutch size was higher during the period with abundant grasshoppers, and fledgling success was slightly higher in the period with more grasshoppers (4.46 birds/nest) than in the period with few grasshoppers (3.85 birds/nest). Nestlings also were larger when

grasshoppers were abundant. This example demonstrates the opportunistic behavior commonly observed with birds, wherein they adapt to increases in resource availability by changing their feeding behavior to take advantage of the additional food. Interestingly, they are slow to adopt new foods, tending to avoid novel items, but once they find an acceptable food source they tend to search for more of the same food type. In this case, reproductive output was enhanced by the abundance of grasshopper food, and the adult sparrow population increased by 42% following the period of high grasshopper abundance.

Similarly, red-winged blackbird, *Agelaius phoeniceus*, responded positively to the mass emergence of periodical cicada, *Magicicada cassini* (Hemiptera: Cicadidae) in a study conducted in Illinois, USA. As shown in Table 4.9, the number of nests, number of successful nests, and number of birds fledging successfully doubled when cicadas were available; among the measures of bird breeding success, only the number of eggs per clutch (not shown) did not benefit from cicada abundance. The marsh-dwelling blackbirds shifted their feeding behavior to take advantage of the superabundance of food. In the years before (1975) and after (1977) the season with peak abundance (1976) the proportion of trips foraging birds made to nearby woodlands to obtain cicadas was about 4%. In contrast, during peak abundance, the proportion of trips to the cicada-inhabited woodland increased to 50%.

The number of residing female blackbirds increased from 143 to 178, and then to 204 during the 3-year period, showing the benefits of abundant food. Likewise, the number of nests per female increased from 0.66 to 0.75, and then to 1.13 during the 3-year

Table 4.8. Proportion (%) of food brought to nestlings by adult Savannah sparrows during years with high grasshopper density and low grasshopper density (adapted from Miller *et al.* 1994).

	High grasshopper density	Low grasshopper density
Arachnida	25.2	25.2
Orthoptera	60.8	0
Neuroptera	0	0.8
Coleoptera		
Adults	0	5.9
Larvae	0	3.7
Lepidoptera		
Adults	1.4	2.2
Larvae	5.6	25.2
Diptera	7.0	16.3
Hymenoptera	0	20.7

Table 4.9. Breeding success of red-winged blackbirds in an Illinois marsh with an adjacent emergence of periodical cicadas in 1976. Periods 1 and 2 are May or June for nest parameters, and May 10–19 or May 20–June 10 for fledged birds (adapted from Strehl and White 1986).

Year	Period	Total nests	Successful nests	No. fledged birds
1975	1	23	20	55
	2	42	18	39
1976	1	71	30	80
	2	102	38	107
1977	1	42	14	37
	2	95	9	59

period of study. Nestlings also were heaviest during 1976. However, the number of nests lost to predators also increased over this three-year period. In 1975, more nests were successful than were lost to predators, but by 1977 there were three times as many nests lost to predators as there were successful nests. The principal predators were northern raccoon, *Procyon lotor*; gray and fox squirrels, *Sciurus carolinensis* and *S. niger*; groundhog, *Marmota monax*; black rat snake, *Elaphe obsoleta*; watersnakes, *Natrix* spp.; and cottonmouth snake, *Agkistrodon piscivorus*. Thus, it appears that bird predators also benefited from the superabundance of cicadas, and that the nutritional benefits of the cicada emergence were spread over at least a two-year period.

HOW INSECTS AVOID BECOMING FOOD FOR WILDLIFE

With insects subject to predation by so many forms of animal life, it should not be surprising that they have evolved many ways to escape being eaten. Some ways of escaping predation are morphological, some behavioral, and yet others are chemical or physiological. They are not mutually exclusive, of course, so some insects display combinations of morphology and behavior, or behavior and chemistry, etc., that enhance their survival. The principal means of avoiding predation can be classified as crypsis, aposematism, mimicry, flight, startle behavior and attack, physical and chemical defenses, group actions, and nocturnal activity.

Crypsis

Probably the most common manner by which insects avoid being eaten is through camouflage, which is also called **crypsis**. Cryptic insects blend into the background by imitating the color or shape of background features. Insects commonly resemble items that might not be edible to a hungry vertebrate predator. This might include leaves, flowers, stems, bark, thorns, lichens, stones, soil, and bird droppings (Fig. 4.14). An example of this might be the resting of green lacewings (Neuroptera: Chrysopidae) on green plant foliage. More dramatic (repugnant, maybe repellent) is the shiny black and white appearance of some swallowtail (Lepidoptera: Papilionidae) larvae, which look like bird droppings. One of the most common forms of crypsis is **disruptive coloration**, an appearance in which the

Fig. 4.14. Insects are masters of camouflage. Some examples of insect crypsis include: (a) a twig-mimicking caterpillar, *Iridopsis defectaria* (Lepidoptera: Geometridae); (b) a moth of *Catocala micronympha* (Lepidoptera: Noctuidae) blends in perfectly with the bark of trees; (c) a bird-dropping mimic, the caterpillar of *Papilio cresphontes* (Lepidoptera: Papilionidae) (photos by L. Buss).

outline of the insect is disrupted by patterns that help it blend into the background. Dark or light streaks or stripes are examples of disruptive coloration.

Crypsis involves more than appearance; the insect must not only look like something other than an insect, but it must appear in the correct location. So resting behavior is critical; for an insect to avoid predation by resembling tree bark, it must frequent tree bark, not flowers or leaves. Orientation can be important in fooling sharp-eyed birds, so some twig mimics angle their bodies so that one end appears to be attached to a stem. Looper caterpillars (Lepidoptera: Geometridae) often anchor themselves with their posterior prolegs, extending the rest of their body to resemble a broken twig. Likewise, movement affects crypsis, most often the absence of movement, but some leaf mimics display 'fluttering' as if they were being blown by the wind. Behavior that eliminates shadows is also helpful in reducing detection by predators. Insects may also display an ability to change their color to match the background, though unlike chameleons, they cannot change quickly, requiring a few days to adapt.

Aposematism

Aposematic coloration is appearance that advertises the presence of the potential prey item. In a sense, this is exactly the opposite of crypsis. Why would insects want to advertise their presence? Warning coloration is used to advertise that the prey are distasteful, toxic, or can inflict pain (see 'Physical and chemical defenses,' below). Aposematic coloration is often red, yellow, or orange accompanied by contrasting background, often black or white (Fig. 4.15). This is a sharp contrast with the more muted colors of cryptic species, which are often green, brown, gray, or black. Often, aposematic insects aggregate or display other behaviors that further demonstrate their presence. They are usually active during the daylight hours, some stridulate or make noises in some manner, and they may contain toxins or feeding deterrents. It is interesting that throughout the Class Insecta, the same warning colors are used, even among insects that have no geographic or ecological overlap. This suggests that most vertebrates, or at least most birds (the most important predators that depend on visual skill for hunting), have an innate hesitation to consume these colors. Not all aposematic insects are, in fact, distasteful or toxic (see mimicry, below).

Fig. 4.15. Monarch butterflies, *Danaus plexippus* (Lepidoptera: Nymphalidae) provide a classic example of aposematic coloration, displaying an orange and black coloration that warns potential predators that they are distasteful. Their well-known aggregation behavior possibly reinforces the effect of the warning coloration, the distastefulness, or perhaps is an example of predator satiation (photo by University of Florida/IFAS).

Mimicry

When one organism resembles another organism, it is called **mimicry**. Insects may mimic or imitate other insects if it confers advantage, as when a harmless species imitates a toxic or dangerous species. The truly dangerous insect is called the 'model' and the harmless insect is the 'mimic'; the mimic derives some benefit (survival advantage) from looking like the model even though it is not dangerous or distasteful (Fig. 4.16). A good example of mimicry is found in flower flies

Fig. 4.16. Insects mimic the appearance of other insects if it confers survival advantage. Bee and wasp mimics are especially common. Shown here is a robber fly (Diptera: Asilidae) that mimics bees (Diptera: Apoidae). Presumably, non-stinging insects such as flies are attacked by predators less often if they resemble stinging insects such as bees (photo by Scott Weihman).

(Diptera: Syrphidae), which frequent blossoms (as do bees (Hymenoptera: Apoidae)) and tend to be black and yellow and rather hairy (as are bees). Thus, the flies resemble bees both behaviorally and morphologically, though they lack the ability to sting. Sometimes mimicry can involve acoustic or chemical signals instead of physical appearance.

There are several types of specialized mimicry, including:

• **Batesian mimicry** is the evolutionary convergence of appearance among palatable species to resemble an unrelated unpalatable species. The idea here is that a predator tastes one or more unpalatable 'models' and learns to avoid insects of a particular appearance. In order for this process to be effective, the mimics must closely resemble the model, and must not be too abundant relative to the model.

• **Müllerian mimicry** is the evolutionary convergence of unrelated unpalatable species to resemble one another. This occurs because it is advantageous for both the prey and the predator if the predator only has to learn to avoid a single appearance, as there will be less consumption (loss) of the co-mimetic species and less waste of time trying to ingest unpalatable prey.

• **Aggressive mimicry** occurs when a predator mimics something that another species considers desirable or attractive. The species being attracted is then eaten by the mimic.

• **Wasmannian mimicry** occurs among non-ant insects that live within ant nests. These mimics may have the appearance of ants and even produce the same chemicals produced by ants, allowing them to take shelter in the nests and to access food from the nest.

Flight and Startle Behavior

Crypsis or disguise is a primary manner of avoiding predation, but when insects are discovered they sometimes produce a startle, fright or flight response that can serve to allow them to escape predation. A startle response can be achieved by **flash coloration**, whereby the insect quickly produces a flash of color that was not exposed previously. This is common in moths (Lepidoptera) and grasshoppers (Orthoptera), and may startle the predator long enough to allow the prey to hop or fly away. Also, **appendage extension**, either of the legs or wings, can be used to make the insect appear larger, and perhaps frighten the predator. This is common among grasshoppers, katydids and wetas (all Orthoptera). Exposure of **eye spots**, or the appearance of large, false eyes on the wings of moths (Fig. 4.17) or the body of caterpillars not only serves to produce a flash of color, but perhaps to startle a predator into thinking that the prey is a well-defended reptile rather than a nearly defenseless insect. Some

Fig. 4.17. When insects suddenly reveal large 'eyes' or what appears to be the forked 'tongue' of a snake, predators may be deterred from attacking, or at least deterred long enough for the insect to escape. Two common examples are shown here: (a) the 'eye spots' on the hind wings of a Polyphemus moth, *Antheraea polyphemus* (Lepidoptera: Saturniidae); and (b) the tongue-like everted osmeterium of a *Papilio cresphontes* larva (Lepidoptera: Papilionidae) (photos by L. Buss).

swallowtail butterfly caterpillars extrude a colorful and malodorous two-pronged structure called an **osmeterium** from the anterior region when they are disturbed. This structure resembles the tongue of a snake.

Flight is a common escape response of insects, and means escape by fleeing, not just by flying. Some species and stages are able to escape by use of their wings, although some are not particularly agile in flight. Flight has many advantages, including allowing some species to escape predation, including predation by birds. For those that can't fly, hopping or running away can provide an effective means of escape by 'flight.' Species capable of hopping away are characterized by greatly expanded hind legs, especially by having well-developed hind femora. Grasshoppers, katydids, crickets (all Orthoptera) and flea beetles (Coleoptera: Chrysomelidae) are examples of taxa that display wonderful hopping reactions. Other species escape by pretending that they are dead, and dropping from the foliage, a behavior called **death feigning** or **thanatosis**. When insects feign death, it is most likely the absence of movement that allows insects to escape detection and therefore consumption. Sometimes insects display a behavior called **dodging**, in which they move around the substrate (usually around a plant stem) in a manner that keeps the substrate between them and the predator. This is a surprisingly effective means of avoiding predation, at least temporarily.

Physical and Chemical Defenses

Another way to startle a predator is to attack it. A bird or small mammal might not expect to be attacked by an insect, thereby startling the predator and allowing the insect time to escape. It is certainly easy to imagine something as large as a mantid (Mantodea) striking back at small predators, but this behavior also occurs among insects such as katydids (Orthoptera: Tettigoniidae) that bite, and many bees (Hymenoptera: Apoidea), wasps (Hymenoptera: Vespidae, Mutillidae and others), and ants (Hymenoptera: Formicidae) that sting.

The insect **sting**, a mechanical device used to inject biologically active substances such as toxins, is a formidable weapon. Sometimes insects that are about to suffer predation give up a portion of their body, a process called **autonomy**. Autonomy, or detachment of body parts, is expressed dramatically when an insect breaks off an appendage, but this also can be extended to include insects shedding spines or hairs to deter predation. Many insects bear **urticating hairs** (Fig. 4.18). These bear barbed spines and are irritating, or are hollow spines that may have poison glands associated with them. Insects such as the two-striped walking stick, *Anisomorpha buprestoides* (Phasmatodea: Pseudophasmatidae) and bombardier beetles (Coleoptera: some Carabidae) eject **irritating spray** into the eyes of attacking vertebrates. Some taxa of insects bleed profusely when disturbed, and the insect blood contains chemicals that are feeding deterrents. This is called **reflex bleeding** or autohemorrhage. In a few insects, the blood is forced out rapidly, producing a popping sound, which is thought to be a form of startle defense.

Toxins and venoms are poisonous substances. They occur widely among insects. **Venoms** are chemicals that are physically injected. Hymenoptera are the only order that possesses a true sting. Other taxa that inject venoms use their mouthparts or venom-containing hairs or spines. The ability to cause pain is useful for defense, but the ability to cause damage or death to the attacking vertebrate organism is even better. Thus, some venoms are quite toxic. Social bee and wasp venoms typically have toxicity levels of 2–6 mg venom per kilogram of body weight. For comparison, rattlesnake venoms average about 4 mg per kilogram of body weight. Thus, social insects and rattlesnakes are about equally venomous. These insects (individually) cannot deliver as much venom as snakes, which is why more people are not killed. However, a few stings, or even as little as one sting, can be enough to kill a small vertebrate insectivore such as a mouse.

The evolution of stinging is particularly interesting. The more primitive wasps (order Hymenoptera) such as family Sphecidae use venoms to subdue prey (other insects). As bees evolved a more vegetarian diet (pollen and nectar) they no longer needed the sting for subduing prey, and they use the sting apparatus only for defensive purposes. The evolution of sociality meant that there was a bonanza of insect food resources available to wildlife that could access nests of colonial species. The evolution of effective venoms that would kill or repel vertebrates was an important factor in allowing the evolution of colonial nesting and sociality. The wasps in the family Vespidae display a range of sting functions depending on their ecology, while spider wasps (Pompilidae) and many ants (Formicidae) use venom and the sting apparatus both to kill prey and in defense. Ants often nest below-ground and forage stealthily. However, ants that are active in the daytime are exposed to predation from lizards, birds and amphibians. These ants, in particular, tend to have powerful venoms.

Toxins are sprayed or are active upon ingestion. Some plant-produced toxins are immediately excreted or metabolized by herbivorous insects, and this is quite common with generalist (polyphagous) species that are not well adapted to ingest chemistry of a particular plant. Some species are better adapted to the host plant's defensive chemistry, and tend to specialize on related hosts. They are considered specialists or even monophagous (feeding on a single genus of plants). Specialist herbivores sometimes acquire (sequester) toxins when they ingest them along with their food. Sequestration allows the insect herbivore to use the plant's defenses for its own chemical defense. Some of the better known toxins used by insects to defend

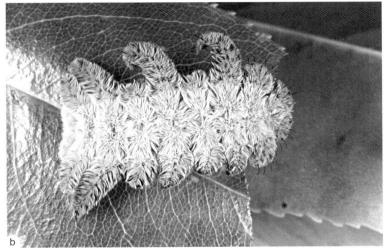

Fig. 4.18. Toxins and venoms are not uncommon among insects. Shown here are two species with urticating hairs; (a) a puss caterpillar, *Megalopyge opercularis* (Lepidoptera: Megalopygidae); and (b) the caterpillar of the hag moth, *Phobetron pithecium* (Lepidoptera: Limacodidae) (photos by L. Buss).

against predation are pyrrolizidine alkaloids, sinigrin, cardenolides, and cyanogens.

Group Actions

The collective, coordinated action of several or many insects can help compensate for their small size. This is evident when many ants, bees, or wasps attack simultaneously, a process also known as **mobbing**. These colonial insects use alarm pheromones to alert members of the colony to danger, stimulating mass attack. The simultaneous emergence of thousands of winged ants, termites, cicadas and other insects is usually considered to be a form of **predator satiation** (Fig. 4.19) because the predators cannot eat all of them if they emerge at the same time, ensuring that some will survive and reproduce, but this also can be a form of startle defense. For species displaying **aposematic coloration**, group displays of color patterns that discourage predation likely are more effective than the same appearances/actions displayed singly.

Fig. 4.19. Insects occurring in groups often have better survival that those occurring individually. Mass emergence (a) of predation-susceptible insects such as eastern subterranean termites, *Reticulitermes flavipes* (Isoptera: Rhinotermitidae), likely provides predator satiation, whereas (b) aposematically colored larvae of io moth, *Automeris io* (Lepidoptera: Saturniidae) likely derive better feeding deterrence due to group display of their warning colors (photos by L. Buss).

Despite the ability to mount collective defenses, ants and termites are not immune to predation. The colonial nature of these small insects makes them particularly attractive to some larger animals that otherwise might ignore them. On an individual basis, ants and termites don't contribute much nutrition; collectively, however, they can be attractive. Animals that specialize on feeding on ants and termites are called **myrmecophages**. Among the well-known myrmecophages are aardvark, *Orycteropus afer*; aardwolf, *Proteles cristata*; sloth bear, *Melursus ursinus*; tamanduas, *Tamandua* spp.; giant anteater,

Myrmecophaga tridactyla; silky anteater, *Cyclopes didactylus*; echnidas or spiny anteaters, family Tachyglossidae; pangolins or scaly anteaters, *Manis* spp.; numbat, *Myrmecobius fasciatus*; and some armadillos, family Dasypodidae. Myrmecophages often have a long, thin tongue, viscous saliva, reduction or loss of teeth, and a muscular stomach structure. A quick and extendable tongue evidently is useful in collecting ants and termites, as this feature is common among mymecophagous vertebrates. Some have structural adaptations that make them less susceptible to the defenses of ants and termites. This includes the carapace of armadillos and the scales of pangolins. Behavioral adaptations include moving to a new nest after a short feeding, and feeding selectively on worker ants trailing to and from a nest, rather than confronting soldier ants. Some vertebrates are also selective in determining their prey, avoiding the more aggressive species. The evolution of these different structures and behaviors among vertebrates demonstrates the effectiveness of the defenses mounted by insects.

Nocturnal Activity

One way for insects to escape predation by most birds is to become active after dark. Nocturnal activity is common for many arthropods, particularly some flies (e.g., midges, mosquitoes), beetles (e.g., scarabs, ground beetles), and moths. Although this allows insects to escape the vision-oriented hunting behavior of some predators, it certainly does not allow total escape from predation. Many mammals hunt at night, and possess good night-vision or depend on their sense of smell to locate food. A few birds have evolved the ability to hunt well into the night, including the owls (Strigiformes), and the nightjars, goatsuckers, or nighthawks (Caprimulgiformes). Mammals such as northern raccoon, *Procyon lotor*, and striped skunk, *Memphitis memphitis*, are principally active at night, and have no trouble locating prized prey such as scarab beetle larvae (white grubs) whether it is day or night. However, the most important night-hunting mammals, at least if you are an arthropod, are the bats (Chiroptera).

Bats are the only mammals that are capable of true flight, and not surprisingly, they have evolved a means to capture flying insects. Or perhaps they capture insects because they have the powers of flight. They fly mostly at night, but sometimes can be seen in the early evening. About 70% of the world's bat species feed primarily on insects. In temperate regions they must either hibernate to survive the insect-free winter months, or migrate to warmer climates. They are renowned for capturing flying insects, which is why their flight often seems erratic. However, some species glean insects from the surface of vegetation, and even from the surface of water. They have good eyesight but also use echolocation to detect and catch insects that are flying at night. Bats emit ultrasonic (high-frequency) sound. When the sound bounces back to the bats it indicates that something (often an insect) is in

the air, allowing the bats to navigate (to avoid trees, for example) and find their prey in the dark. Moths are common prey for many species, and it has been demonstrated that some moths can hear the ultrasonic echolocation sensing system used by bats. At least 14 families of moths have ears that are adapted to detect the echolocation signals of bats. Hearing organs are found in the labial palps of the sphingid (Lepidoptera: Sphingidae) moths, on the metathorax of the noctuids (Lepidoptera: Noctuidae), and on the first abdominal segment of the geometrids (Lepidoptera: Geometridae) and pyralids (Lepidoptera: Pyralidae). When the signals indicate that a bat is nearby, moths will suddenly change their direction of flight, perform spirals and loops, or even stop flying and plummet toward the ground. Moths flying more than 5 m from bats usually just change directions, whereas those less than 5 m tend to spiral or dive to escape predation. Such evasive tactics can allow moths to escape, but this is not guaranteed, as the bats are also agile fliers, plucking moths from the air with their mouths, or even scooping the insects up with their wings and then grabbing them with their mouth. Research has shown that 87% of moths captured by bats were those that did not exhibit defensive maneuvers. Some moths, particularly arctiids (Lepidoptera: Arctiidae) have developed the ability to produce sounds, including bat-like sounds. These are thought to signal to bats that the moths are unpalatable, or in the case of bat-mimics, to confuse the bats into thinking that the moth is actually another bat. Arctiids larvae are relatively unpalatable to most birds, and display aposematic coloration, so the (lack of) palatability hypothesis seems quite logical.

SUMMARY

• Amphibians and reptiles are largely carnivorous, and smaller species and younger animals are largely insectivorous.
• Many mammals are strictly herbivorous, although carnivorous and omnivorous species often feed on insects, especially the smaller species of mammals. A few taxa are specialized to feed exclusively on insects.
• Birds are among the most important insect predators, although some carnivorous species feed exclusively on other animal prey, or on plants. Many species adjust their foraging behavior to feed insects to their actively growing young despite granivorous tendencies in the adult stage.

• Fish generally are carnivorous, and most fresh water-inhabiting species feed on insects, at least early in life. Even ocean-dwelling species may feed on insects.

• Wildlife having improved access to insects for food often display increased fitness, including faster growth rates, larger body size, higher survival rates, and higher rates of reproduction.

• Insects avoid being consumed by wildlife through camouflage (crypsis), advertising their distastefulness (aposematism), resembling distasteful or toxic species (mimicry), by fleeing from predators (flight and startle behavior), by possessing physical or chemical defenses, by acting in concert (group actions), or by being active at night (nocturnal activity) when they are less apparent.

REFERENCES AND ADDITIONAL READING

Aldous, S.E. (1941). Food habits of chipmunks. *Journal of Mammalogy* **22**, 18–24.

Almenar, D., Aihartza, J., Goiti, U., Salsamendi, E., & Garin, I. (2007). Diet and prey selection in the trawling long-fingered bat. *Journal of Zoology* **274**, 340–348.

Altig, R., Whiles, M.R., & Taylor, C.L. (2007). What do tadpoles really eat? Assessing the trophic status of an understudied and imperiled group of consumers in freshwater habitats. *Freshwater Biology* **52**, 386–395.

Atsalis, S. (1999). Diet of the brown mouse lemur (*Microcebus rufus*) in Ranomafana National Park, Madagascar. *International Journal of Primatology* **20**, 193–229.

Bailey, V. & Sperry, C.C. 1929. Life history and habits of grasshopper mice, genus *Onychomys*. *USDA Technical Bulletin* 145. 19 pp.

Bairlein, F. (1997). Food choice in birds and insect chemical defenses. *Entomologia Generalis* **21**, 205–216.

Ballinger, R.E. & Ballinger, R.A. (1979). Food resources utilization during periods of low and high food availability in *Sceloporus jarrovi* (Sauria: Iguanidae). *Southwestern Naturalist* **24**, 347–363.

Bartonek, J.C. & Hickey, J.J. (1969). Food habits of canvasbacks, redheads, and lesser scaup in Manitoba. *The Condor* **71**, 280–290.

Basabose, A.K. (2002). Diet composition of chimpanzees inhabiting the montane forest of Kahuzi, Democratic Republic of Congo. *American Journal of Primatology* **58**, 1–21.

Baumann, P.C. & Kitchell, J.F. (1974). Diel patterns of distribution and feeding of bluegill (*Lepomis macrochirus*) in Lake Wingra, Wisconsin. *Transactions of the American Fisheries Society* **103**, 255–260.

Beal, F.E.L. (1896). The meadow lark and Baltimore oriole. *Yearbook of the US Department of Agriculture for 1985*, 419–430.

Beal, F.E.L. (1897). The blue jay and its food. *Yearbook of the US Department of Agriculture for 1986*, 197–206.

Beal, F.E.L. (1900). Food of the bobolink, blackbirds, and grackles. *USDA Division of Biological Survey Bulletin* 13. 77 pp.

Beal, F.E.L. (1911). Food of the woodpeckers of the United States. *USDA Division of Biological Survey Bulletin* 37. 64 pp.

Beal, F.E.L. (1912). Food of our more important flycatchers. *USDA Division of Biological Survey Bulletin* 44. 67 pp.

Beal, F.E.L. (1915). Food of the robins and bluebirds of the United States. *USDA Bulletin* 171. 31 pp.

Beal, F.E.L. (1915). Food habits of the thrushes of the United States. *USDA Bulletin* 280. 23 pp.

Beal, F.E.L. (1918). Food habits of the swallows, a family of valuable native birds. *USDA Bulletin* 619. 28 pp.

Beal, F.E.L. & McAtee, W.L. (1912). Food of some well-known birds of forest, farm, and garden. *USDA Farmers' Bulletin* 506. 35 pp.

Bennett, L.J., English, P.F., & Watts, R.L. (1943). The food habits of the black bear in Pennsylvania. *Journal of Mammalogy* **24**, 25–31.

Bent, A.C. (1927). Life histories of North American shore birds, Order Limicolae (Part 1). *Smithsonian Institution, U.S National Museum Bulletin* 142. 420 pp.

Blancher, P.J. & Robertson, R.J. (1987). Effect of food supply on the breeding biology of western kingbirds. *Ecology* **68**, 723–732.

Blum, M.S. (2008). Allelochemicals. In Capinera, J.L. (ed.), *Encyclopedia of Entomology*, 2nd edn., pp. 121–133. Springer Science & Business Media B.V., Dordrecht, The Netherlands.

Bradley, W.G. (1968). Food habits of the antelope ground squirrel in southern Nevada. *Journal of Mammalogy* **49**, 14–21.

Brigham, R.M. & Saunders, M.B. (1990). The diet of big brown bats (*Eptesicus fuscus*) in relation to insect availability in southern Alberta, Canada. *Northwest Science* **64**, 7–10.

Burger, J.C., Patten, M.A., Rotenberry, J.T., & Redak, R.A. (1999). Foraging ecology of the California gnatcatcher deduced from fecal samples. *Oecologia* **120**, 304–310.

Carrera, R., Ballard, W., Gipson, P., *et al.* (2008). Comparison of Mexican wolf and coyote diets in Arizona and New Mexico. *Journal of Wildlife Management* **72**, 376–381.

Chessman, B.C. (1986). Diet of the Murray turtle, *Emydura macquarii* (Gray) (Testudines: Chelidae). *Australian Wildlife Research* **13**, 65–69.

Clark, C.F. (1943). Food of some Lake St. Mary's fish with comparative data from Lakes Indian and Loramie. *American Midland Naturalist* **29**, 223–228.

Clark, D.B. & Gibbons, J.W. (1969). Dietary shift in the turtle *Pseudemys scripta* (Schoepff) from youth to maturity. *Copeia* **1969**, 704–706.

Cobb, V.A. (2004). Diet and prey size of the flathead snake, *Tantilla gracilis*. *Copeia* **2004**, 397–402.

Coop, J.D., Hibner, C.D., Miller, A.J., & Clark, G.H. (2005). Black bears forage on army cutworm moth aggregations in the Jemez Mountains, New Mexico. *The Southwestern Naturalist* **50**, 278–281.

Corbin, G.D. & Schmid, J. (2005). Insect secretions determine habitat use patterns by a female lesser mouse lemur (*Microcebus murinus*). *American Journal of Primatology* **37**, 317–324.

Corn, J.L. & Warren, R.J. (1985). Seasonal food habits of the collared peccary in south Texas. *Journal of Mammalogy* **66**, 155–159.

Dearing, M.D. & Schall, J.J. (1992). Testing models of optimal diet assembly by the generalist herbivorous lizard *Cnemidophorus murinus*. *Ecology* **73**, 845–858.

Deblauwe, I., Dupain, J., Nguenang, G.M., Werdenich, D., & Van Elsacker, L. (2003). Insectivory by *Gorilla gorilla gorilla* in southeast Cameroon. *International Journal of Primatology* **24**, 493–502.

Delany, M.F. (1990). Late summer diet of juvenile American alligators. *Journal of Herpetology* **24**, 418–421.

Dickman, C.R. & Huang, C. (1988). The reliability of fecal analysis as a method for determining the diet of insectivorous mammals. *Journal of Mammalogy* **69**, 108–113.

Dixon, J. (1925). Food predilections of predatory and fur-bearing mammals. *Journal of Mammalogy* **6**, 34–46.

Drobney, R.D. & Fredrickson, L.H. (1979). Food selection by wood ducks in relation to breeding status. *Journal of Wildlife Management* **43**, 109–120.

Edmunds, M. (2008). Crypsis. In Capinera, J.L. (ed.), *Encyclopedia of Entomology*, 2nd edn., pp.1122–1128. Springer Science & Business Media B.V., Dordrecht, The Netherlands.

Engel, S. (1976). Food habits and prey selection of coho salmon (*Oncorhynchus kisutch*) and cisco (*Coregonus artedii*) in relation to zooplankton dynamics in Pallette Lake, Wisconsin. *Transactions of the American Fisheries Society* **5**, 607–614.

Fay, J.M. & Carroll, R.W. (1994). Chimpanzee tool use for honey and termite extraction in Central Africa. *American Journal of Primatology* **34**, 309–317.

Freed, A.N. (1982). A treefrog's menu: selection for an evening's meal. *Oecologia* **53**, 20–26.

Freeman, P.W. (1979). Specialized insectivory: beetle-eating and moth-eating molossid bats. *Journal of Mammalogy* **62**, 166–173.

Freeman, P.W. (1981). Correspondence of food habits and morphology in insectivorous bats. *Journal of Mammalogy* **60**, 467–479.

Gadsden, H., Palacios-Orona, L.W., & Cruz-Soto, G.A. (2001). Diet of the Mexican fringe-toed lizard (*Ulma exsul*). *Journal of Herpetology* **35**, 493–496.

Georges, A. (1982). Diet of the Australian freshwater turtle *Emydura krefftii* (Chelonia: Chelidae), in an unproductive environment. *Copeia* **1982**, 331–336.

Greenwood, R.J. (1981). Foods of prairie raccoons during the waterfowl nesting season. *Journal of Wildlife Management* **45**, 754–760.

Gursky, S. (2000). Effect of seasonality on the behavior of an insectivorous primate, *Tarsius spectrum*. *International Journal of Primatology* **21**, 477–495.

Hamilton, Jr., W.J. (1933). The insect food of the big brown bat. *Journal of Mammalogy* **14**, 155–156.

Hamilton, Jr., W.J. (1935). Notes of food of red foxes in New York and New England. *Journal of Mammalogy* **16**, 16–21.

Hamilton, Jr., W.J. (1936). Seasonal food of skunks in New York. *Journal of Mammalogy* **17**, 240–246.

Hamilton, Jr., W.J. (1948). The food and feeding behavior of the green frog, *Rana clamitans* Latreille, in New York state. *Copeia* **1948**, 203–207.

Hartman, G.D., Whitaker, Jr., J.O., & Munsee, J.R. (2000). Diet of the mole *Scalopus aquaticus* from the coastal plain region of South Carolina. *American Midland Naturalist* **144**, 342–351.

Hirai, T. & Matsui, M. (2001). Feeding ecology of *Bufo japonicus formosus* from the montane region of Kyoto, Japan. *Journal of Herpetology* **36**, 719–723.

Hisaw, F.L. & Emery, F.E. (1927). Food selection of ground squirrels, *Citellus tridecemlineatus*. *Journal of Mammalogy* **8**, 41–44.

Holmen, J., Olsen, E.M., & Vøllestad, L.A. (2003). Interspecific competition between stream-dwelling brown trout and Alpine bullhead. *Journal of Fish Biology* **62**, 1312–1325.

Holycross, A.T., Painter, C.W., Prival, D.B., *et al.* (2002). Diet of *Crotalus lepidus klauberi* (Banded rock rattlesnake). *Journal of Herpetology* **36**, 589–597.

Hooks, C.R.R., Pandey, R.R., & Johnson, M.W. (2003). Impact of avian and arthropod predation on lepidopteran caterpillar densities and plant productivity in an ephemeral agroecosystem. *Ecological Entomology* **28**, 522–532.

Hynes, H.B.N. (1950). The food of fresh-water sticklebacks (*Gasterosteus aculeatus* and *Pygosteus pungitius*) with a review of methods used in studies of the food of fishes. *Journal of Animal Ecology* **19**, 36–58.

Johnson, B.K. & Cristiansen, J.L. (1976). The food and food habis of Blanchard's cricket frog, *Acris crepitans blanchardi* (Amphibia, Anura, Hylidae), in Iowa. *Journal of Herpetology* **10**, 63–74.

Jones, J.C. (1940). Food habits of the American coot with notes on distribution. *United States Department of the Interior, Bureau of Biological Survey, Wildlife Research Bulletin* **2**, 52 pp.

Jones, N.E., Tonn, W.M., & Scrimgeour, G.J. (2003). Selective feeding of age-0 Arctic grayling in lake-outlet streams of the Northwest Territories, Canada. *Environmental Biology of Fishes* **67**, 169–178.

Karasov, W.H. (1990). Digestion in birds: chemical and physiological determinants and ecological implications. *Studies in Avian Biology* **13**, 391–415.

Kaspari, M. & Joern, A. (1993). Prey choice by three insectivorous grassland birds: reevaluating opportunism. *Oikos* **68**, 414-430.

Kauhala, K. & Auniola, M. (2001). Diet of raccoon dogs in summer in the Finnish archipelago. *Ecography* **24**, 151–156.

Keevin, T.M., George, S.G., Hoover, J.J., Kuhajda, B.R., & Mayden, R.L. (2007). Food habits of the endangered Alabama sturgeon, *Scaphirhynchus suttkusi* Williams and Clemmer, 1991 (Acipenseridae). *Journal of Applied Ichthyology* **23**, 500–505.

Kilgore, Jr., D.L. (1969). An ecological study of the swift fox (*Vulpes velox*) in the Oklahoma Panhandle. *American Midland Naturalist* **81**, 512–534.

Knutsen, J.A., Knutsen, H., Ghostaeter, J., & Jonsson, B. (2001). Food of anadromous brown trout at sea. *Journal of Fish Biology* **59**, 533–543.

Krakauer, T. 1968. The ecology of the neotropical toad, *Bufo marinus*, in south Florida. *Herpetologica* **24**, 214–221.

Kreivi, P., Muotka, T., Huusko, A., Mäki-Petäys, Huhta, A., & Meissner, K. (1999). Diel feeding periodicity, daily ration and prey selectivity in juvenile brown trout in a subarctic river. *Journal of Fish Biology* **55**, 553–571.

Lampe, R.P. (1982). Food habits of badgers in east central Minnesota. *Journal of Wildlife Management* **46**, 790–795.

Landry, Jr., S.O. (1970). The Rodentia as herbivores. *The Quarterly Review of Biology* **45**, 351–372.

Lee, Y-F. & Severinghaus, L.L. (2004). Sexual and seasonal differences in the diet of Lanyu scops owls based on fecal analysis. *Journal of Wildlife Management* **68**, 299–306.

Lee, Y-F. & McGracken, G.F. (2005). Dietary variation of Brazilian free-tailed bats links to migratory populations of pest insects. *Journal of Mammalogy* **86**, 67–76.

Legler, J.M. (1977). Stomach flushing: a technique for chelonion dietary studies. *Herpetologica* **33**, 281–284.

Legler, J.M. & Sullivan, L. (1982). The application of stomach flushing to lizards and anurans. *Herpetologica* **35**, 107–110.

Lemos-Espinal, J.A., Smith, G.R., & Ballinger, R.E. (2003). Diets of three species of knob-scaled lizards (genus *Xenosaurus*) from Mexico. *The Southwestern Naturalist* **48**, 119–122.

Lima, E.M. & Ferrari, S.F. (2003). Diet of a free-ranging group of squirrel monkeys (*Saimiri sciureus*) in eastern Brazilian Amazonia. *Folia Primatologica* **74**, 150–158.

Linzey, D.W. (1967). Food of the leopard frog, *Rana p. pipiens*, in central New York. *Herpetologica* **23**, 11–17.

Long, J.L. (1984). The diets of three species of parrots in the south of Western Australia. *Australian Wildlife Research* **11**, 357–371.

Martin, A.C. & Uhler, F.M. (1939). Food of game ducks in the United States and Canada. *USDA Technical Bulletin* **634**, 156 pp.

Martin, A.C., Zim, H.S., & Nelson, A.L. (1961). *American Wildlife and Plants. A Guide to Wildlife Food Habits*. Dover Publications, New York. (reprint of 1951 edition by McGraw-Hill Book Company). 500 pp.

Martins, M.M. & Setz, E.Z.F. (2000). Diet of buffy tufted-eared marmosets (*Callithrix aurita*) in a forest fragment in southeastern Bazil. *International Journalof Primatology* **21**, 467–476.

Mathur, D. (1972). Seasonal food habits of adult white crappie, *Pomoxis annularis* Rafinesque, in Conowingo Reservoir. *The American Naturalist* **87**, 236–241.

Maury, M.E. (1995). Diet composition of the greater earless lizard (Cophosaurus texanus) in central Chihuahuan Desert. *Journal of Herpetology* **29**, 266–272.

McAuslane, H.J. (2008). Aposematism. In Capinera, J.L. (ed.), *Encyclopedia of Entomology*, 2nd edn., pp. 239–242. Springer Science & Business Media B.V., Dordrecht, The Netherlands.

McAuslane, H.J. (2008). Mimicry. In Capinera, J.L. (ed.), *Encyclopedia of Entomology*, pp. 2397–2401. Springer Science & Business Media B.V., Dordrecht, The Netherlands.

McAney, C., Sheil, C., Sullivan, C., & Fairley, J. (1997). Identification of arthropod fragments in bat droppings. *The Mammal Society Occasional Publications* **17**, 56 pp.

McAtee, W.L. (1905). The horned larks and their relation to agriculture. *USDA Division of Biological Survey Bulletin* **23**, 35 pp.

McAtee, W.L. (1908). Food habits of the grosbeaks. *USDA Division of Biological Survey Bulletin* **32**, 92 pp.

McAtee, W.L. (1917). Life and writings of Professor F.E.L. Beal. *The Auk: a Quarterly Journal of Ornithology* **34**, 243–264.

Messick, J.P. & Hornocker, M.G. (1981). Ecology of the badger in southwestern Idaho. *Wildlife Monographs* **76**, 3–53.

Miller, C.K., Knight, R.L., McEwen, L.C., & George, T.L. (1994). Responses of nesting Savannah sparrows to fluctuations in grasshopper densities in interior Alaska. *The Auk* **111**, 962–969.

Mittermeier, R.A. & van Roosmalen, M.G.M. (1981). Preliminary observations on habitat utilization and diet in eight Surinam monkeys. *Folia Primatologica* **36**, 1–39.

Nakano, S., Kawaguchi, Y., Taniguchi, Y., et al. (1999). Selective foraging on terrestrial invertebrates by rainbow trout in a forested headwater stream in northern Japan. *Ecological Research* **14**, 351–360.

Nekaris, K.A.I. & Rusmussen, D.T. (2003). Diet and feeding behavior of Mysore slender lorises. *International Journal of Primatology* **24**, 33–46.

Nelson, A.L. (1933). A preliminary report on the winter food of Virginia foxes. *Journal of Mammalogy* **14**, 40–43.

Oplinger, C.S. (1967). Food habits and feeding activity of recently transformed and adult *Hyla crucifer crucifer* Wied. *Herpetologica* **23**, 209–217.

Pearson, T.G. (ed.). (1936). *Birds of America*. Garden City Books, Garden City, New York. 876 pp.

Pennak, R.W. (1978). *Fresh-Water Invertebrates of the United States*, 3rd edition. John Wiley & Sons, Ltd, New York.

Plummer, M.V. (1981). Habitat utilization, diet and movements of a temperate arboreal snake (*Opheodrys aestivus*). *Journal of Herpetology* **15**, 425–432.

Powders, V.N. & Tietjen, W.L. (1974). The comparative food habits of sypatric and allopatric salamanders *Plethodon glutinosus* and *Plethodon jordani* in eastern Tennessee and adjacent areas. *Herpetologica* **30**, 167–175.

Ramírez-Bautista, A. & Lemos-Espinal, J.A. (2004). Diets of two syntopic populations of frogs, *Rana vaillanti* and *Rana brownorum*, from a tropical rain forest in southern Veracruz, México. *The Southwestern Naturalist* **49**, 316–320.

Reynolds, H.C. (1945). Some aspects of the life history and ecology of the opossum in central Missouri. *Journal of Mammalogy* **26**, 361–379.

Rikardsen, A.H., Amundsen, P.-A., Bjørn, P.A., & Johansen, M. (2000). Comparison of growth, diet and food consumption of sea-run and lake-dwelling Arctic charr. *Journal of Fish Biology* **57**, 1172–1188.

Rocha, C.F.D. (1989). Diet of a tropical lizard (*Liolaemus lutzae*) of southeastern Brazil. *Journal of Herpetology* **23**, 292–294.

Rocha, C.F.D. (2004). Diet of the lizard *Mabuya agilis* (Sauria; Scincidae) in an insular habitat (Ilha Grande, RJ, Brazil). *Brazilian Journal of Biology* **64**, 135–139.

Sadzikowski, M.R. & Wallace, D.C. (1977). A comparison of the food habits of size classes of three sunfishes (*Lepomis macrochirus* Rafinesque, *L. gibbosus* (Linnaeus) and *L. cyanellus* Rafinesque). *American Midland Naturalist* **95**, 220–225.

Sanz, C., Morgan, D., & Gulick, S. (2004). New insights into chimpanzees, tools, and termites from the Congo Basin. *The American Naturalist* **164**, 567–581.

Schmidt, J.O. (2008). Venoms and toxins in insects. In Capinera, J.L. (ed.), *Encyclopedia of Entomology*, 2nd edn., pp. 4076–4089. Springer Science + Business Media B.V., Dordrecht, The Netherlands.

Schulz, M. (2000). Diet and foraging behavior of the golden-tipped bat, *Kerivoula papuensis*: a spider specialist? *Journal of Mammalogy* **81**, 948–957.

Schwenk, K. (2000). *Feeding. Form, Function, and Evolution in Tetrapod Vertebrates*. Academic Press, San Diego, California, USA. 537 pp.

Seaburg, K.G. & Moyle, J.B. (1964). Feeding habits, digestive rates, and growth of some Minnesota warmwater fishes. *Transactions of the American Fisheries Society* **93**, 269–285.

Sealy, S.G. (1980). Reproductive responses of northern orioles to a changing food supply. *Canadian Journal of Zoology* **58**, 221–227.

Shewchuk, C.H. & Austin, J.D. (2001). Food habits of the racer (*Coluber constrictor mormon*) in the northern part of its range. *Herpetological Journal* **11**, 152–155.

Shine, R. (1977). Habitats, diets, and sympatry in snakes: a study from Australia. *Canadian Journal of Zoology* **55**, 1118–1128.

Sikes, R.S., Heidt, G.A., & Elrod, D.A. (1990). Seasonal diets of the nine-banded armadillo (*Dasypus novemcinctus*) in a northern part of its range. *American Midland Naturalist* **123**, 383–389.

Simmen, B., Hladik, A., & Ramasiarisoa, P. (2003). Food intake and dietary overlap in native *Lemur catta* and *Propithecus verreauxi* and introduced *Eulemur fulvus* at Berenty, southen Madagascar. *International Journal of Primatology* **24**, 949–968.

Souza, L.L., Ferrari, S.F., & Pina, A.L.C.B. (1997). Feeding behaviour and predation of a bat by *Saimiri sciureus* in a semi-natural Amazonian environment. *Folia Primatologica* **68**, 194–198.

Spence, R., Fatema, M.K., Ellis, S., Ahmed, Z.F., & Smith, C. (2007). Diet growth and recruitment of wild zebrafish in Bangladesh. *Journal of Fish Biology* **71**, 304–309.

Sperry, C.C. (1940). Food habits of a group of shorebirds: woodcock, snipe, knot, and dowitcher. *United States Department of the Interior, Bureau of Biological Survey, Wildlife Research Bulletin* **1**. 36 pp.

Stafford, P.J. (2005). Diet and reproductive ecology of the Yucatán cricket-eating snake *Symphimus mayae* (Colubridae). *Journal of Zoology* **265**, 301–310.

Starkov, V.G., Osipov, A.V., & Utkin, Y.N. (2007). Toxicity of venoms from vipers of *Pelias* group to crickets *Gryllus assimilis* and its relation to snake entomophagy. *Toxicon* **49**, 995–1001.

Stocks, I. (2008). Reflex bleeding (autohemorrhage). In Capinera, J.L. (ed.), *Encyclopedia of Entomology*, 2nd edn., pp. 3132–3139. Springer Science & Business Media B.V., Dordrecht, The Netherlands.

Strehl, C.E. & White, J. (1986). Effects of superabundant food on breeding success and behavior of the red-winged blackbird. *Oecologia* **70**, 178–186.

Stuewer, F.W. (1943). Raccoons: their habits and management in Michigan. *Ecological Monographs* **13**, 203–257.

Suazo-Ortuño, I., Alvarado-Diaz, J., Raya-Lemus, E., & Martinez-Ramos, M. (2007). Diet of the Mexican marbled toad (*Bufo marmoreus*) in conserved and disturbed tropical forest. *The Southwestern Naturalist* **52**, 305–309.

Tevis, Jr., L. (1953). Stomach contents of chipmunks and mantled squirrels in northeastern California. *Journal of Mammalogy* **34**, 316–324.

Tolonen, A., Kjellman, J., & Lappalainen, J. (1999). Diet overlap between burbot (*Lota lota* (L.)) and whitefish (*Coregonus lavaretus* (L.)) in a subarctic lake. *Annales Zoologici Fennici* **36**, 205–214.

Trajano, E. (1997). Food and reproduction of *Trichomycterus itcarambiensis*, cave catfish from south-eastern Brazil. *Journal of Fish Biology* **51**, 53–63.

Tumlinson, R., Cline, G.R., & Zwank, P. (1990). Prey selection in the Oklahoma salamander (*Eurycea tynerensis*). *Journal of Herpetology* **24**, 222–225.

Twigg, L.E. How, R.A., Hatherly, R.L., & Dell, J. (1996). Comparison of the diet of three sympatric species of *Ctenotus* skinks. *Journal of Herpetology* **30**, 561–566.

Valdez, R.A. & Helm, W.T. (1971). Ecology of threespine stickleback *Gasterosteus aculeatus* Linnaeus on Amchitka Island, Alaska. *BioScience* **21**, 641–645.

Vitt, L.J. & Cooper, Jr., W.E. (1986). Foraging and diet of a diurnal predator (*Eumeces laticeps*) feeding on hidden prey. *Journal of Herpetology* **20**, 408–415.

Vogt, R.C. (1981). Food partitioning in three sympatric species of map turtle, genus *Graptemys* (Testudinata, Emydidae). *American Midland Naturalist* **105**, 102–111.

Waters, D.A. (2003). Bats and moths: what is there left to learn? *Physiological Entomology* **28**, 237–250.

Webb, J.K. & Shine, R. (1993). Dietary habits of Australian blindsnakes (Typhlopidae). *Copeia* **1993**, 762–770.

Welker, T.L. & Searnecchia, D.L. (2003). Differences in species composition and feeding ecology of catostomid fishes in two distinct segments of the Missouri River, North Dakota, USA. *Environmental Biology of Fishes* **68**, 129–141.

Whitaker, Jr., J.O. (1963). Food, habits and parasites of the woodland jumping mouse in central New York. *Journal of Mammalogy* **44**, 316–321.

Whitaker, Jr., J.O. (1963). Food of 129 *Peromyscus leucopus* from Ithaca, New York. *Journal of Mammalogy* **44**, 418–419.

Whitaker, Jr., J.O. (1963). Summer food of 220 short-tailed shrews from Ithaca, New York. *Journal of Mammalogy* **44**, 419–420.

Whitaker, Jr., J.O. (1966). Food of *Mus musculus, Peromyscus maniculatus bairdi* and *Peromyscus leucopus* in Vigo County, Indiana. *Journal of Mammalogy* **47**, 473–486.

Whitaker, Jr., J.O. & Mumford, R.E. (1972). Food and ectoparasites of Indiana shrews. *Journal of Mammalogy* **53**, 329–335.

Wight, H.M. (1928). Food habits of Townsend's mole, *Scapanus townsendii* (Bachman). *Journal of Mammalogy* **9**, 19–33.

Willig, M.R., Camilo, G.R., & Noble, S.J. (1993). Dietary overlap in frugivorous and insectivorous bats from edaphic cerrado habitats of Brazil. *Journal of Mammalogy* **74**, 117–128.

Chapter 5

INSECTS IMPORTANT AS FOOD FOR WILDLIFE

In Chapter 4 we saw that a great diversity of insects was consumed by vertebrate wildlife. Nearly all the orders listed in Chapter 1 appear in the diets of wildlife; most that do not tend to be small or occur relatively infrequently. A few groups, particularly mantids (Mantodea) and stick and leaf insects (Phasmida) are not uncommon insects, but seem not to be very important as food items, possibly because they are so cryptic that they often escape detection by predators. Also, a few (e.g., mantids) are quite capable of mounting a defense against predation, or contain alleochemicals that deter predation (e.g., some leaf beetles [Coleoptera: Chrysomelidae]), so it is likely that there are several factors that determine whether or not insects are important prey.

The taxa that commonly are eaten by vertebrate wildlife are listed below. Some occur only in aquatic environments, but other taxa include both terrestrial and aquatic species. Note that although mayflies (Ephemeroptera), stoneflies (Plecoptera), dragonflies and damselflies (Odonata), and caddisflies (Trichoptera) generally are viewed as important food resources for fish, and thus are listed as aquatic insects, the adult forms are aerial and sometimes serve as important food resources for terrestrial vertebrates, particularly avifauna.

Order Ephemeroptera – the mayflies
Order Odonata – the dragonflies and damselflies
Order Plecoptera – the stoneflies
Order Blattodea – the cockroaches

Order Isoptera – the termites
Order Orthoptera – the grasshoppers, katydids, and crickets
Order Dermaptera – the earwigs
Order Psocoptera – the barklice or psocids
Order Hemiptera – the bugs
Order Megaloptera – the alderflies, dobsonflies, and fishflies
Order Neuroptera – the lacewings, antlions, and mantidflies
Order Coleoptera – the beetles
Order Trichoptera – the caddisflies
Order Lepidoptera – the butterflies and moths
Order Diptera – the flies
Order Hymenoptera – the wasps, ants, bees, and sawflies

AQUATIC INSECTS

Mayflies (Order Ephemeroptera)

These are the most primitive winged insects. The adults do not feed. Although usually bearing two pairs of wings, the hindwings may be greatly reduced in size (Fig. 5.1). The wings are held vertically at rest. These insects often are associated with running water in temperate regions, and on windless nights they tend to produce large swarms of males, with females entering the swarms to mate. Mating occurs principally while in flight. Eggs usually are deposited in the water. The immatures develop in the water, but mature nymphs come to the surface to emerge, sometimes crawling out on a rock. They display an unusual form of hemimetabolous development. Curiously, the translucent-winged 'adult' that emerges is not really the adult, but

Insects and Wildlife: Arthropods and Their Relationships with Wild Vertebrate Animals, 1st edition. By J.L. Capinera. Published 2010 by Blackwell Publishing.

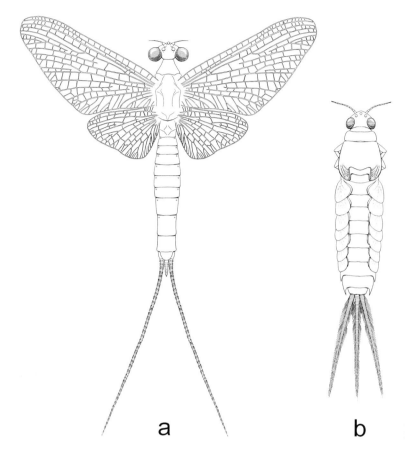

Fig. 5.1. Typical mayflies: (a) adult (Ephemeroptera: Ephemeridae); and (b) immature (Ephemeroptera: Siphlonuridae).

a 'subadult' that often is called a 'dun' and which greatly resembles the adult. The subadult is a transitional stage, however, and usually flies off to rest on a twig where it molts again in about 24 h into the true transparent-winged adult stage. The true adults are similarly short-lived, and the female deposits her eggs in the water, sometimes in long gelatinous filaments. Nymphs have elongate bodies, relatively long antennae, large compound eyes, tracheal gills, and (usually three) long caudal filaments. Nymphs can be found in either the quiet waters of ponds and lakes or the running waters of streams and rivers. They are herbivorous, detritivorous, or occasionally carnivorous. Over 2100 species are known worldwide.

Stoneflies (Order Plecoptera)

These primitive insects usually are associated with running water. The adults are long-lived as compared to mayflies, often surviving 3-4 weeks. The two pairs of wings fold over the body at rest. Usually two cerci and long antennae are present (Fig. 5.2). They are weak fliers and may be diurnal or nocturnal. Adults of both sexes can produce species-specific sounds, called drumming, which they use for locating the opposite sex. Unlike mayflies, they usually mate on the ground. Eggs normally are deposited in the water, usually in masses and sometimes buried in the streambed. The eggs are highly varied in appearance and have diagnostic value.

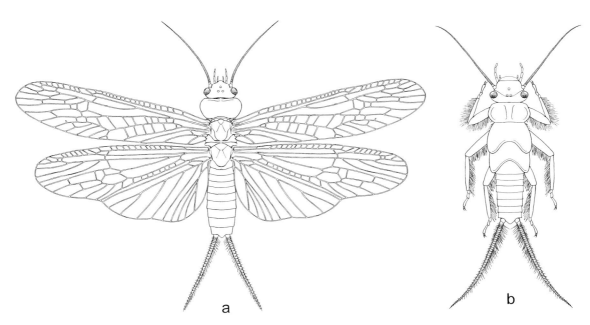

Fig. 5.2. Typical stoneflies (Plecoptera: Perlidae): (a) adult; and (b) immature.

Nymphal development can be prolonged, lasting 2–3 years while they undergo 12–23 molts. The pattern of development is hemimetabolous. The nymphs feed on detritus, algae and other insects. They have tracheal gills that are found in various locations on the body, though a few species lack them, with passive ventilation occurring through their cuticle. Stoneflies generally are limited to cold (less than 25 °C) oxygen-rich water. This, combined with their inability to survive in water with high levels of organic enrichment, has led to their routine use as indicators of water quality. The adults usually emerge in the fall or winter. Some species feed as adults, but others do not. Some families are predominantly predators, whereas others feed on detritus or algae. However, food selection tends to develop later in nymphal development, with young of all stoneflies tending to feed on fine particles of organic matter. About 2000 species occur worldwide.

Dragonflies and Damselflies (Order Odonata)

Odonates have two pairs of wings that are very similar in size, and with numerous wing veins arranged in a net-like pattern. They also have three-segmented tarsi, short antennae, large compound eyes, and chewing mouthparts. Their development is hemimetabolous. Adults, but not the nymphal instars, are often brightly colored. Nearly all odonates are placed into the suborders Anisoptera (the dragonflies), or Zygoptera (the damselflies). Over 5500 species are known, most from the tropics.

Dragonflies (Fig. 5.3) are distinguished from damselflies by having hindwings that are slightly broader at the base and held horizontally and outstretched when at rest. They are strong fliers. Their eyes do not project strongly from the head. Their nymphs are robust (Fig. 5.3), and bear rectal gills. They can expel water through their rectum, causing a jet-propelled escape response. The eggs are usually laid on the surface of the water or on the surface of aquatic plants.

In contrast, in damselflies the front and hind wings are virtually equal in size, and are folded and held vertically when at rest. The adults are weak fliers. The eyes are bulbous and prominent. They nymphs are slender, and bear paddle-like gills at the tip of the abdomen (Fig. 5.4). These gills assist in swimming in addition to ventilation, but they lack the 'jet-propulsion' capabilities of dragonfly nymphs. The eggs are inserted in the stems of aquatic plants.

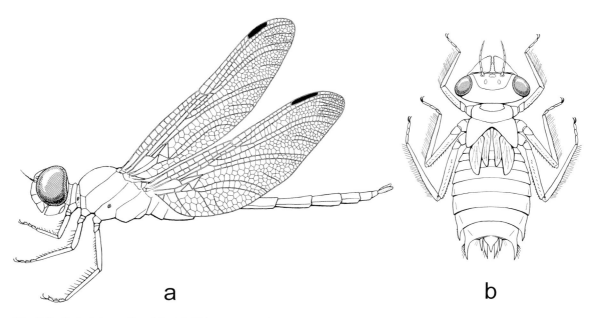

a   b

Fig. 5.3. Typical dragonflies: (a) adult (Odonata: Gomphidae); and (b) immature (Odonata: Libellulidae).

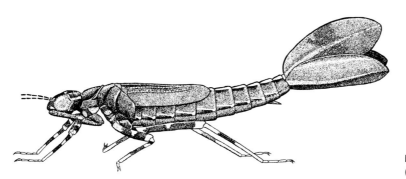

Fig. 5.4. An immature damselfly
(Odonata: Coenagrionidae).

Dragonflies and damselflies are predators in both the adult and immature stages. Adults often display thermoregulation behavior on cool mornings, orienting their bodies to gain heat from the sun, and enabling them to fly. Although these are considered to be primitive insects, they have amazing powers of flight, including the ability to hover and to fly backwards. Dragonflies generally capture their prey while in flight, whereas damselflies tend to capture prey that are resting. The nymphs of dragonflies bear an unusually long and extensible labial structure that is used for capturing prey. Initially, the nymphs feed on small organisms such as the larvae of flies, but as they grow larger they feed on larger invertebrates and even vertebrates such as immature fish and amphibians. They often take 1 or 2 years to complete their development, with nymphs undergoing about 10–15 molts. The nymphs climb from the water when the adult is ready to emerge.

One of the most unusual aspects of odonate biology is the mating behavior. Initially, the male tends to grasp a female behind her head with the tip of his abdomen. Then the female swings the tip of her abdomen forward to come into contact with the second or third segment of the male's abdomen, where a sperm-storing seminal vesicle occurs, and the male pumps in his sperm after emptying the female's spermatheca (sperm receptacle) of any sperm deposited there by any previous mating. Often the male and female can be seen flying in tandem (the male grasping the female behind the head) prior to sperm transfer, or after sperm transfer but before oviposition. By holding onto the female, the male eliminates the possibility that another male will come along and supplant his sperm.

Bugs (Order Hemiptera)

A great number of different families of the order Hemiptera live on or in water and serve as food *for* aquatic wildlife. However, in a few cases, they feed *on* aquatic wildlife. Although they are very diverse, with over 100,000 species described worldwide, they all have piercing-sucking mouthparts. Wing length is variable between and among species, but generally the forewing is smaller than the hind wing, and often the forewing is thickened basally (a basally thickened wing is called a **hemelytron**) (Fig. 5.5). Most have a triangular plate called a **scutellum** that is present at the base of the forewings. The legs tend to be long, and some are flattened or have hairs that increase the surface area and allow them to function as oars when they swim on or in water (Fig. 5.6). Bugs are hemimetabolous, lacking a pupal stage. Normally, they have five nymphal instars.

Aquatic bugs are almost entirely predatory (corixids can be an exception, often ingesting algae), and many feed on other insects or spiders that fall on the surface, or emerge at the surface of the water. They may locate their prey visually, but often use the ripples created by the motion of insects at the surface to detect potential meals. Among the common surface-feeding insects are:

Water striders (family Gerridae)
Broad-shouldered water striders (family Veliidae)
Water treaders (family Mesoviliidae)
Water measurers (family Hydrometridae)

Subsurface predators are often found resting near the surface, but once they locate a prospective prey item

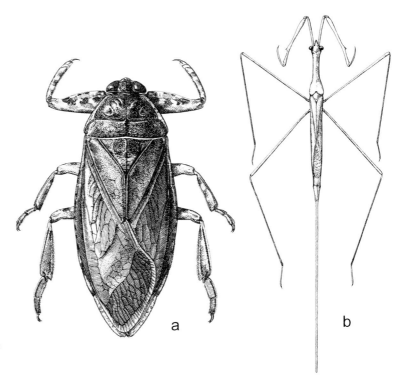

Fig. 5.5. Some large aquatic bugs: (a) a giant water bug, *Lethocercus americana* (Hemiptera: Belostomatidae); a waterscorpion, *Ranatra* sp. (Hemiptera: Nepidae).

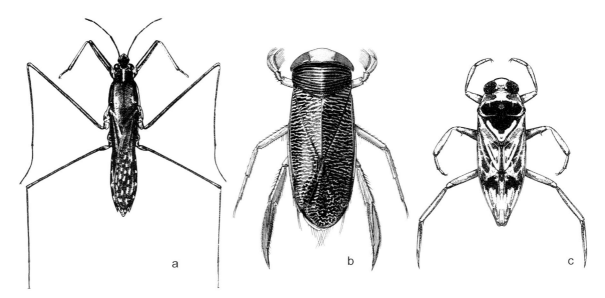

Fig. 5.6. Some small aquatic bugs: (a) a water strider, *Gerris* sp. (Hemiptera: Gerridae); (b) a water boatman, *Hesperocorixa* sp. (Hemiptera: Corixidae); (c) a backswimmer, *Notonecta* sp. (Hemiptera: Notonectidae).

they chase and subdue it. The forelegs of such predators are adapted for grasping prey, often bearing stout spines and teeth. Some bugs, especially giant water bugs, will even feed on small vertebrates. Once captured, the prey item is pierced by the bug's mouthparts and quickly subdued. Among the common subsurface predators are:

Giant water bugs (family Belostomatidae)
Creeping water bugs (family Naucoridae)
Water scorpions (family Nepidae)
Backswimmers (family Notonectidae)
Water boatmen (family Corixidae)

Alderflies, Dobsonflies, and Fishflies (Order Megaloptera)

There are two families, the Sialidae (alderflies) and the Corydalidae (dobsonflies and fishflies) in this order. They occur mostly in temperate regions of the globe. The Corydalidae are large, with the wingspan reaching 16 cm in some species; the sialids are smaller. Some adult corydalids have robust mouthparts that are elongated to form tusks (Fig. 5.7). The Megaloptera have two pairs of wings and long antennae. This small order contains only about 200 species, and often is grouped together with other related insects into the order Neuroptera. Megaloptera might not be worth mentioning, except that they can be large and have a strikingly different appearance relative to the other aquatic insects. The large (up to 90 mm long) larvae have large gills on the first seven to eight abdominal segments. They are predators, feeding on a number of aquatic organisms, including other insects.

Beetles (Order Coleoptera)

This most speciose of all insect orders (over 370,000 species) is well represented in aquatic environments. However, only a few families occur exclusively in water, and even fewer are common enough to be important prey for wildlife. Adult beetles are most easily distinguished by the hard forewings (*elytra*) that meet in the middle and (generally) cover the abdomen (Figs. 5.8, 5.9). The beetles have holometabolous development so the immature stages differ greatly in appearance, and sometimes habits, from the adults.

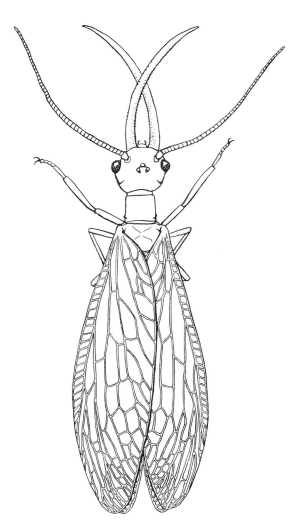

Fig. 5.7. Typical adult male alderfly (Megaloptera: Corydalidae).

Many of the aquatic beetles are predatory, but some are herbivores or scavengers. Some beetles, especially predaceous diving beetles, will feed on vertebrates. Among the common aquatic beetles are:
Predaceous diving beetles (family Dytiscidae)
Whirligig beetles (family Gyrinidae)
Crawling water beetles (family Haliplidae)
Water scavenger beetles (family Hydrophilidae)
Riffle beetles (family Elmidae)
Water-penny beetles (family Psephenidae)

Flies (Order Diptera)

This is another large group of holometabolous insects, with over 100,000 species described worldwide. Flies are distinguished by having only one pair of wings in the adult stage (Figs. 5.10, 5.11). The membranous wings are attached to the mesothorax, and a pair of club-like halteres are found on the metathorax. Mouth and antennal characteristics are variable, and the larval stage is often primitive and cylindrical (Figs. 5.10–5.12). Flies, particularly midges, are the most important sources of nutrition in some aquatic systems. Only the larval forms of flies are found in the water, and adults fly off to feed on nectar or blood. Fly larvae may be detritivores, herbivores, and even predators, but they gain most of their food by filtering out particulate matter from the water. Some species of crane flies are terrestrial. Among the most important aquatic flies are:
Crane flies (family Tipulidae)
Mosquitoes (family Culicidae)
Black flies (family Simuliidae)
Midges (family Chironomidae)
The mosquitoes and black flies are important pests and disease vectors, so they are treated more fully in Chapter 10, Arthropods as parasites of wildlife.

Caddisflies (Order Trichoptera)

The holometabolous caddisflies are closely related to moths, so it is not surprising that the adults look similar to moths, the larvae closely resemble caterpillar larvae, and the larvae spin silk. The wings may be similar in size, or the hindwings smaller (Fig. 5.13). Often the wings bear numerous hairs. This is not a large group of insects, with perhaps 7000 species known. They are quite important to fish, however.

The larvae (Fig. 5.13) are aquatic, and although they are found in all types of aquatic environments they are most common in cool running waters. The larvae feed mostly on algae, diatoms, and detritus and associated microorganisms, but a few are predatory. It is their case-making behavior that makes them quite unusual, and the appearance of the case is different for each species. Cases (Fig. 5.14) are cemented together or held together with silk. Case construction can be categorized as follows:
• ***Free living forms.*** Mostly predators, the larvae move about freely and do not construct a case until

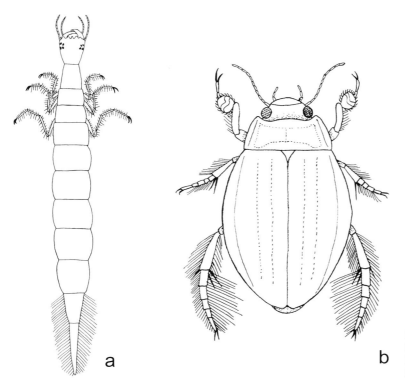

Fig. 5.8. Typical predaceous diving beetles (Coleoptera: Dytiscidae) (a) larva of *Cybister* sp.; (b) adult of *Dytiscus* sp.

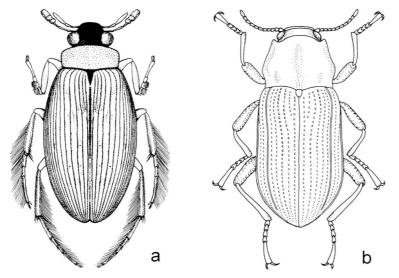

Fig. 5.9. Some additional water beetles (Coleoptera): (a) an adult water scavenger beetle, *Berosus* sp. (Hydrophilidae); (b) an adult riffle beetle, *Stenelmis* sp. (Elmidae).

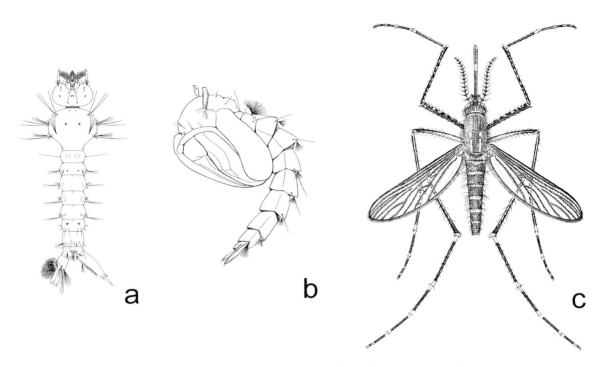

Fig. 5.10. Some stages of typical mosquitoes (Diptera: Culicidae): (a) larva; (b) pupa; and (c) adult.

they are ready to pupate. The pupation chamber is fastened to an underwater stone.

• *Saddle-case makers.* The larvae construct portable cases resembling tortoise shells from rock fragments. They feed on diatoms and detritus. The larvae fasten the dome-shaped structure to a rock or log at pupation.

• *Purse-case makers.* These free-living larvae construct portable purse-shaped or barrel-shaped cases only when they are ready to pupate. They feed on algae and diatoms.

• *Net-spinners.* These are usually sedentary larvae, constructing fixed retreats with capture nets to collect food particles from the flowing water. The retreats can be constructed from organic materials or mineral fragments, and the silken net is placed near the opening. The retreat takes various forms.

• *Tube-case makers.* These larvae make portable tubular cases covered with sand, twigs or bark. They are highly variable in their feeding habits. They fasten their case to the substrate when they are ready to transform to the adult.

TERRESTRIAL INSECTS

Termites (Order Isoptera)

Termites are hemimetabolous insects that have two pairs of nearly equally sized wings during the dispersal portion of the adult stage. They have chewing mouthparts. The antennae are moderately long and thread-like, and not elbowed (Fig. 5.15), as is the case with the similar appearing ants (Hymenoptera: Formicidae). They are soft-bodied and often pale in color. The wings break off after dispersal and the males and females mate and create new colonies. Emergence of the flying adults is synchronous, and such swarms are quite attractive to avian predators. Also, some mammalian and reptilian predators dig up termite mounds, which may contain hundreds of thousands of individuals. About 2900 species occur worldwide. They are less common in cold-weather areas.

Termites are renowned for having reproductive, worker and soldier castes that differ markedly in form and function, and with a highly organized social

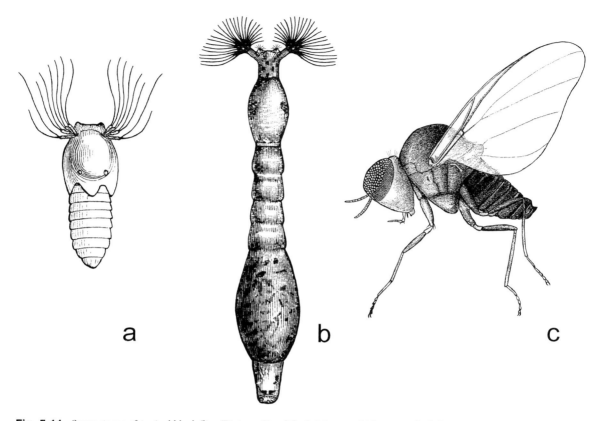

Fig. 5.11. Some stages of typical black flies (Diptera: Simulidae): (a) pupa; (b) larva; and adult.

system. The eggs are tended by workers, and the large abdomen of the female is specialized for prodigious egg production. Solders may be mandibulate, bearing oversized mandibles, or nasute, possessing the ability to secrete or squirt chemical deterrents. Many, but not all, termites are sensitive to low humidity so their colonies are found in contact with the soil, or enclosed in paper-like material called 'carton.'

Termites are among the most ecologically important insects, functioning to assist in the breakdown of cellulosic materials, primarily dead trees, but also soil humus, detritus, grass, and dung. The termites are sometimes divided into two groups: lower (more primitive) and higher (more advanced). The lower termites possess celluolytic protozoa in the hindgut as an aid in digestion of cellulosic materials, whereas the higher termites lack protozoa. Both higher and lower termites possess other microbial symbionts in their guts,

however, particularly bacteria, and both secrete endogenous (originating in the termite rather than in the microorganism) cellulases.

Cockroaches (Order Blattodea)

These hemimetabolus insects are primarily tropical and, like termites, often are quite important in breakdown of organic matter and nutrient cycling. Nocturnal species are common and are dull in color. However, brightly colored species also are common in tropical areas and are active during the day. They have long legs and antennae, and winged species have two pairs of wings (Fig. 5.16). The adults of some species, and of course the nymphs of all species, lack wings. They have chewing mouthparts. Cockroaches are not good fliers despite their large wings. Although many species can

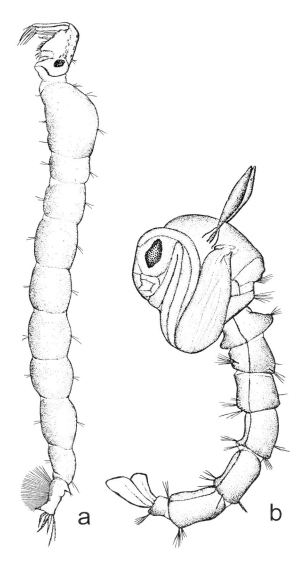

Fig. 5.12. The (a) larval and (b) pupal stage of a typical chironomid midge (Diptera: Chironomidae).

female carries the egg capsule about until hatching occurs. About 4000 species occur worldwide.

In earlier times, the cockroaches, stick insects, earwigs, crickets, grasshoppers, and related insects were grouped together into the order Orthoptera. Although most authorities now separate these groups into individual orders, the older literature can be confusing, and even some recent literature considers cockroaches to be in the order Orthoptera, obscuring the relative importance of this taxon. It is best to distinguish between 'orthopteroids' (primarily mantids, cockroaches, walkingsticks, earwigs, grasshoppers, katydids and crickets) and true Orthoptera (grasshoppers, katydids, and crickets), as it provides additional insight into dietary habits.

Grasshoppers, Katydids, and Crickets (Order Orthoptera)

These common insects are found nearly everywhere, and because they often are abundant and fairly large in size, they are often consumed by wildlife. They are hemimetabolous, and have chewing mouthparts. Most bear four wings, though some are wingless. Normally, the forewings (called tegmina) are slightly thickened and pigmented, and the hindwings are wider, entirely membranous, and often transparent. They vary in their flight capabilities. The antennae are moderately long (grasshoppers, Fig. 5.17) or quite long (katydids and crickets, Fig. 5.18). The hind legs are enlarged relative to the other legs, and adapted for leaping, but normally they move by walking. The eggs may be deposited singly or in clusters (in grasshoppers they are called **egg pods**). **Tympana**, or hearing organs, are commonly found in these insects. In grasshoppers, the tympana are found on the abdomen, but in katydids and crickets they are found on the front tibia. Sound production is used to facilitate mate location. Grasshoppers tend to favor sunny habitats rich in herbaceous vegetation. Katydids may sometimes be found in such environments, but also are abundant in shrubs and trees. Crickets prefer dense vegetation such as forests. Grasshoppers and katydids tend to be diurnal, with crickets nocturnal. There are about 25,000 species of Orthoptera worldwide. Among the most important groups are:

Grasshoppers (family Acrididae)
Lubber grasshoppers (family Romaleidae)
Katydids (family Tettigoniidae)

fly, they usually disperse by running. Some can be quite large in size, attaining 10 cm in length. They are flattened, which allows them to seek shelter beneath bark and in other cryptic locations. Some species burrow in soil. Generally, cockroaches favor dark, humid locations, and they are very abundant on the forest floor of the tropics. The eggs are encapsulated into structures called oothecae, and sometimes the

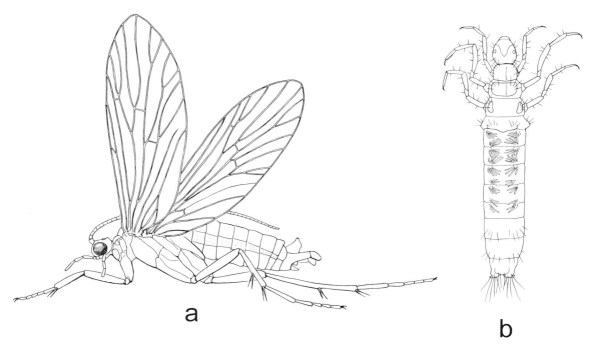

Fig. 5.13. A typical caddisfly (Trichoptera: Limnephilidae): (a) adult; and (b) larval stage (case not shown).

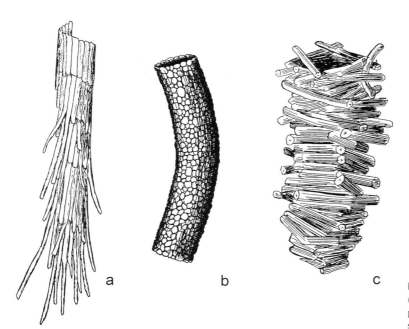

Fig. 5.14. Representitive caddisfly (Trichoptera) larval cases: (a) a phryganeid case; (b) case of *Psilotreta* sp.; (c) case of *Oecetis* sp.

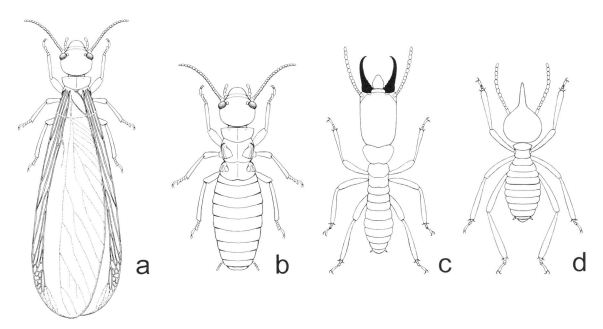

Fig. 5.15. Typical termites (Isoptera): (a) adult bearing wing; and (b) adult after the wings have been shed; (c) mandibulate soldier; (d) nasute soldier.

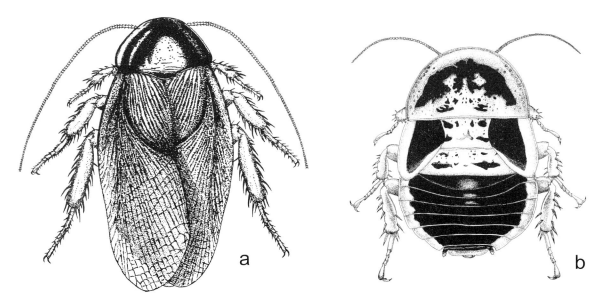

Fig. 5.16. Typical cockroachs: (a) *Pycnoscelus surinamensis*; (b) *Sibylloblatta panesthoides* (both Blattodea: Blaberidae).

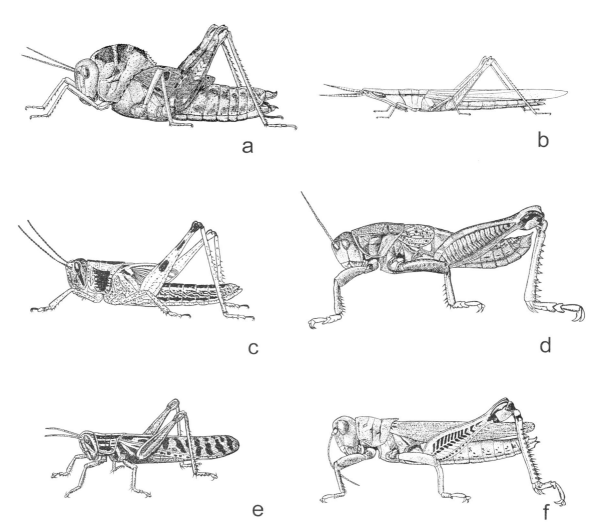

Fig. 5.17. Some representative grasshoppers (Orthoptera: Acrididae and Romaleidae): (a) *Tropidolophus formosus*; (b) *Achurum sumichrasti*; (c) nymph of *Schistocerca americana*; (d) *Brachystola magna*; (e) *Schistocerca americana*; (f) *Melanoplus differentialis*.

Crickets (family Gryllidae)
Mole crickets (family Gryllotalpidae)
Jerusalem crickets (family Stenopelmatidae)

Earwigs (Order Dermaptera)

This is a fairly small taxon, consisting of about 1500 species worldwide. They are distinctive because most bear a pair of terminal forceps-like cerci. They have chewing mouthparts, moderately long antennae, and (when present) two pairs of wings. The forewings are shortened and thickened (tegmina) (Fig. 5.19). The hindwings are broad and membranous, but fold up beneath the short forewings when not in use. They are nocturnal, often hiding in crevices and beneath stones and bark during the daylight hours. Earwigs are omnivorous, but a few are important predators. They

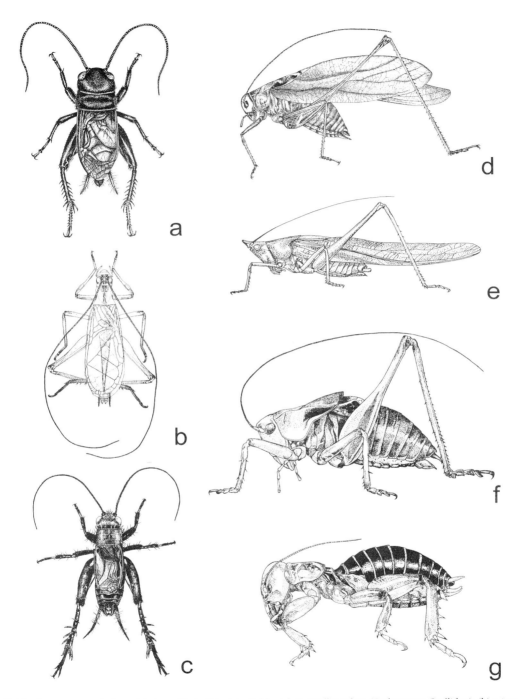

Fig. 5.18. Some representative crickets and katydids: (a) a field cricket, *Gryllus veletis* (Orthoptera: Gryllidae); (b) a tree cricket, *Oecanthus nigricornis* (Orthoptera: Gryllidae); (c) a ground cricket, *Alloneobius griseus* (Orthoptera: Gryllidae); (d) a false katydid, *Amblycorypha oblongifolia* (Orthoptera: Tettigoniidae); (e) a coneheaded katydid, *Neoconocephalus ensiger* (Orthoptera: Tettigoniidae); (f) a shield-backed katydid, *Atlanticus monticola* (Orthoptera: Tettigoniidae); (g) a Jerusalem cricket, *Stenopelmatus fuscus* (Orthoptera: Stenopelmatidae).

are hemimetabolous and may display some primitive sociality, with mothers providing care for eggs and young nymphs.

Barklice or Psocids (Order Psocoptera)

These hemimetabolous insects are commonly found on trunks of trees (hence the name barklice), but also in

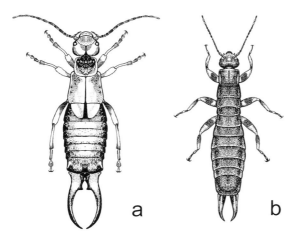

Fig. 5.19. Representative earwigs: (a) male European earwig, *Forficula auricularia* (Dermaptera: Forficulidae); (b) female red-legged earwig, *Euborellia annulipes* (Dermaptera: Carcinophoridae).

animal nests, on vegetation and within detritus. They feed on microflora (algae, lichens, fungi) and organic debris. Barklice are small insects, rarely exceeding 10 mm in length, but despite their small size they are often attractive to predators such as birds. They have two pairs of membranous wings, with the forewings slightly larger than the hindwings (Fig. 5.20). However, often they are wingless. The antennae are moderately long. Some species are gregarious and construct webs resembling dense spider webs on tree trunks. This is a moderately small taxon numbering about 4000 species worldwide.

Bugs (Order Hemiptera)

The Hemiptera are very diverse and quite a large order, with over 100,000 species described worldwide. These insects can be very numerous, so they have considerable importance as a food resource, not only in the aquatic environment (discussed above) but even more so in terrestrial environments. All bugs have piercing-sucking mouthparts. Wing length is variable within and among species, but generally the forewings are smaller than the hindwings. The forewings of some Hemiptera are thickened basally, forming wings called **hemelytra** (literally half elytra or half-thickened) (Figs. 5.21, 5.22). Others have membranous forewings and hind wings. A triangular plate called a **scutellum** may be present at the base of the forewings. Many of these insects are wingless, at least during part of their

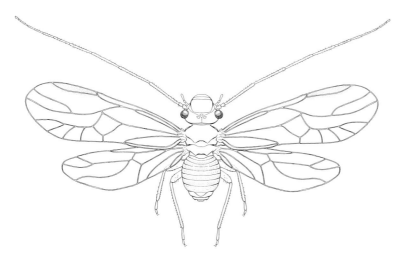

Fig. 5.20. A representative barklouse (Psocoptera).

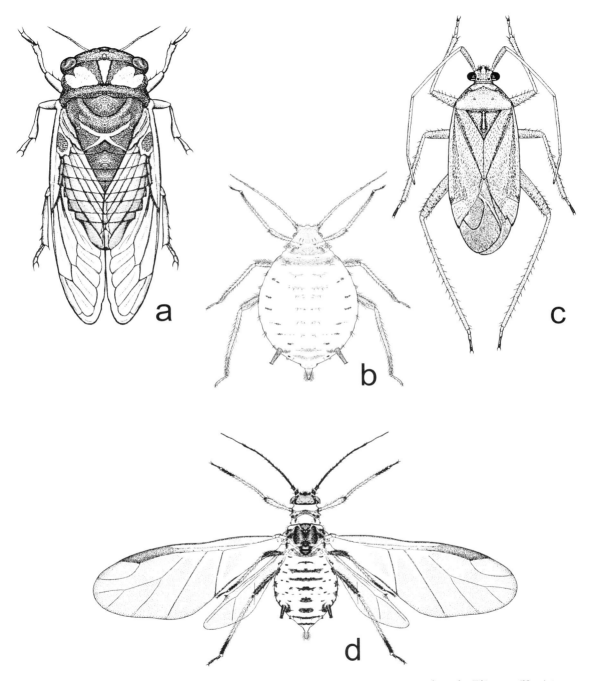

Fig. 5.21. Some representatives of the order Hemiptera, bugs and their allies: (a) an annual cicada, *Tibicen* sp. (Hemiptera: Cicadidae); (b) wingless adult female and (d) winged adult female of the aphid *Aphis fabae* (Hemiptera: Aphididae); (c) a plant bug, *Adelphocoris lineolatus* (Hemiptera: Miridae)

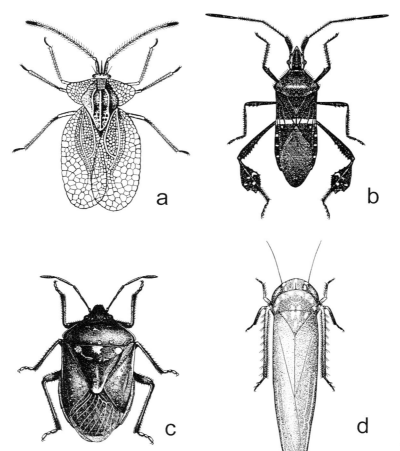

Fig. 5.22. Some representatives of the order Hemiptera, bugs and their allies: (a) a lace bug, *Gargaphia solani* (Hemiptera: Tingidae); (b) a leaffooted bug, *Leptoglossus phyllopus*; (c) a stink bug, *Chlorochroa sayi* (Hemiptera: Pentatomidae); (d) a leafhopper, *Empoasca fabae* (Hemiptera: Cicadellidae).

life cycle. Bugs are hemimetabolous, lacking a pupal stage (there are a few with unusual development, however). Bugs have several nymphal instars, and although most resemble the adults, in other cases they are quite different in appearance. Bugs are mostly plant feeders, and some induce abnormal growth such as leaf curling or swellings (**plant galls**) that allow the insects to dwell within a cavity surrounded by plant tissue. Because piercing-sucking mouthparts are conducive to spread of microbial organisms, bugs often transmit plant diseases. Some terrestrial bugs transmit disease to animals, and are discussed more fully in Chapter 10, Arthropods as parasites of wildlife. Among the terrestrial bugs important to wildlife are:

Plant bugs (family Miridae)
Seed bugs (family Lygaeidae)
Stink bugs (family Pentatomidae)
Cicadas (family Cicadidae)
Treehoppers (family Membracidae)
Leafhoppers (family Cicadellidae)
Aphids (family Aphididae)

Lacewings, Antlions and Mantidflies (Order Neuroptera)

The high frequency of veins in the wings of these insects is the basis for the order name, which means

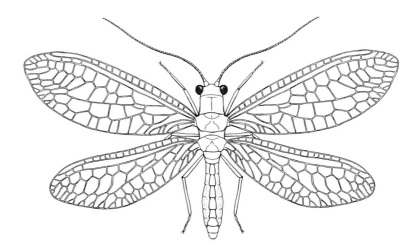

Fig. 5.23. A typical lacewing (Neuroptera: Chrysopidae).

nerve-winged. Although commonly there are many veins and cells (Fig. 5.23), sometimes venation is quite reduced. They are holometabolous, and bear large biting mouthparts. However, the mandible and maxilla of these biting jaws is grooved to form a tube so that liquid foods can be ingested. The wings and abdomen of antlions are narrow and elongate, causing them to resemble damselflies (Odonata). However, the long antennae of adult antlions distinguishes them from damselflies. Nearly all the neuropterans are predatory, though the mantidflies (mantispids) are parasites of spider eggs. Two families, the Mantispidae and the Berothidae, have raptorial forelegs. The neuropterans construct silken cocoons for pupation using a spinneret found near the tip of the abdomen. One family, Sisyridae, consists of species with aquatic larvae that feed on sponges, but they are not common. The antennae are moderately long. About 5000 species are known worldwide. The most common neuropterans are:

Green lacewings (family Chrysopidae)
Brown lacewings (family Hemerobiidae)
Antlions (family Myrmeleontidae)

Beetles (Order Coleoptera)

The largest of all insect orders (Coleoptera, with over 370,000 species) beetles and weevils occupy nearly all terrestrial niches, and often are quite abundant. Thus, they can be quite important food resources, especially for birds and mammals. They range in size from about 1 mm to over 100 mm. Adult beetles are most easily distinguished by the hard forewings (**elytra**; singular, elytron) that meet in the middle and (generally) cover the abdomen (Figs. 5.24, 5.25). The membranous hindwings usually are longer than the elytra, and fold under the forewings for protection. The antennae are variable, the legs usually short. Beetles have chewing mouthparts. They have holometabolous development, so the immature stages differ greatly in appearance, and sometimes habits, from the adults. The larvae usually have pronounced heads and legs, but unlike the adults, which tend to have a rigid exoskeleton, the larvae are soft and more digestible. They lack the prolegs found on caterpillars (Lepidoptera); the weevil larvae lack both true legs and prolegs, usually having little more than ridges to help propel them. Some of the beetles that potentially are very important food resources tend to escape predation from wildlife by feeding within the soil (white grubs, Scarabaeidae), beneath bark (bark beetles, Curculionidae) or within wood (long-horned beetles, Cerambycidae). Others are often found on the soil surface but are nocturnal (ground beetles, Carabidae, and darkling beetles, Tenebrionidae). In contrast, the leaf beetles feed on foliage during daylight hours but often are defended by toxic allelochemicals, and so are avoided by many potential predators. Among the more important beetles for wildlife are:

Ground beetles (family Carabidae)
Long-horned beetles (family Cerambycidae)

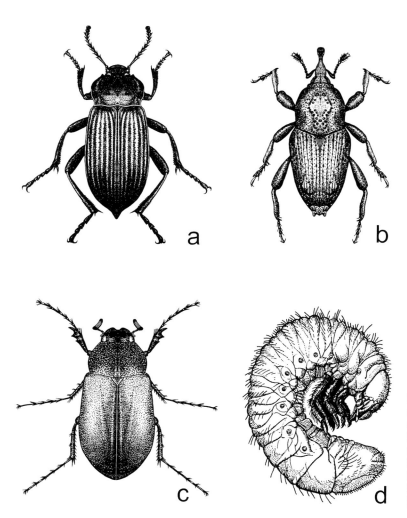

Fig. 5.24. Representative beetles: (a) a darkling beetle, *Eleodes suturalis* (Coleoptera: Tenebrionidae); (b) a billbug, *Sphenophorus maidis* (Curculionidae); (c) a May or June beetle, *Phyllophaga* sp. (Coleoptera: Scarabaeidae); (d) a white grub, or larva of a June beetle, *Phyllophaga* sp. (Coleoptera: Scarabaeidae).

Leaf beetles (family Chrysomelidae)
Scarab beetles (family Scarabidae)
Darkling beetles (family Tenebrionidae)
Weevils (family Curculionidae)

Moths and Butterflies (Order Lepidoptera)

This is a large taxon, over 100,000 species. They are holometabolous. The larvae of Lepidoptera, called caterpillars (Fig. 5.26), have chewing mouthparts and feed mostly on plants. Nearly all adults have siphoning mouthparts that are specialized for drinking nectar from flowers. The adults are distinguished by having two pairs of scaly wings, with the forewings often larger (Fig. 5.27). The adults, but not the larvae, have large compound eyes. The caterpillars have three pairs of segmented, true legs, but also bear up to five pairs of unsegmented leg-like structures called **prolegs** on their abdomen. The prolegs have small hooks called **crochets** so the prolegs assist the larva in walking and anchoring to the substrate. The larvae have well-developed silk glands and many produce silk for constructing a tent, webbing together a shelter of leaves,

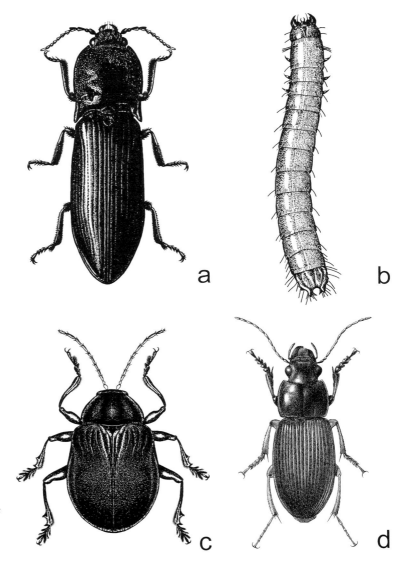

Fig. 5.25. Representative beetles: (a) a click beetle, *Ctenicera glauca* (Coleoptera: Elateridae); (b) a wireworm, or larva of the click beetle, *Ctenicera glauca* (Coleoptera: Elateridae); (c) a leaf beetle, *Typophorus nigritus* (Coleoptera: Chrysomelidae); (d) a ground beetle, *Harpalus* sp. (Coleoptera: Carabidae).

dispersing in the wind, and spinning a cocoon for protection during pupation. Some caterpillars are quite hairy or spiny, whereas others are almost naked. The spines sometimes have poison glands associated with them for protection from predators (usually wildlife). Many caterpillars escape predation by crypsis, but some display aposematic coloration and likely are toxic to predators. The pupal period is spent in a cocoon, often below-ground or in another cryptic location, as they generally are quite susceptible to predation. Adults also are fed upon readily by wildlife, though the wings are usually removed and discarded. Some moths have thoracic hearing organs (**tympana**) that allow them to hear the echolocation calls of bats and avoid predation by these insect specialists.

The Lepidoptera are divided into two suborders, the moths and the butterflies, based mostly on antennal characteristics. The moths lack a terminal antennal

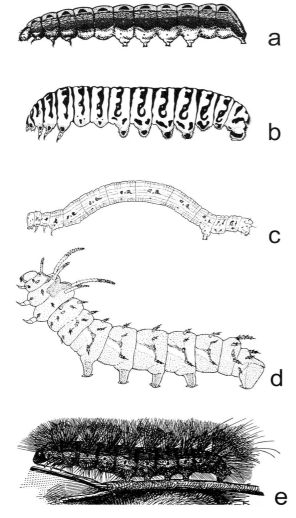

Fig. 5.26. Some representative caterpillars: (a) an armyworm or cutworm, *Spodoptera dolichos* (Lepidoptera: Noctuidae); (b) a swallowtail, *Papilio polyxenes* (Lepidoptera: Papilionidae); (c) a looper, *Cingilia catenaria* (Lepidoptera: Geometridae); (d) hickory horned devil, *Citheronia regalis* (Lepidoptera: Saturniidae); (e) a woollybear, *Spilosoma virginiana* (Lepidoptera: Arctiidae).

matic. Among the more important lepidopteran sources of food for wildlife are:
Loopers/geometer moths (family Geometridae)
Webworms/snout moths (family Pyralidae)
Cutworms/noctuid moths (family Noctuidae)
Hornworms/sphinx moths (family Sphingidae)
Tent caterpillars (family Lasiocampidae)
Giant silkworms (family Saturniidae)
Swallowtails (family Papilionidae)
Pierids (family Pieridae)

Flies (Order Diptera)

The Diptera are a large group of holometabolous insects, with over 100,000 species described world-wide. There are few habitats that are not exploited by flies. They tend to be relatively small and soft-bodied. Adult flies have only one pair of wings (Fig. 5.28). The membranous wings are attached to the mesothorax. A pair of club-like structures that assist in balance, called **halteres**, is found on the metathorax in place of the hindwings. The membranous wings usually are transparent, but some are pigmented. Mouth and antennal characteristics are variable, but generally are either piercing-sucking or sponging. Adults of terrestrial flies often feed on nectar or honeydew. The larval stage (Fig. 5.29) is often primitive and cylindrical, but in terrestrial flies the mouthparts often include **mouth hooks**, structures for tearing the food source. Larvae may be detritivores, herbivores, parasites or predators. Larvae often are called 'maggots.' Among the flies important as food for terrestrial wildlife are:
Crane flies (family Tipulidae)
Mosquitoes (family Culicidae)
Midges (family Chironomidae)
Black flies (family Simulidae)
Robber flies (family Aslidae)
Large fruit flies (family Tephritidae)
Leaf-miner flies (family Agromyzidae)
Muscid flies (family Muscidae)
Blow flies (family Calliphoridae)
Flesh flies (family Sarcophagidae)

Wasps, Ants, Bees, and Sawflies (Order Hymenoptera)

Insects in the order Hymenoptera have two pairs of membranous wings, with the hindwings smaller than

expansion (a knob), tend to be nocturnal, and often are dull in coloration. The butterflies have knobbed antennae, tend to be diurnal, and often are colorful. Likewise, the moth caterpillars tend to be dull or cryptic; the butterfly larvae may be cryptic, but often are apose-

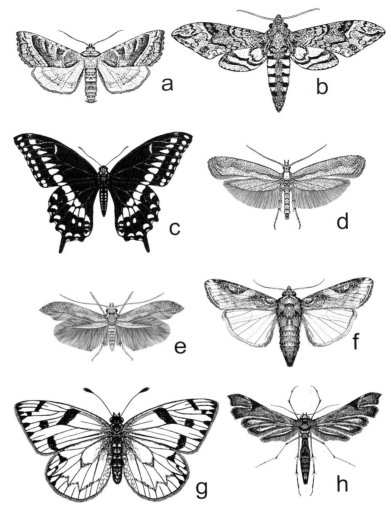

Fig. 5.27. Some representative moths and butterflies: (a) a stem borer, *Hydraecia immanis* (Lepidoptera: Noctuidae); (b) a sphinx or hawk moth, *Agrius cingulatus* (Lepidoptera: Sphingidae); (c) a swallowtail, *Papilio polyxenes* (Lepidoptera: Papilionidae); (d) diamondback moth, *Plutella xyllostella* (Lepidoptera: Plutellidae); (e) a leafminer, *Bedellia orchilella* (Lepidoptera: Lyonetiidae); (f) an armyworm, *Spodoptera latifascia* (Lepidoptera: Noctuidae); (g) checkered white, *Pontia protodice* (Lepidoptera: Pieridae); (h) a plume moth, *Platyptilia carduidactyla* (Lepidoptera: Pterophoridae).

the forewings (Fig. 5.30). This is one of the largest orders, with over 120,000 species described world-wide. Not surprisingly, it is quite diverse. Generally, the mouthparts are adapted for chewing, but a few are adapted to take up liquid. The antennae are moder-ately long to quite long. Sometimes the ovipositor is well developed, and sometimes a modified ovipositor, the **sting**, is present for defense. Hymenoptera are holometabolous. The larvae have well-developed heads (unlike most flies), but the legs are variable. Some Hymenoptera have maggot-like, legless larvae, whereas others (sawflies) have both true legs and more than five pairs of prolegs. It is difficult to generalize about the importance of insects of this order. Bees and parasitic wasps are often highly regarded as beneficial to humans and wildlife, ants can be highly regarded (as predators and as food) or viewed as a nuisance (tending honeydew-producing insects or invading homes), and sawflies generally are regarded as plant pests but can be important wildlife food. This order is best known as the group with the most advanced insects, as they display the highest degree of sociality. Among the Hymenoptera important to wildlife as food are:

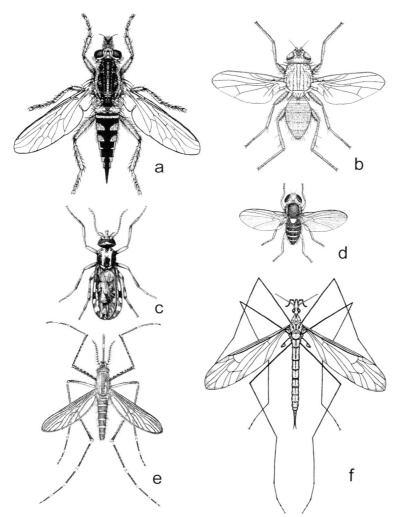

Fig. 5.28. Some representative flies: (a) a robber fly, *Cyrtopogon lateralis* (Diptera: Asilidae); (b) an anthomyid fly, *Pegomya betae* (Diptera: Anthomyiidae); (c) a biting midge, *Culicoides furens* (Diptera: Ceratopogonidae); (d) a leafminer, *Liriomyza trifolii* (Diptera: Agromyzidae); (e) a mosquito, *Culex tarsalis* (Diptera: Culicidae); (f) a crane fly, *Tipula paludosa* (Diptera: Tipulidae).

Sawflies (family Tenthredinidea and others)
Sphecids (family Sphecidae)
Bees (family Apidae and others)
Paper wasps (family Vespidae)
Ants (family Formicidae)

SUMMARY

• The aquatic insects most important as food for wildlife include: mayflies (Ephemeroptera), stoneflies (Plecoptera), dragonflies and damselflies (Odonata), bugs (Hemiptera), alderflies and allies (Megaloptera), beetles (Coleoptera), flies (Diptera), and caddisflies (Trichoptera).

• The immature (nymphal/larval/pupal) stages are often most important to fishes, but the adult stage of aquatic or terrestrial species may also be quite important as a food resource. Birds also may take advantage of the emergence of the adults of aquatic species.

• The terrestrial insects most important as food for wildlife include: termites (Isoptera), cockroaches (Blattodea), grasshoppers and allies (Orthoptera), earwigs

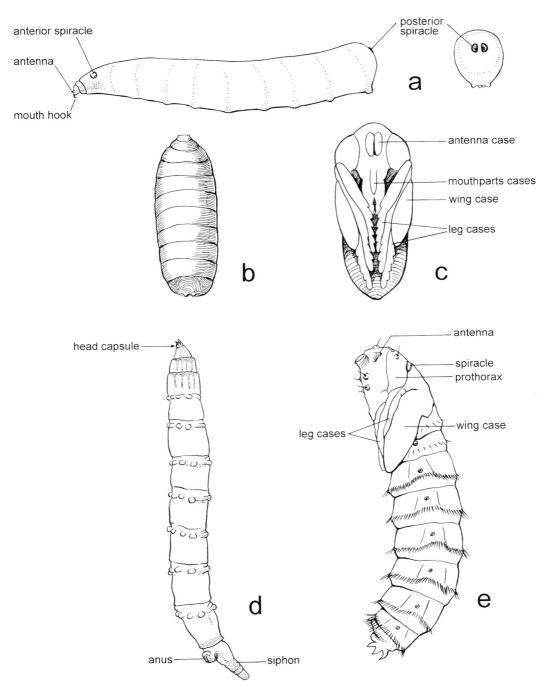

Fig. 5.29. Immature life forms of some flies: (a) larva, (b) puparium, and (c) pupa of house fly, *Musca domestica* (Diptera: Muscidae); (d) larva and (e) pupa of a horse fly (Diptera: Tabanidae).

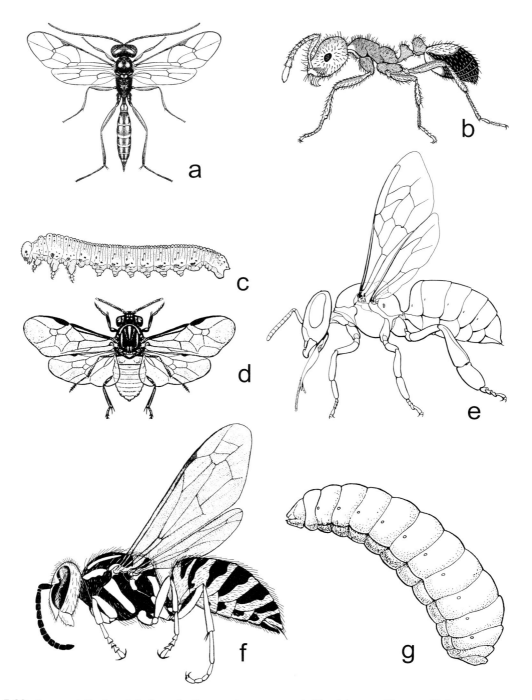

Fig. 5.30. Representative insects in the order Hymenoptera, wasps and allies: (a) a parasitic wasp, *Olesicampe* sp. (Hymenoptera: Ichneumonidae); (b) red imported fire ant, *Solenopsis invicta* (Hymenoptera: Formicidae); (c) larva and (d) adult of a sawfly, *Sterictophora cellularis* (Hymenoptera: Argidae); (e) adult and (g) larva of a honey bee, *Apis mellifera* (Hymenoptera: Apidae); (f) a paper wasp, *Vespula squamosa* (Hymenoptera: Vespidae).

(Dermaptera), bark lice or psocids (Psocoptera), bugs (Hemiptera), lacewings and allies (Neuroptera), beetles (Coleoptera), moths and butterflies (Lepidoptera), flies (Diptera), and wasps and allies (Hymenoptera), particularly ants.

REFERENCES AND ADDITIONAL READINGS

Arnett, R.H. (2000). *American Insects: A Handbook of the Insects of America North of Mexico*, 2nd edn. CRC Press, Boca Raton, Florida, USA.

Arnett, R.H. & Thomas, M.C. (2001). *American Beetles*, vol. 1. CRC Press, Boca Raton, Florida, USA.

Arnett, R.H., Thomas, M.C., Skelley, P.E., & Frank, J.H. (2002). *American Beetles*, vol. 2. CRC Press, Boca Raton, Florida, USA.

Capinera, J.L., Scott, R.D., & Walker, T.J. (2004). *Grasshoppers, Katydids, and Crickets of the United States*. Cornell University Press, Ithaca, New York, USA.

Cranshaw, W. (2004). *Garden insects of North America*. Princeton University Press, Princeton, New Jersey, USA.

Commonwealth Scientific and Industrial Research Organization (CSIRO) (1991). *The Insects of Australia*, 2nd edn. Cornell University Press, Ithaca, New York, USA.

Dindal, D.L. (ed.) (1990). *Soil Biology Guide*. John Wiley & Sons, Ltd, New York, USA.

Henry, T.J. & Froeschner, R.C. (eds.)(1988). *Catalog of the Heteroptera or True Bugs of Canada and the Continental United States*. E.J. Brill, Leiden, The Netherlands.

Johnson, N.F. & Triplehorn, C.A. (2004). *Borror and DeLong's Introduction to the Study of Insects*, 7th edn. Thomson Brooks/Cole, Belmont, California, USA.

Merritt, R. W. & Cummins K.W. (1989). *An Introduction to the Aquatic Insects of North America*, 2nd edn. Kendall/Hunt Publishing, Dubuque, Iowa, USA.

McAlpine, J.F. (ed.) (1981). *Manual of Nearctic Diptera*, vol. 1. Canadian Government Publishing Center, Toronto, Canada.

McAlpine, J.F. (ed.) (1987). *Manual of Nearctic Diptera*, vol. 2. Canadian Government Publishing Center, Toronto, Canada.

McAlpine, J.F. (ed.) (1989). *Manual of Nearctic Diptera*, vol. 3. Canadian Government Publishing Center, Toronto, Canada.

Mullen, G. & Durden, L. (eds.) (2002). *Medical and Veterinary Entomology*. Academic Press/Elsevier, Amsterdam, The Netherlands.

Pennak, R.W. (1978). *Fresh-Water Invertebrates of the United States*, 3rd edn. John Wiley & Sons, Ltd., New York.

Stehr, F.W. (1991). *Immature Insects*. vol. 2. Kendall/Hunt Publishing, Dubuque, Iowa, USA.

Stehr, F.W. (2007). *Immature Insects*, vol. 1. Kendall/Hunt Publishing, Dubuque, Iowa, USA.

Wagner, D.L. (2005). *Caterpillars of Eastern North America*. Princeton University Press, Princeton, New Jersey, USA.

Ward, J.V. (1992). *Aquatic Insect Ecology. 1. Biology and Habitat*. John Wiley & Sons, Ltd., New York, New York, USA.

Williams, D.D. & Feltmate, B.W. (1992). *Aquatic Insects*. CAB International, Wallingford, UK.

Chapter 6

INSECTS AND ECOSYSTEMS

An ecosystem is a fundamental unit of ecological organization. It is usually defined as the community of living organisms, as well as the abiotic (nonliving) environment, occurring in an area. The scale of an ecosystem is variable, so 'area' can be fairly small or quite large. Ecosystems do not operate independently; rather, they are interconnected, as are the elements operating within the ecosystem. Ecosystems have several functional groups, or levels of feeding, which are usually called **trophic levels**. The interaction of the functional groups is often portrayed by **food webs** (sometimes called **food chains** or **trophic pyramids**) which are diagrams showing the feeding links between elements of an ecosystem (see Fig. 6.1 for an example). When the abundance or activity level of one element is modified, it affects other elements. When a predator depresses the activity of its prey, this may release a lower tropic level from predation, allowing it to increase in abundance; this is called a **trophic cascade** because the effects can cascade through the food web. Each ecosystem has primary producers (**autotrophs**) and several consumers or secondary producers (**heterotrophs**). Autotrophs usually are plants, which create large quantities of matter via photosynthesis. Heterotrophs can be divided into several types, depending on their source of food. Primary consumers eat primary producers (usually plants), secondary consumers eat primary consumers, tertiary consumers eat secondary consumers, and quaternary consumers eat tertiary consumers. Insects are usually primary or secondary

consumers, but also function as decomposers or detritivores by breaking down dead plant and animal matter. Wildlife can function at all levels of consumption. Sometimes the same animal can function at more than one level, depending on its behavior. For example, when a blue jay, *Cyanocitta cristata*, feeds on acorns it is a primary consumer; when it feeds on oak leaf rollers such as *Archips semiferana* (Lepidoptera: Tortricidae) it is a secondary consumer; when it feeds on the eggs of house wren, *Troglodytes aedon*, it is a tertiary consumer.

Ecosystems display both primary and secondary productivity. **Primary productivity** is the production of organic compounds from carbon dioxide, normally via photosynthesis. This is accomplished by primary producers, or autotrophs, which generally are plants. **Net primary productivity (NPP)** is the rate of conversion of carbon dioxide from the atmosphere or water into plant tissue less the costs of respiration or maintenance of the plants. This is an important measure of ecosystem production because it is this activity that determines what is available for use or consumption by animals. The rate of ingestion of food and conversion of food into energy by consumers is called **secondary productivity**, and is accomplished by both herbivores and carnivores, or heterotrophs. Insects and other invertebrates are important and energetically efficient heterotrophs. It is noteworthy that about 20% of the energy assimilated by invertebrates is reflected in secondary production (growth), whereas only about 2% of assimilated energy in mammals and birds is reflected in growth. The difference is that warm-blooded animals (**endotherms**) expend more energy in maintenance costs (e.g., temperature regulation) than do cold-blooded animals (**ectotherms** or **poikilotherms**). So ecosystems dominated by ectotherms (invertebrates,

Insects and Wildlife: Arthropods and Their Relationships with Wild Vertebrate Animals, 1st edition. By J.L. Capinera. Published 2010 by Blackwell Publishing.

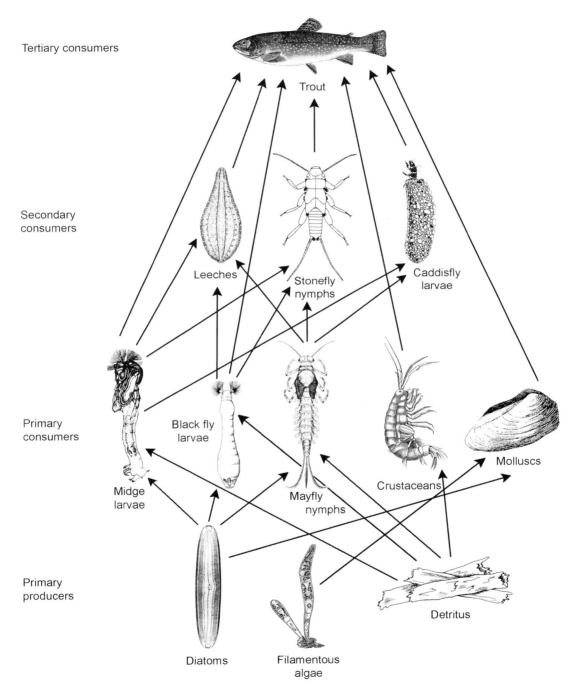

Tertiary consumers

Trout

Secondary consumers

Leeches

Stonefly nymphs

Caddisfly larvae

Primary consumers

Black fly larvae

Midge larvae

Mayfly nymphs

Crustaceans

Molluscs

Primary producers

Diatoms

Filamentous algae

Detritus

Fig. 6.1. A representative aquatic food web showing some of the important links among producers and consumers.

amphibians, reptiles, and fish) are more productive, and can pass along more energy to higher trophic levels.

Products of ecosystems include energy, water, nutrients, detritus, and living organisms. The availability of these products helps to define the characteristics of the ecosystem, though typically they are recognized principally by the dominant vegetation (e.g., grassland, coniferous forest) or type of water (e.g., fresh-water pond, marine estuary) that is present. Various processes occur within and between ecosystems, mostly nutrient and energy **fluxes**, and these fluxes can be affected by organisms, including insects. Nutrient and energy fluxes are the amount of nutrients or energy flowing through an area per unit of time. Here we are concerned mostly with how insects affect ecosystem structure and function, and thereby wildlife populations. The roles of insects in ecosystems are often overlooked or underestimated. There are several reasons for this, including:
• the total biomass of insects is small compared to some vertebrate animals and to plants;
• the average amount of net primary productivity consumed by insects is relatively small (usually 10% or less);
• ecosystem-level effects can only be quantified in long-term studies, but this is rarely done.
Despite the lack of appreciation for insects as important determinants of ecosystem function, insects do have important roles. In Chapter 3 we saw how many forms of wildlife benefit from the conversion of plant material into more nutritious animal protein, so nutrient conversion is an important function of insects. Among the other important roles are decomposition, nutrient cycling, herbivory, pollination and seed dispersal. Finally, anyone who has witnessed insect 'outbreaks' understands how under some circumstances insects can entirely change the landscape and drastically affect the ecosystem. During **outbreaks**, insects become abnormally abundant and can consume up to 100% of the standing crop, resulting in large and immediate effects on nutrient fluxes. **Outbreak insects** are species that become very abundant, sometimes for several years, and feed extensively, often modifying the ecosystem. Outbreaks typically cause nutrient and energy **pulses**, or the relatively instantaneous change in the normal level of fluxes. Although appearing destructive, the result of insect feeding can be quite positive by causing significant increases in nutrient availability and light, allowing growth of younger or early-succession plants. Examples include spruce budworms, *Choristoneura* spp. (Lepidoptera: Tortricidae), in spruce and fir forests, pine beetles, *Dendroctonus* spp. (Coleoptera: Curculionidae: Scolytinae), in pine forests, and grasshoppers, various species (Orthoptera: Acrididae), on grasslands. Outbreaks are most common when susceptible hosts become dominant in the ecosystem. Most outbreaks occur on a local scale, and escape much notice, but outbreaks are very common phenomena. By suppressing, eliminating, or changing the age structure of plant communities, such insects have profound influences on ecosystems. Even if they do not cause permanent changes in ecosystem flora, they may have significant temporary impact on the resident wildlife fauna.

Aquatic insects are responsible for a large portion of the secondary productivity that occurs in aquatic environments and wetlands. Insects feed on green plant material, detritus, zooplankton, and other insects, and in turn are eaten by fish, amphibians, and water-dwelling birds (Fig. 6.1). Aquatic insects help to maintain water quality by consuming algae and aquatic vegetation, but also are sensitive to the quality of the environment. Water quality and cleanliness directly affect the abundance of certain aquatic insects, so they are a valuable index of water quality, and are often used to judge the health of water bodies and wetlands.

Less obvious but also very important to consumers are the population increases of aquatic insects, though they are not called 'outbreaks'. They can result from various changes in hydrology and nutrient stimuli. One of the most interesting is the post-reproductive death of anadromous fishes such as salmon, *Oncorhynchus* spp. Such fish are born in fresh water, live a portion of their lives in salt water, and return to fresh water to spawn and die. Their decaying bodies are a tremendous nutrient input for streams and rivers where they occur. Many organisms, including some aquatic insects (mostly midges [Diptera: Chironomidae] and baetid mayflies (Ephemeroptera: Baetidae), respond to such nutrient fluxes with higher reproductive rates, leading to greater abundance. Similarly, the bodies and excrement (feces) of insects also provide a nutrient pulse in the aquatic environment.

Spawning salmon are prey for bears, *Ursus* spp., which remove large numbers of fish from streams during their feeding frenzy and deposit portions of fish or entire fish carcasses in adjacent terrestrial habitats. In British Colombia, Canada, more than 90% of such carcasses were colonized by flies, mostly blow flies

(Diptera: Calliphoridae). Along the streams studied, these carcasses produced about 196–265 grams of blowfly larvae per meter of stream length. The flies developing from salmon carcasses supplemented the diet of at least 16 vertebrate and 22 invertebrate species in the immediate area. Thus, flies also are an important vector for salmon-derived nutrients, much of which were ultimately derived from the marine environment, but which support growth and survival of terrestrial organisms not obviously linked to the fish.

In many areas, aquatic midge larvae and emerging adults are important food for fish, but most midges escape fish predation and emerge in very large numbers. The adults then become prey for predators such as birds, bats, spiders and insects in terrestrial areas near the emergence sites. However, the bodies of dead midges also are an important food source for decomposers and provide nutrients, particularly nitrogen and potassium, for plants. Thus, the entire ecosystem benefits from midge-derived nutrient flux.

The effects of continuous, low-level occurrence of insects are less apparent and less well documented, but probably equally important. In many cases, the activities of insects such as cicadas, ants and termites are punctuated by synchronized emergence that provides yet additional nutrient pulses as insects move from the belowground to the above-ground environment. Though not as impressive as the actions of outbreak species, or as fish-derived aquatic nutrient pulses, they nevertheless can be significant at the local level. Wildlife, especially birds, are quick to notice and take advantage of the sudden availability of numerous winged insects.

INSECTS AND DECOMPOSITION

The process of decomposition is basically the degradation of a dead organism or its remains into the parts or elements that comprise it. Eventually the organism cannot be recognized and its complex organic molecules are broken down into inorganic elements. Sometimes decomposition is divided into two parts, the initial destruction phase and the subsequent degradation phase. In the **destruction phase**, mechanical breakdown of the dead organism occurs, resulting in smaller-sized particles. In the **degradation phase**, the particles are degraded further into molecules, releasing mineral salts, carbon dioxide, and water. Insect **decomposers** play a major role in the decomposition process, but

especially in the destruction phase. The arthropod taxa involved in decomposition are numerous, but consist principally of millipedes (Diplopoda), woodlice (Isopoda), mites (Acari), springtails (Collembola), beetles (Coleoptera, especially Scarabaeidae, Geotrupidae, Silphidae), and flies (larvae of Diptera, especially Calliphoridae, Sarcophagidae, Muscidae, Faniidae). When a full complement of decomposers exists, and the weather is suitable, decomposition proceeds rapidly. When elements are missing, however, bottlenecks in recycling can develop, resulting in disruption of ecosystems. A classic example of a bottleneck is the introduction of sheep and cattle to Australia, which lacked native hooved mammals. The excrement (dung) produced by millions of grazing mammals lacked appropriate entomofauna for rapid decomposition, and the dung accumulated. Not only did dung matter accumulate, smothering vegetation and reducing availability of plant matter to wildlife and livestock, but the dung was suitable for the production of large numbers of dung-breeding flies, which became a serious nuisance. Introduction of dung-feeding insects from South Africa and elsewhere helped to speed up the process of dung decomposition, boosting forage production and decreasing the number of dung-breeding flies.

Insects play important roles in both belowground and aboveground decomposition, and the above- and belowground decomposers affect each other. In many ecosystems, the most above-ground net primary production enters the soil as plant litter, fueling large populations of **detritivores** (animals that feed on plant litter, or detritus). The quality of the litter thereby affects the belowground processes. Often overlooked is the fact that high levels of net primary productivity occur belowground, thus providing a significant energy source for consumers. Although the nutritive value of root and mycorrhizae (plant-dwelling fungi) tissue is often less than that of foliar tissues, in some systems (grasslands, for example) there is more biomass belowground. Soil-dwelling arthropods not only facilitate breakdown of organic matter, but affect its distribution in the soil profile, water holding capacity, and nutrient availability.

Decomposition of Plant Remains

All sources of carbon are not equally digestible to decomposers. Dead plant material contains high proportions of lignocellulose (lignin, cellulose, and hemicellulose) which are generally indigestible by animals but digestible by microbial organisms. A major function of insects is to break up organic matter and make it accessible to soil-inhabiting microbes. Decomposition of wood, for example, is enhanced by the attack of bark beetles and other wood boring insects that penetrate the outer bark, vector decay fungi and other microorganisms, and initiate log fragmentation. Fungi, particularly basidiomycetes, are the primary agents of wood decay in most forests. However, some termites (Isoptera) and cockroaches (Blattodea) possess enzymes capable of degrading cellulose and hemicellulose, and termites and cockroaches also possess symbiotic relationships with microorganisms that dwell in their digestive tract, facilitating digestion of wood and litter. Lignin is less digestible by insects, but they serve an important role by fragmenting wood and increasing the surface area to make it more accessible to microorganisms that can digest it. Whether or not degradation of lignin by insects is direct or indirect can be debated, but certainly lignin is affected as it is passed through termites, so they must be considered important. Increasingly, there is evidence that termites produce lignases.

Digestibility-reducing compounds are sometimes found in plant foliage. Some leaf material contains phenolic compounds and tannins, for example, which resist herbivore digestibility and microorganism decomposition. Herbivores sometimes encounter difficulty in obtaining energy via digestion of such plants. Many sources of carbon cost more energy to digest than is obtained by feeding. Often, foliage that is difficult for herbivores to digest is also difficult for decomposers to digest, resulting in some leaves that persist for long periods whereas others break down quickly. Woody plants tend to have higher levels of digestibility-resistant substances than do herbaceous plants. The less digestible plants often are avoided by insects, which in turn ties up nutrients in insect and decomposer-resistant leaf litter. When only the more readily consumed and digested plants are consumed, this can result in significant change of the floral composition of plant communities, sometimes making it less suitable for wildlife browsers. On the other hand, early-succession plants are both more palatable and digestible to herbivores and more digestible to decomposers. So when insects radically modify a plant community, as by killing trees, or not so radically, as by altering the competitiveness of plants, they can dramatically alter the suitability of the vegetation to both invertebrate and vertebrate heterotrophs, sometimes enhancing suitability to wildlife.

In some ecosystems, millipedes (Diplopoda) and woodlice (Isopoda) are important as decomposers, specializing in leaf litter decomposition. Although symbiotic micro-organisms may be present within some of these arthropods, others take advantage of extraintestinal digestion that occurs when the organic matter is attacked by fungi. Insects such as *Dendroctonus* spp. bark beetles, and *Xylosandrus* spp. and *Platypus* spp. ambrosia beetles (Coleoptera: Curculionidae) as well as *Atta* spp. and *Acromyrmex* spp. fungus-tending ants (Hymenoptera: Formicidae) are examples of insects that take advantage of the digestive powers of extraintestinal microorganisms, though the insects are responsible for transport of the fungus. Conversely, wood ants (*Formica rufa*) create large mounds of organic material in northern forests. Decomposition and mineral release are lower in mounds maintained by these ants, likely due to the dryer conditions of the mound structure. In aquatic systems, the microbial organisms present in decaying leaf detritus not only help to degrade the complex carbon sources into more digestible products but are an important source of nutrition for fish.

The association of microorganisms with insects extends to fruit-feeding insects as well. In this case, it usually is yeasts that cause fermentation of the fruit, allowing fruit flies (Diptera: Drosophilidae and Tephritidae), wasps (Hymenoptera), and sap beetles (Coleoptera: Nitidulidae) to use this resource. These insects ingest the microbes in addition to the fermenting fruit, which has nutritional advantages for the insects.

Decomposition of Excrement (Dung)

The dung of herbivores, being partly digested material, is a rich source of nutrients for decomposers, and so it usually attracts a large number of insects. It is rich both in organic matter and moisture. Although the dung of carnivores might be expected to be an even richer food source than herbivore dung for dung-feeding insects due to the nutritious food consumed by carnivores, it is actually less suitable for the development of decomposers. This is because carnivores are more efficient at extracting nutrients from their food than are herbivores. Dung feeding insects are said to be **coprophagous**.

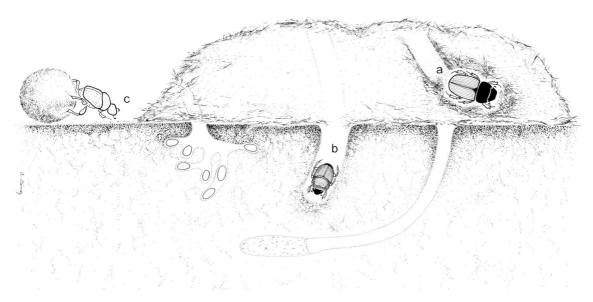

Fig. 6.2. Behavior of some dung beetles (Coleoptera: Scarabaeidae). Dung beetle behavior is both specialized and diversified, allowing for various methods of dung resource use. Beetles in the subfamily Aphrodiinae do not make nests, dwelling instead in the dung pat (a). Beetles in the subfamily Geotrupiinae and portions of Scarabaeinae produce tunnels (b). Some Scarabaeinae are dung rollers (c), making a ball of dung that they roll away from the dung pat (from Galante and Marcos-Garcia 2008; used with permission of the authors).

Dung is a transient resource. It dries quickly in most environments, and is rapidly exploited by several types of arthropods, often disappearing within a few days. Thus, it is important that dung feeders be able to sense the presence of fresh dung and move quickly to take advantage of this fleeting resource. Dung feeders also must be very competitive. The strategy of many flies is to develop rapidly and to convert the dung into fly bodies quickly. In contrast, the strategy of dung beetles is to carry away or bury the dung before other insects can eat it. Both strategies can work quite well, though neither guarantees that there will be adequate food for the offspring of these insects.

The first insects to arrive at freshly deposited dung typically are flies (Diptera), particularly those in the families Muscidae and Scatophagidae, but also Faniidae and Calliphoridae. Often flies arrive at dung within minutes of it being deposited. The flies feed and oviposit (or larviposit) on the dung, but many adults also feed on nectar or blood. Most fly larvae found on dung feed on this resource, but some larvae are predatory on other flies in the dung. Following the arrival of flies, next to be attracted are beetles (Coleoptera), particu-

larly the family Scarabaeidae. Among the most important are the dung beetles. Dung beetles bury portions of the dung pat or roll pieces of dung into balls that are buried in the vicinity of the pat (see Fig. 6.2). The piece of dung is inoculated with an egg by the female, and the larvae develops to adulthood on this fecal resource. Within some groups of dung beetles, the parents provide protection and care to their young. Dung beetles can be very important and numerous. For example, a 1.5 kg (3.3 lb) heap of elephant dung in East Africa attracted 16,000 dung beetles that ate, buried, or rolled away the dung in two hours! Of course, Africa is known for it huge herds of large mammals, which produce large amounts of dung as they graze the grasslands, so dung-feeding insects would be expected to be abundant there. On the other hand, in places like South America that lack large native herbivorous animals, the dung beetle population is much more limited. Another taxon to exploit dung is the mites (Acari). Several families of mites are involved, and many function as predators of flies, but some also feed on fungi and bacteria on the dung, or on the remains of the dung pat. In some areas of the

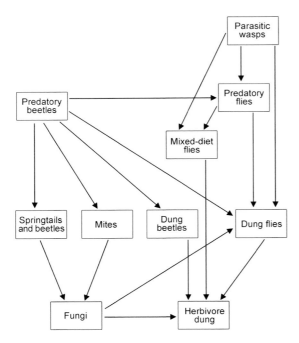

Fig. 6.3. The food web of herbivore dung in England, showing only the most important arthropod interactions (adapted from Hanski and Cambefort 1991).

world such as the steppe (shortgrass) region of North America, termites exploit the dung resources, tunneling to the surface to degrade the remains. The interrelationships of some insects associated with dung are shown in Fig. 6.3.

Decomposition of Carrion

Like dung, the bodies of dead animals (carrion) are a rich but unpredictable and transient resource. An ecologically significant portion of energy, possibly 40%–64% of energy transfer in ecosystems, may pass through carrion. There is intense competition among vertebrates, arthropods, and microbes for carrion, especially in warm climates, where microbes and arthropods develop quickly. Microbes use toxins and substrate decomposition to inhibit consumption by other scavengers, but microbes are relatively immobile so they benefit from the transport provided by arthropods. Indeed, arthropods increase the rate of carcass

decomposition by microbes by tunneling into tissue and allowing access by microbes, and through aeration of the carrion. When arthropods are experimentally excluded from a dead animal, bacteria cannot fully metabolize them, and the carcass will mummify rather than decompose. As carrion decomposes, it emits odors (e.g., hydrogen sulfide and putrescine) that initially are attractive to decomposers. As it decomposes, it eventually produces a variety of amines and sulfur compounds that signal to vertebrate scavengers that it is no longer edible. Arthropods are not deterred by even the late stages of decomposition, though the species using the resource change as decomposition proceeds.

Perhaps even more than with dung, carrion-feeding insects can sense carrion and move rapidly to exploit the resource. A number of chemicals associated with decay are attractive to insects, and help determine the sequence of arriving species. Some flesh flies (Diptera: Sarcophagidae) deposit larvae instead of eggs, a rather clever way to give their young a competitive advantage over egg laying insects. An alternative approach of 'edging out' the offspring of competitors is seen in some *Nicrophorus* beetles (Coleoptera: Silphidae) that transport small mites as they move about. The mites leave their host when new carrion is available and move onto the dead animal to feed on the eggs and young larvae of flies. Remarkably, after the beetles have completed brood rearing and are ready to seek new carrion, these symbiotic mites climb back aboard their hosts and are ready to exploit a new source of food.

As is the case with dung, the sequence of insects exploiting the animal resource begins with flies (Diptera), usually in the families Calliphoridae and Muscidae (see Fig. 6.4). Often a second wave consisting of flies in the family Sarcophagidae, and additional species of Calliphoridae and Muscidae, follow a few hours later. Predatory beetles and even predatory flies soon start to arrive, primarily to take advantage of the availability of fly larvae as food. The families of beetles (Coleoptera) represented are Staphylinidae, Histeridae, and Silphidae, although they also feed on the carrion (see Fig. 6.5). A third wave of decomposers begins as the dead animal exhibits significant decay, and is represented by flies in the families Phoridae, Drosophilidae, and certain Syrphidae. A fourth wave of insects again consists of flies, including cheese skippers (Piophilidae), and then the fifth wave, or final arrivals, typically consisting of beetles (Coleoptera:

Fig. 6.4. Adult and larva of carrion-feeding blow fly (Diptera: Calliphoridae) on a dead lizard. Blow flies constitute the first wave of arthropods exploiting carrion (from Galante and Marcos-Garcia 2008; used with permission of the authors).

Dermestidae, Trogidae, Cleridae) and tineid caterpillars (Lepidoptera: Tineidae) clean up the dry tissues, hair, and feathers.

Despite the large number of insects attracted to carrion, it is principally the calliphorid fly larvae that perform the critical role of opening up the body, and by secreting enzymes, facilitate liquifaction and growth of microorganisms (Fig. 6.6a). Some of the microorganisms developing in carrion produce toxins that affect wildlife, but do not harm insects. Thus, although carrion is an important food resource for some wildlife, it can also be a source of disease-causing organisms or toxins that affect wildlife. The rapid decomposition of dead animals by insects (Fig. 6.6b) can prove to be useful by quickly eliminating disease-infected wildlife, perhaps helping to reduce the spread of disease or poi-

soning of vertebrate scavengers. They also eliminate bodies that would otherwise smother vegetation, and return organic matter to the soil. In the case of carrion beetles (Coleoptera: Silphidae), their tunneling increases aeration and porosity of soil, and takes nutrients deep into the soil. Importantly, the conversion of cadaver tissue to insect tissue means that the nutrients are preserved and available to insectivorous wildlife. Many animals, but particularly birds, consume the maggots and other insects found on or near carrion. Occasionally, this can be a problem (e.g., avian botulism, Chapter 8), but usually this is a rich source of nutrition.

An example of the sequence of carrion-frequenting arthropods is shown in Table 6.1. Here you can see the pattern of attraction of insects to carcasses of herring

Fig. 6.5. Male and female *Nicrophorus* burying (or carrion) beetles (Coleoptera: Silphidae) burying a dead mouse beneath the soil. These insects are specifically adapted for living in carrion. They produce secretions that are applied to the surface of the carrion to deter the development of microorganisms while they proceed with the burial, as this limits the development of the decay odor that would attract other insects, and produce competition for their young. The adults also provide care for the larvae, including feeding them with regurgitated food (from Galante and Marcos-Garcia 2008; used with permission of the authors).

gull, *Larus argentatus*, and black-backed gull, *Larus marinus*, in two separate but nearby habitats along the seacoast of Maine, USA. In most respects, the pattern of attraction is typical, initially commencing with flies (Diptera: mostly Calliphoridae) but soon followed by beetles (Coleoptera: Staphylinidae, Histeridae, Dermestidae). However, you can also see some site-specific differences, namely the abundance of ants (Hymenoptera: Formicidae) and woodlice (Isopoda) in the wooded sites, and earwigs (Dermaptera) at the beach sites. Ants, and particularly woodlice, are common residents of mesic woodlands. The occurrence of earwigs might seem unusual, but the species involved (*Anisolabis maritima*), commonly is found beneath accumulated seaweed, and forages on marine carrion.

NUTRIENT CYCLING

Insects involved in decomposition affect principally carbon and nitrogen cycling. When the subject of nutrient cycling and insects is raised, most people think first of termites. Indeed, termites are critically important in the recycling of ligno-cellulosic material bound up in trees into soil components, including nutrients. Termites produce both carbon dioxide (CO_2) and methane (CH_4) as products of digestion, so they can release considerable amounts of carbon. Insects are less important in releasing nitrogen, despite the vast amounts of food that they process, largely because they are so effective at capturing nitrogen from their food and incorporating it into their bodies. What this means, however, is that, when insects emerge from the soil (e.g., flights of termites, ants, or cicadas), there is a major redistribution of nitrogen from belowground to above-ground. A parallel process occurs in aquatic environments, with detritivores capturing energy and nutrition and converting it into the bodies of aquatic insects. Most aquatic insects are quite mobile in the adult stage, so they are capable of moving nutrients a considerable distance from water, providing nutrient pulses to terrestrial environments.

Conversion of wood into soil, though well appreciated, is only part of the nutrient cycling process involving insects. Less apparent but also important

Fig. 6.6. Decomposition of carrion by blow flies (Diptera: Calliphoridae): (a) after initially feeding on the head (left of photo), and leaving little but bones and teeth, the maggots have moved to the front shoulders of a hog. The maggot mass will eventually gain entrance to the interior portions of the cadaver, where they will consume, with the assistance of other insects that attack later in the sequence of decomposition, nearly all edible material. Thus, after perhaps two months and depending on weather, there is little but (b) bones and perhaps some skin and hair remaining. The thousands of maggots that developed on the carrion have been transformed into insects, which then become a source of food for wildlife (photos by S. Gruner).

is consumption of leaves, roots, and detritus. The mechanisms by which insect herbivory can cause changes in nutrient cycles include:

• herbivory can result in deposition of large quantities of excrement (insect fecal matter, usually called frass when produced in a dry form) onto litter and soil, increasing the rate of return of nitrogen from plants to soil;

• nutrients returned to the soil in insect cadavers are more easily decomposed than leaf litter, and during insect outbreaks a large number of insect cadavers are available for decomposition. Their presence may also stimulate litter decomposition;

• herbivory affects the chemistry and nutrient level of precipitation that drips from the foliage onto the soil. The water, known as 'throughfall', carries dissolved nutrients leaching from the leaves and from herbivore excrement on the foliage;

• herbivory can change the quantity and quality of the leaf litter falling to the soil surface by affecting

Table 6.1. Average abundance of carrion-infesting arthropods collected from gull carcasses in Maine, USA, in thicket and cobble beach habitats. The stages of decomposition are shown in the bottom line of this table. Calliphorid fly life stages are indicated by E for eggs, L for larvae, and P for pupae (adapted from Lord and Burger 1984).

	Postmortem interval (days)																
	0	2	4	6	8	10	12	14	16	18	20	22	24	26	28	30	32
Coastal thicket habitat																	
Calliporidae	12	28	19	27	16	4											
immatures	E	L	L	L	L	P	P	P	P	P							
Sarcophagidae		1	1														
Piophilidae				5	6	5	3	5	2		3	7	5	3		1	
Staphylinidae				9	22	9	45	28	15	12	8	3		5	3	2	2
Histeridae			3	15	2	1	8	7	2	1	3	3	1				
Dermestidae				10	8	15	26	11	19	20	23	28	41	26	20	20	20
larvae									2		40	25	41	20	29	31	34
Nitidulidae				5	4		27	17	8	6	5	13	10	9	7	6	1
Formicidae	104	98	73	81	115	160	127	117	57	63	75	63	56	62	50	41	22
Isopoda		36	48	93	108	139	82	46	41	44	67	62	63	63	56	50	24
Cobble beach habitat																	
Calliporidae	4			8	2												
immatures	E	L	L	L	L	L	P	P	P	P							
Sarcophagidae			1														
Piophilidae				1	1												
Staphylinidae			2	2	9	4	6	1	1								
Histeridae			1	8	5			1									
Dermestidae			2	19	23	25	44	44	22	41	29	27	29	19	19	5	
larvae				14	21	25	19	38	50	39	43	42	79	87	87	54	
Nitidulidae					4	2	16	9	3	6	9	6	3	1			
Formicidae														1			1
Dermaptera		2															
Stage of decomposition	Fresh						Active						Advanced			Dry	

the timing of leaf fall and the chemistry of the foliage;
• herbivory can affect root exudates and the interactions of exudates with soil microorganisms;
• herbivory can affect the structure of plant canopies, influencing the amount of sunlight reaching the soil, and thereby the temperature and moisture levels of the soil;
• herbivory may cause changes in plant community composition, thereby affecting nutrient uptake from the soil and the contribution of leaf litter to the soil.

Large-scale defoliation of plants results in increases in litter and insect excrement (frass) accumulations on the soil surface earlier in the season than would occur normally, though there is not necessarily a change in the overall level of nutrients. Litter and intact leaves do not decompose as fast as insect frass, leading to slower nutrient release. Spatial redistribution of nutrients also occurs, as when leaf-cutter ants (Hymenoptera: Formicidae) move foliar biomass to their underground fungus gardens. Organic leaf litter material that is digested by fungi is often grazed by springtails (Collembola), which often occur at very high densities. The fecal material of springtails is an excellent source of nutrients that can leach back into the soil as the fecal materials are digested by microbes. Feeding by springtails on mycorrhizae can sometimes be stimulatory, but at higher levels of feeding it can be deleterious to host plants.

The pattern of nutrient cycling attributable to insects is variable, and nutrients may flow in both directions. Decomposition processes occurring belowground affect the availability of carbon, the rate of plant respiration, and the occurrence of soil organisms. Aboveground feeding usually is considered to be deleterious to belowground activities, but herbivory sometimes seems to stimulate belowground processes. Herbivory also can affect the quality of the detritus being fed upon at the soil surface and belowground. Herbivory often increases nitrogen concentration in foliage, for example, thereby improving the quality of decomposing plant materials. Conversely, herbivory can sometimes cause increases in the concentration of phenolic compounds, retarding decomposition. Aboveground feeding also results in production of insect frass, sometimes at high levels. Increased availability of insect excrement (frass) has been shown to boost net primary production. However, it can also stimulate production of soil-dwelling microbes and thereby immobilization of nitrogen because it is incorporated in their bodies.

Overall, we must conclude that although insect herbivory can accelerate nutrient cycling and enhance net primary productivity (NPP), this does not always occur. The level of herbivory, the plant selection behavior of the insects, and the decomposition rates of the plants selected or not selected by the insects interact to make the outcome less predictable. Because the outcome is not independent of the environment (moisture, light, mineral nutrients, competition, etc.) it is difficult to generalize about the role of insects in nutrient cycling. Any significant changes in nutrient levels can be expected to affect nutrient levels in plants and structural characteristics of vegetation, however, both factors that will affect wildlife.

Forest insects occurring at low (nonoutbreak) levels of abundance seem to have only modest effects on forest ecosystems. Excrement from caterpillar larvae (Lepidoptera) and aphids (Hemiptera: Aphididae) that rains down on plant leaves and needles will enhance growth of epiphytic microorganisms (bacteria, yeast, fungi) growing there. Also, insect-infested trees will have higher concentrations of dissolved potassium, sulfur, calcium, organic carbon, and nitrogen leaching from leaves and needles relative to uninfested trees. However, the chemistry of the soil on the forest floor is not significantly affected by these inputs, demonstrating the great buffering capacity inherent in many ecosystems.

The degradation of grass litter by springtails (Collembola) and terrestrial isopods (Crustacea: Isopoda) affects the levels of nutrients in the soil. Feeding by these invertebrates causes increased levels of calcium, potassium, magnesium and nitrogen in soil, and the penetration of the nutrients into the soil profile is increased by the presence of these animals.

The construction of **termitaria**, or termite mounds, also affects wildlife. Some termites, especially African species in the genus *Macrotermes* (Isoptera: Termitidae), produce large above-ground mounds. Some *Macrotermes* mounds reach 6 m in height, 30 m in diameter, and consist of up to 27 m^3 of soil. Soil associated with the mounds is mineral-rich, containing high levels of nitrogen, calcium, magnesium, potassium and phosphorus. These soils also have higher levels of organic matter, higher pH, and moisture. Most important, unlike the mounds of most species, *Macrotermes* mounds support vegetation. Thus, areas populated by termites have greater biodiversity. For example, a study conducted in Zimbabwe found that more than 25% of the plants found on termite mounds were not present in nearby termite-free areas. This biodiversity extended to animal life as well, as mounds supported phytophagous insects, termite-feeding vertebrates, and granivorous and frugivorous animals attracted to the vegetation associated with mounds. This vegetation provided both food and shelter, and often supported evergreen species that provided islands of green vegetation throughout the year. The biomass of small vertebrates associated with termite mounds was about twice that found in surrounding areas, so mounds may serve as significant sources of food for predatory wildlife.

Grazing animals are also attracted to termitaria. Nutrient availability is often a limiting factor in mammalian productivity, and foraging behavior by ungulates is related to soil fertility. The woody vegetation growing on termite mounds not only has higher nutrient concentrations, but has lower levels of digestibility reducing substances such as tannins and lignins than nearby vegetation. Wildlife that graze preferentially on mound vegetation include impala, *Aepyceros melampus*; Burchell's zebra, *Equus burchelli*; Defassa waterbuck, *Kobus ellipsiprymnus*; topi, *Damaliscus lunatus*; bushbuck, *Tragelaphus scriptus*; warthog, *Phacochoerus africanus*; and elephant, *Loxodonta africana*. Although termite mounds comprise only 5%–7% of most areas, the vegetation associated with these mounds may provide over 25% of the vegetation consumed by elephants. This clearly demonstrates the enhanced palatability of vegetation associated with termitaria.

It is also worth mentioning that termitaria (Fig. 6.7) provide more than improved nutrition to wildlife. The mounds are exploited by many animals for nesting and resting sites. For example, in Panama the males of the bat *Lophostoma silvicolum* excavate cavities in termite mounds for roosting. Also, some Peruvian birds construct nests within the nests of arboreal termites, including *Brotogeris* spp. parakeets and black-tailed trogons, *Trogon melanurus*, as do some Australian parrots, *Psephotus* spp. Similarly, many lizards, amphisbaenians and snakes use termitaria for oviposition/egg incubation sites. For some lizards, the relationship seems to be obligate, but for other reptiles it seems to be facultative. Among monitor lizards, *Varanus* spp., an obligate relationship with termites is especially common. At the other extreme, an iguana in the Bahamas, *Cyclura cychlura cychlura* (known as Andros iguana) is the only iguanine that lays its eggs in termite mounds. Mounds are warmer and less variable in temperature and humidity, resulting in good hatching

success, but the habitat inhabited by these iguanas is also soil-deficient, so this may also help explain the unusual behavior. Andros iquanas are endangered, so the termites should be considered an essential element in conservation management. Andros iguanas, iguana eggs, and termitaria are shown in Fig. 6.8.

Ant mounds are also attractive oviposition sites for some reptiles, particularly amphisbaenians and snakes. Leaf-cutter ants, *Acromyrmex* spp. (Hymenoptera: Formicidae), are especially prone to construct environments that provide the high, relatively constant humidities that favor reptile eggs. Thermoregulation of nests is a common attribute of social insects, and leaf-cutter ants are well known to maintain a meticulous, hygienic environment, including destruction of alien fungal spores. Thus, there are several advantages for reptiles to occupy ant mounds. However, it may be difficult for some young reptiles to escape from ant (and termite) mounds once they hatch, so small species or species with sharp claws are favored. Several observations have been made to indicate that female monitor lizards will dig into mounds and create a tunnel, facilitating escape of her offspring, once hatching has occurred.

HERBIVORY BY INSECTS

Herbivory is the consumption of any plant part by animals; an alternative name is **phytophagy**. Insect herbivores have various ways to feed on plants, including consumption of foliage, blossoms, and fruit, or consumption of most of the leaf tissues but not the veins of the plant (this is usually called skeletonizing); sap removal, including sucking xylem or phloem; gall formation; tunneling in stem tissue, roots, or fruit (called boring); or beneath the epidermis of leaf tissue (called leaf mining). It is difficult to measure most impacts of herbivory. Defoliation and plant death are usually fairly apparent and easily quantified, but the effects of stem boring, root feeding and sap sucking require controlled experimentation to assess properly, so often we do not have good estimates of these effects.

The Importance of Herbivory

Herbivory is a major factor in ecosystem function. Herbivory influences the availability of food and habitat resources for heterotrophs such as insects and wildlife

Fig. 6.7. Examples of termitaria (termite mounds): (a) arboreal nest of *Microcerotermes* sp. (Termitidae: Nasutitermitinae) in a dry forest of Brazil; (b) Nest of *Nasutitermes* sp. (Termitidae: Nasutitermitinae) in an Amazonian grassland; (c) arboreal nest of *Constrictotermes cyphergaster* (Termitidae: Nasutitermitinae) in a Cerrado grassland; (d) nests of *Cornitermes cumulans* (Termitidae: Nasutitermitinae) in a Cerrado grassland; (e) nest of *Syntermes wheeleri* (Termitidae: Nasutitermitinae) in a pasture; (f) nest of *Macrotermes* sp. (Termitidae: Macrotermitinae), Ivory Coast. (photos a–e by Reginaldo Constantino; f by Rudi Scheffrahn).

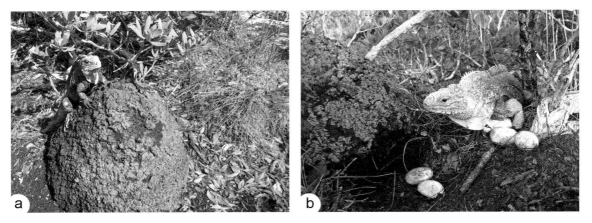

Fig. 6.8. Use of termitaria by the Andros iguana, *Cyclura cychlura cychlura* in the Bahamas: (a) adults climb and perch on the mound of *Nasutitermes ripertii* (Isoptera: Termitidae); (b) adults deposit eggs at the entrance to termite mounds, then push them into the central mound area, and finally cover the damage with back-fill consisting of carton material, soil, pine needles and other debris. The termite colony will soon encapsulate the chamber and repair the damage, providing a good environment for incubation of the eggs (photos by C.R. Knapp).

in addition to influencing nutrient cycling and plant productivity. In the Neotropics, for example, leaf-cutter ants (Hymenoptera: Formicidae) remove about 17% of the total leaf biomass from forests, more than any other taxon. In the Serengeti of Africa, insect herbivores remove about 6% of annual plant production whereas vertebrate herbivores (which are infinitely more visible) consume about 7%. Insect herbivory usually is relatively inconspicuous, but rarely absent. Levels of plant defoliation are variable, of course, but following are some additional examples of estimates from around the world: 21%–23% loss in herbaceous vegetation due solely to grasshoppers in western North America, 4%–8% loss in tropical grasslands in East Africa, 14% loss in tropical savanna in southern Africa, 15% defoliation of eucalyptus forests in Australia, 7%–10% loss in temperate forests of Europe, 2%–10% loss in temperate forests in North America, and 9%–12% loss in tropical evergreen forest of Southeast Asia. A survey of forest defoliation found that herbivory in tropical forests averaged 10.9%, whereas in temperate areas it was significantly lower, about 7.5%. Some have suggested that the impact of insect herbivores on ecosystem processes is far greater than that of vertebrate herbivores.

Does it seem illogical that insects and the herbivory they provide sometimes seems detrimental to ecosys-

tem functioning, when at other times it seems beneficial? How can it be that consumption by insect herbivores reduces plant abundance but at other times speeds up nutrient cycling, increasing plant abundance and fitness? As mentioned previously, the level or timing of herbivory, the plant community, and environmental factors interact to influence the outcome of herbivory. Following are some explanations for the apparent contradictions related to the benefits and detrimental effects of insect herbivory on ecosystem function.

The costs and benefits of herbivory have a strong temporal component. *When herbivory occurs* is important to the outcome. First, consider an example involving leaf and fruit production by an individual tree. Leaf removal by caterpillars from a fruit-producing tree occurring early in the growing season may inhibit the production of fruit because the foliage produces carbohydrates that are stored as fruit. Loss of foliage translates into less carbohydrate available to be stored in the fruit. On the other hand, if the leaf removal happens sufficiently late in the season, the fruits may already be fully formed, so fruiting may not be affected. However, the residual effect of leaf removal, whether occurring either early or late in the growing season, may be the reduction of fruiting the following year because the storage organs of the tree have been depleted by the previous season's defoliation. So our perception of the importance of herbivory is affected by the timing of the herbivory and the timing of assessment. Secondly, consider tree death due to attack by bark beetles, and the effects on the forest ecosystem. Initially, the destruction of mature trees reduces both biomass and net primary productivity in the forest ecosystem. However, eventually the decomposition of the wood by insects and fungi provides a large nutrient input. Young trees growing where old trees once grew are now freed from the water, nutrient, and shading constraints that existed in the mature forest, and can grow rapidly. So net primary productivity not only increases again, but may surpass that which existed in the forest when the trees were senescent. Again, the importance of herbivory is a function of when the process is examined.

Also influencing the assessment of herbivory is how the outcome is measured: *what outcomes are important to you*. In the example of tree death due to bark beetles, foresters may lament the loss of timber but wildlife managers may welcome the renewal of the forest.

The costs and benefits of herbivory are also affected by the *level of defoliation*. Net primary productivity (NPP) and biomass are normally considered to be inversely related to the level of herbivory. However, when experimental data are collected over a sufficiently wide range of insect feeding levels, and the data are analyzed properly, a non-linear relationship is often seen (see Fig. 6.9). Low levels of herbivory often do not result in decreased NPP or plant growth due to the compensatory abilities of the plants. **Plant compensation** is the ability to repair or recover from injury (see below). Plants that are fed upon by herbivores often replace their foliage or produce additional shoots, often causing a change in plant architecture but not necessarily inhibiting plant growth, and perhaps even stimulating growth (overcompensation for feeding injury). Affected plants may also respond chemically via induction of chemical antiherbivore defenses. These chemical defenses further limit the ability of the insects to feed, enhancing the apparent compensation abilities of the plants.

Insect feeding damage to plants is not always trivial, however, and the impact is not limited to the host. Severe defoliation or plant death can significantly affect the environment. For example, higher fuel load in forests following tree death can result in more fire damage. Plant loss can also produce changes in heat loss and light reflection, higher surface temperatures, elevated carbon dioxide levels, lower

Grazing optimization

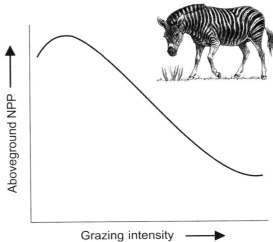

Fig. 6.9. A conceptual model of the relationship of herbivory to plant net primary productivity (NPP) or biomass, showing the non-linear relationship of plant response to herbivory and the small increase in NPP that is the basis for the grazing optimization hypothesis.

humidity and lower rainfall. Thus, herbivory is not strictly a local phenomenon because it can affect many ecosystem processes, including some on a regional basis.

Plant Compensation

To better understand the phenomenon of plant compensation, consider a scenario involving the mowing of turfgrass. If you mow your lawn occasionally, it continues to grow vigorously, and it grows thicker. In contrast, if you do not mow your lawn, it soon senesces, produces seeds, and stops growing. It is tall but not very thick. If you harvest the latter lawn you discover that you have produced less biomass per unit of area by letting the lawn mature early. In a sense, insect herbivory functions like a lawnmower, changing the physiology of the plant and keeping it in a more juvenile and rapidly growing condition. The net result, of course, is that over time proper mowing doesn't harm the turfgrass, and actually increases grass production. Similarly, low levels of insect herbivory not only fail to inhibit plant productivity, they may stimulate it. This explains the small elevation of NPP associated with low levels of herbivory shown in Fig. 6.9.

The concept that herbivory does not have a completely negative effect on plant productivity, and that plant productivity can be enhanced by low levels of herbivory, is called the **grazing optimization hypothesis** (also known as the herbivore optimization hypothesis). Basically, plant growth is thought to be optimized by some low level of herbivory (grazing). What that 'optimal' level might be is determined by several factors, including the nature of the plant, the timing of herbivory, and environmental conditions. Plants that are most capable of compensation tend to experience herbivory early in the season, experience little competition, grow quickly, have multiple meristems (growing points), and have adequate access to water, light, and nutrients. Plants less likely to compensate for herbivory tend to experience herbivory late in the season, suffer from high competition, grow slowly, have meristem limitations, and have inadequate access to water, light and nutrients. Thus, annual and biennial plants are especially resilient and tolerate grazing, whereas long-lived woody perennials are less tolerant.

While some have argued that plants have evolved to *benefit* from low levels of herbivory, hence the notion of 'optimization,' others have argued that plants simply are very good at compensating for injury, or resilient, and did not really evolve to benefit from herbivory. The controversy is due partly to misunderstanding or disagreement about what is important. Whether we are concerned about the growth of an individual plant or the function of the plant community determines whether herbivory can be considered beneficial. Similarly, whether we are considering growth or fitness (reproductive output) influences our assessment of herbivory. The question of plants benefiting from herbivory remains unresolved, but it is clear that low levels of feeding by insects usually do not damage plants or ecosytems. Thus, the average levels of herbivory in various ecosystems mentioned earlier likely are not damaging, though a robust population of insects is supported by such herbivory, in turn supporting wildlife populations.

Insect Outbreaks

Sometimes herbivores attain levels of abundance well above the long-term mean. Such occurrences are called **outbreaks** by entomologists. Outbreaks of phytophagous insects can induce significant changes in ecosystem nutrient flow and ecosystem structure. Outbreaks occur for several reasons, including:

• favorable environmental conditions allow unusually high insect survival or reproduction;
• environmental perturbations such as drought weaken plant defenses;
• excessively high plant density leads to competition for inadequate levels of nutrients and water, weakening plant antiherbivore defenses;
• plant senescence weakens plant antiherbivore defenses;
• plant homogeneity makes it easy for specialist herbivores to find food and reproduce, allowing population increase;
• invasion by non-native species that, lacking the full complement of their natural enemies, attain unusually high densities.

Outbreaks can cause partial or complete loss of foliage, or damage to the stems, trunk, or roots of plants. Often, feeding by insects causes leaf abscission, reducing plant productivity. The ratio of belowground to aboveground plant parts may be altered. Fungi and other plant pathogens may be better able to invade plants weakened by insect feeding.

Importantly, repeated defoliation may cause mortality among plants, and changes in availability of plant resources for wildlife. Plants usually are quite resistant to such damage. However, when defoliated for 3–5 years, especially if defoliation is accompanied by drought, plants may perish. This can result in decreased ecosystem productivity in the short run, but as plants are thinned and older plants are removed, the survivors may benefit and afterwards an increase in growth rate may occur. For example, following an outbreak of mountain pine beetle, *Dendroctonus ponderosae* (Coleoptera: Curculionidae: Scolytinae), the change from a mature even-aged pine forest to a stand of dead mature pines with an emerging understory of aspen, birch, and immature pines is a dramatic change for humans to behold. However, it is even more dramatic if you are an animal that is dependent on the old overstory (e.g., loss of pine cones for squirrels) or benefit from release of the understory (e.g., increased availability of browse for deer). In summary, herbivory by insects can have little or no effect on plant growth and ecosystem productivity, or it can induce quite important changes in structure and function of ecosystems, or something intermediate. As there is a continuum of herbivory from little to much, there is a continuum of effects from few to many.

As suggested above, herbivory also affects habitat suitability for wildlife. By altering the composition of the plant community, herbivory can make the habitat more or less favorable for vertebrates. For example, forest stand characteristics determine the species composition and abundance of birds in forest ecosystems. Early-succession conifer forests have very diverse and abundant bird communities because they have a dense shrub understory (see Fig. 6.10). As the forest matures, shrubs disappear due to shading, and bird populations then decline. The reduction in bird abundance and the increased stand density and maturity can result in increases in conifer-feeding insects such as spruce budworms, *Choristoneura* spp. (Lepidoptera: Tortricidae). However, if the insect population is unchecked, the budworms can kill the trees, allowing shrubs to reinvade, and the forest once again supports a rich community of birds. A less dramatic but interesting effect of insect feeding involves the herbivory of pinyon pine trees by the caterpillar of *Dioryctria albovitella*

(Lepidoptera: Pyralidae). Destruction of terminal shoots and cones of pinyon trees not only modifies the architecture of the tree, making it shorter and thicker, but destroys the cone production, effectively making the trees male only. Trees lacking the female function are not attractive to the pinyon jay, *Gymnorhinus cyaocephalus*. The birds abandon groves of trees affected by such herbivory. On the other hand, grazing of grasslands by black-tailed prairie dogs, *Cynomys ludovicianus*, keeps the protein levels of plants elevated. This is not only suitable for the prairie dogs, but proves attractive for other grazers such as American bison, *Bison bison*, and pronghorn, *Antiolocapra americana*, which graze there preferentially.

PLANT DISEASES AND INSECTS

In addition to the direct effects of feeding insects on plants, insects and mites can affect plants indirectly by transmitting plant pathogens (plant disease agents). If the disease is sufficiently pathogenic it may kill the plant. The loss of individual plants may not greatly impact wildlife because they are mobile, but if key resource-providing species or major elements of the ecosystem are eliminated, wildlife will be seriously affected via reduction of food and **harborage**.

Insects often facilitate movement of pathogens, or entry of pathogens into plants, which then may result in expression of disease. Several types of plant pathogens can be transmitted by insects to plants, primarily viruses, bacteria, fungi, protozoa, and nematodes. Pathogens can be carried accidentally by external contamination of the insect's body (many bacteria and fungi), or internally by insects that are adapted to do so (all protozoa, many viruses, some fungi, bacteria, and nematodes). It has been estimated that 30–40% of plant disease results from the involvement of insects and mites.

In terms of ecosystem disruption by insect-transmitted plant pathogens, all types can be involved, but often fungi are most important. This might seem surprising to some, because viruses and bacteria are important pathogens of plants. However, from the perspective of natural ecosystem disruption, or elimination of dominant ecosystem flora, fungi tend to be involved more frequently. Consequently, beetles (Coleoptera) are commonly involved because although beetles are not well adapted to vector some diseases, they are often associated with fungi.

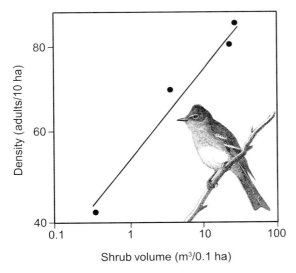

Fig. 6.10. The density of foliage-gleaning birds (adults/10 ha) as a function of shrub volume (cubic meters/0.1 ha) in a conifer forest in Washington, USA (adapted from Safranyik, L. (ed.) 1983). Insectivorous birds like this ash-throated flycatcher, *Myiarchus cinerascens*, are important consumers of foliage-feeding insects, often keeping insect populations from attaining damaging densities.

Most of the important fungal pathogens of plants are transmitted by bark (Curculionidae: Scolytinae) and ambrosia (Curculionidae: Platypodinae) beetles. However, the pine-killing woodwasp, *Sirex noctilio* (Hymenoptera: Siricidae) is an important exception. *Sirex* generally attacks pine plantations that are 10–12 years old. *Sirex* injects a toxic mucus as well as spores of the tree-pathogenic fungus *Amylostereum areolatum*, when feeding. After a few weeks or months the tree dies. The *Sirex* larvae feed on the fungi, a symbiotic relationship. Although mostly a pest in the southern hemisphere (Australia, New Zealand, South Africa, South America), *Sirex* has recently been introduced to North America.

Mountain pine beetle, *Dendroctonus ponderosae* (Coleoptera: Curculionidae: Scolytinae), is indigenous to the forests of western North America. Periodic outbreaks of this insect kill millions of trees, drastically altering the landscape. Mountain pine beetles affect pine trees, particularly ponderosa, lodgepole, Scotch, and limber pine by feeding on the phloem. Initially the trees that are attacked tend to be old, stressed or damaged trees, as these are most susceptible to attack. However, as the beetle population grows even healthy trees succumb to the beetles. Mountain pine beetles transport the spores of blue-staining fungi (mostly *Ceratocystis* spp.) on their bodies and in a special structure associated with their integument called a mycangium. Once inoculated into the tree by the tunneling activities of the beetle, the fungi develop and spread throughout the sapwood, interrupting the flow of water to the crown. The fungus also reduces the tree's flow of pitch, a resinous plant defense mechanism, thus increasing the ability of the beetle to overwhelm the tree. It is the combined action of both beetles and fungus that kills the trees. Other bark beetles have a similar relationship with blue stain fungi, including southern pine beetle, *Dendroctonus frontalis*; spruce beetle, *Dendroctonus rufipennis*; black turpentine beetle, *Dendroctonus terebrans*; and several engraver beetles, *Ips* spp.

Many different fungi can be transported by bark and ambrosia beetles, and the mycangia can be found in several different locations on the insect's body, depending on the species of insect. As noted previously, wood is a rather poor food resource, but partly decayed wood and the fungi themselves are more nutritious. Thus, it is advantageous for the beetles to transmit the fungi and to feed on it, and they are well suited to do so. The beetle-fungus relationship is therefore considered to be a case of mutualism.

The most important insect-vectored nematode affecting ecosystems is likely pinewood nematode, *Bursaphelenchus xylophilus*, the causative agent of pine wilt disease. This problem is mostly an issue in the pine forests of Asia, where the nematode is vectored by adults of the longhorned wood borer *Monochamus alternatus* (Coleoptera: Cerambycidae) which is commonly known as the Japanese pine sawyer. The nematodes are carried in the trachea of *M. alternatus* adults. When the adults feed on young pine twigs, nematodes leave the beetle and enter the tree through the feeding scars. The conductive tissues of the pine trees infected with nematodes become plugged, resulting in wilting and eventual death of the tree. Trees can be sprayed with antihelminth chemicals, but the high cost of such treatments limits its use to high-value sites.

Hardwood trees as well as pines are severely affected by beetle-borne fungi. Oak wilt fungus, *Ceratocystis fagacearum*, is a serious problem in the eastern USA. It affects oaks, of course, but also chestnuts and other trees. It is spread by sap beetles (Coleoptera: Nitidulidae), and possibly by other beetles, but also underground from tree to tree via root grafts. Dutch elm disease, caused by *Ophiostoma* spp. fungi, similarly is spread by both beetles (mostly *Scolytus* spp. bark beetles) and root grafting. It has virtually eliminated American elm from the North American and European landscapes. Chestnut blight fungus, *Cryphonectria parasitica*, has eliminated one of North America's most important trees, the American chestnut, from eastern North America. It can be spread by various means, including wind, birds, rain, and insects, and also affects oaks. Chestnut was not only a dominant overstory tree in much of eastern North America, but its large crop of chestnut fruit was an important food resource for many forms of wildlife, including white-tailed deer, *Odocoileus virginianus*; black bear, *Ursus americanus*; wild turkey, *Meleagris gallopavo*; gray squirrel, *Sciurus carolinensis*; and ruffed grouse, *Bonasa umbellus*. In most areas it was replaced by oak, hickories, and maples, which provide a lower level and less stable production of fruits (called 'mast' by foresters). Thus, the overall carrying capacity of affected forests is believed to have been lowered by the demise of the chestnut.

Palm trees are affected by many plant diseases, including some that are transmitted by insects. One particularly damaging disease is called lethal yellowing. A bacterium (phytoplama) that is transmitted by the planthopper *Myndus crudus* (Hemiptera) is responsible for the disease, which results in the death of many types of palms in the Caribbean region. In the same geographic region, weevils ovipositing on palms transmit the coconut palm nematode, *Bursaphelenchus cocophilus*, which can cause red-ring disease.

POLLINATION AND SEED DISPERSAL BY INSECTS

Arthropods have been associated with the reproductive parts of plants since at least the Devonian Period, 417 to 354 million years ago, although initially they were involved mostly in consuming the spores of primitive plants and did not develop an intimate relationship with plants until the radiation of the Angiospermae during the Cretaceous Period 144–65 million years ago. Since then, their relationship has evolved beyond consumption to now include transport of pollen to facilitate fertilization, which we call **pollination**. The evolution and dominance of the Angiospermae, the flowering plants, is possible only because insects were present to serve as pollinators. Not all modern plants are insect pollinated, of course, as the Poaceae or Graminae (the graminoids, or grasses and close relatives) manage quite nicely based on wind-based transport of reproductive materials. In warmer climates, however, the insect-pollinated plants predominate. Insect pollination allows plants to obtain the genetic benefits of outcrossing, and allows angiosperms to exist as widely separated individual plants because the pollen can be carried with winged pollinators instead of being carried passively by wind. This dispersion likely allows plants to minimize predation and disease, which occur most commonly at high densities.

Pollination is essential for the reproduction of flowering plants. In flowering plants, insects seeking nectar accidentally transfer pollen grains from the male anther to the female pistil, usually the sticky upper portion of the pistil called the stigma. The pollen grain produces a germ tube that grows down the pistil to reach the ovule in the base of the pistil. The nuclei of the pollen fuse with the nuclei of the ovule, accomplishing fertilization and resulting in production of seed. Most plant fruits will not develop if pollen transfer is not accomplished. Plants can be self-pollinated or cross pollinated. In **self-pollination** the pollen of a plant fertilize flowers of the same plant. In **cross-fertilization**, pollen from one plant fertilizes another plant. Cross-pollination results in plants that are stronger and more vigorous. Many plants are self-sterile (self-pollination fails), and others will set fruit but are much more fruitful if cross-pollinated. Although insects are the most important pollinators, some birds (e.g., hummingbirds) and bats also can be effective pollinators.

Flowering plants advertise their presence with attractive odors or visual cues. The goal of flowers is to attract potential pollinators with some attractive feature, leading to the inadvertent transfer of pollen from flower to flower. Vision is useful for attracting insects, but only for short-range orientation. Insects possess color vision, although the colors they normally see are ultraviolet (UV), blue, and green. Humans do not see UV. Most birds, but few insects, perceive red, so bird-pollinated plants often have red flowers. Bat-pollinated plants may only open at night to escape feeding by day-flying insects, and often have white flowers that are easily seen in the dark. Floral scents, on the other hand, are useful in both close and long-range attraction of insects. Day-pollinated flowers tend to have delicate floral scents and night-pollinated species (which are generally white) tend to have heavy scents. Birds are not very good at perceiving scents.

Pollinating insects are rewarded with **nectar**, a sugary drink that also contains some amino acids. Some pollinators such as bees also eat nutrient-rich pollen when they visit flowers. **Pollen** produces sperm cells for fertilization, but also contains protein, lipids, amino acids, vitamins, and other important food elements. Bees also may collect pollen to feed to their young. Other plant products may be collected from blossoms, including floral oils, resins and gums. Insects such as some beetles and thrips feed on pollen but not nectar; many also provide fertilization services to plants.

Pollination is mutually beneficial, as the pollinators obtain food while the plants are fertilized. Plants often evolve specific mechanisms to maximize pollination, and because of this, very specific relationships may develop wherein only one insect or one type of insect is capable of pollinating a particular plant species. These insects are called **pollination specialists**. Some bees and moths fall into this category. Other species feed more generally, not developing behavioral or morphological modifications. These general feeders are called **pollination generalists**, and include some beetles, thrips, and flies.

Most angiosperms must not only attract pollinators to produce viable seeds, but they must attract seed dispersers in an effort to get their seeds dispersed to locations where their seeds will germinate and their seedlings will survive. It is an advantage for the plants if their seed can be distributed widely to new and potentially better habitats. It is also beneficial if the seeds are widely dispersed instead of simply dropped to the soil where the young plants will suffer from competition by the parent. Plants have evolved attractive, nutritious 'packaging' for seeds, that we call fruit (berries). Fruit ensures that a large animal, usually a bird or mammal, but sometimes a reptile, will feed on the fruit and then defecate the seeds elsewhere, thereby providing transport for the seeds. Fruit-feeding animals are called **frugivores**. Some animals eat the seeds, destroying them; these are called **seed predators**. Examples of seed predators that consume the entire seed include birds, rodents and ants, although other insects develop within the seed. Other animals eat some seeds, but move others to storage elsewhere; these are called **seed cachers**. Examples of seed cachers include rodents, ants, and rarely birds. **Seed vectors** are animals that transport seed by contact, usually because the seed is clinging to their fur or feathers. Insects are not good vectors, though birds and mammals are effective.

Ants (Hymenoptera: Formicidae) are particularly good seed predators and cachers, particularly the 'harvester ants,' which collect and return seeds to a central nest site and cache the seeds underground in granaries. The small size and low energy requirements of ants allows them to search for small seeds, whereas the larger size and higher energy requirements of vertebrates forces seed-eaters such as birds and rodents to concentrate on the energetic rewards of larger seeds. It is tempting to infer that due to the seed-eating habits, ants are detrimental to ecosystems. This is not the case, and ecosystems are usually well adapted to their complement of ants. In fact, if ant populations are disturbed there can be adverse consequences for the plant community because the failure of ants to harvest seeds may lead to an increase in the abundance of plants that produced the ant-preferred seed. Similarly, non-indigenous (exotic) ant species can disrupt plant communities by selectively predating certain seeds and changing the relative abundance of flora.

Ants also can be important in the dispersal of seeds. Ants seem to be most important in seed dispersal in areas with infertile soils such as Mediterranean climates, and where there is high fire frequency. Benefits include providing seeds with shelter from fire and from consumption by larger seed predators, and by providing dispersal to high-nutrient sites (ant mounds).

Thus, insects provide vital pollination and seed dispersal services that foster angiosperm diversity and habitat vitality. In a sense, however, insects are just part of the 'team' as they often need vertebrate wildlife to help to disperse the seeds resulting from their pollination activities. Although some insects also exploit seeds and fruit, the benefits of fruit production to wildlife are immense. As shown in Tables 4.3–4.5, fruit is a common element of the diet of many birds and mammals. Particularly in the winter months, birds often depend heavily on dried fruit, which tend to be sugar-rich, thus providing considerable calories.

Pollination is rarely as simple as an insect transporting pollen from plant to plant. As suggested above, pollinators are just one part of food webs, and their behavior is subject to modification by trophic cascades, as are all other components of food webs. Predatory insects can interfere with pollinators and pollination, so wildlife can have indirect effects on their habitat and food supply by feeding on insects. One good example comes from ponds in Florida, USA, some of which have resident fish, whereas others lack fish. We might not ordinarily expect fish to affect pollination, but fish feed on dragonfly larvae within the ponds, and fish-containing ponds have fewer adult dragonflies nearby. Dragonflies feed on flying insects, including pollinators, so the presence of more dragonflies results in fewer pollinators and fewer pollinator visits to flowering plants. Thus, the presence of vertebrate predators, in this case fish, can promote angiosperm pollination and biodiversity. More importantly, this example demonstrates how seemingly small effects in predator-prey relationships can reverberate through ecosystems, having unexpected effects. This phenomenon of obtaining unexpected results from small individual changes sometimes is called 'emergent properties' because the final product is much more than the sum of the individual effects.

INVASIVENESS OF INSECTS

Invasion by organisms of environments where they do not naturally occur has always been a major source of environmental problems, often leading to disruption of environmental functions and loss in biodiversity among naturally occurring organisms. Insects are among the most effective invaders of new environments. Once established in a new area, they often effectively exploit the resources (usually plants) of their new environment, become abundant, and are seen as pests. Indeed, in many areas of the world the principal pests are invaders from another area, usually another continent. This is not to say that invaders or pests cannot be local or regional in origin. For example, agricultural pests commonly move from native flora or weeds to crops, and can be considered invaders of the crop. Likewise, biting insects often disperse short distances from undisturbed natural areas to places inhabited by people or livestock, although these are better considered dispersers than invaders. Sometimes highly dispersive pests fly thousands of kilometers each year to invade areas where they cannot survive the winter. However, here we discuss nonindigenous (adventive, alien) invaders, which are considered to be pests that have moved a considerable distance, rather than locally, to permanently establish populations where they have not occurred previously. Invading insects can affect wildlife directly by biting or stinging, or by transmitting new or existing diseases. They can also affect wildlife indirectly by affecting food plant resource ability, or by modifying the structural characteristics of their plant habitat.

Not all invading organisms become pests, of course, and some simply blend into the environment and cause little disruption. Interestingly, it is this uncertainty about how important an invader will be that creates something of a problem. Often we don't know whether or not to be alarmed by the arrival of an invader, and if we should attempt to eradicate a potential pest, because we can't anticipate its impact. By waiting, the invader becomes established and well dispersed, and then the option of relatively simple eradication is lost and we must adapt to the presence of the newly established invader.

The terminology surrounding invasiveness of organisms is often used incorrectly or inconsistently. The biggest problem is the incorrect use of the term 'endemic,' which actually is the opposite of epidemic, but often is used to mean indigenous or native. Following is some clarification of terms:

• *Invasive.* These organisms have a tendency to spread. Invading (adventive) organisms differ in their invasive potential, but have a tendency to be more invasive if they find a particularly suitable habitat or are free of the suppressive agents occurring in their home environment. However, naturally occurring (precinctive) organisms can also become invasive, as

might occur when habitat is altered, making it more suitable for the precinctive organism.

• *Indigenous.* These are organisms that are present, occur naturally, and evolved in this location. Indigenous = 'native.' The opposite of indigenous is 'nonindigenous,' and, although 'adventive' is a preferable term relative to nonindigenous, the latter is used widely because it is unambiguous.

• *Precinctive.* These organisms are indigenous or native, but only found in a certain area. It is also possible for an organism to be indigenous but not precinctive, as it may be found widely. The term 'endemic' is often used in place of precinctive, but it has other meanings so should not be used with respect to distribution.

• *Adventive.* These organisms came from elsewhere. Adventive = not native, or nonindigenous, or alien. Adventive organisms may be introduced or immigrants.

• *Introduced.* These are organisms that are not native to an area but that have arrived as a result of purposeful introduction. Introduced species sometimes are quite invasive and escape beyond the region or purpose of their release, so this must be guarded against.

• *Immigrant.* These are organisms that were accidentally introduced. An immigrant is an organism that is not native to an area (it is adventive), and has arrived as the result of hitchhiking (traveling) or by its own means of dispersal (usually by flight or rafting). Immigrants are often called 'invaders' or 'invasives,' though their invasive properties vary considerably once they have arrived.

Pathways of Invasion

Natural dispersal of insects (dispersal without assistance of human activities) is frequent, and the natural range of most insects was established by natural dispersal long before humans were navigating the globe. Even remote locations such as the Hawaiian Islands and the Galapagos Islands had a modestly rich insect fauna before the arrival of humans. Rafting on floating debris, hitchhiking on birds, and direct flight are the likely modes of transport, of course, with rafting probably most important. However, since regular movement by humans to these remote locations began about 300 years ago, the fauna has changed considerably. Most significantly, the pathways have changed through the use of ships and aircraft. Such rapid forms of transport allow short-lived organisms to hitchhike to new locations and to survive the trip. For example, the number of insect species in Hawaii has increased immensely with the addition of about 2300 invaders to the approximately 3750 indigenous species, and in the Galapagos by the addition of 295 nonindigenous species to the 1550 indigenous species. At these island locations, nearly all of the increase in species richness is due to the inadvertent introduction of hitchhiking arthropods, although there have been some deliberate introductions. This is not to say that natural dispersal over such long distances is not possible. The relocation of large numbers of desert locusts, *Schistocerca gregaria*, to the Caribbean and northern South America from Africa with the assistance of weather fronts in 1988 stands out as an example of how natural, long-distance transport remains possible. And although the dispersing *S. gregaria* did not establish successfully in the western hemisphere during this dispersal episode, the ancestors of the numerous *Schistocerca* species inhabiting the western hemisphere may have originated from a similar episode thousands of years ago.

Although insects often possess the power of flight, human-assisted dispersal now is the principal means by which nonindigenous insects gain access to new territory. Sometimes relocation of nonindigenous species is deliberate. European honey bee, *Apis mellifera* (Hymenoptera: Apidae), for example, has been relocated to everywhere that humans exist except for polar regions. Other examples of deliberate relocation include silkworm, *Bombyx mori* (Lepidoptera: Bombycidae), for silk production, and the many predators and parasites that attack plant pests of crops grown in greenhouses. Nowadays, these relocations are usually accompanied by appropriate regulatory oversight, and there is minimal risk of ecological and economic problems.

Accidental dispersal of insects is much more likely to result in problems. In earlier times, dispersal of nonindigenous insects was often associated with sailing ships, or more specifically, with disposal of ballast carried in sailing ships. The unloading of ship's ballast on the seashore of ports often resulted in inoculation with nonindigenous organisms. Ballast is weight needed to stabilize sailing ships so that they do not tip excessively. The traditional ballast for sailing ship in the 1600s–1800s was stones, gravel, and rubble. Local ordinances in ports forbade the dumping of ballast in the water (where it would soon clog the

harbor), so it was unloaded on the shore. Many arthropods, molluscs, and nonindigenous plants were introduced to port cities via ballast, which then spread from these sites of introduction. Interestingly, ships traveling from Europe to North America often contained human passengers or refined goods, which did not constitute a great deal of weight, hence the need for ballast. However, on the return journey, the cargo consisted of heavy, raw materials such as lumber, hides, grain and tobacco, so ballast was not needed. This perhaps explains the overwhelming movement of pests from Europe to North America, and the relatively slight movement from North America to Europe.

As sailing ships were replaced by steam-powered ships, the importance of ballast as a source of inoculum was diminished because steamships were less susceptible to tipping. However, the proliferation of steamships in the late 1800s and early 1900s was coincidentally accompanied by a strong interest in botanical gardens, and the introduction of new plants for food and forage crops with the aim of improving agriculture. Travel by steam-powered ships was more reliable and often faster than by wind-powered vessels, so shipments of plants (and their hitchhiking arthropod pests) were more likely to arrive at their destination alive. Thus, accidental transport of insects increased rather than decreased.

This time period was also accompanied by the proliferation of canals and railroads. Thus, not only was it more likely for pests to traverse the world's oceans, but upon introduction to new territory the pests were quickly moved far inland from the ports by canal boats and railroad cars. Thus, the inland environments as well as the coastal environments were inoculated with nonindigenous organisms.

In the latter half of the 1900s, the movement of products by sea in sealed containers became dominant, and travel by air became commonplace not only for people, but for expensive and perishable goods such as flowers and fruit. This allowed short-lived organisms easier access to new environments. Also, the nature of ballast had changed, and liquid (seawater or riverwater) rather than solid ballast predominated. Thus, marine and freshwater invertebrates, algae, and vascular plants became important items of concern, and arthropods were less likely to be associated with ballast. However, arthropods remain important as contaminants of sea containers, perishable goods and other agricultural commodities, and as contaminants of handbags carried by the ever-increasing numbers of tourists. Infestation of wood used to construct crates

Table 6.2. Probable pathways of introduction of insects accidentally introduced into Japan during 1917–1999 (adapted from Kiritani and Yamamura 2003. Exotic insects and their pathways for invasion, pp. 44–67 in Ruiz and Carlton 2003).

Pathway of dispersal	Number of species
Flight (air currents, typhoons, etc.)	4
Plant material (nursery stock, bulbs, etc.)	52
Military transportation	10
Hay (dried fodder)	9
Fruits and vegetables	8
Grain	8
Hitchhiking	3
Wood and packing materials	2
Cut flowers	1
Seeds	1
Total	98

also became an issue as tree-infesting pests were moved around the world, and ants proved especially adept at entering shipping containers and being relocated with the expanded international commerce.

A less-important but notable method of introduction of invasive nonindigenous organisms is contamination of military equipment. Within the context of war, concern about contamination of equipment and material by pests is replaced by concerns about safety and success of the military operation. Thus, in the wake of major conflicts of the twentieth century there were some relocations of pests. But even during relatively peaceful times, movement of military equipment is a potentially important source of pest inoculum. By way of example, Table 6.2 highlights some of the pathways of invasion of Japan by immigrant insects.

Ecological and Taxonomic Patterns of Invasion

In addition to changes in the method of transport of nonindigenous organisms, another change that has occurred over time has been the nature of the pests being relocated. In North America, many of the pests initially introduced were stored products pests. This was a period of relatively slow transport, when perishable goods did not survive the many months at sea necessary for long distance travel, so long-lasting grains and legume seed were disproportionately used as food for sailors and transported to support colonists who were struggling to establish new economies. As the rate of transport was enhanced, vegetable and fruit pests were moved around the globe, as such foodstuffs could survive the weeks of transport from location to location. More recently, ornamental plants are being moved with greater frequency, and the pest complex has changed again. Often, it is plant propagative materials that are being moved from low-labor cost areas to higher labor cost areas in an effort to reduce the overall costs associated with ornamental plant production. Also, cut flowers are commonly produced in relatively small areas favored by climate or labor costs, and then redistributed around the world. Homogenization of ornamental plant pests, especially plant pests of greenhouse crops, seems inevitable. Most recently, the transport of material in wood crates has resulted in the introduction of wood boring insects, sometimes with devastating effects on North America's trees. When forest pests are introduced to new areas, there is risk of major habitat change, with resulting impacts on wildlife.

The taxonomic shift is as significant as the ecological changes among North American invaders. Early in the settlement of North America, beetles (Coleoptera) were the dominant invaders. By the late 1800s, however, Lepidoptera (moths) and Hemiptera/suborder Stenorrhyncha (aphids, whiteflies, scales, and mealybugs) assumed dominance. In the 1900s, Hymenoptera (principally ants, Formicidae) became quite important. Most recently, contaminants of wood such as bark beetles (Coleoptera: Curculionidae: Scolytinae), longhorn beetles (Coleoptera: Cerambycidae) and other wood borers have become more prominent.

In Japan, a slightly different pattern of introduction emerged. There, the island nation was largely protected from invasive pests until about 1868 when the national isolation policy ceased. Immediately thereafter, new fruit tree varieties were imported, and scale insects and mealybugs were introduced along with nursery stock. Following the cessation of World War II in 1945, massive quantities of grain were imported to compensate for food shortages caused by the war. Not surprisingly, the grain insects that earlier had been moved around the world were finally introduced to Japan. Then, during the 1960s to the mid-1980s, a number of weevils were introduced that affected food crops, ornamental trees, and turf. Finally, at the close of the twentieth century, Japan saw the introduction of the many greenhouse pests that were being redistributed elsewhere in the world, including thrips, whiteflies, mealybugs, leafminers, and aphids. Not surprisingly, the principal arthropods associated with nursery stock and other plant propagative materials were aphids, thrips, whiteflies and mites; the contaminants of cut flowers were aphids, thrips and mites; the contaminants of fruits and vegetables were fruit flies, weevils and leaf beetles; the contaminants of grain and other seeds were bean weevils and khapra beetles; and the contaminants of wood were bark beetles, longhorn beetles and other wood borers.

Establishment and Spread

Introduction of nonindigenous species is usually determined by human activities. Hand luggage is often cited as a pathway of introduction, and indeed, many travelers naïvely or deliberately bring fresh produce (mostly fruits and vegetables) into new areas, and sometimes these are infested with pests. Even smuggling of wildlife or pets is risky, because ticks and other parasites have been introduced with such unlikely hosts as tortoises and snakes. Although some pests probably are introduced in this manner (large fruit flies (Diptera: Tephritidae) are perhaps the most likely to accompany smuggled fruit and vegetables from tropical areas) the legal international shipment of propagative plant materials now seems to be the principal route of entry for most pests, with contamination of wood crates or other packing material also an increasing problem.

Certain patterns of introduction of immigrant insects, or potential introduction, usually emerge when the history of invasion is studied. For example, in the USA, 70% of the interceptions of immigrants are made at only three ports: Miami, New York, and Los Angeles. In Miami, the vast majority of detections originate in Central America, whereas in New York, Europe and China predominate as sources, and in Los Angeles most interceptions originate in China, Thailand and Central America. Table 6.3 shows the top sources of invading insect detections found on plant materials intended for propagation that were imported in recent times. These data show that the sources of pests do not necessarily reflect trade with the 'top' trading partners of the USA; rather, they reflect the

Table 6.3. Arthropod interceptions on plant material intended for propagation in USA (top 12 countries for 1990–1999) (after National Research Council 2002).

Origin	Number of interceptions
Costa Rica	4723
Guatemala	1456
Mexico	1193
Netherlands	902
Honduras	886
Thailand	770
China	588
Singapore	498
Japan	339
Belize	232
South Africa	217
Dominican Republic	200

unique circumstances of the plant propagation trade, and point out the need to fully understand the unique pathways of international trade. In contrast, as China has become a major supplier of manufactured goods to North America, there has been the predictable increase in the number of wood boring insects entering from China.

The increase in the 'free trade' movement throughout most of the world in recent years has resulted in vast increases in material moving by truck, boat and aircraft. Only a small proportion of the material crossing international boundaries is inspected, and even then it is difficult to carefully examine many commodities. In New Zealand, for example, 20% of sea containers are inspected, and during the period of 1999–2000, 24.8% of the containers were found to be contaminated and require quarantine action. Unfortunately, the incidence of contamination by pests has been increasing steadily since 1993–1994, when only 6.4% of the containers held pests. In contrast, only 1%–2% of shipments to the USA are inspected, despite the fact that 10.4% of the aircraft arriving at Miami International Airport in 1998–1999 contained insects. Aircraft from Central American countries had a higher incidence of infestation, 23%. These alarmingly high levels of infestation clearly indicate that insects are routinely transported in commercial cargo, and that inadequate efforts are being made to prevent such transport. Shippers and retailers of imported products often are unwilling or unable to curtail contamination, which is best handled by prevention before shipment, not inspection after shipment.

Establishment of pests in new environments is related to several factors, including:

• *Frequency of inoculation.* Some pests are a greater hazard because they more frequently are transported to new environments. Pests that are more cryptic, that infest perishable produce which must be transported with little delay, or that infest material that is imported in high volume are more likely to be transported.

• *Climatic suitability of the new environment.* Transport of insects between like climates is most likely to result in successful establishment of the immigrant. Thus, movement of insects between Europe and North America has a higher potential for successful inoculation of a new pest than transport between North America and South America, because in the former case the seasonality is about the same, whereas in the latter case the opposite seasons prevail. Related to this is the fact that benign climates are more likely suitable for insect invaders than are severe climates.

• *Biological characteristics of the new environment.* The availability of suitable hosts is critical for successful establishment of invaders. Obviously, insects with a narrow host range are more at risk of not arriving in a location where a suitable host is present, but even insects with a broad host range have hosts that vary in suitability for survival, so their survival is not guaranteed. Insects that are preadapted (that have coevolved or are already well adapted) have an extraordinarily high rate of success in tracking their host plants as those host plants are moved around the world, whereas insects that encounter 'new' hosts may or may not be successful in exploiting these new resources. The likelihood of an invading organism establishing successfully is positively related to the size of the area (e.g., number of hectares) supporting a suitable host. Disturbed habitats, and habitats with few competing species, are especially favorable for establishment of invaders.

• *Biological characteristics of the potential invaders.* Certain biological characteristics predispose insects to establish successfully. Invaders that are small enough to escape detection, that have good powers of flight, that reproduce parthenogenetically, that have a high rate of reproduction, and that are preadapted for the new environment have an above average chance of establishment. The likelihood of

establishment also increases when the invader arrives with a large founding population.

The actual pattern of invasion varies with the nature of the introduction, the environment into which the invader is introduced, and the biological characteristics of the invading organism. The invasion and spread of eastern North America by imported cabbageworm/small white cabbage butterfly, *Artogeia rapae* (Lepidoptera: Pieridae) (Fig. 6.11) serves to illustrate several points about invaders. *Artogeia rapae* was first found in North America at Montreal, Canada (top, right of the map) about 1860. After this primary invasion event, it spread by natural means in all directions, but in some cases it apparently was assisted in its dispersal by movement of host material. Thus, in a secondary invasion event it was detected in the area of New York City in 1868, before it occurred in surrounding areas. Likewise it was introduced to Charleston, South Carolina about 1873 before the adjacent areas were infested, etc. It is not surprising that port cities would be points of introduction (including the inland lake port of Chicago, Illinois in 1875). The inoculation of central Indiana in 1874 before the surrounding areas were infested could be due to movement of infested produce to this area, an unusual weather event depositing winged adults, or an artifact of early detection.

Weather can play a significant role in assisting dispersal, but climate often sets boundaries on long-term establishment. In the case of *A. rapae*, the entire potentially inhabitable region of North America has been infested with this adaptable species. In contrast, there are many examples where climate limits distribution of immigrant species. Imported red fire ant, *Solenopsis invicta* Buren (Hymenoptera: Formicidae), has successfully invaded North America only in the southern USA north to Tennessee, whereas Japanese beetle, *Popillia japonica* (Coleoptera: Scarabaeidae), has invaded the northern USA south to Georgia, and gypsy moth, *Lymantria dispar* south to Virginia. The occurrence of host plants limits the distribution of species with a narrow host range, but usually the geographic distribution of the host plants exceeds that of the arthropod herbivore, emphasizing the importance of climate.

Latency Among Invaders

The period of time between the introduction of nonindigenous species and their detection has been called the **latent period**. Normally, an invader becomes

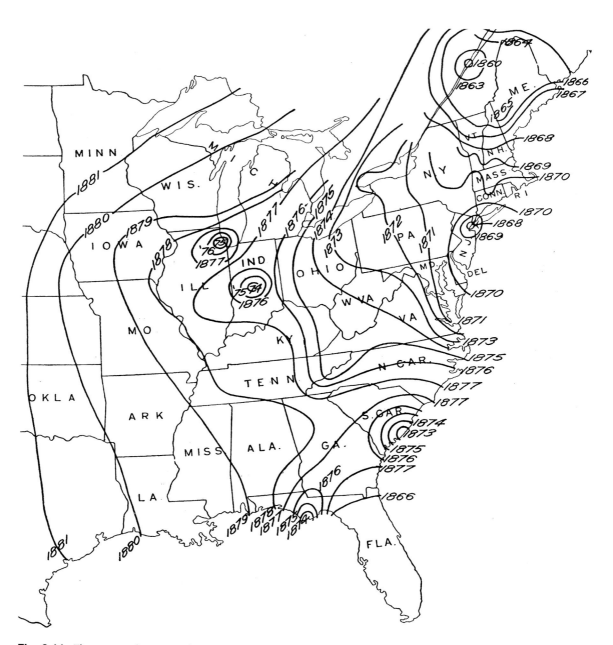

Fig. 6.11. The pattern of invasion of eastern North America by the butterfly *Artogeia rapae* (Lepidoptera: Pieridae) following its accidental introduction about 1860 (from Chittenden 1916).

numerous before it is detected. It is usually impossible to pinpoint the exact time that an invader arrives, though sometimes it can be deduced with a degree of accuracy. One exception to this generalization is the accidental release of gypsy moth, *Lymantria dispar* (Lepidoptera: Lymantriidae) in Medford, Massachusetts in 1869. In this case the date was known precisely, but because the city administration did not care, nothing was done until this species became abundant and damaging 20 years later, in 1889. This might be viewed as a surprisingly long period for such a potentially damaging insect to be latent and unnoticed. However, it is not uncommon for gypsy moth to be at a low level of abundance for 10 or more years, only to suddenly increase in abundance, presumably due to some shift in weather.

In Japan, an analysis of the latent period of invading arthropods showed a range of from 0.5 to 80 years, and a mean of 11.8 years. However, the mean value is inordinately affected by a few 'statistical outliers.' In fact, most species (25 of the 35 species) had a much shorter period of latency, and of this group, the mean latent period was only 3.8 years. Extremely damaging species, and perhaps species that affect certain crops, are more likely to be detected earlier. Thus, because greenhouse crops tend to be high value and inspected regularly, we would expect pests of such crops to be noticed soon after invasion. On the other hand, tree pests that inflicted little damage might go undetected for a considerable period. Often, the likelihood of detection is a function of morphology or behavior; things that look strikingly different or behave differently are apt to be noticed.

The latent period for detection of invaders is almost never known precisely, but in most cases it is a considerable period of time. Thus, there is a good possibility that new invaders will have time to establish and spread before detection, making eradication difficult or impossible. Most countries make efforts to prevent establishment of potential invaders not only by inspection of incoming materials, but also by trapping at ports and other routes of entry.

Often there is a population outbreak displayed by immigrants following establishment and adaptation. The outbreak (and the lack of sizable populations during latency) is likely due to genetic shifts in the invading organisms as they adapt to new hosts and new weather, or as they await opportunity to disperse to an optimal environment. Once conditions are suitable, and especially if natural enemies of the pests have been left behind, populations can increase. Often they attain alarmingly high numbers, but then subside. Subsidence may be due to a shift of indigenous natural enemies to new hosts; a shift in the genetics of the host, induced resistance in the hosts; change in weather; or a change in the management practices by people in the affected area.

Why Invasive Species become so Abundant

The major pests throughout the world often are nonindigenous. In some cases, it may be the newness of the invader that explains the severity of the problem, and once people in the new (invaded) environment learn to manage the invader the problem dissipates. Other times, however, the nonindigenous species becomes a continuing problem. Several explanations have been advanced to explain why nonindigenous insects are more severe pests, including:

• ***Enemy release.*** This is the explanation that is most commonly advanced to explain the inordinate abundance of invaders. When invaders enter a new area, usually they do so without their natural enemies, including predators, parasites and disease agents. With this in mind, attempts commonly are made to introduce from the area of origin of the invader natural enemies that are selective in their action in an attempt to establish biological suppression of the invader. Sometimes this is successful, and the process is called 'classical biological control' or 'introduction biological control.'

• ***Competitive advantage.*** Not all niches are occupied by indigenous species, and even if the niche is occupied, some invaders display superior fitness. Thus, the invaders displace indigenous species and may become quite abundant.

• ***Novelty.*** The newness or novelty of the invader may require several generations to allow adaptation of indigenous natural enemies. Eventually, the indigenous species adapt and provide suppression, but during the initial phase of invasion the abundance of the new invader may be quite high.

• ***Preadaptation.*** If invaders are well adapted for a particular resource or environment they can become unusually abundant when they gain access. This is most obvious when plants are moved to new continents and then insects adapted to those plants gain access later, especially if specialist natural enemies are left behind.

Impacts of Invaders

The effects of nonindigenous arthropods are difficult to assess accurately. In the case of impacts on crops and livestock, it is possible to estimate the costs of pesticides used for their protection, and to add to this the loss experienced by the commodity even in the presence of pesticide use. These data often are not very reliable. Even more problematic are aesthetic costs (e.g., damage to the appearance of lawns and gardens) and environmental effects (e.g., water quality, habitat degradation, loss of wildlife). Nevertheless, nonindigenous insects often account for disproportionately large amounts of damage. For example, of the total crop losses attributable to arthropods in Hawaii, Florida, and California, the proportions due to nonindigenous species are estimated to be 98%, 95%, and 67%, respectively. Although these data may be imperfect, they clearly demonstrate how costly nonindigenous insects can be to societies throughout the world.

The impacts of nonindigenous species on wildlife are evident from a qualitative perspective, but difficult to assess quantitatively. However, if natural ecosystems are significantly modified by the addition of new insects, or outbreaks of nonindigenous insects affect resource availability, wildlife will be affected. Examination of the effects of gypsy moth (*Lymantria dispar*; Lepidoptera: Lymantriidae) invasion can provide some insight into the nature of ecological disruption following insect invasion or population increase. Gypsy moth defoliates numerous species of hardwood trees in the northeastern USA and southeastern Canada. Outbreaks typically last 1–5 years at intervals of 4–12 years. When defoliation by caterpillars exceeds 60%, trees usually refoliate, producing an additional flush of leaves during the year of defoliation. However, during the year of defoliation the growth of the trees is reduced, and reduced growth rates may continue into subsequent years. Vigorous trees survive repeated defoliation for several years, though tree energy reserves decline and subsequent environmental stress (e.g., drought, frost) may kill the trees. Defoliated trees also are more susceptible to attack by wood-boring insects and root-attacking fungi. Areas of forest experiencing defoliation by gypsy moth experience changes in composition of the bird assemblage. For example, a study conducted in Pennsylvania, USA, found that defoliated areas had more house wrens *Troglodytes aedon*, though not a change in overall abundance of bird life. Defoliated areas also had fewer black-capped chickadees, *Parus atricapillus*; wood thrushes, *Hylocichla mustelina*; blue jays, *Cyanocitta cristata*; and northern oriole, *Icterus galbula*.

Importantly, defoliation affects seed production by oaks (*Quercus* spp.), which are among the trees most favored by this insect. The caterpillars will consume the oak flowers, induce acorn abortion, and disrupt flower bud initiation. The high carbohydrate content of acorns provides the energy necessary for winter survival of many species of wildlife. Loss of acorns greatly diminishes the food resources available in many forest ecosystems and causes large-scale reductions in the quality of wildlife habitats. Surviving trees, stressed by defoliation, may require up to 10 years to regain normal acorn production. Repeated defoliation and mortality of oaks also encourages change in tree stand composition, fostering increases in red maple (*Acer rebrum*) and black birch (*Betula lenta*) abundance. Even low levels of gypsy moth abundance have been shown to diminish oak tree growth while intermixed sugar maple (*Acer saccharum*) and white ash (*Fraxinus americana*) trees grew unimpeded by the insects because of differential preference by gypsy moth larvae for the oak foliage. Thus, the presence of gypsy moths can affect ecosystem stability by modifying forest stand composition and short and long term availability of food resources, often to the detriment of wildlife. Presently, gypsy moth is found in only about 25% of the USA where it can survive, and feeds on about 26% of the available host trees; thus, its impact will only increase with time.

The example of gypsy moth is by no means unusual. Among the many invasive insect species that have disrupted natural ecosystems are hemlock woolly adelgid, *Adelges tsugae* (Hemiptera: Adelgidae); emerald ash borer, *Agrilus planipennis* (Coleoptera: Buprestidae); Asian longhorn beetle, *Anoplophora glabripennis* (Coleoptera: Cerambycidae); Africanized honey bee, *Apis mellifera scutellata* (Hymenoptera: Apidae); cactus moth, *Cactoblastis cactorum* (Lepidoptera: Pyralidae); red imported fire ant, *Solenopsis invicta* (Hymenoptera: Formicidae); yellow crazy ant, *Anoplolepis gracilipes* (Hymenoptera: Formicidae); and Argentine ant, *Linepithema humile* (Hymenoptera: Formicidae). The effect of one of these insect invaders (red imported fire ant) on wildlife is discussed more fully in Chapter 13.

SUMMARY

• Ecosystems are complex, interconnected ecological organizations. Plants normally are the primary producers (autotrophs) in ecosystems. Insects are consumers and secondary producers (heterotrophs). Although often overlooked due to their small size, insects are very important in ecosystems due to their roles in nutrient conversion (plant to animal matter), decomposition, nutrient cycling, herbivory, and pollination and seed dispersal.

• The importance of insects is quite evident during insect outbreaks, when they become unusually abundant and have significant effects on the plant components of ecosystems. Insects also are central to many aquatic ecosystems. Less obvious but also important are the insect-based nutrient pulses that move nutrients, such as the emergence of belowground insects into the aboveground environment.

• Decomposition involves the breakdown of dead organisms into its component parts, and eventually into inorganic elements. Often, decomposers break complex carbohydrates into smaller units that fungi are able to degrade further. Among the important decomposers are termites, cockroaches, woodlice, and millipedes. Flies and dung beetles are important in degradation of dung. Carrion is degraded principally by flies and beetles.

• Insects have important roles in nutrient cycling, producing nutrient pulses as a result of herbivory, and affecting both belowground and aboveground productivity. Although it is usually inconspicuous, herbivory is an important influence on plant productivity, and herbivory by insects is more important than herbivory by vertebrates in some ecosystems. Herbivory can have deleterious effects on plants, but often it has no effects or can even be stimulatory due to plant compensation. The concept that some low level of grazing optimizes plant productivity is called the grazing optimization hypothesis.

• Insect outbreaks result from increased availability of susceptible hosts, increased susceptibility of hosts due to age or weather events, unusually high insect survival or reproduction rates, and invasion by nonindigenous insects. By affecting the structure and availability of plants, insect herbivory affects the suitability of ecosystems for wildlife.

• Insects and mites can affect plants indirectly by transmitting plant pathogens (plant disease agents). If the disease is sufficiently pathogenic it may kill the plant. The loss of individual plants may not greatly impact wildlife because they are mobile, but if key resource-providing species or major elements of the ecosystem are eliminated, wildlife will be seriously affected. Several types of plant pathogens can be transmitted by insects to plants, primarily viruses, bacteria, fungi, protozoa, and nematodes. Fungi, transmitted mostly by beetles, are particularly disruptive to natural ecosystems.

• Pollination by insects is essential for the reproduction of many plants. Plants advertise their presence and reward pollinators with nectar. Some insects, mostly ants, collect and redistribute seeds. The occurrence of wildlife can affect pollination and plant biodiversity.

• Insects are important invaders, sometimes dispersing from continent to continent naturally by flight or by rafting, but more often with the inadvertent assistance of humans. Formerly, nonindigenous insects often gained access to new environments by hitchhiking on food, but over time the route of introductions changed, and most recently the invading insects have been hitchhiking on plant propagative materials and on wood used for crates. Establishment of nonindigenous species is related to the frequency of inoculation, climatic suitability of the new environment, the biological characteristics of the new environment, and the biological characteristics of the invaders. Often there is a period of latency or delay after introduction, and then the nonindigenous species attain high levels of abundance and are detected. Nonindigenous species may become especially abundant due to release from natural enemies, competitive advantage, novelty, and preadaptation. In agricultural systems, nonindigenous species dominate the pest assemblage of most crops. Effects of nonindigenous insects have been difficult to assess in natural ecosystems, but are clearly a problem.

REFERENCES AND ADDITIONAL READING

Agrios, G.N. (2008). Transmission of plant diseases by insects. In Capinera, J.L. (ed). *Encyclopedia of Entomology*, 2nd edn., pp. 3853–3884. Springer Science & Business Media B.V., Dordrecht, The Netherlands.

Barbosa, P. & Schultz, J.C. (eds.) (1987). *Insect Outbreaks*. Academic Press, San Diego, CA, USA.

Bardgett, R.D. & Wardle, D.A. (2003). Herbivore-mediated linkages between aboveground and belowground communities. *Ecology* **84**, 2258–2268.

Barker, J.S. (2008). Decompositon of Douglas-fir coarse woody debris in response to differing moisture content and initial heterotrophic colonization. *Forest Ecology and Management* **255**, 598–604.

Beaver, R.A. (1989). Insect–fungus relationships in the bark and ambrosia beetles. In Wilding, N., Collins, N.M., Hammond, P.M., & Webber, J.F. (eds.) *Insect–Fungus Interactions*, pp. 121–143. Academic Press, London, UK.

Bedding, R.A. (2009). Controlling the pine-killing woodwasp, *Sirex noctilio*, with nematodes. In Hajek, A.E., Glare, T.R., and O'Callaghan, M. (eds.). *Use of Microbes for Control and Eradication of Invasive Arthropods.*, pp. 213–235. Springer Science & Business Media B.V., Dordrecht, The Netherlands.

Belovsky, G.E. & Slade, J.B. (2000). Insect herbivory accelerates nutrient cycling and increases plant production. *Proceedings of the National Academy of Sciences, USA* **97**, 14412–14417.

Brightsmith, D.J. (2000). Use of arboreal termitaria by nesting birds in the Peruvian Amazon. *The Condor* **102**, 529–538.

Capinera, J.L. (2002). North American vegetable pests: the pattern of invasion. *American Entomologist* **48**, 20–39.

Capinera, J.L. (2008). Invasive species. In Capinera, J.L. (ed.). *Encyclopedia of Entomology*, 2nd edn., pp. 2025–2040. Springer Science & Business Media B.V., Dordrecht, The Netherlands.

Chittenden, F.H. (1916). *USDA Farmers' Bulletin 766*.

Davidson, D.W. (1993). The effects of herbivory and granivory on terrestrial plant succession. *Oikos* **68**, 23–35.

DeGraff, R.M. (1987). Breeding birds and gypsy moth defoliation: short-term responses of species and guilds. *Wildlife Society Bulletin* **15**, 217–221.

DeVault, T.L., Rhodes Jr., O.E., & Shivik, J.A. (2003). Scavenging by vertebrates: behavioral, ecological, and evolutionary perspectives on an important energy transfer pathway in terrestrial ecosystems. *Oikos* **102**, 225–234.

Diamond, S.J., Giles Jr., R.H., Kirkpatrick, R.L., & Griffin, G.J. (2000). Hard mast production before and after the chestnut blight. *Southern Journal of Applied Forestry* **24**, 196–201.

Domisch, T., Ohashi, M., Finér, L., *et al.* (2008). Decomposition of organic matter and nutrient mineralization on wood ant (*Formica rufa* group) in boreal forest of different age. *Biology and Fertility of Soils* **44**, 539–545.

Fleming, P.A. & Loveridge, J.P. (2003). Miombo woodland termite mounds: resource islands for small vertebrates? *Journal of Zoology, London* **259**, 161–168.

Fonte, S.J. & Schowalter, T.D.(2005). The influence of a neotropical herbivore (*Lamponius portoricensis*) on nutrient cycling and soil processes. *Oecologia* **146**, 423–431.

Frost, C.J. & Hunter, M.D. (2004). Insect canopy herbivory and frass deposition affect soil nutrient dynamics and export in oak mesocosms. *Ecology* **85**, 3335–3347.

Galante, E. & Marcos-Garcia, M.A. (2008). Decomposition by insects. In Capinera, J.L. (ed.). *Encyclopedia of Entomology*, 2nd edn., pp. 1158–1169. Springer Science + Business Media B.V., Dordrecht, The Netherlands.

Hanski, I. & Cambefort, Y. (eds.) (1991). *Dung Beetle Ecology*. Princeton University Press, Princeton, New Jersey, USA. 481 pp.

Hartley, S.E. & Jones, T.H. (2004). Insect herbivores, nutrient cycling and plant productivity. In Weisser, W.W. & Siemann, E. (eds.). *Insects and Ecosystem Function*, pp. 27–52. Springer, Berlin, Germany.

Hocking, M.D. & Reimchen, T.E. (2006). Consumption and distribution of salmon (*Oncorhynchus* spp.) nutrients and energy by terrestrial flies. *Canadian Journal of Fisheries and Aquatic Sciences* **63**, 2076–2086.

Holdo, R.M. & McDowell, L.R. (2004). Termite mounds as nutrient-rich food patches for elephants. *Biotropica* **36**, 231–239.

Howard, F.W., Moore, D., Giblin-Davis, R.M., & Abad, R.G. (2001). *Insects on Palms*. CABI Publishing, Wallingford, UK.

Hunter, M.D. (2001). Insect population dynamics meets ecosystem ecology: effects of herbivory on soil nutrient dynamics. *Agricultural and Forest Entomology* **3**, 77-84.

Huntly, N. 1991. Herbivores and the dynamics of communities and ecosystems. *Annual Review of Ecology and Systematics* **22**, 477–503.

Kalko, E.K.V., Ueberschaer, K., & Dechmann, D. (2006). Roost structure, modification, and availability in the white-throated round-eared bat, *Lophostoma silvicolum* (Phylostomidae) living in active termite nests. *Biotropica* **38**, 398–404.

Kevan, P.G. (2008). Pollination and flower visitation. In Capinera, J.L. (ed.) *Encyclopedia of entomology*, 2nd edn., pp. 2960–2971. Springer Science & Business Media B.V., Dordrecht, The Netherlands.

Knapp, C.R. & Owens, A.K. (2008). Nesting behavior and the use of termitaria by the Andros iguana (*Cyclura cychlura cychlura*). *Journal of Herpetology* **42**, 46–53.

Knight, T.M., McCoy, M.W., Chase, J.M., McCoy, K.A., & Holt, R.D. (2005). Trophic cascades across ecosystems. *Nature* **437**, 880–883.

Lord, W.D. & Burger, J.F. (1984). Arthropods associated with herring gull (*Larus argentatus*) and great black-backed gull (*Larus marinus*) carrion on islands in the Gulf of Maine. *Environmental Entomology* **13**, 1261–1268.

Mobæk, R., Narmo, A.K., & Moe, S.R. (2005). Termitaria are focal feeding sites for large ungulates in Lake Mburo National Park, Uganda. *Journal of Zoology, London* **267**, 97–102.

Naidoo, R. & Lechowicz, M.J. (2001). Effects of gypsy moth on radial growth of deciduous trees. *Forest Science* **47**, 338–348.

National Research Council (2002). *Predicting Invasion of Nonindigenous Plants and Plant Pests*. National Academy Press, Washington DC. 194 pp.

Pieper, S. & Weigmann, G. (2008). Interactions between isopods and collembolans modulate the mobilization and transport of nutrients from urban soils. *Applied Soil Ecology* **39**, 109–126.

Pimentel, D. (ed.). (2002). *Biological Invasions. Economic and Environmental Costs of Alien Plant, Animal, and Microbe Species*. CRC Press, Boca Raton. 369 pp.

Putman, R.J. (1983). *Carrion and Dung. The Decomposition of Animal Wastes*. Edward Arnold, London, UK.

Reed, M.A. & Tidemann, S.C. (1994). Nesting sites of the hooded parrot *Psephotus dissimilis* in the Northern Territory. *Emu* **94**, 225–229.

Riley, J., Stimson, A.F., & Winch, J.M. 1985. A review of Squamata ovipositing in ant and termite nests. *Herpetological Review* **16**, 38–43.

Ruiz, G.M. & Carlton, J.T. (eds.) (2003). *Invasive Species. Vectors and Management Strategies*. Island Press, Washington. DC. 518 pp.

Safranyik, L. (ed.) (1983). *The Role of the Host in the Population Dynamics of Forest Insects*. Canadian Forest Service, Victoria, British Colombia, Canada.

Sailer, R.I. (1978). Our immigrant insect fauna. *Bulletin of the Entomological Society of America* **24**, 3–11.

Schädler, M, Alphei, J., Scheu, S., Brandl, R., & Auge. H. (2004). Resource dynamics in an early-successional plants community are influenced by insect exclusion. *Soil Biology and Biochemistry* **36**, 1817–1826.

Scharf, M.E. & Tatar, A. (2008). Termite digestomes as sources for novel lignocellulases. *Biofuels, Bioproducts, and Biorefining* **2**, 540–552.

Schowalter, T.D. (1996). *Insect Ecology. An Ecosystem Approach*. Academic Press, San Diego. 483 pp.

Shaimazu, M. (2009). Use of microbes for control of *Monochamus alternatus*, vector of the Invasive pine wood nematode. In Hajek, A.E., Glare, T.R., & O'Callaghan, M. (eds.). *Use of Microbes for Control and Eradication of Invasive Arthropods*, pp. 213–235. Springer Science & Business Media B.V., Dordrecht, The Netherlands.

Shivik, J.A. (2006). Are vultures birds, and do snakes have venom, because of macro- and microscavenger conflict? *BioScience* **56**, 819–823.

Stadler, B., Solinger, S., & Michalzik, B. (2001). Insect herbivores and the nutient flow from the canopy to the soil in coniferous and deciduous forests. *Oecologia* **126**, 104–113.

Stowe, K.A., Marquis, R.J., Hochwender, C.G., & Simms, E.L. (2000). The evolutionary ecology of tolerance to consumer damage. *Annual Review of Ecology and Systematics* **31**, 565–595.

Wardle, D.A. & Bardgett, R.D. (2004). Indirect effects of invertebrate herbivory on the decomposer subsystem. In Weisser, W.W. & Siemann, E. (eds.) *Insects and Ecosystem Function*, pp. 53–69. Springer, Berlin, Germany.

Webber, J.F. & Gibbs, J.N. (1989). Insect dissemination of fungal pathogens of trees. In Wilding, N., Collins, N.M., Hamond, P.M., & Webber, J.F. (eds.) *Insect–Fungus Interactions*, Academic Press, London, UK, pp. 161–193.

Weisser, W.W. & Siemann, E. (2004). The various effects of insects on ecosystem functioning. In Weisser, W.W. & Siemann, E. (eds.) *Insects and Ecosystem Function*, pp. 3–24. Springer, Berlin, Germany.

Whitham, T.G., Maschinski, J., Larson, K.C., & Paige, K.N. (1991). Plant responses to herbivory: the continuum from negative to positive and underlying physiological mechanisms. In Price, P.W., Lewinsohn, T.M., Fernandes, G.W., & Benson, W.W. (eds.) *Plant–Animal Interactions: Evolutionary Ecology in Tropical and Temperate Regions*, pp. 227–256. John Wiley & Sons, Ltd., New York, USA.

ARTHROPODS AS DISEASE VECTORS AND PESTS

Chapter 7

TRANSMISSION OF DISEASE AGENTS TO WILDLIFE BY ARTHROPODS

Like most branches of science, disease biology has generated a somewhat unique language to describe its elements and effects. For example, **infectious diseases** are considered to be caused by small (micro) organisms, whereas **parasitic diseases** are caused by larger (macro) organisms. Microparasites causing infectious disease include viruses, bacteria, and fungi. Scientists who study such microparasites are called **microbiologists**. In contrast, macroparasites causing parasitic diseases are larger, and consist of protozoa, helminths (worms), and arthropods (insects and their close relatives). Scientists who study these larger parasites are called **parasitologists**. The difference between the infectious diseases and parasitic diseases may seem arbitrary, but the separation between the two groups is based on more than just size, as some other factors also serve to separate the two groups. **Microparasites** are characterized by:
• being relatively small;
• lacking a special infective stage;
• having a shorter generation time;
• having a short life-span relative to the host;
• producing a shorter-lived but more severe infection resulting in either death or immunity in the host;
• immunity, when it occurs, usually is life-long.

In contrast, **macroparasites** are characterized by:
• being relatively large;
• possessing a free living infective stage that transmits the organism between generations;
• having a relatively long generation time;
• causing chronic infection that usually does not result in host mortality;
• producing only transient immunity.

Parasites living and replicating inside or on hosts are said to **infect** hosts. However, when arthropods are present on a host they are usually said to **infest** hosts. Despite the confusing terminology, however, it is clear that both microparasitic and macroparasitic organisms can affect wildlife.

ARTHROPOD FEEDING BEHAVIOR

Host location behavior differs according to the degree of association that the blood-feeding arthropods have with their host. Some have essentially a permanent relationship (e.g., lice), riding around on the host; this allows them to feed almost any time. Others are temporary parasites (e.g., mosquitoes, black flies), and feed periodically. This latter type of parasite has more obstacles to overcome in trying to locate an appropriate host. Such host location requires the presence of more sophisticated or better-developed host seeking physiology. Thus, sucking lice (Phthiraptera) have about 10–20 antennal receptors and fleas (Siphonaptera) about 50, but the far-ranging stable fly, *Stomoxys calcitrans* (Diptera: Muscidae), possesses nearly 5000 antennal receptors. Comparing bugs (Hemiptera), bed

Insects and Wildlife: Arthropods and Their Relationships with Wild Vertebrate Animals, 1st edition. By J.L. Capinera. Published 2010 by Blackwell Publishing.

bugs (*Cimex* spp., family Cimicidae) have only about 50 antennal receptors whereas kissing bugs (*Triatoma* spp., family Reduviidae) have nearly 3000.

Both **olfaction** and **vision** are important in host location. Olfaction is important in essentially all blood-feeding insects, though in some cases it operates only at relatively close range. Among the important olfactory stimuli are carbon dioxide, lactic acid, ammonia, octenol, acetone, butanone, fatty acids, indole, phenolic components of urine, and others. The failure of these chemicals to attract as many insects as live animals, however, suggests that many other odors remain undiscovered. Perhaps the most important chemical, or at least the most frequent attractant, is carbon dioxide. Many insects display upwind flight when exposed to elevated carbon dioxide levels. Antennal receptors in flies respond to elevated carbon dioxide levels as low as 0.02%–0.05%. Carbon dioxide acting alone, however, is not as effective an attractant as when other chemicals are present. Species differ in their attraction to chemical stimuli, so there is no universally effective attractant, and these differing chemical responses are part of the mechanism that allows partitioning of the host resources. Generally, it is thought that insects fly upwind in a zig-zagging flight, seeking the point of origin of the attractive chemical, as a means of host location. Odor concentration is not thought to be important, only the presence or absence of the stimulus.

Vision can be quite important for host location by diurnal (day-active) insects. Such insects often are drawn to dark silhouettes. However, generally it is thought that host searching is initiated by host odor. Odor then is used to find the general location of the host, followed by visual responses that lead to host attack. Visual stimuli include form, pattern, color, contrast, light intensity, and movement. Blood-feeding insects often are attracted to blue, green, and sometimes red, but not attracted to yellow. This is quite different from plant-feeding species, which often are attracted to the yellow region of the spectrum. Note that color alone is not the entire story on visual attractiveness; intensity contrast, or the ability of the host to stand out from the background, also is important. Movement also can play a role in attraction, and insect eyes are especially well adapted to detect movement. Large, dark objects that move are especially attractive to many blood-feeders such as tsetse flies, *Glossina* spp. (Diptera: Glossinidae), and horse and deer flies (Diptera: Tabanidae).

In most cases, transmission of disease agents occurs only if arthropods feed successfully on their hosts, but many are well adapted to feed on wildlife. Arthropods tend to be much smaller than their host, which helps them to avoid detection. The shape of the parasite (often flattened) may allow freedom of movement on the host and the ability to evade the host's grooming activities. Some are equipped with modified feet, combs and other structural modifications that facilitate attachment and movement on the host, or make it difficult for the host to detach the parasite. Many, like kissing bugs (*Triatoma* spp. [Hemiptera: Reduviidae]), are stealthy and secrete a salivary anesthetic, so despite their relatively large size they usually escape detection. Their shape and secretive behavior also allows the insects the ability to hide within the host's nest or shelter.

Arthropods do not occur randomly or uniformly on most host animals. The arthropods are affected by the covering of the animal (e.g., fur, feathers, scales), the behavior of the host, and sometimes by the temperature of the host. Permanent ectoparasites tend to gravitate to locations where the host body temperature favors the feeding of the parasite. This behavior is compromised by the grooming behavior of the host, and also by features of the body covering such as length and density of hairs. Similarly, nonpermanent ectoparasites such as mosquitoes tend to feed where hair density is less and hair length is shortest. Thus, study of mosquito feeding on eastern chipmunk, *Tamias striatus*, and grey squirrel, *Sciurus carolinensis*, showed that feeding was disproportionately high on the ears, eyelids, and feet of the hosts. Even the thickness of the host's skin affects feeding behavior. In general, underlying skin is thinner where the overlying covering is thicker. Different species of blood-sucking flies will characteristically attack different regions of a host animal's body, depending on the size of the insects' mouthparts and the proximity of vascular tissue. Areas of the body that bear the greatest number of blood vessels tend to be warmer, so it is no surprise that insects not only use heat to find animal hosts, but also seek the warmest portions of the body to feed.

Large animals may receive hundreds or even thousands of bites per day from insects during peak periods of insect activity. This is sometimes called **insect harassment**. It is not surprising, then, that animals may display **defensive behaviors** to reduce such nuisance, pain and blood loss. Animals often display head

Fig. 7.1. Aggregation of feeding sticktight fleas, *Echidnophaga gallinacea*, around the eye of a bird. Parasites often are found on the head, ears, and neck because it is difficult to groom such regions. Unlike most fleas, sticktight flea has a wide host range and affects various forms of wildlife, both birds and mammals (photo by Phil Kaufman).

shaking, bill rubbing, foot scratching, wing flapping, ear flicking, neck shuddering, and mane and tail swishing as anti-insect behavior. Also, they may seek environments such as trees, windy hillsides, and ice patches, or attempt to run away from biting insects, submerge themselves in water, or cover themselves with dust or mud to escape biting. Biting insects, on the other hand, will feed most frequently where they are least disturbed (see Fig. 7.1). Wildlife display many behaviors designed to disrupt feeding or to kill ectoparasites. Grooming can often restrict parasitic insects to a small portion of the body, as the insects are crushed or eaten by the host. Mutual grooming is even more effective, because it allows better access to the back of the neck, back, and other hard-to-reach places. Mutual grooming is not limited to higher animals such as primates; birds and even rodents display mutual grooming. Group or herd formation reduces the likelihood of being bitten, at least for the animals in the center of the herd. How effective are such behaviors? Research in Zimbabwe showed that sedated animals were 15 times more likely to be fed upon by tsetse flies (*Glossina* spp. [Diptera: Glossinidae]) than unsedated animals, so these actions greatly reduce the frequency of biting. Insect harassment of wildlife is also discussed in Chapter 10.

DISEASE IN WILDLIFE

Whether they are infectious or parasitic, disease agents of wildlife share a common evolutionary pattern: all disease-causing organisms extract nutrients from their hosts. However, if they extract too much too quickly they jeopardize their own survival and ability to reproduce. For their progeny to survive, the parasite has to mature, reproduce, and either be transmitted to another host or be put in an environment where they (or their progeny) can likely find another host. Natural selection favors organisms that are successful in this pursuit; they are most fit.

When disease becomes unusually abundant, this is usually called an **epizootic**, **epidemic**, or **outbreak**. An exceptionally widespread epidemic is called a **pandemic**. In contrast, when the disease is at a low or normal level (not readily observable) it is said to be **enzootic** or **endemic**. Unfortunately, conservation biologists and biogeographers often use the term 'endemic' to refer to organisms that are native to an area, so this term has more than one meaning, a confusing situation. Native organisms are better referred to as **indigenous**. When diseases are capable of spreading from one individual to another, they are said to be **contagious**. Diseases that are new or increasing in prevalence are called **emerging pathogens**. For example, West Nile virus (see also discussion of West Nile virus, Chapter 8) first attained the western hemisphere in 1999. Because it was new, it was viewed as an emerging pathogen. Initially, it caused epizootics among some forms of wildlife, particularly some avifauna. In most areas where it has occurred for a few years it has fallen to an enzootic state due to mortality among the most susceptible hosts and development of resistance among others. However, it will certainly cause an epizootic again at some time in the future as resistance diminishes, a new more virulent form of the virus evolves, or as the number of vectors increases. West Nile virus may be viewed as an emerging pathogen as it gains access to areas where it has not occurred previously, but it will never be an indigenous species in the western hemisphere.

Disease is often considered to consist of the effects of infective and parasitic organisms, but in fact disease is much more encompassing than that. The definition of disease can be expanded to include nearly anything that causes an impairment of the host animal. This includes (in addition to parasites) environmental factors like nutrition, toxicants, weather, inherited abnormalities, and combinations of these factors. Thus, when assessing disease in wildlife, it is important to consider that:

• disease is not measured by deaths of individuals, but by impairment of performance. Wildlife populations can go into decline due to decreased reproduction and longevity, but with no noticeable increase in mortality.
• disease is caused not only by extrinsic factors such as parasites, but also by intrinsic factors such as inherited physiological processes.
• disease may be the result of factors acting individually, or in concert with other factors. Indeed, several functions may be impaired simultaneously, none very noticeably, but collectively very important to fitness of the organisms. Often, parasitic organisms are held in check by immune responses or general vigor of the host, but when stressed by lack of food or cold weather, the animals will succumb to the parasite.

In addition, it is useful to understand that disease can be brought about by the host's physiological response to a disease agent, not simply by the direct effect of the disease agent. For example, invasion of a host by a microbial parasite may eventually cause injury, but disease develops initially as the host's immune system recognizes the presence of the foreign bodies and responds with elevated body temperatures and increased numbers of white blood cells. The host animal may be less alert and unable to feed or hunt while responding metabolically to this invasion. Thus, although many animals have immune systems designed to thwart invasion or minimize feeding, it comes at a physiological cost.

VIRULENCE

The ability of a disease agent to cause impairment or dysfunction in an animal is called **virulence** or **pathogenicity**. These terms are synonymous, and can be used to describe the effects of any type of disease, but generally are used in the context of infectious diseases. Virulence and pathogenicity can be measured in terms of host mortality, reproduction, and altered life history. High virulence can occur as a coincidental by-product of infection, or it can be adaptive (beneficial) to the infectious agent.

Coincidental virulence offers no adaptive advantage to the causative agent; thus, virulence may be viewed as simply an accident. Coincidental virulence is commonly seen in new or novel associations. Infectious diseases that have coevolved a relatively benign relationship with wildlife, for example, can be highly virulent to domestic animals or newly introduced wildlife because they have not had opportunity to evolve a relationship that benefits both the host and parasite. One-sided relationships favoring either the host or disease agent are not advantageous to the disease agent. In cases where the host is favored, the disease agent may be fully suppressed and not able to reproduce and spread to new hosts. On the other hand, relationships favoring the disease may result in premature death of the host, also resulting in failure of the disease to spread to new hosts. An example of coincidental virulence occurs with elaeophorosis, a nematode disease of wildlife (see discussion on elaeophorosis, Chapter 9). The nematodes are transmitted to wildlife by the bite of horseflies (Diptera: Tabanidae) but usually cause no harm to their normal host in western North America, mule deer (*Odocoileus hemionus*). The same nematode, when transmitted to abnormal hosts such as elk, *Cervus elaphus*; moose, *Alces alces*; white-tailed deer, *Odocoileus virginianus*; or bighorn sheep, *Ovis canadensis*; can cause disease. High virulence of infectious diseases originating with wildlife but adversely affecting humans and livestock is most often coincidental.

Adaptive or **beneficial virulence** increases the fitness of the infectious disease agent in some way, rather than being a coincidence. For example, if the disease alters the behavior of the host in some manner, thereby allowing hosts to be more easily captured and passed on to a predator (perhaps serving as an additional host, or a means of enhanced dispersal), fitness of the disease agent may benefit. For example, birds that are infected with encephalitis viruses commonly become less active. This reduces their tendency to avoid biting mosquitoes, increasingly the likelihood that mosquitoes will ingest an adequate viremia (presence of virus in the blood), and spread the virus to new, uninfected birds.

As suggested earlier, hosts and disease agents have the potential to co-evolve. Development of resistance by an animal population to a disease is beneficial to the prospective hosts, but comes at a metabolic cost. Resistance usually requires an expenditure of metabolic resources or energy that affects some aspect of life history; resistant individuals may grow more slowly or produce fewer offspring, for example. So, while natural selection will preserve the presence of more resistant individuals when disease is present and causing mortality among the non-resistant members of the population, resistance will dissipate when selective pressure (mortality) no longer occurs among the more disease-susceptible but otherwise more fit hosts. This sets the stage for establishment of a continuing shift in the balance or homeostasis between disease virulence and host resistance, the mechanics of which we attribute to co-evolutionary processes. The situation is never static; there are constant shifts between an ever more virulent disease caused by occurrence of new strains of the disease agent, and the ever more effective resistance in the hosts. The importance of co-evolution is perhaps best seen in the susceptibility of rabbits in Australia to myxomatosis virus (see discussion on myxomatosis, Chapter 8), wherein a disease that was initially highly virulent became less virulent after causing high initial mortality, but after becoming less virulent became more virulent, etc.

The shifting pattern of disease susceptibility in wildlife is shown diagrammatically in Fig. 7.2. Here you can see how a naïve population of susceptible hosts diminishes in abundance over time as they are exposed. Simultaneously, the number of infected hosts increases. Assuming that the infected hosts recover from infection, the population of recovered hosts increases to attain the same level of abundance formerly held by

Typical pattern of disease outbreak

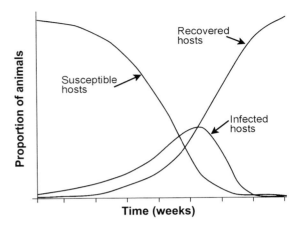

Fig. 7.2. The conceptual pattern of disease spread among a population of wildlife. The naïve (previously unexposed) population of susceptible hosts diminishes in abundance after exposure, to be replaced by infected and then recovered animals (adapted from Wobeser 2006).

the susceptible hosts. On the other hand, if the disease is often fatal, or predisposes the infected animals to predation, the number of recovered hosts might be considerably lower than the earlier number of susceptible hosts. This would result in a rather precipitous decline in animal abundance. The arboviruses (e.g., eastern equine encephalitis) are good examples of insect-transmitted diseases that display such patterns, though for many arboviruses there is considerable mortality among certain species of birds.

It is easy to underestimate the effects of disease on wildlife. One rarely observes animals in nature suffering from disease; they often hide. If an animal dies, its carcass often is quickly destroyed by scavengers. If the affected animal does not perish, it may be even more difficult to detect the presence of disease. The effects of subtle changes in behavior, less efficient digestion or assimilation of nutrients, reduced rate of growth, increased energy use for thermoregulation or fever, and partial reproductive impairment typically become obvious only after several generations, and even then it is difficult to ascribe a cause. Also, the effect of both biotic and abiotic agents may be realized only when a threshold level is surpassed; low levels may not have detectable influences whereas higher levels result in impairment. It is important to recognize that even when the host is successful at fending off infection or parasitism, it comes at a metabolic cost that may affect performance and fitness.

During the time that the animal is dealing with the disease agent it must also contend with the environment. The environment often interacts with the disease-causing agent to influence the outcome for the host animal. Wildlife biologists describe this interaction as a **disease triangle**, with the host animal, the disease agent, and the environment collectively determining the epidemiology of the disease. Plant pathologists subscribe to the same concept, with almost identical terminology: host plant, pathogen, and environment. Insect pathologists seem less impressed with of the role of host animal or plant on the insect's health, though readily acknowledging that there is an interaction with environment. Typically, insect population biologists ascribe population regulation to the interplay of **biotic** (predation and parasitism) and **abiotic**

(weather) elements, as well as **competition** for food (resource quantity and quality). So whatever the discipline, the role of environment is acknowledged to be of great significance in determining disease.

Wildlife often are host to a large number disease agents and parasites in addition to being subject to stresses caused by abiotic conditions and competition. For example, Table 7.1 shows the occurrence of disease agents and parasites in some wildlife found in Florida, USA. Nearly all groups of disease agents and parasites have several members that can affect wildlife. Arthropods are among the most diverse of the disease agents and parasites, averaging 12 different species affecting the mammals shown here. Mites and ticks dominate the arthropod diversity. However, numerical abundance is not the same as impact. Although arthropods are numerous, they or the diseases they cause are not always a major cause of morbidity or mortality in Florida's mammalian wildlife. The major arthropod-associated problems are tularemia in cottontail and marsh rabbit; dirofilariasis (heartworms) in gray fox,

Table 7.1. Number of species in various taxa of diseases and parasites found affecting some mammalian wildlife in Florida, USA (adapted from Forrester 1992).

Wildlife species	Disease agent or parasite					
	Viruses	Bacteria	Fungi	Protozoa	Helminths	Arthropods
Virginia opossum, *Didelphis virginiana*	0	4	2	2	18	11
Nine-banded armadillo, *Dasypus novemcinctus*	2	19	2	2	1	0
Eastern cottontail rabbit, *Sylvilagus floridanus*	0	2	1	0	7	15
Marsh rabbit, *Sylvilagus palustris*	2	1	0	1	0	9
Gray squirrel, *Sciurus carolinensis*	5	10	19	5	9	22
Fox squirrel, *Sciurus niger*	1	1	0	1	2	7
Hispid cotton rat, *Sigmodon hispidus*	13	2	0	1	27	26
Marsh rice rat, *Oryzomys palustris*	3	0	0	2	45	14
Cotton mouse, *Peromyscus gossypinus*	4	2	0	1	17	15
Round-tailed muskrat, *Neofiber alleni*	1	1	0	0	15	13
Raccoon, *Procyon lotor*	12	36	1	8	42	10
River otter, *Lutra canadensis*	2	2	0	0	10	1
Black bear, *Ursus americanus*	0	0	0	1	20	6
Bobcat, *Lynx rufus*	5	2	0	3	25	12
Gray fox, *Urocyon cinereoargenteus*	6	2	0	3	21	13
Red fox, *Vulpes vulpes*	0	0	0	0	12	3
White-tailed deer, *Odocoileus virginianus*	14	26	0	10	31	29

Table 7.2. Occurrence of ectoparasitic arthropods on round-tailed muskrat (*Neofiber alleni*) in Florida, USA, based on a sample of 25 animals (adapted from Smith *et al.* 1988).

Arthropod	Proportion infested (%)	Mean no. per host
Mites		
Listrophorus caudatus	96	770
Listrophorus laynei	92	734
Prolistrophorus birkenholzi	84	236
Laelaps evansi	92	45
Androlaelaps fahrenholzi	72	8
Euschlongastia splendens	28	16
Garmania sp.	28	2
Radfordia sp.	20	1
Macrocheles mammifer	12	1
Tyrophagus sp.	12	1
Fleas		
Polygenis gwyni	4	1

red fox and coyote; sarcoptic mange in red fox; and hemorrhagic disease in white-tailed deer. These data do not really capture the impact of nuisance and harassment due to biting by flies, which is hard to document. The importance of the various disease agents and parasites varies regionally and temporally, of course, though arthropods and arthropod-borne diseases are often a component of the disease triangle.

Often wildlife are host to several disease agents and parasites simultaneously. Table 7.2 shows data from round-tailed muskrats in Florida and serves as an example of how several arthropods can occur simultaneously. Note that four of the mite species occurred on over 80% of the animals. Often parasites have their own particular niche, occupying only a portion of the animal's body, thereby reducing competition among them. The simultaneous occurrence of several parasites may weaken a host and exacerbate the effect of a particular problem, but often the diseases or parasites are of minor importance. This is especially true with mites (Acari) and lice (Insecta: Phthiraptera), which generally are not too debilitating.

DISEASE HOSTS

There are several types of hosts for disease agents. Understanding the different types is important for understanding disease biology and epidemiology. A **reservoir host** (usually simply called 'reservoir') is an animal that harbors a disease agent. Reservoir hosts must be able to support the disease in the absence of other species, providing a means of long-term persistence. The reservoir host must be able to provide the disease agent to other species, allowing spread of the disease. The reservoir host usually is not seriously affected by the disease, though there may be signs of infection. **Primary** or **definitive hosts** are species in which the disease agent passes the adult, sexual, or multiplicative stage of the life cycle. The primary hosts can also be the reservoir hosts. **Intermediate hosts** are animal species that the disease agent passes through during the immature or nonsexual phase of the disease life cycle. **Amplifier hosts** are animals in which the disease agent abundance is increased without severely affecting the host. **Incidental**, **aberrant**, or **unnatural hosts** are not the normal hosts for a disease, and although in some cases they are not susceptible, in other cases they are extremely susceptible, displaying overt signs of infection. These are also called **dead-end hosts** or **dilution hosts** because they may perish or not support high concentrations of the disease agent, proving to be unsuitable for uptake of the disease agent by a vector.

Diseases normally have an environment or habitat where they persist in a relatively stable way. Here the environment (or host) and the disease agent are co-evolved, allowing both to co-exist and neither to eliminate the other. The modification of habitat, or the movement of new wildlife, livestock or humans into an environment where a disease has evolved a stable relationship with its host can upset the balance and allow establishment of new disease-host relationships. In such situations, wildlife, livestock and humans can prove to be various types of host, and arthropods can prove to be vectors and/or intermediate hosts.

DISEASE TRANSMISSION

Wildlife disease agents can be transmitted vertically or horizontally. **Vertical transmission** refers to transmission of a disease agent from parent to offspring. **Horizontal transmission** refers to transmission of a disease agent from animal to animal, independent of their parental relationship. Horizontal transmission is the most common method of disease spread. It can result from several actions, including:
• skin to skin contact between members of the same species;
• airborne transmission of droplets containing disease agents;
• contact with secretions and excretions, including residual fecal materials;
• contact with genital and sexual materials;
• discharge from lesions;
• contact with infected carcasses;
• ingestion of contaminated water and food;
• transmission by other species.
Here we discuss only the latter means of disease transmission, transmission by other species, and specifically by arthropods. Because another species is involved in the transfer from host to host, it is considered to be a form of **indirect transmission**. Not surprisingly, **direct transmission** is defined as the transfer of a disease agent (pathogen) from one host to another without the involvement of another species. Transmission of an infective stage of a disease agent to a host can occur in various ways. **Passive transmission** occurs when the host is contaminated or infected accidentally through ingestion of food, water, or an infected arthropod; this occurs with many nematodes. **Active transmission** occurs when the disease agent actively penetrates the bodies of their host after gaining contact with them; this occurs with hookworms. Finally, **inoculative transmission** occurs when a vector such as a mosquito injects the disease agent into the new host

during the process of blood feeding, as occurs with the protozoa causing malaria.

The importance of arthropod transmission as a means of disease spread varies greatly among diseases. For some wildlife diseases it is the only means or most important form of transmission, but for other diseases it is less important or arthropod transmission does not occur. It is useful to note that transmission of disease agents by arthropods is not completely independent of the other routes of horizontal transmission. Arthropods can be associated with food and carrion, for example, and wildlife could contract a disease from an insect or by feeding on contaminated food. Nevertheless, because an arthropod (serving as a 'vector') is involved in the transmission or transport process, such diseases are called **vector-borne**.

Vectors differ in their ability to acquire disease agents and to infect hosts (**vector competence**). Insects usually are short-lived, and acquire and transmit disease agents quickly, with the incubation period in the vector lasting perhaps a few days. Insects may be quite mobile, and typically take many small blood meals. In contrast, ticks are more long-lived, and incubation may require months. Ticks lack wings, so they are much less mobile, and normally take few but large blood meals.

Differences in vectorial capacity or vector competence are due to several factors. There are genetic variants among a single species of disease agent, for example, that differ in their ability to be acquired or transmitted. The vectors also differ in susceptibility to oral infection and efficiency of transmission, population structure (density, longevity, etc.), host preference, and geographic distribution. Also, the vertebrate hosts differ in susceptibility, which may be manifested in the ability of the host to develop concentrations adequate to infect the vector, population structure (availability of susceptible stages), immune status (prior exposure may confer immunity), and overlap in space and time with the vectors. Often, the suitability of a host animal to produce adequate concentration of the disease agent, or the ability of the disease agent to replicate in the vector, determines the ability of the vector to acquire and transmit the agent. Disease concentration thresholds seemingly exist, especially with mosquitoes, below which transmission does not occur. Thus, certain hosts or vectors are more important in disease transmission cycles. With tick vectors, however, the concentration of disease agent within the host seems to be less important, as ticks can infect one another while feeding together on the same host. Such 'co-feeding' infection occurs with only minimal or incomplete systemic infection, and usually the 'donor' and 'recipient' ticks must be feeding in proximity.

CAUSES OF DISEASE

It can be quite difficult to determine the cause of a disease. Factors that are thought to be causative factors of disease are called **putative** factors because their occurrence is often correlated with presence of the disease, but eventually evidence of their association is necessary before a cause and effect relationship is confirmed. The cause of a disease is often called the **etiologic agent**. It is also common to discuss the **sign** or **clinical sign** to indicate objective evidence that a disease is present, such as hair loss. It is unusual to use the term **symptom** with respect to diseases of animals, as this is a subjective assessment such as pain or lack of energy, and is something that cannot be communicated to humans by animals. It is interesting to note that entomologists have developed slightly different terminology than have wildlife biologists when describing diseases; entomologists use 'sign' to indicate physical manifestations of disease, and 'symptom' to indicate changes in behavior or function attributed to disease.

In some cases, criteria have been established to define whether or not a cause and effect relationship exists between a disease and a putative cause. Often, healthy animals are exposed to a putative cause; if the disease develops, causation is considered to be proved. For infectious disease agents, it is necessary to prove **Koch's Postulates** to establish the relationship between a disease and its putative cause. The rules of Koch's postulates are:
• the putative agent must be found in every case of the occurrence of disease;
• the putative agent must not be associated with the absence of disease;
• the putative agent must multiply in the host when provided the opportunity;
• the putative agent must be capable of being re-isolated from experimental inoculated individuals.
Though Koch's postulates are generally acceptable in ascribing cause and effect relationships, in many cases disease results from many stressors, including some abiotic factors. Thus, disease may be expressed only when a combination of events or factors occurs. Commonly, disease has primary and secondary causes, or

situations that predispose an animal to disease. For example, it is easy to imagine how poor diet or adverse weather could interact with microbial pathogens by weakening the host's ability to fend off infection. Some diseases are **latent**, which means that the potential for expression of the disease is present, but expression is suppressed and the animal is not contagious until triggered by something in the animal's biology or environment. Indeed, it is not uncommon for an animal to be infected (to have the disease-causing organism within its body), but for there to be no apparent dysfunction (lacking disease). The lack of detectable dysfunction can be due to inherent resistance (the same pathogen in another population or species induces expression of disease) or by the condition or vigor of the animal, which suppresses disease expression.

For **noninfectious diseases** (those lacking an infectious agent that can be spread from host to host) such as exposure to pesticides, it is not possible to apply Koch's postulates. In these cases, we must be content with exposing healthy animals to the stressor and observing the animal's response, or by searching for chemical residues in animals suffering from disease. This is not always satisfactory, as it is difficult to know the appropriate dose to test. Also, though it is easy to assess **acute toxicity** caused by high doses because the animal's response is usually rapid, it can be difficult to assess the effects of **chronic toxicity** caused by low doses applied over a long period. Low doses can interact with host metabolism, such as the hormonal system of the animal, or with other agents such as microbial pathogens, and it can be difficult to identify the true factor responsible for poor animal performance.

THE NATURE OF PARASITISM

Organisms that live in association with another animal are usually referred to as a host and symbiont, and the relationship is referred to as **symbiosis**. In the case of wildlife disease, the microparasites (e.g., viruses, bacteria, and fungi) or macroparasites (e.g., protozoa, helminths, and arthropods) are the **symbionts**, and the wildlife (but in some cases livestock and humans) are the final or definitive hosts. In the examples here, arthropods are often intermediate hosts. The nature of symbiosis varies, with the most common types:
• *Mutualism*. The host and symbiont are dependent on one another, and the relationship is mutually

beneficial. The microbial flora found in the digestive system of ruminants is a good example of mutual dependency.

• *Commensalism*. The host provides the habitat and food for the symbionts, which live without benefit or harm. In this case, the symbionts are dependent on the host, but the host does not depend on the symbiont. Many microbial parasites of wildlife have co-evolved with their normal hosts and cause no measurable harm, so under normal conditions they qualify as commensals. However, this relationship can shift when the host is stressed, and under these conditions may cause injury.

• *Parasitism*. The host supplies physiological support for the symbiont by providing habitat and sustenance, and often transport, and the symbiont is harmful to the host. The symbionts discussed here are parasites of wildlife to some degree, and sometimes to livestock, pets or humans. Their effect on wildlife may be severe or mild, and sometimes nearly imperceptable.

There are several forms of parasitism, with parasitism classified according to where the parasites are found, their temporal occurrence, and their relationship with their host. Some major forms of parasitism include:

• *Ectoparasites*. These organisms live on the external surface of the host, or cavities that open directly onto the surface. Foremost among the ectoparasites are the arthropods, such as lice, mites, and ticks.

• *Endoparasites*. These live within the body of the host, including the digestive system, lungs, liver, tissues, cells and freely in the body cavity. Viruses, bacteria, fungi, protozoa, tapeworms, and nematodes are good examples of endoparasites.

• *Temporary parasites.* These visit the host only briefly, usually for food. Bloodsucking arthropods are a good example of such parasites.

• *Stationary parasites*. These spend a definite period of development in association with the host, either on or in it. Some are **periodic parasites**, leaving to spend another portion of their life in a non-parasitic mode. Examples of this include ticks and botflies. Others are **permanent parasites**, spending all of their life on a host except for a brief period when transferring from host to host. The viruses, bacteria, protozoa and helminths are examples of this (although in some cases they may persist for long periods, in the absence of a suitable host, in a resting stage).

• *Incidental* or *aberrant parasites*. These are parasites that occur only occasionally in a host, usually because there are behavioral or ecological barriers that keep the parasite from the prospective host.

• *Obligate parasites*. These parasites cannot continue with their life cycle unless they have access to a certain host (or hosts). Most of the parasites discussed here have a certain host or a restricted range of hosts to which they must gain access.

• *Facultative parasites*. These are parasites that do not require a certain host, but sometimes are found in association with it.

Parasites display many adaptations that make it possible to have a symbiotic relationship with their host. Although they have the same basic nutritional requirements as the host, the symbiont must have adaptations that allow them to enter or attach to a new host after they have escaped from their old host. The tarsal adaptations of lice that allow them to cling to the body hairs of their host are a readily visible adaptation, as are the hooks and suckers of helminths living in association with the host's digestive tract. The symbiont must also escape the host, or at least its progeny must escape, in order to find new hosts. The avenue of escape may be *direct*, as when eggs, immature stages of nematodes, or body segments (proglottids) of tapeworms are released from the host's anus. Escape may also be *indirect*, as when an arthropod feeds on the blood of a host, takes up the symbiont, and transports it. Symbionts escaping both directly and indirectly may have one or more additional hosts that provide transport, maintenance, or an opportunity for propagation.

Organisms with direct escape from the old host and direct transmission to the new host can be said to have a **direct life cycle**. Those with indirect escape and indirect transmission can be said to have an **indirect life cycle**. The latter condition is more complicated, as host, symbiont, vector, and environment must all coincide for infection and development to occur. Here we are concerned mostly with disease organisms with an indirect life cycle, and specifically those that involve arthropods.

PARASITE-INDUCED CHANGES IN HOST BEHAVIOR

Parasites often have an intimate, highly developed relationship with their host that formed over evolutionary time. Both definitive and intermediate hosts display an array of defense mechanisms allowing prospective hosts to avoid, eliminate or tolerate parasites, while parasites display ingenious ways to find and exploit their host. Among the most intriguing aspects of parasitism is the modification by parasites of host behavior when the parasite has successfully infected the host. This is usually presented as an adaptation on the part of the parasite to enhance its fitness.

An **adaptation** is a genetically determined feature that becomes prevalent in a population because it confers a selective advantage to the organisms bearing the feature. In the case of symbiotic organisms, one species may benefit, but not the other. In the case of parasitic organisms, the changes in host behavior are believed to enhance parasite transmission from host to host. Arthropod behavior seems to be readily manipulated by diseases and parasites. For example, caterpillars that are infected with insect-pathogenic viruses often climb to the top of their host plant before they die, where they attach with their prolegs. After they die, their integument ruptures, allowing their liquefying body contents to drip down. Thus, foliage at lower levels of the plant is contaminated with virus, and other caterpillars ingest the virus as they feed. This assists in the spread of the virus from insect to insect. Similarly, flies and grasshoppers infected with fungal disease climb to the top of vegetation and attach, the fungus grows out of the dead insects and produces spores, and the elevated position of the cadaver facilitates the spread of fungal spores by wind or rain. Birds infected with arboviruses often display reduced activities, including reduced preening and other antimosquito behavior, allowing other bird-feeding mosquitoes better opportunity to feed on the infected host, and to acquire the virus for further spread. A classic example of modified host behavior involves the Lancet fluke, *Dicrocoelium dendriticum*, and *Formica* ants. When ants are infected, instead of returning to their nest during the evening and cool periods like uninfected ants, they attach to foliage where they can be consumed by grazing animals. Once ingested, the flukes can infect their vertebrate host, completing the parasite's life cycle (see Chapter 9).

It has become commonplace to infer that modified behavior in parasitized hosts is an adaptation induced by parasites, but often the evidence is lacking. Rarely is an improvement in the functioning of the parasite actually demonstrated. Usually we can observe only the present state of a character, not its evolutionary history, so caution is needed before drawing conclusions about the adaptive value of a trait. For example,

although the tendency of insects that are infected with viruses and fungi to perish in elevated locations is well documented, and logically this behavior will enhance dispersal of the parasites, there is another reason for insects to ascend vegetation. Infected insects will climb vegetation and bask in the sun, elevating their body temperatures to levels that are unsuitable for the parasites that infect them, and optimizing the temperature at which the insect's immune system functions. This phenomenon is called **behavioral fever**, a form of self-medication that can result in elimination of the parasites due to exposure of ectotherms such as insects to high temperature. Thus, there are multiple explanations for a single behavior, and it pays to be cautious about concluding adaptive significance to behavior without adequate experimental evidence.

SUMMARY

• To transmit disease agents effectively, insects must be able to locate and feed on their host animals. Some have a permanent relationship with their host, and are normally found hitchhiking. Others, however, feed only periodically, so host location is a more important issue. Both olfaction and vision are important in host location.

• Host animals display many defense behaviors that help to deter feeding by arthropods, including grooming, head shaking, wing flapping, ear flicking, and tail swishing.

• Infectious diseases of wildlife are caused by microorganisms (e.g., prions, viruses, bacteria, fungi), and are studied by microbiologists. Parasitic diseases are caused by larger organisms (e.g., protozoa, helminths, insects), and are studied by parasitologists.

• Diseases occurring at low or normal levels in animal populations are called enzootic or endemic, whereas when the disease becomes abundant it is called an epizootic or epidemic. The disease agent, the host animal, and the environment interact to determine disease expression.

• The ability of disease agents to impair the functioning of an animal is called virulence or pathogenicity. It is easy to underestimate the importance of disease in wildlife populations. Sick animals often are killed and consumed by predators or scavengers.

• Wildlife differ in their relationship with disease agents. They may serve as reservoirs, or as primary, intermediate, amplifier, and incidental hosts.

- Insects (arthropods) provide a means of indirect transmission of disease agents among wildlife; such diseases are called vector-borne. Arthropod vectors differ in their ability to contract and transmit diseases.
- The effect of infectious and parasitic organisms on their hosts is variable, ranging from complete lack of effects to serious impairment. Among the types of parasitism found in wildlife are ectoparasites, endoparasites, and temporary, stationary, incidental, obligate, and facultative parasites.
- Parasites can cause not only physiological effects, but also can modify the behavior of their hosts. The modified behavior sometimes benefits the parasite, but sometimes benefits the host.

REFERENCES AND ADDITIONAL READING

Chowdhury, N. & Aguirre, A.A. (2001). *Helminths of Wildlife*. Science Publishers, Inc., Enfield, New Hampshire, USA.

Forrester, D.J. (1992). *Parasites and Diseases of Wild Mammals in Florida*. University Press of Florida, Gainesville, Florida, USA.

Hudson, P.J., Rizzoli, A., Grenfell, B.T., Heesterbeek, H., & Dobson, A.P. (2002). *The Ecology of Wildlife Diseases*. Oxford University Press, Oxford, UK.

Lehane, M. (2005). *The Biology of Blood-Sucking in Insects*, 2nd edn. Cambridge University Press, Cambridge, UK.

Moore, J. (2002). *Parasites and the Behavior of Animals*. Oxford University Press, Oxford, UK.

Olsen, W.O. (1986). *Animal Parasites. Their Life Cycles and Ecology*. Dover Publications, New York, New York, USA. (reprint of 3rd edn., 1974)

Poulin, R. (1995). 'Adaptive' changes in the behaviour of parasitized animals: a critical review. *International Journal for Parasitology* **25**, 1371–1383.

Smith, M.A., Whitaker Jr, J.O., & Lane, J.N. (1988). Ectoparasites of the round-tailed muskrat (*Neofiber alleni*) with special emphasis on mites of the family Listrophordae. *American Midland Naturalist* **129**, 268–275.

Wobeser, G.L. (2006). *Essentials of Disease in Wild Animals*. Blackwell Publishing, Oxford, UK.

Chapter 8

INFECTIOUS DISEASE AGENTS TRANSMITTED TO WILDLIFE BY ARTHROPODS

The groups of organisms that typically cause infectious disease in wildlife are listed below, along with some of their characteristics, and examples of diseases that are caused by these insect-transmitted agents. Of considerable interest is the fact that wildlife can be hosts of some of the diseases affecting humans (and our livestock and pets). This should not be surprising, because humans, livestock, and pets are animals too, just like wildlife, so it is to be expected that the organisms affecting wildlife might also affect us. But this means that wildlife, by serving as a reservoir or intermediate host of animal diseases, can be a threat to the health of humans and domestic animals. Of course it also means that unmanaged insect populations can be a threat to wildlife, humans, and domestic animals, and that correct management of insect vectors has the potential to alleviate diseases affecting wildlife and humans.

The relationships among wildlife, diseases affecting wildlife, other hosts (humans and domestic animals), and insects are varied and complex. In some cases, insects are important vectors of disease agents to wildlife, whereas in other cases they are incidental. Sometimes the diseases are maintained almost exclusively in wildlife, but in other cases they spill over to other hosts. Sometimes wildlife have co-evolved with the disease and are immune to disease that affects other hosts, but in other cases wildlife are severely affected. In some cases, insects and disease affect wildlife indirectly, via food poisoning (toxins). The examples of diseases discussed below were selected to show some of the different types of relationships known among insects, disease, and wildlife.

VIRUSES

Viruses are infectious agents that are unable to grow or reproduce outside the cells of their host; they are obligate parasites. Each virus particle, or **virion**, consists of genetic material (either DNA or RNA) within a protective protein coat, or **capsid**. Viruses infect all forms of life, including animals, plants and bacteria. Most are very small and cannot be seen using ordinary (light) microscopes. Viruses are a very simple form of life, and some have argued that they are non-living because they do not meet all the criteria of the definition of life, such as the presence of cells. However, viruses have genes and evolve by natural selection, so it is best to consider them life forms. Inside a host, the virus uses the host cell's metabolic processes to produce more virus. Viruses are found nearly everywhere, but not all viruses cause disease. Some viruses are very hardy, being very resistant to the external environment, but in general they do not survive well outside the animal. Some cause important diseases of wildlife, of insects, or are transmitted by insects to wildlife or to humans, perhaps using wildlife as one host. Undoubtedly, viruses are important disease-causing agents of wildlife and humans, and insects are often involved in the disease cycle. Table 8.1 shows

Insects and Wildlife: Arthropods and Their Relationships with Wild Vertebrate Animals, 1st edition. By J.L. Capinera. Published 2010 by Blackwell Publishing.

Table 8.1. Some viruses that affect wild animals, or involve wild animals and humans, and are transmitted by insects.

Virus family	Disease	Animal hosts	Vector	Human host
Poxviridae	Myxomatosis	rabbits	fleas, mosquitoes	no
Poxviridae	Avian Pox	birds	mosquitoes	no
Asfarviridae	African Swine Fever	warthogs, swine	soft ticks (*Ornithodoros*)	no
Flaviviridae	West Nile Virus	birds, reptiles, mammals incl. rodents	mosquitoes	yes
Flaviviridae	Dengue Fever	primates	mosquitoes	yes
Flaviviridae	Yellow Fever	primates	mosquitoes	yes
Flaviviridae	St. Louis Encephalitis	birds, some mammals incl. rodents	mosquitoes	yes
Flaviviridae	Japanese Encephalitis	water birds, swine, cattle, horses	mosquitoes	yes
Flaviviridae	Murray Valley Encephalitis	water birds	mosquitoes	yes
Flaviviridae	Louping III Virus	deer, hares, rodents, grouse, domestic animals	hard ticks (*Ixodes*)	yes
Flaviviridae	Omsk Hemorrhagic Fever	muskrats, rodents	hard ticks	yes
Togaviridae	Eastern Equine Encephalitis	birds, mammals incl. horses, reptiles	mosquitoes	yes
Togaviridae	Western Equine Encephalitis	birds, mammals incl. horses, reptiles	mosquitoes	yes
Togaviridae	Venezuelan Equine Encephalitis	birds, mammals incl. horses and rodents	mosquitoes	yes
Togaviridae	Fort Morgan Virus	birds	cimicid bugs	no
Togaviridae	Ross River Virus	birds, mammals incl. horses and rodents	mosquitoes	yes
Reoviridae	Bluetongue Disease	wild and domestic ruminants	midges (*Culicoides*)	no
Reoviridae	Epizootic Hemorrhagic Disease	ruminants, especially deer, bighorn, pronghorn, cattle, sheep	midges (*Culicoides*)	no
Calciviridae	Rabbit Hemorrhagic Disease	wild and domestic rabbits	fleas, mosquitoes	no
Bunyaviridae	Uukuniemi Virus	birds	hard ticks (*Ixodes*)	no
Rhabdoviridae	Flanders Virus	birds	mosquitoes	no
Parvoviridae	Colorado Tick Fever	rabbits, hares	hard ticks (*Dermacentor*)	yes

some examples of virus diseases that involve wildlife and insects.

Viruses cause important diseases of wildlife, but also are important because domestic animals and humans sometimes are affected by the viruses, even though the virus may not thrive in these abnormal hosts. Domestic animals and humans are often called **dead-end hosts** for some viruses because they do not develop significant viremia (presence of virus in the blood) and do not contribute to the transmission cycle. Although

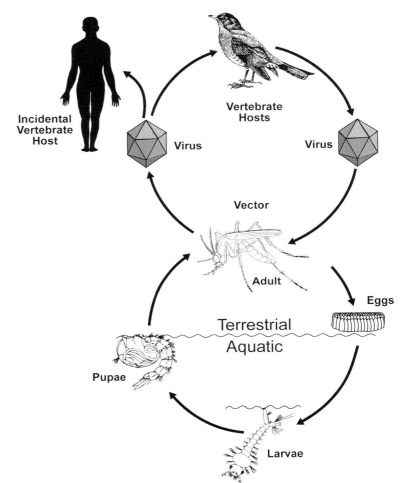

Fig. 8.1. The typical arbovirus transmission cycle. The virus is typically harbored in wild animals and transmitted from animal to animal by an arthropod, most often a mosquito. The wildlife may or may not be affected adversely, but usually develop tolerance and may serve as reservoirs for infecting arthropods. Incidental hosts such as humans and livestock are sometimes infected, and although sometimes the infection is serious or fatal, these incidental hosts rarely develop a high enough titer in their blood to allow further transmission by an arthropod, thus being a 'dead end' for the virus.

infection by the virus can be debilitating or even lethal to wildlife, commonly it is followed by rapid recovery and long-lasting immunity. Mosquitoes and ticks are the most common vectors of arthropod-borne viruses, which often are called **arboviruses**. A very large number of arboviruses occur worldwide, but many are poorly known. Many arboviruses cause **encephalomyelitis**, inflammation of the brain and spinal cord; **febrile illness**, fever; **hemorrhagic fever**, bleeding and fever; or **febrile myalgia and arthralgia**, fever with muscle and joint pain or arthritis in humans and some other hosts. Sometimes, wildlife will be adversely affected by a virus, whereas in other cases they remain

unaffected. Thus, the role of wildlife is that of host or reservoir for most arboviruses. The typical arbovirus life cycle is shown diagrammatically in Fig. 8.1.

Arboviruses typically are managed by vector suppression, either by killing the insects directly with insecticides, or by preventing their survival or reproduction through habitat management. It is difficult to eliminate all vectors, of course, so use of repellents (materials that keep insects away, or repel them) or feeding deterrents (materials that reduce the tendency of insects to feed) are useful, at least for protection of humans. In some cases, disease monitoring is used to acquire advance information on possible epizootics, and to initiate public health campaigns aimed at preventing disease from being dispersed from wildlife to humans and domestic animals. Sentinel animals such as caged domestic chickens, which are known as **sentinel chickens**, are a handy way to assess the occurrence of disease transmission by mosquitoes because they can provide a source of blood that can be examined for the presence of the disease-causing arboviruses.

Myxomatosis

This viral disease of New World rabbits is transmitted by mosquitoes, black flies, fleas, lice, and mites. The virus adheres to the mouthparts of the arthropods as they probe the epidermis. Once the mouthparts are contaminated, the virus can be inoculated into another host during a later probe. The virus does not replicate in the vector. The natural hosts of the myxoma virus are the jungle rabbit, *Sylvilagus brasiliensis*, in South and Central America and *S. bachmani* in California, USA. It has only mild effects on New World rabbits, causing cutaneous fibroma (relatively harmless wart-like skin blemishes) but not a systemic disease. However, the myxoma virus is quite deadly to Old World rabbits, *Oryctolagus* spp. Infected European rabbits display a bedraggled appearance, with partially or completely swollen eyelids, discharge from the conjunctiva and nose, and swollen ears and head. Nodules measuring up to 1 cm in diameter may be found anywhere on the body. In addition to being almost blind, affected animals frequently are in respiratory distress.

In 1950, myxoma virus was introduced to Australia to reduce the population of European rabbits that had been deliberately but regrettably introduced, and where they had few natural enemies. The rabbits in Australia have caused the extinction of numerous animals and plants, and caused extensive soil erosion due to overgrazing. The virus quickly reduced the population of rabbits from 600 million to 100 million, but the rabbit populations have recovered partly as disease resistance has developed. Recently another virus, rabbit calicivirus, has been introduced to attempt population suppression in Australia. Myxomatosis was also introduced to France, where it spread rapidly throughout Europe, causing significant decreases in rabbit populations, and populations of some of the predators that fed upon them. Overall, the reduction in rabbit abundance is estimated at 30%–40% in Australia to 50%–70% in Great Britain. Vaccine is used in Europe to provide protection of rabbits from the virus; both live attenuated virus vaccines and inoculation with rabbit fibroma virus (which is innocuous to European rabbits) are used. In many respects, rabbit fibroma virus can generally be considered the eastern North American equivalent of myxoma virus, and is often seen in eastern cottontails, *Sylvilagus floridanus*, but with few serious effects.

Avian Pox

This widespread disease, also known as fowl pox and bird pox, affects a large number of bird families and is found throughout the world. Infection causes localized proliferation of epithelial cells and production of lesions. In many cases it is only a mild affliction, and does not result in death. However, when the eyelids or mucous membranes of the oral or respiratory cavities are affected, mortality among infected birds can be high. It is a greater problem when population densities of birds are high, as among flocking birds and in zoos. There are a number of strains involved in this disease. It is most abundant in warm and moist regions, and occurs most commonly when mosquitoes are numerous, as they vector the disease.

Distribution and prevalence of this disease are determined by a number of factors, including weather, vector abundance, host abundance, and the strains present. It is a slow-developing disease that is manifested in the presence of pox lesions (wart-like growths) on the unfeathered portions (legs, feet, and skin around the eyes and beak) of the bird. Affected birds also display weakness, vision and breathing problems, difficulty in swallowing, emaciation, and reduced egg production. Secondary infections are common. Among wild birds, it occurs most commonly in upland game birds, songbirds, marine birds, occasionally in raptors and rarely in waterfowl. Avian pox is transmitted mechanically by mosquitoes and other biting arthropods, by direct contact among birds, and by inhalation of virus particles. In remote island locations where exposure to avian pox is limited, exposure of naïve indigenous fauna has resulted in high levels of bird mortality. Also, among wild populations of northern bobwhite, *Colinus virginianus*, and wild turkey, *Meleagris gallopavo*, in the southeastern USA, avian pox is regarded as a serious problem. For many bird species, it is not a major issue. Management involves eliminating mosquitoes and their breeding sites, eliminating infected birds, and in the case of domestic birds, by disinfecting feeders, watering devices, and cages.

West Nile Virus

West Nile Virus has long been present in Africa, the Middle East, southern Europe, India and southern Asia. It was introduced to the western hemisphere in 1999, in New York, USA, and quickly spread through much of North America, attaining the Pacific coast in 2002. The source of the virus in North America is unknown, but several species of Eurasian birds occasionally migrate to North America, especially to the eastern seaboard of North America, and some may have been infected while in Europe. Among the occasional visitors to North America are Eurasian wigeon, *Anas penelope*; green-winged teal, *Anas crecca*; ruff, *Philomachus pugnax*; little gull, *Larus minutus*; and black-headed gull, *Larus ridibundus*. Alternatively, some seabirds are carried by tropical storms annually from Africa to North America, so this is a possible route of introduction. Birds such as cattle egret, *Bubulcus ibis*; black-headed gull, *Larus ridibundus*; yellow-legged gull, *Larus cachinnans*; little egret, *Egretta garzetta*; and gray heron, *Ardea cinerea* are possibilities. Finally, pet, zoo and domestic birds routinely pass through commercial airports such as J.F. Kennedy International Airport in New York, and could be a source of the virus. Epidemics in humans have occurred in Africa and Europe in addition to North America, and infection can cause fever, myalgia, rash, and encephalitis in some victims. In recent years it has become the most important arbovirus in North America. Its importance is likely to diminish as wildlife and humans develop resistance following exposure to the virus.

West Nile Virus is transmitted to birds principally by ornithophilic (bird-feeding) mosquitoes, although it has been found in other animals, and of course in humans. It also can be transmitted by organ transplant, blood transfusion, transplacental, and transmammary. Wild birds are the primary hosts, with *Culex* spp. being especially important vectors, and with humans and horses being accidental or 'dead-end' hosts that do not contribute to continued transmission. Domestic birds, except for geese, generally do not develop sufficient viremia to allow transmission. Domestic geese develop the necessary viremia to amplify transmission, and also suffer mortality. Migra-

tory birds are important in the dissemination of West Nile Virus in temperate regions, and in North America the virus can persist during the winter in southern climates from Florida to California and then be reintroduced to northern areas annually.

In North America, West Nile Virus has become a major mortality factor of corvid birds. American crows, *Corvus brachyrhynchos*, are especially likely to perish when infected. Crows, blue jays, *Cyanocitta cristata*, black-billed magpie, *Pica pica*, and other species of Corvidae account for about 80%–90% of the infected birds in most dead-bird surveys. Other birds that seem to have suffered significant declines in abundance include American robin, *Turdus migratorius*; tufted titmouse, *Baeolophus bicolor*; house wren, *Troglodytes aedon*; chickadee, *Poecile* spp.; and eastern bluebird, *Sialia sialis*. However, some other birds seem to be quite susceptible, including various owls and raptors. The mortality of birds in the western hemisphere is much higher than typically occurs in the eastern hemisphere, likely due to the increased virulence in the naïve birds of the western hemisphere. It appears that most of the birds adversely affected are peridomestic (species that are often found in association with human habitations).

Although corvid birds regularly test positive for West Nile Virus, and often die, this does not mean that they are the most important species in maintaining the virus or enhancing dispersal in North America. American crows, for example, though very susceptible, typically comprise a small proportion of the bird community in an area, rarely exceeding 10%. Also, birds differ greatly in their attractiveness to mosquitoes, which affects the transmission potential. Field studies in the eastern USA have shown that American robin, *Turdus migratorius*, is highly preferred, being fed upon 16 times more often than if mosquitoes displayed no preference, and that virus antibodies are present in over 40% of the robins, documenting exposure to the virus. Importantly, it was estimated that nearly 60% of the infected mosquitoes became infected by feeding on virus-infected robins. In contrast, house sparrow, *Passer domesticus*, is avoided by mosquitoes and less than 20% showed evidence of exposure. Fish crow, *Corvus ossifragus*, also is highly selected by mosquitoes and fed upon more often than would be expected. However, their low level of abundance, like that of American crow, precludes them from being important in the epidemiology of the virus. Thus, variability in host-feeding behavior by mosquitoes, host suitability for virus amplification, and bird host abundance all interact to affect disease potential.

An assessment of birds important in introducing West Nile Virus to Europe from Africa, amplifying the virus, and enhancing its spread, suggested that perhaps 100 of the 300 bird species considered may play a role in West Nile epidemiology. Among the potentially important species are cattle egret, *Bubulcus ibis*; gulls, *Larus* spp.; barn swallow, *Hirundo rustica*; house martin, *Delichon urbica*; common swift, *Apus apus*; common nightingale, *Luscinia megarhynchos*; crows, *Corvus* spp.; blackcap, *Sylvia atricapilla*; and European starling, *Sturnus vulgaris*. Further research will certainly show that not all are equally important. As in North America, *Culex* spp. mosquitoes seem most important in disease transmission.

West Nile Virus is not limited to birds, though it is best known as a pathogen of birds. Among other animals infected are eastern fox squirrel, *Sciurus niger*; black bear, *Ursus americanus*; white-tailed deer, *Odocoileus virginianus*; and American alligator, *Alligator mississippiensis*. It is likely that the list of non-bird hosts will grow as the disease becomes further established in North America.

Yellow Fever

Yellow Fever is found in the Americas, Europe and Africa, but is strangely absent from Asia. Formerly, it was quite important in both North and South America, including some cool-weather cities. Currently, it seems to be increasing in importance in Africa, making it not an emerging pathogen, but a re-emerging pathogen. This disease originated in Africa, and likely was introduced to the western hemisphere by the slave trade. In the 1600s–1800s it caused epidemics in North American cities, including New Orleans, Philadelphia, New York, and Boston in the USA, and Halifax in Canada. Epidemics also occurred in Ireland, Wales, Spain, Uruguay, and Chile. It decimated Napoleon's army in Haiti in 1802, causing the French to scale back their territorial ambitions in the New World. Yellow Fever virus was the first virus that was shown to be transmitted by a mosquito.

Yellow Fever cycles in both urban and forested (sylvan) environments, though the important hosts and vectors vary among locations. Primates (including grivet, *Chlorocebus aethiops*; mangabeys, family Cercopithecidae; bush babies, family Galagidae; baboons,

Papio spp.; and chimpanzees, *Pan* spp.) are the important wild hosts, and yellow fever can be fatal to howler monkeys, *Alouatta* spp; squirrel monkeys, *Saimiri* spp.; spider monkeys, *Ateles* spp.; and owl monkeys, *Aotus* spp. Typically, infection of monkey populations causes reduction in their density and collapse of the yellow fever epidemic, so the disease moves slowly through the population over a large area, returning after the population is replenished. Mortality among monkeys is much less common in Africa than in the Americas. Other animals such as the anteaters *Tamandua tetradactyla* and *Cyclopes didactylus*; kinkajou, *Potos flavus*; and various rodents may be involved as hosts, but their importance is uncertain.

Affected animals display fever, vomiting, pain, dehydration, and prostration, and sometimes hemorrhage including renal failure and death. Human infection produces similar maladies, including liver damage (resulting in jaundice, hence the name 'yellow fever'), and up to 75% of affected humans die.

The mosquito *Aedes aegypti* is often called the yellow fever mosquito due to its association with yellow fever, particularly in urban situations. *Aedes aegypti* is particularly dangerous due to its affinity for houses, where it enters and feeds while the inhabitants sleep. Although eliminated from South America in the 1930s and 1940s, *A. aegypti* has recovered and now poses a significant threat to South American urban areas due to its association with yellow fever. In sylvan (forested) sites of the Americas, other *Aedes* spp., *Haemagogus* spp., and *Sabathes chloropterus* are the important vectors. In Africa, other *Aedes* spp. and *Dicermyia* spp. transmit the virus among monkeys and to people. Antimosquito programs have brought yellow fever under control in many urban areas, but in rural areas it remains a threat. It is sometimes referred to as 'woodcutter's disease' due to its association with forested environments and the people who work there.

St. Louis Encephalitis

This virus is common throughout the western hemisphere, and until West Nile Virus was introduced to New York in 1999 it was the mosquito-transmitted disease agent of greatest medical importance in North America. St. Louis Encephalitis is transmitted by *Culex* mosquitoes in North America. Birds are generally considered to be the most important hosts for amplification of the virus, with the disease transmission cycle typically described as mosquito-bird-mosquito. Nestling birds are considered to be particularly susceptible to infection, possibly because they are confined to the nest and less able to defend themselves from biting mosquitoes. They also display less immunity and display higher levels of viremia than do adults of the same species. Among the birds known to be suitable hosts of St. Louis Encephalitis are northern cardinal, *Cardinalis cardinalis*; American robin, *Turdus migratorius*; northern bobwhite, *Colinus virginianus*; blue jay, *Cyanocitta cristata*; northern mockingbird, *Mimus polyglottos*; and mourning dove, *Zenaidura macroura*. In urban environments birds such as the house sparrow, *Passer domesticus*; house finch, *Carpodacus mexicanus*; and European starling, *Sturnus vulgaris*, are important in virus amplification. Birds inoculated with the virus by mosquitoes quickly develop viremia, allowing other mosquitoes to contract the disease by feeding on the blood of the infected bird, although the virus disappears equally quickly as the birds recover from infection. Thus, the 'window of opportunity' for mosquitoes is quite short, and disease spread occurs mostly when mosquitoes are very abundant. Outbreak of the disease dissipates as the birds become resistant to the virus. Unlike some other mosquito-borne arboviruses such as Eastern Equine Encephalitis and West Nile Virus, St. Louis Encephalitis does not adversely affect the bird hosts.

Other vertebrates are suspected to be hosts in the southeastern USA, where antibodies were found in Virginia opossum, *Didelphis virginiana*; northern raccoon, *Procyon lotor*; cotton mouse, *Peromyscus gossypinus*; and nine-banded armadillos, *Dasypus novemcinctus*. Elsewhere, antibodies were found in big brown bat, *Eptisicus fuscus*; little brown bat, *Myotis lucifugus*; coyote, *Canis latrans*; red fox, *Vulpes vulpes*; striped skunk, *Mephitis mephitis*; jackrabbits, *Lepus* spp.; deer mouse, *Peromyscus maniculatus*; yellow-bellied marmot, *Marmota flaviventris*; and many other vertebrates. The presence of antibodies, though suggestive, does not by itself prove that an animal is an important host because the concentration of virus may be inadequate for successful infection of mosquitoes, or the host may not be attractive to mosquitoes that are important in the spread of disease to humans. A few mammals have been shown experimentally to develop viremia adequate for spread of the virus, including wood rat, *Neotoma mexicana*; Audobon's cottontail rabbit, *Sylvilagus audubonii*; young (but not old) cotton

rat, *Sigmodon hispidus*; and least chipmunk, *Eutamias minimus*.

In Central and South America, the disease is transmitted by at least seven genera of mosquitoes. As in North America, birds seem to be the most important hosts, but the St. Louis Encephalitis virus has also been found in other vertebrates such as vesper mouse, *Calomys musculinus*; grass mouse, *Akodon arviculoides*; a rice rat, *Oryzomys nigripes*; southern opossum, *Didelphis marsupialis*; howler monkey, *Alouatta nigerrina*; a spider monkey, *Atles panisicus*; a three-toed sloth, *Bradypus tridactylus*; various bats, and others. Whether or not these non-bird vertebrates are important in disease cycling is not evident, but at the very least these animals provide mosquitoes with blood meals, enhancing mosquito survival and reproduction.

Many vertebrates have become viremic, some consistently, but disease is not apparent in wildlife. In humans, the virus ranges from unapparent infection to coma and fatal encephalitis. However, only about 1% of infections are clinically apparent and most remain undiagnosed. Different strains of the virus exist, and they differ in their ability to cause disease. The young and elderly are more often affected, but typically it is a more severe ailment in the elderly than in the young. St. Louis Encephalitis inflicts neurological damage such as memory loss, paralysis, and deterioration of fine motor skills in humans. The damage is limited to the brain and spinal cord.

Hemorrhagic Disease

This disease is caused both by bluetongue virus and epizootic hemorrhagic disease virus, with the results of infection referred to collectively as hemorrhagic disease. These are similar orbiviruses in the family Reoviridae, and exist as many strains that differ in pathogenicity. Biting midges (Diptera: Certatopogonidae) of the genus *Culicoides* are the only known vectors of this disease, though many different species are involved around the world.

The name 'bluetongue' is based on the tendency of the tongue and mucosal membranes to turn blue, a result of cyanosis. This also can occur with epizootic hemorrhagic disease infection. Other common signs of infection include an arched back, lameness, and painful or cracked hooves in affected animals. Affected animals can appear disoriented, weak and staggering, or asymptomatic but followed by sudden death. Death may result from congestion of the lungs, necrosis, and internal hemorrhaging. The digestive tract and the organs are typically involved, including the kidneys, thymus, spleen, and lymph nodes. Reproduction is also affected, resulting in aborted or stillborn animals, particularly livestock. Because the viruses can be transmitted in semen, mandatory testing of semen is required prior to artificial insemination. Thus, economic loss results from the disease in livestock, the inability to export semen to countries that lack the disease, and from reduced abundance of wildlife.

Hemorrhagic disease affects wild and domestic ruminants in many parts of the world, including temperate and tropical areas. In the USA, disease is most prevalent in the western and southeastern states. Cattle and sheep are especially susceptible to infection, especially in North America. Most sheep and cattle remain asymptomatic, however. Hemorrhagic virus is reported to cause clinical disease in such North American species as white-tailed deer, *Odocoileus virginianus*; pronghorn, *Antilocapra americana*; bighorn sheep, *Ovis canadensis*; elk, *Cervus elaphus*; mountain goat, *Oreamnos americanus*; and American bison, *Bison bison*. Hemorrhagic disease is particularly lethal to white-tailed deer and pronghorn. Though much less of a problem thus far, many African species seem to be susceptible to infection, including greater kudu, *Tragelaphus strepsiceros*; muntjac, *Muntiacus reevsi*; Grant's gazelle, *Gazella granti*; gemsbok, *Oryx gazella*; sable antelope, *Hippostragus niger*; African buffalo, *Syncerus caffer*; ibex, *Capra ibex*; and others. Antibodies to bluetongue virus have been found in carnivores in Africa, including wild dog, *Lycaon pictus*; lion, *Panthera leo*; cheetah, *Acinonyx jubatus*; spotted hyena, *Crocuta crocuta*; and others. It is possible that these carnivorous animals were infected by preying on infected ungulates.

BACTERIA

Bacteria are small, but not nearly so small as viruses. They are single-celled organisms. Bacteria are **prokaryotes**, which means that they do not have a cell nucleus or other membrane-bound organelles. They contain a single chromosome with double-stranded DNA. Some bacteria are saprophytic, and others are parasites. Some survive for years under adverse environmental conditions. Traditionally, they are divided into one of two groups based on their staining characteristics when exposed to the 'gram stain.' Bacteria that stain blue are

said to be gram-positive; bacteria that stain pink are said to be gram-negative. Bacteria multiply rapidly in infected organisms, but can be eliminated by the host's immunological system. **Rickettsiae** are small, gram-negative, intracellular bacteria that cannot live outside the cell, so unlike many bacteria they cannot be cultured on artificial media. They are variable in form but often rod-shaped, have DNA, RNA, and cell walls. Due to their small size, they were once thought to be viruses or positioned somewhere between viruses and bacteria. They are found in one family (Rickettsiae) and several genera (e.g., *Rickettsia*, *Ehrlichea*, and *Anaplasma*). They are natural parasites of certain arthropods (lice, fleas, ticks, mites) and mammals, and cause many important diseases in humans and animals. **Spirochaetes** are also gram-negative bacteria, but are motile and free-living. Table 8.2 shows some bacteria that involve wildlife and insects.

The role of insects in transmitting bacteria is quite important for some diseases, but not for others. For example, avian cholera is a very important disease of birds, affecting over 190 species around the world, and sometimes causing massive death in both wild and domestic birds. About 35 species of ducks, geese and swans are susceptible, though this is the most susceptible group, with fewer than 10 species known to be susceptible in each of the following: wading birds, gallinaceous birds, doves and pigeons, woodpeckers, shorebirds, hawks and eagles, and owls. Transmission occurs principally by inhalation of aerosols (droplets), or ingestion of contaminated food or water. However, insects such as chewing lice, poultry mites, soft ticks, and tabanid flies have been shown to harbor and transmit the bacterium. Likewise, avian tuberculosis can be found in fowl tick (*Argas persicus*) and its feces, so it is implicated in mechanical transmission, though this is thought to be of little consequence as compared to direct contact among birds and contamination of food and water. In some cases, such as with heartwater, long-distance relocation of wildlife by humans can result in movement of the disease, and the ticks that vector it effectively, because the host animals are not sufficiently screened prior to transport.

Tularemia

The bacterium *Francisella tularensis* affects primarily lagomorphs (rabbits and hares) and rodents in the northern hemisphere, especially western North

Table 8.2. Some bacteria that affect wild animals, or involve wild animals and humans, and are transmitted by insects.

Bacterial species	Disease	Animal host	Vector	Human host
Anaplasma spp.	Anaplasmosis	wild and domesticated ruminants	hard ticks, tabanids	no
Cowdria ruminantium	Heartwater	wild and domesticated ruminants	ticks (*Amblyomma*)	no
Borrelia burgdorferia	Lyme Disease	mammals incl. rodents and rabbits, reptiles, birds	hard ticks (*Ixodes*)	yes
Borrelia anserina	Avian Spirochetosis	birds	soft ticks (*Argas*)	no
Borrelia spp.	Tick-Borne Relapsing Fever	rodents, squirrels, chipmunk	soft ticks (*Ornithodoros*)	yes
Francisella tularensis	Tularemia	mammals incl. rodents, birds	ticks, fleas, tabanids, mosquitoes	yes
Yersinia pestis	Plague	rodents, primates, cats, dogs, American badger	fleas	yes
Pasteurella multocida	Avian Cholera	birds, especially waterfowl	soft ticks, chewing lice, mites	no
Mycobacterium avium	Avian Tuberculosis	wild and domesticated birds, domesticated animals	soft ticks (*Argas*)	rare
Clostridium botulinum	Avian Botulism	birds esp. waterfowl and shorebirds	blowflies	no
Rickettsia rickettsii	Rocky Mountain Spotted Fever	mammals incl. rodents	hard ticks (*Dermacentor*)	yes

America, central Europe, and the former USSR, but also in Asia and Africa. However, it has a broad host range, affecting 190 species of mammals and 23 species of birds in addition to a few amphibians and numerous invertebrates. Among the more important hosts in North America are hares, *Lepus* spp.; New World rabbits, *Sylvilagus* spp.; water voles, *Arviocola* sp.; muskrat, *Ondatra zibithecus*; American beaver, *Castor canadensis*; lemmings, *Lemmus* spp.; voles, *Microtus* spp.; hamster, *Cricetus cricetus*; and red-backed voles, *Myodes* spp. In Europe, the disease is common among European brown hare, *Lepus europaeus*; varying hare, *Lepus timidus*; common vole, *Microtus arvalis*; house mouse, *Mus musculus*; common shrew, *Sorex araneus*; and others. Birds are relatively resistant, though gallinaceous game bird species such as grouse, as well as hawks, owls, and some waterfowl are susceptible.

Francisella tularensis is a highly infectious agent that enters the body in several ways, but primarily via inoculation by blood-feeding arthropods. Other routes of infection include inhalation of aerosols or handling and ingestion of contaminated water or meat. Among the important arthropod vectors are mosquitoes, fleas, tabanid flies, and ticks. A number of ticks from several genera are associated with transmission. Tularemia is primarily an acute disease, infecting the blood and causing inflammation and necrosis in wildlife. The liver, spleen, bone marrow, and lungs are affected.

In western North America, various host and vector systems are evident. In the eastern USA, cottontail rabbit, *Sylvilagus floridanus*, is infected by ticks and biting flies, with tularemia serving as a regulatory mechanism for rabbit populations. In Canada and northern USA, however, muskrat, *Ondatra zibethicus*, and American beaver, *Castor canadensis*, are most affected, with the disease apparently water-borne. In both cases, humans are also at risk, both from tick bites and from handling rabbits. The effect on wildlife populations, especially beaver, is significant in North America. In Europe, a different pattern occurs, with the vertebrate host primarily hares and the vector primarily mosquitoes.

Tularemia is rarely a problem for domesticated animals, though cats, dogs, horses, and sheep are occasionally affected. However, it is an issue for humans in areas where tularemia occurs. In some cases, it is advisable to delay the hunting season until cold weather reduces the density of ticks, and the probability of human infection. Also, trappers handling wildlife need to be aware of the risk. Finally, wildlife managers need to be sensitive to the population shifts due to incidence of tularemia, and may need to adjust harvest quotas in furbearers.

Anaplasmosis

This disease is caused by rickettsiae in the genus *Anaplasma*, with *A. marginale* the best known, and found throughout the tropical and subtropical regions of the world. It is an infectious, but not contagious, disease of ruminant animals that is transmitted mostly by ticks, but to a much lesser degree by tabanid flies. The host range of *Anaplasma* is considerable, including deer, *Odocoileus* spp.; elk, *Cervus elaphus*; giraffe, *Giraffa camelopardalis*; pronghorn, *Antilocapra americana*; American bison, *Bison bison*; cape buffalo, *Syncerus caffer*; Asian water buffalo, *Bubalus bubalis*; wildebeest, *Connochaetes* spp.; bighorn sheep, *Ovis canadensis*; and many African antelope species in addition to domestic cattle, sheep, and goats.

Anaplasma infect erythrocytes (red blood cells) exclusively, causing anemia and reduced hemoglobin concentrations due to damage to the erythrocytes. However, although domesticated ruminants are quite susceptible to infection, wild ruminants are quite resistant. In either type of ruminant, survivors that have been exposed to acute anaplasmosis usually regain normal hematologic parameters, and possess antibodies for some time.

Cattle, and to a lesser degree sheep, are affected by anaplasmosis. Wild ruminants, though substantially immune, play a major role in maintenance and spread of the disease. As noted previously, ticks are largely responsible for inoculating domesticated livestock, and

when wild and domesticated animals share the same pastures, the wild animals become a problem for ranchers. Only where effective tick vectors are absent is the problem diminished.

Lyme Disease

Lyme Disease, the clinical manifestation of Lyme borreliosis, is a chronic, debilitating disease of humans, and presently infects more humans than any other tick-borne infection. Lyme borreliosis is caused by the spirochaete *Borrelia burgdorferi* and is transmitted by hard ticks in the genus *Ixodes*. The important vector species vary geographically: *I. scapularis* and *I. pacificus* are the important species in the eastern and western regions of North America, respectively, whereas *I. ricinus* and *I. persulcatus* are the important species in Europe and Asia, respectively. Lyme disease now occurs in most temperate areas of the Holarctic region, and also from Australia and occasionally from South America and Africa. Lyme borreliosis has become increasingly important to humans in recent years due to increased contact between ticks borne on wildlife, and humans. Due to its increased prevalence, it is often designated an emerging pathogen.

The host range of Lyme borreliosis is impressive, with over 50 species of wild mammals known to support the pathogen. The important mammalian reservoirs in North America are eastern cottontail, *Sylvilagus floridanus*; wood rats, *Neotoma* spp.; kangaroo rats, *Dipodomys californicus*; white-footed mice, *Permyscus leucopus*; eastern chipmunks, *Tamias striatus*; black rat, *Rattus rattus*; and Norway rat, *Rattus norvegicus*. In Europe, the important mammalian reservoirs include mountain hare, *Lepus timidus*; squirrels, *Sciurus* spp.; wood mice, *Apodemes sylvaticus*; yellow-necked mice, *Apodemes flavicollis*; bank voles, *Clethrionomys glareolus*; and Eurasian hedgehog, *Erinaceus europaeus*. Also, a number of birds support the bacterium, including ring-necked pheasant, *Phasianus colchicus*; wild turkey, *Meleagris gallopavo*; Atlantic puffin, *Fratercula arctica*; house wren, *Troglodytes aedon*; robin, *Erithacus rubcula*; yellowthroat, *Geothlypis trichas*; song sparrow, *Melospiza melodia*; house sparrow, *Passer domesticus*; and orchard oriole, *Icterus spurious*. Migratory birds are important in spreading the disease and ticks. Adult ticks feed on larger animals, and various deer are favored. Although the deer are not good hosts for the bacteria, they are important because they are the principal source of nutrition for ticks that are already infected with *B. burgdorferi*. Often overlooked is the potential role of lizards and other reptiles to support ticks. However, research in Hungary showed that lizards were common hosts for *Ixodes ricinus*. Further, *Borrelia burgdorferi* was found in several species of lizards. The Lyme disease cycle is shown diagrammatically in Fig. 8.2.

Vectors of this disease are said to be **three-host ticks** because they take three blood meals, each from a separate vertebrate host. They feed as a larva, again as a nymph, and then as an adult. Fully fed (replete) female ticks detach from their final host to oviposit, and larvae hatch from the eggs. Typically, larvae ingest the bacteria during their first blood meal but are not capable of transmitting disease, whereas nymphs and adults developing from these infected larvae can infect new hosts. Nymphs and larvae usually feed on the same hosts; thus, nymphs are important in continuing the cycle because they inoculate vertebrate hosts that are fed upon by uninfected larval ticks. Rodents and some birds are persistently infected and fed upon by a high proportion of the immature tick population. In contrast, lizards and deer do not become persistently infected so do not act as such important reservoirs of the disease.

The primary site of infection is typically the skin, and a small inflammatory response at the site is normal. The spirochaetes soon spread to the lymph, blood, heart, bone and other organs. Signs of infection generally are absent from naturally infected wildlife, and most wildlife seem to tolerate the infection, though some mice display neurological disturbances. In general, there is no impact of *B. bergdorferi* on wildlife populations. In contrast, skin rash, arthritis and cardiac disease are prevalent in humans following infection.

The key to protecting humans from Lyme disease is personal protection. Hikers should seek to avoid tick infested habitats, wear light clothing to enable detection of ticks, wear long pants and sleeves to minimize skin exposure, tuck pants into boots to keep ticks away from skin, use insect repellents, and search for and remove ticks regularly. Tick control is of limited value. Habitat modification is useful if it decreases contact of humans with wildlife, and the ticks borne by wildlife.

Plague

Yersinia pestis affecting mammals is called plague, and is renowned for causing three pandemics that killed about 200 million people in past centuries. This disease, and its hosts and vectors, have had profound effects on human history. The first pandemic, known as the **Justinian Plague**, affected the Byzantine Empire, specifically Egypt, the Middle East, and Mediterranean Europe. It lasted from about 540 to 700 A.D. and caused about a 50%–60% reduction in human populations in the affected area. The second pandemic, called the **Black Death**, occurred in the 1300s and killed about 75 million people, or about one-third of the people in Europe. The third (unnamed) pandemic started about 1855 in China, but was soon spread to seaports in many countries around the world. India was particularly affected, realizing about 12.5 million deaths in 20 years. It was during this outbreak that the bacterial cause of the disease was discovered. It was not until 1908 that the involvement of rats was established, though it was suspected for centuries because massive deaths of rats typically preceded the occurrence of disease in humans. Technically, the third pandemic has not been terminated as there are continuing cases, particularly in Vietnam, where up to 250,000 cases are estimated to have occurred in the second half of the 1900s. Thus, plague remains an occasional problem, especially in less developed countries. Plague outbreaks typically subside when most susceptible hosts perish, or surviving hosts develop immunity.

Among humans, this bacterium causes three forms of plague, based on the nature of the infection. **Bubonic Plague** is transmitted by fleas to the skin, swellings called **buboes** occur, and lymph nodes are commonly affected. **Septicemic Plague** occurs when the flea injects the bacilli directly into a capillary vein, and the infection becomes general rather than localized in the lymph system. **Pneumonic Plague**, is caused by transmission of the bacilli through coughing or sputum following involvement of the lungs.

Plague normally resides in wildlife, specifically rodents, and is transmitted by fleas. Virtually all mammals can become infected with the pathogen, though susceptibility varies widely, and nearly all non-rodent and non-lagomorph species are considered to be incidental hosts. Maintenance of the plague at relatively low levels in wild rodents is sometimes referred to as **Sylvatic Plague** or **Campestral Plague**. In the USA, it persists at low levels in arid western rural environments but occasionally erupts to attain higher levels, even affecting humans in urban environments as far east as New York City. In the enzootic cycle the reservoir consists of 30–40 species of rodents, with

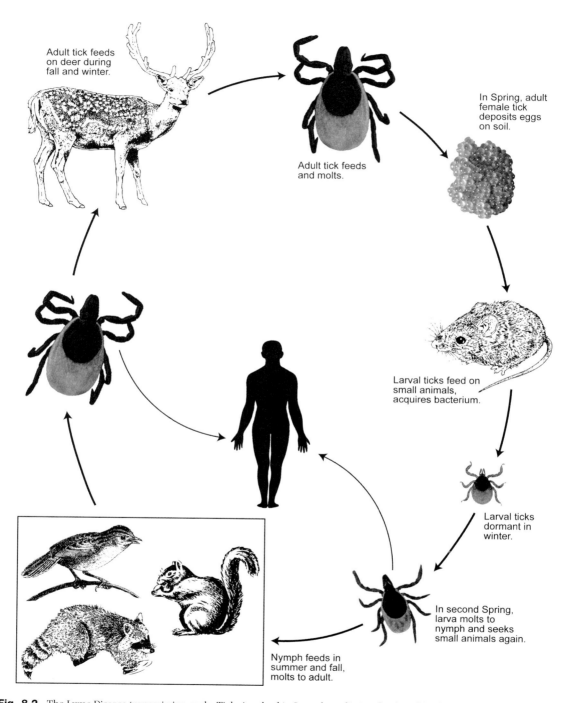

Fig. 8.2. The Lyme Disease transmission cycle. Ticks involved in Lyme borreliosis take three blood meals on the way to maturity, normally feeding on larger and larger hosts as they grow. Wildlife is not normally affected, but humans are incidental hosts that can be severely affected by infection with the spirochaete *Borrelia burgdorferi*.

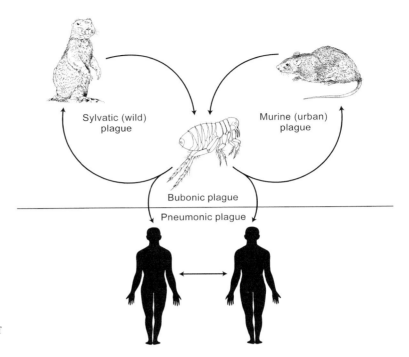

Sylvatic (wild)
plague

Murine (urban)
plague

Bubonic plague

Pneumonic plague

Fig. 8.3. Pathways of plague transmission, showing the different forms of plague and the involvement of fleas and wild animals.

birds, rabbits, carnivores, and primates unaffected. In urban areas, outbreaks usually involve urban rats. Outbreaks based on urban rat hosts is referred to as **Murine Plague**. Around the world, more than 200 species of rodents can be affected. The process of plague transmission is shown diagrammatically in Fig. 8.3.

Plague apparently was introduced to North America about 1900, and it remains mostly a problem in the western USA. During the enzootic stage, not all rodents perish from infection, thus perpetuating the disease in the animal population and allowing fleas to acquire the bacterium. Examples of hosts serving as a reservoir during the enzootic period are grasshopper mouse, *Onychomys leucogaster*; California vole, *Microtus californicus*; and rock squirrel, *Spermophilus variegates*. Some populations of these animals display some resistance, which is not surprising considering that it has been affecting susceptible animals in North America for over 100 years. For plague to emerge from the enzootic stage and attain epidemic status, high densities of susceptible hosts must be present. In the western USA, this role is often filled by prairie dogs, *Cynomys* spp. Plague has devastated populations of this once-abundant

animal. Not only are prairie dogs quite susceptible to the disease agent, but the problem is excerbated by the colonial nature of prairie dogs, which live in large colonies, as fleas and the disease agents they carry can easily be moved from animal to animal.

Among the non-rodent animals affected by plague are domestic cat, *Felis domesticus*; bobcat, *Felis rufus*; mountain lion, *Felis concolor*; black-footed ferret, *Mustela nigripes*; coyote, *Canis latrans*; and the foxes *Vulpes vulpes* and *Urocyon cinereoargenteus*. Reintroduction of Canada lynx, *Lynx canadensis*, to Colorado, USA has been affected by the susceptibility of lynx to plague, which they apparently contract by feeding on infected rodents and rabbits or hares. Oral introduction of the plague bacilli is thought to be the primary route of infection in this case, rather than flea bites. Black-footed ferret populations, which are almost exclusively dependent on prairie dogs as a source of food, were driven nearly to extinction by plague. The ferrets also become infected by eating infected prairie dogs.

Over 1500 species of fleas are potential vectors, though only a few are actually very important. The Oriental rat flea, *Xenopsylla cheopis*, is the most important vector on a worldwide basis, but others are more important locally. Interestingly, in a study of swift foxes, *Vulpes velox*, in Texas, USA, fleas moved from plague-infected prairie dogs, *Cynomys ludovicianus*, to the foxes. Thus, fleas that would not normally be associated with the foxes temporarily acquired a new host and gained additional mobility, and thus could move long distances from dog town to dog town as the foxes hunted. Arthropods other than fleas are insignificant in vectoring plague.

The bacterium affects the flea vector as well as the aforementioned hosts. Specifically, when fleas feed on blood from *Y. pestis*-infected animals, the bacteria may plug up the digestive tract of the flea, a malady called **blocking**. When fleas try to feed again, the incoming blood is mixed with bacilli in the area of blockage and because it cannot be fully ingested and processed the flea will regurgitate some blood and bacilli back into the host animal. The flea, which becomes increasingly hungry so long as the blockage remains, feeds and regurgitates again and again, potentially infecting other hosts. Sometimes fleas are able to clear the blockage, and typically blockage does not occur at temperatures above 28 °C. Infection of new hosts also can occur when an infected host is eaten or bacteria are inhaled. *Xenopsylla cheopis* is more prone to blockage of its intestinal tract than many other fleas, perhaps explaining why it is such an effective vector of plague.

Highly susceptible hosts such as mice and prairie dogs die quickly, often due to collapse of their circulatory system. Lesions (buboes) are formed on some animals, including humans. These swellings often are associated with the lymph nodes. Internally, the liver and spleen also develop lesions. Infected humans can be treated with antibiotics, though wildlife is not. In some cases, prairie dog towns have been treated with insecticide to kill fleas, though usually not out of concern for the prairie dogs, but to protect the endangered black-footed ferrets that feed upon prairie dogs. Research is being conducted on food-based vaccines to protect the prairie dogs, and hence the rare ferrets. Control of rodents near human habitations is often recommended to forestall infection of humans during epizootics.

Avian Botulism

Although bacteria are often transmitted when insects feed on their wildlife hosts, transmission of the toxin that causes avian botulism can occur in a different manner. Avian botulism is the most significant disease of migratory birds, especially waterfowl and shorebirds. Not surprisingly, it was formerly called 'duck sickness.' The bacterial complex known as *Clostridium botulinum* produces protein neurotoxins that cause **food poisoning** when animals (or humans) eat toxin-laden food. There are several different neurotoxins produced by *Clostridium*, with type C_1 being most common among birds, though loons and gulls are typically affected by type E toxin. Although over 250 species of birds have been found to experience botulism poisoning (scavengers such as vultures and crows seem to be resistant), it is filter feeding, dabbling, and fish-eating birds that are especially likely to ingest toxin. Apparently, the poison originates with invertebrates and plants living under anaerobic (depleted of oxygen) conditions in marshes and mud flats, because these bacteria only develop under anaerobic conditions. Decomposing vertebrate carcasses also support high levels of toxin production, producing a secondary form of poisoning called the **carcass-maggot cycle of botulism**. Decomposing animals are a very suitable substrate for growth of *Clostridium*, as decomposition also generates anaerobic conditions and the high temperatures that favor bacterial growth and toxin production. The dead animals are not attractive to waterfowl, but these birds will readily ingest any mature fly larvae (maggots) that disperse from the carcass, and most maggots do move a considerable distance as they leave the carcass and search for dry pupation sites.

Maggots developing in carcasses can have very high levels of toxin, and cause death to birds that feed upon them. Thus, all that is needed to initiate a prolonged cycle of botulism poisoning is initial death of some animals following a storm, collision with power transmission lines, algal poisoning, or feeding on invertebrates or decaying plant material rich in *Clostridium*. The bacteria normally found within the bodies of animals in wet habitats are then free to multiply in their carcasses, which are then assimilated into blowflies, which are fed upon by other birds, which then die and become available to more blowflies. Thus, outbreaks of botulism can originate in several ways, and persist for long periods. It is important to note that the animals are not killed because they became infected with the *Clostridium botulinum* bacterium, they die after ingesting their chemical metabolites, which happen to be poisonous neurotoxins. Thus, Avian Botulism is not really an infectious disease, but a biotoxin, and not quite comparable to most bacterial diseases. The avian botulism cycle is shown diagrammatically in Fig. 8.4.

How important is avian botulism? It is found throughout the world in 30 countries and all continents except Antarctica. Among the birds affected by type C botulism are hawks, owls, geese and ducks, herons and egrets, pelicans, cormorants, grebes, coots and rails, stilts and avocets, sandpipers, terns, ibis and spoonbills, and others. Those affected by type E botulism include ducks, geese, swans, mergansers, gulls, pelicans, grebes, and sandpipers. Some major wildlife die-offs caused by type C botulism include approximately 600,000–1,000,000 waterfowl in the Caspian Sea, USSR, in 1982; 100,000 in Alberta, Canada, in 1995; 117,000 in Manitoba, Canada, in 1996; 1,000,000 in Saskatchewan, Canada, in 1997; and 514,000 in Great Salt Lake, USA, in 1997.

FUNGI

Fungi are **eukaryotes**, which means that they have a cell nucleus, membrane-bounded organelles, and cells organized into complex structures. Thus, they are similar to all higher forms of life, and different from bacteria and viruses. Most fungi have a weak association with animals, and are opportunistic parasites of weakened or debilitated individuals. They may attack wounds or the lungs of animals, for example, usually causing a chronic disease. Wildlife that are stressed by high densities or poor environment are more prone to infection by fungi. Stress gives the appearance that the fungi are important pathogens, but the principal issue is the poor health of the animals, which can lead to infection by fungi.

Insects do not normally vector fungi to animals. This is surprising, because some insects (e.g., bark beetles, leafcutting ants) are well adapted to inoculate plants with fungi, and then to benefit from the association. These insects have evolved complex morphology or behavior to allow transport and inoculation of plants with their fungal associates. With the diversity of insects, and their ability to exploit nearly all environments and resources, it is surprising that they do not use fungi to exploit animal hosts as well. Nevertheless, there is an important wildlife–fungus association that is influenced by insects, the production of **mycotoxins** in plants. Mycotoxins can be produced by a number of fungi, though the most important mycotoxin affecting

wildlife is **aflatoxin**, which is produced by *Aspergillus* fungi, particularly *A. flavus* and *A. parasiticus*. Actually, there are several poisonous metabolites produced by the fungi that differ in their toxicity and etiology, with the metabolite AFB_1 being the most prevalent and toxic. Animals are affected when they ingest food that is contaminated with aflatoxin. As was the case with botulism mentioned earlier, animals perishing from aflatoxin poisoning are not actually infected by the fungus.

Aflatoxin Poisoning

Aflatoxin poisoning, like botulism poisoning, is a case of **food poisoning**. However, it is slightly different than the carcass-maggot cycle of botulism poisoning, wherein the wildlife are affected by feeding on toxin-containing insects. In the case of aflatoxin poisoning, the wildlife ingest plant products that contain high levels of aflatoxin. Like botulism, this disease is not truly infectious, but caused by a biotoxin.

So what is the role of insects in this disease? *Aspergillus* fungi develop in cereal grains when plants are stressed or damaged. Hot, dry weather is stressful to plants, as is damage by insects, hail, and early frost. The products normally affected by aflatoxin are cereal grains and oil seeds, particularly corn, rice, sorghum, cottonseed and peanuts. Insects also serve to move the fungus around from plant to plant in the field, or within stored grains as they feed. Even aflatoxin-free grain can become heavily contaminated if it is not stored properly after harvest. Insect infestation in stored grain causes increase in moisture levels within grain. Increased moisture levels facilitate growth of fungi including *Aspergillus*, and production of heat, which promotes more growth of *Aspergillus*, as optimal growth of this fungus occurs at 26–32 °C.

How important is aflatoxin poisoning to free-ranging waterfowl? Following are some examples of poisoning in the USA. In the winter of 1977–1978, approximately 7500 ducks and geese were killed in Texas following ingestion of crop residue, with corn and rice recovered from the digestive tract of the dead birds containing 500 ppb of AFB_1. Similarly, in 1998–1999 approximately 10,000 waterfowl died in Louisiana after eating affected corn. Corn intended for consumption by mature poultry is not allowed to have over 200 ppb of aflatoxin, so clearly these wild birds ingested too much aflatoxin. Survey of residual corn in farmers' fields, or planted for wildlife, has shown that

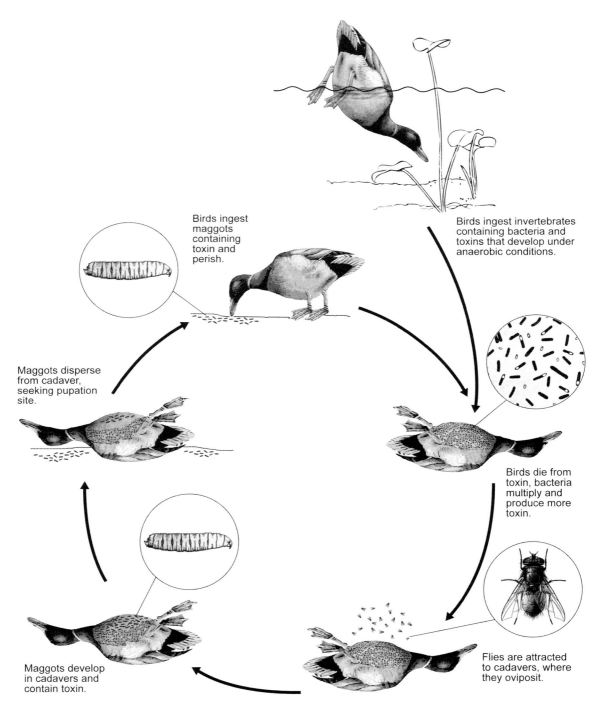

Fig. 8.4. The avian botulism transmission cycle. This disease involves the bacterium *Clostridium botulinum* and affects mostly waterfowl and shorebirds. Birds, but not flies, are affected by the bacterial toxins, so fowl become poisoned when they feed on carrion-feeding maggots bearing the toxin.

such grain may contain high levels of aflatoxin. For example, 32% of the corn growing near the Mississippi Sandhill Crane National Wildlife Refuge was found to contain over 200 ppb of AFB_1, and corn grown in southern Georgia and northern Florida to provide food for quail and other wildlife was found to contain up to 1200 ppb of AFB_1. Corn used to feed deer in North Carolina and South Carolina was found to contain 750 ppb of AFB_1, and even birdseed sold for home bird feeding has been found to contain aflatoxin.

Considerable potential exists for wildlife to be poisoned unintentionally by cultivated crops. Indeed, it is estimated that 25% of the world's food supply of grains is contaminated with aflatoxin, with contamination being higher in developing than industrialized countries. Also, the rate of aflatoxin infection is higher in warm, wet production areas such as the southeastern USA, where phytophagous insects abound. Numerous species have been poisoned by aflatoxins, including fish, birds, and mammals. Young animals are more susceptible than old animals, and animals such as rats and rabbits that metabolize the metabolites rapidly are more susceptible than those that metabolize the aflatoxin slowly. Serious poisonings have involved snow geese, *Chen caerulescens*; Ross's geese, *Chen rossi*; greater white-fronted geese, *Anser albifrons*; mallards, *Anas platyrhynchos*; northern pintails, *Anas acuta*; and ring-necked pheasants, *Phasianus colchicus*, among others.

SUMMARY

• Infectious diseases of wildlife can also affect humans, pets, and livestock, and wildlife often serve as a reservoir for such diseases even though they are not adversely affected.
• Viruses are minute intracellular organisms that use the host cell's metabolic processes to produce more virus. The viruses transmitted by arthropods are called arboviruses. Examples of arboviruses associated with wildlife and arthropods are:

Myxomatosis, which is spread by several insects to rabbits;
Avian Pox, spread mostly by mosquitoes to birds;
West Nile Virus, spread by mosquitoes mostly to birds;
Yellow Fever, spread by mosquitoes to mammals;
St. Louis Encephalitis, spread by mosquitoes to birds and mammals;

Hemorrhagic Disease, spread by biting midges to mammals.
Among the arboviruses that affect humans are West Nile Virus, Yellow Fever, and St. Louis Encephalitis.
• Bacteria (including rickettsiae and spirochaetes) are small, single-celled organisms that possess a cell wall. Bacteria often are transmitted by inhalation of aerosols (droplets), but insects can be important vectors. Among the bacterial diseases involving wildlife and arthropods are:

Tularemia, transmitted by several insects to mammals and a few birds;
Anaplasmosis, spread mostly by ticks to mammals;
Lyme Disease, tick-borne, and occuring in both mammals and birds;
Plague, transmitted by fleas, and normally occurring mostly in rodents;
Avian botulism, unlike most diseases considered here, is not transmitted by the feeding behavior of arthropods. It is a food-borne disease contracted by birds that eat maggots containing bacterial toxin.
Among the bacterial diseases of wildlife that affect humans are Lyme Disease, Tularemia, and Plague.
• Fungi are less intimately involved with vectors, but sometimes infect animals that are weakened due to environmental factors or wounding. Fungi commonly produce toxins, called mycotoxins, that can affect animals. A biotoxin causing disease in wildlife is aflatoxin:

Aflatoxin is a food-borne disease caused by *Aspergillus* fungi developing in grain crops that are stressed while growing, or by poor storage conditions after harvest. Insects are implicated in spreading *Aspergillus* fungi and creating conditions favorable to the disease.
Humans also are susceptible to aflatoxin poisoning.

REFERENCES AND ADDITIONAL READING

Barrett, A.D.T. & Higgs, S. (2007). Yellow fever: a disease that has yet to be conquered. *Annual Review of Entomology* **52**, 209–229.

Beard, C.B., Gordon-Rosales, C., & Durvasula, R.V. (2002). Bacterial symbionts of the Triatominae and their potential use in control of Chagas disease transmission. *Annual Review of Entomology* **47**, 123–141.

Biggins, D.E. & Kosoy, M.Y. (2001). Influences of introduced plague on North American mammals: implications from ecology of plague in Asia. *Journal of Mammology* **82**, 906–916.

Brown, R.N. & Burgess, E.C. (2001). Lyme borreliosis. In Williams, E.S., and Barker, I.K. (eds.). *Infectious Diseases of Wild Mammals*, 3rd edn., pp. 435–454. Iowa State University Press, Ames, Iowa, USA.

Davidson, W.R. & Goff, W.L. (2001). Anaplasmosis. In Williams, E.S. & Barker, I.K. (eds.). *Infectious Diseases of Wild Mammals*, 3rd edn., pp.455–466. Iowa State University Press, Ames, Iowa, USA.

Day, J.F. (2001). Predicting St. Louis encephalitis virus epidemics: lessons from recent, and not so recent, outbreaks. *Annual Review of Entomology* **46**, 111–138.

Földvári, G., Rigó, K., Majláthová, V., Majláth, I., Farkas, R., & Pet'ko, B. (2009). Detection of *Borrelia burgdorferi* sensu lato in lizards and their ticks from Hungary. *Vector-Borne and Zoonotic Diseases* **9**, 331–336.

Forrester, D.J. & Spalding, M.G. (2003). *Parasites and Diseases of Wild birds in Florida*. University Press of Florida, Gainesville, Florida, USA.

Friend, M. & Franson, J.C. (eds.). (1999). *Field Manual of Wildlife Diseases: General Field Procedures and Diseases of Birds*. US Department of the Interior, US Geological Survey, Washington, DC, USA.

Gage, K.L. & Kosoy, M.Y. (2005). Natural history of plague: perspectives from more than a century of research. *Annual Review of Entomology* **50**, 505–528.

Gasper, P.W. & Watson, R.P. (2001). Plague and ersiniosis. In Williams, E.S. & Barker, I.K. (eds.). *Infectious Diseases of Wild Mammals*, 3rd edn., pp. 313–329. Iowa State University Press, Ames, Iowa, USA.

Hoogland, J.L., Davis, S., Benson-Amram, S., Labruna, D., Goossens, B., & Hoogland, M.A. (2004). Pyraperm kills fleas and halts plague among Utah prairie dogs. *Southwestern Naturalist* **49**, 376–383.

Howerth, E.W., Stallknecht, D.E., & Kirkland, P.D. (2001). Bluetongue, epizootic hemorrhagic disease, and other orbivirus-related diseases. In Williams, E.S. & Barker, I.K. (eds.). *Infectious Diseases of Wild Mammals*, 3rd edn., pp. 77–97. Iowa State University Press, Ames, Iowa, USA.

Kilpatrick, A.M., Daszak, P., Jones, M.J., Marra, P.P., & Kramer, L.D. (2006). Host heterogeneity dominates West Nile virus transmission. *Proceedings of the Royal Society B* **273**, 2327–2333.

LaDeau, S.L., Kilpartick, A.M., & Marra, P.P. (2007). West Nile virus emergence and large-scale declines of North American bird populations. *Nature* **447**, 710–713.

Lenghaus, C., Studdert, M.J., & Gavier-Widén, D. (2001). Calcivirus infections. In Williams, E.S. & Barker, I.K. (eds.). *Infectious Diseases of Wild Mammals*, 3rd edn., pp.280–291. Iowa State University Press, Ames, Iowa, USA.

Marquardt, W.C., Demaree, R.S., & Grieve, R.B. (eds.). (2000). *Parasitology and Vector Biology*. Academic Press, San Diego, California, USA.

McGee, B.K., Butler, M.J., Pence, D.B., *et al.* (2006). Possible vector dissemination by swift foxes following a plague epi-

zootic in black-tailed prairie dogs in northwestern Texas. *Journal of Wildlife Diseases* **42**, 415–420.

McLean, R.G. & Ubico, S.R. (2007). Arboviruses in birds. In Thomas, N.J., Hunter, D.B., & Atkinson, C.T.(eds.) *Infectious Diseases of Wild Birds*, pp. 17–62. Blackwell Publishing, Oxford, UK.

Mörner, T. (2007). Tularemia. In Thomas, N.J., Hunter, D.B., & Atkinson, C.T. (eds.). *Infectious Diseases of Wild Birds*, pp.352–359. Blackwell Publishing, Oxford, UK.

Mörner, T. & Addison, E. (2001). Tularemia. In Williams, E.S. & Barker, I.K. (eds.) *Infectious Diseases of Wild Mammals*, 3rd edn., pp. 303–312. Iowa State University Press, Ames, Iowa, USA.

Mullen, G. & Durden, L. (2009). *Medical and Veterinary Entomology*, 2nd edn. Academic Press/Elsevier, Amsterdam, The Netherlands.

Norris, D.E. (2008). Lyme borreliosis. In Capinera, J.L. (ed.). *Encyclopedia of Entomology*, 2nd edn., pp. 2250–2253. Springer Science & Business Media B.V., Dordrecht, The Netherlands.

Olsen, B. (2007). Borellia. In Thomas, N.J., Hunter, D.B., & Atkinson, C.T. (eds.). *Infectious Diseases of Wild Birds*, pp. 341–351. Blackwell Publishing, Oxford, UK.

Robinson, A.J. & Kerr, P.J. (2001). Poxvirus infections. In Williams, E.S. & Barker, I.K. (eds.). *Infectious Diseases of Wild Mammals*, 3rd edn., pp.179–201. Iowa State University Press, Ames, Iowa, USA.

Quist, C.F., Cornish, T., & Wyatt, R.D. (2007). Mycotoxicosis. In Thomas, N.J., Hunter, D.B., & Atkinson, C.T. (eds.). *Infectious Diseases of Wild Birds*, pp. 417–430. Blackwell Publishing, Oxford, UK.

Rocke, T.E. & Bolliner, T.K. (2007). Avian botulism. In Thomas, N.J., Hunter, D.B., & Atkinson, C.T. (eds.). *Infectious Diseases of Wild Birds*, pp. 377–416. Blackwell Publishing, Oxford, UK.

Rothschild, M. & Clay, T. (1953). *Fleas, Flukes and Cuckoos. A Study of Bird Parasites*. Readers Union, London.

Samuel, M.D., Botzler, R.G., & Wobeser, G.A. (2007). Avian cholera. In Thomas, N.J., Hunter, D.B., & Atkinson, C.T. (eds.). *Infectious Diseases of Wild Birds*, pp. 239–269. Blackwell Publishing, Oxford, UK.

Van Riper III, C. & Forrester, D.J. (2007). Avian pox. In Thomas, N.J., Hunter, D.B., & Atkinson, C.T. (eds.). *Infectious Diseases of Wild Birds*, pp.131–176. Blackwell Publishing, Oxford, UK.

Wild, M.A., Schenk, T.M., & Spraker, T.R. (2006). Plague as a mortality factor in Canada lynx (*Lynx canadensis*) reintroduced to Colorado. *Journal of Wildlife Diseases* **42**, 646–650.

Wobeser, G.L. (2006). *Essentials of Disease in Wild Animals*. Blackwell Publishing, Oxford, UK.

Yuill, T.M. & Seymour, C. 2001. Arbovirus infections. In Williams, E.S. & Barker, I.K. (eds.). *Infectious Diseases of Wild Mammals*, 3rd edn., pp. 98–118. Iowa State University Press, Ames, Iowa, USA.

Chapter 9

PARASITIC DISEASE AGENTS TRANSMITTED TO WILDLIFE BY ARTHROPODS

The groups of organisms that typically cause parasitic disease in wildlife are listed below, along with some of their characteristics, and examples of disease that are caused by these insect-transmitted agents. As is the case with infectious diseases, wildlife can be hosts of some of the parasitic diseases affecting humans (and our livestock and pets), and livestock can be a host of some parasites that infect wildlife. By virtue of their ability to transport or cause parasitic diseases, unmanaged insect populations can be a threat to wildlife, humans, and domestic animals. In some cases, proper management of insect vectors has the potential to alleviate diseases affecting wildlife and humans.

As with infectious diseases, insect-borne parasitic diseases affecting wildlife, humans and domestic animals are varied in severity, and complex in life cycle. The examples of diseases discussed below were selected to show some of the different types of relationships known among insects, parasitic disease, and wildlife.

PROTOZOA

The protozoa are a large and diverse group of eukaryotes. They are unicellular but mobile. Protozoa have organelles, but not organ systems. Classification of Protozoa is not stable, so it can be quite confusing. Protozoa have complex life cycles requiring more than one host. Only about 20% are parasitic on vertebrate or invertebrate animals; most are free-living in aqueous environments. Some of the most important diseases affecting humans are caused by protozoa, including malaria, amoebic dysentery, and trypanosomiasis, and these diseases can be quite important to wildlife as well. Like bacteria, they are small and reproduce rapidly in their host; sometimes they induce immunity. Some protozoa that involve wildlife and insects are shown in Table 9.1.

American Trypanosomiasis

American Trypanosomiasis, also called **Chagas** (or Chagas') **Disease** is named after its discoverer, the Brazilian physician Carlos Chagas. It is caused by the trypanosome parasite *Trypanosoma cruzi*, a blood-inhabiting protozoan. This pathogen is transmitted to animals and people by kissing bugs (Hemiptera: Reduviidae: Triatominae). The disease occurs in the western hemisphere from Mexico to Argentina. It is estimated that 15–18 million people in Mexico, Central America, and South America have American Trypanosomiasis. Interestingly, most of these victims do not know they are infected because the acute symptoms fade away. If the disease remains unrecognized and untreated, however, there can be serious consequences later in life and the disease ultimately can be life threatening. About 50,000 people die each year from complications caused by American Trypanosomiasis.

Insects and Wildlife: Arthropods and Their Relationships with Wild Vertebrate Animals, 1st edition. By J.L. Capinera. Published 2010 by Blackwell Publishing.

Table 9.1. Some protozoa that affect wild animals, or involve wild animals and humans, and are transmitted by insects.

Protozoan species	Disease	Animal host	Vector	Human host
Trypanosoma cruzi	American Trypanosomiasis	domestic and wild vertebrates	kissing bugs	yes
Trypanosoma rhodesiense	East African Trypanosomiasis	domestic and wild ungulates, others	tsetse flies	yes
Trypanosoma cervi	Trypanosomiasis	deer, elk, moose reindeer	horse flies	no
Besnoitia besnoiti	Besnoitiosis	kudu, wildebeest, impala, cattle	biting flies, ticks	no
Besnoitia tarandi	Besnoitiosis	musk-ox, caribou, reindeer, mule deer	biting insects	no
Toxoplasma gondii	Toxoplasmosis	ocelots, jaguar, American bobcat, tiger, New World monkeys, Australian marsupials	flies, dung beetles, cockroaches	yes
Plasmodium relictum	Avian Malaria	passerine birds	mosquitoes (*Culex*)	no
Plasmodium spp.	Rodent Malaria	rodents	mosquitoes (*Anopheles*)	no
Leishmania spp.	Leishmaniasis	rodents, dogs	sand flies	yes
Cytauxzoon spp.	Cytauxzoonosis	African antelope	hard ticks	no
Babesia spp.	Babesiosis	deer, caribou, elk, wild sheep	hard ticks	no
Haemoproteus spp.	Haemoproteosis	birds, lizards, turtles, snakes	biting midges, louse flies, mosquitoes	no
Leucocytozoon spp.	Leucocytozoonosis	birds	black flies	no
Hepatozoon spp.	Hepatozoonosis	reptiles, mink, martens, weasel, wild and domestic cats and dogs, northern raccoon, giraffe, impala, deer, hares and rabbits, rodents	hard ticks, mites	no

There are two phases of disease development in humans: the acute and chronic phases. The acute phase occurs in the first few weeks or months of infection. Often it is symptom-free, or exhibits only mild symptoms and signs that are not unique to American Trypanosomiasis. The symptoms noted by the infected individual may include fever, fatigue, aches of the head or body, rash, loss of appetite, diarrhea, and vomiting. The signs of infection include mild enlargement of the liver or spleen, swollen glands, and local swelling. The most easily recognized indication of acute American Trypanosomiasis is called **Romaña's sign**, which includes swelling of the eyelids on the side of the face near the bite wound or where the bug feces were deposited or accidentally rubbed into the eye. Symptoms, if present, usually fade away after a few weeks or months. However, if untreated, the disease persists. Occasionally young children die from severe inflammation of the heart muscle or brain. The acute phase is more severe in immuno-compromised people. The chronic phase of infection may result in no signs of infection for decades or even for life. However, some people develop heart problems including an enlarged heart, heart failure, altered heart rate or rhythm, and cardiac arrest. Also, intestinal problems can occur, including an enlarged esophagus or colon, which can lead to difficulties with eating or defecation. A toxin is responsible for the destruction of tissues.

Although American Trypanosomiasis occurs primarily in the rural areas in Latin America, the movement of large numbers of people from rural to urban areas, and to other regions of the world, has increased its geographic distribution. It is a growing threat in the United States and the Caribbean region, where American Trypanosomiasis is not endemic; it is estimated that about 500,000 unknowingly infected immigrants now live in these more northern countries. In these areas, where kissing bugs are not common or not likely to bite humans, American Trypanosomiasis management should focus on preventing transmission from blood transfusion, organ transplantation, and mother-to-baby (congenital) transmission. Nevertheless, there are ominous reports of *Trypanosoma cruzi* occurring in wild mammal (northern raccoon, *Procyon lotor*; Virginia opossum, *Didelphis virginiana*) populations as far north as North Carolina, USA. For example, a study conducted in Alabama, USA found 13.5% of the opossums and 14.3% of the raccoons, and 6% of the triatomine bugs to be infected with *T. cruzi*. The triatomine bugs indigenous to North America are capable of transmitting American Trypanosomiasis. The failure of American Trypanosomiasis to be a problem in the USA is attributed to the low prevalence of infected bugs and vertebrate hosts, but also to the habit of the northern species not to defecate immediately after feeding. Even if humans are bitten, the North American bugs are less likely to defecate on humans than are the related species found in Central and South America, and thus are less likely to spread the disease (see below).

In American Trypanosomiasis-endemic areas, the principal route of infection generally is considered to be associated with the feeding of triatomine bugs, primarily *Triatoma* spp., and especially *T. infestans*. The bugs contract the typanosomes by feeding on infected animals or people. The bite of kissing bugs typically produces little or no pain, so the sleeping host is usually unaware of the blood-feeding episode. Once infected, the bugs pass *T. cruzi* parasites in their feces. The bugs frequently aggregate in barns, sheds, or houses if they

are not insect proof. In rural areas, many houses are made from materials such as mud, adobe, straw, bamboo, and palm thatch, and lack window screening and, in some cases, doors. Thus, there is nothing to prevent entry of the bugs. During the day, the bugs hide in crevices in the walls and roofs. If the roof is straw or another material with cracks and crevices, the bugs can hide easily and are not readily detected. Then, during the night, when the inhabitants are sleeping, the bugs emerge to feed. (Bugs may feed during the day if their hosts are nocturnal, such as bats.) Because they tend to feed on people's faces, triatomine bugs have also come to be known as **kissing bugs**. The common vectors, in addition to the *Triatoma* spp. mentioned previously, include *Panstrongylus* spp., and *Rhodnius prolixus*. After they bite and ingest blood, they commonly defecate on the person. The person becomes infected if *T. cruzi* parasites in the bug feces enter the victim's body through mucous membranes or breaks in the skin. The unsuspecting, sleeping person may accidentally scratch or rub the feces into the bite wound, eyes, or mouth, facilitating entry of the trypanosomes. Although this has long been considered to be the principal form of *T. cruzi* entry into humans, disease transmission also occurs through consumption of uncooked food contaminated with feces from infected bugs, congenital transmission, blood transfusion, organ transplantation, and (rarely) accidental laboratory exposure. Animals can contract the disease by consuming infected prey, as by consumption of mice by household cats. Recent reports from Brazil indicate that increasing proportions of humans contract Chagas by means other than being bitten by triatomine bugs, and contaminated food is implicated.

The natural reservoirs of *Trypanosoma cruzi* are nearly 200 wild mammal species, although the hosts most commonly infected tend to be opossums, *Didelphis* spp. and *Philander opossum*; armadillos, *Dasypus* spp.; and rodents such as mice (e.g., *Peromyscus* and *Heteromys* spp.); porcupines, *Coendou* spp.; and agoutis, *Dasyprocta* spp. These natural hosts do not develop pathologies. However, *T. cruzi* is also harbored in infected humans, and in domestic animals like cats, dogs, rabbits, and guinea pigs. Canine trypanosomiasis is of veterinary importance in Latin America. In nature, kissing bugs are found in hiding places such as caves, tree holes, hollow trees, fallen logs, palm fronds, and epiphytes. Though kissing bugs feed on birds, and the bugs can be quite numerous in chicken houses on farms, birds seem to be immune to infection and therefore are not considered to be a *T. cruzi* reservoir. The different vectors of American Trypanosomiasis have differing host preferences, so the disease transmission cycle varies accordingly.

The generalized infection cycle follows, and is shown in Fig. 9.1. The infected triatomine bug ingests blood and defecates feces containing trypomastigotes near the feeding site. The victim, irritated by the bite, scratches the area, thereby rubbing the trypomastigote-containing feces into the wound or into intact but susceptible mucosal membranes, such as the conjunctiva. Once inside the host, the trypomastigotes invade cells, where they differentiate into intracellular amastigotes. The amastigotes multiply by binary fission and differentiate into trypomastigotes, which are released into the bloodstream. Cells from a number of different tissues are susceptible to infection by the trypomastigotes, and once inside, they transform into intracellular amastigotes at new infection sites. Intracellular amastigotes destroy tissues such as neurons of the autonomic nervous system in the intestine and heart, leading to digestive and heart problems, respectively. Replication resumes only when the parasites enter another cell or are ingested by another vector. The final step in the cycle is infection of the vector, which occurs when the bug feeds on an infected host containing trypomastigotes. Once ingested by the bug, the ingested trypomastigotes multiply and differentiate (amastigote and epimastigote forms) in the midgut and transform into infective metacyclic trypomastigotes in the hindgut. The feces may contain thousands of metacyclic trypomastigotes.

Management of bugs to disrupt disease transmission is possible. In many areas of Latin America, the interiors of houses are treated with pyrethroid insecticides to eliminate kissing (*Triatoma*) bugs. Several South American countries have rid themselves of the problem in this manner. However, complete control is difficult to accomplish, and houses are often reinfested because some bugs escape the insecticide within the homes by hiding, or because the homes are reinvaded from nearby, untreated areas. Animal shelters are often good harborages for *Triatoma*, and if they are not adequately treated and are near homes, the bugs quickly disperse to the treated home. A distance of at least 1500 m is considered necessary as a buffer zone between untreated and treated harborages. An economic and effective alternative to chemical control is bed nets, though the edges of the net must be tucked beneath the mattress to assure that bugs cannot enter.

Also, the net and mattress must be large enough that the human sleeping beneath the net does not come into contact with the bed netting, because there is risk that bugs will feed through the netting if the human is immediately adjacent. The logistics and cost of meeting these requirements often prove daunting to poor people, so insecticide treatment is more practical. Likewise, improved housing and sanitation would virtually eliminate this problem, but achieving this is presently not possible due to the cost.

Treatment of infected people is difficult because medication is usually effective only if administered during the acute phase, and many victims are unaware of their condition. This issue is compounded by the resistance of the trypanosome to medication in some regions, and by the toxicity and side effects of the drugs. During the chronic phase of infection, treatment consists mostly of treating the symptoms, and both heart replacement and intestinal surgery are practiced. Treatment in either the acute or chronic phase does not guarantee a cure. A vaccine was developed in the 1970s and it proved fairly successful, but implementation is constrained by the high cost of production.

The effect of American Trypanosomiasis is measured less by mortality and more by morbidity. Chronically infected people often suffer decades of weakness and fatigue that effectively removes them from the workplace and prevents them from enjoying a normal life. This results in considerable social disruption and economic loss. A survey of the prevalence of American Trypanosomiasis revealed that about 12% of the residents of Chile and Paraguay were infected, and in Argentina and Bolivia the prevalence was 8%.

African Trypanosomiasis

African Trypanosomiasis is found in sub-Saharan Africa. The trypanosome disease agents are transmitted to wildlife, livestock, and humans principally by the bite of tsetse fly, *Glossina* spp. (Diptera: Glossinidae). The parasites may be acquired by the flies from either humans or animals. These blood-inhabiting trypanosome parasites are in the genus *Trypanosoma*, and several forms and species are known.

When African Trypanosomiasis occurs in humans it is called **Sleeping Sickness**. There are two forms (species or subspecies, depending on the author) important in human disease. *Trypanosoma gambiense* or *T. brucei gambiense* is found in central and western

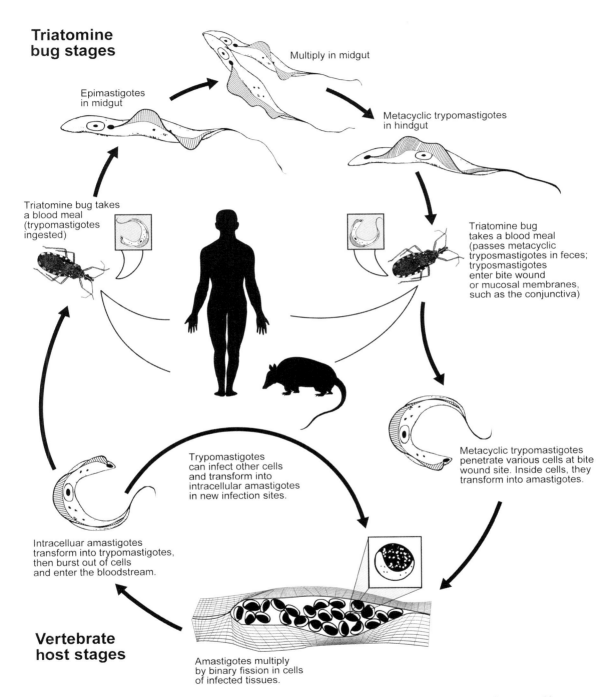

Fig. 9.1. The American Trypanosomiasis (Chagas Disease) transmission cycle in triatomine bugs and their animal hosts. Kissing bugs vector protozoans to various wildlife that remain symptomless, but humans suffer greatly from this debilitating disease, sometimes beginning several years after infection. Domesticated animals also are susceptible to infection.

Africa. It is named after the country of Gambia in western Africa. *Trypanosoma rhodesiense* or *T. brucei rhodesiense* is found is southern and eastern Africa. It is named after the former country of Rhodesia, now Zimbabwe, in eastern Africa. Not only are the two trypanosome forms separated geographically, but they differ in their hosts and in symptoms of infection. Morphologically, however, they are indistinguishable.

Trypanosoma gambiense or West African trypanosomiasis is the more common of the two disease agents, representing over 90% of the known cases in humans. It also tends to cause a chronic infection, and a person can be infected for months or years without seeing symptom expression. However, once the disease is expressed, the disease is far advanced, and the nervous system is adversely affected. Humans are the only important reservoir for this disease, with the trypanosome passed only between humans and vectors. Tsetse of the *palpalis* group are the vectors, often *Glossina palpalis*, *G. tachinoides*, and *G. fuscipes*. These flies inhabit riverine areas and attack people when they collect water, wash clothing, bathe in the rivers, or collect firewood in these areas. The vectors blood-feed on some domestic animals and wildlife, however, including pigs, dogs, and reptiles.

Trypanosoma rhodesiense or East African trypanosomiasis is much less frequent in humans, but occurs in an acute form, with symptoms expressed in weeks or months. As in the case of the *gambiense* form, it affects the central nervous system, but of the disease progresses more rapidly. Tsetse of the *morsitans* group are the principal vectors, including *G. morsitans*, *G. pallidipes*, and *G. swynnertoni*. Cattle and wildlife are the principal reservoir of *rhodesiense*. Among the wild animals serving as hosts for East African trypanosomiasis are antelope (bushbuck, *Tragelaphus scriptus*; duiker, *Sylvicapra grimmia*; wildebeest, *Catoblephas gnu*; impala, *Aepyceros melampus*; hartebeest, *Alcelaphus buselaphus*); bushpig, *Potamochoerus larvatus*; elephant, *Loxodonta* spp.; giraffe, *Giraffa camelopardalis*; lion, *Panthera leo*; warthog, *Phacochoerus africanus*; and zebras, *Equus* spp.

Trypanosomiasis occurs only in sub-Saharan Africa in regions where there are tsetse flies that can transmit the disease. For reasons that are not yet understood, there are many regions where tsetse flies are found, but sleeping sickness is not. Only certain species transmit the disease. They are mainly found in vegetation by rivers and lakes, in forests, and in the savannah, but the different species have different habitat preferences.

When African Trypanosomiasis occurs in livestock it is called **Nagana**. Unlike the human form of the disease, which is not found everywhere tsetse flies occur, Nagana occurs throughout the range of tsetse flies. Nagana is much more common than sleeping sickness. It is basically the same as the tsetse fly-transmitted disease transmitted to humans, but is caused by other trypanosomes such as *Trypanosoma congolense*, *T. suis*, *T. simiae*, *T. uniforme* and *T. vivax* in addition to *T. brucei*. These species do not induce clinical disease in humans.

The disease in domestic animals, and particularly cattle, has been a major obstacle to the economic development of rural central and southern Africa. It is estimated that three million cattle die annually due to this disease. For thousands of years, the ability of tsetse flies to transmit trypanosomes has denied the people of central and southern Africa the use of large domestic animals such as oxen and horses in their agriculture. Also, it has led to shortage of animal protein in their diet. Offsetting this, perhaps, is the benefit of tsetse-borne disease in preventing the southward migration of Arab invaders because their horses and camels were susceptible. Also, the presence of tsetse-borne diseases probably has helped preserve the natural environment of Africa, particularly the abundant wildlife, which are not affected by the disease. There is evidence that some other flies play a minor role in mechanical transmission of the protozoan. Flies in the families Tabanidae and Muscidae also may transmit the disease to animals, and perhaps to humans, though this is of minor importance compared to tsetse transmission.

In most years, 40,000–50,000 cases are reported in humans, but this is likely a serious underestimate of the total number of people affected. Trypanosomiasis is more commonly a problem where public health services are inadequate, in rural habitats and among poor people, and in these situations accurate reporting is infrequent. Rural populations usually depend on agriculture, fishing, animal husbandry or hunting, and in these pursuits they are often exposed to the bite of the tsetse fly and therefore to the disease. Displacement of people following war or famine can exacerbate the problem. Sometimes the trypanosomes cross the placenta and infect the fetus, resulting in mother-to-child infection. Blood transfusions can also be a source of the disease.

A tsetse fly bite is often painful and can develop into a red sore, called a **chancre**. Upon transmission of the disease by the bite of a fly, the trypanosomes multiply in the subcutaneous tissues, blood and lymph of the host. This first stage of the disease is known as the hemolymphatic phase. Associated with this phase are periods of fever and sweating, headache, swelling of the lymph nodes, joint pain, and itching. With time, the trypanosomes cross the blood–brain barrier and infect the central nervous system. This second stage, or neurological phase, has associated with it confusion, mood swings, sensory disturbances, weakness, and poor coordination. As the 'sleeping sickness' designation suggests, another symptom of the disease is disturbance of sleep. Ultimately, without treatment, sleeping sickness often leads to coma or convulsions, and death. The long asymptomatic first stage of the West African form of the disease, and the generalized symptoms that occur when any ailment is noticed, make diagnosis difficult. Serological and cerebro-spinal tests usually are required. The drugs used to treat the disease during the first stage of the disease are not too toxic, easy to administer, and fairly effective. Unfortunately, if early diagnosis is not made, the drugs used to treat the second stage are toxic and difficult to administer.

Development of this trypanosome in the host is complex. Trypomastigotes (one major form of the trypanosome) circulate in the blood and the tsetse fly becomes infected with bloodstream trypomastigotes when taking a blood meal on an infected mammalian host. In the fly's midgut, the parasites transform into procyclic trypomastigotes, multiply by binary fission, and leave the midgut. Eventually they migrate to the proboscis and attach to the epithelium of the salivary glands where they transform into epimastigotes (the other major form of the trypanosome). The epimastigotes continue multiplication by binary fission. The cycle in the fly takes approximately three weeks to progress, and then the trypanosomes transform back to trypomastigotes; at this point they are called metacyclic trypomastigotes and are infective to humans and animals. Once infected, the fly remains infective during its entire 6–12 month life-span. An infected tsetse fly injects metacyclic trypomastigotes into skin tissue when taking a blood meal, and the parasites enter the host's lymphatic system and pass into the blood. There, they transform into bloodstream trypomastigotes and are carried to other sites throughout the body, reach other blood fluids (e.g., lymph, spinal fluid), and continue the replication by binary fission. The transmission cycle is shown in Fig. 9.2.

Prevention and control of the disease usually focuses on elimination of the vector, tsetse flies. Persistence of flies is favored by brush, which provides the flies with cover, so clearing of brush is an important management technique. The flies do not disperse long distances, so it is possible to create barriers around inhabited areas by removing such cover. Land used for crops or grazed periodically by livestock similarly is not very suitable for survival of tsetse. Insecticide application, both ground and aerial applications, has long been used for tsetse suppression. Area-wide management approaches can be implemented using release of sterile insects, but presently is too costly for widespread and continued use. Use of traps (blue or black cloth sprayed with insecticide, or cattle sprayed with insecticides) has had some success, as has traps baited with chemical attractants. Care should be taken to avoid being bitten. Bed netting and thick, long-sleeved clothing are helpful. Avoidance of brushy habitats is also helpful. Vaccines and preventative drugs are not available. It also is desirable to screen communities for occurrence of sleeping sickness to identify foci of disease outbreak and to apply curative medications to affected individuals. Certain breeds of cattle such as N'dama are tolerant of trypanosome infection, so this is an active area of research.

Avian Malaria

Many species of *Plasmodium* are responsible for avian malaria, although *Plasmodium relictum* is most important. In birds, *P. relictum* reproduces in red blood cells. If the parasite intensity is sufficiently high, the red blood cells are destroyed, causing anemia. Blood cells are critical for moving oxygen about the body, so loss of these cells leads to progressive weakness and, eventually, to death of the infected animal. Malaria is found most commonly in the perching birds (order Passeriformes), gallinaceous birds such as grouse, pheasants, and turkeys (Galliformes), and doves and pigeons (Columbiformes). In birds that have co-evolved with the disease, infection prevalence can be high but mortality is absent or occurs at a very low rate. When the disease attains locations where it has not been found previously (e.g., many islands in the Pacific Ocean) or reaches hosts not previously exposed (e.g., penguins), mortality among birds can be high.

The accidental introduction of *P. relictum* and the mosquito *Culex quinquefasciatus* to the Hawaiian

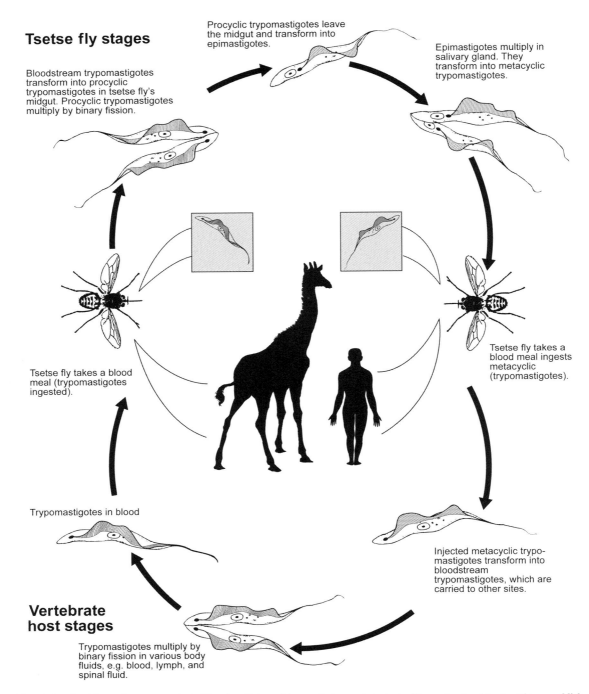

Tsetse fly stages

Bloodstream trypomastigotes transform into procyclic trypomastigotes in tsetse fly's midgut. Procyclic trypomastigotes multiply by binary fission.

Procyclic trypomastigotes leave the midgut and transform into epimastigotes.

Epimastigotes multiply in salivary gland. They transform into metacyclic trypomastigotes.

Tsetse fly takes a blood meal (trypomastigotes ingested).

Tsetse fly takes a blood meal ingests metacyclic (trypomastigotes).

Trypomastigotes in blood

Injected metacyclic trypomastigotes transform into bloodstream trypomastigotes, which are carried to other sites.

Vertebrate host stages

Trypomastigotes multiply by binary fission in various body fluids, e.g. blood, lymph, and spinal fluid.

Fig. 9.2. The African Trypanosomiasis (Sleeping Sickness) transmission cycle in tsetse flies and animal hosts. African wildlife are not adversely affected, but humans and domesticated animals are greatly affected by this protozoan disease that is transmitted by tsetse flies. When affecting livestock, the disease is called Nagana.

Table 9.2. Natural vectors of species of *Plasmodium* from birds (adapted from Atkinson 2009).

Plasmodium species	Locality	Mosquito vector
Plasmodium relictum	California, USA	Culex stimatosoma
	California, USA	Culex tarsalis
	Hawaii, USA	Culex quinquefasciatus
Plasmodium gallinaceum	Sri Lanka	Mansonia crassipes
Plasmodium circumflexum	Sri Lanka	Mansonia crassipes
	New Brunswick, Canada	Culiseta morsitans
Plasmodium rouxi	Algeria	Culex pipiens
Plasmodium juxtanucleare	Malaysia	Culex sitiens
		Culex annulus
	Brazil	Culex saltanensis
Plasmodium hermani	Florida, USA	Culex nigripalpus
Plasmodium elongatum	Maryland, USA	Culex pipiens
		Culex restuans
Plasmodium (Novyella) sp.	Venezuela	Aedeomyia squamipennis
Plasmodium (Giovannolaia) sp.	Venezuela	Aedeomyia squamipennis

Islands is a particularly severe case of the impact of a disease on a population lacking resistance. Of the more than 70 endemic species or subspecies of forest-dwelling birds found there at the end of the 1700s, at least 23 now are extinct and 30 species or subspecies are endangered. Although this phenomenal loss in biodiversity is not due solely to avian malaria, this disease is a major factor restricting recovery of the avifauna in areas where habitat remains. A hopeful sign is that one species, Hawaii amakihi (*Hemignathus virens*), seems to be evolving some resistance and its population is increasing.

This protozoan disease agent is transmitted by *Culex*, *Mansonia*, and other mosquitoes (Table 9.2). Mosquitoes acquire the protozoa when they take a blood meal from the reservoir (bird) host, ingesting gametocytes. These develop into oocysts in the gut of the mosquito. The oocysts then produce numerous sporozoites, which migrate to the salivary gland of the mosquito. These are the infectious agents, and are injected into the blood of the bird during the next blood meal, initiating a complex life cycle in the bird involving infection of the blood cells as well as organs such as the spleen, lung, liver, heart, and kidney. Infected birds experience two cycles of infection: a tissue phase and a blood phase. It is the release of toxins from the ruptured red blood cells that cause the chills and fever so characteristic of malaria infection.

Plasmodium relictum infects hundreds of species of birds, whereas other *Plasmodium* species seem restricted to a few hosts. *Plasmodium* is closely related to some other parasites such as *Haemoproteus* and *Leucocytozoon*. *Haemoproteus* generally is a less serious ailment in birds. Several species of *Leucocytozoon* cause significant population decline in pigeons, doves, waterfowl, and raptors. Some consider infection by these other parasites also to be forms of malaria. Signs of infection in birds include loss of appetite, fever, depression, weakness, and shortness of breath. Captive birds, such as those in zoos, can be treated with antimalarial drugs. This is not feasible for wild birds, of course. Thus, mosquito suppression is sometimes attempted to manage malaria in avifauna.

Toxoplasmosis

The causative agent of toxoplasmosis is *Toxoplasma gondii*, a tissue-inhabiting protozoan. Domestic and wild cats, including jaguar, *Panthera onca*; American bobcat, *Lynx rufus*; Bengal tiger, *Panthera tigris*, and other felids are the definitive hosts of *T. gondii*. However, about 200 species of animals and birds from throughout the world have been infected with this protozoan, thus serving as intermediate hosts. Among wildlife severely affected are marsupials in Australia, monkeys in the New World, Hawaiian crow (*Corvus hawaiiensis*), canaries (*Serinus* spp.), and finches (Fringillidae). Humans also are affected. Surveys from the USA and UK show that 9–40% of people are infected, and in Central and South America and continental Europe the level of infection is 40%–80%. Infected humans may experience headache, fever, and joint and muscle pain, or more serious problems. Although arthropods are not an essential part of the life cycle, invertebrates such as flies, cockroaches and dung beetles often play an important role in transporting the oocysts.

T. gondii is infective during three stages of its life cycle: the tachyzoites (in clusters or groups), the bradyzoites (in tissue cysts), and the sporozoites (in oocysts). Commonly, oocysts are ingested, excyst, and invade the epithelial cells of the intestine. The tachyzoite stage multiplies rapidly in the host's body, and other tissues may be invaded. They may persist as bradyzoites within tissue for months. Host animals shed oocysts, which may be ingested to start new infections, or transported by invertebrates and then ingested. Intermediate hosts also pick up the oocysts, and the tissue from intermediate hosts provides stages of *T. gondii* that are infective to carnivores. Consumption of infected meat is likely a primary route of infection, and is thought to be the principal route of infection for carnivorous birds. In contrast, ground-feeding birds are thought to become contaminated primarily by ingesting oocysts from contaminated soil. Oocysts can survive for years in soil. The transmission cycle is shown in Fig. 9.3.

Animals ingesting *T. gondii* may not show signs of infection, though many organs can be affected. Hosts may display necrosis, resulting in lesions on the intestine, liver, lungs, brain, and other organs, and eventually perish. The prevalence of *T. gondii* is quite high in numerous wild animals, including wild game. Although there are few data available on the impact of toxoplasma infection on wildlife, approximately 20% of the Hawaiian crow population perished from this pathogen during an attempt to restore this bird species to its former habitat in Hawaii. Thus, the pathogenicity of this protozoan is not trivial. Table 9.3 shows some data on the prevalence of *T. gondii* in the USA.

It is not surprising that hunters contract toxoplasmosis from eating undercooked wild game, but care must be exercised with pelt removal from fur-bearing animals as well, even from animals not showing clinical signs of infection.

Table 9.3. Serologic prevalence of *T. gondii* in some wild mammals and birds (adapted from Dubey and Odening 2001, and Dubey 2008).

Species	% positive
Black bear, *Ursus americanus*	15–80
Grizzly bear, *Ursus arctos*	18–25
White-tailed deer, *Odocoileus virginianus*	30–60
Feral pig, *Sus scofa*	13–34
Moose, *Alces alces*	15–23
American bison, *Bison bison*	2
Raccoon, *Procyon lotor*	13–70
Coyote, *Canis latrans*	26–62
Red fox, *Vulpes vulpes*	86–90
Gray fox, *Urocyon cinereoargenteus*	25–75
Striped skunk, *Mephitis mephitis*	50
Virginia opossum, *Didelphis virginianus*	13
American bobcat, *Lynx rufus*	18–73
Cougar, *Felis concolor*	9–100
Wild turkey, *Meleagris gallopavo*	10–71
Barn owl, *Tyto alba*	11–27
Rock pigeon, *Columba livia*	2–100
Black-headed gull, *Larus ridibundus*	1–16
Mallard, *Anas platyrhynchos*	5–12

Toxic or clinical toxoplasmosis has also been recovered from marine animals such as sea lion, *Zalophus californianus*; Atlantic bluenose dolphin, *Tursiops truncatus*; and ungulates such as Saiga antelope, *Saiga tatarica*; pronghorn, *Antilocapra americana*; and gazelles, *Gazalle* spp. Marsupials are susceptible, especially in zoos, but also under natural conditions. They often die without exhibiting clinical signs of infection. Wallabies are more prone to problems than kangaroos.

Meat from wild game is a major source of infection, though in some areas of the world meat from domestic animals such as lamb, pork and rabbit also is commonly infected. Cattle and horses are more resistant. Uncooked meat is dangerous, so knives, counter tops and other surfaces must be cleaned carefully if they come in contact with raw meat. Meat should be cooked thoroughly, attaining a temperature of 67 °C before being consumed.

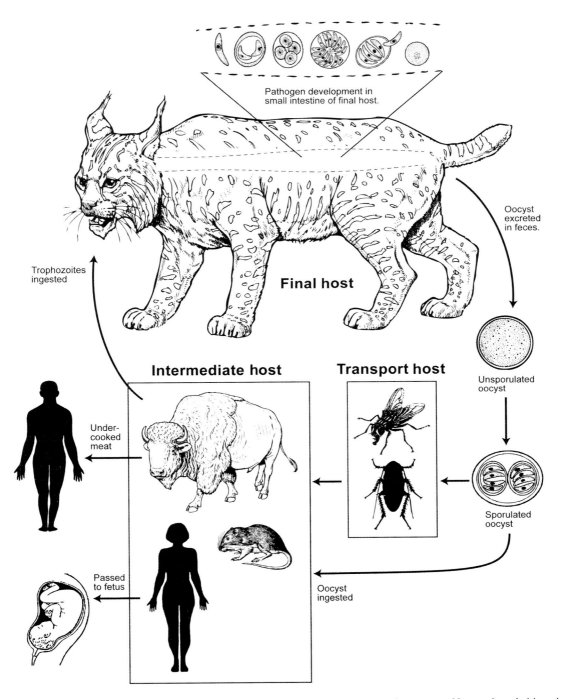

Pathogen development in small intestine of final host.

Oocyst excreted in feces.

Final host

Trophozoites ingested

Intermediate host

Transport host

Under-cooked meat

Unsporulated oocyst

Sporulated oocyst

Passed to fetus

Oocyst ingested

Fig. 9.3. The Toxoplamosis transmission cycle. The protozoan *Toxoplasma gondii* is found in many wild animals and although the disease may or may not affect wildlife, infection can be quite damaging to humans. Insects can be involved it its transmission, though their role is optional. Humans commonly contract the disease from eating undercooked game or through contact with the feces of cats. Unborn children can contract the disease from their mother.

In many areas, domestic cats are the greatest threat of human infection. Cats should be prevented from hunting, and not fed raw meat. Cat feces are a significant hazard, especially for pregnant women. During pregnancy, the mother rarely has symptoms of infection, but the placenta develops lesions and the fetus becomes infected. Later, the infection may localize in the central nervous system, causing disease. Mildly infected children may display visual impairment, whereas severely infected children may develop blindness and mental retardation.

HELMINTHS

Helminth parasites of wildlife are worm-like muticellular organisms in the phyla Nemathelminthes (Nematoda), Platyhelminthes (Trematoda and Cestoda), and Acanthocephala. They normally are visible without magnification, and are relatively long lived. As with the other parasites, the severity of the disease induced by these animals is proportional to the level of infection/infestation. Helminths are endoparasites. Some diseases of wildlife caused by helminthes are shown in Table 9.4.

Nematodes are cylindrical, unsegmented animals. They are long and thread-like, with a buccal (mouth) cavity at the anterior end and a rectal (anal) opening near the posterior end, connected by a complete digestive tract. The posterior end commonly is more pointed. Nematodes, also known as nemas, usually have a relatively simple life cycle. The egg stage is normally followed by several (usually five) developmental stages consisting of four larval stages and an adult stage. The worm molts between each of the five stages, and often the third stage larvae are infective in the primary or definitive host. The nematodes affecting wildlife require intermediate hosts, sometimes arthropods.

Trematodes are diverse but relatively short, flat, unsegmented animals. Often they are called flukes. They have a head region but an incomplete digestive system. Their excretory system is a primitive arrangement made up of flame cells, and is concerned with water and salt balance. Any food that cannot be digested must be excreted through the mouth. Snails are an important part of trematode life cycles, and almost always serve as one of the hosts. The life cycle is complex when arthropods are involved in transmission, and a second host such as an arthropod is normal.

The third (primary) host of a typical trematode is, of course, the vertebrate host. In some, trematode eggs are passed from the primary host in its feces, with the eggs hatching into a miricidium once they reach water. Infection by the miricidium eventually results in production of numerous cercariae. In turn, the cercariae may be incorporated into mucous-like 'slime balls', which are consumed by arthropods. Alternatively, they may penetrate an intermediate host or encyst on vegetation. Eventually, the cercariae give rise to metacercariae once they are ingested by the primary host. The metacercariae then develop into young and then adult flukes.

Cestodes are called tapeworms because they have a long, thin, but flattened body. Unlike trematodes, cestodes have no head, though there is an attachment organ (the scolex) at the anterior end of a chain of body segments (proglottids) that form a chain (strobila). A mouth and digestive system also are lacking. Cestodes occur in the digestive tract of hosts, and generally are the most prevalent internal parasites of wildlife. Their impact on the host is not often thought to be significant, however. Each proglottid absorbs nutrients through its external surface and produces eggs. The egg develops into a larva called an oncosphere; the oncosphere is usually ingested by an invertebrate intermediate host, where it develops into a metacestode. The metacestode is consumed along with the invertebrate by a vertebrate animal, where the metacestode matures.

Acanthocephalans are a small group of cylindrical, unsegmented parasitic worms that are characterized by a thorny retractable proboscis at the anterior end of the body. They are also called 'spiny-headed worms' because of this spiny anterior structure. The body consists of two regions, the presoma (the proboscis and neck) and the trunk. The most notable feature is the abundance of hooks covering the proboscis; these are used for attachment to the host's alimentary canal. Like cestodes, the acanthocephalans lack a digestive tract, instead absorbing nutrients through their body wall. The female worms retain eggs until they are embryonated. Embryonated eggs are released but do not hatch until eaten by insects or crustacean intermediate hosts, which may be aquatic or terrestrial. The vertebrate host is infected when it consumes an intermediate host bearing a fully developed acanthocephalan. More than any other group, acanthocephalans seem to be capable of inducing changes in the behavior of hosts.

Table 9.4. Some helminths that affect wild animals, or involve wild animals and humans, and are transmitted by insects.

Helminth species	Disease	Animal host	Vector or intermediate host	Human host
Nematodes (worms)				
Habronema muscae	Habronemiasis	horses, donkeys, zebras	higher flies (Cyclorrapha)	no
Draschia megastoma	Drachiasis	horses, donkeys, zebras	higher flies (Cyclorrapha)	no
Spirocerca lupi	Spirocercosis	foxes, coyote, jackal, jaguar, American bobcat	scarab beetles	no
Oxyspirura mansoni	Tropical Eyeworm	wild and domestic birds	cockroaches	no
Physaloptera phrynosoma	a spiruroid nematode	horned toad	ants (Pogonomyrmex)	no
Foleyella bachyoptera	a filarioid nematode	leopard frog	mosquitoes	no
Dirofilaria immitis	Dirofilariasis	dogs, other mammals	mosquitoes	rare
Elaeophora schneideri	Elaeophorosis	deer, elk, moose, sheep	horse flies	no
Setaria cervi	Setariosis	deer, domestic livestock	mosquitoes, stable flies	no
Pelecitus scapiceps	Pelecitosis	rabbits, hares	mosquitoes	no
Diplotriaena bargusinica	Filariasis	willow thrush, red-winged blackbird	grasshoppers	no
Pelecitus fulicaeatrae	Avian Filariasis	aquatic birds	lice	no
Sarconema eurycerca	Sarconema or heartworm	swans and geese	lice	no
Cheilospirura spinosa	gizzard worm	grouse, pheasant, quail, partridge, turkey	grasshoppers	no
Pelecitus roemeri	Pelecitosis	kangaroos	horse flies	no
Monanema martini	Monanemosis	African rodents	ticks	no
Brugia malayi	Lymphatic Filariasis	monkeys, wild cats and dogs, viverrids, pangolins	mosquitoes	yes
Loa loa	Loiasis	African primates	deer flies (Chrysops)	yes
Trematodes (flukes)				
Dicrocoelium dendriticum	Lancet (Liver) Fluke	rodents, lagomorphs, camels, primates, deer, wild and domestic sheep, pigs, cattle, horses	snails and ants	rare
Haematoloechus medioplexus	Frog Lung Fluke	frogs and toads	snails and dragonflies	no
Prosthogonimus macrorchis	Oviduct Fluke	wild and domestic birds	snails and dragonflies	no
Crepidostomum cooperi	Pyloric Caeca Fluke	fresh-water fish	clams and mayflies	no
Cestodes (tapeworms)				
Dipylidium caninum	Dog Tapeworm	wild and domestic dogs and cats	fleas, lice	yes
Raillietina cesticillus	Fowl Tapeworm	pheasant, chicken	flies, beetles	no
Raillietina loeweni	a tapeworm	black-tailed jackrabbit	harvester ants (Pheidole)	no
Moniezia benedeni	a tapeworm	American bison, other ruminants	oribatid mites	no
Acanthcephalans (thorny-headed worms)				
Macracanthorhynchus hirudinaceus	Giant Thorny-Headed Worm	wild and domestic mammals	scarab beetles	rare
Moniliformis moniliformis	a thorny-headed worm	rodents, dogs, cats	cockroaches, beetles	yes

Spirocercosis

This disease is caused by *Spirocerca lupi* nematodes (worms or nemas), and affects predatory mammals including the foxes *Vulpes vulpes* and *Urocyon cinereoargenteus*; coyote, *Canis latrans*; jackals, *Canis* spp.; wolf, *Canis lupus*; American bobcat, *Felis rufus*; and jaguar, *Panthera onca*; but also domestic dogs. It occurs principally in warmer climates, and can be common. A study conducted in Texas, USA, for example, found that 82% of the coyotes, 35% of the bobcats, and 20% of the gray foxes were affected. Infection can cause an inflammatory reaction in blood vessels, resulting in an aneurysm of the blood vessel. In the esophagus, abnormal tissue growth is induced, which can result in blockage of the lumen and digestive problems. Interestingly, *S. lupi* is one of a very few helminths that can induce carcinogenic reactions. Dogs develop malignant tumors in the esophagus following years of infection.

Beetles and sometimes other organisms have a role in the life cycle of *S. lupi*. The worms reside in pockets in the esophagus of the mammal host. Eggs produced by the worms are transported out with feces and are then consumed by dung-feeding beetles (Coleoptera: Scarabaeidae) or incorporated into dung balls bearing beetle larvae. The young or old beetles, now containing nematodes, may be eaten directly by a large carnivore. Alternatively, the beetles may be eaten by small carnivores such as rodents, birds and reptiles that in turn are eaten by large carnivores. In any event, the nematode has been transported to another large host through consumption of beetles, completing the cycle. The worms move from the intestine of the carnivore to the aorta, and then eventually to the esophagus, where they mature and produce more eggs. The transmission cycle of Spirocercosis is shown in Fig. 9.4.

Dirofilariasis

Dirofilaria nematodes are rather long and robust. Except for *D. immitis*, the cause of dog heartworm, the *Dirofilaria* spp. are found subcutaneously in mammals, principally carnivores and primates. They are transmitted by mosquitoes (Diptera: Culicidae) except for the case of *D. ursi*, which is transmitted by black flies (Diptera: Simuliidae). The infective microfilarial stage of all species occurs in the blood, allowing blood-feeding insects to access the worms. There are many species of *Dirofilaria*, most associated with a particular host. For example, in North America *D. tenuis* affects northern raccoon, *Procyon lotor*; *D. ursi* affects bears, *Ursus* spp.; *D. subdermata* affects porcupine, *Erethizon dorsatum*; and *D. striata* affects Canada lynx, *Lynx canadensis*. Because they normally are simply subcutaneous parasites, they generally cause little injury. *D. immitis* is an important exception.

In *D. immitis*, the microfilariae are transmitted by about 100 species of mosquitoes (Diptera: Culicidae), including *Culex*, *Aedes*, *Anopheles*, *Psorophora* and *Mansonia* spp. The larvae initially invade the subcutaneous tissues, but as they mature they invade the thoracic cavity, and mature worms are found in the heart. The transmission cycle is shown in Fig. 9.5. The adult worms are long, slender, and white. Males measure 12–16 cm long and females are 25–30 cm. Adult worms cause disease by clogging the heart and main blood vessels, reducing the blood supply to the other organs of the body, particularly the lungs, liver, and kidneys. The heart may become enlarged. The signs of infection and extent of the disease are related to the abundance of worms. Initially, most animals tolerate the worms, even when they are numerous. However, as blood deprivation affects the other internal organs the health of the affected animal deteriorates. Loss of stamina, nervousness, weight loss, and shortness of breath are common signs of infection. The prevalence of *D. immitis* in wildlife is not trivial; in the southeastern USA about 8% of black bears, *Ursus americanus*, were found affected. Among coyotes, *Canis latrans*, 66% were infected in Arkansas, and up to 27% in northern California, and 12.5% in Indiana. In gray foxes, *Urocyon cinereoargenteus*, prevalence was 16% in Alabama and Georgia, but only 3.7% in Indiana.

D. immitis is a cosmopolitan parasite of dogs, especially in warmer regions. Because it is nearly impossible to prevent all mosquitoes from having contact with dogs, preventative chemical therapy is usually recommended. Dogs are commonly provided with microfilaricidal chemicals to prevent development of the adult worms. It is possible to kill the adult worms, but the products used to kill the mature worms carry a greater risk of toxicity to the dog, and the death of numerous, mature worms in the host carries with it the potential for complications. Occasionally, *D. immitis* and other *Dirofilaria* spp. are recovered from humans and other abnormal hosts.

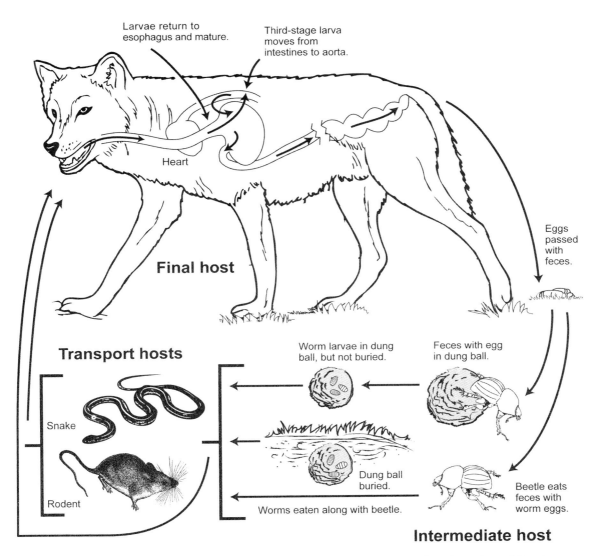

Larvae return to
esophagus and mature.

Third-stage larva
moves from
intestines to aorta.

Heart

Final host

Eggs
passed
with
feces.

Transport hosts

Worm larvae in dung
ball, but not buried.

Feces with egg
in dung ball.

Snake

Dung ball
buried.

Beetle eats
feces with
worm eggs.

Rodent

Worms eaten along with beetle.

Intermediate host

Fig. 9.4. The Spirocercosis transmission cycle. The nematode *Spirocerca lupi* causes a number of vascular and digestive disorders in predatory animals, and the incidence is high in warm climates. Dung beetles (Scarabaeidae) serve as intermediate hosts.

Elaeophorosis

Elaeophora schneideri nematodes infect various ruminants in North America, and usually are located in the arteries and heart. The microfilariae are found in capillaries, primarily of the skin of the forehead and face, and also are found in the insect vector for the first stages of development. Transmission is accomplished by horse flies (Diptera: Tabanidae) when they feed. The parasites are common in mule deer, *Odocoileus hemionus*, in western North America, and less common in white-tailed deer, *Odocoileus virginianus*. Elk, *Cervus canadensis*, and moose, *Alces alces*, are also infected, as are domestic sheep. Exotic deer and

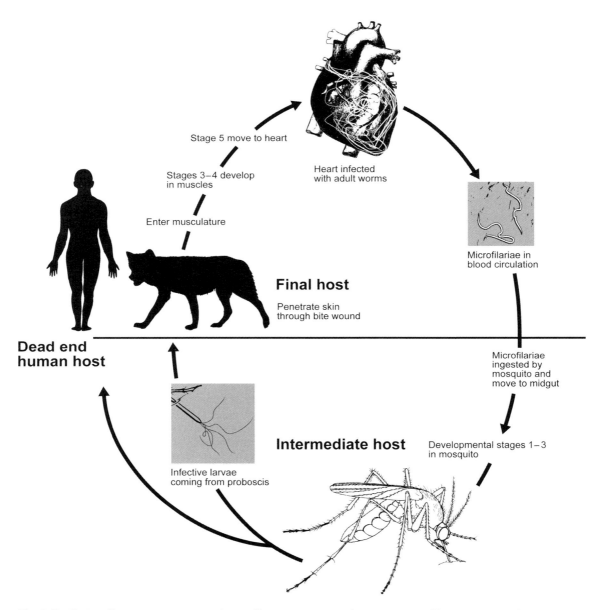

Stage 5 move to heart

Stages 3–4 develop
in muscles

Heart infected
with adult worms

Enter musculature

Microfilariae in
blood circulation

Final host

Penetrate skin
through bite wound

**Dead end
human host**

Microfilariae
ingested by
mosquito and
move to midgut

Infective larvae
coming from proboscis

Intermediate host

Developmental stages 1–3
in mosquito

Fig. 9.5. The Dirofilariasis transmission cycle. *Dirofilaria immitis* nematodes are transmitted by mosquitoes to foxes, coyotes and bears. The nematodes infect muscles, including the heart, where they interfere with blood flow and damage organs that depend on adequate blood supply.

sheep introduced to game farms also are susceptible to infection.

In well-adapted hosts such as mule deer and black-tailed deer, there is little or no consequence to infection by this parasite. However, elk and moose suffer blindness, abnormal antler development, emaciation and other maladies. Sheep and goats suffer similar problems.

Lancet Fluke

The lancet liver fluke, *Dicrocoelium dendriticum*, affects a number of herbivorous mammals on most continents, though its origin is likely Eurasia. Among the numerous hosts are wild and domestic cattle, pigs, sheep, horses, deer, rodents, rabbits and hares, and camels. Eggs from adult flukes are deposited in the bile ducts of the primary host, and pass into the intestine where they are voided with the feces. Hatching occurs only after the eggs have been consumed by snails, producing a form known as a miracidium. In the snail, miracidia migrate through the gut wall, where reproduction occurs in the digestive gland. Cercariae are produced, and migrate to the respiratory chamber where they mix with mucus. At low temperatures, the mucus and cercariae are expelled in the form of slime balls. The slime balls may be discovered by *Formica* spp. ants that forage on the vegetation favored by the snails. Ants will carry the slime balls back to the nest, where worker ants consume them and become infected. In infested areas, up to 35% of the adult ants have been found to be infected. The cercariae transform into metacercariae in the ant. Once infected, the behavior of the ants changes. They are stimulated to climb vegetation, where they attach using their mandibles, and remain torpid so long as the temperature is low (less than 20 °C). Grazing animals ingest the ants along with vegetation, and metacercariae are introduced to the digestive system of the grazing animal. Soon, young flukes are produced that enter the liver and gall bladder. Within a few weeks, they produce eggs to begin the cycle again. The Lancet Fluke transmission cycle is shown in Fig. 9.6.

The effect of the liver flukes can be significant. Up to 7000 flukes have been recovered from a single gall bladder, and 50,000 from a liver. Although generally considered mostly a serious pest of domestic sheep, this parasite affects large numbers of wild animals. For example, in Sweden it was found in 22% of roe deer, *Capreolus capreolus*; 16% of moose, *Alces alces*; 3% of European hares, *Lepus europaeus*; and 11% of mountain hares, *Lepus timidus*. Very similar life cycles are found associated with liver flukes of many different birds in North America, although the species of flukes and of arthropods are different.

Dog Tapeworm

This tapeworm occurs in the small intestine of wild and domestic canids (dogs) and felids (cats). Occasionally it infects humans, particularly children. It is more correctly known as double-pored tapeworm, *Dipylidium caninum*, because it has two genital pores. However, it is a common parasite of dogs, hence the alternate designation.

These tapeworms develop in the intestine of carnivorous vertebrates. They release proglottids containing egg packets. The proglottids can be found on the animal in the anal region, on its bedding or nest area, and in the feces. After the eggs are released by the proglottid, the eggs may be consumed by larval fleas (especially dog flea, *Ctenocephalides canis*, or cat flea, *C. felis*) but also chewing lice such as dog louse, *Trichodectes canis*. Ingested eggs produce oncospheres that penetrate the intestinal wall of the insect and transform into cysticercoids. The cysticercoids live in the body cavity of the insect, growing quickly in immature lice but more slowly in flea larvae. In fleas, there is little growth during the larval stage, considerable growth in the pupal stage, and maturation in the adult. The definitive or primary host is infected when it eats the insects, where the cysticercoids escape into the intestine and produce mature tapeworms in 3–4 weeks. Adult tapeworms are up to 60 cm long and 3 cm wide. Humans acquire the tapeworm in the same manner, by ingesting fleas. The Dog Tapeworm transmission cycle is shown in Fig. 9.7.

Giant Thorny-headed Worm

Macracanthorhynchus hirudinaceus, also known as Giant Thorny-headed Worm, is a well-known parasite of domestic pigs, but also is associated with wild boars, hyenas, moles, squirrels, dogs, and likely other mammals. The large adult worm (up to 9 cm long in males and 65 cm long in females) is yellowish white or yellowish orange, and remains attached to the intestine of its primary (vertebrate) host using a rasp-like structure found at the anterior end of the worm (see Fig. 9.8). Eggs leave the host's body with feces, where

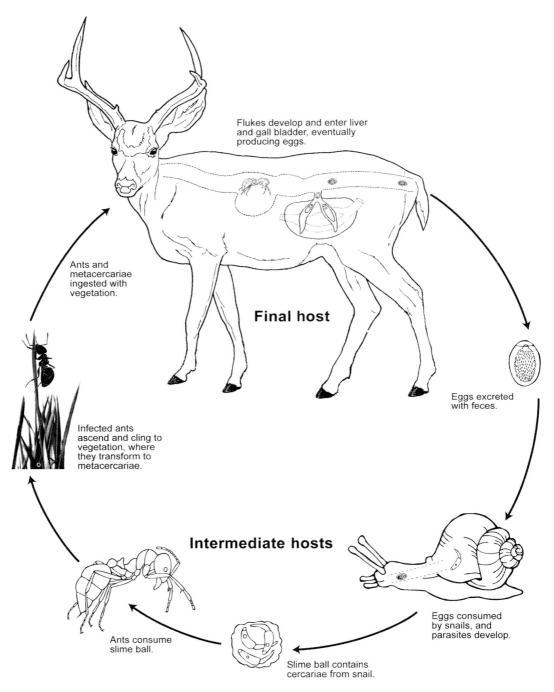

Flukes develop and enter liver and gall bladder, eventually producing eggs.

Ants and metacercariae ingested with vegetation.

Final host

Eggs excreted with feces.

Infected ants ascend and cling to vegetation, where they transform to metacercariae.

Intermediate hosts

Eggs consumed by snails, and parasites develop.

Ants consume slime ball.

Slime ball contains cercariae from snail.

Fig. 9.6. The Lancet Fluke transmission cycle. Liver flukes such as *Dicrocoelium dendriticum* affect a number of herbivorous mammals. The flukes pass through snails and ants before being consumed by the final host. Infection by the fluke causes changes in the behavior of *Formica* ants wherein they climb vegetation and cling to it with their mandibles, facilitating ingestion by grazing animals.

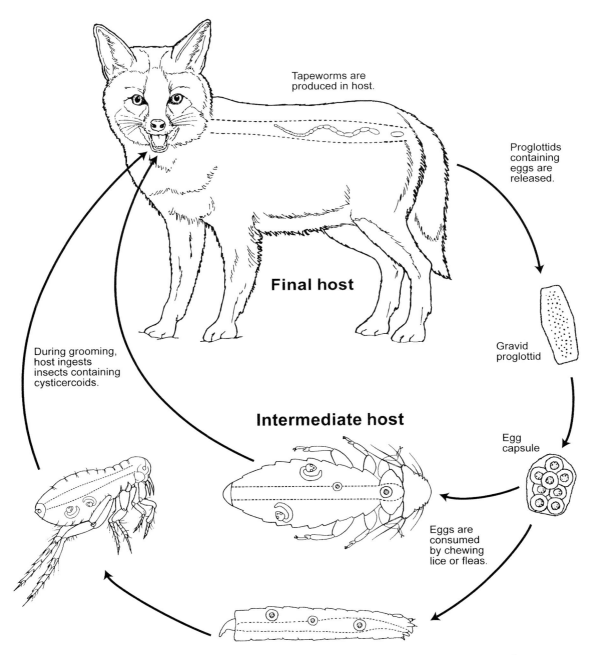

Tapeworms are
produced in host.

Proglottids
containing
eggs are
released.

Final host

Gravid
proglottid

During grooming,
host ingests
insects containing
cysticercoids.

Intermediate host

Egg
capsule

Eggs are
consumed
by chewing
lice or fleas.

Fig. 9.7. The Dog Tapeworm transmission cycle. The tapeworm *Dipylidium caninum* attaches to the intestine of carnivorous vertebrates. Fleas and chewing lice serve as intermediate host by consuming the tapeworm eggs. Wildlife eat these parasitic insects while grooming, completing the transmission cycle.

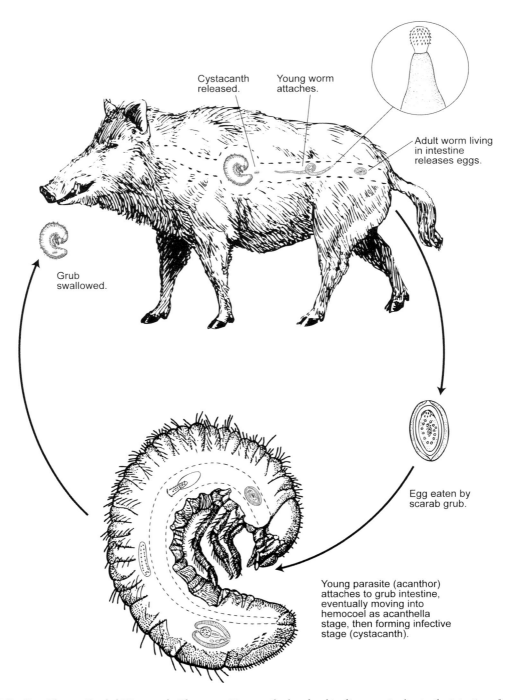

Fig. 9.8. Giant Thorny-Headed Worm cycle. The worm *Macracanthorhynchus hirudinaceus* attaches to the intestine of vertebrate hosts and releases eggs into the host's feces. White grubs developing in the fecal material ingest the eggs and support growth of the worm until they, too, are eaten by vertebrates.

they can be ingested by immatures (white grubs) of scarab beetles (Coleoptera: Scarabaeidae) that live in the soil, especially soil high in organic matter. *Phyllophaga* and *Cotinus* spp. are common grub hosts in North America. Once swallowed by the grub, eggs hatch quickly and an infective stage called the acanthor penetrates the insect's gut wall into the body cavity and attaches to the wall of the intestine. As the acanthor grows it becomes detached from the intestinal wall, becoming another stage called the acanthella, which is a juvenile worm. The final form of the acanthella is called the cystacanth, and is infective to vertebrate hosts that ingest the grubs. Maturity is attained in the primary host within 2–3 months. Females can produce over 250,000 eggs per day for a period of 10 months.

SUMMARY

• As was the case with infectious diseases, parasitic diseases of wildlife can also affect humans, pets, and livestock, with wildlife often serving as a reservoir for such diseases even though they may not be adversely affected. Parasitic diseases vary in severity, and often have very complex life cycles involving intermediate hosts.

• Protozoa are mobile, unicellular organisms. Most are free living, and not associated with animal hosts. However, some of those that have developed a parasitic association have developed into significant disease agents. Among the protozoan diseases that involve wildlife and insects are:

American Trypanosomiasis, also known as Chagas Disease, is transmitted by reduviid bugs to mammals. This is an important disease of humans in South and Central America.

African Trypanosomiasis is transmitted to mammals by tsetse flies in sub-Saharan Africa. It also is known as Sleeping Sickness when it affects humans and Nagana when infecting livestock.

Toxoplasmosis is normally associated with wild and domestic cats, but can infect a large number of mammals and birds. Insects are not an essential part of the cycle, although they are often involved.

American Trypanosomiasis, African Trypanosomiasis, and Toxoplasmosis affect humans as well as wildlife.

• Helminths are worm-like multicellular organisms from several phyla. Nematodes, trematodes, cestodes, and acanthocephalans are examples of helminths associated with wildlife and insects. Among the insect-transmitted diseases affecting wildlife are:

Spirocercosis is found in predatory mammals. Dung-feeding beetles and other insects acquire the eggs of the nematode, which are consumed along with the insect, resulting in infection of the insect predator.

Dirofilariasis is a nematode-caused disease; this condition is also known as dog heartworm. Generally, mosquitoes are the vector.

Elaeophorosis is a nematode disease transmitted by black flies. Elk, moose and some livestock are adversely affected by these nematodes.

Lancet fluke infects the liver of several wild mammals and livestock. It is transmitted to mammals when ants containing an infective stage are inadvertently consumed, but snails are an essential intermediate host.

Dog tapeworm infects wild and domestic dogs and cats. Tapeworm eggs are consumed by flea larvae and chewing lice, and then are ingested by the mammals when grooming.

Giant thorny-headed worm infects several mammals, attaching to their intestine. Worm eggs leave the host's body with feces, where they are consumed by scarab beetle larvae. If a mammal eats a beetle larva, the worm eggs hatch and a new infection occurs.

Among the aforementioned helminth diseases, only dog tapeworm commonly infests people. Humans occasionally are a dead-end host for dirofilariid nematodes.

REFERENCES AND ADDITIONAL READING

Anderson, R.C. (2000). *Nematode Parasites of Vertebrates. Their Development and Transmission* (2nd edn.). CABI Publishing, London, UK.

Araújo, C.A.C., Waniek, P.J., & Jansen, A.M. (2009). An overview of Chagas disease and the role of triatomines on its distribution in Brazil. *Vector-borne and Zoonotic Diseases* **9**, 227–234.

Atkinson, C.T. (2009). Avian malaria. In Atkinson, C.T., Thomas, N.J., & Hunter, D.B. (eds). *Parasitic Diseases of Wild Birds*, pp. 35–53. Wiley-Blackwell, Ames, Iowa, USA.

Atkinson, C.T., Thomas, N.J., & Hunter, D.B. (eds) (2009). *Parasitic Diseases of Wild Birds*. Wiley-Blackwell, Ames, Iowa, USA.

Bartlett, C.M. (2009). Filarioid nematodes. In Atkinson, C.T., Thomas, N.J., & Hunter, D.B. (eds.). *Parasitic Diseases of Wild Birds*, pp. 439–462. Wiley-Blackwell, Ames, Iowa, USA.

Brenner, R.R. & Stoka, A.M. (eds.) (1988). *Chagas Disease Vectors*, vols. **1–3**. CRC Press, Boca Raton, Florida, USA.

Chowdhury, N. & Aguirre, A.A. (2001). *Helminths of Wildlife*. Science Publishers, Inc., Enfield, New Hampshire, USA.

Coura, J.R., Junqueira, A.C.V, Fernandes, O., Valente, S.A.S., & Miles, M.A. (2002). Emerging Chagas disease in Amazonian Brazil. *Trends in Parasitology* **18**, 17–176.

Craig, T.M. (2001). *Hepatozoon* spp. and hepatozoonosis. In Samuel, W.M., Pybus, M.J., & Kocan, A.A. (eds.). *Parasitic Disease of Wild Mammals*, 2nd edn., pp. 462–468. Iowa State University Press, Ames, Iowa, USA.

Crompton, D.W.T. & Nickol, B.N. (1985). *Biology of the Acanthocephala*. Cambridge University Press, Cambridge, Massachusetts, USA.

Dias, J.C.P. & Schofield, C.J. (1999). The evolution of Chagas disease (American Trypanosomiasis) control after 90 years since Carlos Chagas discovery. *Memorias do Instituto Oswaldo Cruz* **94**, 103–121.

Dubey, J.P. (2008). Toxoplasma. In Atkinson, C.T., Thomas, N.J., &Hunter, D.B. (eds.) *Parasitic Diseases of Wild Birds*, pp. 204–222. Wiley-Blackwell, Ames, Iowa, USA.

Dubey, J.P. & Odening, K. (2001). Toxoplasmosis and related infections. In Samuel, W.M., Pybus, M.J., & Kocan, A.A. (eds.). *Parasitic Diseases of Wild Mammals*, 2nd edn., pp. 478–493. Iowa State University Press, Ames, Iowa, USA.

Févre, E.M., Picozzi, K., Jannin, J., Welburn, S.C., & Maudlin, I. (2007). Human African trypanosomiasis: epidemiology and control. In Molyneux, D.H. (ed.). *Control of Human Parasitic Diseases*, pp. 167–221. Elsevier, Amsterdam.

Friend, M. & Franson, J.C. (eds.) (1999). *Field Manual of Wildlife Diseases: General Field Procedures and Diseases of Birds*. US Department of the Interior, US Geological Survey, Washington, DC, USA.

Huffman J.E. (2009). Trematodes. In Atkinson, C.T., Thomas, N.J., & Hunter, D.B. (eds.). *Parasitic Diseases of Wild Birds*, pp. 225–245. Wiley-Blackwell, Ames, Iowa, USA.

Kirchhoff, L.V. (1993). American trypanosomiasis (Chagas disease) – a tropical disease now in the United States. *New England Journal of Medicine* **329**, 639–644.

Leek, S.G.A. (1998). *Tsetse Biology and Ecology: Their Role in the Epidemiology and Control of Trypanosomosis*. CABI, New York, USA.

Leighton, F.A. & Gajadhar, A.A. (2001). *Besnoitia* spp. and besnoitiosis. In Samuel, W.M., Pybus, M.J., & Kocan, A.A. (eds.). *Parasitic Disease of Wild Mammals*, 2nd edn., pp. 468–478. Iowa State University Press. Ames, Iowa, USA.

Marquardt, W.C., Demaree, R.S., & Grieve, R.B. (2000). *Parasitology and Vector Biology*. Academic Press, San Diego, California, USA.

McLaughlin, J.D. (2009). Cestodes. In Atkinson, C.T., Thomas, N.J., & Hunter, D.B. (eds.). *Parasitic Diseases of Wild Birds*, pp. 261–276. Wiley-Blackwell, Ames, Iowa, USA.

Molyneux, D.H. (ed.) (2007). *Control of Human Parasitic Diseases*. Elsevier, Amsterdam.

Nayar, J.K. (2008). Dirofilariasis. In Capinera, J.L. (ed.). *Encyclopedia of Entomology*, 2nd edn., pp. 1224–1229. Springer Science & Business Media B.V., Dordrecht, The Netherlands.

Olsen, W.O. (1986). *Animal Parasites. Their Life Cycles and Ecology*. Dover Publications, New York, New York, USA (reprint of 3rd edn., 1974).

Pybus, M.J. (2001). Liver flukes. In Samuel, W.M., Pybus, M.J., & Kocan, A.A. (eds.). *Parasitic Diseases of Wild Mammals*, 2nd edn., pp. 121–149. Iowa State University Press, Ames, Iowa, USA.

Richardson, D.J. & Nickol, B.B. (2009). Acanthocepthala. In Atkinson, C.T., Thomas, N.J., & Hunter, D.B. (eds.). *Parasitic Diseases of Wild Birds*, pp. 277–288. Wiley-Blackwell, Ames, Iowa, USA.

Rothschild, M. & Clay, T. (1953). *Fleas, Flukes and Cuckoos. A Study of Bird Parasites*. Readers Union, London.

Seed, J.R. & Black, S.J. (eds.) (2001). *The African Trypanosomes*. Kluwer Academic Press, Dordrecht, The Netherlands.

Sterner III, M.C. & Cole, R.A. (2009). *Diplotriaena, Serratspiculum*, and *Serratospiculoides*. In Atkinson, C.T., Thomas, N.J., & Hunter, D.B. (eds.). *Parasitic Diseases of Wild Birds*, pp. 434–438. Wiley-Blackwell, Ames, Iowa, USA.

Van Cleave, H. (1953). *Acanthocephala of North American Mammals*. University of Illinois Press, Urbana: Illinois Biological Monographs.

Wobeser, G.L. (1997). *Disease of Wild Waterfowl*, 2nd edn. Plenum Press, New York, New York, USA.

Wobeser, G.L. (2006). *Essentials of Disease in Wild Animals*. Blackwell Publishing, Oxford, UK.

Zeledón, R. & Rabinovich, J.E. (1981). Chagas' disease: an ecological appraisal with special emphasis on its insect vectors. *Annual Review of Entomology* **26**, 101–133.

Chapter 10

ARTHROPODS AS PARASITES OF WILDLIFE

Previously we discussed arthropods as vectors of micro- and macroparasites, but some are quite capable of causing disease independent of symbiotic disease agents. Most arthropod parasites of wildlife are **ectoparasites**, living on the surface of the host. Some, however, are **endoparasites**, living within the host for at least a portion of their developmental period. The arthropods (phylum Arthropoda) that are considered to be important parasites of wildlife are limited to a few taxa, as follows:

Class Arachnida – scorpions, spiders, mites, ticks, etc.
  Subclass Acari or Acarina – mites and ticks
    Order Mesostigmata – mites
    Order Prostigmata – mites
    Order Astigmata – mites
    Order Ixodida – ticks
      Family Argasidae – soft ticks
      Family Ixodidae – hard ticks
Class Insecta – insects
  Order Phthiraptera – chewing and sucking lice
  Order Hemiptera – bugs
    Family Reduviidae – assassin bugs, subfamily Triatominae – kissing or blood-sucking conenose bugs
    Family Cimicidae – bed bugs, swallow bugs, and bat bugs
  Order Diptera – flies
    Suborder Nematocera – lower (primitive) flies
      Family Culicidae – mosquitoes
      Family Simuliidae – black flies
      Family Ceratopogonidae – biting midges
      Family Psychodidae – sand flies
    Suborder Brachycera (Cyclorrapha) – higher (advanced) flies
      Family Tabanidae – horse and deer flies
      Family Glossinidae – tsetse flies
      Family Muscidae – muscid flies
      Family Calliphoridae – blow flies
      Family Sarcophagidae – flesh flies
      Family Oestridae – bot and warble flies
      Family Hippoboscidae – louse flies
  Order Siphonaptera – fleas

There are other arthropod taxa associated with wildlife, of course. Some insect taxa contain a small number of species that exploit wildlife, such as the earwigs (Dermaptera), or certain families of beetles (Coleoptera) such as Leptinidae, Platypsyllidae, and Staphylinidae, but they are not considered to be very important. Also, among the flies (Diptera) are two families (Nycteribiidae, Streblidae) that, while species-rich, are limited to feeding on bats, and are not well known. Table 10.1 shows the distribution of ectoparasitic species within class Insecta that affect mammals and birds, as determined by Adrian G. Marshall in his book *The Ecology of Ectoparasitic Insects*. Though instructive, this list leaves out such important parasitic groups as mosquitoes (Diptera: Culicidae) and black flies (Diptera: Simuliidae). Although these are not as intimately associated with wildlife (they are transient ectoparasites, feeding periodically) as are Marshall's selections, they are nevertheless important parasites, so they are included in the discussion that follows. Marshall also included some taxa of little importance to wildlife, such as Pyralidae (Lepidoptera) and Scarabaeidae (Coleoptera). Though it is interesting to know that a few species within taxa that are normally considered to be

Insects and Wildlife: Arthropods and Their Relationships with Wild Vertebrate Animals, 1st edition. By J.L. Capinera. Published 2010 by Blackwell Publishing.

Table 10.1. Insects ectoparasitic on mammals and birds (adapted from Marshall 1981).

Order	Family	Number of species	Hosts
Dermaptera	Arixeniidae	5	mammals (bats)
	Hemimeridae	11	mammals (rodents)
Phthiraptera	(all)	3250	birds and mammals
Hemiptera	Cimicidae	89	mammals and birds
	Polyctenidae	32	mammals (bats)
Lepidoptera	Pyralidae	3	mammals
Coleoptera	Leiodidae	1	mammals (lagomorphs)
	Leptinidae	6	mammals (rodents)
	Platypsyllidae	2	mammals (rodents)
	Staphylinidae	57	mammals (rodents)
	Languriidae	2	mammals (rodents)
	Scarabaeidae	3	mammals (edentates)
Diptera	Carnidae	2	birds
	Mystacinobiidae	1	mammal (bat)
	Hippoboscidae	197	birds and mammals
	Nycteribiidae	256	mammals (bats)
	Streblidae	221	mammals (bats)
Siphonaptera	(all)	2018	mammals and birds

plant or detritus feeders have adapted to use animals as hosts, they do not warrant further discussion here. A few additional taxa have some importance to wildlife because they are occasional mechanical vectors of disease, because they are occasional intermediate hosts in disease cycles, or because they occasionally bite or sting wildlife, but they are mentioned only briefly at the end of this chapter.

Wildlife can support diverse assemblages of arthropods, some of which can be important. As an example of arthropod diversity on wildlife, Table 10.2 provides a list of arthropods known to be associated with selected wild birds in Florida, USA. Note that there is little overlap among arthropod species assemblages on the various birds, even when comparing the two species of pelicans. Ducks are a well-known exception, however, in that these closely related birds often share parasites. Only 14 species of parasitic arthropods are known to affect ducks in Florida. Mixed infestations of mites and chewing lice are common, and more likely to occur on juvenile birds than adults. Only when the health of the ducks is compromised by some other factor (usually weather, food, or injury) are these ectoparasites thought to be important.

There are two important exceptions to the notion that arthropod parasites are not often important unless their health is compromised by another factor. The exceptions involve **nest re-use** and **colonial nesters**. Nests of birds and some mammals become infested with arthropods, especially ectoparasites. If the host animal continues to inhabit the nest, the population of parasites builds to levels that may affect the fitness of the adult, and certainly damages the health of the juveniles. A number of blood-feeding arthropods are involved in nest re-use problems, including mites and ticks (Acari), bugs (Hemiptera), lice (Phthiraptera), fleas (Siphonaptera), and flies (Diptera). Colonial (group-nesting) birds and mammals may be subject to the same problems even if they do not occupy a previously occupied nest, however, because the arthropods can disperse short distances and locate the new, uninfested nests nearby.

Arthropods differ in their propensity to infest nesting animals. Mites have a short development time, sometimes only a few days, so their population can increase very rapidly. Two to several generations can develop on young birds while they are in the nest, with some mites remaining on the chicks when they fledge, and others lingering in the nest where they will become dormant until the next period of nesting, even if it is a year later. Studies conducted in Europe of barn swallows, *Hirundo rustica*, showed that 45% of the first

Table 10.2. Arthropods associated with selected birds in Florida, USA (adapted from Forrester and Spalding 2003).

Bird host	Arthropod group	Arthropod species
American white pelican, *Pelecanus erythrorhynchos*	Chewing lice	*Colpocephalum unciferum*
		Pectinopygus tordoffi
		Piagetiella peralis
	Mites	*Alloptes* sp.
		Scutomegninia sp.
Brown pelican, *Pelecanus occidentalis*	Chewing lice	*Colpocephalum occidentalis*
		Pectinopygus occidentalis
		Piagetiella bursaepelicani
	Mites	*Neottialges apunctatus*
		Phalacrodectes pelicani
		Phalacrodectes punctatissimus
		Scutomegninia sp.
	Ticks	*Argas radiatus*
	Flies	*Olfersia sordida*
		Olfersia spinifera
		Wohlfahrtia vigil
Wood stork, *Mycteria americana*	Feather mites	*Analloptes* sp.
		Ingrassia sp.
		Mycteralges mesomorphus
		Taeniosikya sp.
	Skin mites	*Ornithonyssus bursa*
		Unidentified chigger
	Subcutaneous mites	*Neottialges eudocimae*
		Neottialges kutzeri
		Neottialges mycteriae
		Pelargolichus sp.
		Phalacrodectes mycteria
	Chewing lice	*Ardeicola loculator*
		Ciconiphilus quadripustulatus
		Colpocephalum mycteriae
		Colpocephalum scalariforme
		Neophilopterus heteropygus
Wood duck, *Aix sponsa*	Feather mites	*Bdellorhynchus* sp.
		Freyana largifolia
		Ingrassia sp.
	Nasal mites	*Rhinonyssus rhinolethrum*
	Chewing lice	*Anaticola crassicornis*
		Anatoecus dentatus
		Holomenopon clauseni
		Trinoton querquedulae
	Cimicid bugs	*Ornithicoris pallidus*
Red-tailed hawk, *Buteo jamaicensis*	Feather mites	*Pseudalloptinus* sp.
	Chewing lice	*Colpocephaum napiforme*
		Craspedorrhynchus americanus
		Degeeriella fulva
		Kurodaia fulvofasciata
	Louse flies	*Icosta americana*
	Fleas	*Echidnophaga gallinacea*

Table 10.2. *Continued*

Bird host	Arthropod group	Arthropod species
Northern bobwhite, *Colinus virginianus*	Skin mites	*Eutrombicula alfreddugesi*
		Neoschoengastia americana
		Neotrombicula whartoni
		Microlichus sp.
		Rivoltasia sp.
		Ornithonyssus sylviarum
	Nasal mites	*Boydaia colini*
		Colinoptes cubanensis
	Feather mites	*Colinolichus virginianus*
		Colinophilus wilsoni
		Megninia sp.
		Dermoglyphus sp.
		Apionacarus wilsoni
	Ticks	*Amblyomma americaum*
		Amblyomma maculatum
		Amblyomma tuberculatum
		Dermacentor variabilis
		Haemaphysalis chordeilis
		Haemaphysalis leporispalustris
		Ixodes minor
		Rhipicephalus sp.
	Chewing lice	*Colinicola numidiana*
		Goniodes ortygis
		Menacanthus pricei
		Oxylipeurus clavatus
	Fleas	*Echidnophaga gallinacea*

brood was infested with tropical fowl mite, *Ornithonyssus bursa*, with a few nests having up to 10,000 mites. Although this abundance is exceptional, nests of small avifauna in Europe are commonly found with mites, swallow bugs (*Oeciancus vicarious*), and fleas (*Ceratophyllus* spp.). Research conducted on an island inhabited by Peruvian booby, *Sula variegata*, showed that an abundance of *Ornithodorus* ticks caused nest abandonment, with the abandonment starting at one end of the island and gradually spreading to the other, presumably due to dispersal of the hungry ticks. This is more an example of coloniality than nest re-use, of course. Burrows inhabited by either birds or mammals may likewise be infested. European badgers, *Meles meles*, build dozens of nest chambers along their burrows and continuously relocate their bed chamber to avoid

sleeping in chambers infested with fleas, ticks, and lice. It should come as no great surprise, then, that cavity nesting birds discriminate against cavities that have been used previously, or that swallows will avoid mite-infested old nests in favor of building a new nest despite the great time and energy investment required to construct an entirely new nest. Swallows will also deposit their eggs in the nests of neighbors if their own nest is heavily infested. Perhaps most interesting is the apparent **nest fumigation behavior** of some birds. Birds from throughout the world collect fresh vegetation that is laid upon, or incorporated into, the nest. Release of volatile secondary compounds from the plant material is thought to be the basis for this behavior. Among both passerines and raptors, species that re-use their nests are more likely to display this behavior.

MITES AND TICKS (ARACHNIDA: ACARI OR ACARINA: SEVERAL ORDERS)

This is a very large group of small (mites) or medium-sized (ticks) arthropods, with a confusing system of classification. For all practical purposes, ticks are simply large mites. Mites and ticks cause direct injury and also transmit disease to wildlife. They often are abundant on mammals and birds, but also sometimes build to high levels on reptiles, where they cause skin lesions and anemia. Their developmental stages are basically: egg > larva > nymph > adult, but the nymphal stage is sometimes subdivided.

Mites

Mites differ from insects in several ways. Instead of the three body regions found in insects, mites have only two, and they are almost imperceptibly merged so there appears to be no body segmentation (Fig. 10.1). The anterior region is called the **prosoma** or **cephalothorax**, and bears five to six pairs of appendages. True antennae are lacking. The first pair of appendages is used for feeding, and called **chelicerae**. The chelicerae are pincher-like, but in some cases they are modified for piercing skin to obtain blood. The second pair of appendages is called the **pedipalps**, and possesses chemical and tactile sensors for detection of food location and environmental cues. The immature mites (larvae) developing from eggs bear three pairs of legs in most species, but older mites (nymphs and adults) bear four pairs of legs. The principal body region is called the **opisthosoma** or **abdomen**.

Development of mites is similar to development of insects. The nymphal stage is typically divided into three instars, the protonymph, deutonymph, and tritonymph. Each stage or instar is separated by a molt. Their respiratory system often contains tracheal ducts and spiracles, like insects, but because of their small size gases often can diffuse adequately through their body surface. Both male and female forms usually occur. Note, however, that many ticks have only a single nymphal stage.

Mites only ingest liquified food, and they may secrete digestive enzymes into the host to liquefy it; this is called **preoral digestion**. A few burrow into the skin of humans or animals, or infest hair follicles and associated dermal glands. A few even invade the lungs or urinary tract of wildlife and humans. Although mites commonly are ectoparasites of wildlife, they usually cause relatively little injury. All birds are infested with mites, some of which are specific to the host but others of which are not host specific. These same mites may be more damaging to domestic animals or abnormal hosts (introduced wildlife). A few are capable of transmitting disease, but this is usually minor as compared to disease transmission by ticks. Table 10.3 provides some examples of mites associated with wildlife.

Many mites fed on the skin on wildlife, particularly mites in the families Dermanyssidae, Macronyssidae, Laelapidae, and Trombiculidae. They can feed on skin and blood and cause skin irritation (dermatitis), anemia, and they may induce scale loss among snakes. Other mites are known as fur mites because they possess special adaptations for living on the hair coat of their mammalian hosts. Among the families containing fur mites are Cheyletidae, Myobiidae, Listrophorida, Atopomelidae, and Myocoptidae. As with the first group of mites, they pierce the skin and feed on fluids, causing skin irritation and thickening, and hair loss in the host animals. *Demodex* spp. (Demodicidae) are thin-bodied mites that burrow down into sebaceous glands and hair follicles. Though not thought of as damaging to wildlife, *Demodex* spp. can be a problem for domestic animals.

Mange Mites

Some mite species cause extreme skin irritation called **mange** among wildlife and domestic animals. For example, mites in the genus *Psoroptes* (Psoroptidae) cause psoroptic mange. *Psoroptes ovis* affects bighorn, *Ovis canadensis*, and *P. cervinus* affects bighorn and elk, *Cervus elaphus*, in western North America. Chorioptic mange caused by *Chorioptes bovis* affects caribou (*Rangifer tarandus*) in Canada and llama (*Lama glama*) and its relatives in South America. Caparinic mange is caused by infection by *Caparinia* spp.; *C. tripilis* causes mange in Eurasian hedgehog, *Erinaceus europaeus*, and *C. erinacei* affects African hedgehog, *Atelerix albiventris*. Mites in the family Sarcoptidae cause especially severe problems, and *Sarcoptes scabiei* is discussed further below. Other important sarcoptid mites are *Notoedres* spp. *Notoedres cati* affects Siberian tiger (*Felis tigris*), American bobcat (*Lynx rufus*), and sometimes foxes. *Notoedres centrifera* affects American squirrels, *Sciurus* spp., and porcupine, *Erethizon dorsatum*. *Notoedres muris* affects wild rodents, marsupials, and hedgehogs.

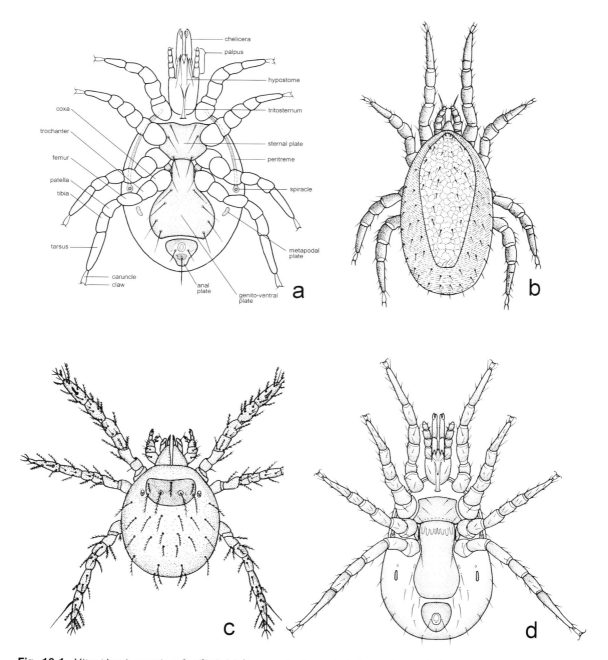

Fig. 10.1. Mites (Acarina: various families): (a) diagram of mite, ventral view, with some body parts labeled; (b) poultry red mite, *Dermanyssus gallinae* (Dermanyssidae), a common pest of wild and cultivated birds; (c) a chigger mite, *Trombicula alfreddugesi* (Trombiculidae), which is commonly found on both birds and mammals; (d) a common bird- and mammal-infesting mite, *Haemolaelaps megaventralis* (Laelapidae).

Table 10.3. Some parasitic mites that affect wild animals, or are associated with wild and domestic animals. Diseases caused or transmitted by these mites also are shown.

Mite	Wildlife host	Domestic animal host	Disease
Dermanyssus gallinae	wild birds	chickens, pigeons, canary	Red fowl mite
Ornithonyssus bursa	wild birds	chickens, ducks, turkeys	Tropical fowl mite
Ornithonyssus sylviarum	wild birds	chickens, pheasants	Northern fowl mite
Larvae of Trombiculidae	snakes, lizards, mammals	rare on livestock, pets	Chiggers
Cheyletiella parasitvorax	European rabbit	rabbits	Cheyletiellosis
Opthalmodx sp.	foxes, rabbits, rodents, bats, lemurs	cattle, sheep, goats, swine	Demodectic mange
Sarcoptes scabiei	foxes, wolf, coyote, deer, elk, fisher, ferrets, gazelle, primates	cattle, sheep, goats, horses, pigs, llama	Sarcoptic mange
Notoedres cati	foxes, tiger, American bobcat, civets	dogs, cats	Notoedric cat mite
Notoedres centrifera	squirrels, porcupine	none	Notoedric squirrel mite
Notoedres muris	rats, marsupials, hedgehogs	none	Notoedric rat mite
Psoroptes cervinus	deer, elk, bighorn sheep	none	Elk scab mite
Psoroptes natalensis	zebu, water buffalo	cattle, horses	A scab mite

Birds are not exempt from mange problems. Skin diseases called bird mange, tassel foot, bumble foot, and scaley leg are examples of mite-induced problems. Many times mange problems show up on captive birds, but epizootics sometimes occur in wild birds as well. For example, epizootics of knemidocoptic podoacariasis were documented in American robin, *Turdus migratorius*, in the eastern and central USA, and in red-winged blackbird, *Agelaius phoeniceus*, common grackle, *Quiscalus quiscala*, and brownheaded cowbird, *Molothrus ater*, in eastern Canada.

Respiratory Mites

Respiratory mites live in the nasal passages and lungs of mammals, birds, and reptiles. Respiratory mites are found in several families, including Ascidae, Entonyssidae, Rhinonyssidae, Cytoditidae, Turbinoptidae, Halarachnidae, Trombiculidae, Lemurnyssidae, Pneumocoptidae, and Ereynitidae. Transmission is thought to occur by direct contact or by sneezing or coughing. Though usually considered benign, they can be a serious problem among captive animals, and hundreds of mites can be found in some specimens.

Ear Mites

Ear mites are so named because they infest the ears of animals. Commonly, it is members of the families Psoroptidae, Trombiculidae, and Raillietidae that are involved. Typically, crusty lesions form inside the ears, accompanied by foul-smelling discharges. Infected animals display ear scratching and head shaking, droopy ears, and sometimes infestation is accompanied by loss of equilibrium. One of the most widespread ear mites is *Psoroptes cuniculi*, known as psoroptic ear mite. It affects numerous animals including white-tailed deer, *Odocoileus virginanus*; mule deer, *O. hemionus*; Nubian mountain goat, *Capra ibex nubiana*; and blackbuck antelope, *Antilope cervicapra*. It is most commonly encountered in domestic animals such as horses, donkeys, mules, and guinea pigs, however. *Otodectes cynotis*, known as otodectic ear mite, affects wildlife such as foxes, ferrets and raccoons, but also is common in cats and dogs. *Raillietia* spp. infest Alpine ibex, *Capra ibex*, in Europe; waterbuck, *Kobus ellipsiprymus*, in Africa; banteng, *Bos javanius*, in Indonesia; wombat, *Vombatus ursinus*, in Australia; and others. *Raillietia auris* is commonly found in cattle everywhere, though it is not normally considered to be a problem.

Bird Mites

Birds are host to several types of mites, and over 2000 species of bird mites are known. Feather mites are found on the plumulaceous down feathers (down mites), vane surfaces of contour feathers (vane mites), inside the quill portion of flight and tail feathers (quill mites), as well as on the surface of the skin (skin mites). Down mites occur in the families Analgidae, Psoroptoididae, and Xolalgidae. Vane mites occur in Alloptidae, Avenzoariidae, Thysanocercidae, Trouessartiidae, and Proctophyllodidae. Quill mites are found in Apionacaridae and Dermoglyphidae. Skin mites of birds occur in Dermationidae, Epidermoptidae, and Knemidocoptidae. The bird feather mites are highly evolved, with many birds having their own species. Feather mites feed mainly on the waxes and fatty acids found on feathers. Interestingly, they sense that feathers are about to be lost when birds molt, and move from the feathers that are about to be molted to those that will be retained. The mechanism that allows the mites to perceive the timing of feather molting is not yet known. Bird mites in the families Ascidae, Rhinonyssidae, Cytoditidae, Turbinoptidae, and Ereynitidae dwell in the nasal passages and tracheal tissues, with mites in the family Cloacaridae invading the lungs. Indeed, mites have been known to invade the alimentary tract, air sacs, and the urinary tract, as well as the lungs. These mites are not uncommon, with studies in Texas, USA and Manitoba, Canada demonstrating that about 20% of birds are infested with nasal mites.

Sarcoptic Mange Mite

The mite *Sarcoptes scabiei* (Sarcoptidae) causes a disease called **Sarcoptic Mange** in animals, and **Scabies** in humans. Sarcoptic mange mite is quite small; males measure 213–285 microns long whereas females are 300–504 microns long. The tarsi have blade-like claws. All stages of the mites live in the skin of the host animal, within burrows created by the female mite. Larvae and nymphs will leave the burrows and crawl on the surface of the skin, leading to accumulations of up to several thousand mites per square centimeter. The life cycle requires about 2 weeks for completion. The mite, and its relatives, are often referred to as 'burrowing mites' or 'itch mites.' The host responds to the burrowing of the mites, and the accumulation of mite fecal material, by forming lesions. The skin becomes reddish, itchy, thickened, crusted with exudates, and

infected. The animal exacerbates the problem by rubbing and scratching. Hair loss is a common manifestation of infection, but is followed by weight loss, reduced food consumption, impaired hearing, blindness, exhaustion, and eventually death. Other mites and other diseases produce similar signs, so skin scrapings must be taken to ascertain the true cause.

Mites are transferred from animal to animal by contact, and normally the disease is limited to intraspecific transmission because different host-specific races are common. Lesions initially form in a limited area, often the face and neck regions. The problem is worst in the winter months. Animals sometimes recover from infection without treatment.

A large number of animal species, both wild and domestic, are susceptible to infection. More than 100 species of mammals are reported to be susceptible to infection. The problem is greatest with canids such as red fox, *Vulpes fulva*; gray fox, *Urocyon cinereoargenteus*; wolf, *Canis lupus*; coyote, *C. latrans*; dingo, *C. lupus dingo*; and also among cervids such as roe deer, *Capreolus capreolus*; red deer, *Cervus elaphus*; elk, *Cervus canadensis*; and reindeer, *Rangifer tarandus*. A few other of the many animals affected include fisher, *Martes pennanti*; ferrets, *Mustela* spp.; black bear, *Ursus americanus*; llamas, *Lama* spp.; Thompson's gazelle, *Gazella thompsonii*; wildebeests, *Connochaetes taurinus*; chamois, *Rupicapra rupicapra*; ibex, *Capra ibex*; chimpanzees, *Pan troglodytes*; and wombat, *Lasiorhinus latifrons*.

Surprisingly little is known about the basis for epizootics of this disease. Prevalence of mange was about 20% of coyotes and 11% of wolves in Alberta, Canada, whereas 80% of coyotes in Texas, USA, were infected. In the Texas study, 70% of the coyotes died, but the population soon recovered. Natural spread of sarcoptic mange mite-infested red foxes from Estonia to Finland, Sweden and Norway resulted in over 50% mortality of fox populations within a few years. It took 20 years for the fox population to recover to its former levels of abundance. In Britain, the fox population was reduced by 95% following accidental introduction of sarcoptic mange mite. Mange is the principal cause of death of chamois, *Rupicapra rupicapra*, and ibex, *Capra* spp., in the European mountains, and is considered a threat to their continued existence.

Exposure to sarcoptic mange mite induces protection to subsequent exposure in some animals, but not in others. Attempts to eliminate the disease from natural populations of wildlife are usually unsuccessful, though there are effective acaricides (miticides) suitable for small-scale treatment, especially of endangered animals. Humans can contract scabies from wildlife, and though it tends to be short-lived and self-limiting, those handling animals should take precautions.

Ticks

As noted previously, ticks are basically large mites, so their structure and development are the same as described above for mites. However, ticks are the most important ectoparasites affecting wildlife. Some important ticks are listed in Table 10.4. In addition to the diseases they transmit, they cause injury through the removal of blood, create irritation at the feeding site, provide sites for secondary infection, and induce tick paralysis. Nearly all ticks belong to one of two families: the Argasidae, called soft-bodied ticks or soft ticks, or Ixodidae, called hard-bodied ticks or hard ticks. Often ticks are referred to as **one-host**, **two-host**, or **three-host ticks**, which simply indicates the number of host animals that a tick feeds upon in order to complete its life cycle. The different hosts may be different individuals of the same species, or different species.

Soft ticks lack a **scutum** (Figs. 10.2, 10.3), or hard plate that covers all or the anterior portion of the abdomen. Instead, they are leathery appearing. Also, when viewed from above, the mouthparts of soft ticks are not visible. They tend to be oval in body shape. The most important genera of soft ticks are *Argas*, *Ornithodoros*, *Otobius*, and *Antricola*.

In contrast, **hard ticks** possess a scutum that covers most of the abdomen (males) or the anterior region (females and young). Also, their mouthparts protrude and their body is teardrop in shape. Among the important genera of hard ticks are *Ixodes*, *Haemaphysalis*, *Boophilus*, *Rhipicephalus*, *Dermacentor*, *Hyalomma* and *Amblyomma*.

The hard and soft ticks differ slightly in their pattern of feeding. In nearly all ticks, each immature (larva, nymph) stage and the adult stage feed once. However, hard ticks have only one nymphal stage, and thus require only one feeding period as a nymph. Soft ticks have two or more nymphal stages, each requiring a blood meal, so soft ticks typically have a feeding pattern that involves taking more blood meals from multiple hosts. Not surprisingly, the soft ticks more typically inhabit the nests, burrows or shelters of hosts, so they

have good opportunity for repeat feeding. Ticks that inhabit the shelters of their hosts are called **nidicolous**. Soft ticks tend to feed relatively rapidly, requiring 30 minutes to several hours to acquire a blood meal, and then leave the host. Usually they are **nocturnal**, or active at night. Development time requires months to years, and the number of instars varies among species. The soft ticks are more tolerant of low humidity and high temperatures than are the hard ticks, and often occur in arid or semiarid environments. The abundance of soft ticks in their shelters can be quite high. In caves inhabited by bats, for example, hundreds of thousands of ticks may be present. Seabirds, which like all animals are reluctant to abandon their offspring, have been documented to abandon their nests during the middle of the breeding season due to the enormous abundance of ticks.

Signs of feeding by soft ticks include bites, lesions, and inflammation, but affected animals may also display itching, head shaking, ear scratching, secondary infections, ear damage, deafness, and convulsions. Infested animals often display nervousness and digestive disorders. Soft ticks also are capable of transmitting several serious diseases involving wildlife, including the bacteria responsible for Relapsing Fever, Tularemia, Avian Spirochetosis, Avian Cholera, and the virus responsible for African Swine Fever.

Hard ticks are more commonly encountered on wildlife than are soft ticks because they have a more protracted period of blood feeding, often 3–20 days. Most hard ticks take three blood meals from different hosts, with the ticks dropping to the ground to molt between feedings. Hence, they are called **three-host ticks**. Often, the immature ticks feed on small animals and the adults on larger species. Some, however, remain on a single host, and are subsequently called **one-host ticks**. Hard ticks become extremely distended when blood feeding, with females becoming as much as 100 times as heavy as when not recently blood-fed. Like soft ticks, the hard ticks often are nidicolous. Hard ticks may require 1–3 years to complete their development. They can be found on their hosts during any portion of the year, though there tends to be a species-specific pattern of abundance for each type of tick. Some ticks are rather specific in their feeding preference. For example, *Ixodes banksi* feeds mostly on beaver (*Castor canadensis*), whereas *Ixodes soricis* feeds on shrews (*Sorex* spp.) and shrew-mole (*Neurotrichus gibbsii*). In contrast, *Ixodes scapularis*, which has been well studied due to its association with Lyme disease,

Table 10.4. Some parasitic ticks that affect wild animals, or are associated with wild and domestic animals. Diseases caused or transmitted by these ticks also are shown.

Tick	Wildlife host	Domestic animal host	Disease
Soft ticks (Argasidae)			
Argas persicus	birds	chickens, turkeys	none
Ornithodoros concanensis	bats	none	none
Ornithodoros hermsi	chipmunk, deer mouse, woodrats	none	Relapsing Fever
Ornithodoros parkeri	rats, mice, rabbits, hares, weasels	none	Relapsing Fever
Ornithodoros turicata	ground squirrels, prairie dogs, woodrat, kangaroo rats, rabbits	cattle, horse, pig	Relapsing Fever, Rocky Mtn. Spotted Fever,
Ornithodoros puertoricensis	rats, rabbits	cats	African Swine Fever
Otobius megnini	deer, rabbits, coyote, elk, mountain goat, mountain sheep	cat, dog, horse, cattle, goat, sheep, pig	Tick Paralysis
Hard ticks (Ixodidae)			
Amblyomma americanum	white-tailed deer, wolf, coyote, mountain lion, American bobcat, skunk, northern raccoon, American badger, otters, Virginia opossum, rabbits, foxes, rats, gophers	cattle, horse, sheep, goat, dog, cat	Tularemia, Babesiosis, Rocky Mtn. Spotted Fever, etc.
Amblyomma tuberculatum	white-tailed deer, rabbit, fox, squirrels	cattle, dogs	none
Boophilis annulatus	white-tailed deer	cattle, horse, goat, sheep	Bovine Babesiosis, Anaplasmosis
Dermacentor andersoni	mule deer, elk, mountain goat, pronghorn, bighorn, American bison, American bobcat, bears, rabbits, prairie dog, mice	cattle, horse, sheep, goat, dog, cat	Tularemia, Colorado Tick fever, Rocky Mtn. Spotted Fever, etc.
Dermacentor halli	peccary	none	none
Dermacentor variabilis	white-tailed deer, wolf, mountain lion, weasels, bears, feral swine, muskrat, Virginia opossum, American badger, northern raccoon, rabbits, rats, mice	cattle, horse, sheep, goat, dog, cat	Babesiosis, Anaplasmosis, Tularemia, Cytauxzoonosis, Rocky Mtn. Spotted Fever, etc.
Haemaphysalis spinigera	birds, monkeys, small and large mammals	cattle	Kyasanur Forest disease
Ixodes pacificus	mule deer, coyote, cougar, wolf, American bobcat, mink, squirrels, ground squirrels, chipmunks	cattle, horse, burro, mule, goat, dog, cat	Tularemia, Lyme Disease, Ehrlichiosis
Ixodes peromysci	deer mouse	none	none
Ixodes scapularis	white-tailed deer, wolf, foxes, feral swine, skunks, northern raccoon, rabbits, squirrels, chipmunks, rats, mice, voles	cattle, horse, sheep, mule, goat, pig, dog, cat	Lyme Disease, Tularemia, Anaplasmosis, Human Babesiosis, etc.

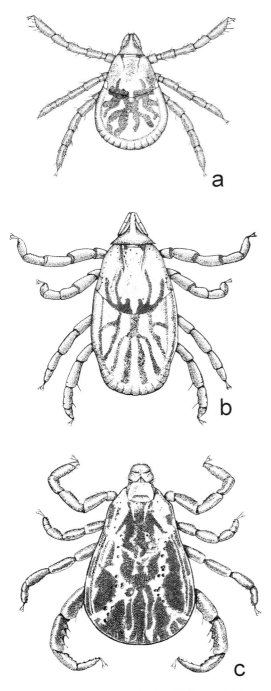

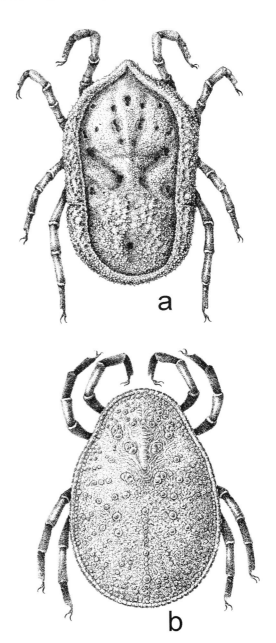

Fig. 10.3. Some representative 'soft ticks': (a) cave tick, *Ornithodoros tholozani* (Acarina: Ixodida: Argasidae): (b) fowl tick, *Argas persicus* (Acarina: Ixodida: Argasidae).

Fig. 10.2. A representative 'hard tick,' the wood tick or American dog tick, *Dermacentor variabilis* (Acarina: Ixodida: Ixodidae): (a) larva; (b) nymph; (c) adult. This is one of the most commonly occurring ticks on mammals.

has been reported from 41 species of mammals, 57 species of birds, and several lizards.

Tick abundance on animals can vary greatly. Recent studies found the number of *Ixodes scapularis* on individual deer to be 2.9–10.7 ticks per deer in Alabama, and 4.0 in Minnesota. Tick numbers were over 40 per host on white-footed mouse, *Peromyscus leucopus*, in Massachusetts but only 4 per host in Connecticut. Prevalence on deer was 75% in Alabama, 98% in Arkansas, 54% in Florida, 76% in Georgia, 99% in Louisiana, 99% in North Carolina, and 56% in Virginia. *Dermacentor albipictus* was found to infest nearly all samples of moose, *Alces alces*; elk, *Cervus canadensis*; deer, *Odocoileus* spp.; and American bison, *Bison bison*, taken in western Canada, and to attain exceedingly high densities, ranging from 0.5 ticks/cm^2 on deer, *Odocoileus* spp., to over 1/cm^2 on moose, *Alces alces*. On moose, this translated to over 50,000 ticks per animal in some cases! Even in the absence of disease, ticks can cause severe debilitation and death among large animals; tick-induced mortality among animals as large as elk and moose are not rare.

Feeding by hard ticks can cause irritation, local inflammation and exudate, blood loss, anemia, hair loss and sometimes paralysis. **Tick paralysis** is a systemic ailment that occurs in a wide range of vertebrates after even a single tick attaches and begins to feed. Tick paralysis is linked to protein toxins injected by female ticks during their feeding, but no microbial pathogen is known. It is a progressive ailment, with severity positively associated with the length of time that the ticks feed. Initially the limbs are affected, but eventually the entire body is affected, including respiratory failure. Prompt removal of ticks arrests the condition, and the animal recovers, sometimes very quickly. It has been associated with over 40 species of ticks worldwide, though the ability to induce paralysis varies considerably. In North America, for example, it is most commonly induced by *Dermacentor andersoni*, *D. variabilis*, and *D. occidentalis*. It has been reported from mule deer, *Odocoileus hemionus*; American bison, *Bison bison*; black bear, *Ursus americanus*; gray fox, *Urocyon cinereoargenteus*; striped skunk, *Mephitis mephitis*; coyote, *Canis latrans*; red wolf, *Canis rufus*; and western harvest mouse, *Reithrodontomys megalotis*. Cattle are particularly susceptible to this problem. Though less frequent, soft ticks can also cause paralysis.

Numerous infectious diseases are associated with hard ticks. Among the more important are Lyme Disease, Tularemia, Powassan Virus, Rocky Mountain Spotted Fever, Human Babesiosis, Canine Babesiosis, Bovine Babesiosis, Anaplasmosis, Equine Ehrlichiosis, Canine Ehrlichiosis, Cytauxzoonosis, and many others.

Taiga Tick

Ixodes persulcatus (Ixodida: Ixodidae) is called taiga tick, a reflection of its occurrence in northern Asia and eastern Europe, in a coniferous forest habitat called 'taiga.' It is one of the world's best-studied ticks due to its medical and veterinary importance. It is the principal vector of tick-borne encephalitis, a viral disease of humans. It also transmits Powassan virus and Kemerovo virus, and only recently was found to vector Lyme borrelioses caused by *Borrelia* spp. It also is associated with other diseases. Its broad distribution seems to be limited in the north of Asia by cold temperatures, but in the south by low humidity.

Taiga tick is a three-host tick. Each stage feeds on a vertebrate host only once. The larva feeds for 3–5 days, the nymph for 3–6 days, and the adult female for 6–10 days. Females consume more than males, and engorged females are 100 to 150-fold heavier than males. Females produce 2000–4000 eggs. The life cycle of a tick is as little as 2 years in the southern portions of its range, to as much as 5 years in the north.

Taiga ticks, like many other ticks, locate their host by questing. **Questing**, or host seeking, is accomplished often using an 'ambush' strategy, wherein unfed ticks cling to the substrate with their front legs extended, awaiting opportunity to attach to a host animal. This also serves to expose chemoreceptors called Haller's organs that are located on the front tarsi. If they attach to the tips of vegetation along trails frequented by animals they have good opportunity to locate a suitable host. Although they will remain in this questing position for long periods, they cannot do so indefinitely as the temperature and humidity in such locations are adverse, so during mid-day they typically retreat to the woodland litter. Ticks can detect host stimuli from distances of 5–10 m, so they can reposition to take better advantage of regular host movements, or they can move directly to a host and attach.

Taiga tick has a very broad host range, with about 100 species of mammals and 175 species of birds, plus a few reptiles recorded as hosts. Larvae feed principally on small mammals such as rodents (mice, rats, squirrels) and insectivores (shrews and moles), nymphs select medium-sized animals such as birds, rodents, lagomorphs (rabbits and hares), and some carnivores, whereas adults select large mammals. The abundance of taiga tick corresponds strongly with the abundance of preferred hosts such as bank vole, *Clethrionomys glareolus*; gray-sided vole, *C. rufocanus*; wood mouse, *Apodemus specious*; striped field mouse, *A. agrarius*; and Siberian chipmunk, *Eutamias sibiricus*.

Wood Tick

The wood tick, *Dermacentor variabilis* (Ixodida: Ixodidae), is widespread in North America, and absent only from the Rocky Mountain and intermountain west and northern Canada. Due to its prevalence on dogs, it is also known as American dog tick. Its host range is very broad, however, with larvae and nymphs feeding on small mammals, and nymphs and adults feeding on larger mammals including livestock and humans. Only the adult stage attacks dogs, livestock, and humans. This tick is the primary vector of Rocky Mountain spotted fever, but also transmits tularemia and can cause tick paralysis. It is most commonly found in tall grass or brushy, shrubby areas where its animal hosts are abundant. It is reddish brown, variable in color pattern, and unfed females measure about 5–6 mm in size.

The life cycle of *D. variabilis* requires three hosts, and requires anywhere from 2 months to 2 years to complete. Eggs hatch in the spring, and larvae occur from March to July. Nymphs occur from June to September. Adults may be found at nearly any time of the year.

Larvae feed for 2–24 days on a small host such as a mouse, then drop from the host to digest the meal (which may require a week or more) and molt to the nymphal stage. Nymphs can persist for up to six months without a blood meal, but then feed on a larger animal such as a raccoon. The nymphs blood-feed for 3–10 days, then drop from the host to digest their meal and molt to the adult stage. The nymphal stage persists for 3 weeks to several months. Adults can persist for up to 2 years without feeding, but actively seek hosts by questing. Once locating a host, the female mates, feeds on blood for 6–13 days, and then drops from the host to deposit eggs. She produces 4000–6500 eggs and dies. Eggs will hatch in about a month, depending on temperature.

The wildlife hosts of *D. variabilis* are numerous. At least ten species of mice and rats, plus several rabbits and squirrels are documented hosts. In addition, *D. variabilis* feeds on coyote, *Canis latrans*; red fox, *Vulpes vulpes*; gray fox, *Urocyon cinereoargenteus*; Virginia opossum, *Didelphis virginiana*; North American porcupine, *Erethizon dorsatum*; striped skunk, *Mephitis mephitis*; American bobcat, *Lynx rufus*; and groundhog, *Marmota monax*.

Blacklegged Tick

The blacklegged tick, *Ixodes scapularis* (Ixodida: Ixodidae) is found widely in eastern North America wherever its primary reproductive host the white-tailed deer (*Odocoileus virginianus*) is found in abundance. Hence, it is also commonly known as the deer tick. *I. scapularis* is a small tick, with unengorged females measuring only about 3 mm in size and males 2 mm. The legs are black, or course, as is the prosoma (cephalothorax) and the scutum. The scutum covers nearly the entire abdomen of the male, so it appears nearly entirely black. The scutum covers only the anterior portion of the abdomen in females and in immature ticks, so they are orange or reddish posteriorly.

I. scapularis is a three-host tick. Adults usually are associated with deer, but the larval and nymphal stages feed on smaller wildlife, mainly rodents or birds. Blacklegged tick is the principal vector of Lyme disease in eastern North America, which was discussed in Chapter 8. Because blacklegged tick vectors the bacterial (spirochaete) pathogen responsible for Lyme disease, *Borrelia burgdorferi*, it has been the subject of considerable research on disease and tick suppression.

Personal protection is the first line of defense against infection from diseases spread by ticks. Avoidance of tick habitats (wooded areas frequented by wildlife), walking in the center of trails (to avoid questing ticks), application of insect repellents (especially to the ankles and legs), and close examination of the body after being in wooded areas are recommended.

Habitat modification can be used to deter wildlife that harbor ticks and disease. Thick brush, wood piles, and debris favor rodents, so these should be eliminated. Burning, mowing or raking vegetation can eliminate shelter for some small animal hosts, and burning will kill some ticks as well. Fencing, deer repellents, and elimination of favored deer browse can discourage the occurrence of deer.

Insecticide or acaricide application is sometimes recommended, though it is hard to get good coverage when the vegetation and leaf litter are thick. Insecticides have been used in conjunction with tick pheromone to maximize contact of the insecticide and ticks. Providing insecticide-treated food and nesting material to small wildlife in another way to increase insecticide contact. Deer feeding stations have been developed that cause self-treatment of deer with systemic insecticides (see Chapter 14, Fig. 14.9).

INSECTS (INSECTA)

Lice (Phthiraptera)

Lice are small, wingless ectoparasites of mammals and birds. As adults, they range in size from 0.4 to 10 mm. They are flattened dorsoventrally and have incomplete (hemimetabolous) development. There are two major groups, often considered to be suborders of Phthiraptera: the sucking lice or Anoplura, and the chewing (biting) lice or Mallophaga. They differ principally in the arrangement of their mouthparts (Fig. 10.4). The **sucking lice** have piercing-sucking mouthparts whereas the **chewing lice** have chewing mouthparts. They can generally be distinguished even without examining their mouthparts, however, because the head of sucking lice generally is much narrower than the body, and even narrower than the thorax, whereas in chewing lice the head is wider than the thorax, and often nearly as wide as the body. Sucking lice are hematophagous, or blood feeding. Nearly all chewing lice feed on the fur, feathers, and skin particles of their host. Chewing lice often elicit allergic reaction in their host, transmit pathogens, and even produce cutaneous wounds, so despite their innocuous-seeming feeding habits, they can be a problem. Both sucking and chewing lice are equipped with large tarsal claws; those of sucking lice are adapted to grasp the hair of their hosts. Eyes are reduced or lacking (Fig. 10.5). Feather-infesting lice often feed preferentially on the downy portion of large feathers. Because they feed on feathers, this does not mean that they do without blood feeding; some puncture the young feathers and take blood from the central pulp that supplies the growing feather. The feather-feeding forms are quite diverse in their body form, and some are quite elongate.

The life cycle of lice is quite short, with a complete generation often occurring in 45 days. Eggs are securely attached to the host's hair or feathers, and hatch in a few days. In the nymphal stage there are three instars, each requiring 2–8 days. The adult can live for up to 30 days. Lice mate on their host, and leave only to move to another host. This generally is accomplished when hosts come into contact with one another, but occasionally lice will crawl a short distance or attach to flies for transport. Lice are quite specific (see Table 10.5 for examples), sometimes feeding only on a single species of host (displaying host specificity), or even on a certain region of the host's body (displaying site specificity). Each species of chewing louse is found on an average of 2.0 bird or 2.6 mammal taxa. Thus, when we examine the number of different species of lice associated with hosts, it is usually fairly limited. For example, the number of chewing louse species found on turkey vulture, *Cathartes aura*, is four; on California condor, *Gymnogyps californianus*, it is three; on American bald eagle, *Haliaeetus leucocephalus*, it is six; and on Chilean flamingo, *Phoenicopterus chilensis*, it is three. Not surprisingly, some lice feed on closely related species (e.g., among squirrels, deer, or ducks), and in such instances are associated with many more host species. Although lice are found continuously on wildlife, their abundance changes, and they generally are more numerous during the winter months.

There are approximately 550 species of sucking lice in the world, but over 2700 species of chewing lice. Probably many more species of chewing lice await description. The disparity in species abundance is related to the greater number of bird species, which are the primary hosts of chewing lice. Nevertheless, about 400 species of chewing lice parasitize mammals. Lice cannot survive long away from their hosts. In the case of sucking lice, survival decreases rapidly if they are off the host for a few hours. In the case of chewing lice, survival diminishes after a few days.

The effect of lice on their hosts is variable. Most animals have a few lice, and both mammals and birds spend considerable time in **host grooming** (mammals) or **preening** (birds), which are forms of lice removal. Low densities are not considered to be threatening to wildlife, but numbers can escalate under conditions of nutritional deficiency, illness, senility, or if they become immuno-compromised (diminished immunity). Lice can cause anemia, itching, and allergic reactions that lead to intense grooming, fur matting or loss, secondary infections, reduced growth rate, and transmission of parasites. So despite the fact that a few lice are 'normal' and not considered to be threatening, when lice become numerous they can affect the health of

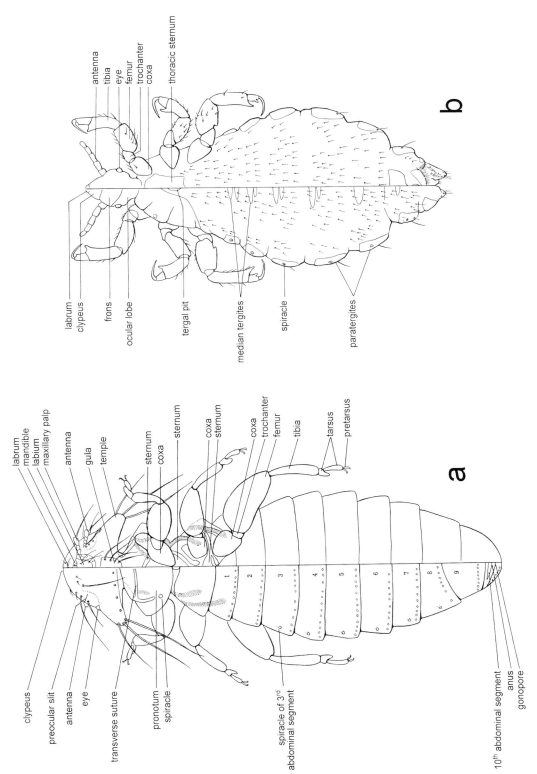

Fig. 10.4. Diagrams of representative lice (Phthiraptera): (a) chewing louse; (b) sucking louse. Each diagram is divided into a dorsal (left half) and ventral (right half) view.

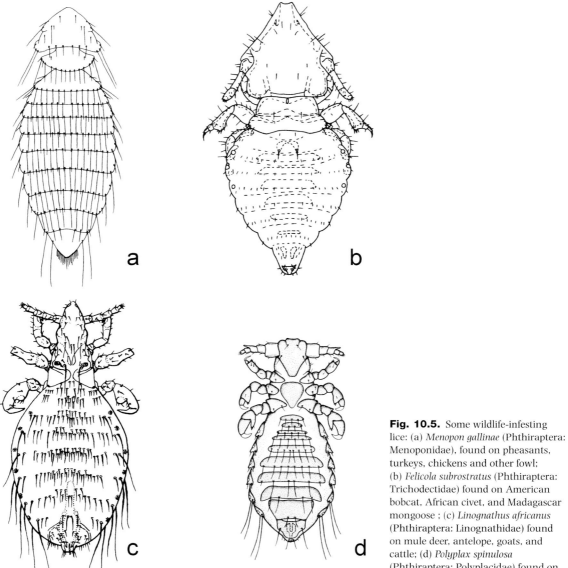

Fig. 10.5. Some wildlife-infesting lice: (a) *Menopon gallinae* (Phthiraptera: Menoponidae), found on pheasants, turkeys, chickens and other fowl; (b) *Felicola subrostratus* (Phthiraptera: Trichodectidae) found on American bobcat, African civet, and Madagascar mongoose ; (c) *Linognathus africanus* (Phthiraptera: Linognathidae) found on mule deer, antelope, goats, and cattle; (d) *Polyplax spinulosa* (Phthiraptera: Polyplacidae) found on rodents.

wildlife. In some cases, nest desertion, decreased mating success, clutch and offspring size, and even death of wildlife is attributed to lice infestations, though there likely are confounding health issues.

Some behaviors displayed by birds may also be indicative of the significance of lice. Wildlife can avoid infestation by choosing resting or nesting sites that lack the parasites, or choosing mates that are louse free. **Bathing** in water and dust, as well as **grooming** or **preening** using either the feet or bill helps to relieve the birds of their parasites. Although we can argue that birds bathe in water to rid them-

Table 10.5. Some parasitic lice that affect wild animals, or are associated with wild and domestic animals. Diseases caused or transmitted by these lice also are shown.

Louse	Wildlife host	Domestic animal host	Disease
Sucking lice			
Linognathous pedalis	mountain goat	sheep	none
Hoplopleura sciuricola	gray, fox and red squirrels	none	none
Neohaematopinus sciuropteri	southern flying squirrel	none	Epidemic Typhus
Haemodipsus ventricosis	wild rabbits	domestic rabbit	Tularemia
Hoplopleura acanthopus	voles and lemmings	none	none
Solenopotes ferrisi	elk, deer	none	none
Solenopotes muntiacus	muntjac	none	none
Haematopinus suis	feral swine	pigs	none
Chewing lice			
Pseudomenopon pilosum	coots, rails, gallinules	none	Avian Filariasis
Trinoton aserinum	geese and swans	none	none
Menacanthus affinis	northern wheatear	none	none
Menacanthus alaudae	skylark	none	none
Pseudomenopon austalis	black-tailed native-hen	none	none
Felicola felis	mountain lion, ocelot	none	none
Bovicola sedecimdecembrii	American bison	none	none
Eutrichophilus setosus	porcupine	none	none
Tricholipeurus parallelus	white-tailed and mule deer	none	none
Heterodoxus spiniger	jackals, red wolf, gray fox, coyote, etc.	dog	none

selves of dirt, this certainly would not explain why birds commonly engage in taking **dust baths**. It is well established that the abrasion provided by dust can be extremely disruptive to the water relations of insects, affecting their survival. Birds also will lay with outstretched wings upon ant nests. This behavior, called **passive anting**, is believed to allow ants to rummage through their plumage to search for prey, namely lice. In a related behavior, called **active anting**, birds will crush ants with their bill, and then rub the crushed ant over their feathers. This is believed to liberate formic acid produced by the ants, killing any parasites dwelling there. Ants are not the only source of louse toxins, of course, and birds also will rub walnuts, berries, and other plant materials on their feathers. The basis for anting has not been proven experimentally, but neither have the antiparasitic ideas been disproved. Another possible mechanism to reduce parasite load is **sunning**, which can cause ectoparasites to overheat, or possibly to make them more vulnerable to preening by moving to escape the excess heat.

Although lice are important vectors of human infectious disease, readily transmitting some diseases among humans, they are not generally considered to be important vectors of wildlife infectious diseases. Nor are they usually considered to be very important in transmission of disease from wildlife to livestock or humans. One disease of interest, however, is epidemic typhus, also known as louse-borne fever. The bacterium *Rickettsia prowazekii* is the causative agent of epidemic typhus. Normally, it is associated only with the human body louse, *Pediculus humanus*. However, the southern flying squirrel, *Glaucomys volans*, has been identified as a reservoir of the bacterium in the eastern and southern USA. Up to 90% of the squirrels are infected during autumn and winter, which corresponds to periods of peak louse abundance. The sucking louse *Neohaematopinus sciuropteri* can maintain and transmit the pathogen to uninfected squirrels. Several cases of human infection followed contact with flying squirrels, especially after squirrels established nests in attics of houses. Thus, humans may be at risk from certain louse–wildlife associations. Also, on occasion, the double-pored (dog) tapeworm, *Dipylidium caninum*, is transmitted to children when lice containing tapeworm eggs are inadvertently consumed.

Bugs (Hemiptera: Reduviidae, Cimicidae, and Polyctenidae)

The order Hemiptera is large and diverse in form and function. However, they all possess piercing-sucking mouthparts, with which they imbibe food in liquid form. Most feed on the xylem (transport tissues that move water and minerals from the roots to the foliage) or phloem (transport tissues that move water containing products of photosynthesis, mostly sugars) of plants. However, the groups of importance to wildlife (other than as food) are blood-feeding. Some transmit diseases. When winged, these insects possess two pairs of wings, with the front pair of wings varying from completely membranous (like the hind wings) to thickened basally but membranous distally, to entirely thickened (but usually showing some evidence of veins). Metamorphosis is incomplete (hemimetabolous).

Assassin Bugs, Subfamily Triatominae – Kissing or Blood-Sucking Conenose Bugs (Hemiptera: Reduviidae)

Assassin bugs comprise a fairly large family of about 4000 species. Most are known for their tendency to attack and feed upon other insects, and in general they are considered to be useful insects because they are valuable for biological suppression of pest insects. However, one subfamily, the Triatominae, is known as kissing bugs. Kissing bugs derive their name from the propensity of these insects to feed (often on the face) on humans at night while they are sleeping. They are also called conenoses, reflecting the tapered shape of the head. Some kissing bugs are important as vectors of **American Trypanosomiasis**, which they spread from wildlife to humans. As noted previously in the section on parasitic diseases, the insects involved in transmission to humans are typically *Triatoma* spp., *Panstrongylus* spp., and *Rhodnius prolixus*, and quite a number of species are vectors. After these insects bite and ingest blood, they often defecate on the person. If the victim scratches or rubs the feeding site, the *T. cruzi* parasites in the bug feces may enter the victim's body

through mucous membranes or breaks in the skin. In addition to the bite wound, the eyes and mouth are entry routes for the trypanosome disease agents. Though not many species of Triatominae vector American Trypanosmiasis, many species feed on wildlife. Wildlife typically are not affected by the trypanosomes.

Triatomines range from about 5 to 45 mm in length, though most are 20–28 mm long. Their coloration is highly variable, but their body shape is fairly consistent (Fig. 10.6). The head is small, but the eyes are pronounced. The antennae are four-segmented and the mouthparts of these bugs (the **rostrum** or 'beak') are three-segmented. The thorax is often triangular in shape, and the front wings (**hemelytra**) are thickened basally but membranous distally. The hind wings, if

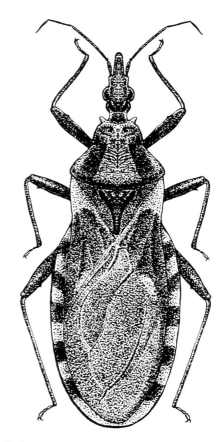

Fig. 10.6. A representative blood-sucking conenose bug (Hemiptera: Reduviidae): *Triatoma sanguisuga.*

present, are membranous. The legs are long but slender.

The triatomine bugs frequent shelters. Natural shelters where triatomines are found include animal burrows, tree holes, nests, rock piles, hollow trees and logs, and dense clumps of vegetation such as the crowns of palm trees. Nearly 200 wild mammal species are fed upon by triatomine bugs, although the hosts most commonly infected tend to be opossums, *Didelphis* spp. and *Philander opossum*; armadillos, *Dasypus* spp.; and rodents such as mice (e.g., *Peromyscus* and *Heteromys* spp.), porcupines, *Coendou* spp., agoutis, *Dasyprocta* spp.; and various south American monkeys. In addition to wild animals, domestic animals such as dogs, cats, and chickens are commonly involved in the disease cycle. Blood loss due to bug feeding is a serious problem with domestic chickens, and though not studied, blood feeding (especially on young birds) may well be a problem for wild birds in habitats favored by these bugs. The birds do not support reproduction of the trypanosomes, however.

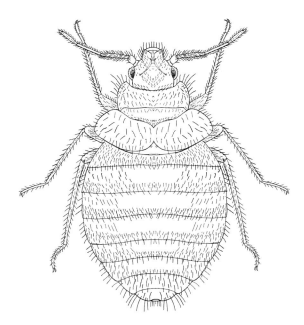

Fig. 10.7. The martin bug, *Oeciacus hirundinis* (Hemiptera: Cimicidae).

Bed Bugs, Swallow Bugs, and Bat Bugs (Hemiptera: Cimicidae and Polyctenidae)

The bed bugs and swallow bugs are in the family Cimicidae. Insects called bat bugs occur in both Cimicidae and the closely related family Polyctenidae. These bugs are flattened and wingless (Fig. 10.7). They are obligate blood-feeding ectoparasites.

There are 12 genera of cimicid bugs associated with bats, and 9 genera found on birds (Table 10.6). In addition, the members of the genus *Cimex* occur on both bats and birds, and three species infest humans: *Cimex hemipterus*, *C. lectularius*, and *Leptocimex boueti*. The species infesting humans also can infest chickens and occasionally some other domestic animals. *Leptocimex boueti*, found in western Africa, infests both bats and humans. Obviously, feeding on humans is something of an anomaly for these bugs, and likely developed from the dwelling in caves by humans earlier in hominid history. The cimicids infesting birds most commonly are associated with swallows, martins, and swifts – birds that commonly nest in caves. Polyctenidae is a small group of about 30 species, and are likely merely highly modified cimicids.

These bugs vary in size, though none are very large; *Cimex* attains a length of 5–7 mm and a width of 2.5–3 mm. They can swell considerably, however, when feeding. In Cimicidae, typically the head is reasonably large, and though eyes are pronounced, ocelli are absent. In Polyctenidae, the head is large, eyes are lacking and the insects are blind. The mouthparts of these bugs are three-segmented. The wings (in adults) are reduced to pads or ridges in these flightless insects. There are three nymphal stages preceding the adult in Polyctenidae, and five recorded in Cimicidae. Cimicids deposit eggs (oviparity). Polyctenids apparently retain their eggs, giving birth to nymphs (viviparity). Both cimicids and polyctenids exhibit **traumatic insemination**, wherein the male bug pierces the female's abdominal integument and injects a large amount of sperm into the hemocoel (body cavity). The sperm migrate to the oviducts and penetrate, then move to the ovarioles and ova to accomplish fertilization. This rather bizarre form of mating is rare in Insecta.

Bedbugs, swallow bugs, and bat bugs are likely very important parasites of colonial (group-nesting) wildlife such as bats and swallows, but data on their effects are scarce. However, the data that do exist show that early-developing swallows have much better survival rates than later-developing birds. This is due to the increase in bug size and abundance as the bird nesting

Table 10.6. Hosts of Cimicidae and the cimicid genera associated with these hosts (adapted from Askew 1971).

Host family	Genera of cimicid parasites
Typical bats (Vespertilionidae)	Bucimex, Propicimex, Cimex, Cacodmus, Aphrania, Stricticimex
Free-tailed bats (Molossidae)	Primicimex, Propicimex, Stricticimex, Loxaspis, Crassicimex
Sac-winged bats (Emballonuridae)	Loxaspis, Leptocimex
Fruit bats (Pteropodidae)	Aphrania, Afrocimex
Fish-eating bats (Noctilionidae)	Latrocimex
Old World leaf-nosed bats (Hipposideridae)	Stricticimex
Horse-shoe bats (Rhinolophidae)	Cimex
Flying lemurs (Cynocephalidae)	Cacodmus
Humans and domestic animals	Cimex, Leptocimex
Swallows, martins (Hirundinidae)	Oeciacus, Paracimex, Ornithocoris, Hesperocimex
Swifts (Apodidae)	Paracimex, Ornithocoris, Cimexopsis, Synxenoderus
Domestic fowl (Phasianidae)	Cimex, Ornithocoris, Haematosiphon
Domestic pigeon (Columbidae)	Cimex
Flycatchers (Muscicapidae)	Cimex
Parrots (Psittacidae)	Psitticimex
New World vultures (Cathartidae)	Haematosiphon
Typical owls (Strigidae)	Haematosiphon
Barn owls (Tytonidae)	Haematosiphon
Eagles, falcons (Falconidae)	Haematosiphon
Oven-bird (Furnariidae)	Gaminicimex

season progresses. Bats cannot groom themselves with their limbs, so they often harbor large populations of ectoparasites – not only cimicids and polyctinids, but also nycteribiids and streblids (Diptera). Disease transmission is not important in these bugs, but blood loss is an important cause of mortality among nestlings of colonial avifauna such as swallows. Fort Morgan virus is a variant of western equine encephalitis virus that affects swallows and has been isolated from the cliff swallow bug, *Oeciacus vicarius*, and *Trypanosoma* spp. protozoa associated with bats have been found in species of Cimicidae, but the importance of these pathogens is not known.

Research was conducted in Texas, USA, on cliff swallow, *Hirundo pyrrhonota*, to assess the effects of natural ectoparasite populations (mostly swallow bug, *Oeciacus vicarious*, but also to a lesser degree two ticks (swallow tick, *Argas cooleyi* (Acari: Argasidae) and *Ornithodoros concanensis* (Acari: Ixodidae)) on swallows. This was accomplished by measuring elements of fitness from birds with natural infestations and comparing this to nearby nests that had been treated with insecticide. Fledgling success was lower when ectoparasites were present, and birds left the nest earlier when

infested with arthropods. Parasitized birds weighed less, and blood chemistry was affected. Perhaps most significantly, nestling survival was much reduced when ectoparasites were present.

Flies (Diptera: Several Families)

Flies are characterized by possessing only one pair of wings, and holometabolous (complete) development. Taxonomists usually consider Diptera to consist of two suborders, Nematocera and Brachycera, but sometimes a third suborder, Cyclorrhapha, is presented. Each group contains some insects of importance to wildlife. Not surprisingly, livestock and humans are also affected by this order of insects. The Nematocera tend to be small, delicate, long-legged flies with long, beaded antennae and aquatic larvae. In the blood-feeding species, females are the blood-feeders. Important groups include mosquitoes, black flies, biting midges, and sand flies. The Brachycera are stout-bodied flies with short antennae, and larvae that are found in either aquatic or semi-aquatic environments. Important groups include deer flies and horse flies, louse flies, tsetse flies, muscid flies, blow flies, bot flies, and others. Flies are well known for transmission of infectious and parasitic diseases to wildlife (Chapters 8 and 9), but biting by adults independent of disease transmission constitutes an important and sometimes overlooked stress factor. The larvae of a few flies infest animals, causing a condition called **myiasis**. Most myiasis-causing flies are incapable of invading healthy animal tissue, but a few are capable of attacking living tissue, which can lead to death of the host. Others infest cavities or organs such as the nasal cavity, eye, ear, and intestine.

Flies also are well known for their ability to harass animals, both wildlife and domesticated animals. **Insect harassment** is feeding or oviposition (larviposition) behavior by insects that results in persistent defensive behavior by animals. Behavioral responses by wildlife to insect harassment include head shaking, body shaking, ear flicking, foot stamping, tail swishing, biting, wing shaking, running, lying, disruption of feeding, herding, relocation, and dust, mud and water bathing. The presence of 'clouds' of mosquitoes (Culicidae) or black flies (Simulidae) most often comes to mind when the subject of harassment is raised. This is because most people have experienced some form of harassment by these insects, and indeed mosquitoes and black flies are quite capable of harassing wildlife. However, other insects that occur at lower density also harass wildlife.

Caribou (reindeer), *Rangifer tarandus*, live in extreme northern climates of North America and Eurasia where mosquitoes and black flies are notoriously abundant during the brief summer periods. Body counts of mosquitoes or black flies of 6000 to 10,000 flies on individual animals have been documented, so efforts have been made to determine the effects of insects on these animals in various studies conducted in Canada and Scandinavia. Weather plays an important role in the ability of insects to harass wildlife. Harassment is greater during warmer periods, sunny weather, and when wind speeds are lower. In the presence of insects, caribou display most of the aforementioned defensive behaviors. Their feeding behavior is markedly affected, and they spend less time feeding, and more time standing, walking or lying. In particular, their use of snow patches increases and their use of shrubby and wooded areas decreases. The insect harassment is deleterious to caribou body condition as expressed by reduced milk production by females, lower calf birth weight, and decreased autumn weight of calves. Interestingly,

however, some studies indicate that oestrids (Diptera: Oestridae; warble and nose bot flies) are more important than other flies. Caribou respond to black flies and mosquitoes, but the response to oestrid fly harassment was distinct and violent. Even a single oestrid is enough to cause the animals to cease feeding and to send caribou running and jumping in a panicked manner. Often, they end their running when they attain windy outcroppings, but in the meanwhile the entire herd could be running for hours. Sometimes, the herd will seek snow patches, where presumably the air temperatures were too cool for the flies. Some might seek forage at the edges of the snow but they soon will be driven back by the attentions of the flies. Only when cloud cover develops or nightfall arrives will most caribou recommence feeding if oestrids are present.

Mosquitoes (Diptera: Culicidae)

Adult mosquitoes are thin bodied, frail-looking insects with long legs (Fig. 10.8a,b). Their antennae are long, their mouthparts (proboscis) are long and slender, and their wing veins and wing margins possess scales. Adults emerge from the pupal stage at the surface of the water. Adults of both sexes feed on nectar from plants but only females feed on blood from animals. Activity, including feeding, varies among mosquitoes, though most are active at dawn and dusk. They use both visual and olfactory cues to locate hosts. Carbon dioxide, lactic acid and octenol are important chemical host attractants. Some mosquitoes can deposit eggs without first taking a meal of blood; this is call **autogeny**. Most, however, must blood-feed prior to developing eggs; this is called **anautogeny**. Adults usually remain in the area where the larvae developed; few disperse more than 2 km from such habitats. Thus, most mosquitoes are found near water. A few species, such as *Aedes sollicitans* and *Aedes vexans*, will disperse greater distances.

Species differ in their preferred oviposition site and larval habitat. In general, they can be categorized as preferring permanent water (e.g., marshes, stream, ponds) or temporary water sources (e.g., pools from melting snow, ditches, tree holes, containers). Temporary water sources often favor mosquito production because they lack predators. Mosquitoes originating in temporary water sources are often characterized as floodwater mosquitoes or container breeding mosquitoes. Mosquitoes normally are found associated with

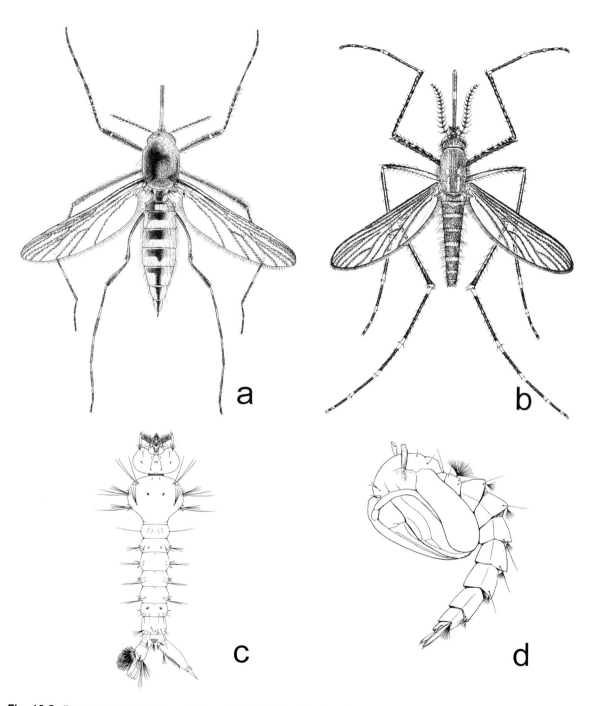

Fig. 10.8. Representative mosquitoes (Diptera: Culicidae): (a) adult *Psorophora confinis*; (b) *Culex tarsalis*; (c) typical mosquito larva; (d) typical mosquito pupa.

fresh water, but some inhabit brackish water (a mixture of fresh and salt water that often occurs in coastal areas).

Mosquitoes also vary in their egg-laying behavior. Some adult female mosquitoes deposit their eggs singly on the surface of the water, or in a location near water such as on emergent vegetation that will be submerged in water at a later time. Others deposit their eggs in clusters called **egg rafts** on the water surface. Some eggs possess floats, or air-filled modifications of the chorion, that help them to float. The eggs are usually elongate.

Mosquito larvae (Fig. 10.8c), called **wigglers** or **wrigglers**, as well as the pupae, called **tumblers**, are aquatic. These names accurately describe their motion when they swim. Larvae possess a number of body hairs, especially in the thoracic and anal regions. The larvae feed on organic matter and small organisms that they collect from the water, often by filtering. Larvae are susceptible to infection by parasites and pathogens, including mermithid nematodes, protozoa, microsporidia, fungi, and viruses, though typically only a small proportion of the larvae are affected.

The pupal stage (Fig. 10.8d) does not feed but remains active. The larvae and pupae of many species obtain oxygen from the water surface, and both stages may have special structures for this purpose, called the **respiratory siphon** in the larval stage and **respiratory trumpets** in the pupal stage. The respiratory siphon is located in the anal region and the larvae rest head-down in the water, with their posterior end exposed to the water surface. The pupal respiratory trumpets, on the other hand, tend to occur just behind the head because the pupae tend to rest head-up. In other mosquitoes such as *Mansonia* spp., however, the siphon and trumpet are saw-like and are used to puncture aquatic vegetation and to obtain oxygen from within plants.

Feeding by adults often causes only minor, local irritation accompanied by formation of localized swellings, or wheals. If mosquitoes continue to feed, however, animals become irritated, nervous, and seek to avoid continued biting by these insects. In environments with very large numbers of mosquitoes such as arctic and subarctic regions, mosquitoes can cause extreme stress on caribou and other mammals, causing the herds to move continuously in an attempt to escape the biting. Under some circumstances, mosquitoes can remove nearly all of their host's blood (called **exsanguination**), which results in death. More commonly, mosquitoes vector infectious or parasitic diseases (Table 10.7). Many 'diseases' have little effect on wildlife, because the host and disease agents have co-evolved. However, some wildlife suffer illness, or even death, due to the presence of infectious or parasitic diseases. Mosquitoes are well known to be vectors of viruses, some of which are quite important to wildlife, particularly avian pox, myxomatosis, epizootic hemorrhagic disease, West Nile virus, and the various encephalitis viruses. When nonindigenous pathogens such as West Nile Virus are introduced, or even when indigenous pathogens such as Yellow Fever Virus become abundant in populations lacking immunity, considerable mortality among wildlife can occur.

Table 10.7. Some mosquitoes that affect wild animals, or are associated with wild and domestic animals. Diseases caused or transmitted by these mosquitoes also are shown.

Mosquito	Wildlife host	Domestic animal host (of disease)	Disease
Culiseta melanura	many birds	horse	Eastern Equine Encephalitis
Culex qinquefasciatus	many birds	rare in horses	St. Louis Encephalitis
Culex tritaeniorhynchus	many birds, some mammals	horse	West Nile Virus
Culex nigripalpus	birds, mammals	horse	Eastern Equine Encephalitis
Ochlerotatus sollicitans	many birds	horse	Eastern Equine Encephalitis
Ochlerotatus triseriatus	gray squirrel, tree squirrel, eastern chipmunk	rare in dogs	Lacrosse Virus
Anopheles quadrimaculatus	white-tailed deer	cattle, sheep, horses	Cache Valley Virus
Aedes vexans	foxes, coyote, bears	dog	Dirofilariasis

Not all mosquitoes display the same feeding behavior. Some mosquitoes feed primarily on one type of host, such as birds or amphibians, but many mosquito species feed opportunistically on a large number of hosts. Sometimes mosquito taxa display important trends in feeding behavior. For example, *Culex* spp. tend to feed on birds whereas *Culiseta* spp. feed primarily on large mammals, and *Aedes* spp. often feed on mammals without regard to size. There are important species-specific differences, however. For example, *Culiseta melanura* is an anomaly in the genus, preferring birds. Seasonal host shifts are common due to climatic changes (e.g., frequent rainfall) that allow them to exploit different habitats, and availability of hosts (e.g., nestling birds are more susceptible than adults). An example of how host availability influences feeding behavior can be seen in a study of *Aedes trivittatus* feeding in different habitats: blood meals were most numerous from dogs (43%), rabbits (24%) and cats (14%) in an urban environment, whereas they were most numerous from horses (37%), deer (29%), northern raccoon (12%), and rabbits (8%) in a suburban location, and from deer (57%) and raccoons (21%) in a rural location.

Mosquito management is usually accomplished by modification or removal of larval breeding sites, normally by reducing the retention time of water suitable for development of larvae. In some cases, tidal water is impounded to maintain a relatively constant level, which can minimize mosquito egg hatch and preserve mosquito predators. However, addition of mosquito-eating fish such as *Gambusia* spp., or reduction of aquatic vegetation (which can provide hiding places for mosquito larvae) can greatly reduce mosquito breeding potential. In many locations, the preferred (most effective or economic) method of mosquito suppression is by insecticidal suppression of larvae, or of adults, often with the aid of aircraft. Both the habitat modification and insecticide application procedures raise concern about environmental impact, especially effects on wildlife. The use of microbial insecticides such as the bacterium *Bacillus thuringiensis israelensis* can be used to kill larvae with minimal negative impacts on the environment. Insect growth regulators, though not as selective and benign as microbial insecticides, are safer than standard chemical insecticides. Also very safe is the use of monomolecular surface film. Most mosquito larvae must move to the surface of the water to obtain oxygen through their ventilatory tube, or respiratory siphon. Application of isostearyl alcohol,

an 18-carbon fatty alcohol, to standing water results in formation of a monomolecular layer at the water surface that modifies the surface tension, interfering with the ability of the larvae to obtain air. Mosquito larvae usually die within 1–3 days, the residue is gone within 2–10 days, and organisms breathing with gills such as fish and other aquatic invertebrates are not affected.

Black Flies (Diptera: Simuliidae)

Black flies (Fig. 10.9), also known as buffalo gnats and turkey gnats, are well known by anyone who has spent much time out-of-doors, as their blood-feeding habits make a strong impression despite the small size of these pests. There are over 1700 species worldwide, and they are found nearly everywhere so long as there is flowing water such as rivers and streams in which to breed. The greatest species richness is found in the Palearctic Region, especially in Russia. The largest genus by far is *Simulium*, which occurs worldwide. In many respects they are like mosquitoes in their impact, ranging from

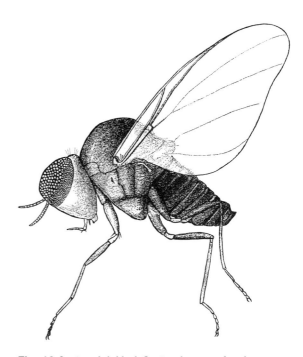

Fig. 10.9. An adult black fly, *Simulium meridionale* (Diptera: Simuliidae).

a mild nuisance to causing exsanguination due to direct feeding, but also transmitting many important diseases. The pests account for only 10%–20% of the species, however, and a few species do not feed as adults.

Eggs of black flies are deposited in batches on the water surface, or on objects, in moving water. All types of flowing fresh water are exploited, though generally they are intolerant of polluted water. The eggs are oval or triangular, with a gelatinous coating. Larvae (Fig. 10.10) are largely sedentary, and attach to stones or other submerged substrate by using minute hooks at the end of the abdomen. The larvae produce an anchor of silk using their salivary glands, and it is this structure that they use for attachment to the substrate. The larvae normally display six or seven instars. They filter-feed, collecting minute organic particles from the moving water with the aid of head fans. They may also scrape food from the surface of stones and other debris in the water. Larvae are susceptible to infection by parasites and pathogens, including mermithid nematodes, protozoa, microsporidia, fungi, and viruses, though typically only a small proportion of the larvae are affected. Pupae (Fig. 10.10) likewise often are attached to substrates, and are housed in cocoons, though some burrow into the sand and silt to pupate. Many produce elongate gills that can be quite diverse in structure. The adults (Fig. 10.9) are small but robust, with humped backs and with cigar-shaped antennae consisting of seven to nine flagellar segments. Adults are small, measuring about 1–5 mm in length. They are variable in color, but as their name suggests, they often are black. Their mouthparts are serrated, and females, but not males, take blood from mammals or birds. The anterior margin of their wings is marked by strong venation.

Black flies are particularly abundant in colder regions. They are more host specific than mosquitoes. Although a great number of mammals and birds are attacked, different black fly species tend to specialize on either mammals or birds. The majority of black flies are not attracted to humans and livestock, and feed only on wildlife. Like mosquitoes, they use carbon dioxide, host odors, and visual cues to locate hosts. They can inject toxins while feeding, resulting in a malady called **Black Fly Fever** in humans and **Simuliotoxicosis** in animals. Affected humans display heavy, erratic breathing and trembling muscles; livestock can suffer from toxemia and anaphylactic shock. Weight loss is a common problem for animals affected by black fly

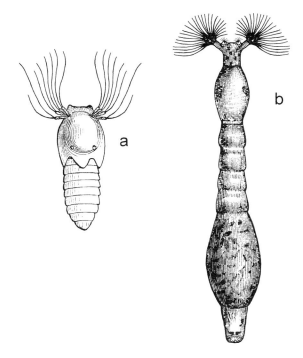

Fig. 10.10. Life stages of the black fly *Simulium piscidium* (Diptera: Simuliidae): (a) pupa; (b) larva.

feeding at more modest levels. The site of feeding is often marked by a reddened, swollen area, which often has a drop of dried blood at the center. Hypersensitivity to black fly feeding is not uncommon in naïve hosts, but adapted animals typically suffer much less. Black flies can vector nematodes, protozoa, and occasionally viruses (Table 10.8), and although these diseases can be an issue for wildlife, they more commonly are a problem for livestock and humans. Some areas are unsuitable for livestock production due to black fly feeding, and they can transmit vesicular stomatitis virus and eastern equine encephalitis, among other diseases.

The haematozoan parasites commonly associated with waterfowl are members of the genera *Haemoproteus*, *Leucocytozoon*, and *Plasmodium*. These are insect-transmitted diseases usually vectored by black flies (Simuliidae), biting midges (Ceratopogonidae), and mosquitoes (Culicidae), respectively. Sexual reproduction occurs in the insect and asexual reproduction in the birds.

Table 10.8. Some black flies that affect wild animals, or are associated with wild and domestic animals. Diseases caused or transmitted by these black flies also are shown.

Black fly	Wildlife host	Domestic animal host (of disease)	Disease
Simulium venustum, others	wild ducks, geese	domestic ducks, geese	*Leucocytozoon simondi*
Simulium areum, others	wild turkey	domestic turkey	*Leucocytozoon smithi*
Simulium vernum, others	owls	none	*Leucocytozoon ziemanni*
Austrosimulium ungulatum	penguins	none	*Leucocytozoon tawaki*
Prosimulium impostor, others	deer, moose	none	*Onchocerca cervipedis*
Simulium venustum	bears	none	*Dirofilaria ursi*

Leucocytozoon simondi affects ducks, geese, and swans, and often is found throughout the world wherever these birds and black flies occur together. Different species of black flies are important vectors, depending on location and the time of the year. The black flies acquire the protozoa by feeding on blood containing gametocytes. Within a few hours, sporozoites can be found in the salivary glands of the flies, which can then be injected during the next episode of blood feeding. Once injected, the sporozoites are carried to the liver where they attack blood cells and mature. The parasites can persist in the host birds for at least two years, and black flies acquire them at many times, though the level of **parasitemia** (level of parasites in the blood) varies seasonally and even during the course of the day.

The effects of infection on the health of the birds are variable. Affected birds are listless, depressed, anemic, and reluctant to feed. Liver and spleen damage also are associated with infection by these parasites. High levels of mortality among ducklings and goslings are well documented, with the youngest birds more likely to succumb. Ill birds may recover following infection, especially from low levels of infection, or following prior exposure. Adults display less severe effects; they may appear listless, but usually do not perish.

Since the 1980s, the preferred means of suppressing black flies has been through the application of the bacterium *Bacillus thuringiensis* var. *israelensis*. This bacterium produces a toxin that is most effective against flies such as mosquitoes and black flies, and so is relatively selective when ingested. When applied to flowing water, it kills mostly black flies (mosquitoes typically avoid flowing water) without affecting other insect species such as stone flies (Plecoptera) and mayflies (Ephemeroptera) that can be so important as a food resource for fish.

Biting Midges (Diptera: Ceratopogonidae)

Biting midges are known by several names, including punkies, no-see-ums, and sand flies. Flies in the family Psychodidae are also called sand flies, a source of considerable confusion. Biting midges (Fig. 10.11) are small flies, measuring only 0.6–5.0 mm, and usually less than 2.5 mm long. Their antennae consist of 15 segments, a large number, though often the antennae do not appear to be very long. Their wings, which are held over the back when at rest, often are spotted. Their mouthparts are not very elongate. There are over 4000 species worldwide, with 600 in North America. The genera *Culicoides*, *Forcipomyia*, and *Leptoconops* are most important, especially *Culicoides*.

Adult midges deposit eggs, often shaped like a banana, singly in moist habitats. Larvae (Fig. 10.12) are elongate and cylindrical, measuring up to 5 mm in length. They may possess or lack abdominal and anal appendages. The whitish larvae are aquatic or semiaquatic, and found in sand, soil, decaying vegetation, manure, and organic debris in the intertidal zone of coastal areas, and in some cases under moss or bark. They shred organic matter or feed on detritus or microorganisms. The pupae (Fig. 10.12) are brown, and possess thoracic respiratory horns. The adults of females have an elongate but short proboscis with biting mouthparts that are used to pierce and obtain blood. Males do not feed. Swarms develop for mating.

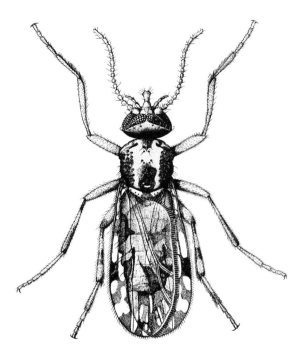

Fig. 10.11. Adult form of the biting midge *Culicoides furens* (Diptera: Ceratopogonidae).

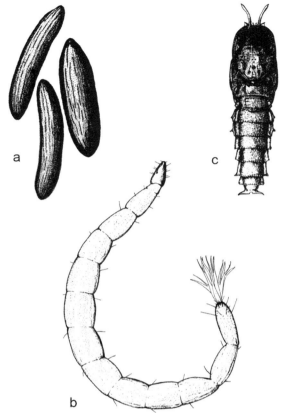

Fig. 10.12. Life stages of the biting midge *Culicoides furens* (Diptera: Ceratopogonidae): (a) eggs; (b) larva; (c) pupa.

Adults are most active at dusk and during the evening, though in some species feeding may occur at any time. High humidity is generally required for flight activity. Some species are autogenous for the first complement of eggs, but further egg production requires a blood meal.

Biting midges affect both wildlife and livestock (Table 10.9). They can be quite troublesome, though the direct effects of feeding vary in severity. Biting midges are not very host specific. Like mosquitoes, they are attracted to hosts by the presence of carbon dioxide and host odors. The same species often attack wildlife such as white-tailed deer, *Odocoileus virginianus*, or bighorn sheep, *Ovis canadensis*, and livestock such as cattle or sheep. Transmission of bluetongue virus to bighorn, white-tailed deer, mule deer, *Odocoileus hemonius*, and pronghorn, *Antilocapra americana*, can result in expression of severe clinical disease called **Hemorrhagic Disease**. In North America, the principal vector of bluetongue virus is *Culicoides variipennis*, a species that occurs fairly widely, but other *Culicoides* spp. are important regionally. Epizootic hemorrhagic disease, a virus very similar to bluetongue, also results in Hemorrhagic Disease, and is more prevalent in wild ruminants, especially deer. Similarly transmitted by *Culicoides* spp., it displays even more severe health effects and is the most important disease of wild deer populations in the USA, causing sporadic periods of significant mortality. Among other diseases transmitted by biting midges are vesicular stomatitis virus, African horse sickness, and arboviruses including eastern equine encephalitis. Biting midges also transmit filarial nematodes, including *Mansonella* and *Onchocerca* spp., and several protozoans. Haemosporidian protozoans in the genera *Hepatocystis*, *Haemoproteus*, and *Leucocytozoon* also are vectored by biting midges to birds, mammals, and reptiles.

Table 10.9. Some biting midges that affect wild animals, or are associated with wild and domestic animals. Diseases caused or transmitted by these biting midges also are shown.

Biting midge	Wildlife host	Domestic animal host (of disease)	Disease
Culicoides variipennis	rabbits, hares	none	Buttonwillow Virus
Culicoides paraensis	sloths, primates (incl. humans)	none	Oropouche Virus
Culicoides spp.	hedgehogs, rats	cattle, sheep	Kotonkan Virus
Culicoides spp.	marsupials	cattle	Mitchell River Virus
Culicoides spp.	antelope	sheep, goats, cattle, camels, buffalo	Rift Valley Fever
Culicoides spp.	white-tailed deer, bighorn, pronghorn	cattle	Hemorrhagic Disease
Culcoides spp.	finches, sparrows, other passerine birds	none	Haemoproteus fringillae
Culcoides downesi	ducks, geese	domestic ducks	Haemoproteus nettionis
Culicoides spp.	squirrels	none	Hepatocystis brayi

Independent of disease transmission, the painful bites caused by midges are annoying to wildlife, livestock, and humans. Their abundance in mangrove-dominated coastal areas in the southeastern USA, and other moist habitats, can make it unbearable for outdoor recreation, and considerable effort has been made to modify some habitats to reduce breeding. Other species of midges more commonly affect humans, especially *C. furens*, *C. hollensis*, *C. melleus*, *C. mississippiensis*, and *C. barbosai*. In the Gulf Coast region, *Leptoconops linleyi* is important. Habitat modification, in addition to being expensive, can have significant deleterious environmental effects. Alternatives are application of insecticide, confinement of livestock to buildings, avoidance of larval habitats, and avoidance of peak feeding periods. Unfortunately, standard window screening is not very effective at excluding these small flies.

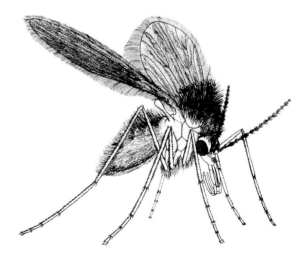

Fig. 10.13. A female phlebotomine sand fly (Diptera: Psychodidae).

Phlebotomine Sand Flies (Diptera: Psychodidae: Phlebotominae)

The subfamily Phlebotominae of family Psychodidae includes biting species known as sand flies. As noted previously, they sometimes are confused with the biting midges (Ceratopogonidae) so it is useful to differentiate the psychodid sand flies by referring to them as psychodid or phlebotomine sand flies. About 600 species occur worldwide, and 380 in the New World. The most important genera are *Phlebotomus* in the Old World and *Lutzomyia* in the New World.

Like other Nematocera (lower or primitive flies), adult phlebotomine sand flies are delicate, tending to have long legs (Fig. 10.13). Also, their antennae are moderately long, with 12–16 segments. The antennae of males and females are similar. The antennae, body, and wings usually bear a dense layer of hairs. Adults are small, normally measuring less than 5 mm in length, and have a decidedly humped appearance. The adult mouthparts form a short proboscis, with knife-like and toothed components for biting and removal of blood from their hosts.

The eggs of phlebotomine sand flies are elongate, dark in color, and shiny. They are deposited in groups, in moist terrestrial habitats. Eggs and larvae are commonly found in leaf litter, crevices of soil, manure, tree hollows, animal burrows and other locations where both organic matter and moisture accumulate. The eggs bear surface markings that are useful for identification. Eggs hatch after a period of 4–20 days, and larvae require 30–60 days for completion of their four instars. The whitish larvae also are elongate, and can attain a length of about 5 mm. Like the other Nematocera, they have a distinct head, which is dark in color. The head and body segments bear setae, and long caudal setae are found posteriorly. They lack true legs but may bear leg-like abdominal structures called **prolegs**. The pupal stage requires only about a week. Pupation occurs within the last larval covering (exuviae).

The adults live for 2–6 weeks and are active fliers, though flights typically are characterized by short, almost hopping movements. Adults feed on nectar, fruit juices and other sugar sources in addition to blood. Females may be autogenous or anautogenous. Autogenous species, which do not require a blood meal before producing their first batch of eggs, require a blood meal before producing subsequent batches. Anautogenous species, which require blood before eggs are produced, may require more than one blood meal per egg batch in this group of insects. Feeding typically occurs at twilight or darkness, though some species feed during the day. The flight range of adults is limited, normally not more that a few hundred meters from where the adults rest during the day. These insects are more abundant in environments with high humidity, such as forests, and during the periods of the year when humidity is higher, such as the rainy season.

Hosts of phlebotomine sand flies include a large number of mammals, and also some reptiles. Burrow-dwelling and nesting animals are their principal hosts, but few mammals escape their attacks where the flies are abundant, as they are opportunistic feeders. Thus, in North America, phlebotomine sand flies typically are associated with such wildlife as rabbits, *Sylvilagus*

spp.; yellow-bellied marmots, *Marmota flaventris*; nine-banded armadillo, *Dasypus novemcinctus*; woodchucks, *Marmota monax*; northern raccoon, *Procyon lotor*; woodrat, *Neotoma albigula*; and rock squirrel, *Citellus variegates*; but also feed on white-tailed deer, *Odocoileus virginianus*; pronghorn, *Antilocapra americana*; American bobcat, *Lynx rufus*; and others.

The biting of these flies produces distinct discomfort. Typically, they make several bites in close proximity and then feed from hemorrhages in surface capillaries. They may concentrate their activities on particular regions of the host such as the belly and genital areas. More important, though, is their ability to transmit diseases (Table 10.10). They vector viruses, bacteria, and protozoans. Among the important diseases transmitted by phlebotomine sand flies are vesicular stomatitis and leishmaniasis. The sand fly *Lutzomyia shannoni* seems to be the most important vector of **Vesicular Stomatitis** in North and South America. However, transmission of this disease is not limited to sand flies, as black flies are also implicated. Short-term illness is characterized by flu-like symptoms: fever, malaise, headache, nausea, and muscular pain are typical of stomatitis in humans and livestock. Reservoirs of stomatitis include opossums, monkeys, porcupine, raccoon, American bobcat, and pronghorn.

Leishmaniasis is a very important disease complex that is transmitted by phlebotomine sand flies to humans. It is caused by several species of protozoa in the genus *Leishmania*, and is found in many (mostly tropical or subtropical) areas of the world. **Cutaneous Leishmaniasis** causes ulcers, especially of the nose, mouth, and pharynx, that may be secondarily affected by bacteria and fungi, or infested by fly larvae. Rodents such as gerbils, *Rhombomys opimus*, and ground squirrels, *Spermophilopsis leptodactylus*, provide important breeding sites, though other wildlife can be important hosts. Wildlife are not much affected by the disease, however. *Lutzomyia* spp. are important vectors in the New World, whereas *Phlebotomus* spp. are important in the Old World. **Visceral Leishmaniasis**, also known as kala-azar, is a more chronic form of the disease, beginning at the site of the insect bite but gradually circulating in the body, and affecting several internal organs and causing wasting and fever. Opossums, rats and dogs are considered to be reservoirs, but wildlife are not much affected by this disease. *Lutzomyia* spp. are important vectors in the New World, whereas *Phlebotomus* spp. are important in the Old World.

Table 10.10. Some phlebotomine sand flies that affect wild animals, or are associated with wild and domestic animals. Diseases caused or transmitted by these sand flies also are shown.

Sand fly	Wildlife host	Domestic animal host (of disease)	Disease
Lutzomyia trapidoi, others	rodents, primates	none	Sandfly Fever (New World)
Phlebotomus papatasi, others	rodents	none	Sandfly Fever (Old World)
Lutzomyia shannoni, others	opossums, porcupine, northern raccoon, American bobcat, pronghorn	horse, cattle, sheep, swine	Vesicular Stomatitis
Lutzomyia anthophora, and many others	opossums, sloths, monkeys, rodents, anteaters, northern raccoon	horse, dog, cat	Cutaneous Leishmaniasis (New World)
Phlebotomus caucasicus, others	rodents, monkeys, hyraxes	dogs	Cutaneous Leishmaniasis (Old World)
Phlebotomus argentipes, others	opossums, canines	none	Visceral Leishmaniasis (New World)
Lutzomyia evansi, others	rats, canines	none	Visceral Leishmaniasis (Old World)

Management of leishmaniasis is based largely on preventing humans from being bitten. Use of repellents and protective clothing or bed netting are foremost, but traps and insecticide can be effective. Other techniques involve elimination of breeding sites, and control of animals that serve as reservoirs. Habitat management, though ecologically disruptive, can be quite effective in affecting phlebotomine sand flies. Deforestation and urbanization leads to reduction in sand fly populations for species that favor these habitats.

Horse Flies and Deer Flies (Diptera: Tabanidae)

Horse (Fig. 10.14) and deer flies are fairly large insects that are well known to inflict painful bites. ***Horse flies*** (several genera) are usually larger, about 9–33 mm long. Horse flies often are green to black in color. ***Deer flies***, *Chrysops* spp., are smaller, about 6–11 mm long. Deer flies tend to be yellow–orange, with darker markings. Deer flies are more likely to have spotted wings than are horse flies. *Diachlorus ferrugatus* is the most common of a group of species known as **yellow flies** due to their yellowish bodies. In most respects they are

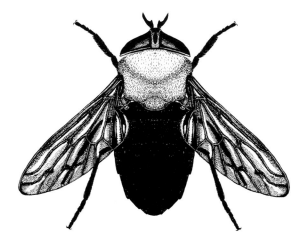

Fig. 10.14. A horse fly, *Tabanus* sp. (Diptera: Tabanidae).

equivalent to the deer flies, and where they are abundant they are are a serious nuisance. **Klegs**, *Haematopota* spp., comprise another group of tabanids. Though klegs are similar to deer flies in appearance, they usually are larger. Klegs are not as important as horse flies and deer flies. All of these flies are stout, but speedy in flight. Their eyes are large and often striped or patterned with metallic green, yellow, orange or violet. Over 4000 species of tabanids occur worldwide, with about 300 species in North America.

The eggs of tabanid flies are 1–3 mm long and typically are deposited in clusters on vegetation near aquatic or semi-aquatic habitat. The eggs are white initially but darken with time. Each cluster may contain 100–800 eggs. Eggs hatch within 2–3 days and move into the mud or water along streams, ponds, and marshes. The larvae are elongate, tapered at both ends and thickest in the middle. They lack true legs in the larval stage, but have abdominal extensions called **pseudopods** that aid in locomotion. They usually possess a respiratory siphon posteriorly. They typically are 15–30 mm long, but some are up to 60 mm long. Some prefer terrestrial habitats in leaf litter beneath trees. Larvae of horse flies, but not necessarily deer flies, are cannibalistic. These flies seem to feed on a mixture of small invertebrates and detritus. When larvae capture prey, they immobilize it quickly, probably by using a toxin. The larvae seek drier nearby soil when they are ready to pupate.

In temperate areas, tabanids usually display only one generation per year, and though some species seem to partition the summer months to reduce competition, others seem to be present for several months. Some species have multiple generations per year. Adults of many species are **autogenous**, not requiring a blood meal before developing a batch of eggs. They also feed on carbohydrates such as floral or extrafloral nectaries, and aphid honeydew. The adults mate while in flight. Males often aggregate in certain locations, including hilltops, where they pursue passing females.

The flies normally are active during the daylight hours. Host color, size, and the presence of carbon dioxide and odors affect the attraction of tabanid flies to host animals. They have toothed, blade-like mandibles that cut the host, causing blood to accumulate. They also inject an anticoagulant to stimulate blood flow. They are persistent feeders, and if dislodged or disturbed while feeding they will return again and again. They are not highly dispersive, tending to remain in a particular locality for long periods. Areas adjacent to marshes or other breeding sites often support large numbers of adults, as do edges of woodlands. It is common for tabanid feeding to cause great discomfort to animals, and for them to seek to escape the fly attacks by fleeing. Formation of herds may benefit animals because the animals at the center are less susceptible to attack by these insects. Tabanids tend to attack larger mammals. The horse flies, in particular, attack large animals such as deer and moose. The smaller deer flies also attack deer, but also feed on smaller mammals such as rabbits, opossums and raccoon, and birds such as crows, ducks, and robins, and even on reptiles.

In addition to the ability of tabanid flies to provide annoyance, a painful bite, and cause considerable blood loss, tabanids transmit several infectious and parasitic diseases (Table 10.11). They are important vectors of tularemia and loiasis, but in most cases they are incidental vectors. *Trypanosoma theileri, T. vivax,*

Table 10.11. Some tabanid flies that affect wild animals, or are associated with wild and domestic animals. Diseases caused or transmitted by these tabanid flies also are shown.

Tabanid fly	Wildlife host	Domestic animal host (of disease)	Disease
Tabanus spp., others	deer, elk, moose	sheep, goats	Elaeophorosis
Chrysops callidus	turtles	none	Trypanosomiasis
Chrysops discalis	rabbits, rodents, other mammals, birds	none	Tularemia
Tabanus nigrovittatus	birds	horse	Western equine encephalitis
Haematopota variegata	carnivores	cat	Dirofilariasis
Dasybasis spp.	kangaroos	none	Dirofilariasis
Besnoitia besnoiti	kudu, wildebeest, impala	cattle	Besnoitiosis
Chrysops spp.	African primates	none	Loiasis

Dirofilaria roemeri, *Elaeophora schneideri*, and some other parasitic diseases develop within the body of the tabanid vector, so they serve as much more than a mechanical vector for some maladies.

Management of horse flies and deer flies is very difficult. The moist habitats favored for oviposition and larval development cannot be manipulated without causing significant ecological disturbance. Application of persistent insecticides can disrupt larval development, but will create a significant ecological disturbance. Similarly, large-scale insecticide application will provide some temporary elimination of adults, but can be very disruptive to other animal life. Repellents are short-lived, and although useful for some livestock and humans, there is no feasible mechanism to apply such protective actions to wildlife. Tabanids are reluctant to enter shelters such as barns, so although such structures are useful for livestock and humans, that approach affords little opportunity for wildlife. Box traps, black ball-shaped sticky traps, or insecticide-treated traps can provide some short-term, localized reductions in population density, but this approach needs further research as there are no traps designed specifically for tabanid flies. Because deer flies are ambush predators that attack moving vertebrate prey as they enter wooded areas, removal of trees and brush has been shown to decrease deer fly problems. Forest clearing may not always be desirable, however. For hunters and other humans in tabanid-infested habitats, protection normally involves wearing long-sleeved clothing and use of an insect repellent. However, additional benefit can be derived by affixing a blue adhesive-coated cylinder or panel on or behind a hat or cap to capture alighting flies. The back of the head is a favored landing place, and a considerable number of flies can be captured on such sticky materials.

Tsetse Flies (Diptera: Glossinidae)

Formerly considered to be a muscid fly (Muscidae), tsetse flies (pronounced set-see or tet-see) recently have been elevated to family status, though the family contains only one genus (*Glossina*) and about 23 species. Their distribution is almost entirely limited to central Africa. Though a small family in number of species and geographic distribution, the importance of this group should not be underestimated. Their ability to transmit **African Trypanosomiasis** (Sleeping Sickness and

Nagana) has had profound influence on the politics, economy, and natural history of central Africa. As noted previously (Chapter 9), three million cattle die annually due to this disease. For thousands of years, the ability of tsetse to transmit trypanosomes has denied the people of central and southern Africa the use of large domestic animals such as oxen and horses in their agriculture. Also, it has led to shortage of animal protein in their diet. However, tsetse-borne disease has prevented the southward migration of Arab invaders because their horses and camels were susceptible to the disease, helping to preserve the culture and autonomy of the affected area. Also, the presence of tsetse-borne diseases probably has helped preserve the natural environment of Africa, particularly the rich wildlife, which are not affected by the disease. About 40,000–50,000 additional cases of sleeping sickness are reported annually in humans, so the impact on the region continues to be great.

Tsetse fly adults (Fig. 10.15) are 9.5–14 mm long, heavy-bodied, and tan to brown. Some bear dark segmental bands on the abdomen, but other species are uniformly colored. The family is distinguished by the swollen, bulbous base of the mouthparts (proboscis). The proboscis extends forward in front of the head when not feeding. The antennae consist of only three segments. The basal segment is small, the second segment is two to four times as long as the base, and the distal segment is much larger and oval in form. This latter segment bears a feathery structure called an arista; the presence of branched hairs on the arista is unique among the aristate flies. The eyes are large. The wings are transparent or dusky.

Tsetse flies largely orient visually, though females produce a pheromone that stimulates mating. After mating, the female produces a single egg that hatches internally, and the larva is provided with nutrition by the female from 'milk glands.' The larva is retained by the female for a few days until mature, when it is 'larviposited' on the soil. The mature larva (Fig. 10.16) is 3–8.5 mm long, depending on species, and bears two black knob-like structures posteriorly that serve as respiratory structures. The larva burrows into the soil and pupates, emerging 30–40 days later. The low reproductive rate of tsetse females is unusual among insects, but is compensated for by the degree of protection provided the larva by the female.

Although occurring over a large geographic area in Africa, the flies actually inhabit only a subset of the environment. The preferred habitat consists of patches

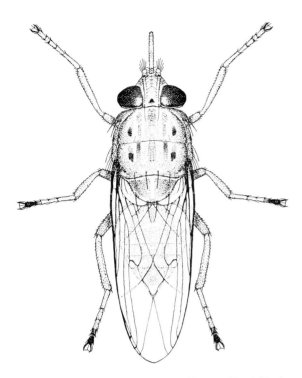

Fig. 10.15. A tsetse fly, *Glossina* sp. (Diptera: Glossinidae)

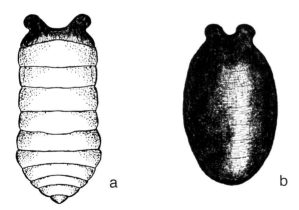

a   b

Fig. 10.16. A tsetse fly, *Glossina* sp. (Diptera: Glossinidae)
life stages: (a) larva; (b) pupa.

of forest and brush where appropriate levels of shade and humidity occur. The flies also are selective in their feeding behavior. A subset of the family called the *palpalis* group (*G. palpalis* and close relatives) feed mostly on reptiles such as crocodiles, *Crocodylus* spp., and monitor lizards, *Varanus* spp., along rivers, though also attacking some mammals and humans. The flies of the *morsitans* group prefer more open habitat, feeding mostly on mammals such as warthog, *Phacochoerus africanus*; giraffe, *Giraffa camelopardalis*; rhinoceros, *Ceratotherium simum*; kudu, *Tragelaphus strepsiceros*; and others. The flies of the *fusca* group favor forests where they feed on some of the same animals as the *morsitans* flies, but also others such as elephant, *Loxodonta* spp.; hippopotamus, *Hippopotamus amphibious*; bushpig, *Potamochoerus larvatus*; aardvark, *Orycteropus afer*; and ostrich, *Struthio camelus*. Flies use both vision and olfaction to locate hosts, and they feed exclusively on blood. When they alight on a host, the toothed labellum and labium are used to penetrate the skin, rupturing a capillary, and anticoagulant is injected, with blood pumped into the fly for 1–10 minutes. A fed fly is not very agile, and is able to fly only poorly until it excretes excess water. Water excretion requires only about 30 minutes and the meal is fully digested within 48 hours. Flies tend to feed at 3–5 day intervals.

Native wildlife is not affected by the trypanosomes transmitted by tsetse flies, though they serve as hosts of these protozoa. Attempts to manage tsetse-transmitted disease usually involve one or more of the following:

• *elimination of wildlife hosts.* Selective reduction of preferred host has been used to minimize the prevalence of disease. This is not often pursued any longer.

• *modification of natural habitat.* Elimination of vegetation reduces the likelihood that flies will inhabit an area. This is not often pursued any longer.

• *insecticide application.* Tsetse flies are quite susceptible to insecticides, and knowledge about where the flies prefer to rest make it possible to target defined resting sites for treatment with persistent insecticide. Alternatively, area-wide treatment, especially multiple applications, can significantly reduce fly numbers.

• *attractants and trapping.* Chemical baits that mimic the odor of host animals, combined with large, dark attractive devices (simulating the profile of wildlife hosts) that are treated with persistent insecticides can be used to greatly reduce the abundance of flies. This approach is now used widely.

The release of sterile male tsetse flies to mate with wild insects and neutralize their reproductive potential has been used to eliminate tsetse flies from the island of Zanzibar. If this technology can be applied broadly in Africa, it could result in elimination of the flies and diseases they transmit with little negative environmental impact. This assumes, however, that in the absence of African trypanosomiasis, human and domestic animal populations would not be allowed to explode.

Muscid Flies (Diptera: Muscidae)

Muscid flies possess no profoundly distinguishing characters; most are drab, ordinary-looking flies. Thus, they tend to be gray–black in color, striped on the thorax and about 4–9 mm long. Muscidae is a large family, with about 700 species just in North America. Although many muscids are associated with animal excrement or decaying plant material, a few are biting or nuisance pests of animals. Adults often feed by sponging liquids. Some, like the face fly, *Musca autumnalis*, feed only on natural (ocular, nasal, salivary) secretions. Others bite and feed on the blood of their victims, including stable fly, *Stomoxys calcitrans* (Fig. 10.17), and moose fly, *Haematobosca alcis*. They also accept plant nectar. The stable fly and the house fly, *Musca domestica* (Fig. 10.18) are implicated in trans-

mitting disease among humans and among livestock, but are of little or no consequence as disease carriers to wildlife. They can be a nuisance to animals, however.

Adults typically deposit eggs in clusters, though some eggs hatch immediately, so perhaps they can be said to larviposit. Larvae of muscids are typical maggots, lacking much in the way of spines or bristles, and also lacking the pseudopods of some other flies. They have strong mandibles called **mouth hooks** anteriorly. The mouth hooks and supporting internal sclerites are collectively called the **cephalopharyngeal skeleton**. They also bear posterior spiracles at the end of their body that are covered by plates bearing slits. The shape and size of the cephalopharyngeal skeleton, and the shape and number of slits of the posterior spiracles, are characteristic of the species and can be used as aids in identification. Larvae of some flies, particularly face fly, develop on animal manure, whereas others such as stable fly may also develop in wet straw. Larvae undergo three instars before pupating within the integument of the third instar (this structure is called a **puparium**, and the process is called **pupariation**). Larvae usually wander away from the larval host before pupating. Adults are active fliers.

Muscid flies are known mostly as associates of cattle. However, face flies (*Musca autumnalis*) affect wildlife such as deer, pronghorn, and American bison. Stable flies also have a broad host range, including moose and bighorn. They also can mechanically transmit *Trypanosoma evansi* and other diseases. Moose flies maintain a close association only with moose, where they typically are found on the hind legs and apparently cause wet, open sores.

A group of about 50 species in the genus *Philornis* seem to be obligate parasites of nestling birds. They occur widely in the tropical regions of the western hemisphere, but some have been found as far north as Texas, USA, where they parasitized ferruginous pygmy-owl, *Glaucidium brasilianum*, and eastern screech owl, *Megascopes asio*, and also in Florida, USA, where they parasitized eastern bluebird, *Sialia sialis*. These flies occur as far south as Argentina in South America. *Philornis* spp. in Puerto Rico are among the most important mortality factors affecting survival of pearly-eyed thrasher, *Margarops fuscatus*, and Puerto Rican Parrot, *Amazona vittata*. *Philornis downsi* Dodge & Aiken first was recorded from the Galápagos Islands in 1964 and now has spread throughout the archipelago where it is causing high nestling mortality among Darwin's finches. The first and early second

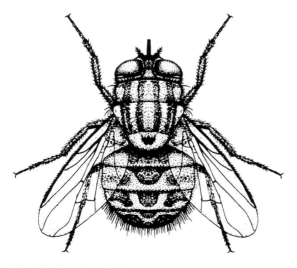

Fig. 10.17. The stable fly, *Stomoxys calcitrans* (Diptera: Muscidae).

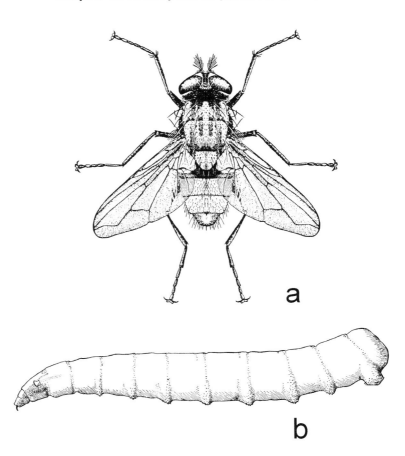

Fig. 10.18. The house fly. *Musca domestica* (Diptera: Muscidae): (a) adult; (b) larva.

larval instars live in finch nostrils and other tissues but the late second and third instars dwell in the nest and feed externally on blood of the nestlings.

Management of muscid flies is difficult. They quickly evolve resistance to insecticides, but then it is not often feasible to apply insecticides to wildlife, anyway. One possible exception is self-application, wherein animals rub on dust bags or trigger sprayers that apply insecticides. Environmental management usually involves eliminating breeding sites (e.g., burning or composting accumulated vegetation), though this also is difficult. Repellents are typically too short-lived to be effective.

Stable Fly

The stable fly, *Stomoxys calcitrans* (Fig. 10.17), also called the dog fly or the biting house fly, is primarily known as a pest of cattle. It is now found throughout the world. In general appearance, it closely resembles a house fly, *Musca domestica*, bearing four black stripes on the thorax and having a tessellated (checkered) abdomen. However, stable fly possesses piercing-sucking mouthparts instead of the sponging mouthparts found in house fly. Both sexes of the stable fly feed on blood from a number of warm-blooded animals, including wildlife and humans. Feeding by flies can produce annoyance in both wildlife and livestock, resulting in avoidance behavior that includes constant stomping of feet, tail switching, and 'bunching,' whereby the herd gathers or bunches together in a tight circle with heads facing inward to protect their front legs from feeding flies. Although normally associated with mammals, stable fly has been found to be a significant pest of American white pelican, *Pelecanus erythrorhynchos*, in the northern Great Plains region of

North America. Because stable flies feed on blood, they are capable of transmitting disease to their hosts. Although most stable flies disperse no farther than needed to obtain a blood meal, this pest can travel distances up to 8 km routinely, and as far as 225 km when carried by storms.

Stable flies deposit eggs on decayed vegetation, or vegetation mixed with animal feces and/or urine. Improperly composted vegetative material, bales of hay, green chop (green vegetation intended for livestock feed), or decomposing grass clippings that remain wet can provide adequate developmental medium for stable flies. Even marine and freshwater grasses that have washed ashore can be suitable for larval development. The females deposit from 60 to 800 eggs. The eggs hatch in 1 to 5 hours and the larvae go through three instars. When the larva is mature, it ceases feeding and often migrates from the developmental area to pupate in drier conditions. Sometimes larvae will enter the ground to the depth of about 2 to 4 cm to pupate. At the time of pupation, the exoskeleton from the last larval instar will harden, forming the puparium in which the pupa will reside. Depending on temperature and other environmental parameters, the development period from egg to adult can require from 10 days to 2 months. Several generations are produced annually and tremendous numbers of biting adults can develop. Adult flies live about 20 days under favorable conditions. Adult stable fly seasonal peaks vary, often being abundant in both spring and fall in warm-weather areas, but most abundant during summer in cooler areas.

Stable flies can be managed by eliminating favored larval habitat, but this is usually impractical in natural areas, where wildlife typically occurs. Stable fly pupae are sometimes heavily parasitized by wasps such as *Spalangia cameroni* Perkins (Hymenoptera: Pteromalidae), but a practical method of manipulating them to attain biological suppression of flies in natural areas is not known. Application of insecticides is sometimes directed to the adult stage. Fly repellents do not give long-term relief for animals or humans against stable fly biting.

House Fly

House fly, *Musca domestica* (Fig. 10.18), originated on the steppes of central Asia, but now occurs on all inhabited continents, in all climates from tropical to temperate, and in a variety of environments ranging from rural to urban. It is commonly associated with animal feces, but has adapted well to feeding on garbage, so it is abundant almost anywhere animals or people live. It is the most common fly to invade homes, and often is the dominant species around livestock and poultry. Despite its inability to bite, it is a nuisance for wildlife and livestock, and a public health problem for humans.

The life cycle of the house fly may be completed in as little as 6 days, but under suboptimal conditions may require up to two months. It breeds continuously in warm climates such as in the tropics and subtropics, producing more than 20 generations annually. Even in more temperate areas they commonly undergo 10 generations. In cooler areas they overwinter as larvae or pupae, but adults perish when exposed to cold. In heated areas in cold climates, however, they breed throughout the winter months if the adults have access to food and larvae to suitable developmental media.

The cylindrical-oval eggs are white, and 1.0–1.2 mm long. Eggs typically are laid in clusters, often numbering 75–150 eggs per cluster. A female normally deposits two to six clusters of eggs during her life span. Maximum egg production occurs at intermediate temperatures, 25–30 °C. Often, several flies will deposit their eggs in close proximity, leading to large masses of larvae and pupae. Eggs must remain moist or they will not hatch. Egg hatch usually occurs within 8–20 h. Due to its rapid development time and high reproductive capacity, the house fly has the capacity to increase in abundance rapidly. Indeed, scientists impressed with its reproductive capacity have calculated that, lacking mortality, a single pair of flies could grow to a population of 191,000,000,000,000,000,000 in only 5 months, enough to cover the entire earth in a layer of flies several meters deep!

The larvae are creamy white, cylindrical, legless, and taper to a point at the head. The head of larval house flies, like most maggots, has dark mouth hooks. There are three larval instars. The optimal temperature for larval development is 35–38 °C and larvae complete their development in 4–13 days at optimal temperatures. The moisture level of the food affects larval survival, and 50%–70% moisture content favors survival. Nutrient-rich substrates such as animal manure provide an excellent developmental substrate. Very little manure is needed for larval development, and sand or soil containing small amounts of degraded manure allows for successful belowground development. At maturity, larvae are 7–12 mm long, and prefer to disperse to a dry location to pupate.

The mature larva pupates within the cuticle of the last larval instar, though the shape of the pupal covering is quite different from the larva, being bluntly rounded at both ends. It is about 8 mm long. The color of the puparium darkens with time, becoming almost black. Pupae complete their development in 2–6 days at 32–37 °C.

The adult is about 4–7.5 mm long. The female is typically larger, and can be distinguished from the male by the relatively wide space between the eyes (in males, the eyes almost touch). The dull, grayish fly bears four narrow dark stripes on the thorax. The abdomen may be gray, but often is yellowish, and has a narrow dark line dorsally. The wings are transparent except for the dark veins. They have sponging mouthparts, which allows them to lap up liquid food. Thus, they cannot bite animals and humans. Flies regurgitate readily, secreting saliva onto solid foods so it can be liquified and ingested. Adults may live up to 2 months, but more typically live 2–3 weeks. They require food before they will copulate, and copulation is completed in as few as two minutes or as long as 15 minutes. Oviposition commences 4–20 days after copulation. Female flies need access to suitable food (protein) to allow them to produce eggs, and manure alone is not adequate. They prefer sunlight, and are active fliers during warm days. They are inactive at night, and commonly can be seen perching on the ceilings of barns or other shelters.

House flies are mechanical vectors of animal and human pathogens, particularly enteric diseases. At times, they have been found contaminated with viruses, bacteria, fungi, protozoa, and nematodes. These microbial organisms adhere to their tarsi, legs, mouthparts, and elsewhere as the flies move about their environment, and then these microbes can be relocated when the flies move to new food sources. Also, when consumed by flies, some pathogens can be harbored in the mouthparts or alimentary canal for several days, and then be transmitted when flies defecate or regurgitate. Among the pathogens commonly transmitted by house flies are *Salmonella*, *Shigella*, *Campylobacter*, *Escherichia*, *Enterococcus*, *Chlamydia*, and many other species that cause illness. These flies are most commonly linked to outbreaks of diarrhea and shigellosis, but also are implicated in transmission of food poisoning, typhoid fever, dysentery, tuberculosis, anthrax, ophthalmia, and parasitic worms.

In addition to disease transmission, house flies can be a severe nuisance. If left undisturbed, the flies may feed on the secretions of the mouth, nose and eyes.

The most important approach for fly management is sanitation. Flies should be deprived of suitable oviposition sites, and larval environments should be eliminated or made too dry for high levels of survival to occur. Natural biological suppression of the house fly results primarily from the actions of certain chalcidoid wasps (Hymenoptera: Pteromalidae), of which many species have been associated with house fly around the world. Among the more important are *Muscidifurax* and *Sphalangia* spp. Ichneumonids and other parasitoids, as well as some predatory insects (especially histerids [Coleoptera: Histeridae] and staphylinids (Coleoptera: Staphylinidae)), also contribute to fly mortality, but under optimal fly breeding conditions the house fly quickly builds to high numbers.

Blow Flies (Diptera: Calliphoridae)

This is a large group of flies, with over 1000 species known worldwide, and four-fifths of the species restricted to the Old World. These blue, green or bronze and metallic-looking flies often are called **bottle flies**, such as 'blue bottle flies' or 'green bottle flies.' They are medium to large in size, measuring about 4–16 mm long, robust, and sometimes male and female flies differ in appearance. Most are oviparous. The larvae are pale yellow to white, cylindrical, tapering anteriorly, and have three instars. Some also are called **screwworms**, apparently due to the belief that they twist to screw their way into the flesh of hosts or because they look like screws imbedded in flesh. The blow flies are not considered to very important as vectors of disease.

Except for the host preference and larval feeding behavior of some species, the biology of blow flies is very similar to muscid flies. Blow flies associated with animals tend to be coprophagous, necrophagous, or myiasis-producing. **Coprophagous** flies ingest the feces of larger animals. **Necrophagous** species feed on carrion, or dead animals. Many blow fly adults oviposit on recently killed animals, but some will infest living hosts. Invasion of a living host animal by fly larvae is called **myiasis**. Several groups of flies are capable of causing myiasis, but blow flies are most important in causing this condition. The **facultative myiasis**-causing flies generally develop in carrion or feces, but are capable of affecting living hosts. Examples of facultative myiasis include infestation by some *Lucilia* spp. green bottle flies, and *Calliphora* spp. and *Protophormia* spp. blue bottle flies. Some flies are capable of **primary facultative myiasis**, which means that they initiate myiasis by attacking living tissues, whereas others display **secondary facultative myiasis**, which means that they invade the host only after primary species have attacked. The **obligate myiasis**-causing flies develop only in living hosts; the *Cochliomyia* and *Chrysomya* spp. screwworms are examples of this. Myiasis can also occur in several areas of the host's body, including dermal, subdermal, nasopharyngeal, intestinal, and urogenital regions.

Several genera contain important myiasis-producing blowflies, including *Auchmeromyia* (found only in Africa), *Calliphora* (these are cosmopolitan), *Chrysomya* (eastern hemisphere only), *Cochliomyia* (western hemisphere only), *Cordylobia* (Africa only), *Lucilia* (cosmopolitan), *Phormia* (western hemisphere), *Protophormia* (Holarctic region), *Bufolucialia* (cosmopolitan), and *Protocalliphora* (Holarctic region). Some obligate myiasis-causing flies affecting wildlife include *Bufolucialia* spp., which attack amphibians, *Protocalliphora* spp. attacking nestling birds, and *Protophormia terraenovae* Robineau-Desvoidy attacking caribou. Probably most important, however, are the primary screwworm flies, *Chrysomya bezziana* Villeneuve, known as Old World Screwworm, and *Cochliomyia hominivorax* (Cocquerel), known as New World Screwworm. Both attack a very large number of vertebrate animals. Although the obligate myiasis-producing flies are usually critical in initiating myiasis, the subsequent attack by facultative myiasis-causing flies is often quite important in causing injury and death.

The importance of myiasis to wild birds has likely been underappreciated. The presence of fly maggots on bird carcasses is often interpreted as a post-mortem phenomenon, but this may not be entirely accurate. The recent discovery that the muscid *Philornis downsi* (mentioned previously) is affecting the abundance of Darwin's finches in the Galapagos Islands has stimulated interest in myiasis as a mortality factor. Myiasis is well known, perhaps because it is better studied, in Europe than in North America. Any nitidicolous (nest-dwelling) bird may be subject to attack, but water and shore birds often seem to be free of attack. The prevalence of flies dwelling in bird nests is high among some bird species, especially colonial species. Intensity of infestation also varies, and large sturdy nests often support more fly larvae.

The occurrence of blowflies is not necessarily detrimental. In the case of *Protocalliphora sialis*, a bird nest-inhabiting species, considerable tolerance to infestation is apparent. This insect infests the nests of many birds, including screech owl, *Otus asio*; downy woodpecker, *Picoides pubescens*; great crested flycatcher, *Myiarchus crinitus*; house wren, *Troglodytes aedon*; European starling, *Sturnus vulgaris*; eastern bluebird, *Sialia sialis*; and several species of swallows but especially tree swallow, *Tachycineta bicolor*. Tree swallows are small insectivores (they weigh about 22 g) that live colonially and lay 4–7 eggs together in feather-lined grass nests within natural and artificial cavities. *P. sialis* blowflies deposit 10–20 eggs in the nest when the nestlings are one-fourth to one-third grown, allowing the flies adequate time to feed and grow before the birds fledge. The fly eggs hatch within 24 hours, and the larvae feed periodically for the next two weeks, retreating to the nesting material between feeding bouts. It is quite interesting that there do not seem to be consistent detrimental effects of feeding by flies on the health of the young birds. Although some authors report negative effects, others report no measurable effects of 30–50 fly larvae per bird. It seems that this is a highly co-evolved parasite host system; such relationships typically do not demonstrate much effect on the host. However, it may be that it simply requires much higher parasite levels to measure significant effects on bird health.

Blowfly populations are often managed best by sanitation. Removal of garbage, fecal material, and dead animals is important, depending on the biology of the blowfly. Trapping and chemical control of adults also proves effective. Area-wide management can be accomplished by chemical suppression of adults, but these flies are susceptible to eradication by release of sterile individuals. This can provide permanent relief from attack by screwworm flies without continued use of chemical insecticides. This has been accomplished in North America and most of Central America, as well as some small areas elsewhere.

New World Screwworm Fly

New World screwworms (Fig. 10.19), the larvae of *Cochliomyia hominivorax*, are extremely important and obligate myiasis-causing flies. These screwworms occur naturally in tropical and subtropical regions of the western hemisphere, although they can extend their range to temperate regions during warm seasons.

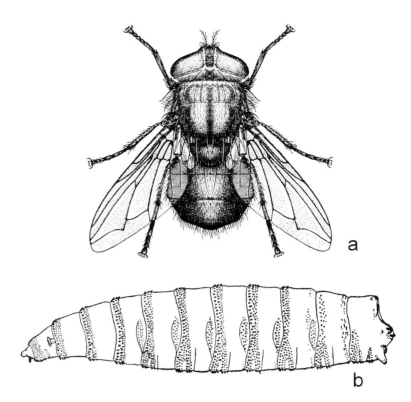

Fig. 10.19. The New World screwworm, *Cochliomyia hominivorax* (Diptera: Calliphoridae): (a) adult; (b) larva.

Animals are very susceptible to infestation during birthing, as are their newborns, via screwworm oviposition on the umbilical cord. Wounds of any type, including tick infestations, can lead to infestation and to death in untreated animals. They affect hundreds of species of animals. They were once the most important factor affecting survival of white-tailed deer, *Odocoileus virginianus*, in the southeastern USA. For example, before eradication, 25%–80% of fawns were lost to this fly annually in southern Texas. After eradication of screwworm from North America, deer populations increased markedly in the southeastern USA.

Although screwworms no longer occur in North America, they remain in Panama and South America (except Chile) and some Caribbean islands, including Cuba, Hispaniola, Trinidad, Tobago, and Jamaica. The elimination of screwworm from North America and Central America can be attributed to use of the sterile insect technique, in conjunction with education and quarantine measures carried out by the United States Department of Agriculture, the Mexico-American Commission, and the national agencies of Belize, Nicaragua, Guatemala, El Salvador, Honduras, and Panama. **Sterile insect technique** involves release of sterilized males, which mate with fertile females, resulting in disruption of reproduction by the fertile, wild population. A principal challenge, of course, is to flood the areas inhabited by fertile flies with sterile flies at a high enough ratio that fertile females are most likely to mate with a sterilized, unfertile male. After several generations of releases, if the releases are successful, the fertile population declines and then disappears.

Screwworm flies are bluish black, with the head orange and the thorax striped. The adults are about 8–10 mm long. Female *C. hominivorax* flies are autogenous, not requiring a protein meal to initiate the first ovarian cycle. However, they require exogenous protein to support subsequent ovarian cycles. The number of eggs produced range from about 250–425 eggs, and some or all may be deposited on a single wound. Females usually become inseminated by day five and single matings are the rule. There is no diapause or quiescent period that allows screwworms to survive winter or dry seasons. Generations continuously overlap. In North America before they were eradicated, overwintering by screwworms in USA was confined to southern Texas and southern Florida during most winters.

The key factor in the natural limitation of screwworm abundance is thought to be the availability of suitable oviposition sites. Indeed, before screwworms were eliminated from North America, inspection of wounds on domestic animals and control of larvae found there was the chief method of preventing screwworm infestation. Screwworm densities, averaged over time and space, often are quite low. However, densities can be quite high locally, but transiently so, when wounded hosts are available. Such clumped distributions have an important bearing in planning sterile fly releases because enough sterilized males must be present to neutralize the reproduction of fertile flies. Thus, sterile fly dispersions must be made continuously wherever screwworms occur until they are eliminated.

The ability to mass-produce flies, and to sterilize flies with ionizing radiation, are important components of the sterile male technique. It is also important to understand the dispersal behavior of males, which affects the mating potential of sterilized males. Given the propensity of screwworm to disperse rapidly and widely, a containment strategy instead of eradication was necessary initially. For this reason, arrangements were made with the Mexican government in 1962 to carry out education, inspection, and sterile fly releases jointly in a 'barrier zone' on the USA–Mexico frontier that extended 160 km into Mexico. Eventually, eradication efforts were extended to southern Mexico north of the Isthmus of Tehuantepec, where a much smaller barrier could be established and maintained. A factory for screwworm production was constructed in Tuxtla Gutierrez, Chiapas State, Mexico, with weekly production capacity of up to 550 million sterile flies. The factory went into production in 1976 and is still operational.

Eventually the sterile insect release program was extended further south to the Darien Gap in Panama, as this is the region of smallest area that could serve as a barrier. It was estimated that such a barrier would require for its maintenance only 50 million sterile flies (half males) weekly. That goal was reached in 2004 and no self-sustaining screwworm populations have

been detected in the barrier zone since then. The barrier is treated with 40 million sterile flies weekly. Sterile flies have since been used successfully in northern Africa for an eradication program in Libya. Screwworms have also been eliminated from Puerto Rico and the Virgin Islands.

The calliphorid *Chrysomya bezziana* (Fig. 10.20) is called the Old World screwworm. It occurs in the Afrotropical and Oriental regions, and is functionally the Old World equivalent of the New World screwworm. Like *Cochliomyia hominivorax*, it is best known for being a pest of livestock but also attacks a wide range of vertebrate hosts, including numerous wildlife species.

Flesh Flies (Diptera: Sarcophagidae)

The flesh flies are similar in appearance to muscid flies during both the larval and adult stages (Fig. 10.21). Adults are black and grey and usually bear a striped thorax and a checkered abdominal pattern. However, they are all larviparous, with the female retaining the eggs until they hatch. They deposit 30–200 larvae, depending on the species of fly. One subfamily of flesh flies, the Miltogramminae, consists mostly of insect parasitoids. The other subfamily, the Sarcophaginae, contains necrophagous flies that usually are associated with carrion or feces. However, some cause facultative or obligatory myiasis in animals. The most important flies from the perspective of wildlife likely are *Wohlfahrtia* spp. (Fig. 10.22), which despite the overall trend of the subfamily Miltogramminae to be insect parasitoids, are vertebrate parasites. The Old World species *W. magnifica* and the New World species *M. opaca* and *M. vigil* are quite invasive. Females larviposit at wounds, of course, but maggots can also penetrate unbroken skin, where they feed gregariously. This is a large group, with over 2000 described species, and most are found in tropical or warm climates.

In Canada, wild ducklings have been found to be susceptible to myiasis caused by *Wohlfahrtia opaca*. The larvae enter the hatched bird via the umbilicus, then burrow subcutaneously in the abdominal region. Birds typically are found with several to many flies, and the birds perish. These flies also are capable of entering the egg through cracked eggshells, and of killing unhatched birds.

Following are some other examples of flesh flies that affect wildlife. Their vernacular names reflect their importance, but they are poorly known: lizard flesh fly,

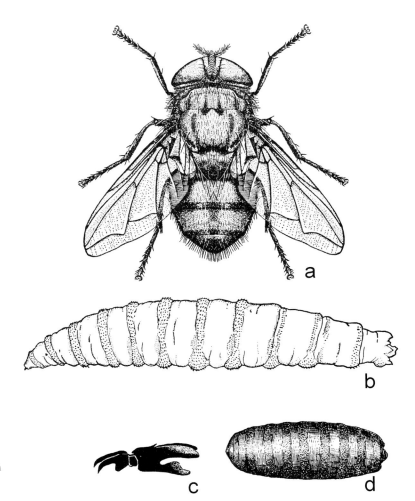

Fig. 10.20. The Old World screwworm, *Chrysomya bezziana* (Diptera: Calliphoridae): (a) adult; (b) larva; (c) cephalopharyngeal skeleton (internal mouthparts), a diagnostic feature of fly larvae; (d) puparium.

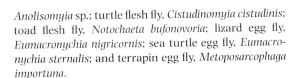

Anolisomyia sp.; turtle flesh fly, *Cistudinomyia cistudinis*; toad flesh fly, *Notochaeta bufonovoria*; lizard egg fly, *Eumacronychia nigricornis*; sea turtle egg fly, *Eumacronychia sternalis*; and terrapin egg fly, *Metoposarcophaga importuna*.

Bot and Warble Flies (Diptera: Oestridae)

Bot fly (Figs. 10.23, 10.24) and warble fly larvae are obligate parasites of mammals, infesting both body tissues and cavities. Body tissue-dwelling larvae feed on connective tissue, serum, and white blood cells. Body cavity-dwelling larvae feed on cellular debris and mucosal secretions. As is the case with infestation of animal hosts by other flies, infestation by bot and warble fly larvae is called **myiasis**.

This is not a particularly species-rich family, consisting of perhaps 150 species. It is most diverse in Africa and central Asia. The African elephant, *Loxodonta* spp., apparently supports the greatest number of bot flies, five species. African and Asian ungulates are often affected, as are rodents and rabbits/hares (see Table 10.12). In North America, the oestrid fauna is not very diverse, with *Cuterebra* parasitizing rodents and rabbits/hares, *Cephenemyia* in deer, and a Holarctic species, *Hypoderma tarandi*, affecting caribou. A few species have been introduced with livestock.

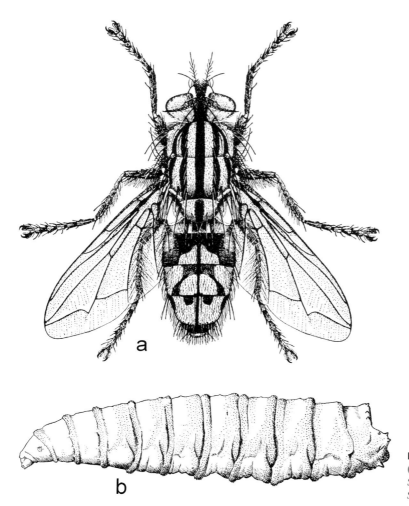

Fig. 10.21. Representative flesh flies (Diptera: Sarcophagidae): (a) adult of *Sarcophaga haemorrhoidalis*; (b) larva of *Sarcophaga crassipalpis*.

This family of flies consists of four subfamilies (formerly treated as families): Oestrinae, the nose and pharyngeal bots; Hypodermatinae, the warbles; Gasterophilinae, the stomach bots; and Cuterebrinae, the rodent bots. Though they have similar life cycles, they also display some marked differences, as can be seen in their oviposition behavior.

• Female **nose and pharyngeal bots** are viviparous. They deposit larvae into the muzzle or eyes of the host, often without landing. The first instars migrate into the eyes, nose or mouth, but the second and third instars are located in the sinus cavities or pharyngeal regions of the host. They are largely Palearctic in distribution, though a single species is known from kangaroos in Australia, and in North America *Cephenemyia* spp. affect deer, *Odocoileus* spp.; moose, *Alces alces*; elk, *Cervus canadensis*; and caribou, *Rangifer tarandus*. Also, *Oestrus ovis* Linnaeus was introduced with sheep.

• Female **warble flies**, *Hypoderma* spp., deposit eggs directly onto the hairs of the hosts. First instars burrow through the skin of the host and then infest connective tissue. Originally found in Africa and Eurasia, some species have been moved elsewhere with cattle, although they remain important pests of Old World deer.

• Females of most **stomach bots** deposit their eggs directly on the hairs of their hosts, the location of which on the host is bot fly-specific. Normally they are

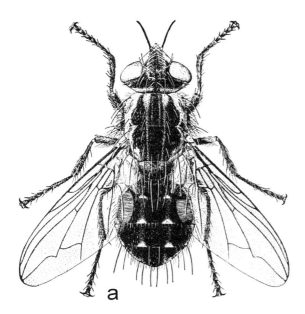

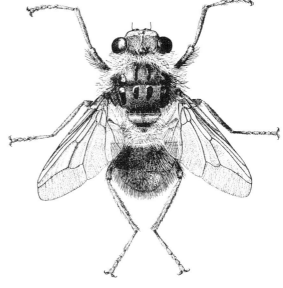

Fig. 10.23. The bot fly *Hypoderma lineatum* (Diptera: Oestridae) also called heel fly and cattle grub, affects deer, reindeer, antelope, and cattle.

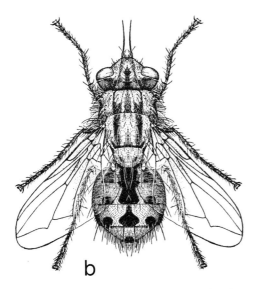

Fig. 10.22. Some common myiasis-producing flesh flies: (a) *Wohlfahrtia vigil*; (b) *Wohlfahrtia magnifica*.

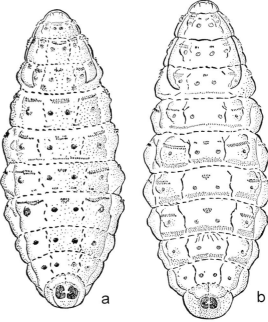

Fig. 10.24. Larvae of two bot flies (Diptera: Oestridae): (a) deer warble, *Hypoderma diana*; (b) heel fly, *Hypoderma lineatum*.

Table 10.12. Bot and warble flies (Diptera: Oestridae), their host relationships, and distribution (adapted from Wood 1987).

Subfamily and genus	Number of species	Hosts	Larval site in host	Geographic range
CUTEREBRINAE – rodent bots				
Alouattamyia	1	howler monkeys	skin	S. America
Andinocuterebra	1	?	?	S. America
Cuterebra	60+	rodents, rabbits, hares	skin	New World
Dermatobia	1	mammals	skin	S. America
Pseudogametes	2	?	?	S. America
Rogenhofera	6	?	?	S. America
Neocuterebra	1	African elephant	skin	W. Africa
Ruttenia	1	African elephant	skin	W. Africa
GASTEROPHILINAE – stomach bots				
Gasterophilus	9	horses, zebras	gut	Eurasia, Africa (some now elsewhere)
Gyrostigma	3	rhinos	gut	Africa, Asia
Cobboldia	3	elephants	gut	Africa, Asia
HYPODERMATINAE – warbles				
Hypoderma	5	deer, caribou, cattle	skin	Eurasia (some now elsewhere)
Oestroderma	1	pikas	skin	C. Asia
Oestromyia	5	mice, marmots, pika	skin	Eurasia
Pallasiomyia	1	saiga antelope	skin	C. Asia
Pavlovskiata	1	goitered antelope	skin	C. Asia
Portschinskia	7	mice, pikas	skin	Eurasia
Przhevalskiana	6	gazelles, goats	skin	C. Asia
Strobiloestrus	3	*Kobus* spp.	skin	Africa
OESTRINAE – nose and pharyngeal bots				
Cephalopina	1	camels	nasal	Asia
Cephenemyia	8	deer, caribou	nasal	Holarctic
Gedoelstria	2	antelope	nasal	Africa
Kirkioestrus	2	antelope	nasal	Africa
Oestrus	6	antelope, sheep, goats	nasal	Eurasia, Africa
Pharyngobolus	1	African elephant	trachea	Africa
Pharyngomyia	2	deer	nasal	Eurasia
Rhinoestrus	11	horses, zebras, pigs, giraffe, sheep, hippos, springbok	nasal	Eurasia, Africa
Tracheomyia	1	kangaroos	trachea	Australia

associated with the mouth initially, then descend to the stomach, duodenum or rectum to complete their larval development. Originally found in Africa and Asia, some species have been moved elsewhere with horses.

• **Rodent bots** also deposit eggs, but not directly on their hosts. Instead, they scatter eggs on vegetation of the soil surface where prospective hosts are likely to be found. The heat from the body of the prospective host causes the eggs to hatch, and the larvae attempt to locate and penetrate the host, often though natural openings. This group is restricted to the New World. Note that although they are often associated with rodents, they have a broader host range (see Table 10.12).

Adults of the bot and warble flies are medium to large in size, 9–25 mm long. They are compact, with short antennae, and often very hairy, resembling heavy-bodied bees. The head is typically broad and flat. The mouthparts are reduced and the adults apparently do not feed, relying instead on fat reserves for energy. Adults are rarely seen, but animals such as caribou can sense them, and will often stampede in their presence. They can be oviparous or viviparous, and the eggs are sometimes modified for attachment to a hair. The whitish, legless larvae often are equipped with bands of spines circling the body, but sometimes plates. By the third (final) instar they are stout. Pupation occurs in the soil.

Males of the nose and pharyngeal bots, and the rodent bots, gather at characteristic aggregation sites with species-specific aggregation characteristics, often hilltops. Warble flies similarly aggregate, but not on hilltops. Adults are very dispersive, with warble flies known to fly hundreds of kilometers when following migrating species. Bot and warble flies generally show a high degree of host specificity. Frequency of infestation is considerable. For example, a survey of white-tailed deer, *Odocoileus virginianus*, in Alberta, Canada, showed that 27% of fawns contained nasal bots, *Cephenemyia* spp. Other studies, such as one in Utah, USA, found that 100% of mule deer, *Odocoileus hemionus*, were infested.

The effect of bot and warble flies is not great, considering the size of the insects. They seldom cause death, unlike myiasis by screwworm, *Cochliomyia hominivorax* (Calliphoridae), or *Wohlfahrtia* spp. (Sarcophagidae). However, avoidance behavior is well developed in wildlife. For example, deer, caribou and howler monkeys expend considerable energy avoiding the flies. The larger larvae of *Hypoderma* produce an immuno-suppressant, predisposing their host to infection by disease, and weight gain is often reduced. For small animals, infestation can be more hazardous; for example, when bot fly levels exceed about seven larvae per *Neotoma* wood rat, mortality exceeds 50% and survivors are emaciated. Irritation is often evident in eastern gray squirrels, *Sciurus carolinensis*, which frequently scratch warbles. Movement may be awkward, and sometimes reproduction is thought to be inhibited

among small hosts infected with fly larvae. In Europe, where deer are cultured for meat and hides, the economic loss is considerable.

Management of bot flies is, at this point, largely limited to use of systemic insecticides applied via food-based bait. This is certainly applicable for intensively managed wildlife, but less so for unmanaged wildlife.

Louse Flies (Diptera: Hippoboscidae)

All the flies in this group are ectoparasites of birds and mammals, and feed on the blood of their hosts. Birds are hosts most commonly: 18 orders of birds are affected, as are five orders of mammals. Hippoboscidae is a small group, with only about 150 species known worldwide. It is closely related to Streblidae and Nycteribiidae, bat-feeding flies, and sometimes these flies are grouped into a single family. It is most diverse in the tropics and subtropics (see Table 10.13).

Louse flies are robust, and most are peculiarly flattened (Fig. 10.25). They range in size from about 1.5 to 12 mm. Due to their flattened condition, the head is oriented forward. The mouthparts are adapted for piercing. The eyes usually are large. Normally, the wings are fully developed, but sometimes they drop off soon after emergence, or are poorly developed. Louse flies are **larviparous**; the egg and larvae of louse flies remain in the female until maturity, instead of the egg being deposited. While in the female, the larvae are nourished by secretions of the milk glands until they reach maturity. The female supports only a single offspring at a time. Upon maturity, the mature larva is extruded, and then usually falls to the substrate, where it transforms into a puparium. Only one generation per year occurs, and the female fly produces perhaps seven or eight larvae.

The importance of louse flies is uncertain. Specificity varies, with mammal-feeding species more specific than bird-feeding species. Occasionally, louse flies are implicated in transmission of trypanosomes, protozoans, and filarial worms, but this appears to be less important than their blood-feeding habits. With up to 80% of deer infested with louse flies in the southeastern USA, and sometimes supporting up to 2300 flies per animal, anemia and other ailments are probably common. They also inadvertently allow other parasitic arthropods such as lice and mites to move from host to host by hitchhiking on the strong-flying flies. There are few options available for control of louse flies on

Table 10.13. Some louse flies that affect wild animals, or are associated with wild and domestic animals. Diseases caused or transmitted by these louse flies also are shown.

Louse fly	Wildlife host	Domestic animal host	Disease
Lipoptena cervi	white-tailed deer and elk in NE North America	none	none
Lipoptena mazamae	white-tailed deer in S North America, Central, and South America	cattle	none
Lipoptena depressa	mule deer, white-tailed deer	none	*Corynebacterium*
Hippobosca longipennis	hyenas	dogs	*Dipetalonema dracunculoides*
Icosta albipennis	wading birds	none	none
Icosta americana	owls, hawks, turkeys, grouse	none	none
Olfersia sordida	pelicans, cormorants	none	none
Stilbometopa impressa	quail	none	*Trypanosoma* sp., *Haemoproteus lophortyx*

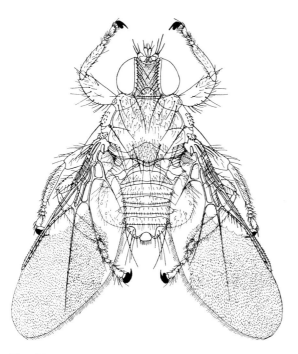

Fig. 10.25. A bird-infesting louse fly, *Ornithoica vicina* (Diptera: Hippoboscidae).

wildlife. Insecticides can be applied in situations where wildlife are intensively managed, but otherwise these parasites must be endured.

Fleas (Siphonaptera)

Fleas, like louse flies, are ectoparasites. More than 2000 species are known. They are small, measuring 1–5 mm long, and usually dark (brown or black) in color. The wingless adults (Figs. 10.26, 10.27) are blood feeders, extracting blood with their piercing-sucking mouthparts while hiding in their host's fur. Blood is required for successful development of eggs. They produce smooth, oval eggs that typically drop off the host into the host's nest or burrow. The larvae that hatch from the eggs have chewing mouthparts and feed on organic matter such as the bedding material and other organic material off the body of the hosts. For larvae of some species, feces of adult fleas (containing host blood) are an important source of nutrients. The larvae are legless and elongate, resembling fly larvae but possessing a distinct head capsule. They display three larval instars. Larvae produce a small silken cocoon and often incorporate sand grains or other debris in its construction as they prepare to pupate. Like the larvae, the pupae are located near but not upon the host. The presence of a host is often stimulus to emerge, and the fully formed adults will remain within the cocoon until a prospective host is sensed. Most species are quite host specific. Rodents are a predominant host; 74% of fleas feed on rodents, and other mammals also are common hosts. In contrast, only 6% of fleas attack birds.

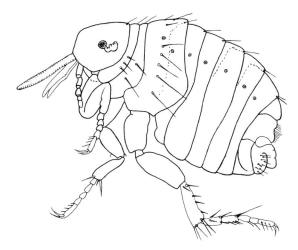

Fig. 10.26. A common bird-infesting flea, *Ceratophyllus gallinae* (Siphonaptera: Ceratophyllidae) also known as hen flea.

Adult fleas are flattened laterally (from side to side), in contrast to lice and louse flies, which are flattened dorso-ventrally (top to bottom). Thus, fleas are well adapted for slipping among the hairs of the host. Their tarsi are adapted for clinging to the host, and they possess tooth-like spines forming combs that help the fleas maintain their presence within fur during host grooming or preening. Thus, hedgehogs (class Mammalia, order Erinaceomorpha, subfamily Erinaceinae), which are covered with spines, often support large populations of fleas because they cannot groom effectively. Fleas also possess an especially tough integument that resists scratching and chewing by their hosts. Their hind legs are adapted for leaping, both for escape and to attain the host, often leaping 30 cm high and 20 cm in distance. Fleas can remain alive for considerable periods away from their hosts, but in the presence of hosts, hungry fleas are quite adept at locating a meal.

Fleas are regarded mostly as a nuisance, but they also vector diseases such as plague and parasites such as tapeworm, so they have considerable importance to wildlife, domestic animals, and occasionally to humans. Reproduction may be related to the reproductive state of the host. Specifically, a classic series of studies showed that the female rabbit flea's ovaries do not develop until the flea feeds on a pregnant female

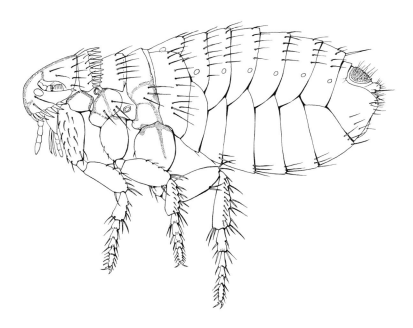

Fig. 10.27. A common mammal-infesting flea, *Ctenocephalides felis* (Siphonaptera: Pulicidae), also known as cat flea.

Table 10.14. Some fleas that affect wild animals, or are associated with wild and domestic animals. Diseases caused or transmitted by these fleas also are shown.

Flea	Wildlife host	Domestic animal host	Disease
European rat flea, *Xenopsylla cheopis*	rodents such as rats, prairie dogs, rock squirrels, ground squirrels; others such as rabbits	dog, cat, chicken	Plague, Murine Typhus
Cat flea, *Ctenocephalides felis*	Northern raccoon, Virginia opossum, rabbit, coyote, fox, monkey	dog, cat, horse, goat, sheep, cattle	Anemia, Dirofilariasis
Dog flea, *Ctenocephalides canis*	coyote, foxes, Virginia opossum, ground squirrels	dog, cat	Anemia, Dirofilariasis
Orchopeas howardi	flying squirrels	none	Q Fever
Ceratophyllus ciliatus	chipmunks, squirrels, wood rat	none	Plague
Hen flea, *Ceratophyllus gallinae*	Great tits, blue tits, blackbirds, bobwhite, burrowing owl, house sparrows, foxes, American bobcat, ground squirrels, American badger, rabbits	chickens, pheasant	none
Sticktight flea, *Echidnophaga gallinacea*	quail, turkey, many wild birds; some mammals	chickens, turkeys, dogs, cats	none
Ceratophyllus celsus	cliff swallows, bank swallows	none	none

rabbit or young rabbits. The rabbit's hormones stimulate reproduction in the flea, assuring that young fleas will have suitable hosts on which to feed.

Fleas are not very host specific, but nearly all affect mammals (Table 10.14). Although most commonly associated with rodents (rats, squirrels, and relatives), they commonly affect lagomorphs (rabbits and relatives) because they share a common habitat, and canids (dogs and relatives) and felids (cats and relatives) because these predators feed on flea-infested animals. As noted previously, only about 6% of fleas specialize on birds.

The hen flea, *Ceratophyllus gallinae*, and the sticktight flea, *Echidnophaga gallinacea*, are important examples of bird-feeding fleas, though like many fleas they also feed on mammals. As we will see in Chapter 11, hen fleas can affect fitness in birds. Hen fleas have a very wide host range, but prevalence of infestation is highest in hole-nesting families of birds. The family Paridae (tits, chickadees, and titmice) are most often infested and harbor the highest populations of fleas.

Study of sticktight fleas affecting the threatened Florida scrub-jay, *Aphelocoma coerulescens*, in Florida, USA, showed that about 15% of the population were infested with these fleas, and that 46% of the flea-infested birds died within a year. Importantly, the hematocrit level (a measure of red cell level in the blood) of infested jays was diminished by 17%, and was negatively correlated with the level of infestation. Sticktight fleas were also found associated with a plague outbreak among ground squirrels, *Citellus beecheyi*, in California, USA. Interestingly, plague-infested sticktight fleas were found on a burrowing owl (*Athene cunicularia*) in the area, and it is postulated that owls may assist in the movement of plague among squirrels. Thus, fleas are clearly an issue with birds, as well as their more normal hosts, mammals.

Other Taxa of Occasional Importance

The taxa that are most commonly associated with wildlife, and which are most likely to affect them negatively, were treated above. Below are a few additional taxa that are more occasional problems or restricted in distribution.

Eye Gnats (Diptera: Chloropidae)

These small flies vary in their feeding habits, though most larvae feed on living or decaying plant material. The adults of some species are attracted to animals and humans. *Hippelates* spp. eye gnats are annoying and persistent feeders, causing localized irritation and transmitting pinkeye and perhaps infectious kerato-conjunctivitis to white-tailed deer, pronghorn, and moose. In Australia, species of *Batrachomia* cause myiasis in the body of frogs, and sometimes induce mortality.

Snipe Flies (Diptera: Rhagionidae)

These long-legged flies often are predatory, feeding on other insects. A few readily attack vertebrates, including humans and livestock as well as wildlife. The bite is painful and these insects can be quite annoying in certain locales. They have not been implicated in disease transmission.

Bees and Wasps (Hymenoptera: Various Families)

The colonial or nest-forming species are most likely to be a threat to animals. These insects sting as a defensive reaction when they are disturbed, injured, or if their nest is disturbed. Bee nests are regularly pillaged by wildlife, particularly bears, accounting for the stinging behavior of these insects. Mostly they are only an annoyance. However, the spread of the aggressive form of the honey bee, known as Africanized honey bee (*Apis mellifera scutellata*) in the western hemisphere has likely presented a new and significant threat to some forms of wildlife. These bees will inhabit tree cavities and animal burrows in the soil, and are well documented to attack pets and livestock when disturbed, so it is to be expected that unwary wildlife are also affected. In Puerto Rico, Africanized bees have usurped bird nesting boxes, greatly reducing the reproductive performance of pearly-eyed thrasher, *Margarops fuscatus*.

The colonial species of wasps, including yellow jackets (Fig. 10.28), paper wasps and hornets, may nest above-ground or belowground. When disturbed, they can become very aggressive, and sting repeatedly. Unlike with bees, wasps do not lose their sting when they attack, so they can sting and inject venom repeatedly.

Ants (Hymenoptera: Formicidae)

Many ants are predatory, and will attack small or helpless vertebrates such as nestling birds. The problem is especially acute with invasive species such as red imported fire ant, *Solenopsis invicta* (Fig. 10.29) which disrupts ecosystems everywhere it invades. These ants also sting, so they may harm larger animals even if they are unable to subdue them. Fire ants seem to affect bird and reptilian populations primarily by destroying the eggs and young. For example, a study in Texas found that predation caused a 92% reduction in the number of waterbird offspring when natural habitats were not treated for ant infestations. Fire ant

Fig. 10.28. A common yellow jacket, *Vespula squamosa* (Hymenoptera: Vespidae), also known as southern yellow jacket.

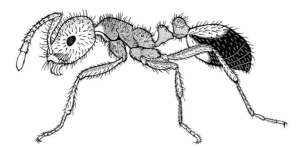

Fig. 10.29. The red imported fire ant, *Solenopsis invicta* (Hymenoptera: Formicidae).

predation on tortoise and reptile hatchlings is also documented, and many ground-nesting birds such as northern bobwhite, *Colinus virginianus*, are thought to be susceptible. This is discussed more fully in Chapter 13.

Dermestids (Coleoptera: Dermestidae)

Dermestids are small, oval beetles that normally are thought of as scavengers, feeding on hair, skin, and cadavers. However, they sometimes inhabit bird nests where they also attack live chicks. Skin lesions, emaciation and death caused by these insects have been reported from Florida and Georgia, USA, in such birds as roseate spoonbill, *Platalea ajaja*; snail kites, *Rostrhamus sociabilis*; wood storks, *Mycteria americana*; and white ibis, *Eudocimus albus*. Infestation levels of up to 18% of nests have been reported.

SUMMARY

• Unlike disease-causing organisms, ectoparasitic and endoparasitic insects are capable of independently attacking wildlife. Like disease-causing organisms, the parasitic insects attacking wildlife also tend to affect livestock and humans.
• The principal ectoparasitic arthropod pests of wildlife include mites, ticks, lice, kissing bugs, swallow and bat bugs, mosquitoes, black flies, biting midges, sand flies, horse and deer flies, tsetse flies, some muscid flies, louse flies, and fleas. Endoparasitic groups include some blow flies, some muscid flies, some flesh flies, and bot and warble flies.

• Though often considered to be of little importance unless the vertebrate host is troubled by other stresses, arthropod parasites can sometimes become very abundant and damaging, especially among animals that re-use nests or nest colonially.
• Mites differ from insects in having two body regions instead of three, by lacking true antennae, and by having four pairs of legs as adults instead of three (the nymphal stages also have four pairs, but the larval stage of mites typically has only three). Mites lack wings. Mites are common ectoparasites, but often cause little measurable injury to wildlife. Sarcoptic mange mite is a common problem for wildlife, and causes a disease called sarcoptic mange. The mites burrow into the skin of the animal, and also live on the surface, causing severe irritation and hair loss. Left untreated, reduced food intake and weight loss are common, followed by death. This ailment is most common among canids such as foxes, and cervids such as deer, but other animals can be affected.
• Ticks are really just large mites, but are very important ectoparasites of wildlife. Soft ticks lack a hard plate (scutum) covering the principal body region; hard ticks have the plate. Ticks may mature after feeding on one host (one-host ticks) or may require multiple hosts (two-host and three-host ticks). Ticks can be very abundant on wildlife, and can cause Tick Paralysis, and increasingly debilitating paralysis caused by toxins secreted by ticks. Ticks transmit many important diseases of wildlife, livestock, and humans.
• Sucking lice are blood-feeding insects whereas chewing lice feed on fur, feathers and skin. The presence of a few lice on wildlife is not considered to be a problem, but when they become numerous they can affect the health of animals. They do not often transmit disease to wildlife.
• Bugs can be important in disease transmission to humans. Kissing bugs transmit American Trypanosomiasis from many mammals to humans, but wildlife seem to be unaffected. Bat bugs and swallow bugs can develop to high levels of abundance on birds and bats, and their presence can reduce host fitness.
• Flies generally are the most important pests of wildlife. Mosquitoes, black flies, biting midges, sand flies, deer and horse flies, and tsetse flies feed externally, extracting blood from wildlife, and transmitting many infectious and parasitic diseases. Muscid flies and blow flies can extract blood from hosts, but they also can be endoparasites, feeding internally. Flesh flies, and bot

and warble flies, are mostly endoparasites, with the bot and warble flies especially well adapted for an internal feeding, and found widely among wildlife. Louse flies are common ectoparasites that affect wildlife mostly though their blood feeding activites, as they are not important disease vectors.

• Fleas are well adapted for living on their hosts, particularly mammals. They transmit diseases in addition to feeding on the blood of their hosts.

• Among insects that are occasional pests of wildlife are eye gnats, snipe flies, bees and wasps, ants, and dermestid beetles.

REFERENCES AND ADDITIONAL READING

Adler, P.H. (2008). Black flies (Diptera: Simuliidae). In Capinera, J.L. (ed.). *Encyclopedia of Entomology*, 2nd edn. Springer Science & Business Media B.V., Dordrecht, The Netherlands, pp. 525–529.

Adler, P.H. & McCreadie, J.W. (2002). Black flies (Simuliidae). In Mullen, G. & Durden, L. (eds.) *Medical and Veterinary Entomology*, pp. 185–202. Academic Press/Elsevier, Amsterdam, The Netherlands.

Allan, S.A. (2001). Ticks (Class Arachnida: Order Acarina). In Samuel, W.M., Pybus, M.J., & Kocan, A.A. (eds.). *Parasitic Disease of Wild Mammals*, 2nd edn., pp. 72–106. Iowa State University Press, Ames, Iowa, USA.

Allan, S.A. (2001). Biting flies (Class Insecta: Order Diptera). In Samuel, W.M., Pybus, M.J., and Kocan, A.A. (eds.). *Parasitic Disease of Wild Mammals*, 2nd edn., pp. 18–45. Iowa State University Press, Ames, Iowa, USA.

Arendt, W.J. (2000). Impact of nest predators, competitors, and ectoparasites on pearly-eyed thrashers, with comments on the potential implications for Puerto Rican parrot recovery. *Ornitologia Neotropical* **11**, 13–63.

Askew, R.R. (1971). *Parasitic Insects*. American Elsevier Publishing, New York, USA.

Bishopp, F.C. & Trembley, H.L. (1945). Distribution and hosts of certain North American ticks. *Journal of Parasitology* **31**, 1–54.

Bornstein, S., Mörner, T., & Samuel, W.M. (2001). *Sarcoptes scabiei* and sarcoptic mange. In Samuel, W.M., Pybus, M.J., & Kocan, A.A. (eds.). *Parasitic Disease of Wild Mammals*, 2nd edn., pp.107–119. Iowa State University Press, Ames, Iowa, USA.

Brenner, R.R. & Stoka, A.M. (eds.) (1988). *Chagas Disease Vectors*, vols. 1–3. CRC Press, Boca Raton, Florida, USA.

Broughton, R.K., Atwell, J.W., & Schoech, S.J. (2006). An introduced generalist parasite, the sticktight flea (*Echidnophaga gallinacea*), and its pathology to the threatened Florida scrub-jay (*Aphelocoma coerulescens*). *Journal of Parasitology* **92**, 941–948.

Catts, E.P. (1982). Biology of New World bot flies: Cuterebridae. *Annual Review of Entomology* **27**, 313–338.

Chapman B.R. & George, J.E. (1991). The effects of ectoparasites on cliff swallow growth and survival. In Loye, J.E. & Zuk, M. (eds.). *Bird–Parasite Interactions*, pp. 69–92. Oxford University Press, Oxford, UK.

Clayton, D.H., Adams, R.J. & Bush, S.E. (2009). Phthiraptera, the chewing lice. In Atkinson, C.T., Thomas, N.J., & Hunter, D.B. (eds.). *Parasitic Diseases of Wild Birds*, pp. 515–526. Wiley-Blackwell, Ames, Iowa, USA.

Clements, A.N. (1992). *The Biology of Mosquitoes*, vols. **1–2**. Chapman and Hall, London, UK.

Colwell, D.D. (2001). Bot flies and warble flies (Order Diptera: Family Oestridae). In Samuel, W.M., Pybus, M.J., & Kocan, A.A. (eds.). *Parasitic Disease of Wild Mammals*, 2nd edn., pp. 46–71. Iowa State University Press, Ames, Iowa, USA.

Colwell, D.D., Gray, D. Morton, K., & Pybus, M. (2008). Nasal bots and lice from white-tailed deer in southern Alberta, Canada. *Journal of Wildlife Diseases* **44**, 687–692.

Currie, D.C. & Hunter, D.B. (2009). Black flies (Diptera: Simuliidae). In Atkinson, C.T., Thomas, N.J., & Hunter, D.B. (eds.). *Parasitic Diseases of Wild Birds*, pp. 537–545. Wiley-Blackwell, Ames, Iowa, USA.

Dabert, J. & Mironov, S.V. (1999). Origin and evolution of feather mites (Astigmata). *Experimental and Applied Acarology* **23**, 437–454.

Dame, D.A. & Jordan, A.M. (2008). Tsetse flies, *Glossina* spp. Diptera: Glossinidae). In Capinera, J.L. (ed.). *Encyclopedia of Entomology*, 2nd edn., pp. 3849–3953. Springer Science & Business Media B.V., Dordrecht, The Netherlands.

Durden, L.A. & Musser, G.G. (1994). The sucking lice (Insecta: Anoplura) of the world: a taxonomic checklist with records of mammalian hosts and geographical distributions. *Bulletin of the American Museum of Natural History*. No. 218.

Durden, L.A. & Musser, G.G. (1994). The mammalian hosts of the sucking lice (Anoplura) of the world: a host-parasite list. *Bulletin of the Society for Vector Ecology* **19**, 130–168.

Fessl, B., Sinclair, B.J., & Kleindorfer, S. (2006). The life-cycle of *Philornis downsi* (Diptera: Muscidae) parasitizing Darwin's finches and its impacts on nestling survival. *Parasitology* **133**, 739–747.

Foil, L.D. (1989). Tabanids as vectors of disease agents. *Parasitology Today* **5**, 88–96.

Foil, L.D. & Hogsette, J.A. (1994). Biology and control of tabanids, stable flies and horn flies. *Revue Scientifique et Technique*. Office International des Epizooites, Paris **13**, 1125–1158.

Forester, D.J. (1992). *Parasites and Diseases of Wild Mammals in Florida*. University Press of Florida, Gainesville, Florida, USA.

Forrester, D.J. & Spalding, M.G. (2003). *Parasites and Diseases of Wild Birds in Florida*. University Press of Florida, Gainesville, Florida, USA. 1132 pp.

Hagemoen, R.I.M. & Reimers, E. (20020. Reindeer summer activity pattern in relation to weather and insect harassment. *Journal of Animal Ecology* **71**, 883–892.

Hansell, M. (2000). *Bird Nests and Construction Behavior.* Cambridge University Press, Cambridge, UK.

Hubbard, C.A. (1947). *The Fleas of Western North America.* Iowa State University Press, Ames, Iowa, USA.

Kettle, D.S. (1995). *Medical and Veterinary Entomology,* 2nd edn. CABI International, Wallingford, UK.

Klassen, W. & Curtis, C.F. (2005). History of the sterile insect technique. In Dyck, V.A., Hendrichs, J., & Robinson, A.S. (eds.). *Sterile Insect Technique. Principles and Practice in Area-Wide Integrated Pest Management,* pp. 3–36. Springer/Verlag, New York.

Krinsky, W.L. (2002). Tsetse flies (Glossinidae). In Mullen, G., and Durden, L. (eds.). *Medical and Veterinary Entomology,* pp. 303–316. Academic Press/Elsevier, Amsterdam, The Netherlands.

Leek, S.G.A. (1998). *Tsetse Biology and Ecology: Their Role in the Epidemiology and Control of Trypanosomiasis.* CABI, New York, USA.

Little, S.E. (2009). Myiasis in wild birds. In Atkinson, C.T., Thomas, N.J., & Hunter, D.B. (eds.). *Parasitic Diseases of Wild Birds,* pp. 546–556. Wiley-Blackwell, Ames, Iowa, USA.

Maa, T.C. & Peterson, B.V. (1987). Hippoboscidae. In McAlpine, J.F. (ed.). *Manual of Nearctic Diptera.* Research Branch, Agriculture Canada, Monograph **28**, Vol. 2, pp. 1271–1281.

Marshall, A.G. (1981). *The Ecology of Ectoparasitic Insects.* Academic Press, London, UK.

McLean, R.G. & Ubico, S.R. (2007). Arboviruses in birds. In Thomas, N.J., Hunter, D.B., & Atkinson, C.T. (eds.). *Infectious Diseases of Wild Birds,* pp. 17–62. Blackwell Publishing, Ames, Iowa, USA.

Meyer, N.L. (1994). *History of the Mexico-United States Screwworm Eradication Program.* Vantage Press, New York, USA. 367 pp.

Moyer, B.R. & Clayton, D.H. (2004). Avian defences against ectoparasites. In van Emden, H. & Rothschild, M. (eds.) *Insect and Bird Interactions,* pp. 241–257. Intercept, Andover, Hampshire, UK.

Mullen, G.R. (2002). Biting midges (Ceratopogonidae). In Mullen, G. & Durden, L. (eds.) *Medical and Veterinary Entomology,* pp. 163–183. Academic Press/Elsevier, Amsterdam, The Netherlands.

Mullen, G.R. & O'Connor, B.M. (2002). Mites (Acari). In Mullen, G. & Durden, L. (eds.) *Medical and Veterinary Entomology,* pp.449–516. Academic Press/Elsevier, Amsterdam, The Netherlands.

Mullen, G. & Durden, L. (2009). *Medical and Veterinary Entomology,* 2nd edn. Academic Press/Elsevier, Amsterdam, The Netherlands.

Norris, K.R. (1965). The bionomics of blow flies. *Annual Review of Entomology* **10**, 47–68.

Olsen, W.O. (1986). *Animal Parasites. Their Life Cycles and Ecology.* Dover Publications, New York, New York, USA (reprint of 3rd edn, 1974).

Pap, P.L., Szép, T., Tökölyi, J., & Piper, S. (2006). Habitat preference, escape behavior, and cues used by feather mites to avoid molting of wing feathers. *Behavior Ecology* **17**, 277–284.

Pence, D.B. (2009) Ascariasis. In Atkinson, C.T., Thomas, N.J., & Hunter, D.B. (eds.). *Parasitic Diseases of Wild Birds,* pp. 527–536. Wiley-Blackwell, Ames, Iowa, USA.

Petersen, J.J. & Greene, G.L. (1989). *Current Status of Stable Fly (Diptera: Muscidae) Research.* Miscellaneous Publication of the Entomological Society of America, **74**. 53 pp.

Price, M.A. & Graham, O.H. (1997). *Chewing and Sucking Lice as Parasites of Mammals and Birds.* United States Department of Agriculture, Agricultural Research Service. Technical Bulletin No. 1849.

Price, R.D., Hellenthal, R.A., Palma, R.L., Johnson, K.P., & Clayton, D.H. (2003). *The Chewing Lice: World Checklist and Biological Overview.* Illinois Natural History Survey Special Publication **24**.

Proctor, H.C. (2003). Feather mites (Acari: Astigmata): ecology, behavior and evolution. *Annual Review of Entomology* **48**, 185–209.

Proctor, H. & Owens, I. (2000). Mites and birds: diversity, parasitism and coevolution. *Trends in Ecology and Evolution* **15**, 358–364.

Rogers, C.A., Robertson, R.J., & Stutchbury, B.J. (1991). Patterns and effects of parasitism by *Protocalliphora sialia* on tree swallow nestlings. In Loye, J.E. & Zuk, M. (eds.) *Bird–Parasite Interactions,* pp. 123–139. Oxford University Press, Oxford, UK.

Rutledge, L.C. & Gupta, R.K. (2002). Moth flies and sand flies (Psychodidae). In Mullen, G. & Durden, L. (eds.) *Medical and Veterinary Entomology,* pp. 147–161. Academic Press/Elsevier, Amsterdam, The Netherlands.

Slansky, F. (2007). Insect/mammal associations: effects of cuterebrid bot fly parasites on their hosts. *Annual Review of Entomology* **52**, 17–37.

Smith, K.G.V. (ed.) (1973). *Insects and Other Arthropods of Medical Importance.* British Museum of Natural History, London, UK.

Soler Cruz, M.D. (2008). Myiasis. In Capinera, J.L. (ed.). *Encyclopedia of Entomology,* 2nd edn., pp. 2517–2527. Springer Science & Business Media B.V., Dordrecht, The Netherlands.

Soulsbury, C.D., Iossa, G., Baker, P.J., Cole, N.C, Funk S.M., & Harris, S. (2007). The impact of sarcoptic mange *Sarcoptes scabiei* on the British *Vulpes vulpes* population. *Mammal Review* **37**, 278–296.

Spicer, G.S. (1987). Prevalence and host–parasite list of some nasal mites from birds (Acarina: Rhinonyssidae, Speleognathidae). *Journal of Parasitology* **73**, 259–264.

Toupin, B., Huot, J., & Manseau, M. (1996). Effect of insect harassment on the behavior of the Rivière George caribou. *Arctic* **49**, 37–382.

Tripet, F. & Richner, H. (1997). The coevolutionary potential of a 'generalist' parasite, the hen flea *Ceratophyllus gallinae*. *Parasitology* **115**, 419–427.

Uspensky, I. (2008). Argasid (soft) ticks (Acari: Ixodida: Argasidae). In Capinera, J.L. (ed.). *Encyclopedia of Entomology*, pp. 283–288. Springer Science & Business Media B.V., Dordrecht, The Netherlands.

Uspensky, I. (2008). Taiga tick, *Ixodes persulcatus* (Acari: Ixodida: Argasidae). In Capinera, J.L. (ed.). *Encyclopedia of Entomology*, 2nd edn., pp. 3687–3690. Springer Science & Business Media B.V., Dordrecht, The Netherlands.

Wheeler, C.M., Douglas, J.R., & Evans, F.C. (1941). The role of the burrowing owl and the sticktight flea in the spread of plague. *Science* **94**, 560–561.

Weladji, R.B. Holand, Ø., & Almøy, T. (2003). Use of climatic data to assess the effect of insect harassment on the autumn weight of reindeer (*Rangifer tarandus*) calves. *Journal of Zoology, London* **260**, 79–85.

Wobeser, G.L. (1997). *Disease of Wild Waterfowl*, 2nd edn. Plenum Press, New York, USA.

Wobeser, G.L. (2006). *Essentials of Disease in Wild Animals*. Blackwell Publishing, Oxford, UK.

Wood, D.M. (1987). Oestridae. In McAlpine, J.F. (ed.) *Manual of Nearctic Diptera*. Research Branch, Agriculture Canada, Monograph **28**, Vol. 2, pp. 1147–1158.

PEST MANAGEMENT AND ITS EFFECTS ON WILDLIFE

Chapter 11

PESTICIDES AND THEIR EFFECTS ON WILDLIFE

In 1962, the biologist and nature writer Rachel Carson authored the landmark book *Silent Spring*, which would make environmental concerns an issue for unprecedented numbers of Americans, change the policies regarding pesticide regulation and use, and eventually lead to creation of the US Environmental Protection Agency. A central theme of *Silent Spring* was the myriad but unforeseen effects by highly toxic and persistent pesticides on wildlife, on other aspects of the environment, and on humans. One of the poignant lines she authored was: '*The question is whether any civilization can wage relentless war on life without destroying itself, and without losing the right to be called civilized*'. Carson's prose was rich with passion for the environment, but she also provided detailed, relevant examples and data. She successfully attacked the dominance of technology and evoked ethical issues such as destruction of nontarget animals by pesticides. In particular, she labeled many seemingly safe insecticides and herbicides as 'biocides', calling into question decades of assurances by the pesticide industry about the safety of their products. Rachel Carson died in 1964, but left an important intellectual legacy that continues to be refined and defended. Some have suggested that she started the entire environmental movement in the USA, and she certainly influenced prominent entomologists who would go on to advance the concept of integrated pest management (IPM). Although the pesticide use situation has improved since Rachel Carson's era, it is not without problems.

Insects and Wildlife: Arthropods and Their Relationships with Wild Vertebrate Animals, 1st edition. By J.L. Capinera. Published 2010 by Blackwell Publishing.

In this section we will review the characteristics of pesticides (particularly insecticides), the effects of pesticides on wildlife, and the alternative pest suppression technologies.

Even though it has been many years since Rachel Carson wrote about the problems of using insecticides, and considerable detailed knowledge has accumulated that documents the problems with using pesticides exclusively to solve insect problems, in many cases insecticides are the first or only method of management considered (see Fig. 11.1). This is an unfortunate reality, and may be due to several factors, including:

• there are no known alternatives to pesticides;
• pesticide application is the least expensive or most convenient technique;
• pesticides are the most reliable technique, serving to minimize risk and pest-related problems;
• due to lack of planning (failure to act in a timely manner to prevent a problem from developing), certain alternatives to pesticides cannot be used.

As we will see in this section, pesticides can cause some serious problems from the perspective of wildlife conservation, so usually it is best to eliminate or minimize pesticide use. However, in most cases it is hard to argue that pesticides are not the most convenient, economic, effective, and reliable technique for pest suppression once the pest situation is out-of-hand (i.e., insects or their damage are so abundant as to cause nuisance or damage). The challenge, then, is to plan in advance for potential problems and to act in advance of the occurrence of pest abundance or damage and to prevent pest-related problems from occurring by using nonpesticide approaches. Failing that, acting early in an upswing of insect abundance or damage in a way that curtails damage and minimizes pesticide use is preferable to extensive and intensive pesticide use after pests

Fig. 11.1. The popularity of insecticides is due not only to the effectiveness of these chemicals for killing insects, but also to the ease and flexibility of use. Among other uses, insecticides can be applied (a) by high-pressure sprayers (Trees can be 25–40 m high, necessitating special high-pressure equipment to treat the foliage in the upper portions of the trees); (b) directly to livestock (The structure in the foreground is called a 'cattle rub'. The burlap wrapped around the rope or chain attached to the pole is impregnated with insecticide. Cattle naturally seek opportunities the scratch their backs, and in rubbing against the treated burlap they transfer insecticide to their bodies); (c) by aircraft (Airplanes and helicopters provide the ability to apply insecticide to large acreages quickly, and without concern for wet soil or compaction. Here you see a large airplane applying insecticide for grasshopper control because vast acreages are often treated for such insect problems. For vegetable and field crop applications, smaller airplanes or helicopters usually are used); (d) with tractor-mounted spray equipment for vegetable and field crop problems (The challenge here often is to attain good coverage of the foliage, so you see insecticide being applied from the sides of the plant as well as from above, in an attempt to control insects on the underside of the leaves) (Fig. 11.1d from USDA–ARS).

have become abundant. Before we discuss using alternatives to pesticides, it is appropriate to discuss pesticides and the problems they cause.

PESTICIDES

Pesticides are an important class of toxicants that, though designed to kill pests, also can affect wildlife. **Pesticides** are a form of poison, or toxic material, used to suppress pests such as insects, plant pathogens, weeds, or vertebrates that cause economic damage or are a health risk or nuisance. The principal types of pesticides are herbicides (used to kill weeds), **insecticides** (for insects), **fungicides** (for fungi), and **rodenticides** (for rats and mice). Other types of pesticides include **acaricides** (for mites and ticks), **bactericides** (for bacteria), **nematicides** (for nematodes), **molluscicides** (for snails and slugs), **avicides** (for pest birds), **piscicides** (for pest fish), **algaecides** (for algae), and other specialty products, including poisons used to kill predators such as coyotes, *Canis latrans*.

In most cases that involve wildlife, the pesticide of concern is an insecticide. However, some pesticides are directed at vertebrates such as prairie dogs or other rodents, or predators such as coyotes. The ability of a pesticide to cause injury or death is called **toxicity**. The active ingredient of most insecticides is called the **toxicant**, and it typically consists of only a small proportion of the formulation that is purchased and mixed (usually with water) for application to the environment to suppress insects (arthropods). Toxicity is dose related; high doses are more toxic. Toxicity is not quite the same as hazard. **Hazard** is the likelihood that a toxicant will cause injury to a non-target organism, and is a function of not only toxicity but also dose of the toxicant, length of exposure, and method of application. Very toxic products can be applied safely if efforts are made to minimize hazard, whereas products with relatively low toxicity can be hazardous if used inappropriately. For example, if insecticide is applied to a flowering crop during the day, high mortality among pollinating insects is probable. However, if applied at dusk, an insecticide with higher toxicity would likely cause much less mortality to pollinating insects because the product would be partly degraded by light and foliar pH before pollinators became fully active the following day.

Toxicants vary greatly in their toxicity, and it is always a good idea to be familiar with the toxicity level of any pesticide that you handle or apply. Toxicants are sometimes grouped into categories, with category I pesticides being most toxic (to the applicator) and category IV being least toxic. Category I chemicals are quite hazardous and should be avoided by anyone without specific training on handling these materials. To purchase category I chemicals, a pesticide applicator license is required. Category II products, in contrast, often consist of readily available materials that farmers and homeowners and others without special knowledge or license can purchase from agricultural supply centers, garden centers, and hardware or discount stores. Category II chemicals, though readily accessible, should be treated with respect because they are capable of harming humans and wildlife. Category III and IV pesticides are less hazardous, but in most cases they also can cause injury and death if misused or misapplied.

Another way to assess toxicity is to obtain the LD_{50} value. The **LD_{50} value** is the dose of active ingredient, expressed in milligrams (mg) of toxicant per kilogram (kg) of test animal, that will kill 50% of the test subjects. Products with a low LD_{50} value are more toxic than those with a high value. Expressing toxicity in this manner adjusts for different sizes (weights) of subjects. This is necessary because a small organism will be more easily killed by a certain dose of toxicant than a large organism that is exposed to the same dose. The route of exposure is also important, with oral (ingestion) being a much more hazardous route of exposure than dermal (skin) exposure. For pesticides that might be inhaled, a slightly different measure of toxicity is used, the **LC_{50} value**; this represents the concentration of toxicant (in milligrams per liter) that induces mortality in 50% of the test subjects.

It may be difficult to obtain detailed toxicity data for some pesticides. One good source of toxicity, handling, and disposal information is the **MSDS (Material Safety Data Sheet)**. The MSDS accompanies all extremely hazardous (restricted use) materials, but also is generally accessible from internet (WWW) databases for all toxicants and hazardous materials. Also, toxicity levels for most products are indicated by signal words found on the label. The **signal word** indicates the degree of hazard, though unfortunately it does not correspond directly to the aforementioned toxicity categories. The signal words are DANGER POISON for toxicity category I, WARNING for toxicity category II, and CAUTION for both categories III and IV (see Table 11.1).

Table 11.1. Toxicity categories of pesticides in relation to hazards.

Hazard indicator	Toxicity category I	Toxicity category II	Toxicity category III	Toxicity category IV
Signal word	Danger Poison	Warning	Caution	Caution
Oral LD_{50}	≤50 mg/kg	50–500 mg/kg	500–5000 mg/kg	>5000 mg/kg
Dermal LD_{50}	≤200 mg/kg	200–2000 mg/kg	2000–20,000 mg/kg	>20,000 mg/kg
Inhalation LC_{50} (dust or mist)	≤2 mg/l	2–20 mg/l	20–200 mg/l	>200 mg/l
Eye effects	irreversible corneal opacity at 7 days	corneal opacity reversible within 7 days, or irritation persisting for 7 days	no corneal opacity, or irritation reversible within 7 days	no irritation
Skin irritation	severe irritation or damage at 72 hours	moderate irritation at 72 hours	mild or slight irritation at 72 hours	no irritation

Toxicity is dose dependent. Low levels of pesticide may be eliminated by an animal from its body without suffering harm, or they may have no measurable effect on the animal. High levels of toxin exposure, of course, are more likely to cause injury. Interestingly, most people fail to appreciate this dose phenomenon, instead either treating all pesticides callously, as if they were not a risk, or overreacting to a perceived risk, as if all pesticides and all exposure rates were extremely hazardous. It is important to treat pesticides as if they were prescription drugs: relatively safe if used according to directions, and hazardous if misapplied. As with prescription drugs, there is some variation among individuals or among species in terms of susceptibility or adverse reaction.

Pesticides vary in toxicity to wildlife. The short-term risk of poisoning, or **acute toxicity**, can be quite different than the risk resulting from prolonged exposure, or **chronic toxicity**. Acute toxicity is mortality resulting from dermal contact, inhalation or ingestion, and typically occurs soon after exposure to recently treated crops or forests. Chronic toxicity often is a more subtle disruption resulting from alteration of biological processes and is manifested by changes in hormone levels, immune responses, reproductive changes, and behavioral modifications. Chronic toxicity typically occurs only after prolonged exposure to low levels of toxicant.

In general, herbicides and fungicides are less toxic to wildlife than insecticides and rodenticides. However, some herbicides and fungicides, while not displaying high levels of acute toxicity, are suspected of being toxic when they move into the water supply or food chain and are ingested over long periods of time. Sometimes there are interactions resulting in greater toxicity of insecticides when other pesticides, including herbicides, are present.

The other components of pesticide formulations, other than the toxicant, are various **adjuvants**, or additives, that enhance the utility or effectiveness of the product by increasing dispersion of the toxicant in the formulation, retention on the target substrate, or in reducing drift. There may also be **inert ingredients** (components that have no active role in effectiveness of the pesticide) present that dilute the other components of the formulation and make it easier to measure and mix the product. Some products are not applied in a water-based spray, especially the pesticides aimed at vertebrates. Often, these are applied as baits, which are a solid material, but even these usually have been pretreated with a liquid formulation of pesticide during the preparation of the bait. The other common nonliquid formulations are granules and dusts. Granules are small pellets of inert clay that are sprayed with toxicant-containing liquid to give them the desired level of toxicant. Dusts are similar, but much smaller in size. Each product type, like the various liquid formulations, has advantages and disadvantages. For example, coverage and penetration of dense vegetation are superior with dusts, but granules are less likely to suffer from drift problems.

Exposure of animals to pesticides usually occurs when animals ingest the product along with their normal food or water, though in some cases they are sprayed directly, with the product being absorbed through the skin. Pesticides that cause toxicity only after they have been eaten are called **stomach poisons**. Those that can affect the insect by external exposure (as by walking on the pesticide residue) are called **contact poisons**. Most modern insecticides are both contact and stomach poisons. Some water-soluble products can be absorbed into the host plant (or animal) and be moved around, contacting the insect wherever it feeds. When applied to plants, these **systemic** pesticides normally are taken up into the plant through the roots or foliar tissue, and move mostly upwards (to the actively growing regions). Systemic pesticides can be translocated in the water-conducting (xylem) and food-conducting (phloem) systems of a plant, especially the xylem. Insects that feed on vascular tissues such as xylem or phloem (insects with piercing-sucking mouthparts) tend to concentrate the insecticide and are especially likely to be poisoned. However, some pesticides move only short distances, as from one side of a leaf to another; these are said to be **translaminar**. Inhalation of toxicants through the insect's ventilatory system is also possible, though not especially common, and such products are called **fumigants**.

Traditionally, pesticide producers have sought fast-acting products because by causing quick mortality damage to the host is minimized, and (not inconsequentially) this usually makes the person who has paid for the product quite happy. However, there are times when fact-acting products are not ideal. This is especially true with social insect species such as ants and termites. It is difficult to contact all the insects in a colony due to specialization of function; some insects may remain deep in the soil, for example, and never venture out to feed on bait or be susceptible to a spray. For such pests it is better to have a slow-acting product that the foragers will eat or take back to the colony, and which will be spread around to nearly all the insects in the colony. Transfer of insecticide can occur when insects have come into contact with insecticide and pass the insecticide along to nest-mates in the course of bumping into one another, grooming, passing along food, or other minor forms of contact. Transfer also can occur when insects cannibalize their dead siblings, feed on dead nest-mates, or feed on feces or regurgitated material. When insecticides are passed from insect to insect, this is sometimes called **horizontal transfer**.

Some products are subject to bioaccumulation. **Bioaccumulation** results when, during the course of an animal's life, the rate of intake of a chemical exceeds the rate of elimination. With persistent pesticides, a long-lived animal can repeatedly acquire small, non-toxic doses, eventually resulting in the accumulation of high levels of toxicant. Bioaccumulation may result from biomagnification or bioconcentration. **Biomagnification** is an increase in the concentration of a chemical at each trophic level of a food chain. For example, persistent chemicals in water may accumulate in algae, then accumulate in algae-eating insects, then accumulate in insect-eating fish, and finally accumulate in fish-eating vertebrates. The net result of the chemical being passed along the food chain can be a very high concentration in the animal at the top of the food pyramid. **Bioconcentration** is also an increase in chemical levels, but occurs independent of trophic levels. Thus, long-lived wildlife can accumulate high levels of pesticides simply by consuming low levels over a long period of time if they are persistent and are not excreted.

Exposure of wildlife to pesticides occurs in various ways. Pesticides usually are localized in distribution because they are applied to a specific location; this type of environmental pollution is called **point source pollution**. Occasionally, extremely long-lived pesticides such as DDT, heptachlor, or chlordane, or persistent environmental pollutants such as mercury become rather pervasive in the environment, residing in the soil or animal tissues for many years and becoming widely distributed; this more general type of pollutant distribution is called **diffuse pollution**.

INSECTICIDE MODE OF ACTION

Pesticides have various ways of killing pests. The manner of causing mortality is called the **mode of action**. The appropriate mode of action depends on the nature of the pest organism and the physiological system within the pest that has been targeted for interference or destruction. Some pesticides are quite selective; some insect growth regulators, for example, affect insects as they molt from stage to stage by interfering with their developmental processes (usually hormones or molting). Vertebrate animals and plants lack many of the developmental systems or physiological processes found in insects, so they are essentially immune to such growth regulator-based insecticides. Thus, some insecticides are considered to be quite selective, and not too hazardous to the environment. They are not perfectly benign, however, because some close relatives of insects such as crustaceans (e.g., shrimp, crabs) can be affected and because beneficial, non-target insects may also be present. The challenge facing pest management specialists who want to use growth regulators (and other toxicants, of course) safely is to ensure that the growth regulators are not applied excessively to aquatic environments where such non-target organisms can be affected. Also, because not all insects are considered to be pests (we want to preserve the valuable predatory, parasitic, and pollinator arthropods) we seek to apply even growth regulators in a manner that minimizes contact with the beneficial arthropods. For example, application of insecticides just before nightfall is a common technique used to minimize exposure of pesticides to honeybees, as they are not active at night. Hopefully, the pesticide will kill any target pests during the night, with the pesticide degraded sufficiently by the moisture and sunlight of the evening and morning periods so that when honeybees become active on the following day there is minimal risk of being affected.

Generally, insecticides and fungicides do not affect plants. Nor do herbicides generally affect insects and plant disease, or fungicides affect insects. There are exceptions, of course, as there are some plants or even varieties of plants that are sensitive to certain chemicals, including the adjuvants mixed with the toxicants, making them unsafe to use in certain situations. A few products are **biocides**, chemicals designed to kill all life. Why would anyone want to kill all life? Well, biocides can be useful when you are trying to be absolutely sure that products such as furniture or grain that are being imported are free of exotic or dangerous organisms. Most biocides are fumigants that are applied to structures to eliminate termite infestations, or to structures containing stored grain. The principal out-of-doors application of biocides is soil fumigation, wherein fumigant is injected into the soil to kill insects, nematodes, plant disease, and weed seeds that threaten a crop. So even though the fumigation occurs out-of-doors, the toxicant is largely retained in the soil, and the toxicity is localized. Needless to say, biocides can be a risk to wildlife if animals are found in these situations, though they typically are not, and the toxicity of the fumigant dissipates quickly. Thus, the hazard to wild-life comes principally from insecticides, or other chemicals directed at animals (nematodes, molluscs, rodenticides), but not often from pesticides directed at plants or plant disease.

The mode of action of most insecticides can be described as **nerve poisons**, because they interfere with normal functioning of the nervous system. The specific mode of nerve inhibition varies among the groups of insecticides (see Table 11.2), but among the organophosphates and carbamates the enzyme **acetylcholine esterase** is inhibited, resulting in excessive neuroexcitation and eventually in salivation, tremors, convulsions, diarrhea, and paralysis in the affected animal. Another common form of toxicity is caused by disruption of nerve axon function due to influx of sodium or chloride ions. As with acetylcholinesterase inhibitors, most insecticides causing disruption of nerve axon function also cause overexcitation. The nerves of both vertebrate and invertebrate animals function similarly, though not identically, so insecticides can pose a risk to wildlife if they are applied in an environment where wildlife occur. Some of the newer classes of nerve toxins take advantage of the differences that exist between the nerves of vertebrates and insects, providing an important measure of safety.

Within the several classes of insecticides, several aspects of nerve transmission are affected, and nearly all classes of insecticides except the insect growth regulators target the nervous system. However, despite their apparent similarities, insecticides differ greatly in persistence, with newer products usually being less persistent and much less likely to accumulate in animals and ecosystems. Also, sometimes the individual products are more or less hazardous to categories of wildlife such as fish or birds, so it is worth investigating the environmental hazards in advance of applying insecticides.

The persistent organochlorines, which were responsible for some serious wildlife problems in the 1950s and 1960s, have been largely eliminated from use (Fig. 11.2). The organophosphates and carbamates, which lack the persistence of the organochlorines but often cause acute poisoning in nontarget organisms, are becoming less common, though they remain widely used. The pyrethroid insecticides are much less hazardous to most wildlife, though fish are quite susceptible. Also, although pyrethroids cause little direct toxicity to vertebrate animals, they can be so effective at killing insects as to potentially induce mortality (indirect toxicity) by depriving wildlife of food. With the newest

Table 11.2. Some major groups of insecticides and their characteristics. Note that nearly all groups affect the insect nervous system.

Class	Examples	Mode of action	Other important characteristics
Insecticides that affect the nervous system			
Inorganic insecticides	boric acid, borates, cryolite, silicates	various, often stomach poisons	persistent, usually low toxicity to mammals
Organochlorines (DDT group)	DDT, dicofol, methoxychlor	prevent nerve repolarization	persistent, metabolically stable, lipophilic, variable toxicity to mammals
Organochlorines (cyclodienes)	dieldrin, chlordane, heptachlor, lindane, endosulfan	GABA receptor antagonist in nerves	persistent, metabolically stable, lipophilic, low toxicity to mammals
Organophosphates	malathion, parathion, diazinon, chlorpyrifos	inhibit acetylcholine esterase in nerve synapses	variable toxicity, unstable, water soluble
Carbamates	carbaryl, methomyl, carbofuran, aldicarb	inhibit acetylcholine esterase in nerve synapses	variable toxicity, water soluble, biodegradable
Pyrethroids	permethrin, esfenvalerate, cypermethrin	prevent nerve repolarization	biodegradable, not very toxic to mammals, very toxic to fish
Nicotinoids (= chlornicotinyls and neonicotinoids)	imidacloprid, acetamiprid, thiamethoxam	activation of nicotinic acetylcholine receptors	low mammalian toxicity, but affect fish and birds
Spinosyns	spinosad	activation of nicotinic acetylcholine receptors	derived from soil bacterium, low toxicity to mammals and birds, moderately toxic to fish
Avermectins	ivermectin, abamectin	activate glutamate chloride receptor	used widely on animals for insect and nematode control but hazardous to mammals and fish
Phenylpyrazoles	fipronil	block GABA chloride receptor	low mammalian toxicity, but affect fish and birds
Insecticides that are insect growth regulators			
Juvenoids	methoprene, hydroprene, fenoxycarb	bind juvenile hormone receptors	keep insects in juvenile form, selective, low toxicity to mammals
Chitin synthesis inhibitors	diflubenzuron, hexaflumuron, buprofezin	inhibit chitin formation	interfere with molting, low toxicity to mammals, birds, and fish

classes of insecticides, such as nicotinoids and spinosyns, the hazards to wildlife are minimized, though indirect effects such as food deprivation remain an issue.

The toxicity of some insecticides is shown in Table 11.3. Note that the toxicity varies considerably among classes of insecticides. The organophosphates and carbamates contain some very toxic materials when they are compared to the pyrethroids. However, even the pyrethroids seem toxic when compared to the insect growth regulators and some of the so-called 'natural products' which are mostly botanical

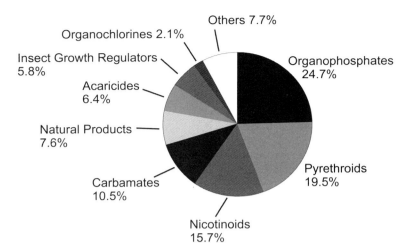

Fig. 11.2. Worldwide use of insecticides (%) by insecticide class in 2004 (adapted from Nauen 2006).

Table 11.3. The toxicity of some common insecticides (products with low values are more toxic).

Insecticide	Toxicity (mg/kg oral LD$_{50}$)	Insecticide	Toxicity (mg/kg oral LD$_{50}$)
Organochlorines		Propoxur	50
DDT	113	**Pyrethroids**	
Aldrin	38	Allethrin	860
Chlordane	250	Bifenthrin	55
Dieldrin	40	Cyfluthrin	500
Endrin	7	Cyhalothrin	56
Heptachlor	40	Cypermethrin	250
Lindane	88	Esfenvalerate	75
Methoxychlor	6,000	Permethrin	430
Toxaphene	49	**Avermectins**	
Organophosphates		Abamectin	11
Acephate	945	Emamectin	93
Chlorpyrifos	135	**Nicotinoids**	
Diazinon	1,250	Acetamiprid	146
Dichlorvos	50	Imidacloprid	424
Dimethoate	387	Thiamethoxam	1,563
Fenthion	250	**Insect growth regulators**	
Fonfos	8	Methoprene	34,600
Malathion	1,375	Hydroprene	5,500
Methyl parathion	3	Fenoxycarb	10,000
Monocrotophos	165	**'Natural products'**	
Parathion	2	*Bacillus thuringiensis*	>15,000
Phorate	1.6	Azadirachtin	5,000
Trichlorfon	250	Rotenone	132
Carbamates		Pyrethrin	200
Aldicarb	0.9	Nicotine	55
Carbaryl	500	Ryania	1,200
Carbofuran	8	Sabadilla	4,000
Methiocarb	20	Spinosad	5,000
Methomyl	17	Petroleum oil	15,000

insecticides but also include a toxin-containing microorganism (*Bacillus thuringiensis*) and a chemical product of soil-dwelling microbes (spinosyns). The toxicity of products within a class of insecticide can also vary. Comparing aldicarb to carbaryl, for example, shows that there is greater than a 500-fold difference in toxicity even within a single class of chemicals. Thus, it is important to know the toxicity of individual products, and to make no assumptions about insecticide safety based on the chemical class. It is also important to note that it is not safe to assume that natural products are inherently safer. *Bacillus thuringiensis* and petroleum oils are practically nontoxic to test mammals, whereas nicotine ranks as a fairly toxic compound.

PERSISTENCE OF INSECTICIDES

Persistence of insecticides is partly a function of the chemistry of the compounds. Organochlorine persistence typically is measured in years, whereas other classes of pesticides may only persist for days or weeks (see Table 11.4). Surely there is no good reason to have products that persist for years in cropping systems. However, for protection of wood structures, there are clear advantages to having wood protected from termites for decades. Within each class of pesticide, however, usually are found some pesticides that persist for months. Indeed, as can be seen from the half-life data on insecticide persistence in soil, there is considerable variation in persistence among different insecticides, even within a single class of insecticides, though persistence of a particular compound is fairly predictable and consistent. Persistence in soil is longer than on foliage.

Placement of the insecticide product also influences persistence. Insecticides applied to plant foliage, the soil surface, or to water are affected by **photodegradation**. Sunlight, and particularly the ultraviolet portion of the spectrum (300–400 nm is the range commonly encountered), causes the pesticide molecule to be broken down into smaller units, causing loss of insecticidal activity. The degradation process in sunlight is also called **photolysis**. Also, insecticides that are mixed into soil may be inactivated by **microbial degradation**, or by **adsorption** to clay minerals and organic particles. The water found in soil may cause water-soluble materials to be released from their bound state to soil, and to become activated and available to affect target organisms. Conversely, should there be

Table 11.4. The half-life (length of time until loss of 50% of toxicity) of some common insecticides applied to soil.

Insecticide	Half-life
Organochlorines	
DDT	3–10 years
Aldrin	1–4 years
Chlordane	2–4 years
Dieldrin	1–7 years
Endrin	4–8 years
Heptachlor	7–12 years
Lindane	20–50 days
Methoxychlor	120 days
Toxaphene	10 years
Organophosphates	
Acephate	3 days
Chlorpyrifos	36–46 days
Diazinon	11–21 days
Dichlorvos	17 days
Dimethoate	2–4 days
Malathion	4–6 days
Methyl parathion	45 days
Monocrotophos	1–5 days
Parathion	180 days
Phorate	2 days
Trichlorfon	140 days
Carbamates	
Aldicarb	30 days
Carbaryl	7–28 days
Carbofuran	26–110 days
Methiocarb	15 days
Propoxur	44–59 days
Pyrethroids	
Bifenthrin	97–156 days
Cyfluthrin	73–95 days
Cyhalothrin	7–21 days
Cypermethrin	4–12 days
Esfenvalerate	75 days
Permethrin	18–23 days
Avermectins	
Abamectin	1 day
Emamectin	1 day
Nicotinoids	
Acetamiprid	1–2 days
Imidacloprid	8–48 days
Thiamethoxam	11–26 days
Insect Growth Regulators	
Methoprene	10 days
Hydroprene	3–10 days
Fenoxycarb	51–75 days

too much water the insecticides may be leached out of the root zone or soil surface regions and become ineffective. Soil pH is also an important determinant of persistence, and alkaline soils generally are deleterious to insecticide persistence. **Hydrolysis** is the principal means of degradation in soil. Hydrolysis is the addition of water to split two chemical bonds. Also, **high temperatures** can lead to rapid degradation, as can high populations of various soil-inhabiting bacteria. Many bacteria can use the pesticides as carbon sources to facilitate growth, and microbe populations can be stimulated to attain high densities when pesticides are applied frequently. Thus, **enhanced biodegradation** occurs in soils containing high densities of pesticide-degrading bacteria, which typically develop following repeated use of pesticides.

Residual activity of organochlorines is usually measured directly, via assessment of pesticide concentrations in animals and substrate. Organophosphates are among the most potent toxicants, causing high levels of acute toxicity, but they are not very persistent, so bioconcentration or biomagnification is unlikely. Therefore, assessment of organophosphate levels in the environment is typically measured indirectly, via assessment of acetylcholinesterase levels in the blood or brains of animals.

ACUTE EFFECTS OF INSECTICIDES

All wildlife can be poisoned by excessive exposure to certain insecticides, but the problem is particularly pronounced with birds. A simple example is the occurrence of seed-feeding flies, *Delia* spp. (Diptera: Anthomyiidae), which damage crop seeds planted in the spring. To prevent damage by such insects while the seed is germinating, coatings containing insecticides are commonly applied to seeds before they are planted. Seed treatment with insecticides (and often fungicides) not only protects the germinating seed, but if the insecticide acts systemically it may also impart protection to young plants, particularly from piercing-sucking insects such as aphids (Hemiptera: Aphididae). Unfortunately, granivorous (seed-eating) birds will often feed on seeds that have been recently planted, and thereby ingest a lethal dose of insecticide. Also, because sand is sometimes consumed by birds to aid in grinding up seeds, birds sometimes feed on granular insecticide, again resulting in bird mortality. Finally, application of liquid insecticides sometimes results in a lethal dose

of insecticide being applied directly to wildlife. Though this seems unlikely, when aircraft are used to apply insecticides an extensive land area is treated quickly, and wildlife may not have adequate time to escape. Not only are crop fields treated, but often adjacent border areas (hedge rows, fence rows, irrigation ditches, road margins) are treated deliberately or inadvertently. In the case of nestling birds, there is no opportunity to avoid exposure, even from tractor-mounted sprayers. The principal locations where insecticide poisonings of birds occurs in the USA, and the seasonality of poisoning, is shown in Fig. 11.3. The high level of poisoning early in the year is due to the early onset of spring in the southern USA.

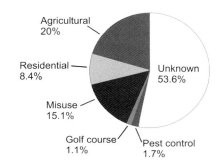

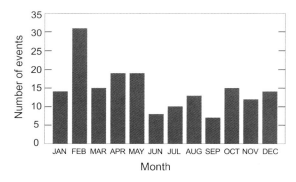

Fig. 11.3. Avian mortality due to application of organophosphate and carbamate insecticides in the USA: (above) where avian mortality (%) is most likely to occur, and (below) seasonality of mortality events. Both data sets are based on data from the U.S. National Wildlife Health Center (1986–1995). Note that in most cases the cause of mortality is unknown. The early occurrence of mortality reflects the early growth of plants and lawns in the southern USA (adapted from Friend and Franson 1999).

Fig. 11.4. This grouse died following the application of insecticide to a large area of rangeland. Such deaths are difficult to observe unless they occur in great numbers because scavengers are quick to consume these food resources.

There is also a problem with birds flying into fields that were recently treated, perhaps to feast on dying insecticide-containing insects, and thereby ingesting a lethal dose of insecticide. For example, studies of horned lark, *Eremophila alpestris*, and McCowan's longspur, *Calcarius mccownii*, in relation to chlorpyrifos-treated wheat fields in Montana, USA, showed that these birds fed on recently killed cutworms. The percent of their stomachs filled with cutworms was 100% and 95.2% for longspurs and larks, respectively, 3 days after insecticide treatment. When examined 9 days after treatment, after most of the susceptible insects would have been killed and scavenged, the cutworm contents had diminished to 71.4% and 70.9%, respectively. Birds collected from untreated areas, in contrast, had only 27.2% and 7.7% cutworms in their stomachs, respectively. Unfortunately, birds scavenging on insecticide-killed cutworms acquired more than an easy meal, as brain cholinesterase activity was inhibited by up to 50%.

Other vertebrates are not immune to such poisoning, but it is most pronounced in avifauna (Fig. 11.3) and fish. With fish, the toxicant is contacted primarily via runoff of water from treated fields. Most insecticide labels prohibit treatment of water bodies, and often require a significant barrier zone or border in an effort to limit the drift of liquid pesticide into water. Nevertheless, contamination of water is not unusual.

How severe is the pesticide poisoning problem? The United States Fish and Wildlife Service estimates that over 670 million birds are exposed to pesticide on farmlands in the USA, and that about 10% die immediately as a result (Fig. 11.4). This does not include those that are sickened and die later, or eggs left unhatched, or nestlings left to starve. Organophosphate and carbamate insecticides are most commonly implicated. Farms are not the only source of pesticide poisoning, of course, as pesticides are a common element of suburban landscape maintenance, too.

One of the most enigmatic forms of acute poisoning is ingestion of insecticide granules by avifauna. Insecticide granules are any large carrier particle to which toxicant will adhere, and from which the toxicant will eventually be washed away by rain or soil water and be made available in the soil to kill pests. Surface application of insecticide and herbicide in the form of granules is common in some crop production systems, especially minimum- or no-tillage systems, because it allows pest control without having to till the soil. Tillage leads to greater energy use, and water and soil loss from cropland, so minimum tillage is generally a highly desirable practice, and application of granular pesticide products is an important component of such crop production systems. However, as noted previously, birds habitually consume small, hard particles (usually silica or 'sand') that lodge in their gizzards, helping to grind up food and enhance digestion. These particles are called grit, and are selected according to size, with larger birds selecting larger grit particles (see Fig. 11.5 for data on the relationship of bird size to grit

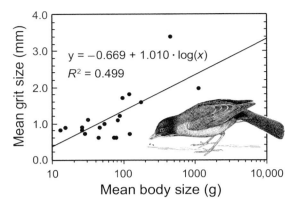

Fig. 11.5. The relationship between grit size selected by birds and their body mass (size). If insecticide granules fall into the size range favored by birds, they likely will be consumed (adapted from Best and Gionfriddo 1991).

size for birds inhabiting cornfields in the midwestern USA). Not only do foraging adults mistake insecticide granules for sand and other forms of grit and eat them, but birds often feed this poisonous 'grit' to nestlings. Grit preferences of birds affect their risk of eating pesticide granules, and of being poisoned by granular insecticides. When size and shape of preferred grit overlap with pesticide granules, they are at greater risk of being poisoned. Availability of granules, color, and granule composition also influence acceptance. In general, greater availability leads to greater consumption, though the relationship is not linear. Light colored grit is often favored over dark for consumption. Not surprisingly, corncob carrier is not as preferred as harder materials. Obviously it would be desirable to formulate insecticide granules in a manner that would discourage ingestion by avifauna. Currently, however, many of the insecticide granules share characteristics with the grit selected by birds. For example, a study of granular insecticide use on the Canadian prairie showed that there was a negative correlation of granular insecticide use and the abundance of several birds, including American robin, *Turdus migratorius*; horned lark, *Eremophila alpestris*; house sparrow, *Passer domesticus*; mourning dove, *Zenaida macroura*; western meadowlark, *Sturnella neglecta*; black-billed magpie, *Pica pica*; European starling, *Sturnus vulgaris*; and killdeer, *Charadrius vociferus*. Thus, although there are advantages to using granular insecticide formulations, there can be environmental consequences as well.

Significant levels of bird mortality are attributed to exposure to granular insecticides, and the threat extends beyond the avifauna that feed directly on the granules to also include their predators. In Canada, potato fields treated with fonofos for suppression of wireworm were entered by ducks and other waterfowl, and the birds often perished following ingestion of the insecticide granules. In turn, the dying and dead birds were fed upon by bald eagles, *Haliaeetus leucocephalus*, which also perished. Not surprisingly, the cadavers of the waterfowl displayed signs of organophosphate poisoning, including an average level of acetylcholinesterase inhibition of 74%. This shows how both primary and secondary mortality is quite possible from acetylcholinesterase inhibitors, despite their relatively brief persistence in the environment.

Exposure of wildlife to insecticides occurs in many ways, not just through the routes already mentioned. Insecticides applied to the soil can wash off the surface during significant rain events and enter ponds and streams. Less apparent, but sometimes quite common, are situations where insecticides leach away from the site of application, become part of the subsurface water supply (ground water or aquifer) and then move into aboveground bodies of water where they can affect animal life. The major considerations related to contamination of water bodies are solubility, distance, persistence and soil characteristics, as described below:

• **Solubility.** Some products dissolve easily (they are more soluble) in water and are more likely to move into water systems.

• **Adsorption.** Some toxicants become tightly attached (strongly adsorbed) to soil particles and less likely to move out of soil into water.

• **Persistence.** Some toxicants break down slowly and remain in the environment for a long time. Sometimes they degrade, but are transformed into other toxic materials, so their persistence can be underestimated.

• **Soil texture.** This soil property is quite important in governing the movement of pesticides, and is a function of the relative proportions of sand, silt, and clay in the soil. Coarse sandy soil allows water to move pesticides more rapidly; it is more permeable.

• **Soil permeability.** This is a measure of how fast soil allows water to move downward. Soils that are more permeable can be a problem, allowing high rates of leaching. Permeable soils are a problem not only from the perspective of environmental contamination, but for crop protection as well. Often insecticides are applied at planting or soon after plant emergence, but protection of the root system is needed for a month or more. In highly permeable soil, it is hard for the insecticide to be retained, making the plants susceptible to injury later in the growing season.

• **Soil organic matter.** The organic matter content influences how much water a soil can hold before the water begins to move downward. High levels of organic matter in soil mean that there is better water retention and less potential for pesticide leaching.

• **Runoff.** This is surface-level movement of pesticide away from the point of application. With excessive rainfall or irrigation, pesticide and soil can move horizontally rather than vertically (downward) and be deposited elsewhere.

• **Leaching.** The downward movement of water through the soil. As suggested above, leaching is determined by the characteristics of the soil and the amount of water applied. However, chemicals differ in their adsorptive properties as well.

• **Back-siphoning.** In commercial agriculture, insecticides are sometimes applied through the irrigation system. This is called insectigation or chemigation. This can be a convenient, economic way to disperse insecticide, and accomplishing this requires that a tank of formulated insecticide is spliced into the water system supplying irrigation. All goes well so long as there are no mechanical failures and the pressure is maintained, expelling the insecticide as planned. If pressure is lost, however, there is risk that the water in the irrigation system, including the contents of insecticide tank, will follow the laws of gravity and flow downward into the water supply. This is called backflow or back-siphoning. This is easily prevented by installing an anti-siphoning device (the water can only flow one way), and normally this is required, and regulatory agencies inspect for this feature. However, this same principle applies to homeowner application of pesticides through hose-end sprayers, which often are connected directly to the drinking water supply, and this is less well regulated.

Despite our best efforts to control the disposition of insecticides, they sometimes become redistributed in unpredictable ways. For example, footwash samples of burrowing owl, *Athene cunicularia*, showed that they were exposed to the insecticide chlorpyrifos, presumably from simply walking on soil treated with this insecticide, which normally is applied to soil for belowground insect suppression. This exposure apparently occurs despite the fact that burrowing owls normally inhabit rangeland, pastures, and other open areas rather than crop fields. Exposure during preening (grooming) is also an issue with avifauna and possibly with other wildlife when aerial applications are made. Clearly, there can be unintended consequences to insecticide use.

Decrease in bird abundance after insecticide treatment is sometimes attributed to dispersal of the birds due to reduction in insect (food) abundance. However, dispersal is not so likely a cause for population decrease of the lizard *Chalarodon madagascariensis* (Iguanidae) in Madagascar because these lizards are highly territorial and have low dispersal capacity. Application of an organophosphate (fenitrothion) and pyrethroid (esfenvalerate) insecticide mixture for control of desert locust (*Schistocerca gregaria* [Orthoptera: Acrididae]) apparently killed young lizards, supporting observations from other areas that lizards were being killed by chlorpyrifos during locust campaigns. The dermal absorption of pesticide deposits on the soil, rather that ingestion of contaminated prey, seems to be the main route of contamination.

More insidious is the intentional poisoning of predators such as coyotes or bears with cholinesterase-inhibiting insecticides. Although illegal, baiting of such wildlife with insecticide-tainted meat is not infrequent in an effort to protect livestock and beehives from coyotes and bears, respectively. Unfortunately, non-target predators and scavengers such as bald eagles, *Haliaeetus leucocephalus*; golden eagles, *Aquila chrysaetus*; black-billed magpies, *Pica pica*; and striped skunks, *Mephitus mephitus*; are also affected when coyotes and other 'less desirable' predators are poisoned.

Organophosphate and carbamate insecticides are cholinesterase inhibitors, interfering with the normal functioning of the nervous system of poisoned animals. Affected animals display hyperstimulation of both the nervous system and muscles. Also, exposure is often followed by salivation, shedding of tears, urination and vomiting. Difficulty breathing and panting behavior may precede respiratory failure. Despite a difficult death, poisoned animals often are found in relatively good condition; or at least they lack wounds and are intact. Such animals should be considered possible poisonings and submitted to laboratories accustomed to dealing with wildlife samples. Diagnosis of poisoning is not a simple task, and standard veterinary laboratories are not often equipped or experienced with wildlife.

SUBLETHAL EFFECTS OF INSECTICIDES

Various behavioral and physiological processes can be affected by pesticides, resulting in disruptive, though sublethal, effects on wildlife. Though not causing direct mortality, these effects nevertheless may indirectly cause increased mortality among wildlife (e.g., greater susceptibility to predation) or reduced reproduction (e.g., hormonal imbalance or eggshell thinning). Disruption of nervous system function, alteration of hormone levels, and inducement of oxidative stress via free radical generation are among the sublethal effects induced by insecticides.

Acetylcholinesterase inhibition in birds is much studied, especially in relation to organophosphate insecticide exposure. In general, when cholinesterase activity drops to 50% of normal or less, behavioral and physiological irregularities are apparent, with death following 80% inhibition or greater. At sublethal levels, affected birds may display impairment of memory and learning and inability to thermoregulate properly, make greater use of cover, exhibit reduction of feeding and flying, and show changes in resting posture. Birds may recover within a few hours, however. Similar responses occur in mammals, fish, reptiles and amphibians. A study of western fence lizard, *Sceloporus occidentalis*, showed that high levels of carbaryl exposure resulted in reduction in arboreal sprint speed and endurance. Such sublethal factors could affect fitness by reducing the ability of these lizards to escape predation. Similarly, earlier studies documented reduced swimming speed and distance in tadpoles of leopard frog, *Rana blairi*.

Not all insecticides are equally disruptive, of course. For example, a study of northern bobwhite (*Colinus virginianus*) foraging in soybean fields of North Carolina found that broods of young quail were present in the soybeans at the time of year when insecticides were applied. Acetylcholinesterase levels and body size were reduced when chicks were exposed to methyl parathion, but not when exposed to methomyl or thiocarb. Methyl parathion is an older product, and its use is now generally banned in the USA.

Probably the best-documented example of sublethal effects is eggshell thinning among predatory birds that was caused by organochlorines. The negative correlation between organochlorine residues in birds and eggshell thickness has been observed for many species, in many areas of the world, causing the eggs to be crushed by the nesting birds. For example, organochlorine residues were negatively correlated with reproductive success in bald eagle, *Haliaeetus leucocephalus*; osprey, *Pandion haliaetus*; peregine falcon, *Falco peregrinus*; Eurasian sparrow hawk, *Accipiter nisus*; American kestrel, *Falco sparverius*; herring gull, *Larus argentatus*; brown pelican, *Pelecanus occidentalis*; and others. This problem is now known to be due to endocrine (hormone) disruption.

Another interesting and well-documented sublethal effect of pesticides is endocrine disruption in American alligator, *Alligator mississippiensis*. Male alligators living in Lake Apopka, Florida, have low testosterone levels. Formerly, the area around Lake Apopka was intensively farmed. Lake Apopka also was the site of a DDT and dicofol (which is closely related to DDT) spill, and the insecticides had estrogen-like effects, resulting in feminization of the males. The penis size of male alligators was reduced by 25%, bone density was affected in females, and egg hatching was reduced. Alligator numbers were reduced by 90% in the years after the pesticide spill. Several organochlorine insecticides in addition to DDT and its closely related compounds, including endosulfan, toxaphene, dieldrin and BHC, have been shown to have potential to disrupt the physiological processes regulated by hormones. DDT is not acutely toxic to most birds and mammals, but long-term exposure is damaging.

DDT and other organochlorine insecticides were widely used before their adverse effects were fully appreciated, and though their use is prohibited in many areas of the world, they remain in use elsewhere due to their effectiveness, persistence, and low cost. Birds that migrate long distances may move into and out of countries where DDT is used, so it remains a continuing threat even where it is not currently used. DDT and DDE (a degradation product of DDT that is not insecticidal) affect enzymes controlling calcium deposition in bird eggshells, so eggshell thinning occurs, disrupting normal egg development. Similarly, long-lived fish and marine mammals continue to be exposed to these pesticides because they wash from the land into the oceans, where they are ingested.

The literature dealing with wildlife poisoning by pesticides has focused for too long on organochlorine effects, despite the fact that they are no longer used widely. In Europe and North America (generally the only comprehensive sources of information on this subject) it is the organophosphate and carbamate acetylcholinesterase inhibitors that now cause most of the problems in wildlife. Among the acetylcholinesterase inhibitors used in the USA, avian mortality often has been associated with famphur (applied to livestock for control of bot flies), carbofuran, diazinon (both agricultural insecticides), and fenthion (used both for agricultural insect control and mosquito control). Bird mortality events often involve famphur and the order Falconiformes (the birds of prey), but nearly as frequent are carbofuran-induced poisonings affecting a large number of bird species. Diazinon-induced deaths have involved mostly waterfowl. The diazinon cases usually involved suburban landscape (turfgrass) environments. Fenthion, like famphur, often involves birds of prey. Interestingly, in recent years the species most commonly reported to be poisoned by acetylcholinesterase inhibitors in the USA is Canada goose, *Branta canadensis*. Canada goose is abundant in both agricultural and suburban habitats, normally is found in flocks, and is large enough to be readily noticed if they perish, so mortality is especially apparent if it occurs. Granular application of diazinon is now greatly restricted in the USA, mitigating this risk.

Evidence that the organochlorine situation has improved, and that these pesticides are less of a threat, is apparent in the wildlife literature from many areas of the world. For example, when osprey, *Pandion haliaetus*, populations were monitored recently along the Columbia River system in Oregon, USA, these predators were found to display increased abundance, higher reproductive rates, and significantly lower egg concentrations of most organochlorine insecticides. As recently as the 1980s and 1990s, organochlorine concentrations were high in the fish preyed upon by ospreys in this area, but the situation has improved markedly. This clearly demonstrates that legislation limiting organochlorine pesticide use has benefited wildlife. In the USA, new restrictions have been placed on the use acetylcholinesterase inhibitors, which replaced organochlorines in most agricultural ecosystems. This should further reduce the risk of wildlife poisoning by insecticides in agricultural areas. However, the newer classes of insecticides are not completely without risk, if only because they deplete insect food resources. Also, use of acetylcholinesterase inhibiting insecticides remains widespread in some areas of the world.

OTHER PESTICIDES

Insecticides are not the only chemical pesticides encountered by wildlife. Rodents are often treated with *blood anticoagulants*. Treated animals can experience spontaneous hemorrhaging or a wound and die from loss of blood due to poor clotting ability of the blood. For many years, a rodenticide known as warfarin was used successfully, but due to the development of resistance in some rodent populations, newer products also are being marketed. Several exposures are typically necessary before warfarin induces death. Warfarin is not too hazardous to scavengers that feed upon warfarin-killed animals. Unfortunately, newer compounds such as brodifacoum and bromadiolone (so-called 'super-warfarins') act after a single exposure, are very persistent, and also affect scavengers that feed on poisoned animals. Like early-generation insecticides, they are lipophilic, tending to accumulate in animals and induce poisoning of both mammals and birds. In France, application of anticoagulant rodenticides against field voles, *Arvicola terrestris*, has resulted in massive poisonings of buzzards, *Buteo buteo*; foxes, *Vulpes vulpes*; and kites, *Milvus milvus* and *M. migrans*. Although herbivores might not be expected to be affected, when rodents are the target, plant materials such as carrots, apples, and grain are often poisoned. Even though the poisoned baits may be buried beneath the soil surface to minimize exposure of non-target animals to the tainted bait materials, a diverse assemblage of animals may be affected. In France, this included hares, *Lepus capensis*; rabbit, *Oryctolagus cuniculus*; wild boar, *Sus scrofa*; and roe deer, *Capreolus capreolus*, in addition to predators. Not all anticoagulant rodenticides are equally hazardous, of course. Chlorophacinone is commonly used for rodent control in both urban and agricultural areas, where it is effective against a number of rodents. The risk to scavengers differs markedly, with mammals being quite susceptible to poisoning, but the risk to birds being minimal to negligible; birds are inherently less affected by this chemical.

Zinc phosphide, an inorganic pesticide but not an anticoagulant, is also used as a rodenticide. Among the target organisms are rats (*Ratus* spp.), mice (*Mus* spp. and others), voles (various), ground squirrels (*Spermophilus beecheyi* and others), prairie dogs (*Cynomys* spp.), nutria (*Myocaster coypus*), muskrats (*Ondatra zibethicus*), and feral rabbits (principally *Oryctolagus* and *Sylvilagus* spp.). Zinc phosphide is formulated most often as a bait, but for indoor use it is also applied as a dust, and mice contact it while walking. In the Great Plains region of the USA, zinc phosphide is used as a toxicant in grain baits that are fed to prairie dogs. Like most pesticides, it has unintended non-target effects on other animals in the area. The endangered black-footed ferret, *Mustela nigripes*, is just one of many species associated with prairie dog colonies. But because these ferrets are quite rare and 90% of their diet consists of prairie dogs, they are particularly at risk when prairie dog colonies are poisoned. Not only mammals, but also birds and fish are commonly poisoned inadvertently by this rodenticide.

INDIRECT EFFECTS OF PESTICIDES ON WILDLIFE

In addition to the direct toxicity caused by pesticides, wildlife may be adversely affected indirectly through deprivation of their primary food source, and these indirect effects may be more important than the direct exposure of wildlife to insecticides. One important indirect effect is the depletion of insect populations caused by insecticide use. Application of broad-spectrum insecticides can cause treated fields to become almost sterile, and if the products are persistent the fields may remain depleted of insect life for weeks. Birds will attempt to compensate for loss of insect food by foraging elsewhere, but there are limits as to how far they can fly and then return regularly to a nest with food for nestlings. If the distance of travel is too great, the nest will be abandoned. Due to the high cost of insecticide development and registration, agrochemical companies favor development of broad-spectrum products because, once registered, they can be used extensively on a large number of crops for numerous pest problems and generate large profits before the patent on the pesticide expires. The nonselective nature of such broad-spectrum products is particularly damaging to bird populations; if only the pests were affected some insect fauna would remain to support bird life, but usually other herbivorous insects, predators, parasitoids, pollinators, and scavengers are also affected, leaving no insects to sustain bird life.

Another indirect effect of pesticides on wildlife is the change in floral diversity (loss of edible weeds, weed seeds, or fungi, and also depletion of habitat or cover) caused by herbicide (and to a lesser degree by fungicide) application. Grass and weed seed can be an important food resource, and clean culture of crops – though beneficial in terms of plant growth efficiency, energy efficiency and water conservation – can greatly reduce food abundance for bird life. The intensification and specialization of agriculture is manifested in the ever-increasing scale (field size) in agriculture, which usually results from merging smaller fields, reducing crop heterogeneity, and in destroying hedge-row and other border area habitat. These practices have negative effects on wildlife because they may have no place to nest, or no place to nest that isn't treated with pesticides, or no source of shelter when crops are harvested, etc. Research in Montana, USA wheat fields showed that herbicide use not only reduced the abundance of broad-leaf weeds, but also the abundance and biomass of insects important to game bird chicks. Weeds also favored the occurrence of carabid beetles, important insect predators in this cropping system. The researchers suggested leaving the edges of fields free of herbicide and insecticide treatments, thereby favoring the survival of both beneficial insects and game birds.

The results of pesticide applications often are dramatic. In Britain, for example, two-thirds of farmland bird species have shown declines in abundance. Even among granivorous birds, which only forage for insects during the breeding season to feed their young, over half of the species have exhibited declining populations. Several taxa of invertebrates are important in the diets of declining farmland birds in Britain, including Coleoptera, Orthoptera, Diptera, Lepidoptera, Hymenoptera, Hemiptera, and Arachnida. Grasshoppers (Orthoptera: Acrididae) and leaf beetles (Coleoptera: Chrysomelidae) are particularly associated with the declining bird species. Not all insects have displayed decreases in abundance, however. Insects in the order Hemiptera, particularly aphids, thrive under intensive farming regimes, and so their abundance has increased. Unfortunately, aphids are not an important food resource for many birds.

In Britain, seeds of many plants were important in the diet of birds, with the seeds of the plant families Graminae (Poaceae), Polygonaceae, Chenopodiaceae, Caryophyllaceae, Asteraceae, Brassicaceae and Fabaceae particularly widespread in the diets of the birds studied. The increased use of herbicides, more frequent harvesting of grasses, more intensive grazing, more efficient harvesting of grains, double cropping, and the decline of traditional rotational practices have resulted in less grass seed being available. Plants in the family Polygonaceae, especially *Polygonum* (knotgrasses and pesicarias) and *Rumex* (docks and sorrels), have high seed capacity and are a rich source of food for birds. However, their abundance in agricultural lands has decreased. Weeds of the family Chenopodiaceae, particularly *Chenopodium* (goosefoots), Caro-phyllaceae such as *Stellaria* (chickweeds) and Brassicaceae (various), also have declined. Other groups are stable or unknown. *Chenopodium*, *Stellaria*, and *Polygonum* are all key food resources for granivorous birds in temperate European farmland, and all are declining in abundance.

The decline in avifauna in Britain is not attributable solely to increased use of pesticides. Loss of hedgerow habitat, more intensive grazing, more efficient harvesting of grain crops, and other cultural practices have perhaps contributed to the decline as well. However, it is clear that the food resource base has been affected negatively, and pesticides are a significant part of the problem.

INSECTICIDES IN THE FOOD CHAIN

Persistent pesticides can accumulate in organisms within a trophic level by intake of food or water, especially when the organisms lack enzymes to degrade them or the ability to excrete them. As noted previously, one form of bioaccumulation is called biological magnification or biomagnification. **Biomagnification** is an increase in the concentration of a chemical at each trophic level of a food chain. For example, if a persistent toxicant is applied to a salt marsh for mosquito control, trace amounts of the toxicant will be incorporated into the tissues of plankton living in the water (this primary level of productivity is often referred to as the 'producers'). The consumers of plankton, such as clams and small fish (collectively called 'primary consumers'), accumulate toxicant at concentrations higher than that occurring in the producers. The primary consumers may be consumed by larger, predatory animals (called 'secondary consumers') such as large fish, which in turn concentrate the toxicant further. The number of steps in the concentration process varies, of course, but eventually the highest level in the trophic pyramid is reached ('tertiary and quaternary consumers'), where very high and damaging concentrations of toxicant may occur (see Fig. 11.6). The final consumers are typically long-lived fish, mammals, and birds. Biomagnification, though generally accepted now, was for many years a very controversial subject. It is very difficult to track compounds like insecticide residues through natural ecosystems.

Convincing evidence of biomagnification was demonstrated through laboratory microcosm (miniature ecosystem) studies by R.L. Metcalf, among others. Laboratory studies of organochlorine concentration in model ecosystems (aquaria) confirmed that the toxicants could be magnified. In these studies, low levels of insecticides entered the water from fecal material produced by caterpillars feeding on insecticide-treated sorghum plants. The chemical toxicants in the water became incorporated into *Oedogonium* algae. The algae, in turn, were consumed by *Physa* snails or *Gambusia* fish. Ecological magnification of the insecticides in Metcalf's studies is shown in Table 11.5. These organochlorine insecticides were shown to accumulate in the tissues of fish and snails to levels from 200 to 84,000 times greater than the concentration in the water of the model ecosystem. The mixed function oxidase system of the caterpillar, which normally protects the organisms against natural toxicants, was ineffective against the insecticides. Also, the water insolubility of the insecticides was directly correlated with their degree of biomagnification, accounting for their transfer from water to blood, and from blood to tissue lipids of the snails and fish.

Organochlorine insecticide residues occur in many ecosystems, but they are not the only toxicants threatening wildlife, and despite their bad reputation, in many cases they no longer occur at injurious levels. For example, a study of toxicant residues occurring in fish that inhabit the Yukon River Basin of Alaska, USA, was conducted to assess the risk of fish consumption to piscivorous (fish-eating) wildlife. The levels of the insecticides DDT, toxaphene, and chlordane in fish did not exceed the toxicity thresholds for fish or piscivorous wildlife. However, some elemental materials were found at threatening levels. Mercury levels exceeded the threshold concentration for hazard to birds and small mammals. Zinc and selenium levels exceeded the threshold concentration for birds. Thus, elemental contaminants (possibly resulting from mining activities) are more of a threat to wildlife in this area than are pesticide residues.

As noted previously, less persistent insecticides, though not necessarily displaying biomagnification can nonetheless cause injury or death in more than one trophic level; this is called **secondary poisoning**. With acetylcholinesterase inhibitors such as organophosphate and carbamate insecticides, the toxin levels in an animal poisoned through contact with these insecticides may be sufficient to poison any predators or scavengers feeding on living or dead animals exposed to such pesticides. Surprisingly few data are

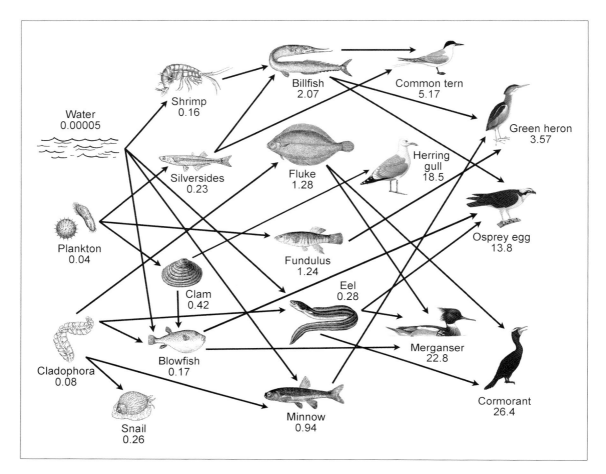

Fig. 11.6. Biological magnification of DDT: pesticide levels (ppm) in water and marine plants, followed by primary, secondary and tertiary consumers as organisms eat organisms and the long-lived pesticide accumulates (data from Woodwell *et al.* 1967). Although insects are the primary target for such insecticides, insects also are the primary consumers in terrestrial and fresh-water food chains, and can contribute to the magnification of long-lived pesticides in the same manner as the DDT magnification shown here.

available on the insecticide residues encountered by wildlife when they feed on insects found within treated crop fields. Recent research has shown that insects and earthworms from alfalfa and citrus fields treated with chlorpyrifos all contained insecticide residues, and that the level of contamination varied among species. Although the authors of this study argued that the residue levels were modest, and less than US Environmental Protection Agency expectations in their calculation of wildlife risk of poisoning, it is difficult to predict the actual impact on predators and scavengers. In

other studies, earthworms living in soil with a sublethal dose of the organophosphate dimethoate proved poisonous to shrews, *Sorex araneus*, that fed upon them. Acetylcholinesterase levels were depressed by 90% in the earthworms and 64% in the shrews. Similarly, in Australia many birds gather to gorge on bands of locusts (Orthoptera: Acrididae), including those locusts that are weakened or dying following exposure to the insecticide fenitrothion. During outbreaks, locusts can be the principal source of food for avian predators for up to 25 days; thus, there is ample oppor-

Table 11.5. Quantitative values for ecological magnification of several organochlorine insecticides in fish and snails when assessed in laboratory microcosms (adapted from Metcalf *et al.*1973).

Insecticide	Water solubility (ppm)	Ecological magnification	
		Gambusia	*Physa*
Aldrin	0.20	3,140	44,600
Dieldrin	0.25	2,700	61,657
Endrin	0.23	1,335	49,218
Mirex	0.085	219	1,165
Lindane	7.3	560	456
Hexachlorobenzene	0.006	287	1,247
DDT	0.0012	84,545	34,545
DDE	0.0013	27,358	19,529

tunity for prolonged exposure to insecticides. In Argentina, over 5000 flocking individuals of Swainson's hawk, *Buteo swainsoni*, were killed in 1996 following application of the organophosphate monocrotophos to alfalfa fields, apparently due to ingestion of dying insects. Invertebrates are not the only route of secondary poisoning, of course. When house sparrows, *Passer domesticus*, were exposed to the organophosphate fenthion, consumption of a single sparrow proved fatal to most American kestrels, *Falco sparverius*, that fed upon them. In British Columbia, Canada, raptors including bald eagle, *Haliaeetus leucocephalus*, and red-tailed hawk, *Buteo jamaicensis*, were found to be poisoned after granular organophosphate and carbamate insecticides were applied for agricultural insect control. Interestingly, in the latter study these products inflicted mortality many months after the products should have degraded. It seems that waterfowl were attracted to fields that had become flooded several months after insecticide use, where they died after contacting residues of the organophosphate insecticides fensulfothion and phorate, and the carbamate carbofuran. The raptors, in turn, scavenged on dead waterfowl and were poisoned.

RISKS OF INSECTICIDES

Given the long history of negative impacts of pesticides on wildlife, it is tempting to indict all pesticides as hazardous, and suspect all wildlife problems as being caused by pesticides. Thus, when yellow-leg frog, *Rana*

mucosa, populations in California, USA, were assessed, it is not surprising that pesticides were one of the factors examined for potentially causing their decline in abundance. Indeed, population declines occurred in portions of the Sierra Nevada Mountains that were often downwind from areas where pesticides were applied. However, the link of pesticides to amphibian population decline, which is a huge worldwide problem, is tenuous at best. Indeed, when the blood and brain of the bullfrog *Rana catesbeiana* were collected from a field of corn and an adjacent pond following application of the organophosphate chlorpyrifos, there was no evidence of an effect on acetylcholinesterase levels.

To demonstrate how complicated the issue of pesticide toxicity really is, consider the results of a study investigating the application of malathion (an insecticide) and glyphosate (an herbicide) on tadpoles of American toads, *Bufo americanus*; leopard frogs, *Rana pipiens*; gray tree frogs, *Hyla versicolor*; and two types of tadpole predators: predatory newts, *Notophthalmus viridescens*, and predaceous diving beetle larvae, *Dytiscus* sp. (Dytiscide). The malathion did not greatly affect the tadpoles directly, but glyphosate did reduce survival. The results are complicated by the presence of predators, especially predatory beetles, because they were more susceptible to the insecticide than were the tadpoles. Thus, the addition of insecticide increased tadpole survival by killing their predators! Also, it seems that the glyphosate, which might not be considered to be toxic to animals because it is registered to kill plants, killed tadpoles because it contained a surfactant formulation additive that was toxic. This is an example of an area where much more research is needed, and it would be premature to attribute such declines in amphibian abundance to pesticides without more research.

Avian raptor populations in agricultural areas have repeatedly been shown to suffer from insecticide poisoning, and as the insecticide use pattern has shifted from being based on organochlorines to acetylcholinesterase inhibitors, it seems reasonable to assume that any recent population declines would be attributable to poisoning by acetylcholine insecticides. In Spain, populations of lesser kestrel, *Falco naumanni*, and Eurasian kestrel, *Falco tinnunculus*, have shown dramatic declines in abundance in recent decades. So assessment of acetylcholinesterase levels in kestrels seemed a logical approach to document the negative effects of organophosphate and carbamate exposure on kestrels. However, research showed that although the kestrel populations were suffering from poor health, the blood levels of acetylcholinesterase did not suggest insecticide toxicity. Also, the blood acetylcholinesterase levels were not different in agricultural areas, where insecticide use is high, and pastoral areas, where insecticide use is low. So pesticides are not considered to be a conservation issue in this case.

There are few places where aesthetics are more important than golf courses. Typically, no expense is spared in an effort to have nearly perfect turfgrass. This usually means that pesticides are used liberally, so one might expect that golf courses would be a particularly inhospitable environment for avifauna. However, assessment of reproductive success in eastern bluebirds, *Sialia sialis*, in Virginia, USA, showed that these insectivores can survive quite well, with survival not differing significantly between bluebirds nesting in golf course environments and non-golf course environments. On golf courses, the number of eggs per nest box was higher (28% greater), as was the number of young birds successfully fledged (17% greater). Overall, pesticide use on golf course did not impose a significant stress on bluebirds, suggesting that pesticide use and wildlife can be compatible.

Pesticides can even prove to be useful in managing wildlife populations. The use of pesticides for rodent and predator control, mentioned earlier, are pertinent examples when the pesticides are used legally and appropriately. In addition, some chemicals (pesticidal and non-pesticidal) serve as wildlife repellents. For example, consumption of corn, rice, and sunflower seed by blackbirds is a significant problem for producers of these crops. In suburban environments, grazing by geese and deer on ornamental plants often creates problems. Several chemicals such as anthraquinone and methyl anthranilate repel birds whereas capsaicin and coyote urine repel deer and elk. However, some insecticides such as methiocarb and chlorpyrifos also reduce consumption of plant materials by birds such as red-wing blackbird, *Agelaitus phoeniceus*. They accomplish this not because they kill the birds, but because the pesticides, even at reduced dose rates, are feeding deterrents.

Sometimes, pesticides can be used to enhance wildlife populations. Often, there is a negative relationship between arthropod abundance and some aspect of survival or reproductive performance. For example, several studies of great tit, *Parus major*, in Europe have found influences of hen flea, *Ceratophyllus gallinae* (Siphonaptera), on aspects of behavior or fitness such as roost site selection, nest site selection, timing of breeding, and body mass of nestlings. Reduction in flea abundance by application of an insecticide inside nesting boxes has been shown to improve elements of fitness, and could be used for other birds troubled by fleas. Animal populations can sometimes be managed for improved health by applying pesticides to them or to food bait to suppress endo- and ectoparasites; this is especially useful where high densities of wildlife are being maintained.

A few comments relative to interpreting the interaction of pesticides and wildlife are warranted here. Assessment of pesticide effects is rarely objective, and results are rarely clear-cut. Qualitative assessment of hazard is fairly sophisticated and reliable; quantitative assessment, however, is often unreliable and controversial. Qualitative assessment will nearly always show that pesticides, particularly insecticides, can harm wildlife. The methodology should always be examined, however, because quantitative assessment is much abused. Unbelievably, some studies are conducted in a no-choice environment, wherein wildlife is presented only with poisoned food (typically poisoned grain), which the animals must consume or perish from starvation. It should come as no surprise that such studies demonstrate that animals tested under these circumstances will eat the poisoned food after several days of avoiding it, and that the poisoned food is toxic. This in no way relates to a natural environment, where most animals are free to select among tainted and pesticide-free food, and are free to relocate.

Similarly, a common practice is to dose test animals with levels of pesticide that are several-fold greater than would be used in the field. As noted earlier, toxicity is a dose-dependent phenomenon, and even materials normally considered to be safe or desirable can be toxic if consumed in massive quantities; aspirin and table salt are pertinent examples for humans. Thus, laboratory studies should be considered a worst-case scenario for wild species. The other common form of data abuse is association; wildlife mortality is sometimes 'associated' or 'correlated' with pesticide use, but no cause and effect is determined. Sometimes this is pure coincidence or hysteria. Wildlife can succumb to a number of diseases, temperature-related stress, or food deprivation, and the cause is not always obvious.

On the other hand, pesticide advocates often attempt to minimize the risk of pesticides to wildlife, or impose unrealistic constraints on use (e.g., 'do not apply within 200 feet of a body of water,' knowing full well that aircraft are not that precise and cannot always control drift, so water will sometimes become contaminated despite label restrictions). Historically, registration of new pesticides has not given adequate consideration to wildlife hazards, and only after a long series of pesticide-related problems has been documented have constraints on use been advanced. Also, the purported economic impact of restricting use of a hazardous material has often been given too much consideration, whereas the effects on wildlife have been given inadequate consideration. Long-term and sublethal effects have been particularly underappreciated because they are hard to measure. Some studies have shown that mortality of birds is greater under field conditions than in confined or laboratory conditions because weakened birds are susceptible to predation in the wild, whereas under confined conditions they can recover without risk of predation.

RESISTANCE TO INSECTICIDES

Insect populations sometimes evolve resistance to pesticides. In nearly all cases, if an insecticide is used regularly, and it is effective at killing a high proportion of the population, eventually the population will no longer succumb to the pesticide. This phenomenon of the pesticide losing its effectiveness is called **insecticide resistance**. It is a heritable loss of sensitivity. In most cases, insecticide resistance is due to widespread occurrence in a population of a pesticide detoxification

mechanism that previously did not occur widely. However, evolution of target site insensitivity is also documented. Where did these physiological mechanisms come from? Well, insects routinely confront toxins in their environment; natural toxins are found in the air, water, and especially in their food. So insects must have the ability to deal with toxins, but typically most have never encountered the synthetic pesticides that we use to kill them. At the biochemical level, destruction (detoxification) or failure to react (target site insensitivity) when confronted with an insecticide is not terribly different than what might occur with a natural toxin. So, chances are that some insects already have the ability to detoxify, or are insensitive to, any insecticide we can formulate. Typically, such insects are rare in the population. Remember, as is the case with people, insects are all slightly different. They differ in their susceptibility to disease, to toxins, to cold and to heat. When you apply an insecticide you select for the survival of the individuals that already possess a mechanism to detoxify the pesticide, or are insensitive to the pesticide. So insecticide resistance is a *population* process. The genetics of the population shifts in response to repeated killing of some members of the population. The survivors of the insecticide application mate with other survivors, and chances are good that their offspring will inherit from their parents the ability to be less affected by the insecticide. This is an example of evolution that we witness routinely; small changes in the biology of organisms eventually translate into significant differences. In this case, the occurrence of insecticide resistance means that we have lost the ability to control the pest insect.

When the loss of insecticide effectiveness begins to occur, the typical response is to believe that the product was not formulated or applied properly, weather interfered with the application, or perhaps the product was 'old' and had lost some of its potency. So the insecticide applicator might apply the product more frequently, or use a higher rate of active ingredient. Often, this gives some immediate but short-term relief, but soon the frequency and rates that are applied are increased higher and higher until the only insects remaining in the population are those that are resistant; all the susceptible members of the population have been killed.

The consequences of insecticide resistance are detrimental to both the applicator (often a farmer) and the environment (both wildlife and humans). When the frequency and rate of application of an insecticide are increased, the expenses associated with producing the crop also go up. Often the new, replacement insecticides are much more expensive than the older materials. This can significantly reduce the profit in growing the crop, and can even put the farmer out of business. Sometimes the producer will lose his entire crop due to pests, which is an economic disaster. Indeed, many farmers do not grow certain crops because pest problems are so severe in their location. On the other hand, working and living around such continuous pesticide application can be detrimental to the health of both humans and wildlife, and the more frequently that insecticides are applied the more likely there will be adverse consequences. So it is very important to not get into a situation where insecticide resistance develops. Keep in mind that this problem is not limited to farms. Resistance occurs among insects living on lawns, flowers, and trees, in homes, and among biting flies such as mosquitoes. This problem affects us all.

There is another form of insecticide resistance that is not so much genetic and population based as it is physiological and individual based. When insects encounter low levels of toxins, their detoxification mechanisms are challenged and stimulated to function. If they again encounter the toxin, their detoxification mechanisms are already operating so they are more likely to resist the toxin. So some argue that low rates of insecticide application are dangerous, and promote development of insecticide resistance. This physiological resistance is limited in scope, however, and is not the basis for widespread insecticide failure. It is, in fact, best to use the minimal effective dose. Sometimes this is even less than the minimal rate recommended by the manufacturer. High rates of insecticide application are more likely to cause the onset of widespread insecticide resistance than are low rates.

How can we prevent insecticide resistance from developing? There are several means of preventing this from happening, or at least minimizing the rate of resistance development in insect populations:

• **Don't use insecticides.** Depend on other means of insect management such as environmental management or natural enemies (biological control).

• **Don't repeatedly use the same insecticide, or insecticides in the same chemical class.** This is because repeated use of the same product or types of products consistently selects for the occurrence of an effective detoxification mechanism in the population.

• **Rotate among different classes of insecticides.** Because different chemical classes of insecti-

cides often have different means of detoxification, rotation lessens the selective pressure for development of a means of detoxifying the insecticides. An organization called the 'Insecticide Resistance Action Committee (IRAC)' serves to advance the education of pesticide users about resistance problems. An important part of their educational program is to provide information on the mode of action of pesticides. This allows users to know if different products truly represent different modes of action, a prerequisite to implementing effective pesticide rotation schemes.

• **Use insecticides only when necessary, and at the minimal effective dose.** Adopt the Integrated Pest Management (IPM) philosophy, which relies on *alternatives* to pesticides to *prevent* pest populations from developing, careful pest population assessment, calculation of the costs and benefits of treating with insecticides, and treatment with insecticides only when and where they are needed. Using insecticides routinely as a preventative measure is, in most cases, unnecessary. Rather, they are used as a corrective measure only if truly needed, and used in a manner that integrates their use with other pest management techniques such as natural enemies of pests.

SUMMARY

• Insecticide application is the principal means of solving most pest problems. As reliable and convenient as this seems, depending on chemical pesticides creates some serious problems for wildlife.

• There are many types of pesticides that affect wildlife. Insecticides are designed to control insect pests, but also affect other animals including wildlife. Other pesticides including avicides, rodenticides, and piscicides may affect wildlife by design, or by accident. However, even chemicals designed to affect plants can have adverse impacts on wildlife, if only indirectly, as by depriving wildlife of food or cover.

• Insecticides differ greatly in their toxicity. A common measure of toxicity is the LD_{50} value, or the concentration of active ingredient expressed in mg/kg of animal body weight, that will kill 50% of the test animals. Information on toxicity can be found on the Material Safety Data Sheet (MSDS) and on the product label.

• Insecticides can cause toxicity after short-term exposure (acute toxicity) or long-term exposures (chronic exposure), and can be toxic in several different ways. Historically, bioaccumulation has been particularly damaging to wildlife. The mode of action of most insecticides is to interfere with nerve function.

• Insecticides degrade in the environment due to photolysis, adsorption, hydrolysis, high temperatures, and microbial actions.

• Poisoning by insecticides is most severe in birds and fish. Exposure often occurs by contact, ingestion of granular formulations, leaching into the water supply, and poisoning of the food supply. Death can also occur by elimination of the food supply, and interference with animal hormones and other metabolic processes. Pesticides used to control nuisance wildlife often are blood anticoagulants.

• Pesticides are not entirely damaging; toxicity is dose-related. Selective applications can be used to manage endo- and ectoparasites for improved wildlife health.

• Insecticide resistance among pests often is a consequence of overuse, and the occurrence of resistance in pest populations tends to stimulate use of more pesticides and more toxic insecticides. Resistance is best managed by eliminating or reducing the use of insecticides. If insecticides must be used, rotation among different classes of insecticides is recommended.

REFERENCES AND ADDITIONAL READING

Allander, K. (1998). The effects of an ectoparasite on reproductive success in the great tit: a 3-year experimental study. *Canadian Journal of Zoology* **76**, 19–25.

Avery, M.I., Evans, A.D., & Campbell, L.H. (2004). Can pesticides cause reductions in bird populations? In van Emden, H.F. & Rothschild, M. (eds.) *Insect and Bird Interactions*, pp.109–120. Intercept, Andover, Hampshire, UK.

Berny, P. (2007). Pesticides and the intoxication of wild animals. *Journal of Veterinary Pharmacology and Therapy* **30**, 93–100.

Berny, P.J., Buronfosse, T., Buronfosse, F., Lamarque, F., & Lorgue, G. (1997). Field evidence of secondary poisoning of foxes (*Vulpes vulpes*) and buzzards (*Buteo buteo*) by bromadiolone, a 4-year survey. *Chemosphere* **35**, 1817–1829.

Best, L.B. & Gionfriddo, G.P. (1991). Characterization of grit use by cornfield birds. *Wilson Bulletin* **103**, 68–81.

Brewer, L.W., McQuillen Jr., H.L., Mayes, M.A., Stafford, J.M., & Tank, S.L. (2003). Chlorpyrifos residue levels in avian food items following applications of a commercial EC formulation to alfalfa and citrus. *Pest Management Science* **59**, 1179–1190.

Carson, R. (1962). *Silent Spring*. Houghton Mifflin Company, Boston.

Cox, C. (1991). Pesticides and birds: from DDT to today's poisons. *Journal of Pesticide Reform* **11** (4), 2–6.

Davidson, C. & Knapp, R.A. (2007). Multiple stressors and amphibian declines: dual impacts of pesticides and fish on yellow-legged frogs. *Ecological Applications* **17**, 587–597.

Dell'Omo, G., Turk, A., & Shore, R.F. (1999). Secondary poisoning in the common shrew (*Sorex araneus*) fed earthworms exposed to an organophosphate pesticide. *Environmental Toxicology and Chemistry* **18**, 237–240.

DeWeese, L.R., McEwen, L.C., Settimi, L.A., & Deblinger, R.D. (1983). Effects on birds of fenthion aerial application for mosquito control. *Journal of Economic Entomology* **76**, 906–911.

DuRant, S.E., Hopkins, W.A., & Talent, L.G. (2007). Impaired terrestrial and arboreal locomotor performance in the western fence lizard (*Sceloporus occidentalis*) after exposure to an ACE-inhibiting pesticide. *Environmental Pollution* **149**, 18–24.

Eason, C.T. & Spurr, E.B. (1995). Review of the toxicity and impacts of brodifacoum on non-target wildlife in New Zealand. *New Zealand Journal of Zoology* **22**, 371–379.

Eason, C.T., Murphy, E.C., Wright, G.R.G., & Spurr, E.B. (2001). Assessment of risks of brodifacoum to non-target birds and mammals in New Zealand. *Ecotoxicology* **11**, 35–48.

Elliott, J.E., Wilson, L.K., Langelier, K.M., Mineau, P., & Sinclair, P.H. (1997). Secondary poisoning of birds of prey by the organophosphorus insecticide, phorate. *Ecotoxicology* **6**, 219–231.

Elliott, J.E., Birmingham, A.L., Wilson, L.K., McAdie, M., Trudeau, S., & Mineau, P. (2008). Fonofos poisons raptors and waterfowl several months after granular application. *Environmental Toxicology and Chemistry* **27**, 452–460.

Fleischli, M.A., Franson, J.C., Thomas, N.J., Finley, D.L., & Riley Jr., W. (2004). Avian mortality events in the United States caused by anticholinesterase pesticides: a retrospective summary of National Wildlife Health Center records from 1980 to 2000. *Archives of Environmental Contamination and Toxicology* **46**, 542–550.

Forbes, A.B. (1993). A review of regional and temporal use of avermectins in cattle and horses worldwide. *Veterinary Parasitology* **48**, 19–28.

Friend, M. & Franson, J.C. (eds.) (1999). *Field Manual of Wildlife Diseases: General Field Procedures and Diseases of Birds*. US Department of the Interior, US Geological Survey, Washington, DC, USA.

Heeb, P., Kölliker, M., & Richner, H. (2000). Bird–ectoparasite interactions, nest humidity, and ectoparasite community structure. *Ecology* **81**, 958–968.

Henny, C.J., Grove, R.A., & Kaiser, J.L. (2008). Osprey distribution, abundance, reproductive success and contaminant burdens along lower Columbia River, 1997/1998 versus 2004. *Archives of Environmental Contamination and Toxicology* **54**, 525–534.

Hinck, J.E., Schmitt, C.J., Echols, K.R., May, T.W., Orazio, C.E., & Tillitt, D.E. (2006). Environmental contaminants in fish and their associated risk to piscivorous wildlife in the

Yukon River Basin, Alaska. *Environmental Contamination and Toxicology* **51**, 661–672.

Hunt, K.A., Bird, D.M., Mineau, P., & Shutt, L. (1991). Secondary poisoning hazard of fenthion to American kestrels. *Archives of Environmental Contamination and Toxicology* **21**, 84–90.

LeClerc, J.E., Che, J.P.K., Swaddle, J.P., & Cristol, D.A. (2005). Reproductive success and developmental stability of eastern bluebirds on golf courses: evidence that golf courses can be productive. *Wildlife Society Bulletin* **33**, 483–493.

Lind, P.M., Milnes, M.R., Lundberg, R., Bermudez, D., Orberg, J., & Guillette Jr., L.J. (2004). Abnormal bone composition in female juvenile American alligators from a pesticide-polluted lake (Lake Apopka, Florida). *Environmental Health Perspectives* **112**, 359–362.

Linz, G.M., Homan, H.J., Slowik, A.A., & Penry, L.B. (2006). Evaluation of registered pesticides as repellents for reducing blackbird (Icteridae) damage to sunflower. *Crop Protection* **25**, 842–847.

McEwen, L.C., Knittle, C.E., & Richmond, M.L. (1972). Wildlife effects from grasshopper insecticides sprayed on short-grass range. *Journal of Range Management* **25**, 188–194.

McEwen, L.C., DeWeese, L.R., & Schladweiler, P. (1986). Bird predation on cutworms (Lepidoptera; Noctuidae) in wheat fields and chlorpyrifos effects on brain cholinesterase activity. *Environmental Entomology* **15**, 147–151.

Metcalf, R.L., Kapoor, I.P., Lu, P-Y., Schuth, C.K., & Sherman, P. (1973). Model ecosystem studies of the environmental fate of six organochlorine pesticides. *Environmental Health Perspectives* **4**, 35–44.

Mullen, G. & Durden, L. (eds.) (2002). *Medical and Veterinary Entomology.* Academic Press/Elsevier, Amsterdam, The Netherlands.

Mineau, P., Downes, C.M., Kirk, D.A., Bayne, E., & Csizy, M. (2005). Patterns of bird species abundance in relation to granular insecticide use in the Canadian prairies. *Ecoscience* **12**, 267–278.

Nauen, R. (2006). Insecticide mode of action: return of the ryanodine receptor. *Pest Management Science* **62**, 690–692.

Palmer, W.E., Puckett, K.M., Anderson, J.R., & Bromley, P.T. (1998). Effects of foliar insecticides on survival of northern bobwhite quail chicks. *Journal of Wildlife Management* **62**, 1565–1573.

Peveling, R. & Nagel, P. (2001). Locust and tsetse fly control in Africa: does wildlife pay the bill for animal health and food security? In Johnston, J.J. (ed.). *Pesticides and Wildlife*, ACS Symposium Series **771**, pp. 82–108. American Chemical Society, Washington, DC, USA.

Poché, R.M. & Mach, J.J. (2001). Wildlife primary and secondary toxicity studies with warfarin. In Johnston, J.J. (ed.) *Pesticides and Wildlife*, ACS Symposium Series **771**, pp. 181–196. American Chemical Society, Washington, DC, USA.

Primus, T.M., Eisemann, J.D., Matschke, G.H., Ramey, C., & Johnston, J.J. (2001). Chlorophacinone residues in rangeland rodents: an assessment of the potential risk of secondary toxicity to scavengers. In Johnston, J.J. (ed.) *Pesticides and Wildlife*, ACS Symposium Series **771**, pp. 164–180. American Chemical Society, Washington, DC, USA.

Relyea, R.A, Schoeppner, N.M., & Hoverman, J.T. (2005). Pesticides and amphibians: the importance of community context. *Ecological Applications* **15**, 1125–1134.

Richards, S.M., Anderson, T.A., Wall, S.B., & Kendall, R.J. (2001). Exposure assessment of *Rana catesbeiana* collected from a chlorpyrifos-treated cornfield. In Johnston, J.J. (ed.) *Pesticides and Wildlife*, ACS Symposium Series **771**, pp. 119–129. American Chemical Society, Washington, DC, USA.

Sanchez-Hernandez, J.C. (2001). Wildlife exposure to organophosphorus insecticides. *Reviews of Environmental Contamination and Toxicology* **172**, 21–63.

Scharf, M.E. (2008). Neurological effects of insecticides and the insect nervous system. In Capinera, J.L. (ed.) *Encyclopedia of Entomology*, 2nd edn., pp.2596–2605. Springer Science & Business Media B.V., Dordrecht, The Netherlands.

Scollon, E.J., Goldstein, M.I., Parker, M.E., Hooper, M.J., Lacer, T.E., & Cobb, G.P. (2001). Chemical and biochemical evaluation of Swainson's hawk mortalities in Argentina. In Johnston, J.J. (ed.) *Pesticides and Wildlife, ACS Symposium Series* **771**, pp. 294–308. American Chemical Society, Washington, DC, USA.

Sepúlveda, M., Del Piero, F., Wiebe, J.J., Rauschenberger, H.R., & Gross, T.S. (2006). Necropsy finding in American alligator late-stage embryos and hatchlings from northcentral Florida lakes contaminated with organochlorine pesticides. *Journal of Wildlife Diseases* **42**, 56–73.

Stafford, T.R. & Best, L.B. (1999). Bird response to grit and pesticide granule characteristics: implications for risk assessment and risk reduction. *Environmental Toxicology and Chemistry* **18**, 722–733.

Story, P. & Cox, M. (2001). Review of the effects of organophosphorus and carbamate insecticides on vertebrates. Are there implications for locust management in Australia? *Wildlife Research* **28**, 179–193.

Taylor, R.L., Maxwell, B.D., & Boik, R.J. (2006). Indirect effects of herbicides on bird food resources and beneficial arthropods. *Agriculture, Ecosystems and Environment* **116**, 157–164.

van Emden, H. & Rothschild, M. (eds) (2004). *Insect and Bird Interactions.* Intercept, Andover, Hampshire, UK.

Vergara, P., Fargallo, J.A., Randa, E., Parejo, D., Lemus, J., & Garcia-Montijano, M. (2008). Low frequency of anti-acetylcholinesterase pesticide poisoning in lesser and Eurasian kestrels of Spanish grassland and farmland populations. *Biological Conservation* **141**, 499–505.

Wilson, J.D., Morris, A.J., Arroyo, B.E., Clark, S.C., & Bradbury, R.B. (1999). A review of the abundance and diversity of invertebrate and plant foods of granivorous birds in northern Europe in relation to agricultural change. *Agriculture, Ecosystems and Environment* **75**, 13–30.

Wobeser, G., Bollinger, T., Leighton, F.A., Blakley, B., & Mineau, P. (2004). Secondary poisoning of eagles following intentional poisoning of coyotes with anticholinesterase pesticides in western Canada. *Journal of Wildlife Disease* **40**, 163–172.

Woodwell, G.M., Wurster Jr.,C.F., & Isaacson, P.A. (1967). DDT residues in an East Coast estuary: a case of biological concentration of a persistent insecticide. *Science* **156**, 821–824.

Yu, S.J. (2008). *The Toxicology and Biochemistry of Insecticides.* CRC Press, Boca Raton, Florida, USA.

Chapter 12

ALTERNATIVES TO INSECTICIDES

There are numerous pest management techniques that provide alternatives to insecticides, but none that can be used for nearly all pest situations, as is the case for insecticides. Alternative approaches often entail advance planning, and rely on preventing problems from developing. Insecticides, in contrast, can be corrective or curative as well as preventative. So when unforeseen problems arise, insecticides can usually be relied upon to 'fix' the problem. Also, in some cases alternatives are not as effective as insecticides, or more costly to implement.

On the other hand, many alternatives to insecticides are safer to humans and wildlife, more economical, and in some cases they are as reliable or convenient. Although using alternatives to insecticides often requires considerable knowledge or advance planning, nonetheless are often preferred over use of insecticides, especially if wildlife conservation is a concern. The following are the principal alternatives to the use of insecticides.

ENVIRONMENTAL MANAGEMENT OR CULTURAL CONTROL

Manipulation of environments or cultural practices can often minimize the suitability of plants or environments for insect pests, thereby reducing or eliminating the need for pesticides. Environmental or cultural practices are normal activities associated with maintaining

Insects and Wildlife: Arthropods and Their Relationships with Wild Vertebrate Animals, 1st edition. By J.L. Capinera. Published 2010 by Blackwell Publishing.

an environment or producing a crop and include activities such as plowing fields, planting a crop, water management, and harvesting (Fig. 12.1). Slight changes in these cultural practices can do much to minimize pest problems, and the need for chemical suppression of pests. Examples of environmental and cultural manipulations include:

• Many mosquitoes (Diptera: Culicidae) typically breed successfully in temporary sources of water such as melt-water from snow, retained overflow from exceptionally high tides along the coastline, excessive rainfall that is stored temporarily, and overflow (tail-water) from incorrect irrigation. More permanent water sources such as ponds, lakes, and even irrigation ditches can be inhospitable to mosquito larvae due to the presence of predatory fish and insects dwelling in these permanent water sources. These permanent water sources may be compromised, however, if emergent or aquatic vegetation are present because the mosquito larvae can hide from the fish. Thus, ensuring that water is present permanently rather than transiently, that predatory fish and insects are present, and that aquatic vegetation is not too abundant, are types of environmental management that can be used to ensure that mosquitoes are not too abundant.

• Larvae of diamondback moth, *Plutella xylostella* (Lepidoptera: Plutellidae), are the most important defoliators of cabbage and related crops. Although they have evolved resistance to many insecticides, due to their small size these insects are quite susceptible to being washed from the plants or drowning. Overhead irrigation greatly reduces the number of these small insects, and the timing and level of irrigation can be manipulated to minimize larval survival.

• The most important pest of alfalfa (lucerne) in many areas of the world is alfalfa weevil, *Hypera postica*

Fig. 12.1. Some environmental management practices that affect insects: (a) minimum tillage (crop residues remaining on the surface of the soil have significant beneficial effects by providing harborage for predatory beetles, or it can be deleterious by providing shelter for slugs); (b) intercropping (crop diversity such as the alternating strips of corn and bean shown here makes it more difficult for specialist herbivores to locate their host plants); (c) reflective mulch (reflective plastic mulch repels flying insects that are seeking to alight, delaying the onset of infestation and the transmission of plant viruses to crops); (d) timing of planting (the crop or a susceptible stage of the crop can be timed to avoid the occurrence of short-lived insects; (e) irrigation (irrigation water can drown young larvae, wash them from the foliage, or raise the humidity in the crop canopy to favor the spread of fungal disease); (f) mowing (plant architecture is important in defining plant suitability for insects; for example, maintaining short vegetation can reduce the suitability of the habitat for grasshoppers around crops or within citrus groves) (Fig. 12.1d from USDA, ARS).

(Coleoptera: Curculionidae). Harvesting of the first cutting of alfalfa crops in some areas can occur before alfalfa weevil reaches the late larval period, a time when it is most voracious and capable of causing the most defoliation. By cutting the alfalfa slightly early and allowing it to dry in the field, the weevil larvae are deprived of suitably fresh foliage and perish. This early harvesting may reduce maximum yield, but by eliminating loss due to weevils not only in the first cutting period but also in subsequent harvests, it more than compensates for the early harvesting.

• Southern chinch bug, *Blissus insularis* (Hemiptera: Blissidae), is more troublesome when lawns are treated with high levels of nitrogen-containing fertilizers because the nitrogen makes their food plants more suitable for chinch bug reproduction. Minimizing nitrogen fertilization and using slow-release fertilizer formulations minimizes the problem caused by chinch bug population growth resulting from sudden increases in nitrogen in lawn grasses.

• Old pine trees, especially those stressed by crowded conditions and drought, are especially suitable hosts for survival of southern pine beetle, *Dendroctonus frontalis* (Coleoptera: Curculionidae). The larvae of these insects tunnel in the inner bark (phloem), disrupting the physiology of the tree. This beetle also transmits a blue-stain fungus, which hastens the killing of the trees. Young and healthy trees are reasonably successful at repelling attacks by beetles. However, once beetle populations build up to very high numbers, even healthy trees are susceptible to attack because the large number of attacking beetles can overwhelm the innate defenses of the trees. Effective management of pine forests that might be susceptible to southern pine beetles involves harvesting some of the trees before they become senescent, and adequate thinning of trees to minimize inter-tree competition for water and nutrients. Careful monitoring of the trees, and rapid removal/harvest of trees and nearby trees that show signs of beetle infestation, also serve to reduce risk to the forest.

• Weed management often is an important means of modifying the environment to eliminate habitat that is suitable for insect pests. Insects such as grasshoppers, aphids, and whiteflies often attain high levels of abundance on weeds growing along fences, roads, irrigation ditches – or in some cases, in fallow (uncultivated) or abandoned crop fields. The insects then may spread from the weedy areas to the crop, infesting the crop with pests and perhaps transmitting plant diseases.

Tilling, burning, cutting, or treating the weedy areas with herbicide may eliminate the threat posed by this harborage.

• Similarly, abandoned or neglected crop fields can serve as a source of insects for nearby crops. Farmers sometimes do not harvest a crop or the entire crop because market conditions do not allow them to make a profit. If they do not treat or destroy the crop promptly it can become a breeding ground for insects that then disperse into adjacent crop fields. This has proven to be a major source of vegetable crop problems in warm-weather areas where whiteflies and the virus diseases they transmit often are harbored in abandoned crop fields. Thus, timely destruction of crop residues is an important management technique.

Environmental management approaches would seem to have great appeal because they can reduce or eliminate pesticide use. However, sometimes there are unintended effects that detract from the value of environmental management. Habitat management for mosquito suppression in coastal areas is a good example of how initial efforts to impose nonchemical control resulted in some undesirable consequences requiring refinement of the technique.

Some salt marsh-dwelling and mangrove swamp-dwelling mosquitoes such as *Ochlerotatus taeniorrhynchus* and *Ochlerotatus sollicitans* will not lay their eggs upon standing water. Rather, they deposit eggs on moist soil, and the eggs hatch when the area is flooded. These (and other) mosquitoes are therefore known as floodwater mosquitoes. Very large numbers of mosquitoes can hatch when these moist coastal soils become flooded and retain water for 5–7 days, allowing the eggs to hatch, complete their immature stages, and emerge as adults. This happens irregularly due to storms and regularly due to 'spring tides.' During the new and full-moon periods, the sun and moon align with the earth, causing exceptionally high and low tides that are called spring tides; thus, approximately twice each month there can be unusually high tides in coastal areas, followed by a sudden burst of mosquito egg hatching and then winged adult mosquitoes.

Initial efforts to prevent floodwater from being retained long enough to allow mosquito emergence along the coast involved construction of parallel ditches (Fig. 12.2) that allowed water to drain quickly and provided access by fish to the temporary pools of floodwater. Mosquito ditching worked reasonably well in areas with high tide amplitude, but not very well in

Fig. 12.2. Some physical and mechanical insect control measures: (a) water management (draining areas of coastal marsh that are flooded only periodically, such as periods when tides are unusually high, eliminates standing water and causes mosquito larvae that are living within this water to perish or to be consumed by fish that swim up the ditches); (b) vacuum equipment (the equipment shown here is good at capturing insects and destroying those that take flight readily in response to disturbance, such as adult plant bugs [Hemiptera: Miridae], but has little effect on more sessile species such as nymphs of whiteflies [Hemiptera: Aleyrodidae]); (c) crop harvesting (early cutting and drying of alfalfa kills larvae of alfalfa weevil and can result in little crop reduction but major insect population suppression); (d) sticky traps (colored traps such as the deer fly trap shown here that attracts flies by the combination of a dark color and movement can physically remove insects from the environment); (e) flaming (burning vegetation can kill insects harbored there); and (f) light traps (a battery-powered light trap, shown here recharged by solar power, captures night-flying insects nightly).

areas with low amplitude. In either case, continual maintenance was required or the ditches would become plugged, causing retention of water and exacerbating the mosquito problem. Thus, this approach was largely abandoned in favor of impounding water.

Mosquito control impoundments are diked areas where water can be maintained at a relatively constant depth, thereby preventing floodwater mosquito eggs from hatching. They also retain water for fish and other aquatic life that can feed on mosquitoes that oviposit in standing water. Impoundments are effective at preventing mosquito populations from developing to high levels, but can have adverse effects on estuary ecology. Impoundments interfere with the flow of nutrients and organisms between estuarine and ocean environments. Because they are relatively closed systems and not replenished by tidal flows, salinity declines and the saltwater marshes can grow to resemble freshwater marshes. The continuous flooding of impoundments and decrease in salinity result in changes in the vegetation and fish abundance and diversity. Species of fishes that use salt marsh vegetation for part of their life cycle are denied access.

The early mosquito control impoundments were more reliable than ditching as effective environmental management tools for alleviating mosquito problems, but they too are being abandoned because they disrupt the ecology of the area, particularly because they impact wildlife. The management of the impoundments can be modified using a technique called **rotational impoundment management** to reduce the impact of impoundments on the local ecology but still achieve mosquito protection. The fundamental concept underlying rotational impoundment management is that water levels need to be strictly controlled only during the mosquito breeding season, and tidal flows can be allowed during the remainder of the season. Thus, even in a warm-weather climate like Florida, USA, where the mosquito breeding season is long, gates can be left open for inflow and outflow of water during the cooler (fall and winter) months. In addition, during the summer months when constant water levels are desired salt water can be pumped in to maintain salinity because they can be equipped with a spillway that allows excess water to flow out, preventing the water level in the impoundments from rising. Thus, with the rotational impoundment management system mosquito management is attained with minimal effect on local ecology.

PHYSICAL AND MECHANICAL CONTROL

Sometimes we can use physical and mechanical means to manage insects (Fig. 12.2). Physical and mechanical methods of insect management often are closely related or overlapping elements of management. **Physical controls** usually involve adverse environmental properties such as heat and cold, irradiation, and modified (controlled) atmosphere. **Mechanical control** typically involves devices, and includes vacuums, barriers, and trapping, including electrocution by 'bug zappers.' Physical and mechanical controls sometimes can be part of environmental or cultural practices, so they can be categorized either way; the examples below involving tillage and weed destruction are appropriate examples of this. Examples of physical and mechanical control include:

• Controlled atmospheres include any deliberately modified deviations from the normal gaseous state of the atmosphere that affect insect survival. Often this means higher levels of nitrogen or carbon dioxide than normal, usually in conjunction with cooler or warmer temperatures, as this provides additional stress to insects. Controlled atmosphere treatments often are used to disinfest fruit of large fruit flies (Diptera: Tephritidae) if they are to be shipped internationally. Controlled atmosphere treatments are usually implemented in storage (warehouses) or in transit to market (in the hold of ships).

• In the case of mosquitoes (Diptera: Culicidae), atmosphere can be regulated under field conditions. Although most aquatic insects use gills to obtain oxygen that is dissolved in water, mosquitoes usually obtain oxygen at the water surface. Thus, periodically they must come to the surface where they use their respiratory siphon to break the water surface and take oxygen directly from the air. Monomolecular surface films are oil-like products that can be applied to the water. So long as the water is standing and not excessively agitated, the films provide a thin barrier at the water surface that prevents mosquitoes from reaching the air above the surface of the water. Mosquitoes usually perish within 1–3 days, and most aquatic invertebrates and fish are unaffected by waters treated with monomolecular films.

• Many plant-feeding insects are attracted to visual stimuli such as color. Although you might expect that green would be the optimal hue for phytophagous (plant-feeding) insects, it turns out that in most cases yellow is more attractive. So yellow traps often are used to attract insects to traps. The trick is to capture them once they are attracted, and typically this is done with a sticky adhesive (a sticky trap) or soapy water (a water pan trap). These traps are most often used as sampling devices, but in a contained environment such as a greenhouse they work quite well in suppressing the abundance of flying pest species such as fungus gnats (Diptera: Mycetophilidae) and whiteflies (Hemiptera: Aleyrodidae).

• Cutworms (Lepidoptera: Noctuidae), which often hide in the soil during daylight hours, often feed at night on seedling plants at the soil surface. Consequently, when gardeners or farmers awaken they find that their plants have been severed by the feeding activities of the cutworms. More alarming, the cutworms may actually drag the severed seedling plant belowground, where they feast upon it during the daylight hours. In this case, the gardeners awaken to discover that the plant has literally disappeared, and only if they dig around in the soil do they find the plant and culprit that is responsible for the 'dirty deed.' On a modest scale, as in a garden, physical barriers can be created to minimize cutting by cutworms. A waxed carton or metal can with both the top and bottom removed can be placed around the plant at the soil surface to serve to block the cutworm from accessing the plant, especially if it is inserted a few centimeters below the soil surface. Another type of barrier can be created by aluminum foil, which is simply wrapped around the stem of the seedling, preventing the cutworm from chewing on the stem of the seedling.

• Insect light traps rely on emission of light at night to disorient insects and cause them to become 'attracted' to the light. This disorientation is often optimized by using 'black-light,' or lights with high levels of emission in the range of nearly invisible (to people) ultraviolet light. The insects, once oriented to the trap, may be physically restrained and captured, or electrocuted (the so-called 'bug-zappers'). Though it may be satisfying to see the piles of dead insects that collect in or below such traps, there are few data to suggest that pest insects, especially mosquitoes, are attracted in great numbers to light-based traps. Thus, they are not recommended by responsible individuals for mosquito control. When combined with odor-based attractants they are more effective for mosquito capture, though such traps are expensive and have a limited range of effectiveness.

• Plowing and disking the soil can injure or crush soil-dwelling insects such as scarab beetle larvae (white grubs) (Coleoptera: Scarabaeidae), caterpillar larvae (cutworms) (Lepidoptera: Noctuidae), and click beetle larvae (wireworms) (Coleoptera: Elateridae). In addition, tillage may bring the insects to the soil surface, exposing them to predators such as birds. Small flocks of birds often can be seen following tractors as they till the soil, where the birds feast on the exposed insects, thereby providing biological suppression.

HOST RESISTANCE

Cultivars, varieties, and breeds are often selected by plant and animal breeders because they differ in their appearance, growth rates, or other performance characteristics. However, such differences also include variation in structure and chemistry beyond the characteristics being assessed by the breeders, and these differences may affect susceptibility to insects. Occasionally, breeders select for insect resistance characteristics, but this is an unfortunately uncommon practice, largely because insecticides are usually available to 'solve' any pest problems should they occur. All too often, breeders select for appearance, taste, texture, and growth rate characteristics without any regard to susceptibility to insects, thus setting the stage for insect problems and greater insecticide use.

Host resistance often is due to lack of attractiveness (due to lack of chemical stimuli, the host is not recognized as a suitable host), but also because feeding or reproductive performance of the insect is inhibited by chemical constituents or by structural characteristics. Examples of the latter include nutritional inadequacy, the presence of **allelochemical compounds** (non-nutritive compounds produced by plants that affect other organisms, in this case insect preference and plant suitability) such as alkaloids, phenolics, resins, glucosinolates, and even **structural characters** such as spines or **trichomes** (Fig. 12.3) that physically impale the insect or entangle it in sticky exudates. It is easy to under-appreciate the significance of a hairy leaf or waxy plant surface, but if you are a small insect trying to alight, search, and feed on these substrates, the physical and chemical attributes of the host can be quite formidable. Think about the difficulties of walking on ice, or navigating through a dense forest, and you have some idea what insects face when trying to feed on a plant. It is easier to appreciate the importance of allelochemicals that affect the taste of a plant because people, like insects, have very definite taste preferences. Using the human analogy again, some people

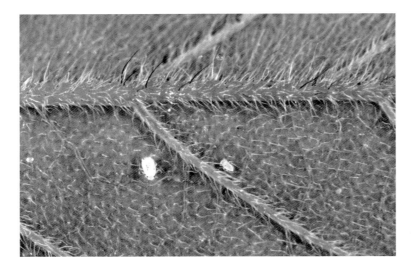

Fig. 12.3. Host plant resistance takes many forms. Here you see the potential impact of trichomes (leaf hairs) on the ability of insects (in this case, aphids) to move and feed, but just as often it is the chemical constituents of plant foliage that are the basis for resistance.

like onion, garlic, pepper or cabbage, but others hate these tastes; insects behave similarly. In fact, the major difference between insect and human feeding preferences is that some insects (monophagous or specialist species that have a restricted host range) are 'addicted' to specific chemical stimulus and will starve rather than eat food lacking this stimulus. Humans tend to be omnivorous, or at least they do not restrict their diet to a single plant species or family. Indeed, unlike insects, humans could not survive on such a restricted diet.

There are several ways to classify the basis of host resistance, but one common approach is to recognize three basic types:

• ***Antixenosis or nonpreference.*** In the expression of antixenosis, the host is not as attractive to the insect as might occur with other strains or varieties of the host. The insect chooses not to select the prospective host, perhaps because the host does not produce the proper chemical stimuli, or the prospective host has physical properties that discourage the insect from attacking the host.

• ***Antibiosis.*** In the expression of antibiosis, the host interferes with some life process of the insect that is attacking the host, keeping the insect from feeding or growing normally, and perhaps even causing death of the insect.

• ***Tolerance.*** In the expression of tolerance, the host has the ability to grow and produce normally despite supporting the feeding of insects. For this to occur, the tissue affected by the insect must not be essential, or

the host has the ability to repair the damage without incurring significant damage.

These classes of resistance portray the various mechanisms that have naturally evolved in hosts to combat the feeding by insects, and give insight into the mechanisms that should be investigated by breeders as they search for non-insecticidal solutions to pest problems.

Most often, breeders identify varieties with resistance that is due to antibiotic properties, probably because these characteristics are easiest to identify. After all, if insects are unable to grow normally or reproduce on a host, this is a dramatic and easily detectable response. Surprising to some, however, is that the antibiosis-based resistance is probably not the ideal form of resistance. When insects that are affected by antibiosis-based resistance fail to reproduce, the genetic structure of the population is at risk of being changed. In other words, the susceptible insects perish without reproducing. Inevitably, it seems, when this kind of dramatic mortality is inflicted on an insect population there will be a few insects that are slightly different genetically, and they will possess the ability to detoxify or overcome in some manner the basis for host resistance. They not only survive but they reproduce, and pass on to their offspring the characteristics that allow them to overcome the host resistance. Thus, in a short time the genetic structure of the population is changed from consisting mostly of individuals that are inhibited by the resistance mechanisms of the host, to a population that mostly can overcome the resistance of the host.

Does this sound familiar? This is the same mechanism we discussed with respect to development of resistance to insecticides. This should not be surprising; a toxin is a toxin, whether natural or synthetic. Thus, if you examine the history of aphid-resistant wheat and sorghum varieties, for example, there is an interesting pattern of new varieties being introduced, followed by adaptation by the aphids that feed on these plants, followed by development of new varieties, etc. So although antibiosis-based host resistance is valuable, it is not perfect. Insect populations possess the ability (in time) to overcome host resistance just as they overcome insecticide toxicity. One important lesson is that the higher the level of insect mortality caused by the resistant host (or the insecticide) the more rapid will be the change in the genetics of the insect and the evolution of mechanisms that overcome resistance. Therefore, it is probably better to select for hosts that are tolerant of insects, rather than possessing antibiosis, because tolerance does not inflict high levels of mortality and cause marked shifts in the genetic structure of insect populations. Unfortunately, it is harder to identify this form of resistance, so it is not often developed. Another important lesson is that less complex forms of resistance (under the control of a single gene, and usually called **monogenic resistance**) are easier to overcome because few mutations are necessary to impart the ability to overcome the resistance mechanism. Thus, more complex forms of resistance (under the control of several genes, and usually called **polygenic resistance**) are more desirable.

A recent trend in the breeding of plants that are resistant to insects is to use molecular techniques to insert into the plant the toxins produced by microbial organisms such as the bacterium *Bacillus thuringiensis*. This insertion confers to the plant the insect-killing properties formerly found only in the bacteria. Mostly this is directed against caterpillar pests, although this technology can also be used to protect against some other insects by using variants of the bacterial toxin. This has resulted in significant reductions in insecticide use in some crops, though by using this toxin widely there is selective pressure on the insect populations that could lead to development of the ability to overcome plant resistance. To forestall this development, crop producers often are required not to plant their entire crop to the genetically modified, bacterial toxin-containing strain, thereby allowing survival of the 'normal' toxin-susceptible individuals. Nevertheless, insects have been able to overcome *Bacillus thur-*

ingiensis-based plant resistance within 7–8 years in some locations. Also, plant breeders now no longer create plant varieties with just a single type of toxin, instead inserting several toxins, thereby making it less likely that the insects can evolve resistance.

SEMIOCHEMICALS

Semiochemicals are chemicals that are used in communication between organisms. For this reason they have also been called 'information chemicals' or 'info-chemicals.' The chemicals that mediate between members of the same species are called **pheromones**. The chemicals that mediate between members of different species are called **allelochemicals**. Both pheromones and allelochemicals have limited use for management of insect pests (Fig. 12.4).

Pheromones are used principally for attraction of adult insects. **Sex pheromones** are used naturally by insects to bring the members of the opposite sexes together for mating, so it is logical for humans to synthesize and use these materials to cause insect attraction to traps, where they are captured. Various pheromone releasers and traps have been devised which provide optimal release of pheromone and capture of insects. This works very well for monitoring the abundance of adults, but is not effective for assessing the abundance of immature insects because they are not attracted and lack much mobility. Sex pheromones are usually not very good for reducing insect abundance because typically they capture only male insects (the females normally produce the chemical scent). On the other hand, if the atmosphere is flooded with pheromone scent, the males get disoriented or confused, and cannot easily find the females. This approach, called 'male confusion,' can result in reduced breeding and population decline. Also, so-called 'attract and kill' technologies use sex pheromones to attract insects to a gel or other matrix containing insecticide; once they contact the toxicant in the matrix, death is assured. **Aggregation pheromones** bring together members of both sexes, commonly to locations that are optimal for both feeding and reproduction. Obviously, aggregation pheromones as well as sex pheromones can be used for attract and kill technologies.

Knowledge of allelochemicals can similarly be used to create traps because some insects use the scent of their host to aid in food location. Floral scents can be

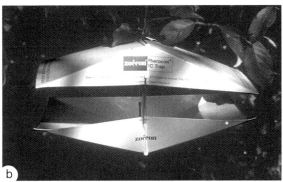

Fig. 12.4. Insect pheromones are used for attracting and capturing insects, either as a population monitoring technique or as a population suppression technique. Traps usually are based on capture of insects with a sticky surface, or with a funnel apparatus as they climb upward or fall downward: (a) attraction of a grape root borer, *Vitacea polistiformis* (Lepidopera: Sesiidae), moth (note the flying moth indicated by the arrow) to a pheromone lure within a universal bucket trap (once attracted to the plastic 'bucket' by a pheromone lure they fall down and are captured in the bucket); (b) a sticky trap (a pheromone lure is suspended within the cardboard trap, and insects attracted to the odor stumble into the sticky bottom half of the trap and are captured); (c) a variation of the bucket trap designed to capture and eliminate gypsy moth, *Lymantria dispar* (Lepidoptera: Lymantriidae), although the females are flightless and only the male moths are attracted to the pheromone; (d) a pheromone-baited water pan trap capturing European corn borer, *Ostrinia nubilalis* (Lepidoptera: Pyralidae), which is more easily captured by a soapy water solution than by a sticky surface; (e) three pheromone traps being assessed for capture of sweetpotato weevil, *Cylas formicarius* (Coleoptera: Brentidae), with (left) a standard bucket trap (the insect falls downward upon entering the trap), (center) a specially designed trap that takes advantage of the weevil's tendency to climb upward on the outside of a metal screen cone and then to fall into a capture device at the center, and (right) a trap designed for boll weevil, *Anthonomus grandis* (Coleoptera: Curculionidae), that captures insects crawling within a screen cone. (Fig. 12.4a by Scott Weihman, 12.4d by David Thompson, 12.4e by Rick Janzen).

used instead of sex pheromones to bait traps and monitor insect population abundance. Allelochemicals usually are not sex specific, so they sometimes can be used for mass capture of insects, but they typically are not very powerful attractants so the range of attraction is limited. Their levels can also be modified in cultivated plants to make the plants less attractive to insect herbivores. Host odors also are part of the basis for capture of mosquitoes in some mosquito traps.

BIOLOGICAL CONTROL

Biological control usually is defined as the action of natural enemies to suppress the abundance of pest species. **Natural enemies** of insects (Figs. 12.5–12.7) are usually predators (mostly insects, but also vertebrates, that quickly kill and consume insects), parasitoids (parasitic organisms larger than microbial organisms, such as other insects and nematodes, that slowly kill their host), and pathogens (microbial disease agents that kill or negatively affect their host). Sometimes the definition of 'biological control' is expanded to include the use of all processes that are fundamentally biological, rather than just natural enemies, and might include host plant resistance or the use of semiochemicals. However, it is better to limit the term 'biological control' to the roles of natural enemies. Biological control by natural enemies can occur naturally, or it can be fostered. There are four principal forms of biological control, reflecting its occurrence:

• *Natural biological control.* This form of biological control reflects the natural state of biological suppression by natural enemies without any habitat or natural enemy manipulations by humans. It is the total suppressive action by various naturally occurring predators, parasitoids, and pathogens. In natural ecosystems, natural biological control often is adequate to keep insect populations from over-exploitation of its food resources, though other factors such as competition and weather also act on insect populations to keep them from attaining a very high level of abundance. Natural biological control agents give the appearance of providing 'balance of nature.' Contrary to the belief by some, there is no **balance of nature**, the idea that in nature there exists an inherent equilibrium founded on the interactions of plants and animals, resulting in a stable, continuing system of life. The so-called 'balance' is continually upset by natural events and by the activities of humans. Outbreaks of insects are an example of the disequilibrium that disproves the notion of 'balance', or at least the notion of stability. However, the **resiliency** (rapid return to equilibrium after perturbation) of both producer and consumer organisms, and the **persistence** (tendency of organisms to persist even when faced with catastrophic changes in abundance), can be construed as supportive of the balance hypothesis, or at least supportive of the idea of continuance or constancy. Ecosystems are mosaics of patches, and random events or biological feedbacks that appear to be destabilizing or catastrophic on a local scale may be much less so when viewed on a larger scale. Clearly, ecological stability is scale-dependent, and stability usually exists only on a very broad scale. The elements of balance in insect populations that do exist, however, are due in large measure to natural enemies.

• *Augmentative biological control.* Augmentation involves release of natural enemies into the environ-

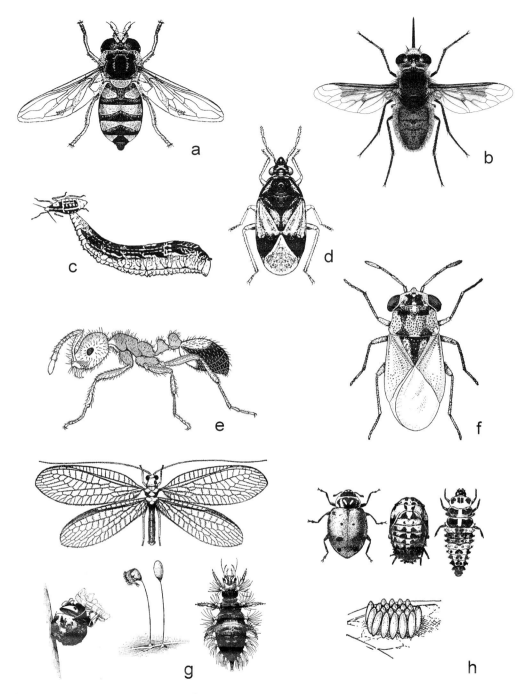

Fig. 12.5. Representative common insect predators: (a) an adult flower fly, *Didea fasciata* (Diptera: Syrphidae); (b) an adult bee fly, *Callostoma fascipennis* (Diptera: Bombyliidae); (c) larva of *D. fasciata* feeding on aphid; (d) a minute pirate bug, *Orius insidiosus* (Hemiptera: Anthocoridae); (e) a predatory ant, *Solenopsis invicta* (Hymenoptera: Formicidae); (f) a big-eyed bug, *Geocoris pallens* (Hemiptera: Lygaeidae); (g) a green lacewing, *Chrysopa microphyta* adult (top), cocoon and shed nymphal cuticle (bottom left), eggs with hatching larva (bottom center), larva (bottom right) (Neuroptera: Chrysopidae); (h) convergent lady beetle, *Hippodamia convergens* adult (top right), pupa (top middle), larva (top right), eggs (below) (Coleoptera: Coccinellidae).

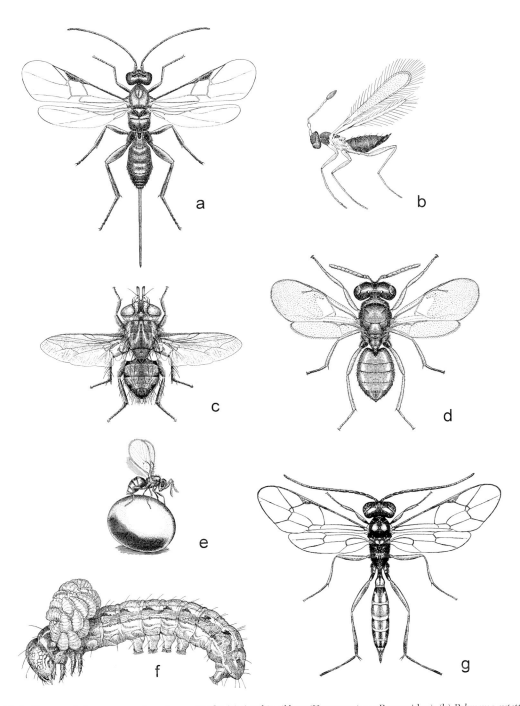

Fig. 12.6. Representative common insect parasitoids: (a) *Agathis gibbosa* (Hymenoptera: Braconidae); (b) *Polynema eutettixi* (Hymenoptera: Mymaridae); (c) *Phorocera claripennis* (Diptera: Tachinidae); (d) *Pteromalus eurymi* (Hymenoptera: Pteromalidae); (e) *Trichogramma minutum* (Hymenoptera: Trichogrammatidae) on moth egg; (f) ectoparasitic larvae of *Euplectrus plathypenae* (Hymenoptera: Eulophidae) on caterpillar; (g) *Olesicampe* sp. (Hymenoptera: Ichneumonidae).

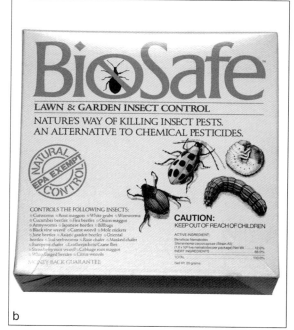

Fig. 12.7. Representative commercially available biological control agents: (a) a bacterial product, based on *Bacillus thuringiensis*, used for caterpillar control; (b) an entomopathogenic nematode product, based on *Steinernema* nematodes, for insects dwelling in soil and other cryptic habitats; (c) a formulation of predatory *Pytoseiulus persimilis* mites, applied to control phytophagous mites; (d) a microsporidian (protozoan) product, based on *Nosema locustae*, used for control of grasshoppers.

ment to supplement naturally occurring biological control. Often it involves artificial culture of natural enemies in insectaries, and periodic release of the natural enemies into closed systems such as greenhouses, where high levels of natural enemies can be maintained, and pests suffer high mortality. This can be accomplished under field conditions as well, but it requires more effort and expense.

There are two forms of augmentative biological control, inoculative and inundative. With **inoculative biological control**, relatively small numbers of biological control agents are released, which then reproduce and spread to suppress the pest species. This approach works well where the environment generally favors the biological control agent but some defect in the system prevents it from occurring continuously. For example, the lady beetle *Cryptolaemus montrouzieri* (Coleoptera: Coccinellidae) is established in California, USA, but can overwinter successfully only along the coast. Annually, beetles are released (inoculated) into citrus groves elsewhere in the state, where they multiply and provide good suppression of mealybugs on the citrus crops. **Inundative biological control** occurs when large numbers of natural enemies can be made available economically, releases of large numbers can be made repeatedly, and perhaps there is need to achieve immediate suppression of the pest. An example of this approach is the use of the wasp parasitoid *Muscidifurax raptor* (Hymenoptera: Pteromalidae) for control of dung-breeding flies in feedlots, poultry operations, and dairy barns. In this case, releases are made weekly during the summer.

• *Conservation biological control.* This involves the implementation of practices that help maintain and enhance the actions of natural enemies. Natural enemies need food, water, shelter, and protection from insecticides if they are to be effective. Maintenance of non-crop plants, especially flowering plants, may enhance survival of natural enemies. Flowering plants can be planted along field margins, or even within crops to enhance natural enemy survival. Reduction in insecticide use, or use of insecticides that possess short residual life, also help to conserve biological control agents.

• *Classical biological control.* This involves importing and establishing natural enemies so that they can suppress the abundance of newly established pest species. It is also known as 'importation biological control.' Classical biological control is based on the assumption that (1) there are effective

natural enemies in the homeland of the invader, and (2) if imported, the natural enemy or enemies will provide effective suppression of the invading pest. However, it can be difficult or impossible to find an appropriate natural enemy from the source of the invading insect. Some potential natural enemies may not thrive in the new environment, or the natural enemies may not be sufficiently specific to allow introduction (generally, specificity is necessary to assure that nontarget insects will not be affected). However, if classical biological control is successful, it often provides spectacular success and permanent, economic suppression of the pest. This approach is an important element of modern insect management because so many serious pests are invaders. For example, nearly 2000 insect pests have successfully invaded the USA and established, often becoming some of the most serious pests. There have been very few instances where insects imported deliberately for biological suppression have caused inadvertent problems, but there is always concern that imported species will not perform as expected. With current risk assessment procedures, it is highly unlikely that imported species will cause harm.

Biological control has many important advantages over most other insect management techniques. From the perspective of wildlife management, the most important is that no insecticides are used, so safety to wildlife and humans is a major advantage. However, habitat modification may not be required, which also can be beneficial to wildlife, as removal of weeds can deprive wildlife of cover and food. From the perspective of the user, sometimes biological control can be implemented very economically, which may be a significant factor in determining how to manage pests. Some examples of successful biological control are shown in Table 12.1.

AREA-WIDE INSECT MANAGEMENT

Pest management is usually performed on a local basis. Individual landowners make their own decisions on what procedures to use and when to deploy them. Their neighbors may do nothing, allowing emigration from neighboring fields to reduce or eliminate the benefits of treatment. An alternative to independent decision-making is to organize a regional approach that optimizes pest control over a large geographic area; this is called **area-wide management**. This is par-

Table 12.1. Examples of completely or partially successful biological suppression of arthropod pests by using natural enemies.

Natural enemy	Pest
Augmentative biological control	
Amblyseius cucumeris (predatory mite)	Thrips
Cryptolaemus montrouzieri (predatory lady beetle)	Mealybugs
Diglyphus isaea (parasitic wasp)	*Liriomyza* leafmining flies
Aphidoletes aphidimyza (predatory gall-midge)	Aphids
Steinernema feltiae (parasitic nematode)	Sciarid fungus gnats
Eretmocerus eremicus (parasitic wasp)	Whiteflies
Amblyseius californicus (predatory mite)	Spider mites
Trichogramma brassicae (parasitic wasp)	Eggs of moths
Orius insidiosus (predatory bug)	Thrips
Classical biological control	
Aphytis holoxanthus (parasitic wasp)	Florida red scale
Amitus herperidum (parasitic wasp)	Citrus blackfly
Eremocerus paulistus (parasitic wasp)	Woolly whitefly
Chrysocharis laricinellae (parasitic wasp)	Birch sawfly
Lathrolestes nigricollis (parasitic wasp)	Birch leafminer
Ageniaspis citricola (parasitic wasp)	Citrus leaf miner
Aphidius smithi (parasitic wasp)	Pea aphid
Apanteles solitarius (parasitic wasp)	Satin moth
Trissolcus basilis (parasitic wasp)	Southern green stink bug
Anagyrus kamali (parasitic wasp)	Pink hibiscus mealybug
Dacnusa dryas (parasitic wasp)	Alfalfa blotch leafminer
Bathyplectes curculionis (parasitic wasp)	Alfalfa weevil
Ooencyrtus kuvanae (parasitic wasp)	Gypsy moth
Ormia depleta (parasitic fly)	Mole crickets

ticularly appropriate when individual landowners are not capable of meeting the challenge from particularly mobile or damaging pests.

Some pest situations lend themselves to regional solutions. Perhaps the most common is mosquito control. Adult mosquitoes are quite mobile, so treating your own property does not ensure relief from mosquito bites and the potential of disease transmission. Therefore, suppression of larval and adult mosquitoes on a neighborhood, city, or county-wide basis is logical and economically effective. The same applies to crop pests, forest pests, and rangeland pests, but rarely do individual landowners band together to manage pests on a regional basis. The successful examples of regional or area-wide pest management tend to involve medical or veterinary pests, especially mosquitoes, and also forest pests, grasshoppers or locusts, and fruit flies. There are exceptions, of course, but most pest problems are dealt with by individual landowners.

Although area-wide management of pests is often the most effective method, if it is not implemented properly it can be quite hazardous to wildlife. Area-wide treatment of rangeland for grasshoppers (Acrididae, various species) and rangeland caterpillar, *Hemileuca oliviae* (Lepidoptera: Saturniidae) has, at times, been shown to cause bird mortality or bird population reductions without mortality being evident. When large areas are treated, the pest insects have no refugia, which is good, but the wildlife similarly cannot escape exposure, which is not good. Further, even if the insecticide is not directly toxic to wildlife, if nearly all the insects are killed due to treatment with a broad-spectrum insecticide, the wildlife are deprived of a food resource, which is nearly as damaging. In the case of birds, they usually leave the area when food is depleted. For less mobile animals, this may not be possible. Thus, careful selection of insecticides to identify those that are least damaging and least residual is important for protection of wildlife. The same problem occurs when wetlands are treated for mosquito suppression.

Area-wide management is a good system for implementing the **sterile insect technique**. Sterile insect technique is a method of suppressing reproduction in an insect population. Reproduction is disrupted by releasing laboratory-reared, sterilized insects into the wild, healthy population. The sterilized insects mate with the wild insects but, of course, no reproduction can occur. At least with some species, females mate only once, so if they mate with a sterile male they are effectively eliminated from the reproducing population. This is also called the 'sterile male technique,' reflecting an emphasis on releasing sterile males. Male insects often mate repeatedly, so it is most desirable to release sterilized males. There are no environmental hazards, and no direct risks to wildlife, associated with the sterile insect technique.

For the sterile insect technique to be effective, many more sterile insects must be present than wild, healthy individuals, as this increases the probability that the wild insects will mate with sterile insects. We usually try to attain a ratio of at least ten sterilized individuals for every healthy individual, so we must be able to mass-produce the insects easily and economically. We also must be able to sterilize them in a manner that does not disrupt their mating competitiveness, as they will be competing with wild males for the attention of wild females. This approach is best used on a large area, so there is minimal possibility of reinvasion by healthy individuals capable of reproduction. Natural barriers such as oceans and mountain ranges often help define the limits of sterile insect releases, and help to prevent reinvasion. When carried out repeatedly, sterile insect releases can be used to drive an insect to extinction, effectively eradicating it from certain geographic locations. This is usually done only for recently invading insects, of course, because their distribution is normally limited to a small geographic area. Once the invading insects spread to the full potential of their geographic range, it may be too expensive to implement a sterile insect program over the entire area.

INTEGRATED PEST MANAGEMENT (IPM)

Integrated pest management (IPM) is both a philosophy and an approach to controlling insects. IPM is based on the notion that the mere presence of insects does not constitute a 'problem'; rather, it is only when insects are abundant enough to be damaging that they are a problem. Sampling or monitoring populations is central to IPM, as only by doing this can we determine when insect abundance or damage levels have reached the 'problem' stage. IPM also promotes the use of a myriad of techniques to prevent insect problems from developing, and as a curative for when problems occur, rather than depending solely on insecticides.

In many cases, the pest status of insects can be defined in an economic context. A central parameter is the concept of the **economic injury level**, a level of insects or insect damage that is equivalent to the cost of controlling them. In other words, at some point it will cost you more to ignore the insect problems than it will cost to reduce them to an acceptable level. Obviously, most agriculturalists don't want to exceed the economic injury level because they are losing money to the activities of insects. Because you cannot get instant control of pests as they attain the economic threshold, you have to plan ahead and initiate suppressive efforts *before* this level is reached; the point at which a response is stimulated is called the **economic threshold** or **action threshold**, and it is typically expressed in abundance of insects or level of insect damage. The significance here is that it is not the mere presence of insects, or their occurrence at low levels, that triggers control activities. Consequently, much less insecticide tends to get applied when agriculturalists adhere to the IPM approach. The threshold concept is more problematic when human or wildlife health is being considered. It is difficult to place health in an economic context. Certainly, we can put dollar figures on work-days lost, and we can estimate revenue lost due to reduced hunting potential, but it is more difficult to express pain and suffering, or reduced abundance of songbirds, in monetary terms.

The emphasis on insect management techniques other than pesticides similarly can result in the reduction of insecticide use. Application of many of the techniques mentioned previously, such as cultural and biological control, can prevent or reduce the need for using insecticides. More commonly, it allows us to restrict insecticide use to a smaller area or a lower frequency of application. In essence, it is the integration of these various insect control techniques, both chemical and non-chemical, that is the core function of IPM, and is the basis for the name '*integrated* pest management.'

There is a general procedure or process for instituting IPM that can be applied to almost any situation. It consists of a sequential series of steps:
• prevent the problem from developing through environmental management or other techniques;
• inspect and monitor for both pests and the natural enemies of pests on a regular basis;
• be certain that you have a correct identification of the pest or problem before proceeding with corrective actions;

• carefully consider all the management options you have available, weighing economic, health, and environmental costs;
• evaluate the effectiveness of whatever management options you selected and implemented;
• follow up with re-inspection and monitoring to be sure that your corrective actions, if implemented, were adequate.

Note that this procedure emphasizes population monitoring (sampling) and impact assessment. IPM is much more than applying an insecticide on a predetermined schedule. It is a knowledge-based approach to population management.

PREVENTING VERSUS CORRECTING PROBLEMS

Many of the wildlife issues associated with insect control stem from our tendency to avoid putting time, energy, and resources into activities that might not be necessary. For example, when farmers select a crop variety to plant, the principal considerations tend to be when the crop can be planted and harvested, the yield potential, the sales potential and, of course, the potential financial return on the investment. In many cases, insects do not present problems every year. So, instead of planning to prevent insect damage (for example, by planting an insect-resistant crop), it may seem reasonable to wait until it appears that insects are threatening, and then act to suppress them. This often leads to insecticide use because these products usually have rapid curative action.

In the long run, preventing pest problems from developing is much more desirable than reacting to problems. Ideally, we would design crop production systems, forest production systems, as well as urban landscapes, and homes and businesses that are resistant or immune to pests. Though this often can be done, protection from pests often is relegated to a secondary issue, with the consequence being dependence on insecticides. As discussed earlier, there are many practices that can be implemented that prevent pest problems from developing:
• if water levels are managed properly, mosquito populations can often be eliminated or minimized;
• installation of door sweeps and weather stripping around doors and windows can practically eliminate the invasion of structures by peridomestic cockroaches, and the need for household pest control;

• window screens are routinely used in most wealthy countries where biting flies such as mosquitoes might be a problem. In poorer areas of the world, however, screens are unaffordable so diseases vectored by mosquitoes remain a serious problem. The simple act of making homes insect-proof would do much to curtail many serious diseases;
• when crop production schedules are adjusted, such as the timing of harvest in alfalfa, pests can be killed without resorting to insecticide;
• plowing down a crop, and destruction of crop residues, is important to keep insects such as whiteflies and aphids from continuing to multiply on the remaining crop and then dispersing from crop to crop;
• selective harvesting and thinning of trees to maintain a variable-aged forest experiencing minimum levels of competition can reduce the potential of bark beetle outbreaks;
• physical barriers, though too expensive and time consuming to use for large scale commercial agriculture, can be quite effective at preventing damage by cutworms in home gardens;
• destruction of weeds along fence rows and in fallow fields to eliminate insect breeding areas can do much to reduce the influx of pests into newly planted crops;
• when alien pests invade a new area and establish, searching their homeland for suitable natural enemies that can be imported into their new home can reduce the pest's density and damage, eliminating need for regular suppression.

We need to do more to promote the design of crop landscapes that resist insect invasion or pest increase, to manipulate crop, forest and rangeland to avoid conditions that lead to pest increase, and to design homes and buildings to inhibit invasion by nuisance pests and wood destroyers. Certainly, we must do a better job at stopping the inadvertent transport of organisms from continent to continent; many invertebrates have hitchhiked across oceans to become serious pests. To do this requires preventative actions, advance planning, and even financial investment when there is no guarantee of financial benefit. Our natural resources, including our wildlife populations, deserve better treatment than a continued assault by chemical insecticides. Preventative practices seem contrary to conventional business and political practices, but perhaps it is time to factor in the benefits of wildlife conservation.

SUMMARY

- Other, non-chemical methods of managing insecticides should be used to decrease dependency on insecticides. Among other useful approaches are:
  - environmental management or cultural control;
  - physical and mechanical control;
  - host resistance;
  - semiochemicals;
  - biological control;
  - area-wide insect management; and
  - integrated pest management.
- Host resistance is usually based on the presence of allochemical compounds or structural characters. Resistance is often classified as being due to antixenosis, antibiosis, or tolerance. The genetic basis for resistance is usually classified as monogenic or polygenic.
- Semiochemicals are often classified as pheromones (providing intraspecific communication) and allelochemicals (providing interspecific communication). The semiochemicals most often used for insect sampling and management are sex pheromones, but aggregations pheromones are sometimes used.
- Biological control generally depends on the use of natural enemies to suppress pest populations. The different forms of biological control include natural biological control, augmentative biological control, conservation biological control, and classical (importation) biological control.
- Integrated pest management approaches often are based on the economic threshold concept, wherein a certain level of pest abundance or damage triggers pest suppressive actions that forestall economic loss. Also inherent in this approach is the blending of different management techniques rather than depending on a single-factor approach, and regular population assessments.
- Environments that experience frequent pest problems should be modified or redesigned to help prevent the development of pest populations, rather than waiting for pest populations to develop and then applying insecticides as a corrective action.

REFERENCES AND ADDITIONAL READING

Blum, M.S. (2008). Allelochemicals. In Capinera, J.L. (ed.) *Encyclopedia of Entomology*, 2nd edn., pp. 121–133. Springer Science & Business Media B.V., Dordrecht, The Netherlands.

Brockmeyer, R.E., Rey, J.R., Virnstein, R.W., Gilmore, R.G., & Earnest, L. (1997). Rehabilitation of impounded estuarine wetlands by hydrologic reconnection to the Indian River Lagoon, Florida. *Journal of Wetlands Ecology and Management* **4**, 93–109.

Dent, D. (2000). *Insect Pest Management*, second edition. CABI Publishing, Wallingford, UK.

Hoy, M.A. (2008). Augmentative biological control. In Capinera, J.L. (ed.) *Encyclopedia of Entomology*, 2nd edn., pp. 327–334. Springer Science & Business Media B.V., Dordrecht, The Netherlands.

Hoy, M.A. (2008). Classical biological control. In Capinera, J.A. (ed.) *Encyclopedia of Entomology*, 2nd edn., pp. 906–923. Springer Science & Business Media B.V., Dordrecht, The Netherlands.

Klassen, W. (2008). Area-wide insect pest management. In Capinera, J.L. (ed.) *Encyclopedia of Entomology*, 2nd edn., pp. 266–282. Springer Science & Business Media B.V., Dordrecht, The Netherlands.

Klassen, W. (2008). Sterile insect technique. In Capinera, J.L. (ed.) *Encyclopedia of entomology*, 2nd edn., pp. 3541–3563. Springer Science & Business Media B.V., Dordrecht, The Netherlands.

McCravy, K.W. (2008). Conservation biological control. In Capinera, J.L. (ed.) *Encyclopedia of Entomology*, 2nd edn., pp. 1021–1023. Springer Science & Business Media B.V., Dordrecht, The Netherlands.

Mullen, G. & Durden, L. (eds.) (2002). *Medical and Veterinary Entomology*. Academic Press/Elsevier, Amsterdam, The Netherlands.

Rey, J.R, Shaffner, J., Tremain, D., Crossman, R.A., & Kain, T. (1990). Effects of re-establishing tidal connections in two impounded tropical marshes in east-central Florida. *Wetlands* **10**, 163–170.

van Emden, H.F. & Service, M.W. (2004). *Pest and Vector Control*. Cambridge University Press, Cambridge, UK.

Thacker, J.R.M. (2002). *An Introduction to Arthropod Pest Control*. Cambridge University Press, Cambridge, UK.

CONSERVATION ISSUES

Chapter 13

INSECT–WILDLIFE RELATIONSHIPS

Insects and wildlife clearly affect each other in myriad ways. Although some wildlife feed regularly on insects, and some insects feed on wildlife, the relationships of insects and wildlife are much more complex and interesting than 'who-eats-whom.' Here I discuss the importance of the insect–wildlife relationship.

HOW WILDLIFE AFFECT INSECT SURVIVAL

Vertebrate animals can affect insects in various ways. Most important is the natural predation of wildlife populations on insects. However, also discussed briefly is the deliberate introduction of vertebrates for the biological suppression of insects.

Naturally Occurring Predation by Wildlife on Insects

Predatory wildlife, particularly birds and bats, are among the principal threats to the survival of insects. As we saw in Chapter 4 (Wildlife diets), a great number of animals take advantage of the nutrition and abundance of insects (arthropods), and consume them. Consequently, many modifications in insect morphology, behavior, and physiology have evolved that assist insects in avoiding predation. Obviously, these adaptations are not completely successful, because many vertebrates are successful in their pursuit of an insect-based

Insects and Wildlife: Arthropods and Their Relationships with Wild Vertebrate Animals, 1st edition. By J.L. Capinera. Published 2010 by Blackwell Publishing.

meal. Table 13.1 shows the taxa that, in general, are most important as food for wildlife. A taxonomic classification of insects is convenient, but perhaps not the best means of characterizing prey. Prey can also be classified by their size, where they are commonly found, or how they generally avoid being eaten. Such ecological classifications may have more relevance when assessing how likely it is that insects will be consumed by wildlife. Prey size and defensive behavior certainly affect wildlife choice and opportunity, and some habitats also affect the ability of wildlife to capture arthropods.

The effect of wildlife on arthropod population densities is variable, but sometimes quite significant. The most important population-level effects of wildlife are in maintaining insect populations at low levels, and in suppressing populations when they begin to increase. In many cases, they prevent potential insect outbreaks from occurring. Wildlife display both functional and numerical responses to increasing numbers of prey. **Functional responses** are behavioral changes, in this case increased predation caused by greater concentration of feeding on the increasing population of insects. Functional responses could be reflected in simple switching of hosts, but could also be manifested in relocation to areas where food is most plentiful. The flocking behavior of birds and schooling of fish are examples of wildlife taking advantage of locally abundant resources. **Numerical responses** are population increases caused by greater survival or reproduction of wildlife following an increase in the food supply. For numerical responses to occur there usually must be a large-scale and protracted increase in prey abundance, and adequate food must be available through most of the year including times when insects might not be numerous. If this is not the case, then wildlife might

Table 13.1. Taxonomic versus ecological categorization of common arthropod prey in wildlife diets (modified from Cooper *et al*.1990).

Taxon	Prey description	Prey length (mm)	Where prey are usually found	Predator avoidance mechanism
Non-insect arthropods				
Araneae	spiders	2–40	various	walking, silking, crypsis, chemical defenses
Isopods	woodlice	0.5–18	leaf litter	nocturnal, cryptic location
Diplopoda	millipedes	10–100	leaf litter	nocturnal, cryptic location, chemical defenses
Insects				
Isoptera	termites	5–15	soil, wood	cryptic location
Ephemeroptera	mayflies			
adult		5–40	air	flight
nymph		5–40	water	crypsis, walking
Odonata	dragonflies and damselflies			
adult		20–90	air	flight
nymph		15–40	water	crypsis, walking
Plecoptera	stoneflies			
adult		5–25	air	flight
nymph		5–25	water	crypsis, walking
Trichoptera	caddisflies			
adult		5–30	air	flight
larva		1.5-25	water	shelter, crypsis, walking
Coleoptera	beetles and weevils			
Carabidae	ground beetles	5–30	soil surface	nocturnal
Cerambycidae	long-horn beetles			
adult		8–30	foliage, bark	falling, flying
larva		8–25	within wood	cryptic location
Chrysomelidae	leaf beetles			
adult		3–15	foliage	aposematism, crypsis, dodging, falling, chemical defenses
larva		3–5	foliage	aposematism, crypsis, regurgitation
Curculionidae	weevils	5–20	foliage, stems	crypsis, falling
	bark beetles	3–8	beneath bark	crypsis
Scarabaeidae	scarabs			
adult	scarabs	5–30	soil, foliage	crypsis, nocturnal
larva	white grubs	5–40	below-ground, within decaying matter	cryptic location
Hemiptera	bugs and allies			
Pentatomidae	stink bugs	8–25	foliage	crypsis, flying, chemical defenses
Coreidae	seed bugs	8–15	foliage	crypsis, chemical defenses

Table 13.1. *Continued*

Taxon	Prey description	Prey length (mm)	Where prey are usually found	Predator avoidance mechanism
Gerridae	water striders	10–30	water surface	crypsis, swimming
Notonectidae	backswimmers	5–20	water surface	crypsis, swimming
Corixidae	water boatmen	5–30	water surface	crypsis, swimming
Membracidae	thorn bugs	5–15	stems	crypsis
Cicadellidae	leafhoppers	3–15	foliage	crypsis, flying
Aphididae	aphids	1–8	foliage, stems	crypsis
Cicadidae	cicadas	25–60	roots, stems	cryptic location, group action
Hymenoptera	wasps, bees, ant, sawflies			
Formicidae	ants	3–12	soil, leaf litter, foliage, bark	walking, cryptic location, biting, stinging
Apidae	bees	8–25	air, blossoms	stinging, aposematism
Vespidae	wasps	10–30	air, foliage	stinging, aposematism
Orthoptera	grasshoppers, katydids, crickets			
Acrididae	grasshoppers	15–70	foliage, soil surface	crypsis, jumping, flying
Tettigoniidae	katydids	20–80	foliage	crypsis, jumping, flying
Gryllidae	crickets	10–40	soil surface, foliage	nocturnal, jumping
Flies				
Asilidae	robber flies	15–30	foliage	flying
Syrphidae	flower flies	8–15	air, blossoms	flying, aposematism
Calliphoridae	blow flies			
adult		8–15	foliage	flying
larva		8–20	carrion, soil	cryptic location
Chironomidae	midges			
adult		1–10	air	flying, nocturnal
larva		1–12	water	cryptic location, swimming
Lepidoptera	moths, butterflies			
	'hairy' caterpillars	10–40	foliage	spines/hairs
	pupae	10–30	soil, foliage	crypsis, cryptic location
	moths	10–40	air	flying, nocturnal, startle
	butterflies	50–160	air, blossoms	flying, mimicry, aposematism, chemical defenses
Noctuidae	cutworms	10–50	foliage, soil	belowground, nocturnal, crypsis
Geometridae	loopers	10–40	foliage	crypsis, silking
Pyralidae	webworms	10–25	foliage	crypsis, webbed foliage
Saturniidae	giant silkworm larvae	25–140	foliage	crypsis, spines, aposematism
Sphingidae	hornworms	20–100	foliage	crypsis
Papilionidae	swallowtail larvae	20–70	foliage	crypsis, startle

perish during these insect-free periods. The long life spans, particularly the long pre-reproductive period of many vertebrates, limit their ability to respond numerically to insect population increases.

Both functional and numerical responses are well documented, though functional responses are much more common. Functional responses to prey availability are most easily observed in animals that aggregate to feed. Aggregations result in increased numbers of predators per unit of area or unit of time. However, aggregation is not the only type of functional response. Change in host selection behavior is also common. Wildlife will often learn to feed selectively on prey that are more readily available; such animals form a 'search image' that allows them to search efficiently. From the perspective of prey survival, functional and numerical responses are equivalent in that there is more predation occurring. The importance of prey abundance on predator abundance can be seen in any numbers of examples, but the following illustrates the behavior of migratory insectivorous birds. Research was conducted on warblers (about six canopy-feeding species) and American redstarts, *Setophaga ruticilla*, overwintering in Jamaica. Bird abundance was found to vary significantly among sites, and density was directly related to arthropod biomass, with birds spending more time where prey were more available. Even within the season, bird abundance shifted in response to food availability. This has conservation implications, of course, and arthropods must be adequately abundant if such wildlife are to take up residence.

Although it is difficult to generalize across all ecosystems, it seems that wildlife can regularly affect insect population abundance, and that wildlife are most effective when prey abundance is low. If weather, food, pesticides, cropping practices or some other factor allows a sudden increase in survival or reproduction of the prey, the predators may not be able to regulate the prey population, and an insect outbreak may ensue. **Population regulation** is a density-related response to greater abundance of a host (insect prey in this case) that causes more mortality (predation) as the host density gets higher, and less mortality when the population decreases. Population regulation keeps the population from increasing to too high a level, and from exhausting its resource base. Thus, it has a stabilizing effect. As you will see from the examples below, in some cases wildlife predators are quite effective at killing insects, but

they may not keep them from attaining high levels of abundance. Predators are perhaps best classified as dampening forces rather than regulatory agents, although when they are part of a multifaceted ecosystem they certainly contribute to population regulation, and sometimes they are capable of completely suppressing insect populations.

Entomologists consider mortality factors such as predators, parasitoids, microbial pathogens, and competition to be **density-dependent factors** because their presence and effectiveness often increase as population density increases. In contrast, weather-related factors are considered to be **density-independent factors** because the mortality caused by these elements often occurs independent of insect density. The distinction between density-dependent and density-independent is not really that clear-cut, however. For example, we tend to consider weather to be a density-independent mortality factor because freezing weather should kill non-diapausing insects whether they are abundant or rare. In reality, insects may benefit from having shelters where they can escape adverse weather, and those shelters could be limiting, so weather may kill different proportions of the population depending on how numerous they are when the adverse weather occurs. Although it may be a simplistic generalization, we consider that both density-independent and density-dependent factors affect insect population size, but only density-dependent factors actually 'regulate' populations in the sense of stabilizing abundance.

Ecologists have long debated the relative importance of the various regulatory processes. Often the debate involves discussion of 'bottom-up' or 'top-down' regulation. **Bottom-up regulation** is imposed by resource limitations, and can be due to either quantity or quality of the resource. **Top-down regulation** is due to predation and related factors such as parasitism and disease. To understand the origin of this terminology, examine a trophic pyramid (Fig. 6.1 or 13.1). At the base of the pyramid we find primary producers or autotrophs, and the abundance of these (normally plants) can affect the occurrence of herbivores that depend upon the plant life, i.e., the organisms directly above them in the trophic pyramid or food chain. Thus, the 'bottom' affects the organisms above them, providing bottom-up regulation. Similarly, tertiary or quaternary consumers normally are the 'top' elements of trophic pyramids, and because they feed on the heterotrophs below, they provide top-

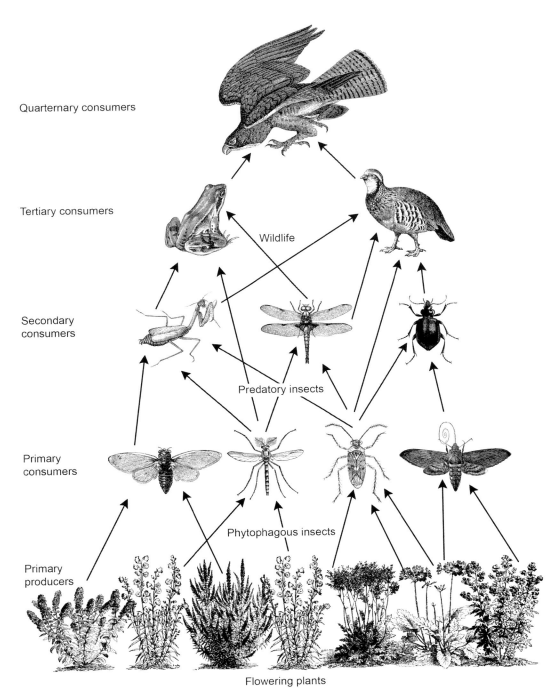

Quarternary consumers

Tertiary consumers

Wildlife

Secondary consumers

Predatory insects

Primary consumers

Phytophagous insects

Primary producers

Flowering plants

Fig. 13.1. A representative terrestrial food web showing some of the important links among producers and consumers.

down regulation. Some ecologists also consider that organisms within a trophic level can affect population abundance, and this is called **lateral regulation**. Lateral regulation occurs when insects cannibalize each other or compete for the same food or other resource, resulting in change in numerical abundance. Research has demonstrated the importance of bottom-up, top-down, and lateral regulation, and while each can assume particular importance at a certain time and place, none is universally most important. Indeed, in most ecosystems, it is the interplay of these different forms of regulatory factors that determine abundance. It is probably safe to conclude that the importance of top-down regulation has been over-estimated in many ecosystems, because it is most apparent. In contrast, lateral regulation has been little studied and so is most often underappreciated.

Western Pine Beetle and Woodpeckers

Western pine beetle, *Dendroctonus brevicomis* (Coleoptera: Curculionidae: Scolytinae), burrows beneath the bark of pines in the western North America, killing trees by girdling them and by transmitting a blue-stain fungus that disrupts the vascular tissues of the trees. Predation of larval western pine beetle was determined using X-rays to assess the presence of larvae beneath the bark. One study in California, USA, found that 32% of larvae were consumed by woodpeckers, and another from the same area reported 58% consumption. In addition to the consumption of beetles by woodpeckers, the bark damage caused by the woodpeckers also made it easier for small parasitoids to insert their ovipositors through the bark, resulting in up to a threefold increase in parasitism of surviving larvae on woodpecker-drilled trees. Drilling by woodpeckers also affected the microhabitat of the insects and reduced their survival rate by increasing desiccation of larvae in the bark.

Spruce Budworm, Birds, and Mammals

Studies of eastern spruce budworms, *Choristoneura fumiferana* (Lepidoptera: Tortricidae), in the northeastern USA showed that birds respond to the abundance of the budworms by being more abundant where budworms were present, and by eating more budworms. Twelve of the fifteen bird species that were studied dis-

played increased budworm consumption. The avifauna that were numerous included red-winged blackbird, *Agelaius phoeniceus*; white-throated sparrow, *Zonotrichia albicollis*; blackburnian warbler, *Dendroica fusca*; black-throated green warbler, *Dendroica virens*; northern parula warbler, *Parula americana*; bay-breasted warbler, *Dendroica castanea*; Tennessee warbler, *Vermivora peregrina*; and golden-crowned kinglet, *Regulus satrapa*. The average number of budworms recovered per bird stomach was 0.4 in endemic (low population), 5.5 in transitional (increasing population), and 7.9 in epidemic (high density) areas. The bird community was estimated to consume 3300 larvae and 2300 pupae per hectare during the endemic phase of budworm outbreak, increasing to 32,000 larvae and 22,000 pupae during the transitional stage, and 52,000 larvae and 36,000 pupae during the epidemic phase. The bay-breasted and Tennessee warblers occurred only in the forest stands with the highest budworm densities, and the authors of the study felt that the birds present during the early phases of the insect outbreak were most important in curtailing population increase. These species were the blackburnian warbler and golden-crowned kinglet. As impressive as all this seems, the birds were not able to keep up with the budworms, and they consumed an estimated 87.2% of the budworms in the endemic populations, but only 23.0% and 2.4% in the transitional and epidemic populations. Birds are not the only important predators of spruce budworms, however, as ants also kill these caterpillars. At least at low densities, ants are able to compensate for birds, and birds for ants, if both are not present. As suggested above, neither can regulate budworm populations if they get very high. Mammals were found to feed on spruce budworms, especially rock vole, *Microtis chrotorrhinus*, and short-tailed shrew, *Blarina brevicauda*, but neither proved to be very important.

Studies of western spruce budworm, *Choristoneura occidentalis*, show essentially the same for population control in western North America, despite the difference in topography, climate, and fauna. When trees with low densities of budworms were protected from birds and ants, there were 10–15 times as many budworms present as compared to trees where predators had ready access. Even when budworm populations were high, there were twice as many budworms present on trees lacking predators. As seen in the East, both the birds and ants could compensate for the absence of the other predator.

Gypsy Moth, Birds, Mammals, and Beneficial Insects

Gypsy moth, *Lymantria dispar* (Lepidoptera: Lymantriidae), like spruce budworm, is susceptible to wildlife predation. Birds are important predators of larvae, but unlike the budworm situation, the important predators are small mammals that feed on pupae, particularly white-footed mouse, *Peromyscus leucopus*. Indeed, when gypsy moth populations were established artificially in Connecticut, USA, control plot mortality was high (resulting in production of only about 6 egg masses per hectare by moths from survivors) at the end of the season, as were plots where birds were excluded (about 3.3 egg masses/ha) by cages. Where mammals were removed by trapping, mortality was low (38.5 egg masses/ha), and where both birds and mammals were absent, mortality was very low, resulting in about 51.5 egg masses/ha. This suggests that predators, but especially small mammals, are important in maintaining gypsy moth populations at low densities. Unlike budworms, gypsy moth larvae descend trees regularly when they are not feeding, and pupate in sheltered locations such as between rocks and beneath debris and overhangs. This makes them quite susceptible to predation by mammals. Like budworm populations, however, once they attain higher densities the vertebrate predators cannot suppress them. Insect parasitoids (Diptera and Hymenoptera) can provide some level of control, and a viral disease (gypsy moth baculovirus) and a fungal disease (*Entomophaga maimaiga*) can have dramatic effects on larval populations, causing large-scale mortality. The baculovirus is effective mostly at high densities. The fungal disease causes epizootics during wet spring weather.

Rangeland Grasshoppers and Birds

Grasshopper (Orthoptera: Acrididae) populations on grasslands in the western USA have been shown to be affected by birds, but the impact varies. Grasslands are less diverse than forests, and most birds share a common resource base. They adjust their feeding according to availability of prey, and usually are not food-limited during the summer months due to the great availability of insects. Studies in Arizona using bird exclosures showed that grasshopper populations could be more than twice as abundant when bird predation was eliminated. At this site, seven of the twelve grasshopper species were more abundant in the absence of birds. In

New Mexico, populations of *Hesperotettix viridis* and *Melanoplus aridis* were 2.2 and 2.9 times higher in bird exclosures, respectively. In Nebraska, the presence of birds depressed grasshopper abundance by about 25%, whereas in North Dakota, it was depressed 26%–37%. Studies conducted in Montana showed that whether or not grasshopper abundance would be affected was determined, in part, by grasshopper size.

It is possible to calculate the potential effect of birds on steppe (shortgrass prairie) in Colorado, USA, based on known consumption rates of grasshoppers by birds, and their average densities. Among the birds consuming grasshoppers in this area are western meadowlark, *Sturnella neglecta*, with an average of 23–48 grasshoppers per bird stomach; horned lark, *Eremophila alpestris*, from 8–42; lark bunting, *Calamospiza melanocorys*, from 1–14; and McCowan's longspur, *Calcarius mccownii* from 1–5. Birds have a high metabolic rate and process their food quickly. Thus, their stomach contents typically represent only a fraction of their daily intake. Research has shown that the daily consumption of these birds may be 64–78 grasshoppers/day, plus hundreds of other insects. The average density of breeding birds was reported by various investigators to be about 2.8, 2.7, and 5.2/ha in this area. Thus, if we assume 3.0 birds/ha, capturing 70 grasshoppers/bird/day, these birds would consume about 25,200 grasshoppers/ha over a 4-month period. Assuming a fairly high density of five grasshoppers/m^2 or 50,000 grasshoppers/ha, the grasshopper population would be suppressed by about 50%. This is a conservative estimate because it is based only on the residential breeding birds, but in reality other far-ranging species are commonly present, and many other mortality factors also come to bear on rangeland grasshoppers. Thus, birds contribute very materially to grasshopper population suppression.

Crop-Feeding Aphids and Birds

The importance of birds as consumers of insects is easily seen in an instance of passerine bird predation of aphids during an aphid outbreak on grain crops on a North Carolina farm. Samples of the more abundant birds showed that aphids comprised 82.7% of the food consumed by American goldfinch, *Carduelis tristis*, whereas with vesper sparrow, *Pooecetes gramineus*, it was 19.5%, with savannah sparrow, *Passerculus sandwichensis*, it was 25.3%, and with chipping sparrow,

Table 13.2. Estimated consumption rate (insects per day) of grain aphids by sparrows on a North Carolina farm (adapted from W.L. McAtee, *Yearbook of Agriculture for 1912*).

	No. birds present	No. eating aphids	Mean no. eaten/bird	Total no. aphids eaten/day
American goldfinch	300	240	132.5	318,000
Vesper sparrow	2,500	1,761	22.5	396,225
Savannah sparrow	70	45	63.5	28,570
Chipping sparrow	245	178	94.7	168,560
Field sparrow	20	10	154.3	15,430
Dark-eyed junco	70	28	14.0	3,920
Song sparrow	6	3	50.0	1,500
			Total	932,205

Spizella passerina, it was 45.1%. Based on the frequency of birds eating aphids, the abundance of aphids in the stomachs of the birds, and the abundance of the birds on the farm, it was possible to estimate the aphid consumption rate, as shown in Table 13.2. Thus, the combined feeding by six bird species (the aforementioned species plus dark-eyed junco, *Junco hyemalis*, and song sparrow, *Melospiza melodi*) consumed nearly one million aphids per day.

Crop-Feeding Caterpillars, Spiders, and Birds

In Hawaii, USA, the colonization of broccoli plants principally by larvae of cabbage butterfly, *Artogeia rapae* (Lepidoptera: Pieridae), but also by cabbage looper, *Trichoplusia ni* (Lepidoptera: Noctuidae) caterpillars, and the resulting herbivory were assessed using removal (spiders) and exclusion (spiders and birds) techniques. The bird species responsible for predation were redcrested cardinal, *Paroaria coronata*, and northern cardinal, *Cardinalis cardinalis*. Several species of spiders were present. The bird-accessible and the bird- and spider-accessible treatments had fewer insects than the spider-accessible treatment and the check (both birds and spiders excluded). Thus, birds were able to suppress the abundance of caterpillars. However, when the weights of the plants were determined, the spider-accessible plant had weights that were not significantly different from the bird-accessible treatments, providing evidence that the spiders provided some benefit to plant growth. Overall, it seems that both predators provided benefit to the plants, but the birds were more beneficial.

Tropical Forest Floor-Dwelling Insects, Lizards, and Birds

In the Netherlands Antilles, exclusion of *Anolis wattsi* and *A. bimaculatus* lizards from tropical forest floor environments caused a two- to threefold increase in the number of adult phorid flies (Diptera: Phoridae) and significant increase in the numbers of small spiders. Large spiders escaped predation. Because the *Anolis* lizards foraged only on the forest floor, vegetation-inhabiting arthropods were not affected. A similar insect depletion on the forest floor was noted in Panama but was caused instead by the checker-throated antwren, *Myrmotherula fulviventris*, which preyed largely on cockroaches (Blattodea: Blattidae). Cockroaches comprised the largest proportion of the available arthropod prey at this study site, and the birds depleted the population by 50% in 6 weeks.

Tropical Forest Insects, Bats, and Birds

It is difficult to find data on the relative contributions of both birds and bats to suppression of arthropods, but recent studies have been quite revealing. In a semideciduous tropical forest in Panama, bird or bat exclosures were manipulated to assess the relative contribution of these predators to arthropod abundance. Control plants (no exclusion of predators) averaged 4.9 arthropods/m^2, whereas plants from which birds were excluded had 8.1 arthropods/m^2 and plants from which bats were excluded had 12.4 arthropods/m^2. The plants also suffered increased herbivory when birds and bats were prevented from gleaning arthropods on the foliage. Control plants had 4.3% of their leaf area removed by insect herbivory, whereas bird exclusion had 7.2% and bat exclusion 13.3%. Thus, both birds and bats were important predators, but bats proved to be more important in this tropical system.

A similar study was conducted in a Mexican shade coffee plantation. In this case, an additional treatment was included: both birds and bats were excluded. Also, the study was conducted in both the dry season and in the wet season. In the dry season, control plots (no exclusion) had 9.9 arthropods/100 leaves, whereas bat exclusion had 10.6, bird exclusion had 12.9, and both bird and bat exclusion had 14.5 arthropods/100 leaves. In this study, birds but not bats removed a significant number of arthropods. In contrast, during the wet season, control plots (no exclusion) had 4.6 arthropods/100 leaves, whereas bat exclusion had 8.5, bird exclusion had 7.3, and both bird and bat exclusion had 9.7 arthropods/100 leaves. Thus, in the wet season, bats were slightly more effective than birds, though both had significant effects on arthropod abundance. In both seasons, the presence of birds and bats had additive effects on suppressing insects, and very marked overall effects, with arthropod densities (averaged over seasons) 46% higher when both birds and bats were absent.

Aquatic Insects, Ducks, and Fish

In northern climates, not all lakes support fish, and ducks can choose between lakes that support or lack fish. In Sweden, aquatic insects such as dragonfly and damselfly larvae (Odonata), mayfly larvae (Ephemeroptera), dytiscids (Coleoptera: Dytiscidae), backswimmers (Hemiptera: Notonectidae) and *Chaoborus* midge larvae (Diptera: Chaoborus) are important food resources both for goldeneye duck (*Bucephala clangula*) and fish (perch, *Perca fluviatilis*, and roach, *Leuciscus rutilis*). Ducks prefer using lakes without fish, apparently because fish reduce the availability of insects for feeding by ducks. Thus, the presence of predatory fish affects the suitability of lakes by removing insect prey, and ducks are displaced. This is clear evidence that insects are an important food resource for ducks, that fish can deplete this insect-based food resource, and that fish can compete effectively with ducks for this resource.

Predation of Animal Ectoparasites by Birds

Birds sometimes enjoy a symbiotic relationship with other wildlife, particularly ungulates. Mammals can be good hosts from some blood-feeding arthropods, but these arthropod pests can serve as a good food resource for avifauna, thereby relieving the animal hosts of their parasites. Perhaps best known among the birds that take advantage of ectoparasites are the African oxpeckers: the red-billed oxpecker, *Buphagus erythrorhynchus* of east Africa and Asia, and the yellow-billed oxpecker, *Buphagus africanus* of most of sub-Saharan Africa. They often are found perching on the backs of large grazing ungulates. They feed on the arthropods associated with these animals, particularly ticks and botflies. Examination of the stomach contents of these birds revealed 16 to 408 ticks per bird, plus various flies and lice. Red-billed oxpeckers eat about 10% more ticks than yellow-billed oxpeckers. They search visually, first choosing the engorged female ticks, then eating the younger and smaller ticks. Yellow-billed oxpeckers prefer to feed on the ectoparasites of buffaloes and rhinos, whereas red-billed oxpeckers prefer foraging on other ungulate hosts. Oxpeckers also prefer to feed on weak animals and will feed repeatedly on individual animals that are heavily infested with ticks. These birds also cry in alarm when they spot danger, giving the grazing animals greater opportunity to escape predation. Oxpeckers gain more than food from the grazers, as they often line their nests with hair from these mammals. Although this is often cited as an example of mutualism, both parties seeming to benefit, close study has revealed that the birds like to feed on animal blood. They obtain blood not only indirectly by consuming bloated ticks, but directly by pecking at the wounds until blood flows. Yellow-billed oxpeckers are more aggressive than red-billed oxpeckers with respect to feeding at wounds, but in either case it is believed to be fairly rare. Thus, this relationship may be more parasitic than mutualistic, though clearly each party gains something from the interactions. Interestingly, the red-billed oxpecker populations in southern Africa recently suffered serious declines due to reduction in the availability of wild hosts and the use of insecticide treatment of cattle for tick control. The birds move freely among wild and domestic grazing animals, but contacted excessive levels of insecticide by their association with cattle. Increase in the number of wild ungulates and the use of safer acaricides can help to preserve these interesting birds.

The well-studied relationship between oxpeckers and ungulates is by no means the only similar bird-mammal relationship. Similar associations are known between fan-tailed raven, *Corvus rhipidurus*, and camel, *Camelus dromedarius*; pale-winged starling, *Onychognathus nabouroup*, and mountain zebra, *Equus zebra*; black-billed magpie, *Pica pica*, and moose, *Alces alces*; yellow-bellied bulbuls, *Alophoixus phaeocephalus*, and klipspringers, *Oreotragus oreotragus*; black caracara, *Daptrius ater*, and Brazilian tapir, *Tapirus terrestris*; and palewinged trumpeter, *Psophia leucoptera*, and gray brocket deer, *Mazama gouazoubira*.

Also common are birds that take advantage of insects flying around ungulates. In Europe and Asia, the yellow wagtail (*Motacilla flava*) is an insectivorous species that plucks flies from the air around grazing animals. Cattle egrets, *Bubulcus ibis*, also feed on insects associated with grazers, including domestic animals. Although grasshoppers are the preferred insect prey of cattle egrets, biting and nuisance flies are also eaten. European starling, *Sturnus vulgaris*, is commonly associated with domestic grazing livestock in Europe and North America, but it is not clear that this association extends to wildlife.

Introduction of Vertebrates for Biological Suppression of Insects

Vertebrates have been relocated in several areas of the world to aid in biological suppression of vertebrate pests. In general, vertebrate predators have been disappointing, and an ecological disaster in some cases. The classical case of biological control malfunctioning is the introduction of Indian mongoose, *Herpestes javanicus*, from India to other tropical countries to control rats, particularly in sugarcane fields. However, mongoose is a typical predator, with broad dietary habits, and soon demonstrated that it would kill native birds, reptiles, and other non-target prey. It has been very disruptive to island fauna, including islands in the Caribbean, Hawaii, and Polynesia. For example, it was responsible for the extinction of five animal species from Jamaica and the virtual extinction of seven ground-nesting birds in Fiji. At least eight endangered birds in Hawaii are prey for mongoose.

Introduction of biological control organisms for *insect* biological control has a much better track record, but generally the introduction of predators has been less successful than the introduction of parasitoids.

Biological control benefits from specificity, and predators tend to have too broad a range of potential hosts to be considered appropriate for biological control. Most of the introductions of *vertebrate* predators were done in the late 1800s and early 1900s, and usually by farmers or other individuals unschooled in the theory and practice of biological control.

One of the few mammal-based biological control successes was attained by the introduction in 1958 of masked shrew, *Sorex cinereus*, into Newfoundland from nearby mainland Canada for suppression of larch sawfly, *Pristophora erichsonii*. Prior to introduction, Newfoundland lacked any native shrews, and ground-level mortality of sawfly cocoons was lacking as compared to other areas of Canada, probably accounting for the extensive damage to trees on this island. The shrews have established and spread, apparently having the desired impact on these tree-feeding insects. On average, the shrews now account for about 40% of the larch sawfly mortality in Newfoundland.

In an effort to improve biological control of mosquitoes, several small fish have been moved around the world, including mosquito fish, *Gambusia affinis*; guppy, *Poecilia reticulata*; dispar topminnow, *Aphanius dispar*; panchax minnow, *Aplocheilus panchax*; goldfish, *Carassius auratus*; and *Epiplatys* spp. *Gambusia* has been especially popular, and is now found throughout the warm weather areas of the world. In addition to inoculating *Gambusia* into new areas, these fish are often maintained and restocked into artificial water containers or ponds that dry out seasonally. Introduction into temporary water is probably the most effective use of this predator. Unfortunately, redistribution was often done without regard for native species, some of which were as effective as *Gambusia*. *Gambusia* can displace native fish species, and prey on other non-target organisms such as immature salamanders.

Amphibians have less often been transported for biological control, although the cane or marine toad, *Bufo marinus*, stands out as an exception, as it has been introduced to several areas of the world. This giant anuran, a native of central and South America, is up to 23 cm long. Like the mongoose, initially it was introduced into sugarcane plantings, where it was thought to feed on white grubs, but subsequently it was released to control other insects. Like most toads, cane toad will eat almost anything it can catch and ingest, but due to its extraordinarily large size there are quite a number of potential prey. Ants, termites, and beetles are their most common prey. However, centipedes, scorpions, spiders, other toads and even snakes may be consumed. In Australia, rainbow bee-eater (*Merops ornatus*), a burrow-nesting bird, loses many of its young and eggs to cane toads. The toads apparently use olfaction to find the burrows. As broad as the diet may be, their most important ecological effect may be indirect. Cane toad is poisonous, so most animals that eat them will succumb to the poison. For example, domestic dogs are reported to perish within 15 minutes of mouthing a cane toad. Research in Australia showed that frogs eating cane toad eggs perished, and that the number of dingoes (*Canus lupus dingo*) in toad-infested areas was reduced. Crocodiles in some dry areas of Australia have been reduced by over 70% due to feeding on these poisonous anurans. Also, frog-eating snakes and monitor lizard suffered from ingesting cane toads. In contrast, turtles were able to eat eggs and tadpoles of cane toad without ill effect. Many of the effects of this toad on indigenous fauna are just becoming known, and the toad continues to expand its range in Australia.

Although birds have often been transported to new regions of the world, few have been released with biological control in mind. One exception is the cattle egret, *Bubulcus ibis*, which was introduced into Hawaii in 1959 for control of flies associated with cattle. Though perhaps eating such flies, they also eat cultivated prawns, interfere with air traffic at airports, and are thought to compete with native birds.

An alternative to introduction of new predators, particularly vertebrate predators, is to maximize the effectiveness of indigenous vertebrates or introduced invertebrate predators by manipulating the environment. Examples of actions that might improve naturally occurring biological regulation of pests include providing nesting boxes for birds and rodents, shelters for bats, and preserving or planting hedgerows and fencerow flora for various wildlife and beneficial insects.

HOW INSECTS AFFECT WILDLIFE SURVIVAL

Insects (arthropods) influence wildlife conservation in ways other than discussed earlier (Chapter 4, Insects in the diet of wildlife; Chapters 7–10, Insects as pests and vectors). Specifically, insects can affect wildlife negatively by preying on them, but some wildlife benefit directly from their association with biting or stinging insects, and some insects can also help to conserve wildlife habitat by deterring hunting, development and habitat destruction. Insects also support wildlife-based recreation. Many of our wildlife watching, fishing and hunting activities would not be possible without insects as a food source, and many more wildlife-rich habitats would likely have been compromised were it not for the occurrence of biting insects.

Predation by Insects on Wildlife

Most people think of insects as small organisms of the appropriate dimensions to be eaten by wildlife, and wildlife as the large predators eating insects and other arthropods. Food chains of ecosystems are typically characterized by phyletic ascent; the more advanced organisms tend to be at the end of the food chain or top of the food pyramid. Thus, plants are presented as producers, insects as primary and secondary consumers, and vertebrates as tertiary consumers. Generally this is accurate, but sometimes there is an inversion of normal ecological relationships, and invertebrates become predators of wildlife. Usually this is an age- or size-related phenomenon. As long as a predaceous insect (arthropod) is larger than the vertebrate, there is possibility that the arthropod will be the predator and the vertebrate the prey. However, if the vertebrate grows, their roles may be reversed, and the wildlife may become the predator. This process of arthropods and vertebrates eating each other at different stages in their lives is called **cross predation**.

Effects on Terrestrial Wildlife

Some terrestrial arthropods feed on small wildlife. The following are some examples of arthropods killing terrestrial vertebrates: the wolf spider *Argiope aurantia* was reported to eat broad-headed skink, *Eumeces laticeps*; the wolf spider *Lycosa ammophila* killed the green anole, *Anolis carolinus*; the giant crab spider *Olios* spp. killed pre-adults of the terrestrial frog *Eleutherodactylus coqui*; and the giant centipede, *Scolopendra gigantia*, killed and ate three species of bats (Chiroptera). Tarantulas of the genus *Avicularia* sometimes are called 'bird spiders' because they tend to be arboreal. These and other tarantulas are reported to capture and eat small vertebrates. The frequency and significance of such interactions are not known, but thought to be rare.

More common is the destruction of wildlife by ants (Hymenoptera: Formicidae). Army ants (especially *Eciton* spp., but also many others) in South America kill arthropods, and are known to affect snakes, lizards and nestling birds, though it is not certain whether they eat them or simply kill them. Driver ants (*Dorylus* spp.) in Africa are renowned for their predatory abilities, swarming over the soil and vegetation at high densities and overwhelming prey in their path. Although often concentrating mostly on insect predation, they will take any animal matter they encounter, tearing it to pieces with powerful mandibles. Also, at least six highly invasive species of ants have been moved widely around the globe, often causing ecological disruption, including impacts on wildlife. These important species are: yellow crazy ant, *Anoplolepis gracilipes*; Argentine ant, *Linepithema humile*; big-headed ant, *Pheidole megacephala*; red imported fire ant, *Solenopsis invicta*; tropical fire ant, *Solenopsis geminata*; and little fire ant, *Wasmannia auropunctata*. While these other species lack the dramatic swarming behavior of army and driver ants, they nevertheless occur at high densities and cause loss to animals they encounter.

Among the invasive ant species, the red imported fire ant is most important in damaging wildlife populations in North America, where it is free from many of the biotic factors that apparently reduce its abundance in its native environment, South America. Red imported fire ant often replaces native ants due to the lack of natural enemies, their high reproductive rate, and aggressive foraging strategy. Also, in the USA fire ants tend to be polygynous (multiple queens per colony) whereas in South America they tend to be monogynous (a single queen per colony). This is important because polygynous colonies have much higher population densities.

Wildlife found in open, sunny locations are most susceptible to disruption because fire ants avoid shaded areas. Turtles are relatively defenseless. Box turtles (*Terrapene* spp.) may close their plastron, but this defense does not work against ants, which find gaps large enough to enter. Gopher tortoise (*Gopherus polyphemus*), Florida red-bellied turtle (*Pseudomys nelsoni*), sliders (*Trachemys* spp.), loggerhead turtle (*Caretta caretta*), and green turtle (*Chelonia mydas*) are just a few examples of turtles that have been killed by red imported fire ant. The turtle hatchlings are most susceptible. The ants cannot penetrate the eggs, but the ants attack as soon as the hatchling begins to break

out of the egg. Estimates of turtle population reduction in ant-infested areas range from 15%–45%.

Other reptiles and amphibians are also susceptible to attack by fire ants. The kingsnake *Lampropeltis getula floridanus* appears to have declined in abundance in areas of its range where fire ants have invaded. More remarkable, the Texas horned toad, *Phrynosoma cornutum*, is adversely affected. This reptile is an ant specialist, though normally feeding on harvester ants, *Pogonomyrmex* spp. If attacked by a few fire ants, horned toads typically will eat them. However, in the presence of numerous ants (20 or more), these toads will flee. The hibernating form and the eggs seem susceptible to attack by ants, but the ants also deprive the toads of their normal prey, harvester ants, which are displaced by these aggressive invaders. Young *Bufo* toads that have recently emerged from the water also may be attacked successfully by ants, whereas older toads at the same location could elude the ants successfully. A similar problem exists with alligator (*Alligator mississippiensis*) eggs and hatchlings.

Fire ants can have dramatic impact on some bird populations. In coastal Texas, USA, ants reduced island-dwelling water bird populations by 92% as compared to insecticide-treated islands. The benefits of ant suppression were not evident early in the season, but during June and July, when the warmer weather allowed the ants to become more active, the effect was very evident. In Mississippi, populations of least tern (*Sterna antillarum*) chicks suffered 33% mortality in the presence of fire ants, but only 6.3% where ants were controlled. Ants have also been implicated in the reduction of breeding success of crested caracara (*Caracara plancus*) in Texas; ground dove (*Columbina passerina*) in South Carolina; black rail (*Laterallus jamaicensis*) in Florida; and blue-gray gnatcatcher (*Polioptila caerulea*), eastern towhee (*Pipilo erythrophthalamus*), northern cardinal (*Cardinalis cardinalis*) and other forest-dwelling birds in Mississippi. Northern bobwhite, *Colinus virginianus*, has been the subject of considerable attention in the southeastern USA due to its general decline in abundance. It is susceptible to attack by fire ants, especially when hatching. Bobwhite populations tend to be much more stable in areas where fire ants are absent as compared to locations where fire ants are present, and abundance declined after invasion by these ants (Fig. 13.2a). Survival of young bobwhite chicks is higher when fire ant populations are suppressed; 21-day survival was much higher when hatching from insecticide-treated bobwhite

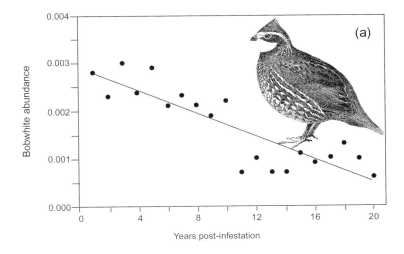

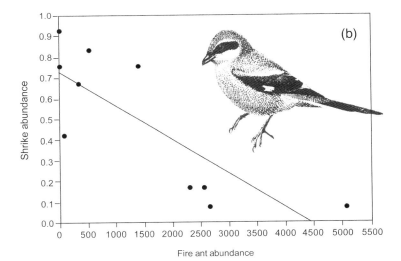

Fig. 13.2. The effect of red imported fire ant (*Solenopsis invicta*) on (a) northern bobwhite (*Colinus virginianus*) and (b) loggerhead shrike (*Lanius ludovicianus*) abundance (after Allen *et al.* 1995 and 2001).

nests (60%) than when hatching from untreated nests (22%). Also, in the case of the endangered black-capped vireo, *Vireo atricapillus*, fire ants accounted for about 31% of the nest predation in Texas, but 92% of the ant visits to nests occurred at night and were detected only by video surveillance. Effects on avifauna may extend beyond direct effects to include disruption of feeding. Along the Gulf of Mexico, loggerhead shrike (*Lanius ludovicianus*) populations were lower after fire ants invaded shrike-inhabited areas. Direct effects of ants on shrike fitness could not be measured, but the

abundance of invertebrates was lower in ant-infested areas (Fig. 13.2b), so researchers concluded that the ants were affecting shrikes indirectly by depleting insect food resources.

Curiously, the effect of red imported fire ants on mammal populations is less certain. Some rodents seem to be adversely affected, and a five-year study of eastern cottontail rabbits, *Silvilagus floridanus*, found that 29% of rabbit litters were destroyed by fire ants. Rabbits less than 4 days old lack the pelage (fur) for adequate protection from fire ant stings. There are some reports that deer suffer from association with fire ants, and in Texas there are reports of fawn survival being twice as high in areas were ants were suppressed, as compared to areas lacking ant control. Foraging on seeds by field mouse, *Peromyscus polionotus*, is suppressed in the presence of fire ants, and cotton rat, *Sigmodon hispidus*, similarly avoids areas infested with these ants. Effects of fire ants on mammals are less well documented than might be expected, however, and more than any other area, research on the effects of ants on mammals is needed.

Overall, it is clear that some vertebrates will suffer negative population-level effects from fire ants. The birthing or hatching periods are particularly vulnerable, and reptile species whose young remain belowground until all the young have hatched prior to emergence are particularly vulnerable. Egg-laying species seem more vulnerable than live-bearing species. Live-bearing species that require nourishment (altricial species) are more vulnerable than those that are largely mobile and independent at birth (precocial species). Species that are ground-nesting or live in open, sun-lit habitats are more susceptible than those living in elevated or canopied habitats due to the preference of fire ants for the former environment. Timing of hatching or birth may be critical because the ants are more active during the warmest periods of the year.

Effects on Aquatic Wildlife

The most important form of insect predation on wildlife takes place in the fresh-water environment. Frogs, salamanders, and fish are often preyed upon by arthropods, particularly by insects. Insects are especially important predators in ephemeral ponds. Among those that are important predators are immature dragonflies (Odonata); caddisflies (Trichoptera); beetles (Coleoptera) in the family Dytiscidae (predaceous diving beetles); bugs (Hemiptera) in the families Belostomatidae (giant water bugs), Notonectidae (backswimmers), Naucoridae (creeping water bugs), and Nepidae (waterscorpions); and some Neuroptera, family Corydalidae (dobsonflies and fishflies). They will attack and eat both eggs and young of the vertebrates. Data on population effects are scarce, but these insects are normally quite common and predation of vertebrates by them is often observed. For example, a reduction in the abundance of pike fry, *Esox lucius*, by 26% in two

weeks was attributed to a mixed population of predatory insects in France. Field studies of the belostomatid *Lethocerus deyrolli* in Japan revealed that frogs, but not fish or aquatic insects, were the important prey items, comprising 86.4% and 78.6% of the diet in spring and summer, respectively. In the USA, eggs of spotted salamander, *Ambystoma maculatum*, were preyed upon by the caddisfly *Banksiola dossuaria*. In this latter study, 62% of the caddisflies in a pond were found on the salamander egg masses, with an average of 11.8 insects per egg mass. Late instars of this caddisfly, but not the co-occurring caddisfly *Platycentropus radiatus*, inflicted mortality on the eggs. Some caddisfly larvae require access to animal tissue before they can mature. Surprisingly, in Israel the late-instar larvae of the mosquito *Culiseta longiareolata* were shown to be a predator of *Bufo* hatchlings. If this proves to be a common but overlooked relationship, predation by insects on frogs and toads may be even more important, as anurans and mosquitoes are commonly found co-inhabiting pools of water.

Symbiotic Relationships Between Insects and Wildlife

As we saw with disease-causing microorganisms, organisms can interact with one another in different ways. When organisms live in association with one another it is usually referred to as **symbiosis**. The nature of symbiosis varies, with the most common types:
• **Mutualism.** The organisms are dependent on one another, and the relationship is mutually beneficial.
• **Commensalism.** The organisms live together but only one benefits from the relationship, without causing harm to the other.
• **Parasitism.** One of the co-existing organisms obtains habitat or sustenance at the expense of the other.
We also discussed symbiotic relationships such as those associated with nutrient cycling (Chapter 6, Insects and ecosystems) and predation of animal parasites (this chapter), but interesting examples of symbiosis are ubiquitous. When examined critically, symbiosis tends to be complex and variable. If environmental conditions change, the relationships of organisms may also change.

Sometimes symbiotic relationships are direct, as when an insect-produced structure is inhabited by wildlife. Other times the relationship is less direct, as when birds place their nests in the vicinity of aggressively biting or stinging insects and derive some protection against predation. Sometimes the relationship is rather indirect, as when plant herbivory by vertebrates affects insect interactions with the host plant, with positive or negative feedback to the vertebrate herbivore. Most often, these insect-wildlife symbiotic relationships appear to be commensal, but some apparently are mutualistic or parasitic.

Reptiles, birds, and mammals often take advantage of insect structures for constructing their nests or shelters by dwelling inside. They normally are tolerated by their host and the vertebrates do not feed on the insects, so are considered to be examples of commensalism. Occasionally, birds will nest within silk structures created by social-living caterpillars or spiders. For example, the northern beardless-tyrannulet, *Campostoma imberbe*, constructs nests within unspecified caterpillar tents in Central America. More commonly, spider webs are used by birds. In Africa, southern double-collared sunbird, *Nectarinia chalybea*, scarlet-chested sunbird, *N. senegalensis*, and amethyst sunbird, *N. amethystine*, nest within spider webs. Apparently, some South American birds such as the tyrannid royal flycatcher, *Onychorhynchus coronatus*, similarly construct nests on debris caught in spider constructions.

More commonly, wildlife take advantage of the more permanent soil or paper nests created by termites (Isoptera), or ants and wasps (Hymenoptera) for nesting sites. The nests may be inhabited or uninhabited by insects during occupancy by the vertebrates. The mound-building termites, particularly *Macrotermes* and *Amitermes* of Africa, Asia, and Australia, create structures that are highly attractive to some birds, bats, and iguanas, possibly because they have a hard crust that impedes digging by predators, or because termites regulate the temperature of their mounds. Occupancy of termite mounds (termitaria) was discussed earlier (Chapter 6, Insects and ecosystems). Also, in both the New World and Old World, *Nasutitermes* spp. construct globular nests in trees from wood and fecal materials. These nests contain small pockets or cells in addition to the papery tissue, and this material is referred to as 'carton'. New World ants of the genus *Azteca* and Old World ants of the genus *Crematogaster* also make large carton arboreal nests. It is not uncommon for vertebrates to use insect carton for nesting. Lastly, the papery nests of vespid wasps sometimes are exploited by birds for a nesting site, though this is not very common. Animals find it particularly easy to excavate carton nests, though why they are tolerated by biting and stinging species is uncertain. Among birds, often it is parrots, kingfishers, and woodpeckers that inhabit nests of termites and ants, but others also display this habit. These also seem to be examples of commensalism.

Constructing a nest near aggressive wasps is a popular strategy for some birds. Why the insects tolerate the birds is not known, though the birds that dwell in proximity to insects often construct covered nests, which may deter attack by the insects. However, tyrant flycatchers (Tyrannidae) are commonly associated with the aggressive wasp genus *Polistes* in the American tropics, and they often construct open nests. Many of the wasp-associated birds are caciques or weaver-type birds that construct suspended nests that are not directly in the foraging path of insects. There are many reports of multiple wasp nests and bird nests in the same tree.

One of the best-studied symbiotic relationships in tropical areas involves ants and acacia trees. The ant-acacia symbiotic relationship varies according to the various species involved, but usually is described as mutualism. Some acacia trees bear large, swollen thorns (Fig. 13.3) that are used by ants as nesting sites. The swellings are plant-produced, and not induced by the insects (unlike **galls**, which are stimulated by arthropod feeding); such arthropod-sheltering structures are called **domatia**. The ants are quite well protected from predation by birds when dwelling within the acacia thorns, as the thorns are thick walled and tough. Not only do the trees provide shelter, but the trees also bear foliar nectaries that provide sugar to the ants, and produce modified leaflet tips called **Beltian bodies**, which constitute the ant's primary protein and oil sources. Thus, the ants have ample reason to dwell on these trees. The trees are susceptible to herbivory, however, so the ants patrol the tree and aggressively attack and kill insect herbivores. They also bite mammalian herbivores that attempt to feed on the acacia trees, and clip any vines or other vegetation that attempt to intrude into the space occupied by the acacia trees.

In Central America, the ant-acacia relationship also can involve birds. Nesting birds, especially rufous-naped wrens (*Campylorhynchus rufinucha*), select nesting sites in acacia trees containing active colonies of *Pseudomyrmex* ants, particularly *P. spinicola* and *P. nigrocinctus*. Presumably, they derive some protection

Fig. 13.3. Domatia (swollen thorns) on *Acacia repanolobium* trees in Tanzania provide shelter for ants (photo by Christine Miller).

from the ants, because the density of birds, especially those with pendant (hanging) nests, is high when aggressive ants are present. Other birds sometimes use these same trees as nesting locations, though less often, or select trees with less active ant species. This is thought to be due to the competition provided by the wrens, which sometimes destroy the eggs of other birds. Overall, there seems to be little benefit derived by the plants or ants to the occurrence of birds, although enhanced dispersal of seeds is a possibility. Thus, most investigators conclude that the bird-ant aspect is a commensal relationship. However, it is an important one, as nest predation is usually the main cause of reproductive failure in birds. Aggressive biting and stinging activities by insects can be an effective deterrent to mammalian and reptile predators. For example, a study in Mexico using artificial eggs showed that nest success was likely greater in ant-protected trees (64.3%) than in ant-free trees (17.8%). So it is very significant that some birds have developed symbiotic relationships with insects, thereby benefiting their reproductive success.

In East Africa, the ant–acacia relationship involves *Crematogaster* ants, and in addition to herbivorous insects the acacia trees are subject to herbivory by large wildlife. Nevertheless, the ants swarm to fend off acacia-feeding herbivores, even if they are large animals like giraffe (*Giraffa camelopardalis*) and African elephant (*Loxodonta africana*). Interestingly, when the large acacia-feeding animals are eliminated from the

African ecosystem there is less need for the plants to be protected by the ants, so the trees produce fewer hollow thorns and nectar producing glands. In turn, the ants compensate for the loss of their plant-produced sugary food resource by culturing more honeydew-producing scale insects. The more aggressive *C. mimosae* ants tend to be replaced by a more docile sibling species, *C. sjostedti*, in the absence of mammal herbivory. This then leads to invasion by a stem-boring beetle (Cerambycidae) that reduces the survival and growth of the acacia trees. So the presence of the large African animals is important in maintaining the African ecosystem, despite their plant-feeding habits. Unlike the avifauna in Central America, the African wildlife seem to have mutualistic relationships with the acacia trees and ants.

The Benefits of Insects for Habitat Conservation

The biting and disease transmission abilities of some insects, particularly mosquitoes, have made some environments unpleasant, sometimes even uninhabitable, for people. Although this is unfortunate for people attempting to live in these environments, it can be quite beneficial for wildlife conservation. The brief periods of summer in arctic regions, for example, allow hordes of mosquitoes to develop, making it unhealthy for livestock and people amidst the enormous numbers

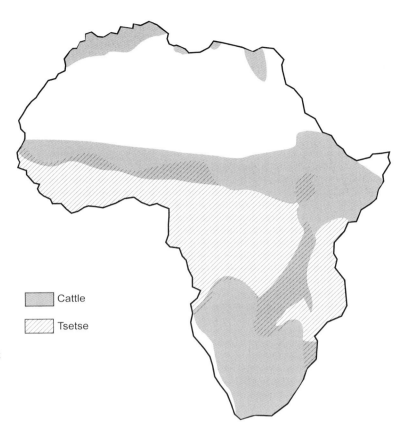

Fig. 13.4. The distribution of cattle grazing areas and tsetse fly-infested areas of Africa. Note the small amount of overlap in their distribution (adapted from the International Laboratory for Research on Animal Diseases (ILRAD) Annual Report, 1989).

Cattle

Tsetse

of blood feeders. This has reduced the development of such areas, making it easier for governments to set aside lands for preservation and wildlife conservation.

Even in the absence of formal protection, insect biting can be sufficiently deterrent to reduce hunting and fishing pressure. For example, several rural counties in northern Florida, USA, desire to develop their natural resources for economic gain, starting with hunting and fishing. However, the abundance of the fiercely biting 'yellow flies' (Diptera: Tabanidae), known elsewhere as deer flies, make it very uncomfortable to be in the outdoor environment during the summer months. This 'problem' has greatly reduced recreation-based economic development, but has helped to protect the wildlife population from exploitation.

The classic example of wildlife conservation and wildlife habitat protection due to insects occurs in central and southern Africa, where tsetse fly-transmit-

ted trypanosomes cause significant mortality among people, and especially among livestock (see African Trypanosomiasis in Chapter 9, Parasitic diseases). The influence of tsetse fly transmission of trypanosomes (resulting in the disease called Nagana) can be seen in Fig. 13.4, which shows the tsetse fly-inhabited regions of sub-Saharan Africa being practically devoid of cattle. The cattle grazing areas crowd the periphery of the tsetse-infested regions to the detriment of these susceptible ungulates, but large areas of central Africa are not much used for grazing because of the high incidence of Nagana found there. Mosquito-borne malaria is also a significant factor limiting human habitation in this area. These diseases are severe hardship for the inhabitants, but have undoubtedly helped with preservation of habitat and animals from exploitation. The white rhino, *Ceratotherium simum*, is an example of wildlife that likely survived extinction because it lived in such inhospitable territory, although in recent years its situation has begun to deteriorate again, especially for the northern subspecies of white rhino. Thus, advancement of technologies of tsetse fly or mosquito suppression, or breeding of livestock that are resistant to trypanosomiasis, must be viewed as a mixed blessing because it will increase human pressure on wildlife, including those in game preserves. Already, in many areas of Africa, wildlife preserves are being illegally used by local residents for harvest of animal food and timber, and for livestock grazing. Growing human populations or the presence of additional domestic grazing animals could prove to be an ecological disaster if left unchecked.

Throughout most of the tropics, and even some of the subtopics, malaria similarly acts to curtail human populations and limit natural resource exploitation. **Malaria** of humans is caused by protozoa in the genus *Plasmodium*. It infects many animals, including reptiles, birds and mammals, but most animals do not display adverse effects from infection. Humans, however, are susceptible to several species. The two most common forms are:

• *Plasmodium falciparum*, which can cause severe, fatal malaria. It may kill 1–3 million people annually, often in Africa, where it is the common form of malaria.
• *Plasmodium vivax*, which is the most common form of malaria because it occurs in Asia, in regions of high population density. It usually does not cause death, but can be incapacitating, resulting in a huge social and economic impact.

The malaria disease transmission cycle is shown in Fig. 13.5. When a female *Anopheles* mosquito feeds on blood, it injects saliva and an anticoagulant into the host. When *Plasmodium*-infected mosquitoes feed, they also inject motile, spindle-shaped cells called sporozoites into the host's blood. The sporozoites travel to the liver where they divide asexually (a process called schizogony) and produce another stage called merozoites. The merozoites affect other liver cells and enter the host's blood, where they invade erythrocytes (red blood cells). Inside these blood cells more asexual reproduction occurs, producing schizonts. The schizonts divide and produce merozoites. The erythrocytes rupture, releasing toxins that cause fever and chills in the host. Some merozoites in erythrocytes develop into gametocytes, which are cells that are capable of producing male or female gametes. The gametocytes must be ingested by mosquitoes, however, before these male and female gametes can form. In the gut of the mos-

quito, fertilization occurs, and zygotes are formed. The zygote becomes attached to the wall of the intestine and oocysts are formed. Sporozoites are then produced within the oocysts, but the sporozoites are released into the hemocoel and eventually move through the insect's blood to the salivary glands of the mosquito where they can be injected to start a new infection cycle. Thus, malaria is a biologically complex disease. It is also deadly, affecting millions of humans annually.

Malaria is transmitted by mosquitoes in the genus *Anopheles*, and only about 30–40 of the over 400 species of *Anopheles* are involved in transmitting the disease. *Anopheles* mosquitoes can be distinguished from most others by their peculiar resting behavior; the tip of their abdomen is pointed up into the air rather than parallel to the substrate on which they rest. Unlike many diseases, where humans are accidental hosts, many species of *Anopheles* have a strong preference for feeding on humans. Many of the vector species prefer to rest in sheltered locations such as houses, so although they have great potential to feed on humans they are easily killed by applying insecticides to the indoor walls of homes. However, insecticide resistance has limited the continued use of many products. Bed netting is often recommended as an alternative to use of insecticides.

As is the case with African Trypanosomiasis, the suppression of malaria is not without ecological costs. For the humans saved from this terrible infliction, it is obviously beneficial, but if malaria suppression allows human populations to increase then there could be terrible impact on wildlife conservation. In many tropical areas where malaria is present the economy remains agrarian, and both population growth and economic development by such societies are typically accompanied by natural resource exploitation.

As is often the case with insects and disease, there is no simple interpretation of the impact. Although I have presented malaria as beneficial to the survival of wildlife, there are other forms of malaria that are more damaging. Foremost is bird malaria, *Plasmodium relictum*. Transmitted by mosquitoes in the genus *Culex*, bird malaria has been devastating to populations of birds lacking a history of exposure, and therefore an appreciable level of resistance. The birds of the Hawaiian Islands, in particular, have been devastated by introduction of this form of malaria into a naïve population.

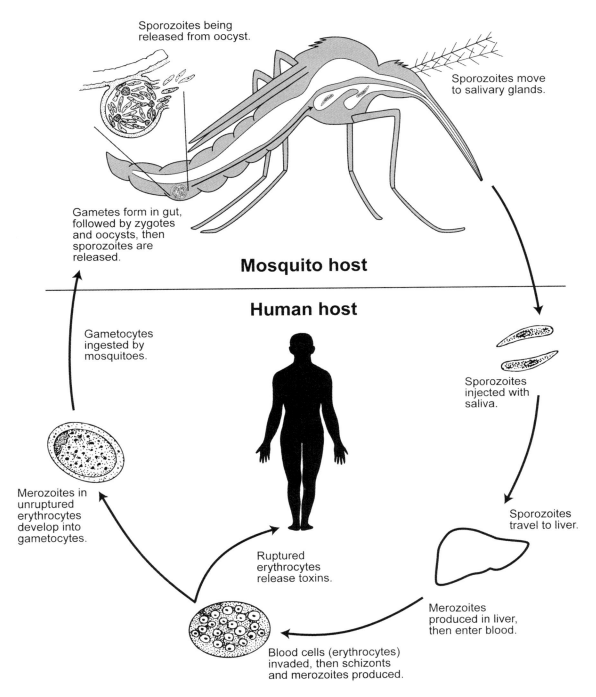

Sporozoites being released from oocyst.

Sporozoites move to salivary glands.

Gametes form in gut, followed by zygotes and oocysts, then sporozoites are released.

Mosquito host

Human host

Gametocytes ingested by mosquitoes.

Sporozoites injected with saliva.

Merozoites in unruptured erythrocytes develop into gametocytes.

Sporozoites travel to liver.

Ruptured erythrocytes release toxins.

Merozoites produced in liver, then enter blood.

Blood cells (erythrocytes) invaded, then schizonts and merozoites produced.

Fig. 13.5. The Malaria disease transmission cycle.

Table 13.3. Insectivory in North American bird species (adapted from Losey and Vaughan 2006).

Order	Common name	No. species	No. primarily insectivorous	No. partially insectivorous	No. not insectivorous
Gaviiformes	Loons	5	0	5	0
Podicipediformes	Grebes	7	5	2	0
Procellariiformes	Tubenoses	6	1	4	1
Pelecaniformes	Pelicans and allies	11	0	5	6
Ciconiiformes	Herons and allies	20	5	12	3
Phoenicopteriformes	Flamingos	1	1	0	0
Anseriformes	Waterfowl	44	19	24	1
Falconiformes	Vultures, hawks, and falcons	31	3	18	10
Galliformes	Quail, grouse, and allies	22	2	20	0
Gruiformes	Cranes and allies	13	8	5	0
Charadriiformes	Shorebirds and gulls	108	51	22	35
Columbiformes	Pigeons and doves	11	0	3	8
Cuculiformes	Cuckoos and roadrunners	6	6	0	0
Strigiformes	Owls	19	6	11	2
Caprimulgiformes	Goatsuckers	8	8	0	0
Apodiformes	Swifts and hummingbirds	20	6	14	0
Trogoniformes	Trogons	1	1	0	0
Coraciiformes	Kingfishers	3	0	1	2
Piciformes	Woodpeckers	22	22	0	0
Passeriformes	Perching birds	285	251	34	0
Total		643	395	180	68
%			61	28	11

The Benefits of Insects for Wildlife-Based Recreation

Insects are quite important as food for wildlife populations that, in turn, support recreational activities such as hunting, fishing, and wildlife viewing. The importance of insects relative to wildlife populations (Table 13.3) is often overlooked, except for freshwater fish, where there is considerable appreciation for the role of insects in supporting the fish population. The importance of wildlife in technologically advanced countries is often limited mostly to recreation, and the analyses reported below (2005 estimates) relate only to such values, and are limited to calculations made only for the USA. However, in some parts of the world wildlife is an important part of the human diet, so the importance of insects could be proportionally greater in such locations.

Small game, but not large game, is often critically dependent on insects for food. Chicks of many popular upland game birds cannot survive without insects as a nutritional resource (see Chapter 4, Wildlife diets). Therefore, the economic effects of insects, based on the proportion of expenditures related to hunting these birds, are about $1.48 billion annually in the USA. The economic effect of insects as food for waterfowl is estimated at $0.56 billion. Other vertebrates that are popular with hunters, such as raccoons, also consume insects as part of their diet, but their contribution to the hunting economy has not been calculated.

Most freshwater fishes are insectivorous, so the entire value of the freshwater fishing (recreational fishing) economy (Table 13.4) can be attributed to insects. This value is $27.9 billion. Although saltwater fishes are not normally thought of as insectivorous,

Table 13.4. Value of commercially landed fishes that rely on insects as a critical nutritional resource (adapted from Losey and Vaughan 2006).

Species	Fish weight (kg landed)	Fish value ($)
Alewife	1,675,935	384,968
Mullet, striped	15,473,230	9,504,673
Mullet, white	509,887	241,064
Mullets	444,000	310,680
Mummichog	4,590	13,221
Perch, white	2,482,006	1,082,354
Perch, yellow	1,714,342	2,914,078
Salmon, chinook	27,345,066	32,633,445
Salmon, chum	92,031,758	16,900,456
Salmon, coho	32,256,133	15,261,440
Salmon, Pacific	176	538
Salmon, pink	334,080,474	24,758,990
Salmon, sockeye	184,505,904	109,897,597
Shad, American	2,074,686	1,190,072
Shad, gizzard	5,306,259	700,916
Shad, hickory	88,339	23,199
Smelt, eulachon	1,081,152	160,842
Smelt, rainbow	489,467	730,685
Smelts	480,212	150,728
Suckers	157,164	45,384
Tilapias	5,482,778	1,223,061
Trout, lake	558,129	228,773
Trout, rainbow	308,306	189,625
Walleye	25,810	42,396
Whitefish, lake	8,604,823	6,048,110
Total ($)		224,637,295

about 25 species spend part of their life in freshwater habitats where they feed on insects. If the proportion of the marine fisheries (commercial fishing) due to these species is calculated, it represents about $0.22 billion.

Wildlife observation is an important form of recreation, and bird-watching is central to wildlife observation. Wildlife viewers often also appreciate opportunities to view small mammals, reptiles and amphibians, which often use insects as well, and many make efforts to view insects. Nevertheless, the benefits of wildlife watching are calculated only on the basis of bird-watching, and only on the basis of species of North American birds that are primarily insectivorous. Thus, this is a very conservative estimate, but still accounts for an economic impact of $19.8 billion.

SUMMARY

• Wildlife can affect insect survival through predation. Most of the major orders of insects are commonly used as food by terrestrial wildlife. Additionally, spiders, isopods and millipedes are frequent food items. Among aquatic wildlife, common prey include mayflies, dragonflies and damselflies, stoneflies, and caddisflies.
• Wildlife often display functional responses to prey availability, and less often show numerical responses. Sometimes wildlife function in population regulation of insects. Wildlife, like most biotic mortality agents, tend to function in a density-dependent manner, whereas environmental mortality factors tend to function in a density-independent manner.
• Wildlife have been relocated to new locations around the world for many reasons, one of which is to provide biologically based suppression of pest insects. Generally this has been a mistake, in large measure because most predatory wildlife are generalists or at least opportunists, and will feed on more than the intended target pests.
• Insects also affect wildlife survival through predation. Small vertebrates, especially helpless young or small aquatic vertebrates, are susceptible to attack by insects. These vertebrates not only can outgrow their susceptibility, but may eventually turn on the insects and make them the prey items. Predation of terrestrial vertebrates by large arthropods such as tarantulas is easy to imagine, but aggressive ant species also are important predators. In the aquatic environment, predatory bugs, beetles and other insects eat eggs and immature stages of amphibians and fish.
• Symbiotic relationships are ubiquitous in nature, and many occur between insects and wildlife. Birds sometimes establish commensal relationships with biting or stinging insects. Their nests appear to be protected by their proximity to aggressive insects, providing some protection for their eggs and nestlings from mammal and reptile predation. In Africa, mutualistic relationships occur between insects and grazing animals.
• Insects can affect wildlife survival by enhancing habitat conservation. When insects make it uncomfortable or unhealthy for humans or their livestock, natural resources (including wildlife) are less likely to be exploited. Mosquitoes and malaria, and tsetse flies and trypanosomiasis, are examples of insects and associated disease that have reduced the rate of natural resource exploitation and helped preserve harborages for wildlife.

• The presence of insects helps foster wildlife-based recreation. Activities such as wildlife-watching, fishing, and hunting would be much more limited were insects not available as a food resource.

REFERENCES AND ADDITIONAL READING

Allen, C.R., Epperson, D.M., & Garmestani, A.S. (2004). Red imported fire ant impacts on wildlife: a decade of research. *American Midland Naturalist* **152**, 88–103.

Allen, C.R., Lutz, R.S., & Demariais, S. (1995). Red imported fire ant impacts on northern bobwhite populations. *Ecological Applications* **5**, 632–638.

Allen, C.R., Lutz, R.S., Lockley, T., Phillips, S.A., Jr., & Damariais, S. (2001). The non-indigenous ant, *Solenopsis invicta*, reduces Loggerhead Shrike and native insect abundance. *Journal of Agriculture and Urban Entomology* **18**, 249–259.

Belovsky, G.E. & Slade, J.B. (1993). The role of vertebrate and invertebrate predators in a grasshopper community. *Oikos* **68**, 193–201.

Blaustein, L. & Margalit, J. (1994). Mosquito larvae (*Culiseta longiareolata*) prey upon and compete with toad tadpoles (*Bufo viridis*). *Journal of Animal Ecology* **63**, 841–850.

Bock, C.E., Bock, J.H., & Grant, M.C. (1992). Effects of bird predation on grasshopper densities in an Arizona grassland. *Ecology* **73**, 1706–1717.

Boland, C.R.J. (2004). Introduced cane toads *Bufo marinus* are active nest predators and competitors of rainbow bee-eaters *Merops ornatus*: observational and experimental evidence. *Biological Conservation* **120**, 53–62.

Buckner, C.H. (1966). The role of vertebrate predators in the biological control of forest insects. *Annual Review of Entomology* **11**, 449–470.

Burger, J.C., Redak, R.A., Allen, E.B., Rotenberry, J.T., & Allen, M.F. (2001). Restoring arthropod communities in coastal sage scrub. *Conservation Biology* **17**, 460–467.

Campbell, R.W. & Sloan, R.J. (1977). Natural regulation of innocuous gypsy moth populations. *Environmental Entomology* **6**, 315–322.

Campbell, R.W. & Sloan, R.J. (1977). Release of gypsy moth populations from innocuous levels. *Environmental Entomology* **6**, 323–330.

Campbell, R.W. & Torgersen, T.R. (1983). Compensatory mortality in defoliator population dynamics. *Environmental Entomology* **12**, 630–632.

Campbell, R.W., Torgersen, T.R., & Srivastava, N. (1983). A suggested role for predaceous birds and ants in the population dynamics of the western spruce budworm. *Forest Science* **29**, 779–790.

Catling, P.C., Hertog, A., Burt, R.J., Wombey, J.C., & Forrester, R.I. (1999). The short-term effect of cane toads (*Bufo marinus*) on native fauna in the gulf country of the Northern Territory. *Wildlife Research* **26**, 161–185.

Chandra, G., Bhattacharjee, I., Chatterjee, S.N., & Ghosh, A. (2008). Mosquito control by larviporous fish. *Indian Journal of Medical Research* **127**, 13–27.

Collins, N.M. & Thomas, J.A. (eds.) (1991). *The Conservation of Insects and their Habitats*. Academic Press, London, UK.

Cooper, R.J., Martinat, P.J., & Whitmore, R.C. (1990). Dietary similarity among insectivorous birds: influence of taxonomic versus ecological categorization of prey. *Studies in Avian Biology* **13**, 104–109.

Crawford, H.S., Titterington, R.W., & Jennings, D.T. (1983). Bird predation and spruce budworm populations. *Journal of Forestry* **81**, 433–435, 478.

Dahlsten, D.L., Copper, W.A., Rowney, S.L., & Kleintjes, P.K. (1990). Quantifying bird predation of arthropods in forests. *Studies in Avian Biology* **13**, 44–52.

Eriksson, M.O.G. (1979). Competition between freshwater fish and goldeneyes *Bucephala clangula* (L.) for common prey. *Oecologia* **41**, 99–107.

Fowler, A.C., Knight, R.L., George, T.L., & McEwen, L.C. (1991). Effects of avian predation on grasshopper populations in North Dakota grasslands. *Ecology* **72**, 1775–1781.

Gardner, K.T. & Thompson, D.C. (1998). Influence of avian predation on a grasshopper (Orthoptera: Acrididae) assemblage that feeds on threadleaf snakeweed. *Environmental Entomology* **27**, 110–116.

Glen, D.M. (2004). Birds as predators of lepidopterous larvae. In van Emden, H.F. & Rothschild, M. (eds.) *Insect and Bird Interactions*, pp. 89–106. Intercept, Andover, Hampshire, UK.

Gradwohl, J. & Greenberg, R. (1982). The effect of a single species of avian predator on the arthropods of aerial leaf litter. *Ecology* **63**, 581–583.

Hirai, T. (2007). Diet composition of the endangered giant water bug, *Lethocerus deyrolli* (Hemiptera: Belostomatidae) in the rice fields of Japan: which is the most important prey item among frogs, fish and aquatic insects? *Entomological Science* **10**, 333–336.

Jachmann, H. (2008). Illegal wildlife use and protected area management in Ghana. *Biological Conservation* **141**, 190–1918.

Janzen, D.H. (1969). Birds and the ant X acacia interaction in Central America, with notes on birds and other myrmecophytes. *The Condor* **71**, 240–256.

Joern, A. (199). Variable impact of avian predation on grasshopper assemblies in sandhills grassland. *Oikos* **64**, 458–463.

Johnson, M.D. & Sherry, T.W. (2001). Effects of food availability on the distribution of migratory warblers among habitats in Jamaica. *Journal of Animal Ecology* **70**, 546–560.

Kalka, M.B., Smith, A.R., & Kalko, E.K.V. (2008). Bats limit arthropods and herbivory in a tropical forest. *Science* **320**, 71.

Kock, R., Kebkiba, B., Heinonen, R., & Bedane, R. (2002). Wildlife and pastoral society – shifting paradigms in disease control. *Annals of the New York Academy of Sciences* **969**, 24–33.

Konvicka, M., Benes, J., Cizek, O., Kopecek, F., Konvicka, O., & Vitaz, L. (2008). How too much care kills species: grassland reserves, agri-environmental schemes and extinction of *Colias myrmidone* (Lepidoptera: Pieridae) from its former stronghold. *Journal of Insect Conservation* **12**, 519–525.

Letnic, M., Webb, J.K., & Schine, R. (2008). Invasive cane toads (*Bufo marinus*) cause mass mortality of freshwater crocodiles (*Crocodylus johnstoni*) in tropical Australia. *Biological Conservation* **141**, 1773–1782.

Losey, J.E. & Vaughan, M. (2006). The economic value of ecological services provided by insects. *Bioscience* **56**, 311–323.

Oliveras de Ita, A. & Rojas-Soto, O.R. (2006). Ant presence in acacias: an association that maximizes nesting success in birds? *The Wilson Journal of Ornithology* **118**, 563–566.

McCormick, S. & Polis, G.A. (1982). Arthropods that prey on vertebrates. *Biological Reviews* **57**, 29–58.

McElligott, A.G., Maggini, I., Junziker, L., & König, B. (2004). Interactions between red-billed oxpeckers and black rhinos in captivity. *Zoo Biology* **23**, 347–354.

McEwen, L.C. (1987). Function of insectivorous birds in a shortgrass IPM system. In Capinera, J.L. (ed.) *Integrated Pest Management on Rangeland. A Shortgrass Prairie Perspective*, pp. 324–333. Westview Press, Boulder, Colorado, USA.

Morris, R.F., Cheshire, W.F., Miller, C.A., & Mott, D.G. (1958). The numerical response of avian and mammalian predators during a gradation of the spruce budworm. *Ecology* **39**, 487–494.

Myers, J.G. (1929). The nesting together of birds, wasps, and ants. *Proceedings of the Royal Entomological Society of London* **4**, 80–88.

Myers, J.G. (1935). Nesting associations of birds with social insects. *Transactions of the Royal Entomological Society of London* **83**, 11–22.

Ohba, S., Miyasaka, H., & Nakasuji, F. (2008). The role of amphibian prey in the diet and growth of giant water bug nymphs in Japanese rice fields. *Population Ecology* **50**, 9–16.

Orrock, J.L. & Danielson, B.J. (2004). Rodents balancing a variety of risks: invasive fire ants and indirect and direct indicators of predation risk. *Oecologia* **140**, 662–667.

Palmer, T.M., Stanton, M.L., Young, T.P., Goheen, J.R., Pringle, R.M., & Karban, R. (2008). Breakdown of an ant–plant mutualism follows the loss of large herbivores from an African savanna. *Science* **319**, 192–195.

Pascala, S. & Roughgarden, J. (1984). Control of arthropod abundance by *Anolis* lizards on St. Eustatius (Neth. Antilles). *Oecologia* **64**, 160–162.

Peres, C.A. (1996). Ungulate ectoparasite removal by black caracaras and pale-winged trumpeters in Amazonian forests. *Wilson Bulletin* **108**, 170–175.

Richter, S.C. (2000). Larval caddisfly predation on the eggs and embryos of *Rana capito* and *Rana sphenocephala*. *Journal of Herpetology* **34**, 590–593.

Rothschild, M. & Clay, T. (1953). *Fleas, Flukes and Cuckoos. A Study of Bird Parasites*. Readers Union, London.

Samish, M., Ginsberg, H., & Glazer, I. (2004). Biological control of ticks. *Parasitology* **129**, S389–S403.

Stapley, L. (1998). The interaction of thorns and symbiotic ants as an effective defence mechanism of swollen-thorn acacias. *Oecologia* **115**, 401–405.

Stout III, B.M., Stout, K.K., & Stihler, C.W. (1992). Predation by the caddisfly *Banksiola dossuaria* on egg masses of the spotted salamander *Ambystoma manulatum*. *American Midland Naturalist* **127**, 368–372.

Toledo, L.F. (2005). Predation of juvenile and adult anurans by invertebrates: current knowledge and perspectives. *Herpetological Review* **36**, 395–400.

Wiens, J.A. & Rotenberry, J.T. (1979). Diet niche relationships among North American grassland and shrubsteppe birds. *Oecologia* **42**, 253–292.

Williams-Guillén, K., Perfecto, I., & Vandermeer, J. (2008). Bats limit insects in a neotropical agroforestry system. *Science* **320**, 70.

Young, B.E., Kaspari, M., & Martin, T.E. (1990). Species-specific nest site selection by birds in ant-acacia trees. *Biotropica* **22**, 310–315.

Chapter 14

INSECT AND WILDLIFE CONSERVATION

Insects are integral components of many ecosystems, and often worthy of conservation. In Chapters 3 and 4 we discussed their importance as food resources for wildlife. In Chapter 6 we discussed the significance of pollination by insects in natural ecosystems, which has great ecological importance not only for plants but also for the animals that feed upon plants. Additionally, insect-based herbivory and decomposition were presented as key elements of ecosystem nutrient dynamics. In Chapter 13 we saw the importance of insects in protecting wildlife habitat, and the value of insects for wildlife-based recreation. However, there remain additional benefits that can be derived from insects and conservation of insects. For example, crop pollination, and honey and silk production, are important economic benefits. Also, maintenance of biodiversity is assuming greater importance in most areas of the world. Although not fully appreciated by some, insects are an important type of wildlife that deserve protection in their own right. Importantly, insect abundance can be manipulated to favor wildlife.

OTHER ECONOMIC BENEFITS OF INSECTS

Pollination

Pollination can be accomplished by insects, vertebrates (birds and bats), and wind, but certain plants are mostly dependent on insects, and bees in particular, for

Insects and Wildlife: Arthropods and Their Relationships with Wild Vertebrate Animals, 1st edition. By J.L. Capinera. Published 2010 by Blackwell Publishing.

successful pollination. Among the crops requiring pollination are most fruits and nuts, many vegetables, and a few field crops. In contrast, the grain crops, including corn (maize), wheat, and rice are wind pollinated, so about the only role insects have in the production of these nutritional staples is one of herbivory.

Wild pollinators (mostly bees and flies) can be quite important for plants requiring insect pollination, and may be completely effective for isolated plants or small fields. However, in modern crop production the high density of crops and the long distance of crops from uncultivated areas may limit the ability of wild pollinators to effectively provide pollination of crop plants. Therefore, hives of bees (usually Hymenoptera: Apidae and specifically *Apis mellifera*) are often moved to the vicinity of crops requiring pollination. In the USA, about 2.5 million hives were rented for pollination services in 1999, clearly indicating the importance of pollination. Nearly 85% of the rentals occurred in only seven crops (in descending order of importance): almond, apple, melons, alfalfa seed, plum/prune, avocado, and blueberry.

Estimates of the dependence of crops on pollination, and the proportion of pollination accomplished by wild pollinators versus domesticated honey bees, are shown in Table 14.1. Note that dependence varies considerably from crop to crop, and even between related crops (e.g., compare grapefruit to lemon, which are both citrus crops). The value of pollination is estimated at over $3 billion in the USA alone (2001 estimate), though there are earlier estimates of $5.7 billion in pollination benefits. Though this is a small value relative to the total value of crops, these insect-pollinated crops account for important diversity in our diet. Imagine subsisting on corn, wheat and barley, spiced up with an occasional potato; a bland diet, indeed! The

Table 14.1. The value of crop production in the USA resulting from pollination, 2001–2003, in relation to the source of pollinators (adapted from Losey and Vaughan 2006).

Crop	Mean annual value ($ millions)	% dependent on pollination	% domesticated nonindigenous bees	% indigenous bees	Mean value from indigenous bees ($ millions)
Fruits and nuts					
Almond	1,120.0	100	100	10	158.51
Apple	1,585.1	100	90	20	4.2
Apricot	30.0	70	80	10	38.24
Avocado	382.4	100	90	10	2.31
Blueberry, wild	23.1	100	90	10	19.29
Blueberry, cultivated	192.9	100	90	10	0.31
Boysenberry	3.9	80	90	10	0.31
Cherry, sweet	290.6	90	90	10	26.15
Cherry, tart	56.3	90	90	10	5.07
Citrus					
Grapefruit	278.4	80	90	10	22.27
Lemon	286.1	20	10	90	51.50
Lime	2.0	30	90	10	0.06
Orange	1713.6	30	90	10	51.41
Tangelo	10.8	40	90	10	0.43
Tangerine	112.0	50	90	10	5.60
Temple	6.1	30	90	10	0.18
Cranberry	159.7	10	90	10	15,097
Grape	2774.8	10	10	90	249.73
Kiwifruit	16.7	90	90	10	1.50
Loganberry	158.0	50	80	20	15.80
Macadamia	31.1	90	90	10	2.80
Nectarine	121.2	60	80	20	14.54
Olive	66.5	10	10	90	5.99
Peach	487.9	60	80	20	58.55
Pear	263.9	70	90	10	18.47
Plum and prune	197.8	70	90	10	13.85
Raspberry	95.8	80	90	10	7.19
Strawberry	1,187.6	20	10	90	213.77
Vegetables					
Asparagus	164.3	100	90	10	16.43
Broccoli	543.4	100	90	10	54.34
Carrot	575.5	100	90	10	57.55
Cauliflower	219.8	100	90	10	21.98
Celery	256.5	100	80	20	51.30
Cucumber	379.5	90	90	10	34.16
Cantaloupe	401.0	80	90	10	32.08
Honeydew	94.1	80	90	10	7.53
Onion	808.0	100	90	10	80.80
Pumpkin	75.5	90	10	90	61.16
Squash	192.3	90	10	90	155.76
Vegetable seed	61.0	100	90	10	6.10
Watermelon	315.9	0.7	0.9	0.1	22.11
Field crops					
Alfalfa hay	7,212.8	100	95	5	360.64
Alfalfa seed	109.0	100	95	5	5.45

Table 14.1. *Continued*

Crop	Mean annual value ($ millions)	% dependent on pollination	% domesticated nonindigenous bees	% indigenous bees	Mean value from indigenous bees ($ millions)
Cotton lint	3,449.5	20	80	20	137.98
Cotton seed	689.3	20	80	20	27.57
Legume seed	34.1	100	90	10	3.41
Peanut	793.1	10	20	80	63.45
Rapeseed	0.3	100	90	10	0.03
Soybean	15,095.2	10	50	50	754.76
Sugar beet	1,057.3	10	20	80	84.58
Sunflower	312.7	100	90	10	31.27
Total ($)					3,074.13

benefits of fruits and vegetables are considerable, both for nutrition and appetite.

However, some caution should be used in interpreting pollination data. Sometimes 'insect pollinated crops' can be *grown* without pollination. Thus, asparagus, carrots and alfalfa are examples of crops can be grown successfully in the absence of pollinators because we do not harvest a pollinated 'fruit' from these crops. They require pollinators for seed production, however. A relatively small area of carrots, if properly pollinated, can produce enough seed for all the carrots grown as vegetables. In the case of alfalfa, the crop can be harvested several times per year, and for several years, before replanting. The extreme case is asparagus, which is normally harvested for a decade before replanting. Thus, although the existence of these crops (and many others) is dependent on insect pollination, the production of any particular field may not require the presence of pollinators. In contrast, for other crops (e.g., apple, avocado, and blueberry) every fruit harvested requires visitation by an insect pollinator.

We take for granted that pastures and prairies will be populated by wildflowers. A walk through woodlands in the spring is a wonderful way to see small herbaceous plants in their full glory. And what would a tropical landscape be without a profusion of flowers? Without insects to perform pollination services, these and many other environments would seem sterile, lacking in the bright colors we normally expect in our landscapes. Most plants that produce colorful flowers do so to attract insects, and without these pollinators the plants would decline or disappear. Birds, mammals, and rodents that depend on the fruits and seeds of such plants would also decline. Biodiversity would decline tremendously, and highly coevolved systems, such as some orchids, would certainly disappear, at least in natural ecosystems. It is difficult to assign economic or even aesthetic values to the loss of insect pollination, but it certainly would provide a very different world from the one we now enjoy.

Honey

Bees are induced to visit flowers by the availability of nectar as food for the bees, and when they feed on nectar the bees may become contaminated with pollen. The pollen may then be inadvertently transferred to the next flower that is visited by the pollinator. Bees collect more nectar than they need for immediate consumption, and store some in the hive as honey (Fig. 14.1). It is stored within wax cells. Thus, honey is an important by-product of pollination. At one time, honey production was more important; farmsteads routinely produced their own supply of sweetener from their own hives. Later, an industry grew up to supply honey, often manned by migratory beekeepers that kept hundreds or thousands of hives and followed the availability of nectar for their bees. While this still occurs to some extent, in many instances the importance of honey has been supplanted by other sweeten-

Fig. 14.1. Honey is normally the product of (a) domesticated honey bees, *Apis mellifera* (Hymenoptera: Apidae) although in some societies it is collected from other colonial bee species. Honey is derived from plant nectar and is stored within (b) bee combs for later consumption by bees, though because the bees are so industrious honey can be extracted from the combs and harvested by humans with enough left to supply the colony (photos by J. D. Ellis).

ers (from sugar cane, sugar beet, or corn). Nevertheless, honey production is an important supplement to pollination services for many beekeepers, and the only product for some. The value of honey produced in the USA was $157 million in 2005, coming from 2.4 million colonies. But about 60% of the honey consumed in the USA was imported from other countries, so this value underestimates its importance. World honey production in 2005 was estimated by the Food and Agriculture Organization (FAO) of the United Nations to be slightly over 1.4 billion kg. Production has increased steadily over the last decade or so, with most of the increases coming from Asia and South America, while honey production in the USA continues to decline, along with the number of beekeepers and colonies of bees. It is difficult to estimate the value of honey throughout the world, but it likely was about (US) $3 billion in 2005.

Silk Production (Sericulture)

Silk is produced by specialized glands (generally modified salivary glands) in the larvae of some Lepidoptera, and by other structures in some immatures and adults of mites and spiders. Only moth larvae, however, have been exploited by humans for their ability to produce silk commercially. The silk naturally serves various functions, such as larval dispersal in the wind, leaf rolling, anchoring of pupae, and construction of cocoons. It is this latter function, which consists of production of a single long strand of silk, that allows the silk to be unwound from the cocoon and harvested for silk production.

In various parts of the world, different species have been used to produce silk, but generally it involves moths of the superfamily Bombycoidea, and particularly *Bombyx mori* L. Silk has been harvested by humans from *Bombyx mori* at least since 2600 BC. It was one of the first, and most valuable, trade commodities between China and Europe. Though originating in China, once the insect was smuggled out of China it was quickly spread around the world, where its cultivation was limited only by the ability to produce mulberry trees, its natural host. Besides China, Japan and India are the most important production centers. Now, silkworms can be produced on artificial diet as well as on mulberry. Silk is a valuable commodity, valued at over $1 billion annually.

Shellac and Lacquer

Several scale insects (Hemiptera: Kerriidae) can be grown on trees in Southeast Asia and used as a source of lac, a resin that is the principal ingredient of shellac. *Laccifer lacca* Kerr is the species most commonly cultured for this purpose, which involves collection and purification of a shiny amber-colored secretion. India produces the largest proportion of the shellac on the world market, though production has fallen greatly due to the availability of synthetic resins. This natural product is also a component of some candy products where a shiny coating is desired.

The giant margarodid scale insect *Llaveia axin* (Hemiptera: Margarodidae), known in Mexico and Central America as aje or niij, respectively, is also a source of a shiny lacquer. However, rather than being a secretion, as is the case with lac insects, it is the fat from the bodies of the niij insects that is collected. These scale insects are about 2.5 cm long and can weigh over 0.5 g each, so they are unusual both for their immense size and the amount of fat that they contain. The extract can be applied alone to produce a

permanent film, or mixed with pigment to produce a finish for pottery, ceramic, leather, metal, wood, and gourds. Not well known outside this region of the Americas, is it important locally in artistry.

Dyes

Commercially important dyes have been extracted from insects. Probably best known is cochineal dye, obtained from the bodies of cochineal scales, *Dactylopius coccus* (Hemiptera: Dactylopiidae) native to Mexico and Central America. Cultivated on *Opuntia* cactus, it can be used to produce a bright red dye. In the seventeenth and eighteenth centuries it was a mainstay of the economy of this region. However, its popularity declined in importance as synthetic materials became available beginning about 1860, and only now is becoming more popular again as a 'natural' dye. It also is useful in cosmetics and food.

Lac dye is a byproduct of shellac production. Lac dye has been used as a skin cosmetic, in medicine, and for dyeing wool, silk, and leather.

Food for Humans and Domestic Animals

Insects are not ordinarily a major source of nutrition for humans in technologically advanced societies. In some societies, however, insects are consumed if they are especially available or provide needed nutrition during times of famine (e.g., locusts in Africa), or as part of cultural tradition (e.g., as a condiment – canned grasshoppers in Southeast Asia or Maguey worms in Mexico). In Amazonia, most of the indigenous peoples are entomophagous. In Africa, Asia, Oceania, and Latin America it is not unusual to find a great diversity of insects in local marketplaces (Fig. 14.2), and in some countries children are especially likely to eat insects. In a few societies, insects have been prized as food. For example, the emperor Montezuma and the Aztec kings who preceded him prized the eggs of aquatic Hemiptera (called 'ahuahutle'). In this society it was the equivalent of caviar, and transported at great effort and expense to Tenochtitlan for ceremonies. Formerly, the indigenous people living in eastern Australia would trek to the Australian Alps to collect bogong moth, *Agrotis infusa* (Lepidoptera: Noctuidae) that aggregated in caves during the heat of the day. After singeing off the wings and legs they would feast on these nutritious

Fig. 14.2. Some examples of insects being used as food by humans: (a) restaurant in Guangzhou, Guangdong Province, southeast China with displays of fresh fish and aquatic insects (in aquaria) and terrestrial insects (in boxes at front of display); (b) close-up of a readily available terrestrial species that is often eaten by humans, pupae of silkworm, *Bombyx mori* (Lepidoptera: Bombycidae) (also shown in the lower left corner of (a)); (c) example of aquatic insects for sale in this restaurant: water scavenger beetles (Coleoptera: Hydrophilidae) and predaceous diving beetles (Coleoptera: Dytiscidae); (d) stir-fried mole crickets (Orthoptera: Gryllotalpidae) and onions from a food stall in Lampong, northern Thailand (photos from China by Theresa Friday and Teresa Olczyk; from Thailand by Bob McGovern).

morsels. Insects are good sources of proteins, lipids, and vitamins, but it is difficult to raise them economically. Thus, they tend to be eaten opportunistically rather than cultured. The economic value of insects as food has not been assessed, but likely is relatively low.

In the Neotropics, ants and termites can be important seasonal resources and are actively collected for food. Leafcutter ant (*Atta* spp.; Hymenoptera: Formicidae) and termite (Isoptera) colonies produce large numbers of dispersive queens at the start of the rainy season. The queens can be collected by hand with probes, trapped as they exit the soil, or induced to come to the surface by channeling water into the nest.

Sometimes insects are cultured for food. In the tropics of Amazonia and New Guinea, palms are cut down to attract ovipositing palm weevils, *Rhynchophorus palmarum* and *Rhinostomus barbirostris* (Coleoptera: Curculionidae). Because *R. palmarum* prefers the base of trees and *R. barbirostris* drills holes along the length of the trunk, both species benefit from this tree cutting. However, if deep cuts are made along the length of the trunk, *R. palmarum* are induced to deposit there, also. The weevils differ in flavor and fat content, so their abundance can be regulated according to 'taste.' These practices can be viewed as a form of 'semi-cultivation.'

Insects often are useful for maintenance of pets, as they are readily accepted and nutritious. Easily cultured insects are most generally used, including house cricket, *Acheta domesticus* (Orthoptera: Gryllidae); mealworm, *Tenebrio molitor* (Coleoptera: Tenebrionidae); waxworms, *Galleria mellonella* (Lepidoptera: Pyralidae); and various flies (Diptera: Muscidae and Calliphoridae). They are most often used as food or a food supplement for amphibians, reptiles and birds, but rodents and some small mammals also accept them. They are sold at pet shops, where they may be found in living, dried, and frozen forms. In recent times, dehydrated insects have been included in some types of wild bird food, although this is still relatively unusual. Also, zoos often seek live insects for their exhibit animals, both due to their nutritional value and also because higher animals often suffer from boredom in a zoo environment, and it is healthy to provide variety in the diet and behavioral diversion in the form of live insects. Thus, large mobile insects such as grasshoppers provide diversion as well as nutrition for some zoo animals, including monkeys. Lastly, but importantly, insects are used as a lure (bait) for fishing, as some fish take live insects preferentially. The same insects used for pet food tend to be used as fish bait.

Insects serve as a food resource for some domestic animals, notably fowl and pigs, when animals are not kept in pens. Turkeys, chickens, ducks and geese readily forage for insects on small farms throughout the world. In fact, in some instances they are herded into crop fields infested with insects to eliminate the pests. Similarly, if pigs are allowed to roam they will readily grub in the soil for scarab beetle larvae, though these are not delicate animals and they have a very broad diet, so they generally are kept away from crops to prevent them from causing crop injury.

Medical Treatment

Though more widely used for medical treatment in ancient societies, insects retain some uses in contemporary treatment of human ailments, most notably in apitherapy (bee venom therapy) and in maggot therapy. **Apitherapy** is sometimes recommended for rheumatic diseases, including arthritis and multiple sclerosis. Traditionally, bees were stimulated to sting the affected area, but injected venom is now more common as a form of treatment. Bee venom is also applied topically in the form of creams, liniments and ointments.

Maggot therapy is more well founded scientifically, and takes advantage of the propensity of some fly larvae (usually *Lucilia* maggots) (Diptera: Calliphoridae) to feed on dead and decaying flesh, but to avoid feeding on living tissue. Thus, live maggots are introduced to wounds of humans or pets to clean out necrotic tissue. Also, maggots excrete products that inhibit growth of microbes leading to infection of living tissue, and stimulate regrowth of healthy tissue. More commonly used prior to the advent of modern antibiotics (the 1940s and 1950s), maggot therapy continues to be used for wounds that display difficulty in healing, and with the emergence of antibiotic-resistant bacteria, shows promise of renewed interest by the medical community.

CONSERVATION OF INSECTS, THE 'SMALLEST WILDLIFE'

Human population expansion and the accompanying natural habitat loss, environmental contamination, and global climate change are severe threats to animal and plant biodiversity. The problem is exacerbated by the accidental (in the case of insects) and deliberate (for plants and fish) introduction of aggressive, **nonindigenous** (from another place, not 'native') species around the globe that displace **indigenous** (naturally occurring in this location, 'native') organisms. The toll in lost biodiversity is impressive. Among insects alone, an estimated 11,200 species have become extinct since 1600, although this is only an approximation because most extinctions of insects go unnoticed. So, the actual count of documented insect extinctions is considerably less. In the USA, the list of extinct or probably extinct insects stands at 162, less than 1%. In Britain, about 11% of the insects are thought to be rare or endangered, and the corresponding levels are 22% and 34% in Austria and Germany, respectively. Overall, about one fourth of all insect species are thought to be in danger of extinction. In Britain, butterflies are being lost more quickly than are plants or birds. For many insects, when their host becomes extinct, be it plant or animal, they also disappear. The problem is more difficult for insects because most are considered to be insignificant, or not worthy of attention. Butterflies are perhaps the exception, but perhaps for the wrong reasons. It is a tragedy when a beautiful insect disappears, but is it less significant when a dung-feeding insect is lost? Insects are animals and (in nearly all cases, because a few can be considered to be 'livestock') insects are wildlife, though too often the term is restricted to vertebrate animals. So, anyone interested in conservation should embrace the idea that insects are important members of any ecosystem.

The principal threats to insect biodiversity are:

• **Loss of natural habitat.** Land transformation, whether for agriculture, reforestation, flood protection, or development of industries or housing has been the principal determinant of species loss. Homogeneous crops and pastures, homes and mowed lawns, roadways, and buildings are not conducive for the continued existence of most insects. Admittedly, some insects prosper under such conditions, and one of the reasons that a few species of ants have become global in distribution is that they are able to exploit such disturbed environments. The development of the city of San Francisco, USA caused the extinction of three species of butterflies that had their distribution limited to the sand dune ecosystem of the San Francisco Bay area. Also, three additional species in the area have had their natural habitat greatly diminished. The loss of tropical forests in Southeast Asia, Africa, and South America is particularly alarming, as that is where

more than half of all insects live. As timber is harvested, or forest is burned to create grazing land, innumerable host-specific insect species disappear. Other less expansive environments especially subject to disruption include islands, where the fauna is often quite vulnerable to invasive organisms, and the delicate environment inside caves, where any disturbance can have irreversible repercussions. On the other hand, open areas such as prairie and pastureland are often needed by flower-feeding insects such as bees and butterflies, and disturbance (especially fire) is essential for maintenance of such areas. Reforestation and protection from fire can lead to growth of dense vegetation that is not suitable for such insects, or for wildlife preferring open habitats.

• **Invasive nonindigenous species.** Invasive nonindigenous species can displace indigenous species and modify ecosystems. This is most apparent when an invasive plant establishes in a new location. For example, the uncontrolled spread of *Melaleuca quinquenervia*, a small tree deliberately introduced to southern Florida, USA, has changed sawgrass-dominated wetlands to dense woodlands that crowd out nearly everything but *Melaleuca*. Invading insects are often less apparent but no less significant. Red imported fire ant (*Solenopsis invicta*) in the southern USA, gypsy moth (*Lymantria dispar*) in the northeastern USA, and argentine ant (*Linepithema humile*) in the western USA are examples of insects that have altered entire ecosystems. However, invaders often have even more radical effects on islands due to the relatively small number of species present. For example, invasion of the Galapagos Islands by red imported fire ant threatens numerous indigenous species.

• **Global climate change.** The warming of the earth has already resulted in tropical species expanding their range, and temperate species having longer periods of activity. It remains to be seen how many species adapt to these environmental alterations, move to new locations, or become extinct. Species with broad geographic distributions are likely to survive with altered distribution. Species with very restricted ranges, especially species with a very narrow host range, are likely to suffer the most.

• **Widespread use of pesticides and other control practices.** Use of pesticides historically has had significant nontarget effects. Agricultural pest control and mosquito control are examples of situations that have had deleterious effects on butterflies, bees, and other nontarget species. Even with the introduction and use of less persistent and less broad-spectrum pesticides in recent years, considerable nontarget impact occurs. Incorporation of *Bacillus thuringiensis* toxin into an increasing number of crop plants poses risks not only to the nontarget insects associated with the crops, but potentially to insects feeding on related plants, because genes can move from species to species. Practitioners of biological control make every effort to ensure that introduced biological control agents are specific, but in the past some have had unforeseen consequences. Pesticides are not as damaging to insect populations as are land modifications, however, because insects can recolonize and recover quickly from population disruption if host plants are available.

It is not unusual for insect populations to vary considerably from year to year. Changes in weather, shifts in the abundance of flowering annual plants, and changes in the occurrence of parasites and disease often account for this variation. Insects generally have a high reproductive capacity and can recover from such population oscillations when their food supply is adequate. However, large-scale changes in habitat such as conversion to agriculture or reforestation of pastures can prevent recovery of insect populations. Figure 14.3 presents an example of how insects are affected, in this case butterfly populations in Switzerland. Note that most species are affected by agriculture, loss of pasture, and reforestation.

Conservation Status

The US **Endangered Species Act**, passed by the United States Congress in 1973, is a far-reaching effort to protect critically imperiled species from extinction as a consequence of economic growth and development. In addition to protecting individual species, the Endangered Species Act is designed to protect the ecosystems upon which they depend. It encompasses plants and invertebrates as well as vertebrates. Because it can impede land development, it has been controversial.

Also created in 1973 was the Convention on International Trade in Endangered Species of Wild Fauna and Flora (**CITES**). CITES is an international agreement restricting international commerce in plant and animal species believed to be actually or potentially harmed by trade. Over-exploitation of any natural resource is not sustainable, and over-collecting of insects is potentially a problem. However, a certain amount of collection is a necessary part of research for preservation purposes, and commercial interest in

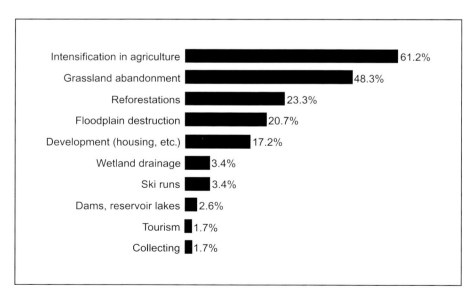

Fig. 14.3. The proportions of alpine butterfly taxa (*n* = 116) in Switzerland that have been affected by changes in land use and other factors (adapted from Erhardt 1995).

specimens can translate into propagation and other actions that ensure survival of threatened species. So, overly strict regulations can be counter-productive, and affect mostly legitimate and necessary movement of specimens. Therefore, CITES has been a controversial agreement, subject to criticism that it is overly restrictive. Currently, there are over 150 parties to this agreement. The US CITES list includes all species protected by the Endangered Species Act, plus species that are vulnerable but not yet threatened or endangered. The **IUCN Red List of Threatened Species** also is a good source of information on conservation status of insects.

The conservation status among insects varies considerably, and at this point relatively few seem to be at risk of extinction, though the number is expected to grow rapidly as more natural lands are developed. **Endangered species** are those considered to be at risk of extinction in a relatively short period of time. Other species experiencing population decline are considered to be at less risk, and are classified as **threatened species**, or are placed in another category to designate that they or their habitat is at risk. The species at greatest risk in the USA are shown in Table 14.2. Note that they represent several orders of insects, but are mostly butterflies or beetles, reflecting popular interest in these groups. Also, the island state of Hawaii is over-represented; throughout the world, island fauna typically is at greater risk.

Advancing the Conservation of Insects

There are several practices that can be used to advance insect conservation. Perhaps the most tangible and attainable is the creation of ecological reserves. Setting aside land that supports indigenous species is helpful, but not all land is equivalent. Reserved land should

Table 14.2. Threatened and endangered insect species in the USA, as established by the United States Fish and Wildlife Service. STATUS: E is endangered; T is threatened; C is candidate taxon, ready for proposal; DA is delisted taxon; PE is proposed endangered; EmE is emergency listing, endangered.

Inverted common name	Scientific name	Listing status	Current range
Beetle, American burying	*Nicrophorus americanus*	E	AR, MA, MI, NE, OH, RI, SD, Canada (Ont.)
Beetle, Coffin Cave mold	*Batrisodes texanus*	E	TX
Beetle, Comal Springs dryopid	*Stygoparnus comalensis*	E	TX
Beetle, Comal Springs riffle	*Heterelmis comalensis*	E	TX
Beetle, delta green ground	*Elaphrus viridis*	T	CA
Beetle, Hungerford's crawling water	*Brychius hungerfordi*	E	MI, Canada
Beetle, Kretschmarr Cave mold	*Texamaurops reddelli*	E	TX
Beetle, Mount Hermon June	*Polyphylla barbata*	E	CA
Beetle, Tooth Cave ground	*Rhadine persephone*	E	TX
Beetle, valley elderberry longhorn	*Desmocerus californicus dimorphus*	T	CA
Beetle, Warm Springs Zaitzevian riffle	*Zaitzevia thermae*	C	MT
Bug, Wekiu	*Nysius wekiuicola*	C	HI
Butterfly, Bahama swallowtail	*Heraclides andraemon bonhotei*	DA	FL
Butterfly, bay checkerspot	*Euphydryas editha bayensis*	T	CA
Butterfly, Behren's silverspot	*Speyeria zerene behrensii*	E	CA
Butterfly, callippe silverspot	*Speyeria callippe callippe*	E	CA
Butterfly, Corsican swallowtail	*Papilio hospiton*	E	France (Corsica), Italy (Sardinia)
Butterfly, El Segundo blue	*Euphilotes battoides allyni*	E	CA

Table 14.2. *Continued*

Inverted common name	Scientific name	Listing status	Current range
Butterfly, Fender's blue	*Icaricia icarioides fenderi*	E	OR
Butterfly, Homerus swallowtail	*Papilio homerus*	E	Jamaica
Buttefly, Karner blue	*Lycaeides melissa samuelis*	E	IL, IN, MI, MN, NH, NY, OH, WI, Canada (Ont.)
Butterfly, Lange's metalmark	*Apodemia mormo langei*	E	CA
Butterfly, lotus blue	*Lycaeides argyrognomon lotis*	E	CA
Butterfly, Luzon peacock swallowtail	*Papilio chikae*	E	Philippines
Butterfly, Mariana eight-spot	*Hypolimnas octucula mariannensis*	C	GU
Butterfly, Mariana wandering	*Vagrans egestina*	C	GU, MP
Butterfly, mission blue	*Icaricia icarioides missionensis*	E	CA
Butterfly, Mitchell's satyr	*Neonympha mitchellii mitchellii*	E	IN, MI, OH
Butterfly, Myrtle's silverspot	*Speyeria zerene myrtleae*	E	CA
Butterfly, Oregon silverspot	*Speyeria zerene hippolyta*	T	CA, OR, WA
Butterfly, Palos Verdes blue	*Glaucopsyche lygdamus palosverdesensis*	E	CA
Butterfly, Queen Alexandra's birdwing	*Troides alexandrae*	E	Papua New Guinea
Butterfly, Quino checkerspot	*Euphydryas editha quino (= E. e. wrighti)*	E	CA, Mexico
Butterfly, Sacramento Mountains checkerspot	*Euphydryas anicia cloudcrofti*	PE	NM
Butterfly, Saint Francis' satyr	*Neonympha mitchellii francisci*	E	NC
Butterfly, San Bruno elfin	*Callophrys mossii bayensis*	E	CA
Butterfly, Schaus swallowtail	*Heraclides aristodemus ponceanus*	E	FL
Butterfly, Smith's blue	*Euphilotes enoptes smithi*	E	CA
Butterfly, Uncompahgre fritillary	*Boloria acrocnema*	E	CO
Butterfly, Whulge checkerspot (= Taylor's)	*Euphydryas editha taylori*	C	No data
Caddisfly, Sequatchie	*Glyphopsyche sequatchie*	C	TN
Cave beetle, beaver	*Pseudanophthalmus major*	C	KY
Cave beetle, Clifton	*Pseudanophthalmus caecus*	C	KY
Cave beetle, greater Adams	*Pseudanophthalmus pholeter*	C	KY
Cave beetle, Holsinger's	*Pseudanophthalmus holsingeri*	C	VA
Cave beetle, icebox	*Pseudanophthalmus frigidus*	C	KY
Cave beetle, inquirer	*Pseudanophthalmus inquisitor*	C	TN
Cave beetle, lesser Adams	*Pseudanophthalmus cataryctos*	C	KY
Cave beetle, Louisville	*Pseudanophthalmus troglodytes*	C	KY
Cave beetle, surprising	*Pseudanophthalmus inexpectatus*	C	KY
Cave beetle, Tatum	*Pseudanophthalmus parvus*	C	KY
Damselfly, blackline Hawaiian	*Megalagrion nigrohamatum nigrolineatum*	C	HI
Damselfly, crimson Hawaiian	*Megalagrion leptodemus*	C	HI
Damselfly, flying earwig Hawaiian	*Megalagrion nesiotes*	C	HI
Damselfly, oceanic Hawaiian	*Megalagrion oceanicum*	C	HI
Damselfly, orangeblack Hawaiian	*Megalagrion xanthomelas*	C	HI

Table 14.2. *Continued*

Inverted common name	Scientific name	Listing status	Current range
Damselfly, Pacific Hawaiian	*Megalagrion pacificum*	C	HI
Dragonfly, Hine's emerald	*Somatochlora hineana*	E	IL, OH, WI
Fly, Delhi Sands flower-loving	*Rhaphiomidas terminatus abdominalis*	E	CA
Gall fly, Po'olanui	*Phaeogramma* sp.	C	HI
Grasshopper, Zayante band-winged	*Trimerotropis infantilis*	E	CA
Ground beetle, [unnamed]	*Rhadine exilis*	E	TX
Ground beetle, [unnamed]	*Rhadine infernalis*	E	TX
Mold beetle, Helotes	*Batrisodes venyivi*	E	TX
Moth, Blackburn's sphinx	*Manduca blackburni*	E	HI
Moth, Kern primrose sphinx	*Euproserpinus euterpe*	T	CA
Naucorid, Ash Meadows	*Ambrysus amargosus*	T	NV
Pomace fly, [unnamed]	*Drosophila aglaia*	PE	HI
Pomace fly, [unnamed]	*Drosophila attigua*	C	HI
Pomace fly, [unnamed]	*Drosophila differens*	PE	HI
Pomace fly, [unnamed]	*Drosophila digressa*	C	HI
Pomace fly, [unnamed]	*Drosophila hemipeza*	PE	HI
Pomace fly [unnamed]	*Drosophila heteroneura*	PE	HI
Pomace fly [unnamed]	*Drosophila montgomeryi*	PE	HI
Pomace fly [unnamed]	*Drosophila mulli*	PE	HI
Pomace fly [unnamed]	*Drosophila musaphila*	PE	HI
Pomace fly [unnamed]	*Drosophila neoclavisetae*	PE	HI
Pomace fly [unnamed]	*Drosophila obatai*	PE	HI
Pomace fly [unnamed]	*Drosophila ochrobsis*	PE	HI
Pomace fly [unnamed]	*Drosophila substenoptera*	PE	HI
Pomace fly [unnamed]	*Drosophila tarphytrichia*	PE	HI
Riffle beetle, Stephan's	*Heterelmis stephani*	C	AZ
Skipper, Carson wandering	*Pseudocopaeodes eunus obscurus*	EmE, PE	EmE = CA, NV, Washoe Co., NV and Lassen Co., CA; PE = CA, NV, Washoe Co., NV and Lassen Co., CA
Skipper, Dakota	*Hesperia dacotae*	C	MN, ND, SD, Canada
Skipper, Laguna Mountains	*Pyrgus ruralis lagunae*	E	CA
Skipper, Mardon	*Polites mardon*	C	CA, OR, WA
Skipper, Pawnee montane	*Hesperia leonardus montana*	T	CO
Tiger beetle, Coral Pink Sand Dunes	*Cicindela limbata albissima*	C	UT
Tiger beetle, highlands	*Cicindela highlandensis*	C	FL
Tiger beetle, northeastern beach	*Cicindela dorsalis dorsalis*	T	CT, MA, MD, NJ, RI, VA
Tiger beetle, Ohlone	*Cicindela ohlone*	E	CA
Tiger beetle, Puritan	*Cicindela puritana*	T	CT, MA, MD, NH, VT
Tiger beetle, Salt Creek	*Cicindela nevadica lincolniana*	C	NE

include transitional land encompassing different ecosystems. This means different topography and floral communities, but also different stages of succession and levels of light penetration, and often involves maintenance through burning or tillage. Thus, landscape heterogeneity is an important element in selection of reserves. Connectivity is also important if insects and other wildlife are to relocate successfully. Thus, it is important to reduce contrast between remnant patches of reserve. Forest-dwelling species, for example, are not as likely to cross over expansive open areas as they are to move through areas with dense vegetation. Corridors or linkages of linear strips of land connecting reserves are preferred over disconnected reserves. Reserves with larger area also have greater buffering capacity, so any disturbance is less likely to prove to be seriously disruptive.

Dealing with invasive pests is more challenging. Reducing the invasion of new areas by nonindigenous species requires the cooperation of crop producers, exporters, importers, tourists, airline and shipping companies, and government regulatory agencies. At one time, transport of goods that might be contaminated with invasive pests or disease was severely restricted. However, the pendulum has swung so far toward the unrestricted transport of goods around the world (so-called 'free trade') that now there are few effective barriers to movement of invasive organisms. In most cases, there is no way to prevent pests from entering, and the principal (and not terribly effective) response to invasion is to detect and eliminate the infestation before it spreads.

Global climate change is seemingly an even more formidable challenge, but because it affects nearly everyone and evidence of the problem has become overwhelming, there is increasing hope of arresting, or at least reducing the rate of, climate change. The principal challenge involves transitioning from dependency on carbon-based fossil fuels to alternative energy sources. This is technically attainable, requiring only the will to change and the willingness to invest the financial resources.

Development and use of pest management procedures that are not based on use of broad-spectrum pesticides remains challenging. The problem is much like that of invasive species: there are too many economic forces aligned with maintaining the status quo, and too few people who have a significant interest in seeing change occur. Also, the feasibility of using alternative technologies remains untested for most pest problems.

Nevertheless, measurable progress has been made and more can be done to minimize pesticide-related impacts on non-target species.

Insect collecting, or more specifically over-collecting of insects, sometimes is implicated as a threat to insect survival. Generally, insect collecting poses no risk to insect populations. However, for species that have very limited distribution it is possible to deplete the population further through uncontrolled collecting. The greatest threat is from collection of rare species. Collectors will sometimes pay very high prices, up to several thousand US dollars per specimen for rare or especially attractive specimens. The eventual disposition of such specimens is museums and wealthy collectors. On the other hand, another threat is derived from the nonselective harvest of insects to make jewelry, ornaments, and artwork. Large numbers of insects are sometimes harvested for this purpose, and although most often they are common species, in some cases rare species are captured. This latter situation may not be as deleterious as imagined because in many cases the insects are collected by baiting, which attracts mostly males, or because males are sometimes more colorful and therefore more desirable. Generally, female survival is considered to be more crucial to the survival of animal species. Lastly, it is worthwhile to mention the 'live-trade' category, in which living insects are used for displays. These are mostly cultured specifically for this purpose, with no collection from the wild.

Conservation of Bumble Bees

Many taxa of insects have experienced declines in recent years, but bumble bees (*Bombus* spp.; Hymenoptera: Apidae: Bombinae) serve as a good example of the issue confronting insect conservation. Bumble bees in both Europe and some areas of North America have declined dramatically in the last half-century. There are pollination-related consequences for both native and cultivated plants, many of which depend considerably on bumble bees. Simultaneously, there are several disease and parasite problems affecting honey bees, *Apis mellifera*, so the loss of the native *Bombus* assemblage of bees is especially troubling.

Several factors account for the decline in bumble bee abundance. A major reason for the decline in bumble bee numbers is the loss or fragmentation of habitat. Improvement of grazing lands by planting improved grasses and application of herbicides, or the planting of homogeneous crops, often deprives bumble bees of flowers for nectar. Bumble bees also require nest sites. Some nest aboveground in dense vegetation, but many nest belowground in abandoned animal burrows. When farmland lacks hedgerows and weedy fencerows or field margins, these bees are less able to find suitable nesting sites. Even better is a mosaic of natural areas and cropland. Application of broad-spectrum insecticides is also a serious problem for bumble bees. Interestingly, some widely used insecticides cause few immediate effects on adults, and hence have gained a reputation for being relatively safe, but affect foraging or colony growth. In North America, commercially produced bumble bee colonies are heavily infected with pathogens. These bumble bees are not always successfully contained in greenhouses, so mixing of bumble bees and bee microbial pathogens likely occurs in the field.

Research conducted in Poland suggests that although bee density may be high in simplified (crop) plant systems, bee diversity is low. Bee diversity is higher in more complex plant communities such as forests and untilled areas. Diverse plant communities function as refugia for bees, allowing them to persist. Annual crops and mown meadows, on the other hand, may be rich sources of food, but the bees forage there only temporarily. Crops attractive to bees include winter rape (canola), sunflower, flax, alfalfa, red clover, yellow lupin, and buckwheat. Red clover and sunflower are especially attractive to bumble bees. Refuge habitats should be free from frequent disturbance, and should constitute about 25% of the landscape.

Conservation of Butterflies

Butterflies have great aesthetic appeal, and so have garnered inordinate interest relative to other insects. They represent only about 20,000 species, or about 0.5% of the total number of insects. Nevertheless, there are legions of amateur lepidopterists that are avidly interested in butterfly species and subspecies, and they have formed numerous organizations to advance knowledge and conservation of butterflies. Conservation efforts have been initiated most often in temperate areas of the world, such as Japan, Europe or North America. However, high rates of habitat loss combined with the much richer diversity of the tropics suggest that there likely is much greater risk of extinction in tropical environments. Habitat loss is usually involved in the decline of butterflies; some examples follow.

Queen Alexandra's birdwing butterfly, *Ornithoptera alexandrae* (Lepidoptera: Papilionidae), is the world's largest butterfly. It exists only in small areas of Papua New Guinea, where it has a limited number of hosts, and where considerable habitat has been damaged by development of oil palm plantations and by volcanic disruption. Because it is so endangered, and trade is prohibited by CITES, commercial production of this insect (butterfly ranching), has been inhibited. Thus, this is a case where regulation designed to preserve a species may actually serve to guarantee the demise of this species. Other birdwing butterflies have benefited from collector interest, and have generated butterfly ranching enterprises in several areas of the tropics.

Homerus swallowtail, *Papilio homerus* (Lepidoptera: Papilionidae), is found in Jamaica. Once occurring widely on that island, its distribution is now much more restricted. Larvae feed only on *Hernandia* spp., and require very high humidity to survive. The existence of this species was especially threatened by reforestation efforts that involved cutting natural forests and replanting with Caribbean pine. The plight of this butterfly was brought to light by a film 'Papilio homerus, the vanishing swallowtail' that generated popular interest in preserving this species. Popularization of this butterfly via school children, and increasing recognition of the commercial value of natural areas and wildlife, provide reasons for optimism concerning its fate.

The Apollo butterfly, *Parnassius apollo* (Lepidoptera: Papilionidae), formerly occurred widely in Europe but populations have been decreasing since the 1930s in most areas. Habitat loss seems to be the major problem, with loss of mountain meadows and other disturbed areas that normally are inhabited by their host plants being a fundamental problem. In some cases, habitat destruction is due to construction, farming or too frequent grazing, but in other cases it is invasion by shrubs and trees. Although patches of its larval host plant and adult nectar sources are available, in many cases they are considerable distances apart, making egg deposition difficult. However, another possible cause of its decline is contamination of its food plant, *Sedum* spp., with heavy metals.

The monarch butterfly, *Danaus plexippus* (Lepidoptera: Nymphalidae), is one of the most abundant and wide-ranging butterflies, but due to its unusual biology it is one of the most susceptible to elimination from

North America. These butterflies are long distance migrants, with adults moving north in the spring and summer months to take advantage of the availability of larval food plants, and then new adults moving south in the autumn months to escape the cold. Butterflies from all over the North American continent gather at only a few overwintering sites in Mexico and California to overwinter in the adult stage. There, thousands cluster together in trees. Only a few sites are suitable for overwintering, so habitat destruction is a significant threat to their existence. Development of land in California, and lumbering and firewood gathering in Mexico, have reduced the extent of suitable habitat. Likewise, agriculture and urbanization, and especially the increased use of herbicides, have reduced the availability of the larval food plant, milkweed (*Asclepias* spp.). Larvae feeding on milkweed acquire chemicals that make them emetic, deterring feeding by birds and other potential vertebrate predators. Not all caterpillars are emetic because the *Asclepias* species, and even different tissues in the same milkweed plant, have different cardenolide concentrations. Individuals of Scott's oriole (*Icterus parisorum*), black-backed oriole (*Icterus abeilleri*), and black-headed grosbeak (*Pheucticus melanocephalus*) feed on some of the butterflies at the overwintering sites, stripping the wings and consuming the abdomens. Not all butterflies contain high concentrations of cardenolides because not all larvae fed on cardenolide-rich plants, and some birds apparently learn to distinguish cardenolide-laden individuals. Thus, the soil surface beneath butterfly aggregations may be littered with hundreds of dead butterflies per square meter, most preyed upon by these birds. Birds tend to feed at the margins of the butterfly aggregations, and mortality is proportionally higher in smaller aggregations. Thus, although some birds have learned to penetrate the cardenolide-based chemical defense, there is benefit for monarch butterflies to aggregate in large numbers, probably explaining their huge numbers at a few sites. However, the small number of overwintering sites and the ability of birds to exploit this food resource make the migratory North American populations susceptible to potential elimination by various natural disasters.

The Atala butterfly, *Eumaeus atala* (Lepidoptera: Lycaenidae), is found in Cuba, the Bahamas, and southeastern Florida, USA. It feeds only on cycads, and in Florida the only host plant is coontie, *Zamia pumila*. This host plant was used by Native Americans as a source of starch, and can be used to make a type of bread. In the early 1900s, several commercial enterprises were begun in southern Florida that depended on collection of wild coontie and extraction of the starch from the roots. The harvest of coontie and the destruction of woodlands in southern Florida quickly led to the near elimination of coontie from the landscape, and the disappearance of Atala butterfly from the USA. During the period from 1937 to 1959 it was absent from Florida. The popularization of coontie as a landscape plant beginning in the 1960s likely set the stage for a comeback. Atala butterfly has re-established a population in southeastern Florida, but it remains limited by the limited availability of its host plant, urbanization, and the use of pesticides in the landscape.

Conservation of Beetles

Several types of beetles (Coleoptera) have displayed significant decreases in abundance. Although habitat destruction explains the decrease in abundance of some, we can only speculate about the causes for others, and unlike some of the aforementioned butterflies we have no reason to be hopeful about their future.

One of the most dramatic decreases in abundance of a beetle is the range decline of the American burying beetle, *Nicrophorus americanus* (Coleoptera: Silphidae). This is the largest carrion-frequenting beetle in North America. As its name suggests, it tends to dig soil from beneath carrion, burying and using the dead animal to support its brood. Formerly found throughout the eastern half of North America, it now is known from only about 5% of its former range. It persists only on an island near Massachusetts, the border of Oklahoma and Arkansas, and central South Dakota and Nebraska, although efforts are being made to reintroduce it elsewhere. Burying beetle feeds on carrion, and requires carrion about the size of a dove or chipmunk. A decline in carrion is not evident, and it has declined in pesticide-free areas such as national and state parks, as well as from agricultural areas, so pesticides are not an obvious cause. Habitat disruption, habitat fragmentation, extinction of the passenger pigeon (*Ectopistes migratorius*), competition with flies, and many other factors have been suggested to explain the disappearance of the beetles, but there are no real explanations.

Tiger beetles (Coleoptera: Cicindelidae) seem to be unusually threatened with extinction. Habitat loss is by far the overriding concern with these species as they are usually quite localized in their distribution, favoring open, sandy areas such as sandbars along rivers, lakes, and marshes. These areas are often affected by development. Among the many species at risk are Salt Creek tiger beetle, *Cicindela nevadica lincolniana*, which inhabits the banks of saline wetlands of Lancaster County in Nebraska, USA; Ohlone tiger beetle, *Cicindela ohlone*, which inhabits native grasslands in limited areas of Santa Cruz County in California, USA; Puritan tiger beetle, *Cicindela puritana*, which inhabits sandy beaches along the Connecticut River in New England, USA, and along the Chesapeake Bay in Maryland, USA; and Highlands tiger beetle, *Cicindela highlandensis*, which inhabits sparsely vegetated scrub areas (relict sand dunes) of two counties in central Florida, USA. Development and recreation are the major threats to these species.

Similarly, there are many reports of declines in the abundance of fireflies (Coleoptera: Lampyridae). Summer evenings in the eastern USA were once punctuated by the flashing of fireflies, especially in the early evenings. This phenomenon now is less often observed in most suburban areas. Even in rural areas, fireflies are less common now. Again, there are hypotheses put forth to explain their decline, most commonly habitat disruption and use of pesticides. However, the abundance of artificial light at night (so-called 'light pollution') may also serve to disorient the fireflies, which use flashing light principally for mate finding, but in some cases for prey selection.

MANAGING INSECT RESOURCES FOR THE BENEFIT OF WILDLIFE

Habitat management is the foundation of wildlife management, with food resources, water, and shelter (cover) widely recognized as important components of preserving and improving wildlife populations. Management techniques such as prescribed fire, thinning, promotion of cover, and uneven stand management or harvesting are commonly used to increase the availability of suitable food resources and cover. The focus of land management for wildlife is almost always plant-based. There seems to be an assumption that if the plant environment is suitable, clean water is reliably available, and hunting and predation are controlled, all wildlife will flourish. Populations of invertebrate animals, though often an integral component of the

food supply, are not part of the management equation. Although insects are acknowledged to be important food, and sometimes a nuisance or vectors of disease, rarely are they actively managed for the benefit of wildlife.

Although insect food resources often are vitally important for wildlife survival and fitness, there are few guiding principles or protocols for managing wildlife via this food source. Wildlife management often promotes manipulation of plant food and cover for better wildlife habitat. For example:

• Leaving snags (dead and decaying trees) to provide nesting sites for cavity-nesting birds is commonly recommended, but rarely is there mention that many wood boring beetles (Coleoptera: Cerambycidae and Buprestidae) also inhabit such trees, or that the death of the tree likely was caused by bark beetles (Coleoptera: Curculionidae); both are important food resources.

• Leaving pasture or fence-row vegetation uncut is often suggested to improve cover for game birds, and of course grass and weed seed are favored by such birds. Rarely is there mention that grasshoppers (Orthoptera: Acrididae) are more abundant in tall vegetation, even though they are prized as a food resource by most birds.

• The occurrence of logs, branches and root masses, aquatic plant beds and rocks in streams are promoted for preservation of fish such as rainbow trout, *Oncorhynchus mykiss*. However, rainbow trout are opportunistic feeders, moving readily to wherever insect prey are found, and feeding frequently on terrestrial insects that fall from overhanging riparian (streamside) vegetation into the water. Manipulation of vegetation to attract or retain terrestrial insects is rarely acknowledged as a fish management technique. Likewise, addition of logs and boulders to streams is proscribed to allow fish a place to seek shelter, without consideration of the usefulness of such substrates for black flies (Diptera: Simuliidae) or other aquatic insects.

• The bats found in North America are almost entirely insectivorous. In temperate areas they either hibernate or migrate south during the insect-free winter months. So how important are insects to bats? The decline in the abundance of bats is usually given as reduced availability of roosting habitat. However, roosting habitats are useless without nearby feeding sites, and of course there must be an abundance of insects.

The poor state of knowledge about insect management for the enhancement of wildlife populations can be easily observed by examining the Fish and Wildlife

Habitat Management Leaflets produced by the United States Department of Agriculture, Natural Resources Conservation Service. These publications acknowledge the importance of insects as food for many forms of wildlife, but offer few suggestions for management. For example, the 'Amphibians and Reptiles' publication (No. 35, dated 2006) lists insects as a habitat requirement for crocodiles and alligators, lizards, snakes, turtles and tortoises, frogs and toads, and salamanders. Among the habitat management recommendations are the usual concerns about availability of water and shelter, invasive species, livestock and vehicles. Avoiding pesticide use is recommended, but the concern is direct toxicity to wildlife, not disruption of the food supply. And not surprisingly, there is no mention of actions that boost the insect-based food supply. This is not mentioned as a criticism of the publication, only a reflection on the state-of-the-art of wildlife management, which unfortunately does not include habitat enrichment via arthropod manipulation.

Principles

There are several fundamental concepts that are applicable to managing wildlife via insect resources, namely:

• *Insects are important wildlife food.* The most important concept is that insects are an important component of the diet of most birds, reptiles, amphibians, freshwater fish, and small mammals. Acknowledgement of this fact is a prerequisite to improving our ability to manage wildlife resources.

• *Wildlife dietary habits vary.* It is important that wildlife managers understand the dietary needs and habits of the wildlife species of concern, and manage specifically for these animals. There are periods when insect food resources are especially important, but many species feed differentially on certain taxa, so feeding preferences should also be considered.

• *Insects can affect wildlife.* Clearly, the presence of insects is not always beneficial. Insects have relationships with wildlife that extend beyond being a source of food. Biting species, and especially disease agent-carrying species, can be a threat to wildlife populations. There are times and places when wildlife benefit from reduction in insect abundance.

• *Pesticides can affect wildlife.* If not used carefully, pesticides can interfere with wildlife survival and reproduction, either directly via toxicity or indirectly via deprivation of food. Pesticide use should be mini-

mized, or pest-specific products and practices employed when management of pests becomes necessary.

• *Insects are not all the same.* It is important to appreciate that insects also have specific diet and environmental requirements, and those needs must be considered if insect populations are to be manipulated successfully.

Practices

Protocols or procedures for enhancing insect abundance to benefit wildlife populations are not well developed. Entomologists have devoted most of the last 200 years to the active pursuit of means to eliminate or suppress insect abundance, not to enriching insect populations. However, enough is known about insects, their interrelationships, and their relationships with their environment to be able to make some recommendations. In particular, the literature on habitat management for the preservation of predatory and parasitic natural enemies, and for conservation and propagation of butterflies, are well developed and can be used as a foundation for fostering abundance of several other types of insects.

Flowering plants are critically important to many insects. From flowers, insects can obtain pollen, which provides amino acids (protein), fat, and minerals. Crude protein levels of pollen range from about 10% to over 30%. More commonly, insects collect nectar, which supplies easily digestible carbohydrates (sugars). A great number of wasps, bees, ants, flies, beetles, and moths and butterflies are attracted to these flower resources. Flowers not only foster reproduction of insect populations but also concentrate insects, making them easier prey for wildlife. Pollination of flowers by insects also results in production of fruits and seeds, which also are important food for wildlife. Thus, the presence of flowering plants, including forbs, bushes and trees, is critically important in boosting both insect and wildlife abundance.

It is important that flower resources be available over a protracted period of time to maximize their effects on supporting insect populations. It would be easy to assume that spring and early summer are the critical periods when insect resources should be maximized because so much bird breeding occurs then. However, some birds produce multiple broods, and many forms of wildlife feed on insects throughout the year, even winter. Also, wildlife about to enter hibernation or to migrate to warmer climates for the winter need high levels of fat in their bodies to carry them through these energetically demanding periods, so insect food resources are valuable in the autumn as well. Maintenance of a diversity of floral resources, including plants that bloom throughout the year, or most of the year (where climate allows) is highly desirable.

One could argue that woodlots, hedgerows, and weedy fence lines already are present in many crop and pasture areas, and that these could provide sustenance to insects, and shelter for wildlife. However, uncultivated or unmanaged land has decreased in most agricultural areas. This trend is due to the availability of effective herbicides, the increased sophistication and use of irrigation systems, the development of larger agricultural equipment, the consolidation of fields into ever-larger units, and the specialization of agriculture. Thus, it is increasingly common to see large fallow fields, bare soil strips bordering crops, pastures without weeds or flowers, and mowed areas along roadsides.

The benefits to wildlife of noncrop plants, and the harm caused by agrichemicals, are particularly well documented in Europe. The decline in hedgerows in Europe late in the 20th century was accompanied by decreases in the abundance of many birds. Reduced abundance of grey partridge, *Perdix perdix*, a popular game bird in Britain, particularly attracted attention to the problems with loss of plant biodiversity and extensive use of agrichemicals. Eventually it was shown that the borders of cultivated fields were heavily used by partridge chicks. However, the use of herbicides eliminated weeds that produced the seeds favored by the birds, and the use of insecticides both eliminated insect prey and killed birds outright. A **conservation headland** program was developed that involved leaving a border of unsprayed cropland, usually 6–9 m wide, around the periphery of the crop fields, thereby providing a relatively safe area for bird foraging. This border area also resulted in a 70%–75% reduction in pesticide drift into adjacent hedgerow areas. Within three years of implementing the conservation headland effort, the density of breeding pairs of partridge increased three-fold. Not surprisingly, the number of butterflies observed in these unsprayed areas also increased. This suggests that relatively minor land use modifications could be used to favor survival of wildlife.

Sometimes agriculturists plant strips of flowers within or alongside crop fields to encourage the presence and activity of insect parasitoids and predators. These strips consist of a large number of different plants so as to provide different periods of bloom, but also because insects are differentially attracted to different flowers, and because different soil and moisture conditions favor different plant species. Wasps and flies respond quite well to the increased availability of flowering plants. This practice is sometimes called **ecological compensation** because it helps to compensate for the lack of diversity, lack of food, and lack of shelter found in the crop. It is usually recommended that these ecological compensation areas should occupy at least 5%–10% of the agricultural landscape, though in practice it is often much less. Mowing or replanting may be necessary periodically to maintain diversity or prevent ecological succession. A variation on planting strips of flowers is the establishment of **beetle banks**. These are within-crop strips of vegetation, initially grass but eventually taller vegetation as succession occurs, that provide shelter for predators, particularly ground beetles (Coleoptera: Carabidae) and rove beetles (Coleoptera: Staphylinidae). The same benefit has been found with diverse crop margins. As with the parasitoids and flowering plants, the predatory beetles disperse into the surrounding crops to affect insect control, and bees disperse to provide pollination. Thus, wasp, bee, fly, butterfly and beetle species composition and abundance are susceptible to floral management (grazing, fertilizer, timing and height of cutting). The practice of ecological compensation could also be adapted to benefit insect-feeding wildlife.

Flowers are not the only source of nectar, of course. **Extrafloral nectaries** occur on many plants. Extrafloral nectaries are nectar-producing glands that are located on the plant in locations other than in flowers. These nectaries can occur on the leaves, petioles, bracts, fruits, and elsewhere (Fig. 14.4). They are found widely among plants, though most common on woody plants and vines. They are more common in tropical environments, and in habitats where ants are common. Nectaries are generally absent from monocots. Nectaries produce mostly sugars, about 95% of which is sucrose, glucose, and fructose, although amino acids and proteins also are present. Importantly, extrafloral nectaries have a longer secretion period than floral nectaries. In temperate areas, extrafloral nectar activity roughly corresponds with the reproductive period of flowering plants, though in tropical areas nectary secretion may occur throughout the year. Extrafloral nectaries likely evolved as part of the

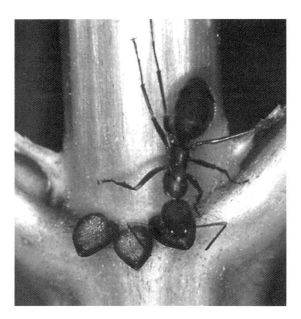

Fig. 14.4. Ant feeding on extrafloral nectary secretion of elderberry, *Sambucus* sp. Extrafloral nectaries are an important source of sugars for predators and parasitoids (photo by R. Mizell).

defenses of plants, and now provide supplemental food to support ants, bugs, beetles, wasps and other insects that help protect plants from predation by phytophagous insects. Thus, the presence of plants bearing extrafloral nectaries complements and extends the attractiveness and suitability of flowering plants, encouraging insect occurrence and reproduction to the benefit of wildlife as well as to the benefit of the nectar-producing plants that require pollination. Naturally occurring, nectary-producing plants could be selectively encouraged as part of habitat management, or they could be planted as part of ecological compensation efforts.

Some insects need **foliar resources** as well as nectar and pollen. This problem is well known by those who encourage the presence of butterflies ('butterfly gardening'). Although butterflies will be attracted to certain flowering plants, in the long run there will be few butterflies unless larval food resources are also available, and larvae almost invariably are foliage feeders. Thus, productive butterfly gardens contain a mixture of adult and larval food resources. Due to the popularity of butterfly gardening, the suitability of plants for butterfly adults and larvae is well defined, and landscape designs for summer-long or year-long production of butterflies are available, depending on the climate. Unfortunately, this does not extend to moths and their larvae, the nocturnal counterparts of butterflies. Nevertheless, the food habits of many moth caterpillars are well known, so it should be possible to enhance their occurrence. Likewise, the dietary habits of the common leaf beetles, weevils, beetles, aphids, leafhoppers, stink bugs, grasshoppers, and katydids are known, so **insect nurseries**, or landscape elements that promote availability of insect food for wildlife, can be designed. As is the case with butterfly gardens (or flower strips or beetle banks mentioned previously), the design elements would vary regionally and temporally and would foster not only insect populations but the wildlife that feed on them.

In the design of insect nurseries, and their incorporation in agricultural, urban, and conservation area landscapes, there are several factors to consider:
• Diversity, and spatial and temporal availability, as well as spatial scale of supplemental plants, are important. For example, revegetation of mine spoils in Wyoming, USA, was accomplished with some elements of the native vegetation found in this sagebrush (*Artemesia* spp.)-dominated shrub-steppe plant community. However, the plants selected for planting were seeded at different rates and in a more homogeneous manner than in the native vegetation, so the grasshopper community that developed on the revegetated area was quite different than the undisturbed areas. Diversity of plants tends to increase as plant succession progresses, and tends to be greater in larger stands than in small. The diversity of insects is correlated with plant diversity. Thus, older and larger areas of flora have more diverse fauna. Trophic relationships also change with succession; younger stands tend to have proportionally more herbivores, whereas older stands have proportionally more scavengers and fungivores.
• Insects can only collect nectar from the types of flowers for which their mouthparts are adapted, so both insect and plant host must be aligned. In other words, some types of biodiversity are more important than other, and the plant resources must match the feeding preferences of the desired insects.
• Perennial plants or annual plants that reseed themselves readily would provide greater economy to nursery plantings. If high levels of maintenance were

required, it would be difficult to promote such designs. Mowing, burning or addition of fertilizer, which are relatively modest forms of maintenance, often stimulate new growth of plants. Young vegetation can be very favorable for some insects, particularly aphids. These maintenance activities also may be necessary to prevent ecological succession, and displacement of the preferred plants by later-successional species.

• Invasiveness or weediness must be considered as well. Although it might be difficult for plants to spread into established ecosystems (e.g., forests, pastures), there is risk that nursery plants seeded along roadsides, fences, irrigation ditches or field edges, or flower strips planted within crops, could provide seed that would invade cropping areas and become weeds. Native plants pose fewer risks, however.

• Origin of the nursery plants would be an issue for some conservationists, and native (indigenous) wild plants would be most desirable. However, nonindigenous plants often have valuable characteristics and might be useful if lack of weediness can be assured. Similarly, crop plants might seem out of character in natural landscape environments, but interplanting of crop plants could prove useful because many are quite susceptible to insect herbivory, because seed is generally available, and because planting and cultural needs are well defined. Interplanting with grain crops, for example, could produce both insects and seed for wildlife.

• Diversity in food plants is vital for maintaining insect diversity, but variation in plant architecture favors both insects and the wildlife that feed upon them. For example, band-winged grasshoppers often prefer open areas with some bare soil whereas slant-faced grasshoppers prefer tall grasses. Some birds only forage in open areas (Fig. 14.5). Rapidly growing, taller plants such as sunflowers provide both shelter and viewing platforms for other birds. Many studies show that wildlife populations are higher in heterogeneous habitats, especially edges of forests, hedges, and edges of fields.

• Availability of seed is an issue for any project that requires extensive planting, and seed of noncrop plants is often unavailable or prohibitively expensive.

• Suitability of the insect nursery plants for production of desirable insects is a major consideration. In some cases, specific wildlife species such as eastern bluebirds, *Sialia sialis*, or northern bobwhite, *Colinus virginianus*, might be targeted for management, but each would require quite different nursery plant elements. Obviously, it would be easier to design insect nursery plantings for such individual species. Sound biological knowledge is a prerequisite to implementing species-specific management efforts. However, the overall

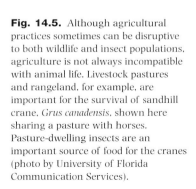

Fig. 14.5. Although agricultural practices sometimes can be disruptive to both wildlife and insect populations, agriculture is not always incompatible with animal life. Livestock pastures and rangeland, for example, are important for the survival of sandhill crane, *Grus canadensis*, shown here sharing a pasture with horses. Pasture-dwelling insects are an important source of food for the cranes (photo by University of Florida Communication Services).

abundance of insects could be boosted with diverse plantings that might benefit a myriad of wildlife.

• Addition of plants to the landscape is not always the ideal approach to fostering insect abundance. Some important groups, particularly grasshoppers and some ants, often thrive best in sunny environments with more limited vegetation. Thus, low-growing vegetation, or even vegetation-free areas, can be important landscape design elements. Wildlife behavior also must be considered. For example, bluebirds prefer hunting insects in open areas. Bobwhites prefer a mix of open and dense vegetation, a mixture of different 'cover' types. Plants that provide appropriate perching opportunities can also foster the occurrence of passerine birds by facilitating their ability to hunt insects.

• Grazing management can also influence availability of insects. Season-long grazing may keep plants preferred by livestock from maturing and producing flowers, whereas rotational grazing (heavy grazing pressure for a short period, then relocation of the livestock to another area) may allow plant maturation. On the other hand, rotational grazing encourages consumption of less preferred plants by livestock, which could prove detrimental if the less-preferred plants are important to insects. Thus, like creation of ecological compensation areas in cropland, there is need for local adaptation.

• In addition to providing food, shelter must also be available. This is especially critical for overwintering of insects, as many species must have appropriate shelter. Perennial plants, especially forests, often provide refugia for insects.

Ecological compensation could also be adapted to benefit important insect groups beyond flower- and foliage-feeders. Termites and wood-boring beetles are important wildlife food, and usually feed on dead or dying trees. Forest management that involves retention of dead trees, even if a small percent of trees had to be girdled annually to create suitable breeding sites, would do much to ensure continuous availability of this food resource for these insects. 'Thinning' of woodlands (a common silvicultural practice) to reduce competition does not have to include complete removal of trees; trees could be killed and left in place to serve as nurseries for wood-breeding insects. Likewise, animal manure can be used to foster development of fly populations. Animal producers often compost manure, or spread it onto crop fields in the winter, or till it into the soil in the spring. Though these are good practices for manure management and soil improvement, some manure could be distributed along field edges where fly breeding could occur. Even carrion that accrues from road-kill, instead of being buried in landfills, could be recycled more naturally by being moved to farmland or conservation areas where scavengers (including insects) could take advantage of this resource.

Because diverse cropping systems (e.g., Figs. 14.6, 14.7) are more favorable to most insects than are monocultural systems, the use of strip cropping, intercropping, cover crops, and smaller crop fields (which have proportionally more 'edge' and often more weedy fence lines or hedgerows) would introduce greater diversity to an area and would foster higher densities and diversity of both insects and wildlife. Research in Florida, USA, for example, not only documented increases in avian abundance and species richness in diverse cropping systems, but found crop diversity to be more important than 'organic' crop production for fostering birds. However, floral composition can influence insect species composition, so it may not be enough to simply introduce diversity without regard for the plants present. For example, management of crop margins in Georgia, USA, to enhance northern bobwhite (*Colinus virginianus*) populations boosted populations of some insects but did not contain adequate floral resources, so wasps were nectar-deficient. In Sweden, widening of field margins and planting with perennial legumes were found to be very successful at attracting insects, though not surprisingly, insect occurrence varied with the type of planting. Clearly, plant food resources can be used to manipulate insects, and hence, their availability as a wildlife food resource.

Obviously, increased abundance of insects is not always beneficial for ecosystem health. For example, steps to increase insect abundance might cause some insects to attain high enough numbers to threaten nearby vegetation. This could easily happen with bark beetles (Coleoptera: Curculionidae: Scolytinae) if plant age, plant health, and weather all favored insect population increase. An interesting example of how increased insect abundance favored wildlife but not the health of the ecosystem can be found in the northwestern USA, and involves spotted knapweed, *Centaurea maculosa*. Spotted knapweed is an invasive weed that gained access to North America from Eurasia, and has become troublesome in semiarid areas. Several insects were introduced in an attempt to provide biological suppression of this weed, including the flies *Urophora affinis* and *U. quadrifasciata* (Diptera: Tephritidae). The flies attack the seeds developing in the flowerheads,

Fig. 14.6. Interplanting of squash with buckwheat, *Fagopyrum esculentum*. Although commonly planted as a grain and cover crop, the flowers of buckwheat are highly attractive to pollinators and parasitic wasps (photo by Teresia Nyoike).

Fig. 14.7. Flower strip sown within a wheat field in Switzerland to introduce diversity, especially nectar-producing plants, into the cropping system (from Abivardi 2008; photo by D. Ramseier).

cause formation of galls, and reduce seed production. However, this plant is not seed limited, and the insects overwintering within the seedheads are suitable food for deer mice, *Peromyscus maniculatus*. The overwintering flies comprise 30%–85% of the diet of deer mice in Montana, USA, increasing winter survival of the mice and allowing the mouse population to be twice as abundant as in areas lacking knapweed-feeding flies. The abundance of deer mice may favor survival of

predators, but this native generalist consumer attained high enough densities to damage the population of native plants through seed removal. Also, the mouse population serves to spread Sin Nombre virus, a hantavirus affecting humans. Thus, there were negative and unforeseen consequences associated with the increased abundance of insects.

Indigenous fauna often are adapted to indigenous plants, so remnant vegetation (remaining undisturbed sections of native plants) is especially valuable for some wildlife. For example, a study in Australia examined the importance of small areas of remnant vegetation versus planted trees for the maintenance of birds, and found that the natural vegetation supported three times the number of species as disturbed areas. They also noted that small patches of vegetation could be quite important. Hedgerow and fencerow areas, or within-crop conservation areas, could prove to be important complements to wildlife as well as to predatory, parasitic and pollinating insects. Is it feasible for insects to locate and populate artificially established plant resources? Encouraging results come from studies such as restoration efforts for coastal scrub communities in California, USA. In these efforts, the arthropod species richness was greater on planted shrubs, and common insect abundance was equivalent to that which was found in undisturbed areas.

Insects associated with water are important food resources for some wildlife. As noted previously (Chapter 4, Wildlife diets), insects support not only many freshwater fish populations, but are important for many birds (e.g., wading birds and ducks), amphibians (e.g., frogs, salamanders), and reptiles (e.g., alligators, turtles). Emerging adult insects also are important prey for birds associated with dense vegetation (e.g., forests) and open spaces (e.g., pastures) in addition to those considered to be water-inhabiting, as most birds

are quick to take advantage of opportunities to feed on abundant insects. For example, research on avifauna attracted to emerging insects along a wooded stream amidst tallgrass prairie in the Great Plains region of the USA showed that several 'flycatchers' (eastern wood-peewee, *Contopus virens*; great crested flycatcher, *Myiarchus crinitus*; eastern kingbird, *Tyrannus tyrannus*) and 'gleaners' (black-capped chickadee, *Parus atricapillus*; common yellowthroat, *Geothlypis trichas*) were taking advantage of this abundant food resource. Not only was there a positive association of emerging insects and birds, but the birds largely partitioned the food resource, with different birds foraging in different areas (Fig. 14.8). In this case, the emerging insects in forested areas were comprised primarily of midges (Diptera: Chironomidae), 53%; mayflies (Ephemeroptera), 25%; and stoneflies (Plecoptera), 12%. In nearby prairie sites the proportions of emerging midges, mayflies, and stoneflies were 94%, 3%, and 1%, respectively. The insect taxa normally associated with water are the mayflies (Ephemeroptera), caddisflies (Trichoptera), stoneflies (Plecoptera), alderflies, dobsonflies and fishflies (Megaloptera), dragonflies and damselflies (Odonata), and some beetles (Coleoptera), bugs (Hemiptera), and flies (Diptera).

Abundance of aquatic insects is often related to availability of a reliable supply of clean water. Water alone is not adequate, of course, so nutrient inputs in

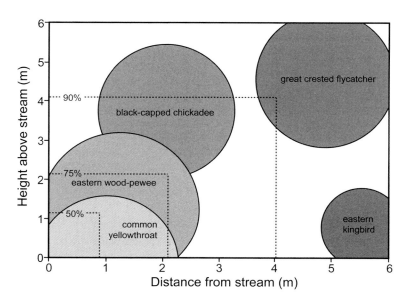

Fig. 14.8. Foraging space of some birds attracted to feed on emerging insects along a stream in Kansas, USA. The enclosed areas represent the space where approximately 80% of the feeding occurred. Dashed lines show the cumulative percentage of aquatic insect biomass present at increasing distance from the stream (adapted from Gray 1993).

the form of leaves and branches are usually necessary, as well as growth of algae and aquatic plants, to feed the populations of invertebrates. Once these plant materials are in place then various invertebrates can prosper, and also can be fed upon by vertebrates. Pesticides, fertilizers, and oxygen supplies often prove to be problems. Both herbicides and insecticides can wash into water supplies, killing plants, invertebrate fish food, and fish or other wildlife. Nutrient run-off from adjacent land can help fuel populations of invertebrates, but too much nutrient flow can cause **eutrophication**, which is stimulation of excessive plant growth and some undesirable environmental side effects. The consequences of this excessive plant growth, sometimes called 'algal blooms,' include:

• The ability of light to penetrate into the water is diminished. This occurs because the algae form mats as a result of being produced faster than they are consumed. Diminished light penetration causes decreases in the productivity of plants living in the deeper waters, and hence their production of oxygen.

• The oxygen supply becomes depleted. When the algae die and decompose, much oxygen is consumed by the decomposers. This, combined with the lack of primary production in the darkened, deeper waters due to poor light penetration causes significant decrease in availability of oxygen.

• The depletion of the oxygen supply results in the death of fish that need high levels of dissolved oxygen, such as trout, salmon and other desirable sport fish. The community composition of the water body can change, allowing fish that can tolerate low dissolved, such as carp (Cyprinidae), to become more abundant. Change in fish populations can affect many aspects of the aquatic ecosystem.

• Some species of algae that produce blooms also produce toxins that render the water unsuitable.

Thus, it is important that some nutrients enter into water supplies, but not too much. Usually, it is excessive nitrates and phosphates from agricultural fertilizers applied to cropland, from animal wastes in feedlots, or from inadequately treated human wastes that cause degradation of water quality. Reduction or elimination in the flow of excess nutrients is a key element in maintaining healthy water supplies for fish and other wildlife associated with ponds, lakes, streams, and rivers. Some of the practices that can help enhance quality and provide for aquatic insects are:

• *Minimal effective use of fertilizers.* Only the amount of fertilizers that will be taken up by the plants should be applied, because excessive fertilization will pollute nearby water supplies or groundwater. A fertilizer-free barrier zone surrounding waterways should be instituted.

• *Minimal use of pesticides.* Pesticides can affect aquatic life directly and indirectly. Pesticides should not be applied directly to water, and should not be allowed to wash into water. Some pesticides are particularly hazardous to fish and should be avoided. As with fertilizers, a pesticide-free zone should be instituted around water.

• *Increased oxygenation of water.* Elevational changes, including dams and spillways that cause water to tumble and become aerated, can enhance aquatic life.

• Preservation or replanting of riparian areas. Vegetation along streams provides several benefits, including shade that keeps the water temperatures lower, roots that reduce the potential for erosion and flow of soil into the stream, insects that inadvertently tumble into the water, and mating and oviposition sites for aquatic insects. In addition, riparian areas provide habitat for animals, especially routes for travel. A band of vegetation 30 m in width is recommended.

• *Unrestricted access of streams by livestock.* This can damage the banks, cause increased erosion, and disrupt the fauna residing in the water. Urine and fecal material produced by livestock will pollute the water supply. It is better to provide water to animals some distance away from natural water sources.

• *Flow and connectivity of water.* For insects to thrive in streams and rivers, continued (but not constant) flow and connectivity of water are important. Water levels can (and should) vary seasonally, but connectivity between pools of water assures better insect and fish survival, just as connectivity between forest stands assures better survival of terrestrial wildlife. Capture and storage of water in reservoirs and pods, combined with timely release, often is key to maintaining water flow and sustainability of animal populations that are dependent on water.

When insects have the potential to be abundant and damaging, nonchemical management of insect populations is safest for wildlife. Environmental manipulations that prevent increase in the abundance of pest populations are the ideal methods. This includes cultural practices such as timing of planting or harvesting crops, planting insect resistant crops, and fostering biological control of pests. These approaches were dis-

Fig. 14.9. White-tailed deer, *Odocoileus virginianus*, at a feeding station. As the deer feed from the grain trough, their ears and neck contact the pads impregnated with insecticide. This modest level of contact with insecticide is adequate to kill the ticks that are commonly found feeding on and in the ears: the lone star tick, *Amblyomma americanum*, and the 'deer tick,' *Ixodes scapularis*. Disruption of tick populations is a key method for reducing the risk associated with Lyme disease (photo by Wayne Ryan, USDA, ARS).

cussed previously in Chapter 8, Pest management and its effects on wildlife. If insecticides must be used in environments where wildlife occurs, it is important to use products that are most selective to the target pest, least persistent, or used in the most targeted manner so as to reduce exposure to wildlife. Innovative approaches to the management of Lyme disease, though directed more to the safety of people than to wildlife, are perhaps good examples of how pesticides can be integrated appropriately. It is possible to reduce tick populations (and thereby disease transmission potential) by selective application of insecticides to wildlife through self-treatment of the animals. Both small animals such as mice and chipmunks, and large animals such as deer, can be induced to contact pads containing insecticide when they feed at bait stations. This treatment kills the ticks found on the animals, and has been found to reduce tick levels by 80%–95%, with little or no exposure to nontarget organisms (Fig. 14.9).

Animal-biting species can be quite damaging to wildlife, though wildlife tend to be more resistant to insects and insect-borne diseases than are domestic animals. In most cases, wild animals and co-occurring biting insects have coevolved, so the existence of the wildlife is not really threatened. The greatest threat comes from introduction of nonindigenous insects or insect-borne diseases. Often there is no inherent resist-

ance to these new pests or diseases, and therefore the risk to wildlife is greater. Therefore, every effort should be made to reduce the risk of importing nonindigenous insects and animal diseases. The movement of animals (exotic wildlife, livestock from other continents, and pets) poses the greatest threat of introducing nonindigenous organisms. Regulations and procedures that prevent movement of insects and diseases deserve strengthening and better enforcement.

In some cases, especially when wildlife species are rare or threatened with extinction, intervention to relieve them of arthropod parasites is warranted. Even the use of chemical insecticides applied systemically in their food, via contact (e.g., dust baths), or into their environment (e.g., bird nesting boxes) may be acceptable, as long as the safety of the insecticides is established.

SUMMARY

• Economic benefits provided by insects include pollination of crops; production of silk, shellac, and natural dyes; use as food for humans and domestic animals; and for medical therapy.

• In some cases, insects warrant protection from extinction. Indigenous species often are at risk, due to competition for resources, if nonindigenous species are accidentally introduced to an area. However, habitat destruction/host loss is the major factor threatening insect biodiversity. Pesticides and climate change also threaten species that occur in limited areas. Some species have been designated as threatened or endangered, and regulations have been established to minimize the possibility of their extinction.

• The concept of habitat management for the improvement of wildlife populations can be expanded to include insects as a key component of wildlife habitat. Insects are an important component of the diet of many forms of wildlife, and these dietary needs of wildlife should be considered as a part of overall wildlife resource needs. Also, pesticides directed at insects can interfere with wildlife, and insects sometimes are pests, or transmit diseases, to wildlife.

• The fundamental concepts governing use of insects for manipulation of wildlife populations are: (1) insects are an important component of the diet of many forms of wildlife; (2) wildlife managers must understand the dietary needs and habits of the wildlife species of concern, and manage specifically for these animals; (4) insects also have specific diet and environmental requirements; (4) pesticides can interfere with wildlife survival and reproduction; and (5) the presence of insects is not always beneficial.

• Specific practices that could foster insect abundance and provide insect food resources (insect nurseries) for wildlife include habitat management such as expanded hedgerows and fencerows, increased diversity of plants within and near crop fields, planting of nectar or foliar resources for insects, and management of ecological compensation areas to favor certain insects. Other food resources that favor insect abundance such as dead trees, animal wastes and carrion, and a reliable and clean supply of water also can be used to foster insect production.

• Excess fertilizer and pesticides can wash into water supplies, leading to depletion of invertebrate populations and decreased fish populations. Protection of riparian areas with barrier zones where pesticides and fertilizers are not applied, fostering growth or replanting of vegetation along the edges of water bodies, keeping animals from direct access to water bodies, and improving aeration of water can enhance aquatic insect populations.

• Insecticides may have a role in wildlife conservation. Selective application can safely protect wildlife, livestock, and people from damaging pests and the diseases they transmit.

REFERENCES AND ADDITIONAL READING

Abivardi, C. (2008). Flower strips as ecological compensation areas for pest management. In Capinera, J.L. (ed.), *Encyclopedia of Entomology*, pp. 1489–1494. Springer Science & Business Media B.V., Dordrecht, The Netherlands.

Banaszak, J. (1992). Strategy for conservation of wild bees in an agricultural landscape. *Agriculture, Ecosystems and Environment* **40**, 179–192.

Barker, A.M. (2004). Insects as food for farmland birds – is there a problem? In van Emden, H.F. & Rothschild, M. (eds.) *Insect and Bird Interactions*, pp. 37–50. Intercept, Andover, Hampshire, UK.

Calvert, W. H., Hedrick, L.E., & Brower, L.P. (1979). Mortality of the monarch butterfly (*Danaus plexippus* L.): avian predation at five overwintering sites in Mexico. *Science* **204**, 84–851.

Carvell, C., Meek, W.R., Pywell, R.F., Goulson, D., & Nowakowski, M. (2007). Comparing the efficacy of agri-environment schemes to enhance bumble bee abundance and diversity on arable field margins. *Journal of Applied Ecology* **44**, 29–40.

Choo, J. (2008). Potential ecological implications of human entomophagy by subsistence groups of the Neotropics. *Terrestrial Arthropod Reviews* **1**, 81–93.

Collins, N.M. & Thomas, J.A. (eds.) (1991). *The Conservation of Insects and their Habitats*. Academic Press, London, UK.

Cunningham, R.B., Lindenmayer, D.B., Crane, M., *et al.* (2008). The combined effects of remnant vegetation and tree planting on farmland birds. *Conservation Biology* **22**, 742–752.

DeFoliart, G.R. (1989). The human use of insects as food and as animal feed. *Bulletin of the Entomological Society of America* **35**, 22–35.

Dennis, R.L.H., Shreeve, T.G., & VanDyck, H. (2003). Towards a functional resource-based concept for habitat: a butterfly biology viewpoint. *Oikos* **102**, 417–426.

Emmel, T.C. & Sourakov, A. (2008). Monarch butterfly, *Danaus plexippus* L (Lepidoptera: Danaidae). In Capinera, J.L. (ed.) *Encyclopedia of Entomology*, pp. 2456–2461. Springer Science & Business Media B.V., Dordrecht, The Netherlands.

Erhardt, A. (1995). Ecology and conservation of alpine Lepidoptera. In Pullin, A.S. (ed.) *Ecology and Conservation of Butterflies*, pp. 258–276. Chapman & Hall, London, UK.

Fiedler, A.K., Landis, D.A., & Wratten, S.D. (2008). Maximizing ecosystem services from conservation biological control: the role of habitat management. *Biological Control* **45**, 254–271.

Goulson, D., Lye, G.C., & Darvill, B. (2008). Decline and conservation of bumble bees. *Annual Review of Entomology* **53**, 191–208.

Gray, L.J. (1993). Response of insectivorous birds to emerging aquatic insects in riparian habitats of a tallgrass prairie stream. *American Midland Naturalist* **129**, 288–300.

Gurr, G.M., Wratten, S.D., & Altieri, M.A. (2004). *Ecological Engineering for Pest Management. Advances in Habitat Manipulation for Arthropods*. Cornell University Press, Ithaca, New York, USA.

Hansell, M. (2000). *Bird Nests and Construction Behavior*. Cambridge University Press, Cambridge, UK.

Hill, D.S. (1997). *The Economic Importance of Insects*. Chapman and Hall, London.

Holland, J.M. (2004). The impact of agriculture and some solutions for arthropods and birds. In van Emden. H.F. & Rothschild, M. (eds.) *Insect and Bird Interactions*, pp. 51–73. Intercept, Andover, Hampshire, UK.

Holzschuh, A., Steffan-Dewenter, I., Kleijn, D., & Tscharntke, T. (2007). Diversity of flower-visiting bees in cereal fields: effects of farming system, landscape composition and regional context. *Journal of Applied Ecology* **44**, 41–49.

Jones, G.A. & Sieving, K.E. (2006). Intercropping sunflower in organic vegetables to augment bird predators of arthropods. *Agriculture, Ecosystems and Environment* **117**, 171–177.

Jones, G.A., Sieving, K.E., & Jacobson, S.K. (2005). Avian diversity and functional insectivory on north-central Florida farmlands. *Conservation Biology* **19**, 1234–1245.

Konvicka, M., Benes, J., Cizek, O., Kopecek, F., Konvicka, O., & Vitaz, L. (2008). How too much care kills species: grassland reserves, agri-environmental schemes and extinction of *Colias myrmidone* (Lepidoptera: Pieridae) from its former stronghold. *Journal of Insect Conservation* **12**, 519–525.

Lagerlöf, J., Stark, J., & Svensson, B. (1992). Margins of agricultural fields as habitats for pollinating insects. *Agriculture, Ecosystems and Environment* **40**, 117–124.

Landis, D.A., Wratten, S.D., & Gurr, G.M. (2000). Habitat management to conserve natural enemies of arthropod pests in agriculture. *Annual Review of Entomology* **45**, 175–201.

Losey, J.E. & Vaughan, M. (2006). The economic value of ecological services provided by insects. *Bioscience* **56**, 311–323.

MacVean, C. (2008). Lacquers and dyes from insects. In Capinera, J.L. (ed.), *Encyclopedia of Entomology*, pp. 2110–2117. Springer Science & Business Media B.V., Dordrecht, The Netherlands.

Mizell III, R.F. & Mizell, P.A. (2008). Plant extrafloral nectaries. In Capinera, J.L. (ed.) *Encyclopedia of Entomology*, pp. 2915–2921. Springer Science & Business Media B.V., Dordrecht, The Netherlands.

New, T.R. (1997). *Butterfly Conservation*, 2nd edn. Oxford University Press, Melbourne, Australia.

New, T.R. (2005). *Invertebrate Conservation and Agricultural Ecosystems*. Cambridge University Press, Cambridge, UK.

Olson, D.M. & Wäckers, F.L. (2007). Management of field margins to maximize multiple ecological services. *Journal of Applied Ecology* **44**, 13–21.

Ouin, A., Aviron, S., Dover, J., & Burel, F. (2004). Complementation/supplementation of resources for butterflies in agricultural landscapes. *Agriculture, Ecosystems and Environment* **103**, 473–479.

Parmenter, R.B., MacMahon, J.A., & Gilbert, C.A.B. (1991). Early successional patterns of arthropod recolonization on reclaimed Wyoming strip mines: the grasshoppers (Orthoptera: Acrididae) and allied faunas (Orthoptera: Gryllacrididae, Tettigoniidae). *Environmental Entomology* **20**, 135–142.

Pearson, D.E. & Fletcher, Jr., R.J. (2008). Mitigating exotic impacts: restoring deer mouse populations elevated by an exotic food subsidy. *Ecological Applications* **18**, 321–334.

Samways, M.J. (1992). Some comparative insect conservation issues of north temperate, tropical, and south temperate landscapes. *Agriculture, Ecosystems and Environment* **40**, 137–154.

Samways, M.J. (2007). Insect conservation: a synthetic management approach. *Annual Review of Entomology* **52**, 465–487.

Sourakov, A. & Emmel, T.C. (2008). Conservation of insects. In Capinera, J.L. (ed.) *Encyclopedia of Entomology*, pp.1025–1035. Springer Science & Business Media B.V., Dordrecht, The Netherlands.

Smith, H.R. & Remmington, C.L. (1996). Food specificity in interspecies competition. Comparisons between terrestrial vertebrates and arthropods. *BioScience* **46**, 436–447.

Woodcock, B.A., Potts, S.G., Pilgrim, E., *et al.* (2007). The potential of grass field margin management for enhancing beetle diversity in intensive livestock farms. *Journal of Applied Ecology* **44**, 60–69.

GLOSSARY

aberrant host Animals that are not the normal hosts for a disease (= incidental or unnatural host).

abdomen The posterior-most major body region of arthropods. The abdomen contains the reproductive organs and most of the digestive organs, but often lacks large appendages. *See* Fig. 2.3.

aberrant parasite Parasites that occur only occasionally in a host, usually because there are behavioral or ecological barriers that keep the parasite from the prospective host (= **incidental parasite**).

absolute sampling measures Techniques that estimate the abundance of organisms in a defined area or volume.

acanthocephalans Cylindrical, unsegmented worms with a thorny proboscis in the phylum Acanthocephala.

acaricide A pesticide used to kill mites (= **miticide**).

acetylcholine esterase An acetylcholine-destroying enzyme that is released after acetylcholine is emitted. This restores the synapse region between nerves to its original state.

acetylcholine A common chemical neurotransmitter that bridges the synapse between nerves, allowing continuation of the impulse from nerve to nerve.

acoustical communication Signaling that depends on the ability to produce (and perceive) vibrations in air, water, or solid substrates.

action threshold The point at which pest suppressive efforts need to be initiated to prevent economic loss from occurring. This is also known as the economic threshold or ET. *See also* **economic injury level**.

active anting Bird behavior in which the bird crushes an ant with its bill, and then rubs the crushed ant over its feathers.

active transmission Transmission of a disease via active involvement of the disease agent, as when the pathogen invades the host.

acute toxicity Poisoning that occurs rapidly after exposure to the toxin.

adipose tissue The fatty tissues of insects (= **fat body**). Although consisting mostly of fatty tissues for energy storage, the fat body tissues also function in storage of waste products and in harboring beneficial microorganisms. *See* Figs. 2.14, 2.15: 'fat body.'

adjuvant An additive to a pesticide formulation that enhances the actions of the toxicant, often by modifying the physical properties of the formulation. Examples of adjuvants are stickers, spreaders, solubilizers, and dispersants.

adventive Organisms that came from elsewhere, not indigenous.

aflatoxin A mycotoxin produced by *Aspergillus* fungi.

aggressive mimicry Resemblance of a predator to something that is attractive to prospective prey.

air sacs Dilations of the trachea. Air sacs serve as an air reservoir and a means for insects to add mass internally despite having an inelastic integument.

alatae (adj., alate) The winged form of aphids. Typically, aphids alternate between winged and wingless (apterous) forms during their life cycle.

algaecide A pesticide used for algae.

allelochemical compounds Non-nutritive compounds (also called allelochemicals) produced by

Insects and Wildlife: Arthropods and Their Relationships with Wild Vertebrate Animals, 1st edition. By J.L. Capinera. Published 2010 by Blackwell Publishing.

plants that affect other organisms (normally herbivorous insects and plant diseases) negatively.

allomones   Allelochemicals that benefit the producing organism.

allopatric speciation   Development of new species due to geographic separation. After long isolation, the genetics of isolated populations change, and if isolated populations then have the opportunity to interbreed they may no longer be able to do so. Allopatric speciation is a principal means by which new species develop. *See also* **sympatric speciation**.

ametabolism   A form of metamorphosis in which there is little or no change in body form as the immature molts and grows.

amplifier host   Animals serving as disease agent hosts in which the disease agent abundance is increased without severely affecting the host.

anadromous   Fish that breed in fresh water but migrate to salt water for part of their life cycle.

anamorphosis   Addition of a body segment at each molt.

anautogeny   The need for blood-feeding insects to have a blood meal prior to production of eggs.

antennae   (singular, antenna): paired sensory appendages, usually elongate in form among adult insects. The antennae provide the ability to detect chemicals and wind speed. *See* Figs. 2.4, 2.5.

anterior   Referring to the front or the end of the insect bearing the head; opposite of posterior.

antibiosis   A type of resistance in which there is some interference with the pest's life cycle.

anticoagulants   Pesticides that interfere with the ability of blood to clot. Animals frequently experience small wounds, and when they have diminished clotting capacity they may bleed to death.

antixenosis   A type of resistance based on the lack of attractiveness of a prospective host to a pest (= **nonpreference**).

apitherapy   Bee venom therapy; treatment of humans with bee venom.

apodemes   Infoldings of the integument, forming ridges that strengthen the integument and provide points of attachment for muscles. The ridges are marked externally by linear depressions called sutures.

apolysis   The first step of molting, the separation of the epidermis from the cuticle.

aposematic coloration   The use of warning coloration to advertise the presence of an organism that is distasteful or harmful.

appendage extension   A form of startle behavior in which the prey extends its appendages to give it the appearance of being larger.

apterae   (adj., apterous) The wingless form of aphids. Typically, aphids alternate between winged (alate) and wingless forms during their life cycle.

arbovirus   An arthropod-borne virus.

area-wide management   a regional approach to pest suppression; pest control over a wide area.

aristate   A type of antenna that bears an arista (a small bristle or hair-like process not normally found at the tip of the antenna). This type of antenna is found on more advanced flies. *See* Fig. 2.5.

armyworm   Larvae of Noctuidae, so-called due to their tendency to occur in large numbers, and sometimes in aggregations. These insects sometimes also are known as cutworms.

augmentative biological control: The release of natural enemies into the environment to supplement naturally occurring biological control.

Australian realm   A biogeographic realm consisting of Australia and nearby islands.

autogeny   the ability of blood-feeding insects to produce a batch of eggs without first feeding on blood.

autohemorrhage   The self-induced seeping of blood from the body (= **reflex bleeding**), usually from leg joints, in response to disturbance. The bleeding releases feeding deterrents and sometimes is accompanied by sound.

autonomy   The shedding of a body part under duress. Some insects readily shed appendages or spines. This likely helps them to avoid capture by predators.

autotroph   A primary producer, normally green plants.

avicide   A pesticide used to kill pest birds.

axon   Elongate, fibrous nervous tissue that transmits nervous impulses electrically. *See* Fig. 2.20.

bactericide   A pesticide used to kill bacteria.

balance of nature   The idea that there is an inherent equilibrium resulting in a stable, continuing system of life.

basement membrane   A membrane found below the epidermal cells and separating the integument from the muscles and internal organs.

Batesian mimicry   Evolutionary convergence of the appearance of an edible organism to resemble an unrelated unpalatable species.

beetle banks   Strips of vegetation, initially grasses but eventually larger and more diverse plants as ecological succession occurs, that provide shelter for predatory insects, particularly ground beetles (Coleoptera: Carabidae).

behavioral fever A form of self-medication that can result in elimination of the parasites due to exposure of ectotherms such as insects to high temperature.

Beltian bodies Modified acacia tree leaflet tips that can provide ants with protein and oil as food.

beneficial virulence High virulence benefits the fitness of the disease agent in some way.

binomial nomenclature A formal system of naming species based on the genus and species. The result is called the scientific name or Latin name because this two-part name is based on Latin terminology. The system was developed by Carl Linnaeus. *See also* **scientific name**.

bioaccumulation Accumulation of pesticide in an animal during the course of its life when the rate of uptake exceeds the rate of elimination. Bioaccumulation occurs via bioconcentration and biomagnification.

biocides Chemicals that kill all forms of life.

bioconcentration An increase in concentration of pesticide in an animal over time, independent of trophic relationships.

biogeographic realms (or regions or provinces). Fauna within a geographic area that share a common phylogeny, and developed in relative isolation from other realms.

biomagnification Increase in the concentration of pesticide at each trophic level in a food chain. The result typically is a very high concentration in the animals at the top of the food chain.

bipectinate A structure, usually a type of antenna, with comb-like teeth on two sides. *See* Fig. 2.5.

Black Death A plague pandemic affecting Europe in the 1300s.

Black Fly Fever A malady caused by injection by black flies of proteins into humans, causing loss of muscle control and erratic breathing.

blocking With respect to plague, this is a malady in which the plague bacteria plug up the digestive tract of the flea, preventing the flea from processing its blood meal and causing it to regurgitate into the host.

bolus The regurgitated undigested matter remaining after the meal of some raptors and other birds (= **pellet**).

bottle flies The metallic flies in the family Calliphoridae (Diptera), also known as blow flies.

bottom-up regulation Population regulation that is imposed by factors lower in the food pyramid or food web. Normally, these are resource limitations, and can be due to either quantity or quality of the resource. Most commonly the resource limitation is food, but it could also be oviposition substrate, mating site, nesting site, or any other vital resource.

brachypterous Referring to the condition of some insects when the wings are reduced in size. This condition is called brachyptery or micropterity.

buboes Lesions or swellings induced by bubonic plague.

Bubonic plague Plague transmitted by fleas to the skin, and characterized by swellings or buboes, particularly swollen lymph nodes.

bulk: the volume, mass or weight of ingested items.

campestral plague Maintenance of low levels of plague in wild rodents (= **sylvatic plague**).

cannibalism Feeding on conspecifics (organisms of the same species).

capitate A structure, usually a type of antenna, bearing a terminal knob-like structure located at the tip. *See* Fig. 2.5.

capsid A protective protein coat surrounding a virion.

carcass-maggot cycle of botulism Fly maggots can successfully develop on carcasses from animals that have died from botulism poisoning. Such maggots contain botulism and may be poisonous to birds that feed on the maggots.

carnivory Feeding on animals.

caudal Pertaining to the anal end of the body.

caterpillars The larvae of Lepidoptera, and sometimes phytophagous sawflies (Hymenoptera: suborder Symphyta).

caudal Referring to the posterior end of the insect. Caudal appendages are typically cerci.

central nervous system The principal system of interconnected nervous tissue. In insects, it consists of a dorsal brain and a ventral nerve cord.

cephalopharyngeal skeleton The mouth hooks of maggots and their supporting internal sclerites. The shape and size of these structures are often used to identify fly larvae.

cephalothorax The chelicerate arthropods (subphylum Chelicerata) such as mites, spiders and crabs, and lobsters have two principal body segments, the cephalothorax and the abdomen. The cephalothorax is the anterior segment, and consists of the fused head and thorax.

cerci (singular, cercus) A pair of lateral appendages found terminally on the abdomen. The cerci apparently have sensory functions and sometimes are involved in mating behavior.

cestodes Tapeworms; long, thin segmented animals in the phylum Platyhelminthes (Cestoda).

Chagas Disease American Trypanosomiasis; a disease caused by the trypanosome *Trypanosoma cruzi*, and transmitted by triatomine (kissing) bugs (Hemiptera: Reduviidae).

chancre A red sore caused by the bite of tsetse flies.

chelicerae The fang-like feeding structure of chelcerate arthropods (subphylum Chelicerata). *See* Fig. 10.1.

chemical communication The emission and perception of odors, resulting in the transmission of information.

chewing lice Lice (Phthiraptera) with chewing mouthparts.

chitin A polysaccharide that is an important component of the cuticle.

chorion The proteinaceous egg covering or egg 'shell.' *See* Fig. 2.29.

chronic toxicity Poisoning that occurs only after long exposure to the toxin.

circulatory system The system of blood flow through the body of the insect. The transport of blood (hemolymph) provides the insect with means of transporting blood cells, nutrients and hormones, but not usually oxygen. The circulatory system of insects is called an 'open' circulatory system because the blood does not flow continuously in vessels, instead often circulating freely in the body cavity. However, the insect has a heart, a large vessel called a dorsal aorta, and veins in the wings. *See* Fig. 2.18.

classical biological control Importation and establishment of new natural enemies that affect nonindigenous invaders.

clavate A club-like structure, usually a type of antenna that thickens gradually away from the base. *See* Fig. 2.5.

clinical sign or sign of disease Objective evidence that a disease is present, often a physical attribute.

coincidental virulence High virulence is a coincidental by-product of infection and provides no adaptive benefit to the infectious agent.

collectors Aquatic organisms feeding on small debris. They can be found nearly anywhere, but usually predominate in the lower reaches of watercourses.

collophore A tube-like structure on the first abdominal segment of springtails; It has various functions, including adhesion to smooth surfaces, water absorption and excretion. *See* Fig. 1.6.

colonial nesters Animals that nest in colonies.

commensalism A form of symbiosis in which the organisms live together but only one benefits from the relationship, and the other is not harmed.

competitive advantage The concept that invasive insects are abundant because they are able to exploit unoccupied niches or are better able to exploit them than indigenous species.

compound eyes The largest and most complex of the insect visual organs, located laterally on the head. There are two compound eyes on almost all insects, providing shape, motion, and color detection. *See* Fig. 2.4.

conservation biological control Practices that maintain or enhance the effectiveness of naturally occurring natural enemies.

conservation headland An area of cropland at the periphery of crop fields that is not treated with pesticides, providing a safe area for bird foraging.

contact poison A pesticide that causes toxicity by external exposure.

contagious The disease organism can readily spread.

coprophagous Feeding on dung, or animal feces.

corpora allata Small endocrine glands associated with the insect brain that produce and store juvenile hormones. *See* Figs. 2.21, 2.23.

corpora cardiaca Small endocrine glands that store and release prothoraciotropic hormones (PTTH). *See* Fig. 2.23.

coxa: the basal leg segment, attaching to the thorax and connecting to the second segment of the leg, the trochanter.

crepuscular Active at dawn or dusk.

crochets Small hooks found on the prolegs of lepidopteran prolegs, but not on sawflies. They assist in attachment to the host plant.

cross-predation The process of different taxa eating each other at different stages in their development. In the case of insects and wildlife, it usually refers to aquatic insects eating small fish, but older fish of the same species then consuming insects.

cross-fertilization Pollen from a plant fertilizes flowers on another plant.

crypsis Becoming less apparent to predators by blending into the background; camouflage.

cultural control The manipulation of environments or cultural practices to reduce the susceptibility of plants or animals to pests. The manipulations typically are based on the modification of normal production practices such as depth of tillage, amount of irrigation, or timing of planting (= **environmental management**).

cuticle The nonliving layers of the integument. See Fig. 2.1.

cutworm Name commonly applied to larvae of Noctuidae due to their plant feeding habits. Those that tend to aggregate often are called armyworms.

dead-end host Host animals that are not suitable for production of high concentrations of a disease agent, or otherwise limit the uptake of the disease by a vector (= **dilution host**).

death feigning The cessation of movement by an animal when disturbed, thereby pretending that it is dead (= **thanatosis**).

decomposition: the degradation of a dead organism into its parts and elements.

deer flies *Chrysops* spp. (Diptera: Tabanidae).

definitive host The animal species in which the disease agent passes the adult, sexual, or multiplicative stage of the life cycle (= **primary host**).

degradation phase of decomposition: degradation follows the destruction phase; in the degradation phase the small particles are further degraded into molecules.

dendrite The end of an insect nerve cell that receives an impulse, which is then transmitted along the axon.

density The number of individuals measured in a sampling unit.

density dependent factors Mortality factors that have greater effect as host population density increases. These are usually biotic factors such as predators, parasitoids, and diseases.

density independent factors Mortality factors that act independent of host density. These are usually abiotic factors such as weather.

destruction phase of decomposition A period in which the dead organism is mechanically degraded into smaller sized particles. This phase is followed by the degradation phase.

detritivores Animals that feed on plant litter, or detritus.

detritophagous Feeding on detritus.

deutocerebrum One of the three principal regions of the brain, and connected to the antennae of insects. *See* Fig. 2.21.

diapause A genetically determined period of arrested development. Typically, diapause allows insects (arthropods) to survive inclement conditions and is induced by environmental stimuli associated with the inclement conditions. Most often, decreasing day-length or decreased nutrient quality of plants cause insects to enter diapause in advance of winter conditions.

diffuse pollution Pollution derived from pesticides originating over a broad area.

digestibility-reducing compounds Chemical compounds that cause plant foliage to be resistant to herbivore digestibility and microorganism decomposition. Not only do insects have difficulty in obtaining energy from such material, but they often avoid these plants.

dilution host Host animals that are not suitable for production of high concentrations of a disease agent, or otherwise limit the uptake of the disease by a vector (= **dead-end host**).

direct flight muscles These muscles pull directly on the wings. Although the direct flight muscles sometimes are adequate to provide lift for the insects, more often they provide auxiliary functions such as wing twisting, leading to propulsion. *See also* **indirect flight muscles**.

direct life cycle With respect to parasitism, this refers to movement of the parasite directly from one host to another.

direct transmission Transmission of a disease from one organism to another without involvement of another (a vector).

disease triangle The interaction of the host, the disease agent, and the environment.

disruptive coloration A form of crypsis in which the outline of the organism is disrupted by a color pattern, helping it to blend into the background.

diurnal Active during the daylight hours.

diversity In biology, this term is used in several ways, and often incorrectly. Diversity is largely a measure of the composition of a community as measured by the number of species present. However, diversity also includes a measure of evenness (a measure of disparity among the number of individuals representing each species present). The number of species measured without considering evenness is called species richness. Diversity is often represented by a diversity index, of which there are several types. Diversity indices reduce the measure of diversity to a single number; although this is convenient for comparisons, considerable information is lost when using such measures. In many ways, measurement of species richness, plus a measure of mean abundance and a variance measurement for each species, is more informative for assessment of diversity.

dodging   An escape behavior in which the insect quickly moves around the stem of a plant, keeping the plant between it and a potential predator.

domatia   Plant-produced swellings that serve as shelters for arthropods.

dorsal   Pertaining to the upper surface of the body; opposite of ventral.

droppings: The fecal deposits or excrement of birds.

dung   The excrement of animals, particularly mammals (= **scats**).

ecdysis   Escape of the insect from its old cuticle (exuviae).

ecdysone   A hormone that stimulates molting in insects, and released by the prothoracic glands.

ecological compensation   Plantings of wildflower strips within crop fields that introduce plant/resource/food diversity, thereby compensating for the lack of diversity associated with the crop monoculture.

economic injury level   A level of pests or damage that is equivalent to the cost of controlling them. This is also known as the EIL.

economic threshold   The point at which pest suppressive efforts need to be initiated to prevent economic loss from occurring. This is also known as the ET, but is also called the action threshold.

ecosystem   A community of living organisms and the abiotic environment of an area.

ectognathous   Arthropods having the mouthparts extruding forward from the head (the insects).

ectoparasites   Organisms that live on the external surface of the host, or cavities that open directly onto the surface.

ectoparasitoids   Parasitoids that develop outside their host, normally living on the surface.

ectothermic   Organisms that do not actively regulate their body temperature, which is governed largely by the environment. Ectotherms are also called 'cold-blooded' or poikilotherms.

egg pods   Clusters of grasshopper eggs, usually produced in conjunction with frothy material.

EIL   An acronym for economic injury level. This is a level of pests or damage that is equivalent to the cost of controlling them.

elytra   (singular, elytron) The thickened forewings of beetles (Coleoptera). They are mostly useful for protection, having little or no value for flight. Generally, they lack veins. *See* Fig. 2.13.

emerging pathogens   Diseases that are increasing in prevalence.

encephalitis   An acute inflammation of the brain. This can be caused by a number of factors, often infection by a virus or bacterium. Symptoms include irritability, fever, headache, confusion, and sometimes seizures.

encephalomyelitis   Inflammation of the brain and spinal cord due to viral infection.

endangered species   Species that are considered to be at high risk of extinction.

endangered species act   Legislation passed by the US Congress in 1973 that legislates protection of species that are threatened with extinction.

endemic: disease occurrence is at a low level (= **enzootic**).

endocrine glands   Glands that release their products to the inside of the insect. The most important of the endocrine glands are the hormone glands including the corpora allata, corpora cardiaca, and prothoracic glands.

endocuticle   The interior layer of the cuticular region of the integument. *See* Fig. 2.1.

endoparasites   Organisms that live within the body of the host, including the digestive system, organs, tissues, cells, and freely in the body cavity.

endoparasitoids   Parasitoids that live inside their host.

endothermic   Organisms that actively regulate their body temperature, and maintain a relatively constant temperature. Endotherms are also called 'warm-blooded.'

enemy release   The idea that invasive insects are abundant because their natural enemies are absent or rare.

enhanced biodegradation   Rapid loss of pesticide effectiveness in the soil due to elevated levels of pesticide-degrading bacteria.

entognathous   Arthropods having the mouthparts sunk into the head (the springtails, proturans, and diplurans).

entomophagous   Feeding on insects (arthropods).

environmental management The manipulation of environments or cultural practices to reduce the susceptibility of plants or animals to pests. The manipulations typically are based on the modification of normal production practices such as depth of tillage, amount of irrigation, or timing of planting (= **cultural control**).

enzootic Disease occurrence is at a low level (= **endemic**).

epicuticle The thin exterior layer of the cuticle, exterior to the exocuticle. *See* Fig. 2.1.

epidemic An unusual abundance of disease (= **epizootic, outbreak**).

epidermal cells The cellular living portion of the integument, located basally, that gives rise to the nonliving layers (cuticle) above the epidermis.

epizootic An unusual abundance of disease (= **epidemic, outbreak**).

ET An acronym for economic threshold. This is the point at which pest suppressive efforts need to be initiated to prevent economic loss from occurring. This is also known as the action threshold.

Ethiopian realm A biogeographic realm consisting of Central and southern Africa.

etiologic agent The cause of a disease.

etiology The study of the causes of diseases.

eukaryotes Organisms that possess a cell nucleus, membrane bounded organelles, and cells organized into complex structures.

eusocial behavior The most advanced form of sociality found in insects, expressed as cooperative care of young by adults, overlap of generations, and division of reproduction among castes.

eutrophication Stimulation of excessive aquatic plant growth by high nutrient levels. This usually is accompanied by environmental effects such as reduced light penetration of the water, reduced oxygen levels, and sometimes by high levels of plant-produced toxins.

exocrine glands Glands that release their products to the outside of the insect. Among the exocrine glands are silk, poison, wax, and pheromone glands.

exocuticle A region of the integument that is exterior to the endocuticle. *See* Fig. 2.1.

exsanguination Removal of an animal's blood by mosquitoes, resulting in the animal's death.

extraoral digestion Digestion of food that occurs outside the digestive system (alimentary canal) resulting from secretion of digestive enzymes into the prey organism. Thus, the insect is able to imbibe partially digested food.

exuviae (singular and plural) The old cuticle once it has been shed by a molting insect.

exuvial space The space created by the separation of the epidermis from the cuticle during apolysis.

eye spots The presence of large, false eyes on the wings that, when exposed, deter consumption by a predator.

facet The hexagon-shaped lens of the ommatidium.

facultative diapause A readily reversible form of arrested development that is terminated whenever favorable conditions appear.

facultative myiasis Infestation by flies that normally develop on carrion or feces, but are capable of affecting living host animals.

facultative parasites Parasites that do not require a certain host, but sometimes are found in association with it.

fat body The fat or adipose tissues of insects. Although consisting mostly of fatty tissues for energy storage, the fat body tissues also function in storage of waste products and in harboring beneficial microorganisms. *See* Figs. 2.14, 2.15.

febrile illness Fever due to viral infection.

febrile myalgia and arthralgia Fever with muscle and joint pain due to viral infection.

femur (plural, femora) The third leg segment, connecting the trochanter and the tibia. The femur is often the largest segment of the leg, bearing the muscles used for leaping or grasping. It is no longer than the tibia, but usually thicker. *See* Figs. 2.3, 2.13.

filiform Long and slender, thread-like in appearance. This term is often used to describe antennae that have relatively uniform diameter segments. *See* Fig. 2.5.

filter chamber A structure that filters out excess water as plant sap is ingested by piercing-sucking insects. Essentially, it is a bypass of part of the midgut to allow excess fluid to be quickly passed through the system. By reducing the amount of water that passes through the midgut, the ingested food is less dilute, and therefore more nutritious.

flabellate A structure with segments laying flat on each other, resembling a fan. Some insect antennae have components that possess fan-like elements. *See* Fig. 2.5.

flagellum The distal portion of the antenna; all the segments beyond the basal scape and pedicel.

flash coloration   A startle or fright behavior displayed by insects that flash unexpected color and surprise a predator.

flux   The amount of an item, usually energy or nutrients, moving through an area in a period of time.

food chains   Diagrams showing the feeding links between elements of an ecosystem (= **food web** or **trophic pyramid**).

food web   The interaction of functional groups, particularly feeding links, in an ecosystem. Food webs are sometimes known as food chains or trophic pyramids.

forbs   An ecological designation for herbaceous flowering plants that are not grasses or ferns. Normally they are broad-leafed.

foregut   The first of the three principal components of the digestive system. This region serves for ingestion and temporary storage of food. *See* Fig. 2.17.

frequency of consumption   The frequency of items ingested, expressed as the number or proportion of each item in the diet.

frequency of occurrence   The proportion of a population that has a food item in its diet (= prevalence).

frugivorous   Feeding on fruit.

fumigants   Pesticides that are inhaled by insects through their ventilatory system.

functional responses   Behavioral responses of wildlife or beneficial insects (mostly predators) that affect survival rates of prey insects.

fungicide   A pesticide used for fungi.

fungivorous   Feeding on fungus.

furcula   A spring-like structure attached near the tip of the abdomen in springtails. This structure propels the springtail with leap. *See* Fig. 1.6.

galls   On plants, abnormal growths and deformities induced by and inhabited by insects (= **plant galls**).

ganglion   (plural, ganglia) A mass of nervous tissue, normally found dorsally as part of the brain or ventrally in association with the nerve cord. *See* Fig. 2.15.

generalist   Having a broad range of feeding habits or occupying a wide niche. For plant-feeding insects, this is usually defined as feeding on more than one family of plants.

geniculate   A structure that possesses an elbow or sharp turn. Some antennae possess such obtuse angles. *See* Fig. 2.5.

genitalia   The structures that are involved in the physical aspects of mating behavior; typically they are found near the tip of the abdomen.

geographic shifts   Changes in behavior related to place (geography).

glycerol   A naturally occurring sugar alcohol found in hemolymph that inhibits freezing in insects.

granivorous   Feeding on seed or grain.

grazers   In aquatic systems, organisms that scrape algae from the surface of submerged items.

grazing optimization hypothesis   The hypothesis that plant growth is optimal when subjected to some level of herbivory (= **herbivore optimization hypothesis**).

guano   The feces or droppings of seabirds. Guano is harvested from coastal areas and used as fertilizer (= **excrement**).

guild   A group of species in a community that are functionally similar, or use resources similarly.

habitat management   Management of food and cover resources for enhancement of wildlife populations.

halteres   Small vibrating structures that take the place of the second pair of wings in Diptera. They aid in balance when the insect flies.

hard ticks   Ticks that possess a scutum, or hard plate, that covers some or all of the body.

hazard   With respect to pesticides, the likelihood that a toxicant will cause injury to a nontarget organism.

head   The most anterior (front) of the three insect tagma (major body regions). The head contains the feeding appendages and most of the important sensory organs, particularly the eyes and antennae. *See* Figs. 2.3, 2.4.

hemelytra   (singular, hemelytron) The forewings of some Hemiptera, thickened basally and membranous distally.

hemimetabolism   A form of metamorphosis in which there is slight change in body form as the immature molts and grows. The adult resembles the immature stages (called nymphs) but also may have fully formed wings and genitalia. This is also called incomplete metamorphosis.

hemocytes   The blood cells of insects, a principal component of the hemolymph.

hemolymph   The blood of insects, so named because it is equivalent to both the blood and lymph systems of vertebrates.

Hemorrhagic fever   Bleeding and fever due to viral infection.

herbivore optimization hypothesis The hypothesis that plant growth is optimal when subjected to some level of herbivory (= **grazing optimization hypothesis**).

herbivory Feeding on plants (= **phytophagy**).

heterotroph A consumer. Heterotrophs may be herbivores or carnivores, depending on their trophic level.

hibernation A reversible state of inactivity and metabolic depression. It is characterized by lower body temperature, slower breathing, and lower metabolic rate. Hibernation allows animals to conserve energy, especially during winter when food is short, instead using energy reserves (body fat) at a slow rate. This term is applied to mammals; the ecologically similar condition in insects is called diapause. (*See also* **diapause**)

hindgut The third of the three principal components of the digestive system, following the midgut. Some digestion occurs in the hindgut, but it principally provides salt and water balance, and excretion. *See* Fig. 2.17.

Holarctic realm A biogeographic region consisting of the Nearctic and Palearctic realms, corresponding approximately to North America, Europe and northern Asia.

holometabolism A form of metamorphosis in which there is radical change in body form as the immature molts and grows. The adult does not resemble the immature stages (called larvae) and is separated from the larval stage by a pupal stage. This is also called complete metamorphosis. *See* Fig. 2.26.

honeydew Liquid excreta passed by some sucking insects, and usually very rich in sugars. It is commonly found on the upper surface of foliage, where it drips from insects feeding above, and serves as a substrate for the growth of sooty mold, a blackish fungus.

horizontal transfer of pesticide Passage of pesticide from insect to insect. This normally occurs due to cannibalism, scavenging, and consumption of feces or regurgitated material.

horizontal transmission Transmission of a disease agent from animal to animal independent of parental relationship.

hormones Chemicals released from endocrine glands that affect another part of the insect. Hormones cause differential expression of genes, and result in development of various larval, pupal, and adult characteristics. In insects, the principal hormones are juvenile hormones, prothoraciotropic (PTTH or brain) hormone, and ecdysone (ecdysteroids).

horse flies Several genera of large tabanids (Diptera: Tabanidae).

host grooming Self-cleaning behavior displayed by mammals.

host resistance Cultivars, varieties, or breeds that are more resistant to pests than other types of the same species.

hydropyle A structure that allows uptake of water.

imaginal discs Cells in holometabolous insects that, once expressed, result in development of adult characters such as wings.

immigrant Organisms that were accidentally introduced.

incidence In parasitology, this refers to the proportion of hosts that become infected.

incidental host Animals that are not the normal hosts for a disease (= **aberrant** or **unnatural host**).

incidental host Parasites that occur only occasionally in a host, usually because there are behavioral or ecological barriers that keep the parasite from the prospective host (= **aberrant parasite**).

indigenous Organisms that are naturally present in an area (= **native**).

indirect flight muscles Muscles that affect flight indirectly, by deforming the integument and using the elastic energy of the integument to provide energy for flight.

indirect life cycle With respect to parasitism, this refers to movement of a parasite indirectly from host to host, usually via a vector.

indirect transmission Transmission of a disease from one organism to another via a third organism, often an insect.

inert ingredients Components of a pesticide formulation that lack significant activity, serving mostly to dilute the formulation to make it easier to handle. Water often serves as the principle inert ingredient.

infection The occurrence and replication of a disease in or on a host.

infectious diseases Diseases caused by microparasites: viruses, bacteria, and fungi. In contrast, diseases caused by protozoa, helminthes, and arthropods are traditionally called parasitic diseases.

infestation The occurrence of arthropods on a host.

inoculative transmission Transmission of a disease agent via a vector.

inoculative biological control   A type of augmentative biological control in which a small number of biological control agents are released, and which then reproduce and spread to suppress the pest.

insect harassment   The presence of inordinate numbers of biting insects that leads to defensive behavior by wildlife.

insect nurseries   Landscape elements, often plants, that favor production of insects as food for wildlife.

insectivores   Animals that feed on insects (arthropods).

instar   The stage of the insect between molts. The instars typically are numbered (e.g., first instar, second instar, etc.).

integrated pest management   The belief that the mere presence of pests is not a serious problem, and that the abundance of pests needs to be monitored and any actions to suppress populations of pests should be based on their potential for loss. This is also known as IPM.

integument   The body wall, which is comprised of both living and nonliving portions. It is also called the exoskeleton, reflecting the importance of the integument for support and protection. *See* Fig. 2.1.

intensity   In parasitology, this refers to the number of parasites in an infected host.

intermediate host   The animal species that the disease agent passes through during the immature or nonsexual stages of the disease life cycle.

intersegmental membranes   Soft, flexible areas of the integument located between the hardened plates.

introduced   Organisms that are not indigenous and were purposefully released.

inundative biological control   A type of augmentative biological control in which large numbers of biological control agents can be released to provide suppression.

invasive   Organisms having a tendency to spread.

invertivores   Animals that feed on invertebrates, usually crustaceans and mollusks in addition to insects.

IPM   An acronym for integrated pest management. IPM is the belief that the mere presence of pests is not a serious problem, and that the abundance of pests needs to be monitored and any actions to suppress populations of pests should be based on their potential for loss.

irritating spray   Liquid or aerosol sprays released by insects to deter predation, they are mostly effective when they come in contact with the eyes of attackers.

Johnston's organ   A mass of nerve cells in the insect antenna (the pedicel) that detects movement of the antenna.

Justinian Plague   A plague pandemic affecting the eastern Mediterranean region from AD 540–700.

juvenile hormones   Endocrine secretions of the corpora allata, causing preservation of immature (larval) characteristics and inhibiting metamorphosis.

kairomones   Allelochemicals that benefit the perceiving organism.

kissing bugs   Several species of triatomine bugs (Hemiptera: Reduviidae) that transmit American Trypanosomiasis, so named because of their tendency to bite sleeping victims on the face.

klegs   *Haematopota* spp. (Diptera: Tabanidae).

Koch's Postulates   A set of rules used to establish the relationship between a disease and its causative agent.

labium   A flap that closes the mouth cavity from below in insects with chewing mouthparts, but modified for other functions in other types of mouthparts. *See* Fig. 2.4.

labrum   A flap that closes the mouth cavity from above in insects with chewing mouthparts, but modified for other functions in other types of mouthparts. *See* Fig. 2.4.

lamellate   Broad and flat, possessing thin sheets, or leaf-like. Some antennae possess such flattened elements. *See* Fig. 2.5.

larva   (plural, larvae) Usually this refers to the immature feeding stage of holometabolous insects. Typically, larvae molt and grow three or more times before pupating, and then transform into an adult. Larvae do not resemble the adult stage. However, in Europe this term sometimes is used to refer to the immatures of hemimetabolous insects (= **nymphs**) as well.

larviporous   Deposition of larvae by females instead of eggs. The eggs are retained in the female, and larvae often receive nutrition from the female prior to being deposited.

latency   Delayed appearance. In the case of invasive organisms, the latent period is the time between introduction and detection. In the case of epidemiology, latent diseases are those that have the potential to be expressed, but are not yet evident.

lateral   Pertaining to the side.

lateral regulation Population regulation that is due to other organisms within the same trophic level. Normally, this is due to intraspecific or interspecific competition, but can also due to cannibalism.

LC_{50} value The concentration of toxicant (in milligrams per liter) that induces mortality in 50% of the test subjects.

LD_{50} value The dose of active ingredient, expressed in milligrams (mg) of toxicant per kilogram (kg) of test animal, that will kill 50% of the test subjects. Products with a low LD_{50} value are more toxic than those with a high value.

learning A change in behavior as a result of prior experience.

legs Structures adapted for walking. The insect legs are attached to the thorax. Each leg normally consists of the (basal) coxa, followed by the trochanter, femur, tibia, tarsus, and the (terminal) pretarsus. Each of the six segments contains muscles and can move independently. *See* Figs. 2.3, 2.13.

lentic An aquatic habitat consisting of standing water.

lotic An aquatic habitat consisting of running water.

lower developmental threshold The lowest temperature limits for growth.

maggot therapy The use of live maggots (Diptera: Calliphoridae) to feed on human wounds and to clean out necrotic tissue. These maggots secrete antimicrobial products and stimulate regrowth of healthy tissue.

Malaria An *Anopheles* mosquito-transmitted disease caused by protozoa in the genus *Plasmodium*.

Malpighian tubules Organs that extract waste products such as salts, amino acids, and nitrogenous wastes from the hemolymph and deposit them into the alimentary canal so they can be excreted.

mandibles Jaw-like feeding structures found on insects and the arthropods other than chelicerates that have chewing mouthparts. They are useful for masticating food as well as grasping. They are modified to provide other functions in some insects. *See* Figs. 2.4, 2.8.

Mange An epidermal (skin) disorder characterized by inflammation, blistering and thickening that is caused by infestation by mites.

Material Safety Data Sheet These provide safety and toxicity information accompanying all extremely hazardous (restricted use) materials, but also are generally accessible from internet (WWW) databases for all toxicants and hazardous materials. They also are called MSDA sheets.

maxillae A secondary pair of jaws in insects with chewing mouthparts, but modified for other functions in other types of mouthparts. *See* Figs. 2.4, 2.7, 2.8.

mechanical controls The use of mechanical devices such as barriers and trapping to disrupt pest populations.

mesothorax The second (middle) segment of the thorax. *See* Figs. 2.3, 2.11.

metabolic heat Heat that is generated from muscle activity.

metamorphosis A change in form or developmental stage.

metathorax The posterior (third) segment of the thorax. *See* Figs. 2.3, 2.11.

microbial degradation of pesticides Breakdown of a pesticide in the soil due to the action of microorganisms.

micron A unit of length equal to one-millionth of a meter.

micropyle A small opening or openings in the chorion through which the sperm enter to fertilize the egg.

midgut The second of the three principal components of the digestive system, following the foregut but preceding the hindgut. This is the principal site of digestion. *See* Fig. 2.17.

mimicry The resemblance of one organism by another. There are many forms of mimicry, the most common being Batesian, Müllerian, aggressive, and Wasmannian mimicry.

mobbing Simultaneous attack by a group. Such group action can compensate for the small size of insects.

mode of action The manner of action of a pesticide; the way that pesticides kill pests.

molluscicide A pesticide used to kill slugs and snails (molluscs).

molting fluid Enzymes that are secreted into the exuvial space and which digest the old cuticle, making the constituents available to construct a new procuticle.

moniliform A structure that is bead-like, consisting of a string of segments with regularly occurring constrictions. Some insect antennae take this form, bearing round or nearly round segments. *See* Fig. 2.5.

monogenic resistance Resistance that is under the control of a single gene.

monophagous Feeding on a single prey species.

mouth hooks The larvae of some insects, particularly flies, have hooks derived from mandibles that are used for rasping the host's tissue prior to it being ingested. The mouth hooks are the most anterior portion of the cephalopharyneal skeleton. *See* Fig. 10.20.

mosquito control impoundments Diked areas of coastal marsh where water can be maintained at a relatively constant depth, thereby preventing floodwater mosquito eggs from hatching.

MSDS An acronym for Material Safety Data Sheet. These provide safety and toxicity information accompanying all extremely hazardous (restricted use) materials, but also are generally accessible from internet (WWW) databases for all toxicants and hazardous materials.

Müllerian mimicry Evolutionary convergence of the appearance of unrelated unpalatable species.

multivoltine Having more than one generation (life cycle) per year.

Murine plague Plague occurring in populations of urban rodents.

mutualism A form of symbiosis in which the organisms are dependent on one another, and the relationship is mutually beneficial.

mycetophagous: feeding on fungus.

mycotoxin A toxin produced by a fungus, normally found in plants or plant products.

myiasis Invasion of a living animal by fly larvae.

myrmecophages Animals that specialize on feeding on ants and termites.

Nagana the name applied to African Trypanosomiasis when it affects livestock.

natural enemies Animals that suppress the abundance of pests; the natural enemies of insects are predators, parasitoids, and pathogens (diseases).

Nearctic realm A biogeographic region consisting of North America except for southern Mexico and Central America.

necrophagous Feeding on dead animals, or carrion.

nectarivorous Feeding on plant nectar.

nematicide A pesticide use to kill nematodes.

nematodes Cylindrical, unsegmented helminths in the phylum Nemathelminthes (Nematoda).

Neotropical realm A biogeographic realm consisting of South and most of Central America, including the Caribbean region.

nerve poisons chemicals that interfere with the normal functioning of nerves.

Net Primary Productivity (NPP) The rate of conversion of carbon dioxide into plant tissue less the cost of maintaining the plant.

nidicolous Inhabiting the nests of other animals.

nocturnal Active at night.

noninfectious diseases Diseases that are not caused by an infectious agent.

nonpreference A type of resistance based on the lack of attractiveness of a prospective host to a pest (= **antixenosis**).

novelty In invasion biology, this refers to the idea that invasive insects are abundant because natural enemies have not yet adapted to exploit this new resource.

numerical responses Changes in the abundance of wildlife or beneficial insects (mostly predators) that affect survival rates of prey insects.

nymph The immature feeding stage of hemimetabolous insects. Typically, nymphs molt three or more times before molting to the adult stage. Nymphs often resemble the adult stage, differing mostly by the lack of fully formed wings. In Europe, immature hemimetabolous insects sometimes are called larvae.

obligate myiasis Species of flies that develop only in living hosts.

obligate parasite Parasites that cannot continue with their life cycle unless they have access to a certain host (or hosts).

obligatory diapause A form of arrested development that is genetically determined to occur, usually requiring a certain period of time or chilling in order to terminate.

oligophagous Feeding on a restricted number of food types. For plant-feeding insects, this is usually taken to mean feeding on plants within a single plant family.

ocelli (singular, ocellus) Very small eyes located on the top of the head, providing mostly perception of light intensity. The dorsal ocellus is most prominent, but lateral ocelli (also called stemmata) occur in some insects. *See* Fig. 2.4.

ommatidium (plural, ommatidia) A visual unit, usually clustered together to form the compound eyes of insects.

omnivory Having broad feeding habits, including both plants and animals.

one-host ticks Ticks that remain on a single host for feeding.

ontogenetic shifts Changes in behavior related to development (maturation).

ootheca A protective case that is extruded by the accessory glands of females to contain and protect their eggs.

opisthosoma The posterior of the two body regions of mites and ticks (= **abdomen**).

optimal foraging theory the concept that animals will optimize their foraging behavior to maximize food intake with respect to energy expended gaining the food.

Oriental realm A biogeographic realm consisting of India and Southeast Asia through Indonesia.

osmeterium A colorful, malodorous two-pronged appendage that is extended from the head region of an insect.

outbreak An abnormal increase in abundance of insects or disease.

ovaries Part of the female reproductive system, origin of the insect ova (eggs). Each ovary usually consists of several individual ova-producing structures called ovarioles. *See* Fig. 2.28.

oviparity Reproduction by production of eggs, which are deposited by the female and which hatch some time after being laid.

ovipositor An egg laying structure.

ovoviviparity Reproduction in which the egg is retained in the female genital tract, but not provided with nutrients from outside the egg, until it hatches. Thus, a larva or nymph is deposited instead of an egg.

Palearctic realm A biogeographic realm consisting of Europe and Asia except for Southeast Asia, and including the Arabian Peninsula and northern Africa.

pandemic An exceptionally widespread epidemic.

parasitemia The level of parasites in the blood of animals.

parasitic diseases Diseases caused by macroparasites: protozoa, helminthes, and arthropods. In contrast, diseases caused by viruses, bacteria and fungi are traditionally called infectious diseases.

parasitism A form of symbiosis in one of the co-existing organisms obtains habitat or sustenance at the expense of the other.

parasitoids Parasites living on or in an insect, and that eventually kill the insect.

parasocial behavior A type of sociality that involved formation of aggregations or colonies in addition to brood care.

parthenogenesis Development of an embryo without fertilization.

passive anting Bird behavior in which the birds rest on ant-hills with their wings outstretched and allow ants to rummage through their feathers searching for lice.

passive transmission Transmission of a disease accidentally, as might occur when food or water is contaminated, and no vector is involved.

pathogenicity The ability of a disease to cause impairment or dysfunction (= **virulence**).

pectinate A structure with comb-like teeth. Some antennae have this form. *See* Fig. 2.5. *See also* **bipectinate**.

pedicel The second segment of the antenna, adjacent to the scape or basal segment.

pedipalps The second pair of appendages on chelicerate arthropods (subphylum Chelicerata). They are used as tactile organs, or in feeding or defense. They sometimes are modified into formidable weapons.

pellet The regurgitated undigested matter remaining after the meal of some raptors and other birds (= **bolus**).

perennial Requiring more than 1 year to complete a generation (life cycle).

periodic parasite A type of stationary parasite that leaves the host to spend another portion of its life in a non-parasitic mode.

peritrophic membrane A semipermeable membrane lining the midgut. The membrane protects the midgut lining from abrasion and invasion by parasitic organisms but allows digested food to pass through.

perikaryon The body of a nerve cell. Projections called axons originate at the perikaryon, and the cell nucleus is found here. The perikaryon is also called a soma. *See* Fig. 2.20.

permanent parasite A type of stationary parasite that spends all of its life on a host except for a brief period when transferring from host to host.

persistence The tendency of organisms to persist after a major decrease in abundance.

pesticide A form of poison, or toxic material, that is used to suppress pests such as insects, plant pathogens, weeds, or vertebrates that cause economic damage or are a health risk or nuisance.

petiole A constriction of the abdomen at the attachment to the thorax. This constriction allows the abdomen to have greater flexibility, which is very useful for depositing eggs or if the insect uses a sting for defense.

phenology   Growth in relation to temperature.

pheromone   A semiochemical that affects organisms of the same species.

photodegradation   Breakdown of a pesticide due to the action of sunlight (= **photolysis**).

photolysis   Breakdown of a pesticide due to the action of sunlight (= **photodegradation**).

phylogeny   The origin and evolution of an organism's lineage through time.

physical controls   The use of adverse environmental properties such as temperature or humidity to disrupt pest populations.

physical gill   A bubble of air that exchanges gasses with dissolved gasses in water. The ability of the insect to retain a physical gill is usually attributable to the presence of a plastron.

phytophagous   Feeding on plants (= **herbivorous**).

picivorous   Feeding on fish.

piercing-sucking mouthparts   Tubular mouthparts that are pointed at the apex for piercing the plant or animal host to imbibe plant sap or blood. *See* Figs. 2.6, 2.7, 2.8.

piscicide   A pesticide used to kill pest fish.

plant compensation   The ability of plants to repair damage or recover from herbivory.

plant galls   Abnormal growths and deformities of plants that are induced by and inhabited by insects (= **galls**).

plastron respiration   Respiration made possible by cuticular structures holding a film or bubble of air (a physical gill) that serves as a gas exchange site when insects are submerged in water.

plastron   Cuticular structures holding a film or bubble of air (a physical gill) that serves as a gas exchange site when insects are underwater.

pleura   (singular pleuron) Sclerites (integumental plates) found laterally (= **pleurites**).

plumose   A structure that branches and divides, resembling a feather or plume. Some antennae have this form. *See* Fig. 2.5.

Pneumonic plague   Plague transmitted by coughing or sputum.

poikilothermic   Organisms that do not actively regulate their body temperature, which is governed largely by the environment. Poikilotherms are also called 'cold-blooded' or ectothermic.

point source pollution   Pollution derived from pesticides applied to a specific location.

pollination generalists   Insects that can pollinate a number of different plants, and lack adaptations for pollinating particular plants.

pollination specialists   Insects that specialize in pollinating a particular plant species, often developing physical adaptations that facilitate pollination.

pollination   The fertilization of plants via movement of pollen. Pollen produces sperm cells that, when coming into contact with the female reproductive structure, results in fertilization. Pollen is also an important food for some insects, particularly bees.

polyembryony   More than one embryo develops from a single egg.

polygenic resistance   Resistance that is under the control of several genes.

polymorphism   A condition of having two or more discrete physical appearances, without intermediates (= **polyphenism**).

polyphagous   Feeding on a broad array of food types. In the case of plant-feeding insects, this is usually taken to mean feeding on plants from more than one plant family.

polyphenism   A condition of having two or more discrete physical appearances, without intermediates (= **polymorphism**).

population regulation   A density-related response to abundance of a host (insect prey, in this case) that causes more mortality (i.e., predation, parasitism or disease) as the host density gets higher, or lesser mortality as the host density decreases. Population regulation keeps the population from increasing to too high a level, and from exhausting its resource base. It likewise keeps the host or prey from being eliminated; total elimination is not to the advantage of the regulatory agent.

posterior   Referring to the back or anal region; opposite of anterior.

precinctive   Organisms that are naturally present (indigenous), but found only in a certain area.

predaptation   The idea that invasive insects are abundant because they are especially well adapted to feed on a particular resource or survive in a particular habitat.

predator satiation   Mass emergence of insects, ensuring that there will be too many for all to be consumed by predators.

preening   Self-cleaning behavior displayed by birds.

preference   Expressed choice of diet that does not simply reflect food availability.

preoral ingestion Secretion of saliva into the host, where some digestion occurs, prior to ingestion.

pretarsus The sixth and terminal leg segment, attached to the tarsus; the pretarsus bears the claws and sometimes other modification that aid in attachment.

prevalence (of infection) In parasitology, this refers to the proportion of the population affected. Computation is based on detection of the parasite, not enumeration of the individual parasites present.

prevalence (of consumption) When considering diet habits, this refers to the proportion (%) of a population that contain a particular food item. (= **frequency of occurrence**).

primary facultative myiasis Flies that are capable of initiating myiasis by attacking living host animals.

primary host The host species in which the disease agent passes the adult, sexual, or multiplicative stage of the life cycle (= **definitive host**).

primary productivity The production of organic compounds from carbon dioxide, normally via photosynthesis.

primer pheromone Semiochemicals that stimulate physiological effects.

prions Infectious proteins causing degenerative disease by aggregating extracellularly within the central nervous system, forming plaques that disrupt the normal tissue structure. Thus far, arthropods are not implicated in transmission of prions, which seem to be transferred only by ingestion.

procuticle The thick interior region of the cuticle beneath the epicuticle that differentiates into the exocuticle and the endocuticle.

prokaryotes Organisms lacking a cell nucleus and other membrane-bound organelles. Most prokaryotes are bacteria.

prolegs Unsegmented, leg-like structures found on the abdomen of some holometabolous larvae. Although not jointed, they assist in attachment and movement. They bear small claw-like structures distally called crochets, and aid in attachment and in walking.

prosoma The anterior of the two body regions of mites and ticks (= **cephalothorax**).

prothoracic glands Relatively large endocrine glands that occur in the thorax of insects and secrete ecdysteroids, including ecdysone.

prothoracicotropic hormone PTTH or brain hormone, a hormone that stimulates the prothoracic glands to secrete ecdysone or a similar ecdysteroid. See Fig. 2.23.

prothorax The anterior (front) segment of the thorax. See Fig. 2.11.

protocerebrum One of the three principal regions, and the largest section of the brain. It is connected to the compound eyes of insects. See Fig. 2.21.

pseudopods Abdominal extensions found on the abdomen of legless larvae that assist in locomotion. Unlike prolegs, the pseudopods lack the ability to grip the substrate, instead providing mostly traction.

PTTH Prothoracicotropic hormone or brain hormone, a hormone that stimulates the prothoracic glands to secrete ecdysone or a similar ecdysteroid.

pulses Sudden increases in the level of nutrients or energy in an area.

pupa A stage in the development of holometabolous insects that follows the larval stage but precedes the adult stage. The pupal stage does not feed; the time is used to reorganize the body into the adult form. The pupal stage often takes place in a protected location such as a cell in the soil or within a cocoon.

pupariation Pupation of a fly larva in the puparium.

puparium The integument of the last instar larva in which fly larvae pupate.

putative Purported or possible, but unproven.

questing Host seeking by ticks wherein ticks ascend vegetation and extend the front legs anticipating that an animal host will come within its grasp and allow attachment.

reflex bleeding The self-induced seeping of blood from the body (= **autohemorrhage**), usually from leg joints, in response to disturbance. The bleeding releases feeding deterrents and sometimes is accompanied by sound.

relative sampling measures Techniques that estimate the relative abundance of organisms (relative to another location or time).

releaser pheromone Semiochemicals that elicit immediate responses in behavior.

reselin A protein found in the cuticle that provides elasticity.

reservoir host An animal that harbors a disease agent.

resiliency The recovery or return of a population to equilibrium after a perturbation.

respiratory siphon   A structure that allows aquatic insects to obtain air while remaining submerged. In most mosquitoes, it is a structure located at the posterior (anal) end of larvae that is used to pierce the water surface and obtain oxygen.

respiratory trumpet   A structure located behind the head of mosquito pupae that is used to pierce the water surface and obtain oxygen.

Rickettsiae   Small gram-negative intracellular bacteria that cannot survive outside a cell.

rodenticide   A pesticide used for rodents.

Romaña's sign   Swelling of the eye region of humans, a sign of infection by American Trypanosomiasis.

rostrum   A beak-like extension of the head of weevils, with the mouthparts located at the tip of the extension. It is also called a 'beak.' *See* Fig. 2.13.

rotational impoundment management   A system of management that reduces the impact of floodwater mosquito impoundments on the local ecology but still achieves mosquito protection by preventing egg hatch. In this system, water levels need to be strictly controlled only during the mosquito breeding season, and tidal flows can be allowed during the remainder of the season.

scape   The basal segment of the antenna. *See* Fig. 2.4.

scats   The fecal deposits or excrement of mammals.

scientific name   The species name based on two Latin names (binomial nomenclature), and consisting of the genus and species. The rules for use include:

Genus name
1. The genus name is written first.
2. The genus name is always underlined or italicized.
3. The first letter of the genus name is always capitalized

Species name
1. The species name is written second.
2. The species name is always underlined or italicized.
3. The first letter of the species name is never capitalized.

Some authors use subspecies designations. These are usually geographic or color variants. Subspecies names occur third, after the species designations, but otherwise follow the aforementioned rules for species. Sometimes the author name, or the name of the person who first described the species, is included as part of the species name. If the genus name has been used in the text, subsequent references may use an abbreviation (usually first letter only, capitalized and followed by a period) to reference the genus (e.g., *D.* for *Diaphania*). If more than one genus begins with the same letter (e.g., *Aedes*, *Anopheles*, a more complex abbreviation is acceptable (e.g., *Ae.*, *An.*, respectively).

sclerites   Hardened, plate-like portions of the integument.

sclerotization   The chemical process that causes the darkening and hardening of the cuticle to form a rigid exoskeleton.

screwworms   Larvae of certain calliphorid flies (Diptera: Calliphoridae) that are parasites of wildlife and livestock. They burrow into flesh when wounds are present, but do not transmit microbial pathogens.

scutellum   A triangular plate found on the back (base of the wings) of many insects in the order Hemiptera.

scutum   A hard plate that covers all or only the anterior portion of the abdomen in ticks. Soft ticks lack an obvious scutum, whereas in hard ticks most of the abdomen typically is covered.s

secondary facultative myiasis   Flies that are capable of attacking host animals, but only after other (primary) flies have already attacked.

secondary plant substances   Non-nutritive allelochemical compounds that affect the preference and suitability of plants to herbivores.

secondary poisoning   Poisoning of nontarget animals, as when predators or scavengers feed on poisoned wildlife and also become poisoned.

secondary productivity   The rate of ingestion and assimilation of energy by consumers, less the expenditure of energy for maintenance by consumer organisms.

seed cachers   Animals that store seeds.

seed predators   Animals that feed on seeds.

seed vectors   Animals that transport seeds by contact, as when seeds cling to their fur.

self-pollination   Pollen from a plant fertilizes other flowers on the same plant.

semiochemical   Exocrine chemicals released by one organism and perceived by another.

sentinel chickens   Caged domestic chickens that are monitored for infection with arbovirus, and serve as an early-warning system of arbovirus outbreaks.

Septicemic plague Plague transmitted by fleas to a vein, and characterized by a general infection rather than localized.

serrate A toothed, notched, or saw-like structure. Some antennae have this form. *See* Fig. 2.5.

setaceous A slender structure that gradually tapers to a point; seta-like. Some antennae have this form. *See* Fig. 2.5.

shredders Aquatic organisms that feed on coarse debris, usually leaf litter. Typically, they are found in the headwaters of streams.

signal words Terminology found on the pesticide label that indicates the hazard of the product. The signal words are DANGER POISON for toxicity category I (the most toxic category), WARNING for toxicity category II, and CAUTION for both categories III and IV (the least toxic category).

Simulotoxicosis A malady caused by injection by black flies of proteins into animals, resulting in toxemia and anaphylactic shock.

siphoning   mouthparts Tubular   mouthparts adapted for ingesting liquid food. These are commonly found on Lepidoptera, and used for feeding on nectar or are water. The mouthparts are coiled when not in use and can be uncoiled to penetrate into the flower to reach the nectar. *See* Fig. 2.9.

Sleeping Sickness The name applied to African Trypanosomiasis when it occurs in humans. It is caused by various trypanosomes and transmitted by tsetse flies, *Glossina* spp. (Diptera: Glossinidae).

sociality Cooperation among individuals of a species that benefits the colony.

soft ticks Ticks that lack a scutum, or hard plate, that covers part of the body of some ticks (the hard ticks).

specialist Having a restricted range of feeding habits or occupying a narrow niche. For plant-feeding species, this is usually defined as feeding on plants within a single plant family, or even within a genus of plants.

species The fundamental unit of taxonomy (biological classification). A species is usually considered to consist of a population capable of interbreeding and producing fertile offspring.

speciose Species-rich; having many species.

spermatheca A portion of the female reproductive system that is used to store sperm until they are needed for fertilization of the eggs. *See* Fig. 2.28.

spines Immovable, rigid, pointed outgrowths of the integument. *See also* **spurs**.

spiracles Openings of the ventilatory structures, the trachea, to the outside of the insect. They are round or oval in shape, and usually bear closing mechanisms internally. Normally they are found laterally in adults. In immature insects that feed in wet environments they may be found caudally (at the back end) which the insect pushes to the surface of the feeding medium so it can accomplish gas exchange.

spirochete: Small gram-negative bacteria that are motile and free-living.

sponging mouthparts Mouthparts that are modified for ingesting liquids without penetrating the host. These are commonly found in Diptera.

spurs Movable spines, most often found on the legs. *See also* **spines**.

stadium (plural, stadia): The interval or period of time between molts. Stadia usually are measured in days.

stationary parasites Organisms that spend a definite period of development in association with the host, either on or in it.

stemmata Small eyes located on the side of the head in some immature insects. They often occur in clusters, and although capable of perceiving form their small number precludes much discrimination. They also are called lateral ocelli.

sterile insect technique A method of suppressing reproduction in an insect population using sterilized insects. Reproduction is disrupted by releasing laboratory-reared, sterilized insects into the wild, healthy population. The sterilized insects mate with the wild insects, neutralizing their reproductive ability.

sterile insect technique An insect suppression tactic that involves release of sterilized insects, usually males, that mate with fertile wild flies and neutralize their reproductive potential.

sterna (singular, sternum) Integumental plates (sclerites) found ventrally (= **sternites**).

sting A structure that is used by insects defensively. The sting is a modified ovipositor, and usually has associated with it a poison gland. Only females possess a sting and can inject biologically active substances (venoms). The sting is not used for oviposition.

stomach poison A pesticide that causes toxicity only after being eaten.

stridulatory apparatus Components of the exoskeleton of insects that provide a file and scraper, allowing the insect to produce sound.

stylate   A structure that possesses a style, a cylindrical or spine-like appendage. Some antennae have this form. *See* Fig. 2.5.

subsocial behavior   A type of primitive sociality expressed as limited brood care.

subspecies   A taxonomic unit that is subordinate to the species. Subspecies are usually geographic or color variants that have little biological significance because they can interbreed with other subspecies within the species.

sucking lice   Lice (Phthiraptera) with piercing-sucking mouthparts. Other lice are called chewing lice because they have different mouthparts and food habits.

sucking mouthparts   Modified feeding structures that allow ingestion of liquid food. Commonly they are piercing-sucking, in which case they are pointed at the apex for piercing the plant or animal host to obtain plant sap or blood. Sucking mouthparts also may be siphoning or sponging, in which case they take liquid food such as nectar that doesn't require that the host be pierced.

sutures   Linear depressions in the integument that correspond to the presence of apodemes. *See also* **apodemes**.

Sylvatic plague   Maintenance of low levels of plague in wild rodents (= **campestral plague**).

symbiosis   The association of one type of organism with another. The nature of the association is variable, and includes mutualism, commensalism, and parasitism.

sympatric speciation   Development of new species due to means other than geographic isolation. The genetic divergence of populations necessary for speciation may result from such factors as specializing on different host plants, separation in time, or the use of different pheromones. *See also* **allopatric speciation**.

symptom of disease   Subjective evidence that a disease is present, often manifested in behavior.

synapse   A microscopic gap or space between insect nerves that is bridged chemically to allow transmission of information along a series of nerves.

tagma   (plural, tagmata) The major body regions of the insect (i.e., head, thorax, abdomen). Each tagma, however, consists of individual body segments. *See* Fig. 2.3.

tagmosis   The coordinated functioning of individual body segments in tagma. See tagma.

tarsus   (plural, tarsi) The fifth segment of the leg, it is usually subdivided into up to five sections called tarsomeres. It is often called the insect 'foot,' as it has that function. It is found between the tibia and pretarsus. *See* Figs. 2.3, 2.13.

taxa   (singular, taxon) Taxonomic units, or groupings of related insects. This term may be applied to any grouping, ranging from species to much larger groupings such as families, orders and classes.

tegmina   (singular, tegmen) The thickened forewings of insects in the order Orthoptera. They are not as thick as the elytra of beetles, bear veins, and are used in flight.

temporal shifts   Changes in behavior related to date or time of day.

temporary parasites   Organisms that visit the host only briefly, usually for food.

tenaculum   A catch that secures the furcula of springtails. *See* Fig. 1.6.

teneral   An individual that has recently molted and is soft and pale in color.

terga   (singular, tergum) Integmental plates (sclerites) found dorsally (= **tergites**). *See* Figs. 2.3, 2.11.

terminal arborization   The end of a nerve cell that receives an impulse traveling along the axon, then translates it into a chemical stimulus by releasing a neurotransmitter into the synapse.

termitaria   Termite mounds. *See* Fig. 2.20.

testes   (singular, testis) Part of the male reproductive system, origin of the insect sperm. *See* Fig. 2.27.

thanatosis   The cessation of movement by an animal when disturbed, thereby pretending that it is dead (= **death feigning**).

thermal constant   A summation, in day-degrees, of time and temperature above the lower developmental threshold that provides adequate time for a phenological event.

thorax   The middle of the three tagma (body regions) in insects, between the head and thorax. The thorax is the center for movement, bearing the legs and wings, if present. The thorax consists of three segments, the prothorax, mesothorax, and metathorax, each with one pair of legs. *See* Figs. 2.3, 2.11.

threatened species   Species that are considered to be at risk of extinction, but at less risk than endangered species.

three-host ticks Ticks that take three blood meals during their development, thus requiring three hosts.

tibia (plural tibiae) The fourth leg segment, connecting the femur and the tarsus. This is one of the long segments, usually thin, and often bearing spines. *See* Figs. 2.3, 2.13.

Tick Paralysis A systemic, progressive paralysis affecting many vertebrates once a tick begins to feed. The ailment is due to injection of toxins and is reversible if the tick is removed promptly.

tolerance A type of resistance in which the host maintains its ability to grow and function in the presence of a pest.

top-down regulation Population regulation that is due to factors higher on a food pyramid or food web. Normally, this takes the form of predation and related factors such as parasitism and disease.

toxicant The active ingredient of a pesticide or poison.

toxicity The ability of a pesticide to cause injury or death.

translaminar Pesticides that move short distances, as from one side of the leaf to the other.

traumatic insemination Mating behavior in which the male penetrates the hemocoel of the female and deposits sperm; the sperm then migrate in the abdomen to the oviducts and fertilize the ova.

trehalose A disaccharide that serves as an energy store or 'blood sugar,' in insects.

trematodes Flatworms or flukes; short, flat, unsegmented animals in the phylum Platyhelminthes (Trematoda).

trichomes Leaf hairs or spines.

tritocerebrum One of the three principal regions of the brain, and connected to the mouthparts of insects. *See* Fig. 2.21.

trochanter The second leg segment, between the coxa and the femur. The trochanter often is often quite small. *See* Figs. 2.3, 2.13.

trophic cascade Change in one trophic level that may affect other trophic levels. For example, suppression of predator abundance may allow increase in abundance of herbivores.

trophic level Level of feeding in an ecosystem.

trophic pyramids Diagrams showing the feeding links between elements of an ecosystem (= **food chains** or **food webs**).

tubular proboscis Tubular mouthparts adapted for ingesting liquid food. These are commonly found on Lepidoptera, and used for feeding on nectar or water. The mouthparts are coiled when not in use and can be uncoiled to penetrate into the flower to reach the nectar.

tumblers Mosquito pupae.

tympanum (plural, tympana) An acoustical organ that functions like the eardrum of vertebrates and allows the insect to hear. It is found in various locations, depending on the taxon.

tympanum A membrane that is stretched over a cavity, allowing some insects to detect sound waves and to hear. *See* Figs. 2.3, 2.11.

univoltine Having one generation (life cycle) per year.

unnatural host Animals that are not the normal hosts for a disease (= **aberrant** or **incidental host**).

upper developmental threshold The highest temperature limits for growth.

urticating hairs Irritating or poisonous spines found on insects.

vector competency The ability of vectors to acquire and transmit diseases.

vector-borne Involvement of an arthropod (the vector) in the transmission of a disease.

venom Biologically active (toxic) substances that are injected by the sting or spines of insects.

ventilatory system The system that provides gas exchange in insects, principally the intake of oxygen and the secretion of carbon dioxide. It consists principally of a system of large and small tubes called trachea and tracheoles, respectively. The ventilatory system opens to the outside of the insect via small openings called spiracles.

ventral Pertaining to the lower surface, or underneath side, of the body; opposite of dorsal.

vertical transmission Transmission of a disease agent from parent to offspring.

virion A virus particle.

virulence The ability of a disease to cause impairment or dysfunction (= **pathogenicity**).

visual communication Signaling that depends on vision.

viviparity Reproduction in which the egg is retained in the female genital tract, and nourished, until it hatches. A larva or nymph is deposited instead of an egg.

voltinism   The number of generations (complete life cycles) that occur annually.

volume   the proportion of items in the diet expressed on the basis of area (displacement).

Wasmannian mimicry   Resemblance of non-ant insects to ants so the non-ants can gain access to ant nests.

wigglers   Mosquito larvae (= **wrigglers**).

wing veins   Tubular braces found within wings that stiffen the membranous wings. They also transport blood, which is used as a hydraulic medium to expand the wings when the insect molts to the adult stage.

wings   Rigid extensions of the integument found on the thorax of some adult insects. They usually are membranous and bear veins, and are used for flight. *See* Figs. 2.3, 2.12.

wrigglers   Mosquito larvae (= **wigglers**).

yellow flies   Some deer flies with yellowish bodies, primarily *Diachlorus ferrugatus* (Diptera: Tabanidae).

INDEX

aardvark, 114, 164, 317
aardwolf, 164
abdomen, 6, 43, 289
abdominal muscles, 47
aberrant host, and disease, 239
abiotic population regulation, 238
absolute sampling measures, 87–91
abundance of insects, determination,
    86–93
Acacia paniculata, 124
Acacia repanolobium, 402
Acanthocephala, 274
acanthocephalans, 274
Acari, 6–8, 285, 289–298
acaricide, 343
Acarina, 6–8, 285, 289–298
Accipiter nisus, 354
Accipitridae, 151
Acer rubrum, 226
Acer saccharum, 226
acetylcholine, 52
acetylcholine esterase, 52, 346
acetylcholine esterase inhibition by
    pesticides, 354
Acheta domestica, 99, 100, 416
Achurum sumichrasti, 184
Achyra rantalis, 68
Acinonyx jubatus, 252
Acipenseriformes, 152, 155
acoustical communication, 60
Acrididae, 181, 184, 393, 425
Acris crepitans blanchardi, 108
Acromyrmex, 202, 210
Actinopterygii, 152
action threshold, 381
actions, group, 163
active anting, 301
active transmission of disease, 240
activity, nocturnal, 165
acute effects of insecticides, 350
acute toxicity of pesticides, 344
acute toxicity of disease, 241
Adelges tsugae, 226
Adelgidae, 226

Adelphocoris lineolatus, 187
adipose tissue, 47
adjuvants, of pesticides, 344
adsorption of pesticides, 349
adventive, 219
Aedeomyia squamipennis, 271
Aedes, 250, 276, 308
Aedes aegypti, 250
Aedes sollicitans, 305
Aedes trivittatus, 308
Aedes vexans, 305, 307
Aegolius arcadius, 151
Aepyceros melampus, 210, 268
aflatoxin, 259
    poisoning, 259, 261
African buffalo, 252
African elephant, 325, 328,402
African hedgehog, 289
African swine fever, 291
African Trypanosomiasis, 266–270,
    316–318, 403–404
African wood owl, 151
Africanized honey bee, 226, 333
Afrocimex, 304
Afroscoricida, 114
Agathis gibbosa, 377
age of insects, 28
Agelaius phoeniceus, 132, 143, 148, 157,
    291, 360, 392
Ageniaspis citricola, 380
agent, etiologic, 241
aggregation pheromones, 373
aggressive mimicry, 160
Agkistrodon piscivorus, 158
agoutis, 265, 303
Agrilus planipennis, 226
Agrius cingulatus, 193
Agromyzidae, 192, 194
Agrotis infusa, 414
air sacs, 51
Aix sponsa, 92, 136, 150, 287
Akodon arviculoides, 251
Alabama sturgeon, 141, 153, 155
alatae, 41

Alaudidae, 142
albatross, 126
Alcelaphus buselaphus, 268
Alces alces, 236, 277, 279, 296, 326, 396
alderfly, 14, 22, 176–177, 432
alewife, 407
Aleyrodidae, 371
alfalfa weevil, 366
algaecides, 343
alimentary canal, 47, 49
alimentary system, 47
allata, corpora, 55
allelochemical compounds, 371
allelochemicals, 55, 61, 373
Alligator mississipiensis, 106, 107,
    112,114, 250, 354, 398
alligator, 106–107, 114, 398
    American, 112, 249, 354
allomones, 61
Alloneobius griseus, 185
allopatric speciation, 5
Alloptes, 287
Alophoixus phaeocephalus, 396
Alouatta, 250
Alouatta nigerrina, 251
Alouattamyia, 328
Alpine ibex, 291
alternatives to insecticides, 366–382
amakihi, Hawaii, 271
Amazona vittata, 318
Amblyomma, 253, 293
Amblyomma maculatum, 287, 294, 434
Amblyomma tuberculatum, 287, 294
Amblyseius californicus, 380
Amblyseius cucumeris, 380
ambrosia beetles, 202, 216
Ambrysus amargosus, 421
Ambystoma maculatum, 400
Ameiurus melas, 153
Ameiurus nebulosus, 141
American alligator, 112, 249, 354
American badger, 291, 332
American bald eagle, 298
American beaver, 253

American bison, 114, 119, 123, 215, 254, 272, 296, 301, 318
American bobcat, 271, 272, 276, 289, 291, 297, 313, 314, 332
American burying beetle, 419
American coot, 138, 150
American crow, 132 249
American dog tick, 295, 297
American goldfinch, 393–394
American kestrel, 151, 354, 359
American redstart, 390
American robin, 130, 149, 249, 251, 291, 352
American shad, 407
American three-toed woodpecker, 128
American toad, 359
American Trypanosomiasis, 263–267, 302–303
American white pelican, 319
American woodcock, 138, 143, 147, 151
ametabolous, 63–64
amethyst sunbird, 401
Amitermes, 401
Amitus herperidum, 380
Ammodramus savanarrum, 98
Ammospermophilus leucurus, 100
Amphibia, 105
amphibian and reptile diets, 105–113
amphibian, 105, 108–113
amphipod, 6, 9
Amphisbaena alba, 106
amphisbaenians, 106
amplifier host, and disease, 239
Amylostereum areolatum, 216
anadromous, 155
Anagyrus kamali, 380
Analloptes, 287
anamorphosis, 12
Anaphe venata, 99
Anaplasma, 252–254
Anaplasma marginale, 254
Anaplasmosis, 254, 291
Anas acuta, 261
Anas crecca, 249
Anas discors, 143, 147
Anas gibberifrons, 136, 150
Anas penelope, 249
Anas platyrhynchos, 261, 272
Anaticola crassicornis, 287
Anatidae, 150
Anatoecus dentatus, 287
anatomy, internal, 45, 48
anautogeny, 78, 305
anchovies, 152
Andinocuterebra, 328
Androlaelaps fahrenholzi, 239
Andros iguana, 212
anemia, 332

angel insect, 14
Anguilliformes, 152
animal ectoparasites, predation by birds, 395–396
Anisolabis maritime, 206
Anisomorpha buprestoides, 162
Anisoptera, 173
Anna's hummingbird, 128, 143
anole, green, 396
Anolis, 394
Anolis bimaculatus, 394
Anolis carolinus, 397
Anolis wattsi, 394
Anolisomyia, 325
Anopheles, 264, 276, 404
Anopheles quadrimaculatus, 307
Anoplolepis gracilipes, 226, 398
Anoplophora glabripennis, 226
Anser albifrons, 261
Anseriformes, 126
ant, 15, 24, 62, 192–193, 195–196, 206, 217, 333–334, 401
   Argentine, 226, 398, 417
   army, 398
   big-headed, 398
   destruction of wildlife by, 397–400
   driver, 398
   fire, 400
   harvester, 398
   leafcutter, 416
   red imported fire, 397–400
   tropical fire, 398
   yellow crazy, 398
ant mound, 210
ant-acacia symbiotic relationship, 401–402
antbirds, 142
anteater, 115, 249, 314, 249, 314
   giant, 164
   scaly, 164
   silky, 164
   spiny, 164
antelope, 254, 268, 312
   blackbuck, 291
   goitered, 328
   Saiga, 252, 272, 328
antelope ground squirrel, 102
antennae, 6, 37, 39, 40
Antheraea polyphemus, 161
Anthomyiidae, 194
Anthonomus grandis, 375
antibiosis, 372
anticoagulants, blood, 355
Antilocapra americana, 215, 252, 254, 272, 311, 313
Antilocapridae, 114
Antilope cervicapra, 291
anting, active, 301

anting, passive, 301
antixenosis, 372
antlions, 15, 22, 188–189
Antricola, 293
antwren, checker-throated, 394
Anura, 105
aorta, 50
Aotus, 250
Apanteles solitarius, 380
ape, 114
Aphanius dispar, 396
Aphelocoma coerulescens, 332
Aphididae, 187–188
Aphidius smithi, 380
Aphidoletes aphidimyza, 380
aphids, 187–188
Aphis fabae, 57, 187
Aphrania, 304
Aphrodiinae, 203
Aphytis holoxanthus, 380
Apidae, 195–196, 226
Apionacarus wilsoni, 287
Apis mellifera, 99, 196, 220, 410, 413, 422
Apis scutellata, 226, 333
apitherapy, 416
Aplocheilus panchax, 396
Apodemes flavicollis, 254
Apodemes sylvaticus, 254
apodemes, 34
Apodemia mormo langei, 420
Apodemia agrarius, 297
Apodemia specious, 297
Apodiformes, 126
Apollo butterfly, 423
apolysis, 35
aposematic coloration, 159, 163
aposematism, 159
apparatus, stridulatory, 60
appendage extension, 160
appendages, caudal, 43
apterae, 41
Apterygota, 14
Apus apus, 250
aquatic collectors, 75
aquatic grazers, 75
aquatic insects, 74, 200
   as wildlife food, 171–179, 200
   as predators, 75
   as shredders, 75
Aquila chrysaetus, 353
Arachnida, 6–8, 285
Araneae, 6–8
arborization, terminal, 52
arboviruses, 245–261
   Avian Pox, 248
   Hemorrhagic Disease, 251
   Hemorrhagic Fever, 247

Myxomatosis, 248
St. Louis Encephalitis, 250
West Nile Virus, 249
Yellow Fever, 250
Archeognatha, 14
Archilochus colubris, 128, 143
Archips semiferana, 198
Arctic charr, 155
Arctic grayling, 152
Arctiidae, 192
Ardea cinerea, 249
Ardeicola loculator, 287
Ardeidae, 151
area-wide insect management, 379–381
Argas, 253, 293
Argas cooleyi, 304
Argas persicus, 252, 252, 294, 296
Argas radiatus, 287
Argasidae, 285
Argentine ant, 226, 398, 417
Argidae, 196
Argiope aurantia, 397
Armadillidium, 9
armadillo, 115, 265, 303
    nine-banded, 118, 120, 123, 238, 251,
        313
army ant, 398
armyworm, 192
Artemesia, 428
Arthopoda, 6
arthralgia, 247
arthropod sampling techniques, 86–91
Artiodactyla, 114
Artogeia rapae, 223–225, 394
Arviocola, 253
Arviocola terrestris, 355
Asclepias, 424
Ash Meadows naucorid, 421
ash-throated flycatcher, 130, 215
Asian longhorn beetle, 226
Asian water buffalo, 254
Aslidae, 192, 194
Aspergillus, 259
Aspergillus flavus, 259
Aspergillus parasiticus, 259
Aspergillus spp. 259, 261
assassin bugs, 285, 302
Astigmata, 285
Atala butterfly, 424
Atelerix albiventris, 289
Ateles, 250
Atelocerata, 6
Athene cunicularia, 151, 332, 353
Atlantic bluenose dolphin, 272
Atlantic puffin, 254
Atlanticus monticola, 185
Atles paniscus, 251
Atta, 106, 202

Atta mexicana, 99
Auchmeromyia, 322
Audubon's cottontail rabbit, 251
augmentative biological control, 375
Australian realm, 32
Austrosimulium ungulatum, 310
autogeny, 78, 305
Automeris io, 164
autonomy, 162
autotrophs, 198
availability, food, 86
Avian Botulism, 258–260
Avian Filariasis, 275, 301
Avian Malaria, 264, 269–271
Avian Pox, 248
avicide, 343
Avicularia, 397
avocet, 259
axon, 52
Aythya affinis, 136, 150
Aythya americana, 136, 150
Aythya valisineria, 136
Azteca, 401

Babesia, 264
Babesiosis, 264
    Bovine, 291
baboon, 249
baby, bush, 249
Bacillus thuringiensis, 308, 310, 349, 373,
        378, 418
backswimmer, 176, 395, 400
bacteria, 252
bacterial diseases, 252–260
    Anaplasmosis, 254
    Avian Botulism, 258–260
    Lyme Disease, 254–256
    Plague, 255–258
    Tularemia, 252–254
bactericides, 343
badger, 114
    American, 119, 123, 291, 332
    European, 288
Baeolophus bicolor, 134, 150, 249
Baetidae, 200
Bahama swallowtail butterfly, 419
Bakerella, 124
balance of nature, 375
bald eagle, 352–354, 359
    American, 298
ballooning, 8
Baltimore oriole, 132
banded rock rattlesnake, 112
bandicoot, 115
bank swallow, 134, 332
bank vole, 254, 297
Banksiola dossuaria, 400
banteng, 291

bark beetle, 202, 216, 221, 425
barklouse, 14, 20, 186
barn owl, 272
barn swallow, 134, 249, 286
barnacle, 9
barracuda, 152
basement membrane, 34–35
bass, largemouth, 140, 153, 155
bat bug, 285, 303–304
bat, 114–115, 122, 125, 394, 425
    big brown, 115, 251
    Brazilian free-tailed, 115, 116, 120,
        125
    golden-tipped, 116
    greater spear-nosed, 116
    hunting by, 165
    little brown, 251
    pale spear-nosed, 116
    Pallas's long-tongued, 116
    Seba's short-tailed, 116
    trawling long-fingered, 120
    white-lined, 115–116
Batesian mimicry, 160
bathing, 300
bath, dust, 301
Bathyplectes curculionis, 380
Batrachomia, 333
Batrisodes texanus, 419
Batrisodes venyivi, 421
bay checkerspot butterfly, 419
bay-breasted warbler, 392
Bdellorhynchus, 287
bear, 114, 125, 276, 353
    black, 119, 120, 125, 238, 249, 276,
        292, 296, 353
    grizzly, 125
    sloth, 164
beardless-tyrannulet, northern, 401
beaver, 115, 293
    American, 253
beaver cave beetle, 420
bed bug, 234, 285, 303–304
Bedellia orchilella, 193
bee, 15, 24, 192–193, 195–196,
        333–334
    Africanized honey, 333
    social, 62
bee-eater, 143
    rainbow, 397
beetle, 15, 22, 176, 189–190, 400
    American Burying, 419, 429
    click, 371
    Coffin Cave mold, 419
    Comal Springs dryopid, 419
    Comal Springs riffle, 419
    conservation of, 425–426
    delta green ground, 419
    Hungerford's crawling water, 419

beetle (*continued*)
    Kretschmarr Cave mold, 419
    Mount Hermon June, 419
    scarab, 371
    southern pine, 368
    Tooth Cave ground, 419
    valley elderberry longhorn, 419
    Warm Springs Zaitzevian riffle, 419
    western pine, 392
    ground, 427
    rove, 427
    threatened and endangered, 419
    wood boring, 425
beetle bank, 427
behavior, defensive, by hosts, 234
    eusocial, 61
    feeding, 233
    flight and startle, 160–162
    nest fumigation, 288
    parasocial, 61
    subsocial, 61
behavioral fever, 243
Behren's silverspot butterfly, 419
Belostomatidae, 175–176, 400
Beltian bodies, 401
beneficial virulence, 237
benefits of insects for wildlife, 156
Bengal tiger, 271
Berosus, 178
Besnoitia besnoiti, 264, 315
Besnoitia tarandi, 264
Besnoitiosis, 264
Betula lenta, 226
big brown bat, 115, 251
big-headed ant, 398
bighorn sheep, 236, 252, 254, 289, 311, 312, 318
bigmouth buffalo sucker, 141, 153
binomial nomenclature, 5
bioaccumulation of pesticides, 345
biocide, 346
bioconcentration of pesticides, 345
biodegradation, enhanced, of pesticides, 350
biogeography, 32
    ecological, 32
    historical, 32
biological control, 375–379
    augmentative, 375
    classical, 379
    conservation, 379
    inoculative, 379
    natural, 375
biology, thermal, 69
biomagnification, 357–358
biomagnifications of pesticides, 345
biotic population regulation, 238
bird diets, 126

bird mite, 292
bird stomach contents, 128–139
bird, perching, 142
    song, 142
birdwing butterfly, Queen Alexandra's, 423
Bison bison, 215, 252, 254, 272, 296
bison, American, 114, 119, 123, 215, 252, 254, 272, 296, 301, 318
biting house fly, 319
biting midges, 194, 285, 310–312
Blaberidae, 183
black bear, 119, 120, 125, 216, 238, 249, 276, 292, 296
black bullhead, 153
black caracara, 396
black crappie, 140, 153
Black Death, 255
black fly, 177, 180, 192, 199, 276, 285, 305, 308–310
Black Fly Fever, 309
black phoebe, 130
black rail, 398
black rat, 254
black rat snake, 158
black tern, 136, 150
black turpentine beetle, 216
black-backed gull, 206
black-backed oriole, 424
black-billed magpie, 249, 352–353, 396
black-capped chickadee, 226, 432
black-capped vireo, 399
black-footed ferret, 257, 356
black-headed grosbeak, 424
black-headed gull, 249, 272
black-tailed native-hen, 301
black-throated green warbler, 392
blackbird, 142, 149
    red-winged, 132, 143, 148, 157–158, 291, 360, 392
    rusty, 132
    yellow-headed, 132
blackbuck antelope, 291
Blackburn's sphinx moth, 421
blackburnian warbler, 392
blackcap, 249
blacklegged tick, 297–298
Blackline Hawaiian damselfly, 420
blacktailed prairie dog, 215
Blanchard's cricket frog, 108
Blarina brevicauda, 116, 120, 392
Blattodea, 14, 16, 180–181, 183
bleeding, reflex, 162
blind snake, 107
Blissus insularis, 368
blood anticoagulant, 355
blood cells, 50
blood-feeding insects, 78

blood-sucking conenose bug, 285, 302–303
blow fly, 192, 285, 321–324
blue bottle flies, 321
blue jay, 132, 143, 148, 226, 249, 251
blue tit, 332
blue-gray gnatcatcher, 398
blue-winged teal, 143, 147
bluebird, 149
    eastern, 130, 249, 318, 322, 360, 429
    mountain, 130
    western, 130
bluegill, 140, 153, 155
boar, wild, 355
bobcat, 238, 257
    American, 271–272, 276, 289, 291, 297, 313, 314, 332
bobolink, 132, 143, 149
bobwhite, northern, 136, 146, 150, 248, 251, 334, 354, 398–399, 429–430
bodies, Beltian, 401
body regions, 35
body fat, 47
bogong moth, 414–415
Boloria acrocnema, 420
bolus, 91
Bombus, 42
Bombycilla cedrorum, 143, 145
Bombycillidae, 142
Bombyx mori, 99, 220, 414–415
Bonaire whiptail lizard, 110
Bonasa umbellus, 143, 146, 216
booby, Peruvian, 288
booklice, 14, 20, 186
Boophilis, 293
Boophilis annulatus, 294
boring insect, 76
Borrelia, 253, 296
Borrelia anserina, 253
Borrelia burgdorferi, 253–256
Borreliosis, Lyme, 254–256
Bos javanius, 291
bot fly, 325–329, 285
bot, nose, 326
    pharyngeal, 326
    rodent, 328
    stomach, 326
bottle fly, 321
    blue, 321
    green, 321
bottom-up regulation of populations, 390
Botulism, Avian, 258–260
Bovicola sedecimdecembrii, 301
Bovidae, 114
Bovine Babesiosis, 291
box turtle, 398
Boydaia colini, 287

brachypterous, 41
Brachystola magna, 184
Bradypodidae, 115
Bradypus tridactylus, 251
brain, 53
Branchiopoda, 9
Branta canadensis, 355
Brazilian free-tailed bat, 115, 116, 120, 125
Brazilian skink, 110
Brazilian tapir, 396
bristletails, 14
broad-headed skink, 110, 396
broad-shouldered water strider, 175
broad-winged hawk, 151
Brotogeris, 210
brown bullhead, 141
brown lacewing, 189
brown lemur, 124
brown mouse lemur, 124
brown pelican, 354
brown thrasher, 134, 149
brown trout, 140, 154
brown-headed cowbird, 132
Brown's leopard frog, 108
brownheaded cowbird, 291
Brugia malayi, 275
Brychius hungerfordi, 419
Bubalus bubalis, 254
Bubonic Plague, 255–258
Bubulcus ibis, 138, 150, 249–250, 396–397
Bucephala clangula, 147, 395
Bucimex, 304
budgets, time, 93
budworm, eastern spruce, 392
    western spruce, 392
buffalo, African, 252
    cape, 254
Bufo 398, 400
Bufo americanus, 359
Bufo japonicus, 108
Bufo marinus, 108, 396
Bufo marmoreus, 108
Bufolucialia, 322
bug, 14, 175–176, 186–188, 302–304, 432
    assassin, 285, 302
    bat, 285, 303–304
    bed, 285, 303–304
    blood-sucking conenose, 285
    cimicid, 287
    cliff swallow, 304
    kissing, 302–303
    martin, 303
    southern chinch, 368
    swallow, 285, 303–304
    Wekiu, 419

bulbul, yellow-bellied, 396
bulk of diets, 93
bullhead, 155
    black, 153
    brown, 141
bumble bee, conservation, 422–423
bunting, 142
    painted, 134, 150
Buphagus africanus, 395
Buphagus erythrorhynchus, 395
Buprestidae, 226, 425
burbot, 154, 155–156
burrowing owl, 332, 353
Bursaphelenchus cocophilus, 217
Bursaphelenchus xylophilus, 216
burying beetle, American, 424
bush baby, 249
bushbuck, 210, 268
bushpig, 268, 317
Buteo albicaudatus, 151
Buteo buteo, 355
Buteo jamaicensis, 287, 359
Buteo lineatus, 151
Buteo platypterus, 151
Buteo swainsoni, 359
butterfly, 15, 24, 190–192
    conservation of, 423–424
    gardening, 428
butterfly, Apollo, 423
    Atala, 424
    Bahama swallowtail, 419
    bay checkerspot, 419
    Behren's silverspot, 419
    cabbage, 394
    callippe silverspot, 419
    Corsican swallowtail, 419
    El Segundo blue, 419
    Fender's blue, 420
    Homerus swallowtail, 420
    Karner blue, 420
    Lange's metalmark, 420
    lotus blue, 420
    Luzon peacock swallowtail, 420
    Mariana eight-spot, 420
    Mariana wandering, 420
    mission blue, 420
    Mitchell's satyr, 420
    Monarch, 423–424
    Myrtle's silverspot, 420
    Oregon silverspot, 420
    Palos Verdes blue, 420
    Queen Alexandra's birdwing, 420, 423
    Quino checkerspot, 420
    Sacramento Mountains checkerspot, 420
    Saint Francis' satyr, 420
    San Bruno elfin, 420
    Schaus swallowtail, 420

    Smith's blue, 420
    Uncompahgre fritillary, 420
    Whulge checkerspot (=Taylor's), 420
buzzard, 355

cabbage butterfly, 394
cabbage looper, 394
cacher, seed, 218
Cacodmus, 304
Cactoblastis cactorum, 226
cactus moth, 226
caddisfly, 15, 22, 177, 179, 182, 199, 400, 432
    Sequatchie, 420
caecilian, 105
caiman, 106
Calamospiza melanocorys, 393
Calcarius mccownii, 351, 393
Calidris acuminata, 138, 150
Calidris canutus, 138, 150
Calidris himantipus, 138, 150
Calidris melanotos, 138, 150
Calidris pusilla, 138, 147, 150
California condor, 298
California quail, 136, 150
California vole, 257
California woodpecker, 128
Callipepla californica, 136, 150
Calliphora, 322
Calliphoridae, 192, 203–204, 206, 285, 431–324
callippe silverspot butterfly, 419
Callithrix aurita, 124
Callophrys mossii bayensis, 420
Callostoma fascipennis, 376
Calomys musculinus, 251
Calypte anna, 128, 143
Calyptra, 78
Cambrian period, 27, 29
camel, 396
Camelidae, 114
camel, 114, 328
Camelus dromedaries, 396
Campestral Plague, 255–258
Campostoma imberbe, 401
Campylobacter, 321
Campylorhynchus rufinucha, 401
Canada goose, 355
Canada lynx, 257, 276
canal, alimentary, 47, 49
canaries
cane toad, 108, 396
Canidae, 114
Canis, 276
Canis latrans, 119, 123, 251, 257, 272, 276, 292, 297
Canis lupus, 276, 292, 296
Canis lupus baileyi, 119, 123

Canis lupus dingo, 292, 397
cannibalism, 79
canvasback duck, 136, 150
Caparinia, 289
Caparinia erinacei, 289
Caparinia tripilis, 289
cape buffalo, 254
Capra, 292
Capra ibex, 252, 291–292
Capra nubiana, 291
Capreolus capreolus, 279, 292, 355
Caprimulgiformes, 126
Caprimulgus vociferous, 126
capsid, 245
capuchin, 124
Carabidae, 189, 191, 427
Caracara cheriway, 151
Caracara plancus, 398
caracara, black, 396
caracara crested, 151
Carassius auratus, 396
Carboniferous period, 29
  Early, 29
  Late, 29
carcass decomposition, 204–208
Carcinophoridae, 186
cardiaca, corpora, 55
cardinal, 142
  northern, 134, 150, 251, 394
  red-crested, 394
Cardinalidae, 142, 150
Cardinalis cardinalis, 134, 150, 251, 394, 398
Cardinalis sinuatus, 134, 150
Carduelis tristis, 393
Caretta caretta, 398
caribou, 114, 289, 305, 325–326, 328–329
Carnivora, 114
carnivory, 85
Carolina chickadee, 146
Carolina wren, 136 143, 145, 150
carp, 433
Carpodacus mexicanus, 134, 150, 251
Carpoides cyprinus, 141
carps, 152
carrion-feeding insects, 204–208
carrion, decomposition, 204–208
Carson wandering skipper, 421
Carson, Rachel, 341
cascade, trophic, 198
cassowaries, 142
Castor canadensis, 253, 293
Castoridae, 115
cat, 114, 123, 257, 265, 291
cat flea, 279, 331
Catasticta teutila, 99
catbird, gray, 134, 149

caterpillars, 190–192
catfish, 152, 155
Cathartes aura, 298
Catharus fuscescens, 130, 143
Catharus guttatus, 132
Cathraus ustulatus, 130
Catoblephas gnu, 268
Catocala micronympha, 158
Catostomus commersoni, 141, 154
cattle, 254, 268
cattle egret, 138, 150, 249, 250, 396
caudal appendage, 43
Caudata, 105
causes of disease, 241
cave beetle
  beaver, 420
  Clifton, 420
  greater Adams, 420
  Holsinger's, 420
  icebox, 420
  inquirer, 420
  lesser Adams, 420
  Louisville, 420
  surprising, 420
  Tatum, 420
cave catfish, 141
cave tick, 295
Cebus, 10
Cebus apella, 124
Cebus nigrivittatus, 124
cedar waxwing, 143, 145
cells, 41
  epidermal, 34
cellulose digestion, 180, 201–202
Cenozoic era, 27, 29
Centaurea maculosa, 430
centipede, giant, 396
centipedes, 6, 11
central nervous system, 52
Centruroides, 8
cephalopharyngeal skeleton, 318
Cephalopina, 328
cephalothorax, 6, 289
Cephenemyia, 325–326, 328–329
Cerambycidae, 189, 221, 226, 425
Ceratitis capitata, 68
Ceratocystis, 216
Ceratocystis fagacearum, 216
Ceratophyllus, 287
Certatophyllus celsus, 332
Ceratophyllus ciliatus, 332
Ceratophyllus gallinae, 331–332, 360
Ceratopogonidae, 194, 285, 310–312
Ceratotherium simum, 317, 403
cerci, 43
Certhiidae, 142
Cervidae, 114
Cervus canadensis, 277, 292, 296, 326

Cervus elaphus, 236, 252, 254, 289, 292
Cestoda, 274
cestode, 274
Chagas Disease, 263–267
chains, food, 198
Chalarodon madagascariensis, 353
chamber, filter, 50, 69
chamois, 292
chancre, 268
Chaoboridae, 395
Chaoborus, 395
Charadriidae, 150
Charadriiformes, 126
Charadrius vociferous, 136, 150, 352
charr, Arctic, 155
checker-throated antwren, 394
checkered white, 193
cheese skipper, 204
cheetah, 114, 252
Cheilospirura spinosa, 275
chelicerae, 6, 289–290
Chelonia mydas, 398
chemical communication, 60
chemical defenses, 162
Chen caerulescens, 261
Chen rossi, 261
Chenopodium, 357
chestnut blight, 216
chewing lice, 14, 20, 279, 287, 298–302
chewing mouthparts, 36–37
Cheyletiella parasitvorax, 291
Cheyletiellosis, 291
chickadee, 142, 249
  black-capped, 226, 432
  Carolina, 146
chicken, prairie, 146
chickens, sentinel, 248
chigger mite, 290–291
Chilean flamingo, 298
Chilopoda, 6, 11
chimpanzees, 124–125, 249, 292
Chinook salmon, 407
chipmunk, 115
  eastern, 234
  least, 118, 251
  lodgepole, 117, 120
  Siberian, 297
  yellow pine, 117, 120
chipping sparrow, 134, 143, 145, 393–394
Chironomidae, 177, 191–192, 200, 432
Chiroptera, 114
chitin, 34
Chlamydia, 321
Chlidonias niger, 136, 150
Chlorocebus aethiops, 250

Chlorochroa sayi, 188
Chloropidae, 333
Chondestes grammacus, 98
Chondrichthyes, 152
chorion, 66
Chorioptes bovis, 289
Choristoneura, 215
Choristoneura fumiferana, 392
Choristoneura occidentalis, 392
chronic toxicity of pesticides, 344
chronic toxicity, of disease, 241
Chrysocharis laricinellae, 380
Chrysomelidae, 190–191
Chrysomya, 332
Chrysoma bezziana, 322, 324–325
Chrysopa microphyta, 376
Chrysopidae, 189
Chrysops, 275, 314, 315
Chrysops callidus, 315
chum salmon, 407
cicadas, 157–158, 187–188
Cicadellidae, 188
Cicadidae, 187–188
Cicindela dorsalis dorsalis, 421
Cicindela highlandensis, 421, 425
Cicindela limbata albissima, 421
Cicindela nevadica lincolniana, 421, 425
Cicindela ohlone, 421, 425
Cicindela puritana, 421, 425
Ciconiiformes, 126
Ciconiphilus quadripustulatus, 287
Cicrocebus murinus, 124
Cimex, 234, 303–304
Cimex hemipterus, 303
Cimex lectularius, 303
Cimexopsis, 304
cimicid bugs, 287
Cimicidae, 285, 302–304
Cingilia catenaria, 192
Cingulata, 115
circulatory system, 50
Circus cyaneus, 151
Cirripedia, 9
Cistudinomyia cistudinis, 325
Citellus beecheyi, 332
Citellus lateralis, 117, 120
Citellus variegates, 313
CITES, 418, 423
Citheronia regalis, 192
classical biological control, 379
classification of Arthropoda, 6–27
classification of insects, 14–27
clay-soil ctenotus skink, 110
Cleridae, 205
Clethrionomys glareolus, 254, 297
Clethrionomys rufocanus, 297
click beetle, 191, 371
cliff swallow, 134, 304, 332

cliff swallow bug, 304
Clifton cave beetle, 420
clinical sign of disease, 241
Clostridium, 258
Clostridium botulinum, 253, 258–260
Clupeiformes, 152
Cnemidophorus murinus, 110
coatis, 114
Cobboldia, 328
Coccyzus americanus, 132, 149
cochineal scale, 414
Cochliomyia hominivorax, 322–324, 329
cockroaches, 14, 16, 180–181, 183, 394
coconut palm nematode, 217
cods, 152
Coenagrionidae, 174
Coendou, 265, 303
Coffin Cave mold beetle, 419
coho salmon, 141, 154, 155, 407
coincidental virulence, 236
Colaptes auratus, 128, 143
Colaptes cafer, 144
Coleoptera, 15, 22, 176, 189–190, 400, 432
Colias eurytheme, 68
Colinicola numidiana, 287
Colinicola virginianus, 287
Colinicola wilsoni, 287
Colinicola cubanensis, 287
Colinus virginianus, 136, 146, 150, 248, 251, 287, 334, 354, 398–399, 429–430
collared peccary, 118
collectors, aquatic, 75
Collembola, 6, 12
collophore, 12
colonial nesters, 286
Colorado Tick Fever, 291
coloration
    aposematic, 159, 163
    disruptive
    flash, 160
Colpocephalum mycteriae, 287
Colpocephalum napiforme, 287
Colpocephalum occidentalis, 287
Colpocephalum scalariforme, 287
Colpocephalum unciferum, 287
Coluber constrictor mormon, 110
Columba livia, 272
Columbiformes, 126
Columbina passerine, 398
Comal Springs dryopid beetle, 419
Comal Springs riffle beetle, 419
commensalism, 241, 400
common goldeneye, 147
common grackle, 132, 291
common nightingale, 249
common shrew, 253

common swift, 249
common vole, 253
common whitefish, 154–156
common yellowthroat, 432
communication, 58
    acoustical, 60
    chemical, 60
    visual, 60
compensation
    ecological, 427
    plant, 213–214
competence, vector, 240
competition-based population regulation, 238
compound eye, 37, 53–55
compounds
    allelochemical, 371
    digestibility-reducing, 202
condor, California, 298
conenose, blood-sucking, 302–303
Connochaetes, 254
Connochaetes taurinus, 292
conservation biological control, 379
    beetle banks for, 427
    headlands for, 427
    of beetles, 425–427
    of bumble bees, 422–423
    of butterflies, 423–424
    of insects, 387–397, 416–434
    of wildlife, 397–407, 425–434
    status, 418–419
constant, thermal, 70
Constrictotermes cyphergaster, 211
consumption, frequency, 93
contact poisons, 345
contagious, 235
contamination of water bodies by pesticides, 353
Contopus cooperi, 130
Contopus virens, 130, 423
control
    biological, 375–379
    cultural, 366–370
    mechanical, 370–371
    natural biological, 375
    physical, 370–371
coontie, 424
coot, 142, 259, 301
    American, 138, 150
copepod, 9
Copepoda, 9
Cophosaurus texanus, 110
coprophagous, 321
coprophagy, 202
Coraciiformes, 126
Coral Pink Sand Dunes, tiger beetle, 421
Cordylobia, 322
Coregonus lavaretus, 154–156

Corixidae, 176
cormorant, 126, 259, 330
Cornitermes cumulans, 211
corpora allata, 55
corpora cardiaca, 55
Corsican swallowtail, butterfly, 419
Cortaritermes silvestri, 99
Corvidae, 142, 149
Corvus, 250
Corvus brachyrhynchos, 132, 249
Corvus hawaiiensis, 272
Corvus ossifragus, 249
Corvus rhipidurus, 396
Corydalidae, 176–177, 400
Corynebacterium, 330
Cotinus, 283
cotton mouse, 238, 251
cotton rat, 251, 400
cottonmouth snake, 158
cottontail, eastern, 254
cottontail rabbit, 238, 253
cougar, 272
cowbird, brown-headed, 132, 149, 291
Cowdria ruminantium, 253
coxa, 36, 41, 46
coyote, 119, 123, 251, 257, 272, 276,
     291–291, 296–297, 301, 332, 353
crab, 6, 9
crane fly, 177, 192, 194
     sandhill, 429
cranes, 142, 406
crappie
     black, 138, 153
     white, 153
Craspedorrhynchus americanus, 287
Crassicimex, 304
crawling water beetle, 177
creeping water bug, 176, 400
Crematogaster, 401–402
Crematogaster mimosa, 402
Crematogaster sjostedti, 402
Crepidostomum cooperi, 275
crested caracara, 151
Cretaceous period, 27, 29
Cricetidae, 115
Cricetus cricetus, 253
cricket, house, 416
crickets, 14, 18, 181, 184–185
crimson Hawaiian damselfly, 420
Crioceris duodecimpunctata, 68
crochets, 190
crocodile, 106, 317
Crocodilia, 106
Crocodylus, 317
Crocuta crocuta, 252
cross predation, 397
cross-fertilization, 217
Crotalus lepidus klauberi, 112

crow, 142, 149, 249
     American, 132, 249
     fish, 249
     Hawaiian, 272
Crustacea, 6, 8–9
crustacean, 6, 8–9
Cryphonectria parasitica, 216
crypsis, 158
Cryptolaemus montrouzieri, 379–380
Cryptotis parva, 117, 120
Ctenicera glauca, 191
Ctenocephalides canis, 279, 332
Ctenocephalides felis, 279, 331–332
Ctenotus grandis, 110
Ctenotus helenae, 110
cuckoo, yellow-billed, 132, 149
cuckoos, 126, 406
Cuculidae, 149, 179
Cuculiformes, 126
Culcoides downesi, 311
Culex, 249–250, 264, 271, 276, 308, 404
Culex annulus, 271
Culex nigripalpus, 271, 307
Culex pipiens, 271
Culex quinquefasciatus, 269, 271, 307
Culex restuans, 271
Culex saltanensis, 271
Culex sitiens, 271
Culex stimatosoma, 271
Culex tarsalis, 194, 271, 306
Culex tritaeniorhynchus, 307
Culicidae, 177, 192, 285, 305–308
Culicoides, 246, 251, 310–311
Culicoides barbosai, 312
Culicoides furens, 194, 311–312
Culicoides hollensis, 312
Culicoides melleus, 312
Culicoides mississippiensis, 312
Culicoides paraensis, 311
Culicoides variipennis, 311
Culiseta, 308
Culiseta longiareolata, 400
Culiseta melanura, 307–308
Culiseta morsitans, 271
cultural control, 366–370
Curculionidae, 190, 425
Cutaneous Leishmaniasis, 313–314
Cuterebra, 325, 328
cuticle, 34
cutthroat trout, 138
cutworm or noctuid moth, 192, 371
Cyanocitta cristata, 132, 143, 148, 198,
     226, 249, 251
Cybister, 178
cycling, nutrient, 206–210
Cyclopes didactylus, 164, 250
Cyclura cychlura cychlura, 210, 212
Cylas formicarius, 375

Cynomys, 257, 355
Cynomys ludovicianus, 215, 258
Cyprinidae, 433
Cypriniformes, 152
Cyrtopogon lateralis, 194
Cytauxzoon, 264
Cytauxzoonosis, 264, 291

Dacnusa dryas, 380
Dactylopius coccus, 414
daddy longlegs, 6, 8
Dakota, skipper, 421
Damaliscus lunatus, 210
damselfly, 14, 16, 173–175, 395, 432
     blackline Hawaiian, 420
     crimson Hawaiian, 420
     flying earwig Hawaiian, 420
     oceanic Hawaiian, 420
     orangeblack Hawaiian, 420
     Pacific Hawaiian, 421
Danaus plexippus, 159, 423
Danio rerio, 154
Daptrius ater, 396
dark-eyed junco, 134, 150, 394
darkling beetles, 190
Daruma pond frog, 108
Dasybasis, 315
Dasyprocta, 265, 303
Dasypus, 265 *303*
Dasypus novemcinctus, 118, 121, 123,
     238, 251, 313
dead-end host, 239, 246
death feigning, 162
Death, Black, 255
decomposer, 201–210
decomposition, 201
     carcass, 204–208
     degradation phase, 201
     destruction phase, 201
     of carrion, 204–208
     of excrement (dung), 202–204
     of plant remains, 201–202
deer, 114, 254, 296, 315, 318, 325,
     326, 328
     gray brocket, 396
     mule, 236, 277, 291, 296, 301, 311,
         329, 330
     red, 292, 279, 292, 355
     white-tailed, 216, 236, 238, 249, 252,
         277, 291, 301, 311–313, 323,
         329–330
deer fly, 285, 314–316
deer mouse, 251, 294, 431
deer tick, 297–298, 434
defense by insects against predation,
     158–165
     aggressive mimicry, 160
     aposematic coloration, 159, 163

aposematism, 159
appendage extension, 160
autonomy, 162
Batesian mimicry, 160
chemical defenses, 162
crypsis, 158
death feigning, 162
disruptive coloration, 158
dodging, 162
eye spots, 160
flash coloration, 160
flight and startle behavior, 160–162
group actions, 163
irritating spray, 162
mobbing, 163
Müllerian mimicry, 160
nocturnal activity, 165
osmeterium, 161
physical defenses, 162
predator satiation, 163
reflex bleeding, 162
sting, 162
thanatosis, 162
toxins, 162
urticating hairs, 162
venoms, 162
Wasmannian mimicry, 160
defensive behaviors by hosts, 234
definitive host, and disease, 239
Degeeriella fulva, 287
degradation phase of decomposition, 201
degradation, microbial, 349
Delhi Sands flower-loving fly, 421
Delia, 350
Delichon urbica, 156, 250
delta green ground beetle, 419
demodectic mange, 291
Demodex, 289
dendrite, 52
Dendroctonus, 202
Dendroctonus brevicomis, 392
Dendroctonus frontalis, 216, 368
Dendroctonus ponderosae, 214, 216
Dendroctonus rufipennis, 216
Dendroctonus terebrans, 216
Dendroica castanea, 392
Dendroica coronata, 134, 150
Dendroica fusca, 392
Dendroica virens, 392
density-dependent factors in population regulation, 390
density-independent in population regulation, 390
Dermacentor, 246, 253, 293
Dermacentor albipictus, 296
Dermacentor andersoni, 294, 296
Dermacentor halli, 294

Dermacentor occidentalis, 296
Dermacentor variabilis, 287, 294–297
Dermanyssus gallinae, 290–291
Dermaptera, 14, 20, 184–186, 206, 285–286
Dermatobia, 328
Dermatobia hominis, 69
Dermestidae, 205–206, 334
Dermoglyphus, 287
desert locust, 220, 353
Desmocerus californicus dimorphus, 419
destruction of wildlife by ants, 397–400
destruction phase of decomposition, 201
detritivore, 201
detritophagous, 79
deutocerebrum, 53
developmental threshold, 70
Devonian period, 27, 29
Diachlorus ferrugatus, 314
diamondback moth, 366
diapause, 71
    facultative, 71
    obligatory, 71
Diatraea grandiosella, 68
Dicermyia, 250
Dicrocoelium dendriticum, 243, 275, 279–280
Dictyssa oblique, 94
Didea fasciata, 376
Didelphimorphia, 115
Didelphis marsupialis, 251
Didelphis, 265, 303
Didelphis virginiana, 118, 121, 123, 126, 238, 251, 265, 272, 297
diets of wildlife
    amphibian and reptile, 105–113
    bird, 126–152
    bulk, 93
    determination of, 91–97
    fish, 152–156
    frequency of consumption, 93
    frequency of occurrence, 93
    mammal, 107, 114–123, 124–126
    mass, 93
    optimal foraging, 97
    prevalence, 93
    time budgets, 97
    volume, 93
digestibility-reducing compounds, 202
digestion, extraoral, 49
digestive system, 47
Diglyphus isaea, 380
dilution host, and disease, 239
dingo, 292, 396
Dioryctria albovitella, 215
Dipetalonema dracunculoides, 330
Diplopoda, 6, 9–11
Diplotriaena bargusinica, 275

Diplura, 6, 13
Dipodomys californicus, 254
Diprotodontia, 115
Diptera, 15, 24, 177, 192, 194–195, 304–330, 333, 432
Dipylidium caninum, 275, 279, 281, 302
direct flight muscle, 45, 49
direct life cycle, 242
direct transmission of disease, 240
Dirofilaria, 276
Dirofilaria immitis, 275–276, 278, 316
Dirofilaria roemeri, 316
Dirofilaria striata, 276
Dirofilaria subdermata, 276
Dirofilaria tenuis, 276
Dirofilaria ursi, 276, 310
Dirofilariasis, 275–276, 278, 332
discs, imaginal, 56
disease
    active transmission, 240
    causes of, 241
    Chagas, 263–267
    clinical sign, 241
    direct transmission, 240
    Hemorrhagic, 251, 311–312
    horizontal transmission, 240
    indirect transmission, 240
    infectious, 245–262
    inoculative transmission, 240
    insects causing disease, 285–338
    Koch's Postulates for establishing cause of, 241
    Kyasanur Forest, 291
    Lyme, 254–256, 291, 296
    noninfectious, 241
    of plants, and ecosystems, 215–217
    parasitic, 233, 263–284
    passive transmission, 240
    putative cause, 241
    sign, 241
    symptom, 241
    transmission by arthropods, 233–244
    vector-borne, 240
    vertical transmission of, 240
disease host
    aberrant, 239
    amplifier, 239
    dead-end, 239
    definitive, 239
    dilution, 239
    incidental, 239
    intermediate, 239
    primary, 239
    reservoir, 239
    unnatural, 239
disease transmission to wildlife, 233
disease triangle, 238
dispar topminnow, 396

disruption, endocrine, by pesticides, 354
disruptive coloration, 158
diving beetles, predaceous, 400
dobsonflies, 14, 22, 176, 400, 432
dodging, 162
dog flea, 279
dog fly 319
dog louse, 279
dog, prairie, 257–258, 332, 355
dog, raccoon, 123
dog tapeworm, 275, 279, 281
dog, wild, 252
dogs, 265, 268, 291
Dolichonyx oryzivorus, 132, 143, 149
dolphin, Atlantic bluenose, 272
domatia, 401–402
donkey, 291
dorsal vessel, 50
Dorylus, 398
double-pored (dog) tapeworm, 302
dove, 126, 269, 406
    ground, 398
    mourning, 251, 352
dowitcher, long-billed, 138, 150
down mites, 292
downy woodpecker, 128, 332
Drachiasis, 275
dragonfly, 14, 16, 173–175, 218, 395, 400, 432
    Hine's emerald, 421
Draschia megastoma, 275
driver ant, 398
droppings, 92
Drosophila aglaia, 421
Drosophila attigua, 421
Drosophila differens, 421
Drosophila digressa, 421
Drosophila hemipeza, 421
Drosophila heteroneura, 421
Drosophila melanogaster, 421
Drosophila montgomeryi, 421
Drosophila mulli, 421
Drosophila musaphila, 421
Drosophila neoclavisetae, 421
Drosophila obatai, 421
Drosophila ochrobsis, 421
Drosophila substenoptera, 421
Drosophila tarphytrichia, 421
Drosophilidae, 204
Dryocopus pileatus, 128, 143
duck, 126, 259, 310
    canvasback, 136, 150
    pink-eared, 136, 150
    redhead, 136, 150
    wood, 136, 150
duiker, 268
Dumetella carolinensis, 134, 149

dung, 92
dung beetles, 203–204
dust baths, 301
Dutch Elm Disease, 216
dyes from insects, 414
Dytiscidae, 177–178, 395, 400
Dytiscus, 178, 359

eagle
    bald, 352–354, 359
    golden, 353
eagles, 126
ear mites, 291
Early Carboniferous, 29
earwigs, 14, 20, 184–186, 206, 285
East African Trypanosomiasis, 264, 268
eastern bluebird, 130, 249, 318, 322, 360, 429
eastern chipmunk, 234, 254
eastern cottontail rabbit, 254, 399
eastern fox squirrel, 249, 329
eastern kingbird, 130, 432
eastern meadowlark, 132, 143–144
eastern mole, 115–116, 120, 125
eastern phoebe, 130, 143, 145
eastern screech owl, 318
eastern spruce budworms, 392
eastern towhee, 398
eastern wood pewee, 130, 432
ecdysis, 35
ecdysone, 56
ecdysteroids, 56
echidna, 164
Echidnophaga gallinacea, 235, 287, 332
Eciton, 398
ecological and taxonomic patterns of invasion, 221
ecological biogeography, 32
ecological compensation, 427
economic benefits of insects, 406–407, 410–416
economic injury level, 381
Economic Ornithology and Mammalogy, Section of, 142
economic threshold, 381
ecosystems, 198
ectognathous, 12, 37
ectoparasites, 242, 285–286
ectoparasitoids, 79
Ectopistes migratorius, 424
ectothermic, 70
ectotherms, 198
eels, 152
egg fly
    lizard, 325
    sea turtle, 325
    terrapin, 325

egg rafts, 307
egg shell, 67
eggs of insects, 66–69
egret, 259
    cattle, 138, 150, 249, 396
    little, 249
Egretta garzetta, 249
Ehrlichea, 252
El Segundo blue butterfly, 419
Elaeophora schneideri, 275, 277, 316
Elaeophorosis, 275, 277–279
Elanoides forficatus, 151
Elaphe obsolete, 158
Elaphrus viridis, 419
Elateridae, 191, 371
elderberry, 428
Eleodes suturalis, 190
elephant, 210, 268, 317, 328
    African, 325, 328, 402
elephant shrews, 114
Eleutherodactylus coqui, 397
elk, 114, 236, 252, 254, 277, 289, 291–292, 296, 315, 326
elf owl, 151
elk scab mite, 291
Elmidae, 177–178
Elopiformes, 152
elytra, 41, 46
Emberizidae, 142, 150
Embiidina, 14
Emerald ash borer, 226
emerging pathogens, 236
Empidonax flaviventris, 130
Empoasca fabae, 188
emus, 142
Emydura krefftii, 112
Emydura macquarii, 112
Encephalitis, St. Louis, 250
encephalomyelitis, 247
Endangered Species Act, 418
endangered species, 419–422
endemic, 235
endocrine disruption by pesticides, 354
endocrine glands, 55
endocuticle, 34–35
endoparasites, 242, 285
endoparasitoids, 79
Endopterygota, 14
endothermic, 70
endotherms, 198
enemies, natural, 375–377
energy fluxes, 200
energy pulses, 200
engraver beetle, 216
enhanced biodegradation of pesticides, 350
Enterococcus, 321
Entognatha, 6, 11–14

entognathous, 12
Entomophaga maimaiga, 393
entomophagous, 79
environmental management, 366–370
enzootic, 235
Eocene epoch, 29
Eon
    Archean, 27
    Hadean, 27
    Phanerozoic, 27, 29
    Proterozoic, 27
Ephemeridae, 172
Ephemeroptera, 14, 16, 171, 432
epicuticle, 34–35
epidemic typhus, 301
epidemic, 235
epidermal cells, 34
epidermis, 35
Epiplatys, 396
epizootic, 235
epoch
    Eocene, 29
    Holocene, 29
    Mississippian, 29
    Miocene, 29
    Oligocene, 29
    Paleocene, 29
    Pennsylvanian, 29
    Pleistocene, 29
    Pliocene, 29
    Tommotian, 29
Eptesicus fuscus, 115, 122, 251
Equidae, 114
Equus, 268
Equus burchelli, 210
Equus zebra, 396
Era
    Cenozoic, 27, 29
    Mesozoic, 27, 29
    Paleozoic, 27, 29
Eremophila alpestris, 132, 149, 351–352,
    393
Erethizon dorsatum, 276, 289, 297
Eretmocerus eremicus, 380
Eretmocerus palustris, 380
Erinaceomorpha, 114
Erinaceus europaeus, 254, 289
Erithacus rubcula, 254
Escherichia, 321
Esociformes, 152, 155
Esox lucius, 154–155, 400
esterase, acetylcholine, 52, 346, 354
estrildid finches, 142
Estrildidae, 142
Ethiopian realm, 32
etiologic agent, 241
Euborellia annulipes, 186
Eudocimus albus, 334

eukaryotes, 259
eulachon smelt, 407
Eulemur fulvus, 124
Eumacronychia nigricornis, 325
Eumacronychia sternalis, 325
Eumaeus atala, 424
Eumeces laticeps, 110, 397
Euphagus carolinus, 132
Euphilotes battoides allyni, 419
Euphilotes enoptes smithi, 420
Euphydryas anicia cloudcrofti, 420
Euphydras editha bayensis, 419
Euphydras editha quino, 420
Euphydras editha wrighti, 420
Euphydras editha taylori, 420
Euplectrus plathypenae, 377
Euplectrus plathypenae, 377
Euproserpinus euterpe, 421
Eurasian hedgehog, 289
Eurasian kestrel, 360
Eurasian sparrow hawk, 354
Eurasian wigeon, 249
European badgers, 288
European brown hare, 253
European hares, 279
European honey bee, 220
European starling, 136, 150, 249, 251,
    322, 352, 396
Eurycea tynerensis, 108
Eurytoma, 68
Euschlongastia splendens, 239
eusocial behavior, 61
Eutamias amoenus, 117, 120
Eutamias minimus, 118, 251
Eutamias sibiricus, 297
Eutamias speciosus, 117, 121
Eutrichophilus setosus, 301
Eutrombicula alfreddugesi, 287
eutrophication, 433
Euxoa auxiliaris, 125
evolution of insects, 15, 27–31
evolutionary success of insects, 31
excrement, decomposition, 202–204
excretory system, 69
exocrine glands, 55
exocuticle, 34–35
Exopterygota, 14
exoskeleton, 34
exsanguination, 307
extension, appendage, 160
extrafloral nectaries, 427–428
extraoral digestion, 49
exuviae, 35
exuvial space, 35
eye gnats, 333
eye spots, 160
eyes, compound, 37
Eyeworm, Tropical, 275

face flies, 318
facets, 54
facultative diapause, 71
facultative myiasis, 321
Fagopyrum esculentum, 431
Falco naumanni, 360
Falco peregrines, 354
Falco sparverius, 151, 354, 359
Falco tinnunculus 360
falcon, 126 (see also, kestrel)
    lesser, 360
    peregrine, 354
Falconiformes, 126, 151
false map turtle, 112
false scorpions, 6–8
fan-tailed raven, 396
Faniidae, 203
fat body, 47
feather mites, 287, 292
febrile illness, 247
febrile myalgia, 247
fecundity, 67
feeding behavior, 233
feigning, death, 162
Felicola felis, 301
Felicola subrostratus, 300
Felidae, 114
felids, 123
Felis concolor, 257, 272
Felis domesticus, 257
Felis rufus, 257, 276
Felis tigris, 289
female reproductive system, 65–66
femur, 36, 41, 46
Fender's blue butterfly, 420
feral rabbits, 355
ferret, 291–292
    black-footed, 257, 356
ferruginous pygmy owl, 151, 318
fever
    African Swine, 291
    Black Fly, 309
    Colorado Tick, 291
    Hemorrhagic, 247
    Q, 332
    Relapsing, 291
    Rocky Mountain Spotted, 291
    Yellow, 250
field sparrow, 145, 394
field vole, 355
Filariasis, 275
    Avian, 275, 301
    Lymphatic, 275
file, 60
filter chamber, 50, 69
finch, house, 134, 150, 251
finches, 142
    estrildid, 142

fire ant, 400
   red imported 196, 223, 334, 397–400, 417
   tropical, 398
fish crow, 249
fish, diets of, 152–156
fish, stomach contents of, 140–141, 153–154
fish, mosquito, 396
fisher, 292
fishes, ray-finned, 152
fishflies, 176, 400, 432
flagellum, 37, 40
flamingo, 126, 142, 406
   Chilean, 298
flammulated owl, 151
flash coloration, 160
flat-headed snake, 107, 112
flatfish, 152
Flatidia coccinea, 124
flea, 15, 24, 257, 288, 330–332
   cat, 279, 331
   dog, 279
   hen, 331–332, 360
   sticktight, 235, 332
   water, 9
flesh fly, 192, 285, 324–326
   lizard, 324
   toad, 325
   turtle, 325
flicker, northern, 128, 143–144
flight and startle behavior, 160–162
Florida red-bellied turtle, 398
Florida scrub jay, 332
flounders, 152
flowering plants, importance to insects, 426–428
fluid, molting, 35
Fluke
   Frog Lung, 275
   Lancet (Liver), 275, 279–280
   Oviduct, 275
   Pyloric Caeca, 275
fluxes, energy, 200
fluxes, nutrient, 200
fly, 15, 24, 177, 192, 194–195, 304–330, 333
   biting house, 319–320
   black, 276, 285, 305, 308–310
   blow, 285, 321–324
   bot, 285, 325–329
   bottle, 321–324
   deer, 285, 314–316
   Delhi Sands flower-living, 421
   dog, 319–320
   face, 318–319
   flesh, 285, 324–326
   fruit, 370

   horse, 285, 314–316
   house, 320–321
   louse, 285, 287, 329–330
   moose, 318–319
   muscid, 285, 318–319
   New World screwworm, 322–324
   Old World screwworm, 322, 324–325
   phlebotomine sand, 312–314
   pomace, 421
   sand, 285, 310, 312–314
   snipe, 333
   stable, 319–320
   tsetse, 285, 316–318, 403–404
   warble, 285, 325–329
   yellow, 314, 403
flycatcher
   ash-throated, 130
   great crested, 322, 432
   olive-sided, 130
   scissor-tailed, 130
   tyrannid royal, 401
   western yellow-bellied, 130
flycatchers, 142
   Old World, 142
flying earwig Hawaiian damselfly, 420
flying squirrel, 332
Foleyella bachyoptera, 275
foliar resources, 428–432
folivorous wildlife, 98
food availability, 86
food chains, 198
food chain, insecticides in, 357–359
food, insects as, 414–416
food poisoning, 259
food preference, 85
food webs, 198
foraging theory, optimal, 93
Forcipomyia, 310
foregut, 49
Forficula auricularia, 186
Forficulidae, 186
Formica, 243, 279–280
Formica rufa, 202
Formicidae, 195–6, 206, 218, 212, 226, 333–334, 398
fowl mite
   northern, 291
   red, 291
   tropical, 291
Fowl Tapeworm, 275
fowl tick, 252, 295
fox squirrel, 238, 301
   eastern, 249
fox, 114, 257, 276, 291, 332, 335
   gray, 119, 120, 158, 238, 272, 276, 292, 296, 297, 301

   red, 119, 120, 158, 238–239, 251, 272, 292, 297
   swift, 119, 258
Francisella tularensis, 252–254
Franklin's gull, 136, 150
frass, 209
Fratercula arctica, 254
Fraxinus americana, 226
frequency of consumption, 93
frequency occurrence, 93
Freyana largifolia, 287
Fringillidae, 142, 150
Frog Lung Fluke, 275
frog, 105, 333
   Blanchard's cricket, 108
   Brown's leopard, 108
   Daruma pond, 108
   gray tree, 359
   leopard, 354, 359
   northern leopard, 108
   southern leopard, 108
   terrestrial, 396
   Vaillant's, 108
   yellow-leg, 359
frugivores, 218
frugivorous wildlife, 98
fruit flies, 192, 202, 370
fruit-feeding insects, 78
Fulica americana, 138, 150
fumigants, 345
functional responses of predators, and population regulation, 387
fungi, 259–261
fungicides, 343
fungivorous wildlife, 98
fungus gnats, 371
fungus-tending ants, 202
furcula, 12
Furnariidae, 142

Gadiformes, 152, 155
gall fly, Po'olanui, 421
gall-forming insects, 76
Galleria mellonella, 99–100, 416
Galliformes, 126
Gallinago delicata, 138, 147, 143, 159
gallinules, 301
galls, 401
Gambusia, 308, 357, 396
Gambusia affinis, 396
Gaminicimex, 304
Gammarus, 9
ganglion, 52
Gargaphia solani, 188
Garmania, 239
Gasterophilus, 328
Gasterosteiformes, 152, 155
Gasterosteus aculeatus, 153

Gaviiformes, 126
Gazalle, 272
Gazalle granti, 252
Gazalle thompsonii, 292
gazelle, 114, 272
    Grants, 52
    Thompson's, 292
Gedoelstria, 328
geese, 126, 259, 301, 310
    Canada, 355
    greater white-fronted, 261
    Ross's, 261
gemsbok, 252
generalists, 75, 85
genitalia, 45
Geocoris pallens, 376
geographic shifts in wildlife feeding, 102
geological time scale, 27–30
Geometridae, 192
Geomyidae, 115
Geothlypis trichas, 254, 432
Geotrupiinae, 203
gerbils, 115, 313
Gerridae, 175
Gerris, 176
giant anteater, 164
giant centipede, 396
giant crab spider, 396
giant margarodid scale, 414
giant pandas, 114
giant silkworms, 192
Giant Thorny-Headed Worm, 275
Giant water bugs, 175–176, 400
gill, physical, 66, 74
Ginkgo biloba, 29
Giraffa camelopardalis, 254, 268, 317, 402
giraffe, 114, 254, 268, 328, 402
Girafidae, 114
gizzard shad, 407
gizzard worm, 275
gladiators, 14
glands, 55
    endocrine, 55
    exocrine, 55
    odor, 55
    pheromone, 55
    poison, 55
    prothoracic, 55
    salivary, 55
    silk, 55
    wax, 55
glandular systems, 55
Glaucidium brasilainum, 151, 318
Glaucidium californicum, 151
Glaucomys volans, 301
Glaucopsyche lygdamus palosverdesensis, 420

Glossina, 234 235, 266, 316–317
Glossinia fuscipes, 268
Glossinia morsitans, 268
Glossinia pallidipes, 268
Glossinia palpalis, 268, 317
Glossinia swynnertoni, 268
Glossinia tachinoides, 268
Glossinidae, 234–235, 285, 316–318
Glossophaga soricina, 116
glycerol, 71
Glyphopsyche sequatchie, 420
gnatcatcher, blue-gray, 398
gnatcatchers, 142
gnat
    eye, 333
    fungus, 371
goat, 254
    mountain, 252, 291, 301
goatsuckers, 126, 165, 406
goitered antelope, 328
golden eagles, 353
golden moles, 114
golden-crowned kinglet, 392
golden-tipped bat, 116
goldeneye duck, 395
goldeneye, common, 147
goldfinch, American, 393–394
goldfish, 396
Gomphidae, 174
Goniodes ortygis, 287
gopher tortoise, 398
gopher, pocket, 115
Gopherus polyphemus, 398
Gorilla beringei beringei, 125
Gorilla beringei grauer, 125
Gorilla gorilla gorilla, 125
gorilla
    Grauer's, 126
    mountain, 126
    western, 126
grackle, common, 132, 291
grackles, 149
grand ctenotus skink, 110
granivorous wildlife, 97
Grant's gazelle, 252
granules, insecticide, 351–352
Graptemys ouachitensis, 112
Graptemys pseudogeographica, 112
grass mouse, 251
grasshopper, 14, 18, 181, 184, 393, 425
    migratory, and sparrows, 156–157
    Zayante band-winged, 421
grasshopper mouse, 115, 117, 257
Grauer's gorilla, 126
gray brocket deer, 396
gray catbird, 134, 149
gray fox, 119, 120, 158, 238, 272, 276, 292, 296–297, 301

gray heron, 249
gray (grey) squirrel, 216, 301
gray tree frogs, 359
grayling, Arctic, 152
grazers, aquatic, 75
grazing optimization hypothesis, 214
great crested flycatcher, 322, 432
great tit, 332, 360
greater Adams cave beetle, 420
greater earless lizard, 110
greater kudu, 252
greater spear-nosed bat, 116
greater white-fronted geese, 261
grebe, horned, 136, 150
grebes, 142, 259, 406
green anole, 396
green bottle flies, 321
green lacewings, 189
green sunfish, 138, 153, 155
green turtle, 398
green-winged teal, 249
grey partridge, 427
grey teal, 136, 150
grey-sided vole, 297
grivet, 249
grizzly bear, 125
grooming, 300
grosbeak, 142
grosbeak, black-headed, 424
grosbeak, rose-breasted, 134, 150
ground beetles, 189, 191, 427
ground dove, 398
ground squirrel, 313, 332, 355
    antelope, 102
groundhog, 158, 297
group actions, 163
grouse, 253, 269, 330, 406
    ruffed, 143, 146, 216
grubs, white, 371
Gruiformes, 142
Grus canadensis, 429
Gryllidae, 184–5
Grylloblattodea, 14
Gryllotalpa africanus, 99
Gryllotalpidae, 184
Gryllus veletis, 185
guano, 92
guinea pigs, 265, 291
gull, 249, 259, 406
    black-backed, 206
    black-headed, 249, 272
    Franklin's, 136, 150
    herring, 206, 354
    little, 249
    yellow-legged, 249
guppy, 396
Gymnogyps californianus, 298
Gymnophiona, 105

Gymnorhinus cyaocephalus, 215
gymnures, 114
gypsy moth, 223, 225–226, 393, 417
Gyrinidae, 177
Gyrostigma, 328

habitat conservation, 402–405
habitat management to benefit wildlife,
    425–434
Habronema muscae, 275
Habronemiasis, 275
Hadean eon, 27
Haemagogus, 250
Haemaphysalis, 293
Haemaphysalis chordeilis, 287
Haemaphysalis leporispalustris, 287
Haemaphysalis spinigera, 294
Haematobosca alcis, 318
Haematoloechus medioplexus, 275
Haematopinus suis, 301
Haematopota, 314
Haematopota variegata, 315
Haematosiphon, 304
Haemodipsus ventricosis, 301
Haemolaelaps megaventralis, 290
Haemoproteus, 264, 271, 309, 311
Haemoproteus fringillae, 311
Haemoproteus lophortyx, 330
Haemoproteus nettionis, 311
hairs, urticating, 162
hairy woodpecker, 128, 143–144
half-life of common insecticides, 349
Haliaeetus leucocephalus, 298,
    352–353, 359
Halictus brachtatus, 68
Haliplidae, 177
halteres, 192, 194
hamster, 253
harassment by insects, 234, 305
hard ticks, 285, 293
hare, 252–253, 257, 312, 325, 355
    European, 253, 279
    mountain, 254, 279
    varying, 253
Harpalus, 191
harrier, northern, 151
hartebeest, 268
harvester ants, 398
harvestmen, 8
Hawaii amakihi, 271
Hawaiian crow, 272
hawk, 126, 253, 259, 330, 406
    broad-winged, 151
    Eurasian sparrow, 354
    red-shouldered, 151
    red-tailed, 359
    Swainson's, 359
    white-tailed, 151

hawkmoth, 193
hazard, of pesticides, 343
head, 37
headland, conservation, 427
heart, 50
heartworm or Sarconema, 275
heat, metabolic, 70
hedgehog, 114, 312, 331
    African, 289
    Eurasian, 254, 289
Helicoverpa zea, 115
helminthes, 274–283
Helotes mold beetle, 421
hemelytra, 41
hemelytron, 175, 186, 302
Hemerobiidae, 189
Hemignathus virens, 271
Hemileuca oliviae, 68, 380
hemimetabolous, 63–65
Hemiptera, 14, 175–176, 186–188, 285,
    302–304, 432
Hemlock woolly adelgid, 226
hemocytes, 50
hemolymph, 50
Hemorrhagic Disease, 239, 251,
    311–312
Hemorrhagic Fever, 247
hen flea, 331–332, 360
Hepatocystis, 311
Hepatocystis brayi, 311
Hepatozoon, 264
Hepatozoonosis, 264
Heraclides andraemon bonhotei, 419
Heraclides aristodemus ponceanus, 420
herbivore optimization hypothesis, 214
herbivory, 76–78, 85
    by boring insects, 76
    by insects, 210–214
    by fruit-feeding insects, 78
    by gall-forming insects, 76
    by leaf-chewing insects, 76
    by leaf-mining insects, 76
    by nectar-feeding insects, 78
    by piercing-sucking insects, 76
    by pollen-feeding insects, 77
    by root-feeding insects, 78
    by seed-feeding insects, 78
    by skeletonizing insects, 76
herbivory, plant compensation for,
    213–214
Hermetia illucens, 100
hermit thrush, 132
Hernandia, 423
heron, gray, 249
herons, 126, 259, 406
Herpestes javanicus, 396
herring gull, 205–206, 354
herrings, 152

Hesperia dacotae, 421
Hesperia leonardus montana, 421
Hesperocimex, 304
Hesperocorixa, 176
Hesperotettix viridis, 393
Heterelmis comalensis, 419
Heterelmis stephani, 421
Heterodoxus spiniger, 301
Heteromys, 265, 303
heterotrophs, 198
Hexapoda, 12
hibernation, 107
hickory horned devil, 192
hickory shad, 407
high temperature and pesticides, 350
Highland tiger beetle, 421, 425
hindgut, 49, 69
Hines emerald dragonfly, 421
Hippelates, 333
Hippobosca longipennis, 330
Hippoboscidae, 285, 329–330
Hippodamia convergens, 376
Hippopotamidae, 114
Hippopotamus amphibious, 317
hippopotamus, 114, 317
hippos, 328
Hippostragus niger, 252
Hirundinidae, 142, 149
Hirundo pyrrhonota, 304
Hirundo rustica, 134, 250, 286
hispid cotton rat, 238
Histeridae, 204, 206
historical biogeography, 32
hogs, 114
Holarctic realm, 32
Holocene epoch, 29
Holomenopon clauseni, 287
holometabolous, 63–65
Holsinger's cave beetle, 420
Homerus swallowtail butterfly, 420, 423
Homo sapiens, 28
honey bee, 196, 422
    Africanized, 226, 333
    European, 220
honey, 412–413
honeydew, 50
hooks, mouth, 37, 318, 325
Hoplopleura acanthopus, 301
Hoplopleura sciuricola, 301
horizontal transfer of pesticides, 345
horizontal transmission of disease, 240
hormone, 55–56
    ecdysone, 56
    juvenile, 56
    PTTH, 56
horned grebe, 136, 150
horned lark, 132, 149, 351–352, 393
horned toad, Texas, 398

hornets, 33
hornworms or sphinx moths, 192
horse flies, 195, 285, 314–316
horses, 114, 291
horseshoe crabs, 6
host
    aberrant, and disease, 239
    amplifier, and disease, 239
    dead-end, and disease, 239, 246
    definitive, and disease, 239
    dilution, and disease, 239
    incidental, and disease, 239
    intermediate, and disease, 239
    primary, and disease, 239
    reservoir, and disease, 239
    unnatural, for disease, 239
host location
    behavior, 233
    olfaction, 234
    vision, 234
host resistance, 371–373
house cricket, 416
house finch, 134, 150, 251
house fly, 195, 320–321
house martin, 249
house mouse, 117, 120, 253
house sparrow, 249, 251, 254, 332, 352, 359
house wren, 136, 150 226, 249, 254, 322
howler monkey, 249, 251, 328–329
human body louse, 301
hummingbird, 126, 406
    Anna's, 128, 143
    ruby-throated, 128, 143
Hungerford's crawling water beetle, 419
Hyaenidae, 114
Hyalomma, 293
Hydraecia immanis, 68, 193
hydrolysis of pesticides, 350
Hydrometridae, 175
Hydrophilidae, 177–178
hydropyles, 66
hyena, 114, 330
    spotted, 252
Hyla versicolor, 359
Hylocichla mustelina, 130, 226
Hymenoptera, 15, 24, 192–193, 195–196, 333–334, 401
Hypera postica, 366
Hypoderma, 326, 328–329
Hypoderma diana, 327
Hypoderma lineatum, 327
Hypoderma tarandi, 325
Hypolimnas octucula mariannensis, 420
hypothesis, grazing optimization, 214
    herbivore optimization, 214
hyraxes, 314

ibex, 252, 292
    Alpine, 291
ibis, 259
    white, 334
Icaricia icarioides fenderi, 420
Icaricia icarioides missionensis, 420
icebox cave beetle, 420
Ichneumonidae, 196
Icosta albipennis, 330
Icosta americana, 287, 330
Icteridae, 142, 149
Icterus abeilleri, 424
Icterus galbula, 132, 156, 226, 424
Icterus spurious, 254
Ictinia mississippiensis, 151
Ictiobus cyprinellus, 141, 153
iguana, 210, 212
illness, febrile, 247
imaginal discs, 56
immigrant, 219
impala, 210, 268, 315
imported cabbageworm, 223–224
impoundments, mosquito control, 370
incidental host, and disease, 239
Indian mongoose, 396
indigenous, 219, 235
indirect effects of pesticides, 356–357
indirect flight muscles, 45, 49
indirect life cycle, 242
indirect transmission of disease, 240
inert ingredients of pesticides, 344
infectious diseases, 233
Ingrassia, 287
ingredients, inert, of pesticides, 344
inhibition, acetylcholinesterase
injury level, economic, 381
inoculative biological control, 379
inoculative transmission of disease, 240
inquirer cave beetle, 420
insect harassment, 305
insect management, area-wide, 379–381
insect nurseries, 428
insect outbreaks, and ecosystems, 214–215
insect outbreaks, causes, 214
insect resources, managing, 425–434
Insecta, 6, 14–33
insecticide granules, 351–352
insecticides, 341–362
    acute effects, 350
    alternatives to, 366–382
    half-life of common, 349
    in the food chain, 357–359
    major groups, 347
    mode of action, 345
    persistence of, 349
    resistance to, 361–362
    risks associated with, 359–361

    sublethal effects, 354–355
    toxicity of common, 348
Insectivora, 114
insectivores, 85
insectivory, 85
insects as food, 414–416
insects, 6, 14–33
    age of, 28
    benefits for wildlife, 156
    carrion-feeding, 204–208
    classification of, 14–27
    transmission of disease agents, 235–236
    eggs of, 66–69
    evolution of, 15, 27–31
    nutrient content of, 97–103
    nutritional value of, 97–103
    outbreak, 200
    parasitic, 79
    predation on wildlife, 397–400
    predatory, 79
    success of, 31
    symbiotic relationships with wildlife, 400–402
insects, angel, 14
insemination, traumatic, 303
instar, 63
integrated pest management, 381–382
integument, 34
intermediate host, and disease, 239
internal anatomy, 45, 48
intersegmental membranes, 34
introduced, 219
invasion, 218–226
    abundance of invaders, 225
    ecological and taxonomic patterns, 221
    establishment and spread, 222
    factors contributing to, 223
    impacts, 226
    latency, 223
    pathways, 219
invasives, 218–226
invertivores, 85
IPM, 381–382
Ips, 216
Iridopsis defectaria, 158
irritating spray, 162
Isopoda, 206
isopods, 6, 9
Isoptera, 14, 18, 179–180, 183, 401
IUCN Red List of Threatened Species, 419
Ixodes, 246, 253–254, 293
Ixodes banksi, 293
Ixodes minor, 287
Ixodes pacificus, 254, 294
Ixodes peromysci, 294
Ixodes persulcatus, 254, 296–297
Ixodes ricinus, 254–255

Ixodes scapularis, 254, 293–298, 434
Ixodes soricis, 293
Ixodida, 285
Ixodidae, 285

jackals, 114, 301
jackrabbit, 251
jaguar, 271, 276
Japanese beetle, 223
Japanese common toad, 108
jay
    blue, 132, 142, 143, 148–149, 198,
        215, 248, 251, 226
    Florida scrub, 332
    pinyon, 215
Jerusalem crickets, 184–5
Johnston's organ, 54
Julus, 9
Junco hyemalis, 134, 150, 394
junco, dark-eyed, 134, 150, 394
jungle rabbit, 248
Jurassic period, 29
Justinian Plague, 255
juvenile hormones, 56

kairomones, 61
kala-azar, 313
kangaroo, 115, 315, 328
kangaroo rat, 254
Karner blue butterfly, 420
katydids, 14, 18, 181, 185
Kerivoula papuensis, 116
Kern primrose sphinx moth, 421
kestrel
    American, 151, 354, 359
    Eurasian, 360
    lesser, 360
killdeer, 136, 150, 352
kingbird, 142
    eastern, 130, 432
    western, 130
kingfishers, 126, 406
kinglet, 142
    golden-crowned, 392
    ruby-crowned, 134, 150
kingsnake, 398
kinkajou, 249
Kirkioestrus, 328
kissing bug, 234, 265, 302–303
kissing bugs, 285
kite, 355
    Mississippi, 151
    snail, 334
    swallow-tailed, 151
klegs, 314
klipspringer, 396
knot, red, 150
koala, 115

Kobus ellipsiprymnus, 210, 291
Koch's Postulates, 241
Krefft's river turtle, 112
Kretschmarr Cave mold beetle, 419
krill, 9
kudu, 315, 317
    greater, 252
Kurodaia fulvofasciata, 287
Kyasanur Forest Disease, 291

labium, 36–37, 40
Laccifer lacca, 414
lace bug, 188
lacewings, 15, 22, 188–189
lacquer from insects, 414
Laelaps evansi, 239
Laguna Mountains skipper, 421
lake trout, 407
lake whitefish, 407
Lama, 292
Lama glama, 289
Lampropeltis getula floridanus, 398
Lancet (Liver) Fluke, 243, 279–280
Lange's metalmark butterfly, 420
Laniidae, 142
Lanius ludovicianus, 128, 143, 399
larch sawfly, 396
large fruit flies, 192
largemouth bass, 138, 153, 155
Laridae, 150
lark, 142
    bunting, 393
    horned, 132, 149, 351–352, 393
Larus, 250
Larus argentatus, 206, 354
Larus cachinnans, 249
Larus marinus, 206, 249
Larus minutes, 249
Larus pipixcan, 136, 150
Larus ridibundus, 249, 272
larvae, 63
larviparous, 329
Lasiocampidae, 192
Lasiorhinus latifrons, 292
Late Carboniferous, 29
lateral regulation of populations, 392
Laterallus jamaicensis, 398
Lathrolestes nigricollis, 380
Latimeria chalumnae, 29
Latrocimex, 304
LD$_{50}$ value, for pesticides, 343
leaf beetles, 190–191
leaf insects, 14, 18
leaf-chewing insects, 76
leaf-miner flies, 192, 194
leaf-mining insects, 76, 193
leafcutter ants, 210, 212, 416
leaffooted bug, 188

leafhoppers, 188
learning, 60
least chipmunk, 118, 251
least shrew, 117, 120
least tern, 397
leg, 36, 41, 46
Leishmania, 264
Leishmaniasis, 264, 313–314
    Cutaneous, 313–314
    Visceral, 313–314
lemmings, 115, 253, 301
Lemmus, 253
Lemna valdiviana, 102
Lemur catta, 124
lemur, 114, 124
    brown, 124
    brown mouse, 124
    lesser mouse, 124
    ring-tailed, 124
lentic, 74
leopard frog, 354, 359
leopards, 114
Lepidoptera, 15, 24, 190–192
Lepomis cyanellus, 140, 153, 155
Lepomis gibbosus, 140, 153, 155
Lepomis macrochirus, 100, 140, 153,
    155
Leptinidae, 285–286
Leptocimex, 304
Leptocimex boueti, 303
Leptoconops, 310
Leptoconops linleyi, 312
Leptoglossus phyllopus, 188
Lepus, 251, 253
Lepus capensis, 355
Lepus europaeus, 253, 279
Lepus timidus, 253–254, 279
lesser Adams cave beetle, 420
lesser kestrel, 360
lesser mouse lemur, 124
lesser pandas, 114
lesser scaup, 136, 150
Lethocerus americana, 175
Lethocerus deyrolli, 400
Leuciscus rutilis, 395
Leucocytozoon, 264, 271, 309, 311
Leucocytozoon simondi, 310
Leucocytozoon smithi, 310
Leucocytozoon tawaki, 310
Leucocytozoon ziemanni, 310
Leucocytozoonosis, 264
levels, trophic, 198
Lewis's woodpecker, 128
Libellula americana, 5
Libellulidae 174
lice, 298–302
    chewing, 14, 20, 279–287, 298–302
    sucking, 14, 20, 298–302

life cycle
  direct, 242
  indirect, 242
lignocelluloses, 201, 206
Limnephilidae, 182
Limnodromus griseus, 138, 150
Limnodromus scolopaceus, 138, 150
Linepithema humile, 226, 398, 417
Linognathous pedalis, 301
Linognathous africanus, 300
Liolaemus lutzae, 110
lion, 114, 252, 268
  mountain, 257, 301
  sea, 272
Lipoptena cervi, 330
Lipoptena depressa, 330
Lipoptena mazamae, 330
Liriomyza, 380
Liriomyza trifolli, 194
Listrophorus caudatus, 239
Listrophorus laynei, 239
little brown bat, 251
little egret, 249
little fire ant, 398
little gull, 249
lizard, 106, 353, 394
  Bonaire whiptail, 110
  greater earless, 110
  Mexican fringe-toed, 110
  Mexican knob-scaled, 110
  monitor, 210, 317
  mountain spiny, 110
  Newman's knob-scaled, 110
  tropical sand, 110
  western fence, 354
lizard egg fly, 325
lizard flesh fly, 324
llama, 114, 289, 292
Llaveia axin, 414
Loa loa, 275
loaches, 152
lobsters, 9
location behavior, host, 233
locust, desert, 353
lodgepole chipmunk, 117, 121
loggerhead shrike, 128, 143, 399
loggerhead turtle, 398
Loiasis, 275
lone star tick, 434
long-billed dowitcher, 138, 150
long-horned beetles, 189, 221
longspur, McCowan's, 351, 393
loons, 126, 406
looper, cabbage, 394
loopers and geometer moths, 192
Lophostoma silvicolum, 210
Loris lydekkerianus, 124
loris, Mysore slender, 124

Lota lota, 154, 156
lotic, 74
lotus blue butterfly, 420
Louisville cave beetle, 420
louse
  dog, 270
  louse flies, 285, 287, 329–330
  human body, 301
Loxaspis, 304
Loxodonta, 268, 317, 325
Loxodonta africana, 210, 402
lubber grasshoppers, 181
Lucilia, 322, 416
Luscinia megarhynchos, 250
Lutra canadensis, 238
Lutzomyia, 312–313
Lutzomyia anthophora, 314
Lutzomyia evansi, 314
Lutzomyia shannoni, 313, 314
Lutzomyia trapidoi, 314
Luzon peacock swallowtail, 420
Lycaeides argyrognomon lotis, 420
Lycaeides melissa samuelis, 420
Lycaon pictus, 252
Lycosa ammophila, 397
Lygaeidae, 188
Lymantria dispar, 223, 225–226, 375, 393, 417
Lyme Borreliosis, 254–256
Lyme Disease, 254–256, 291
Lymphatic Filariasis, 275
Lynx canadensis, 257, 276
Lynx rufus, 238, 271–272, 289, 297, 313
lynx, Canada, 257, 276

Mabuya agilis, 110
mackerels, 152
Macracanthorhynchus hirudinaceus, 275, 279–283
Macrocheles mammifer, 239
macroparasites, 233
Macroscelidae, 114
Macrotermes, 125, 209, 401
Macrotermes subhyalinus, 99
maggot therapy, 416
Magicicada cassini, 157
magpie, black-billed, 249, 252, 353, 396
Malacorhynchus membranaceus, 136, 150
Malacostraca, 9
Malaria
  Avian, 264, 269–271
  of humans, 404–405
  Rodent, 264
male reproductive system, 65
mallard, 261, 272
Malpighian tubules, 69

mammal diets, 107, 114–126
mammal stomach and scat contents, 116–122
Mammalogy, Section of Economic Ornithology and, 142
management
  environmental, 366–370
  pesticide resistance, 362
  resistance, 362
  rotational impoundment, 370
managing insect resources, 425–434
mandibles, 6, 36–37
mandibular palps, 37
Manduca blackburni, 421
mangabey, 249
mange mites, 289–291, 292–293
Mange, Demodectic, 291
Mange, Sarcoptic, 291–293
Manis, 164
Mansonella, 311
Mansonia, 271, 276, 307
  crassipes, 271
mantidflies, 15, 22, 188–189
mantids, 14, 18
mantispids, 189
mantled squirrel, 117, 120
Mantodea, 14, 18
Mantophasmatodea, 14
Mardon skipper, 421
Margarops fuscatus, 318, 333
Mariana eight-spot butterfly, 420
Mariana wandering butterfly, 420
marine toad, 396
marlins, 152
marmosets, 124
marmot, yellow-bellied, 251
Marmota flaviventris, 251, 313
Marmota monax, 158, 297, 313
marmots, 115, 328
  yellow-bellied, 313
marsh rabbit, 238
marsupials, 115
martens, 114
Martes pennant, 292
martin bug, 303
martin
  house, 249
  purple, 132, 149
masked shrew, 396
mass of diets, 93
mast, 216
Material Safety Data Sheet (MSDS), 343
maxillae, 36–37
maxillary palps, 37
mayflies, 14, 16, 171, 199–200
mayflies, 432
Mazama gouazoubira, 396
McCowan's longspur, 351

meadowlark, 149
  eastern, 132, 143–144
  western, 352, 393
mealworm, 416
mechanical control, 370–371
Mecoptera, 15
Megalagrion leptodemus, 420
Megalagrion nesiotes, 420
Megalagrion nigrohamatum nigrolineatum, 420
Megalagrion oceanicum, 420
Megalagrion pacificum, 421
Megalagrion xanthomelas, 420
Megaloptera, 14, 22, 176–177, 432
Megalopyge opercularis, 163
Megascops asio, 151, 318
Megascops choliba, 151
Megascops kennicotti, 151
Megninia, 287
Melaleuca, 417
Melanerpes carolinus, 128
Melanerpes erythrocephalus, 128, 143, 144
Melanerpes lewis, 128
Melanerpes uropygialis, 128
Melanoplus aridis, 393
Melanoplus differentialis, 184
Melanoplus sanguinipes, 68, 156
Meleagris gallopavo, 216, 248, 254, 272
Meles meles, 287
Melospiza melodia, 394, 254
Melursus ursinus, 164
Membracidae, 188
membrane
  basement, 34–35
  peritrophic, 49
  intersegmental, 34
Menacanthus affinis, 301
Menacanthus alaudae, 301
Menacanthus pricei, 287
Menopon gallinae, 300
Mephitidae, 114
Mephitis mephitis, 118, 123, 126–127, 251, 272, 296–297, 353
mergansers, 259
Meropidae, 143
Merops ornatus, 397
Merostomata, 6
Mesostigmata, 285
mesothorax, 41, 44
Mesoviliidae, 175
Mesozoic era, 27, 29
metabolic heat, 70
metamorphosis, 56, 63
metathorax, 41, 44
Metoposarcophaga importuna, 325
Mexican fringe-toed lizard, 110
Mexican knob-scaled lizard, 110
Mexican marbled toad, 108

Mexican wolf, 119, 123
mice, 265, 303, 355
  deer, 431
  grasshopper, 115, 117
  New World, 115
  white-footed, 254
  wood, 254
  yellow-necked, 254
Micrathene whitneyi, 151
microbial degradation of pesticides, 349
microbiologists, 233
Microcebus rufus, 124
Microcerotermes, 211
Microlichus, 287
microparasites, 233
Micropterus salmoides, 140, 153, 155
micropyle, 66
Microtus, 253
Microtus arvalis, 253
Microtus chrotorrhinus, 392
Microtus californicus, 257
midge, 177, 191–192, 199, 395, 432
  biting, 285, 310–312
midgut, 49
migratory grasshopper, and sparrows, 156–157
millipedes, 6, 9–11
Milvus migrans, 355
Milvus milvus, 355
mimicry, 159–160
  aggressive, 160
  Batesian, 160
  Müllerian, 160
  Wasmannian, 160
Mimidae, 142
Mimus polyglottos, 134, 143–144, 149, 251
minnows, 152
  panchax, 396
Miocene epoch, 29
Miridae, 187–188
mission blue butterfly, 420
Mississippi kite, 151
Mississippian period, 27, 29
Mitchell's satyr butterfly, 420
mite, 6–8, 285, 287, 289–293
  bird, 292
  chigger, 290
  down, 292
  ear, 291
  elk scab, 291
  feather, 287, 292
  mange, 289–291, 292–293
  nasal, 287
  Notoedric cat, 291
  Notoedric squirrel, 291
  poultry red, 290
  quill, 292

respiratory, 291
  sarcoptic mange, 292–293
  skin, 287, 292
  subcutaneous, 287
  vane, 292
mobbing, 163
mockingbird, 142
  northern, 134, 143–144, 149, 251
mode of action, insecticide, 345
mold beetle, Helotes, 421
mole, 114–115, 122, 125
  eastern, 115–116, 120, 125
  golden, 114
  Townsend's, 115–116, 122
mole crickets, 184
molluscicides, 343
Molothrus ater, 132, 291
molting, 35
molting fluid, 35
Monanema martini, 275
Monanemosis, 275
Monarch butterfly, 423–424
mongoose, 396
  Indian, 396
Moniezia benedeni, 275
Moniliformis moniliformis, 275
monitor lizard, 210
monkey, 114, 303
  howler, 251, 328, 329
  owl, 249
  spider, 249, 251
  squirrel, 124, 249
Monochamus alternatus, 216
monogenic resistance, 373
monophagous, 76
moose, 114, 236, 277, 279, 296, 315, 318, 326
moose flies, 318
mosquito, 177, 179, 192, 194–195, 276, 285, 305–308, 370
mosquito control impoundments, 370
mosquito fish, 396
Motacilla flava, 396
moths, 15, 24, 190–193
  Blackburn's sphinx, 421
  bogong, 414–415
  diamondback, 366
  gypsy, 393, 417
  Kern primrose sphinx, 421
Mound
  ant, 210
  termite, 209–212
Mount Hermon June beetle, 419
mountain bluebird, 130
mountain goat, 252, 291, 301
mountain goat, Nubian, 291
mountain gorilla, 126
mountain hare, 254, 279

mountain lion, 257, 301
mountain pine beetle, 216
mountain sheep, 291
mountain spiny lizard, 110
mountain zebra, 396
mourning dove, 251, 352
mouse, cotton, 238, 251
mouse
    deer, 251, 294
    field, 400
    grass, 251
    grasshopper, 117, 257
    house, 117, 120, 253
    prairie deer, 117, 120
    vesper, 251
    western harvest, 296
    white-footed, 117, 120, 296, 393
    wood, 297
    woodland jumping, 117, 120
mouth hooks, 37, 192, 195, 318, 325
mouthparts, 37
    chewing, 36–37
    ectognathous, 37
    piercing-sucking, 37, 40, 41
    siphoning, 37, 42
    sponging, 37, 43
    sucking, 37, 40–42
Moxostoma macrolepidotum, 141, 154
MSDS (Material Safety Data Sheet), 343
mule deer, 236, 277, 291, 296, 301,
    311, 329–330
Müllerian mimicry, 160
mullet, 407
    striped, 407
    white, 407
multivoltine, 71
mummichog, 407
Muntiacus reevsi, 252
muntjac, 252, 301
Murgantia histrionic, 68
Muridae, 115
Murine Plague, 255–258, 332
Murray turtle, 112
Mus, 355
Mus musculus, 117, 120, 253
Musca autumnalis, 100, 318
Musca domestica, 195, 318–321
Muscapidae, 142
muscid flies, 192, 194, 285, 318–319
Muscidae, 192, 195, 203–204, 285,
    318–319
Muscidifurax, 321
Muscidifurax raptor, 379
muscles, 45
    abdominal, 47
    direct flight, 45, 49
    indirect flight, 45, 49
    visceral, 47

muscular system, 45, 48
muskoxen, 114
muskrat, 115, 238, 253, 355
Mustela, 292
Mustela nigripes, 257, 356
Mustelidae, 114
mutualism, 241, 400
Myadestes townsendi, 130, 143
myalgia, febrile, 247
mycetophagous, 79
Mycetophilidae, 371
Mycobacterium avium, 253
mycotoxins, 259
Mycteralges mesomorphus, 287
Mycteria americana, 287, 334
Myiarchus cinerascens, 130, 215
Myiarchus crinitus, 322, 432
myiasis, 305, 321, 325
    obligate, 322
    primary facultative, 322
    secondary facultative, 322
Myndus crudus, 216
Myocastor coypus, 355
Myodes, 253
Myotis capaccinii, 120
Myotis lucifugus, 251
Myrmecobius fasciatus, 164
Myrmecophaga tridactyla, 164
myrmecophages, 164
Myrmecophagidae, 115
Myrmeleontidae, 189
Myrmotherula fulviventris, 394
Myrtle's silverspot butterfly, 420
Mysore slender loris, 124
Myxomatosis, 248

Nagana, 268, 316
name, scientific, 5
Napaeozapus insignus, 117, 120
nasal mites, 287
Nasutitermes, 211, 401
Nasutitermes corniger, 99
Nasutitermes ripertii, 212
native-hen, black-tailed, 301
Natrix, 158
natural biological control, 375
natural enemies, 375–377
nature, balance of, 375
naucorid, Ash Meadows, 421
Naucoridae, 176, 400
Nearctic realm, 32
necrophagous, 321
nectar, 217
nectar-feeding insects, 78
nectaries, extrafloral, 427–428
Nectarinia chalybea, 401
Nectarinia senegalensis, 401
Nectarinia amethystine, 401

nectarivorous wildlife, 98
Nemathelminthes, 274
nematicides, 343
Nematoda, 274
nematodes, 274
Neoconocephalus ensiger, 185
Neocuterebra, 328
Neofiber alleni, 238–239
Neohaematopinus sciuropteri, 301
Neonympha mitchellii francisci, 420
Neonympha mitchellii mitchellii, 420
Neophilopterus heteropygus, 287
Neoptera, 14
Neoschoengastia americana, 287
Neotoma, 254, 329
Neotoma albigula, 313
Neotoma mexicana, 251
Neotrombicula whartoni, 287
Neotropical realm, 32
Neottialges apunctatus, 287
Neottialges kutzeri, 287
Neottialges mycteriae, 287
Nepidae, 175–176, 400
nerve poisons, 346
nervous system, 48, 52
nest fumigation behavior, 288
nest reuse, 286
nesters, colonial, 286
net primary productivity (NPP), 198
Neuroptera, 15, 22, 188–189
Neurotrichus gibbsii, 293
New World mice, 115
New World rabbits, 248, 252–253
New World screwworm fly, 322–324
New World warblers, 142
Newman's knob-scaled lizard, 110
newts, 359
Nicrophorus, 204, 206
Nicrophorus americanus, 419, 424
nidicolous, 293
nighthawks, 126, 165
nightingale, common, 249
nightjars, 126, 165
nine-banded armadillo, 118, 120, 123,
    238, 251, 313
Nitidulidae, 216
no-see-ums, 310
Noctuidae, 192–193, 371
nocturnal activity, 165
nomenclature, binomial, 5
noninfectious diseases, 241
North American porcupine, 297
northeastern beach tiger beetle, 421
northeastern blind snake, 112
northern beardless-tyrannulet, 401
northern bobwhite, 136, 146, 150,
    248, 251, 334, 354, 398–399,
    429–30

northern cardinal, 134, 150, 251, 394
northern flicker, 128, 143–144
northern fowl mite, 291
northern harrier, 151
northern leopard frog, 108
northern mockingbird, 134, 143–144,
    149, 251
northern oriole, 226
northern parula warbler, 392
northern pike, 154–155
northern pintails, 261
northern pygmy owl, 151
northern raccoon, 118, 120, 123, 158,
    165, 251, 265, 276, 291, 313–314,
    332
northern saw-whet owl, 151
northern slimy salamander, 108
northern spring peeper, 108
northern wheatear, 301
Norway rat, 254
nose bots, 326
Nosema locustae, 378
Notochaeta bufonovoria, 325
Notoedres, 289
Notoedres cati, 289, 291
Notoedres centrifera, 289, 291
Notoedres muris, 289, 291
notoedric cat mite, 291
notoedric rat mite, 291
notoedric squirrel mite, 291
Notonecta, 176
Notonectidae, 176, 395, 400
Notophthalmus viridescens, 359
Nubian mountain goat, 291
Nudaurelia oxemensis, 99, 100
numbat, 164
numerical responses of predators, and
    population regulation, 387
nurseries, insect, 428
nuthatches, 142
nutria, 355
nutrient
    content of insects, 97–103
    cycling, 206–210
    fluxes, 200
    pulses, 200
nutritional value of insects, 97
Nyctereutes procyonoides, 121, 123
Nycteribiidae, 285–286
nymphs, 63
Nysius wekiuicola, 419

Oak Wilt fungus, 216
obligate myiasis, 322
obligatory diapause, 71
occurrence, frequency of, 93
oceanic Hawaiian damselfly, 420
ocelli, 37, 54

ocelot, 301
Ochlerotatus sollicitans, 307, 368
Ochlerotatus taeniorrhynchus, 368
Ochlerotatus triseriatus, 307
Odocoileus, 254, 296, 326
Odocoileus hemionus, 236, 277, 291, 296,
    311, 329
Odocoileus virginianus, 216, 236, 238,
    250, 250, 272, 277, 291, 297, 311,
    313, 323, 329, 434
Odonata, 14, 16, 173–175, 395, 400,
    432
Odontophoridae, 150
odor glands, 55
Oecanthus nigricornis, 185
Oecetis, 182
Oeciacus hirundinis, 303–304
Oeciacus vicarious, 287
Oedogonium, 357
Oestridae, 285, 325–329
Oestroderma, 328
Oestrus, 328
Oestrus ovis, 326
Ohlone tiger beetle, 421, 425
Oklahoma salamander, 108
Old World flycatchers, 142
Old World rabbits, 248
Old World rats, 115
Old World Screwworm, 322, 324–325
Olesicampe, 196, 377
Olfersia sordida, 287, 330
Olfersia spinifera, 287
Oligocene epoch, 29
oligophagous, 76
Olios, 397
olive-sided flycatcher, 130
ommatidium, 53, 54
omnivory, 85
Onchocerca, 311
Onchocerca cervipedis, 310
Oncorhyncus, 200
Oncorhyncus kisutch, 141, 154
Oncorhyncus mykiss, 152, 425
Ondatra zibethicus, 253, 355
one-host ticks, 293
ontogenetic shifts in wildlife feeding,
    102
Onychognathus nabouroup, 396
Onychomys, 115
Onychomys leucogaster, 117, 257
Onychomys torridus, 117
Onychorhynchus coronatus, 401
Ooencyrtus kuvanae, 380
ootheca, 66, 181
Opheodrys aestivus, 107, 112
Ophiostoma, 216
Opiliones, 8
opisthosoma, 289

opossum, 265, 303, 314
    southern, 251
    Virginia, 118, 120, 123, 126, 238,
        251, 265, 272, 291, 297, 332
Opthalmodx, 291
optimal foraging theory, 93
Opuntia, 414
orangeblack Hawaiian damselfly,
    420
orchard oriole, 254
Orchopeas howardi, 332
Ordovician period, 27, 29
Oreamnos americanus, 252
Oregon silverspot butterfly, 420
Oreotragus oreotragus, 396
organ
    Johnston's 54
    tymbal, 60
Oriental rat flea, 257
Oriental realm, 32
oriole, 149
    Baltimore, 132
    black-backed, 424
    northern, 226
    orchard, 254
    Scott's, 424
Orius insidiosus, 376, 380
Ormia depleta, 380
Ornithicoris, 304
Ornithicoris pallidus, 287
Ornithodoros, 246, 253, 287, 293
Ornithodoros concanensis, 294, 304
Ornithodoros hermsi, 294
Ornithodoros parkeri, 294
Ornithodoros puertoricensis, 294
Ornithodoros tholozani, 296
Ornithodoros turicata, 294
Ornithoica vicina, 330
Ornithology and Mammalogy, Section
    of Economic, 142
Ornithonyssus bursa, 287, 291
Ornithonyssus sylviarum, 287, 291,
    423
Orthoporus dorsovittatus, 10
Orthoptera, 14, 18, 181,184–185
Orthopteroidea, 14
Orycteropus afer, 114, 164, 317
Oryctolagus cuniculus, 355
Oryctolagus, 248, 355
Oryx gazelle, 252
Oryzomys nigripes, 251
Oryzomys palustris, 238
osmeterium, 161
osprey, 354, 355
Osteichthyes, 152
ostrich, 142, 317
Ostrinia, 293

Ostrinia nubilalis, 375
Ostrinia megnini, 294
Otodectes cynotis, 291
otter, 114
    river, 238
Otus asio, 322
Otus elegans, 151
Otus flammeolus, 151
Ouachita map turtle, 112
outbreak, 235
    causes of, 214
    of insects, 200
    of insects, and ecosystems, 214–215
ovaries, 65–66
ovenbirds, 142
Oviduct Fluke, 275
oviparity, 66
ovipositor, 45
Ovis canadensis, 236, 252, 254, 289, 311
ovoviviparity, 66
owl, 126, 151, 165, 253, 259, 330, 406
    monkey, 249
    African wood, 151
    barn, 272
    burrowing, 332, 353
    eastern screech, 151, 318
    elf, 151
    ferruginous pygmy, 151
    flammulated, 151
    northern pygmy, 151
    northern saw-whet, 151
    screech, 322
    tawny, 151
    tropical screech, 151
    western screech, 151
Oxidus gracilis, 10
oxpeckers
    red-billed, 395
    yellow-billed, 395
Oxylipeurus clavatus, 287
Oxyspirura mansoni, 275

Pacific Hawaiian damselfly, 421
Pacific salmon, 407
paddlefish, 152
painted bunting, 134, 150
Palaemontes paludosus, 102
pale spear-nosed bat, 116
pale-winged starling, 396
Palearctic realm, 32
Paleocene epoch, 29
Paleoptera, 14
Paleozoic era, 27, 29
palewinged trumpeter, 396
Pallas's long-tongued bat, 116
Pallasiomyia, 328

palm weevils, 416
Palos Verdes blue butterfly, 420
palps, 36–37, 40
    mandibular, 37
    maxillary, 37
Pan, 124, 250
Pan troglodytes, 292
panchax minnow, 396
panda
    giant, 114
    lesser, 114
pandemic, 235
Pandion haliaetus, 354–355
pangolins, 114, 164
Panstrongylus, 265, 302
Panthera leo, 252, 268
Panthera onca, 271, 276
Panthera tigris, 271
paper wasps, 195–196, 333
Papilio chikae, 420
Papilio cresphontes, 158, 161
Papilio homerus, 420, 423
Papilio hospiton, 419
Papilio polyxenes, 192, 193
Papilionidae, 192
Papio, 250
Paracimex, 304
parakeets, 210
paralysis, tick, 291, 296
parasite-induced changes in host
        behavior, 242–243
parasitemia, 310
parasite
    aberrant, 242
    facultative, 242
    incidental, 242
    obligate, 242
    periodic, 242
    permanent, 242
    stationary, 242
    temporary, 242
    types, 242
parasitic diseases, 233
parasitic insects, 79
parasitism, 241–241, 400
parasitoids, 79
    ectoparasitoids, 79
    endoparasitoids, 79
parasitologists, 233
parasocial behavior, 61
Paridae, 142, 150
Parnassius apollo, 423
Paroaria coronate, 394
parrot, 126, 210
    Puerto Rican, 318
    red-capped, 138, 151
parthenogenesis, 66
partridge, grey, 427

Parula americana, 392
Parulidae, 142, 150
Parus atricapillus, 226, 432
Parus major, 360
passenger pigeon, 424
Passer domesticus, 249, 251, 254, 352, 359
Passerculus sandwichensis, 156, 393
Passeriformes, 142
Passerina ciris, 134, 150
passive anting, 301
passive transmission of disease, 240
Pasteurella multocida, 253
pathogenicity, 235
pathogens, emerging, 236
pathways of invasion, 219
Pavlovskiata, 328
Pawnee montane skipper, 421
pearly-eyed thrasher, 318
Pecari tajacu, 118, 123
peccaries, 114
peccary, collared, 118
Pectinopygus occidentalis, 287
Pectinopygus tordoffi, 287
pectoral sandpiper, 138, 150
pedicel, 37, 40
Pediculus humanus, 301
pedipalps, 6, 289–290
peeper, northern spring, 108
Pegomya betae, 68, 194
Pelecaniformes, 126
Pelecanus, 92
Pelecanus erythrorhynchos, 287, 319
Pelecanus occidentalis, 287, 354
Pelecitosis, 275
Pelecitus capiceps, 275
Pelecitus fulicaeatrae, 275
Pelecitus roemeri, 275
pelican, 126, 259, 330, 406
    American white, 319
    brown, 354
pellet, 91
penguins, 126
Pennsylvanian epoch, 27, 29
Pentatomidae, 188
Peramelemorphia, 115
Perca flavescens, 140, 153, 155
Perca fluviatilis, 395
perch, 152, 395
    white, 407
    yellow, 138, 153, 155, 407
perching birds, 142
Perciformes, 152
Perdix perdix, 427
peregrine falcon, 354
perennial, 71
perikaryon, 52

period, Cambrian, 27, 29
  Carboniferous, 27, 29
  Cretaceous, 27, 29
  Devonian, 27, 29
  Jurassic, 27, 29
  Mississippian, 27, 29
  Ordovician, 27, 29
  Permian, 27, 29
  Quaternary, 27, 29
  Silurian, 27, 29
  Tertiary, 27, 29
  Triassic, 27, 29
periodical cicada, and blackbirds,
    157–158
Perissodactyla, 114
peritrophic membrane, 49
Perlidae, 173
Permian period, 27, 29
Peromyscus, 265, 303
Peromyscus leucopus, 254
Peromyscus gossypinus, 238, 251
Peromyscus leucopus, 117, 120, 296,
    393
Peromyscus maniculatus, 117, 120, 251,
    431
Peromyscus polionotus, 400
persistence of insecticides, 349
persistence of insects, 375
Peruvian booby, 288
pest management, integrated, 381–382
pesticides, 341–362
  acaricides, 343
  algaecides, 343
  avicides, 343
  bactericides, 343
  contamination of water bodies, 353
  fungicides, 343
  indirect effects, 356–357
  insecticides, 343
  mode of action, 345
  mode of application, 345
  molluscicides, 343
  nematicides, 343
  piscicides, 343
  poisoning, 350–356
  resistance management, 362
  resistance to, 361–362
  risks associated with, 359–361
  rodenticides, 343
  sublethal effects, 354–355
  toxicity categories, 344
petiole, 43
petrels, 126
Petrochelidon pyrrhonota, 134
pewee, eastern wood, 130
Phacochoerus africanus, 210, 268, 317
Phaeogramma, 421
Phalacrocorax, 92

Phalacrodectes mycteria, 287
Phalacrodectes pelicani, 287
Phalacrodectes punctatissimus, 287
Phanerozoic eon, 27, 29
pharyngeal bots, 326
Pharyngobolus, 328
Pharyngomyia, 328
Phasianus colchicus, 254, 261
Phasmida, 14, 18
pheasant, 126, 269
  ring-necked, 254, 261
Pheidole, 275
Pheidole megacephala, 398
phenology, 70
pheromone glands, 55
pheromones, 55, 61, 373
  aggregation, 373
  primer, 61
  releaser, 61
  sex, 61, 373
Pheucticus ludovicianus, 134, 150
Pheucticus melanocephalus, 424
Philander opossum, 265, 303
Philomachus pugnax, 249
Philornis, 318
Philornis downsi, 318, 322
Phlebotominae, 312–314
phlebotomine sand flies, 312–314
Phlebotomus, 312–313
Phlebotomus argentipes, 314
Phlebotomus caucasicus, 314
Phlebotomus papatasi, 314
Phobetron pithecium, 163
phoebe, 142
  black, 130
  eastern, 130, 143, 145
  Say's, 130
Phoenicopteriformes, 126
Phoenicopterus chilensis, 298
Pholidota, 114
Phoridae, 204
Phormia, 322
Phorocera claripennis, 377
photodegradation of pesticides, 349
photolysis of pesticides, 349
Phrynosoma cornutum, 398
Phthiraptera, 14, 20, 285, 298–302
Phyllophaga, 190, 283
Phyllophaga rugosa, 99
Phyllostomus discolor, 116
phylogeny of insects, 15, 27–31
Physa, 357
Physaloptera phrynosoma, 275
physical and chemical defenses, 162
physical controls, 370–371
physical defenses, 162
physical gill, 66, 74
phytophagous, 79

phytophagy 76–78, 85
  by insects, 210–214
Piagetiella bursaepelicani, 287
Piagetiella peralis, 287
Pica pica, 249, 352–353, 396
Piciformes, 126
picivorous, 152
Picoides borealis, 128, 143
Picoides dorsalis, 128
Picoides pubescens, 128, 322
Picoides villosus, 128, 143, 144
piercing-sucking insects, 76
piercing-sucking mouthparts, 37, 40,
    41
Pieridae, 192–193
pierids, 192
pigeon, 126, 269, 406
  passenger, 424
  rock, 272
pig, 268
  guinea, 265, 291
pika, 328
pike, 152, 400
  northern, 154–155
pileated woodpecker, 128, 143
Pilosa, 115
pinewood nematode, 216
pink salmon, 407
pink-eared duck, 136, 150
pintails, northern, 261
pinyon jay, 215
Piophilidae, 204
pipefishes, 152
Pipilo erythrophthalamus, 398
piscicides, 343
Plague, 255–258, 332
  Bubonic, 255–258
  Campestral, 255–258
  Justinian, 255
  Murine, 255–258
  Pneumonic, 255–258
  Septicemic, 255
  Sylvatic, 255–258
plant bugs, 187–188
pleant compensation, 213–214
plant diseases, and ecosystems, 215–217
plant galls, 188
plant remains, decomposition, 201–202
plant resistance, 371–373
Plasmodium, 264, 269, 271, 309, 404
Plasmodium circumflexum, 271
Plasmodium elongatum, 271
Plasmodium falciparum, 404–405
Plasmodium gallinaceum, 271
Plasmodium hermani, 271
Plasmodium juxtanucleare, 271
Plasmodium relictum, 264, 269, 271,
    404

Plasmodium rouxi, 271
Plasmodium vivax, 404–405
plastron respiration, 74
plastron, 66
Platalea, 130
Platalea ajaja, 334
Platycentropus radiates, 400
Platyhelminthes,274
Platypodinae, 216
Platypsyllidae, 285–286
Platyptilia carduidactyla, 193
Platypus, 202
Plecoptera, 14, 16, 172, 432
Pleistocene epoch, 29
Plethodon glutinosus, 108
Plethodon jordani, 110
Pleuronectiformes, 152
Pliocene epoch, 29
plume moth, 193
Plutella xyllostella, 193, 366
Pneumonic Plague, 255–258
Po'olanui gall fly, 421
pocket gophers, 115
Podiceps auritus, 136, 150
Podicipedidae, 150
Podicipediformes, 142
Poecile, 249
Poecile carolinensis, 146
Poecilia reticulate, 396
Pogonomyrmex, 275, 398
poikilotherms, 198
point source pollution, 345
poison glands, 55
poisoning
    aflatoxin, 259, 261
    food, 259
    pesticide, 350–356
    secondary, 357
    contact, 345
    nerve, 346
    stomach, 345
Polctenidae, 303–304
Polioptididae, 142
Polioptila caerulea, 398
Polioptila californica californica, 93, 96
Polistes, 401
Polites mardon, 421
pollack, 152
pollen-feeding insects, 77
pollen, 217
pollen generalists, 217
pollen specialists, 217
pollination, 410–413
pollination and ecosystems, 217–218
pollution, point source, 345
Polyctenidae, 302
polyembryony, 66
polygenic resistance, 373

Polygenis gwyni, 239
Polygonum, 356, 357
polymorphism, 57–58
Polynema eutettixi, 377
Polynema eutettixi, 377
polyphagous, 75
Polyphylla barbata, 419
Polyplax spinulosa, 300
pomace fly, 421
Pomoxis, 155
Pomoxis annularis, 153
Pomoxis nigromaculatus, 140, 153
Pontia protodice, 193
Pooecetes gramineus, 393
Popillia japonica, 223
population regulation, 238, 387, 390
    abiotic, 238
    biotic, 238
    competition-based, 238
    density-dependent, 390
    density-independent, 390
porcupine, 265, 276, 297, 301, 303,
    314
Portschinskia, 328
Porzana carolina, 143, 146
possums, 115
Postulates, Koch's, 241
Potamochoerus larvatus, 268, 317
Potos flavus, 250
poultry red mite, 290
Pox, Avian, 248
prairie chicken, 146
prairie deer mouse, 117, 120
prairie dog, 215, 257–258, 332, 355
precinctive, 219
predaceous diving beetle, 177–178, 400
predation by insects on wildlife, 397–400
predation by wildlife on insects, 387–397
predation of animal ectoparasites by birds,
    395–396
predation, cross, 397
predation, defense against, by insects,
    158–165
predator satiation, 163
predators, aquatic, 75
predators, seed, 218
predatory insects, 79
preening, 300
preference, food, 85
preoral digestion, 289
pretarsus, 36, 41
prevalence, 93
primary facultative myiasis, 322
primary host, and disease, 239
primary productivity, 198
primary screwworm, 322
primates, 114, 312, 315

primer pheromones, 61
Primicimex, 304
Pristophora erichsonii, 396
proboscis, tubular, 37, 42
Procellariiformes, 126
procuticle, 34
Procyon lotor, 118, 121,123, 158, 165,
    238, 251, 265, 272, 276, 313
Procyonidae, 114
production, sound, 60
productivity, net primary (NPP) , 198
    primary, 198
    secondary, 198
Progne subis, 92, 132, 149
prokaryotes, 252
prolegs, 41, 190, 192
Prolistrophorus birkenholzi, 239
pronghorn, 114, 215, 252, 254, 272,
    291, 311–314, 318
Propicimex, 304
Propithicus verreauxi, 124
Prosimulium impostor, 310
prosoma, 289
Prosthogonimus macrorchis, 275
Prostigmata, 285
Proteles cristata, 164
Proterozoic eon, 27
prothoracic glands, 55
prothorax, 41, 44
Protocalliphora, 322
Protocalliphora sialis, 322
protocerebrum, 53
Protophormia, 322
Protophormia terraenovae, 322
protozoa, 263–274
protozoan diseases, 263–274
    African Trypanosomiasis, 266–270
    American Trypanosomiasis, 263–267
    Avian Malaria, 269–271
    Chagas Disease, 263–267
    East African trypanosomiasis, 268
    Nagana, 268
    Sleeping Sickness, 266–270
    Toxoplasmosis, 271–274
    West African trypanosomiasis, 268
Protura, 6, 13
proturans, 6, 13
Przhevalskiana, 328
Psephenidae, 177
Psephotus, 210
Pseudacris crucifer crucifer, 108
Pseudalloptinus, 287
Pseudanophthalmus caecus, 420
Pseudanophthalmus cataryctos, 420
Pseudanophthalmus frigidus, 420
Pseudanophthalmus holsingeri, 420
Pseudanophthalmus inexpectatus, 420
Pseudanophthalmus inquisitor, 420

Pseudanophthalmus major, 420
Pseudanophthalmus parvus, 420
Pseudanophthalmus pholeter, 420
Pseudanophthalmus troglodytes, 420
Pseudemys scripta, 103, 107
Pseudocopaeodes eunus obscurus, 421
Pseudogametes, 328
Pseudomenopon austalis, 301
Pseudomenopon pilosum, 301
Pseudomyrmenigrocinctus, 401
Pseudomyrmespinicola, 401
Pseudomyrmex, 401
Pseudomys nelson, 398
pseudopods, 315
Pseudoscorpiones, 6–8
pseudoscorpions, 6–8
Psilotreta, 182
Psittacidae, 151
Psittaciformes, 126
Psitticimex, 304
psocids, 14, 20, 186
Psocoptera, 14, 20, 186
Psophia leucoptera, 396
Psorophora confinis, 306
Psorophora, 276
Psoroptes, 289
Psoroptes cervinus, 289, 291
Psoroptes cuniculi, 291
Psoroptes natalensis, 291
Psoroptes ovis, 289
Psychodidae, 285, 312–314
Pteromalus eurymi, 377
Pterophoridae, 193
Pterygota, 14
PTTH, 56
Puerto Rican parrot, 318
puffin, Atlantic, 254
pulse, energy, 200
pulse, nutrient, 200
pumpkinseed, 138, 153, 155
punkies, 310
pupal stage, 65
pupariation, 318
puparium, 318
Puritan tiger beetle, 421, 425
purple martin, 132, 149
Purpureicephalus spurius, 138, 151
putative cause of disease, 241
Pycnogonida, 6
Pycnoscelus surinamensis, 183
pygmy-owl, ferruginous, 318
Pygosteus pungitius, 141
Pyloric Caeca Fluke, 275
Pyralidae, 192, 215, 226
pyramids, trophic, 198
Pyrgus ruralis lagunae, 421
pyrrhuloxia, 134, 150
Pytoseiulus persimilis, 378

Q fever, 332
quail, 126, 330, 332, 354, 406
    California, 136, 150
Quaternary period, 29
Queen Alexandra's birdwing butterfly,
    423
Quercus, 226
questing, 296
quill mites, 292
quillback, 141
Quino checkerspot butterfly, 420
Quiscalus quiscula, 132, 291

rabbit, 332, 355, 265, 312–325
    Audubon's cottontail, 251
    cottontail, 238, 253, 399
    feral, 355
    jungle, 248
    marsh, 238
    New World, 248, 252–253
    Old World, 248
raccoon, 114, 272
    dog, 120, 123
    northern, 118, 120, 123, 158, 165,
        238, 251, 265, 276, 291, 313–314,
        332
racer, western yellow-bellied, 110
Rachel Carson, 341
Radfordia, 239
rafts, egg, 307
rail, black, 398
Raillietia, 291
Raillietia auris, 291
Raillietia cesticillus, 275
Railietia loeweni, 275
rails, 142, 259, 301
rainbow bee-eater, 397
rainbow smelt, 407
rainbow trout, 138, 152, 407, 425
Rallidae, 150
Ramphotyphlops, 107
Ramphotyphlops australis, 112
Ramphotyphlops polygrammicus, 112
Rana brownorum, 108
Rana catesbeiana, 359
Rana mucosa, 359
Rana pipiens, 108, 359
Rana porosa brevioda, 108
Rana sphenocephala, 108
Rana vaillanti, 108
Ranatra, 175
Rangifer tarandus, 289, 292, 305, 326
Raphidioptera, 15
raptors, 151
rat, 355
    black, 254
    cotton, 251, 400
    hispid cotton, 238

    kangaroo, 254
    Norway, 254
    Old World, 115
    rice, 251
    wood, 251, 254, 329
rat flea, Oriental, 257
rat snake, black, 158
rattlesnake, banded rock, 112
Rattus, 355
Rattus norvegicus, 254
Rattus rattus, 254
raven, fan-tailed, 396
ray-finned fishes, 152
realm, Australian, 32
    Ethiopian, 32
    Holarctic, 32
    Nearctic, 32
    Neotropical, 32
    Oriental, 32
    Palearctic, 32
recreation, wildlife-based, 406–407
red deer, 292
red fowl mite, 291
red fox, 119, 120, 158, 238–239, 251,
    272, 292, 297
red imported fire ant, 334, 297–400, 417
red knot, 138, 150
red squirrel, 301
red wolf, 296, 301
red-backed voles, 253
red-bellied woodpecker, 128
red-billed oxpecker, 395
red-capped parrot, 138, 151
red-capped parrot, 151
red-cheeked salamander, 110
red-cockaded woodpecker, 128, 143
red-crested cardinal, 394
red-eared turtle, 103, 107
red-headed woodpecker, 128, 143–144
red-shouldered hawk, 151
red-tailed hawk, 359
red-winged blackbird, 132, 143, 148,
    157–158, 291, 360, 392
redhead duck, 136, 150
redstart, American, 390
Reduviidae, 285, 302
reflex bleeding, 162
regions, body, 35
regulation, population, 238, 387, 390
    abiotic, 238
    biotic, 238
    bottom-up, 390
    competition-based, 238
    lateral, 392
    top-down, 390
Reguliidae, 142, 150
Regulus calendula, 134, 150
Regulus satrapa, 392

reindeer, 292, 305
Reithrodontomys megalotis, 296
Relapsing Fever, 291
relative sampling measures, 86–91
releaser pheromones, 61
remains, decomposition of plant, 201–202
reproductive system, 65
reptiles, 106–113, 268
reservoir host, and disease, 239
resiliency of insects, 375
resilin, 34
resistance management, insecticide, 361–362
resistance, host plant, 371–373
   monogenic, 373
   polygenic, 373
resources, foliar, 428–432
respiration, plastron, 74
respiratory mites, 291
respiratory siphon, 52, 307
respiratory trumpets, 307
Reticulitermes flavipes, 59, 164
Reticulitermes hageni, 59
reuse, nest, 286
Rhadine exilis, 421
Rhadine infernalis, 421
Rhadine persephone, 419
Rhagionidae, 333
Rhagoletis, 66
Rhaphiomidas terminatus abdominalis, 421
rhino, white, 403–404
rhinoceros, 114, 317, 328
Rhinocerotidae, 114
Rhinoestrus, 328
Rhinonyssus rhinolethrum, 287, 416
Rhipicephalus, 287, 293
Rhodnius prolixus, 265, 302
Rhombomys opimus, 313
Rhynchocephalia, 106
Rhynchophorus palmarum, 99, 100, 416
rice rat, 251
Rickettsia, 252
Rickettsia prowazekii, 301
Rickettsia rickettsii, 253
rickettsiae, 252
riffle beetle, 177–178
   Stephan's, 421
ring-necked pheasant, 254, 261
ring-tailed lemur, 124
Riparia riparia, 134
risks of insecticides, 359–361
river otter, 238
Rivoltasia, 287
roach, 395
roadrunners, 406
robber flies, 192, 194

robin, American, 130, 149, 249, 251, 254, 291, 352
rock crawlers, 14
rock pigeon, 272
rock squirrel, 257, 313, 332
rock vole, 392
Rocky Mountain Spotted Fever, 291
rodent bots, 328
Rodent Malaria, 264
Rodentia, 114–115, 122
rodenticides, 343
rodents, 114–115, 122, 257, 325
roe deer, 279, 292, 355
Rogenhofera, 328
Romalea microptera, 36
Romaleidae, 181, 184
Romaña's sign, 264
root-feeding insects, 78
rose-breasted grosbeak, 134, 150
roseate spoonbill, 334
Ross's geese, 261
Rostrhamus sociabilis, 92, 334
rostrum, 302
rotational impoundment management, 370
rough green snake, 107, 112
rove beetles, 427
ruby-crowned kinglet, 134, 150
ruby-throated hummingbird, 128, 143
ruff, 249
ruffed grouse, 143, 146, 216
rufous-naped wren, 401
Rumex, 356
Rupicapra rupicapra, 292
rusty blackbird, 132
Ruttenia, 328

Sabathes chloropterus, 250
sable antelope, 252
Sacramento Mountains checkerspot butterfly, 420
Saguinus midas, 124
Saiga antelope, 272, 328
Saiga tatarica, 272
Saimiri, 259
Saimiri sciureus, 124
Saint Francis'satyr butterfly, 420
salamander, 105
   northern slimy, 108
   Oklahoma, 108
   red-cheeked, 110
salivary glands, 55
Salmo clarkii, 140
Salmo gairdneri, 140
Salmo trutta, 140, 154–155
salmon, 152
   Chinook, 407
   chum, 407

coho, 141, 154–155, 407
   Pacific, 407
   pink, 407
   sockeye, 407
Salmonella, 321
Salmoniformes, 152, 155
Salt Creek tiger beetle, 421, 425
Salvelinus alpinus, 155
Sambucus, 428
sampling measures
   absolute, 87–91
   relative, 86–91
sampling techniques, arthropod, 86–91
san Bruno elfin butterfly, 420
sand flies, 285, 310, 312–314
sandhill crane, 429
sandpiper, 259
   pectoral, 138, 150
   semipalmated, 138, 147, 150
   sharp-tailed, 138, 150
   stilt, 138, 150
sap beetles, 202, 216
sapsucker, yellow-bellied, 128
Sarconema eurycercav, 275
Sarconema or heartworm, 275
Sarcophaga crassipalpis, 326
Sarcophaga haemorrhoidalis, 326
Sarcophagidae, 192, 204, 285, 324–326
Sarcoptes scabiei, 289, 291–293
sarcoptic mange mite, 289, 291–293
Sarcoptic Mange, 289, 291–293
satiation, predator, 163
Saturniidae, 192
Savannah sparrow, 156–157, 393–394
sawfly, 15, 24, 192–193, 195–196, 333–334
   larch, 396
Say's phoebe, 130
Sayornis nigricans, 130
Sayornis phoebe, 130, 143
Sayornis saya, 130
scabies, 292
scale, giant margarodid, 414
Scalopus aquaticus, 115–116, 120, 125
scaly anteater, 164
Scandentia, 114
Scapanus townsendii, 115–116, 122
scape, 37
Scaphirhynchus suttkusi, 141, 153, 155
scarab beetle, 190, 371
Scarabaeidae, 190, 203, 371
Scarabaeinae, 203
scarlet-chested sunbird, 401
scat contents of mammals, 116–120
Scatophagidae, 203
scats, 92
scaup, lesser, 136, 150
scavengers, 73

Sceloporus jarrovi, 110
Sceloporus occidentalis, 354
Schaus swallowtail butterfly, 420
Schistocerca americana, 5, 184
Schistocerca gregaria, 220, 353
scientific name, 5
scissor-tailed flycatcher, 130
Sciuridae, 115
Sciurus, 254, 289
Sciurus carolinensis, 158, 216, 234, 238, 329
Sciurus niger, 158, 238, 250
sclerites, 34
sclerotization, 35
Scolopacidae, 150
Scolopax minor, 138, 143, 147
Scolopendra, 11
Scolopendra gigantia, 397
Scolytinae, 200, 216, 221
Scolytus, 216
scorpionflies, 15
Scorpiones, 6–8
scorpions, 6–8
    false, 6–8
    whip, 8
    wind, 6–8
Scott's oriole, 424
scraper, 60
screech owl, 322
screwworm
    New World, 322–324
    Old World, 322, 324–325
    primary, 322
scutellum, 175, 186
Scutomegninia, 287
scutum, 293
sea lion, 272
sea spiders, 6
sea turtle egg fly, 325
seahorses, 152
Seba's short-tailed bat, 116
secondary facultative myiasis, 322
secondary plant substances, 75
secondary poisoning, 357
secondary productivity, 198
Sedum, 423
seed bugs, 188
seed cachers, 218
seed dispersal, and ecosystems, 217–218
seed predators, 218
seed vectors, 218
seed-feeding insects, 78
self-pollination, 217
semiochemicals, 55, 60, 373
semipalmated sandpiper, 138, 147, 150
sentinel chickens, 248
Septicemic Plague, 255
Sequatchie caddisfly, 420

sericulture, 413–414
Serinus, 272
Setaria cervi, 275
Setariosis, 275
Setophaga ruticilla, 390
sex pheromones, 61, 373
shad
    American, 407
    gizzard, 407
    hickory, 407
sharp-tailed sandpiper, 138, 150
sheep
    bighorn, 236, 252, 254, 289, 311, 312, 318
    mountain, 291
shell, egg, 66
shellac from insects, 414
shifts in wildlife feeding
    geographic, 102
    temporal, 100
Shigella, 321
shorebirds, 126
short-billed dowitcher, 138, 150
short-tailed shrew, 116, 120, 392
shorthead redhorse sucker, 141, 154
shredders, aquatic, 75
shrew, 114, 293, 359
    common, 253
    elephant, 114
    least, 117, 120
    masked, 396
    short-tailed, 116, 120, 392
    tree 114
shrew-mole, 293
shrike, 142
    loggerhead, 128, 143, 399
shrimps, 6, 9
    mantis 9
    seed, 9
    tadpole, 9
Sialia, 149
Sialia currucoides, 130
Silia mexicana, 130
Sialia sialis, 130, 249, 318, 322, 360, 429
Sialidae, 176
Siberian chipmunk, 297
Siberian tiger, 289
Sibylloblatta panesthoides, 183
Sickness, Sleeping, 266–270, 316
Sigmodon hispidus, 238, 251, 400
sign of disease, 241
sign, Romaña's, 264
signal word, safety of pesticides, 343
silk glands, 55
silk production, 413–414
silkworm, 192, 413–415, 220
    giant, 192

silky anteater, 164
Silphidae, 204–205
Silurian period, 27, 29
Siluriformes, 152, 155
silverfish, 14
Silvilagus floridanus, 399
Simuliidae, 177, 180, 192, 285, 308–310
Simuliotoxicosis, 309
Simulium, 309
Simulium areum, 310
Simulium meridionale, 308
Simulium piscidium, 309
Simulium venustum, 310
Simulium vernum, 310
Siphlonuridae, 172
siphon, respiratory, 52, 307
Siphonaptera, 15, 24, 330–332
siphoning mouthparts, 37, 42
Sirex noctilio, 216
Siricidae, 216
Sittidae, 142
skeleton, cephalopharyngeal, 318
skeletonizing, 76
skin mites, 287, 292
skink
    Brazilian, 110
    broad-headed, 110, 396
    clay-soil ctenotus, 110
    ctenotus clay-soil, 110
    grand ctenotus, 110
skipper
    Carson wandering, 421
    Dakota, 421
    Laguna Mountains, 421
    Mardon, 421
    Pawnee montane, 421
skunk, 291
skunk, 114, 123
    striped, 118, 123, 126–127, 165, 251, 272, 296–297, 353
skylark, 301
Sleeping Sickness, 266–270, 316
sliders, 398
sloth, 115, 312, 314
    bear, 164
    three-toed, 251
small white cabbage butterfly, 223–224
smelling, 37
smelt, 407
    eulachon, 407
    rainbow, 407
Smith's blue butterfly, 420
snail kites, 334
snake, 106
    black rat, 158
    cottonmouth, 158
    flat-headed, 107, 112

northeastern blind, 107, 112
rough green, 107, 112
southern blind, 112
water, 158
Yucátan cricket-eating, 112
snakeflies, 15
snipe flies, 333
snipe, Wilson's, 138, 143, 147, 150
snout moths, 192
snow geese, 261
sociality, 61
sockeye salmon, 407
soft ticks, 285, 293
Solenopotes ferrisi, 301
Solenopotes muntiacus, 301
Solenopsis geminate, 398
Solenopsis invicta, 196, 223, 226,
    333–334, 376, 398, 399, 417
solifugids, 6–8
solitaire, 143
    Townsend's, 130
Somatochlora hineana, 421
song birds, 142
song sparrow, 254, 394
sora, 143, 146
Sorex, 293
Sorex araneus, 253, 358
Sorex cinereus, 396
Soricomorpha, 114
sound production, 60
southern blind snake, 112
southern chinch bug, 368
southern double-collared sunbird, 401
southern flying squirrel, 301
southern leopard frog, 108
southern opossum, 251
southern pine beetle, 216, 368
space, exuvial, 35
Spalangia cameroni, 320
sparrow, 150
    chipping, 134, 143, 145, 393–394
    field, 145, 394
    house, 249, 251, 254, 332, 352, 359
    savannah, 156–157, 393–394
    song, 254, 394
    vesper, 393–394
    white crowned, 134, 392
sparrows, American, 142
specialists, 76, 85
speciation, allopatric, 5
speciation
    allopatric, 5
    mechanisms of, 5
    sympatric, 5
species
    definition, 5
    number of, 4
spectral tarsier, 120

spermatheca, 66
Spermophilopsis leptodactylus, 313
Spermophilus beecheyi, 355
Spermophilus townsendii, 123
Spermophilus variegates, 257
Speyeria callippe callippe, 419
Speyeria zerene behrensii, 419
Speyeria zerene hippolyta, 420
Speyeria zerene myrtleae, 420
Sphalangia, 321
Sphecidae, 195
sphecids, 195
Sphenarium histrio, 100
Spheniciformes, 126
Sphenophorus callosus, 68
Sphenophorus maidis, 190
Sphingidae, 192–193
sphinx moths, 192–193
Sphyrapicus varius, 128
spider, 6–8
    giant crab, 396
    sun, 6–8
    wolf, 396
spider monkey, 249, 251
spiderling, 8
Spilosoma virginiana, 192
spines, 35
spiny anteater, 164
spiracles, 43, 46,51
Spirocerca lupi, 275, 276, 277
Spirocercosis, 275–277
spirochaetes, 252
Spizella passerina, 134, 143–145, 394
Spizella pusilla, 145
Spodoptera dolichos, 192
Spodoptera exigua, 68
Spodoptera frugiperda, 68, 115
Spodoptera latifascia, 193
sponging mouthparts, 37, 43
spoonbill, 126, 259
    roseate, 334
spots, eye, 160
spotted hyena, 252
spotted knapweed, 430
spotted salamander, 400
spray, irritating, 162
springbok, 328
springtails, 6, 12
spruce beetle, 216
spurs, 35
Squamata, 106
squirrel, 115, 254, 312
    flying, 301, 332
    fox, 238, 301
    gray or grey, 216, 234, 301, 329
    ground, 313, 332, 355
    mantled, 117, 120
    monkey, 124, 249

red, 301
rock, 257, 313, 332
St. Louis Encephalitis, 250
stable fly, 233, 319–320
stadium, 63
Staphylinidae, 204, 206, 285–286,
    427
starling, 142
    European, 136, 150, 249, 251, 352,
        396
    pale-winged, 396
Steinernema 378
    feltiae, 380
Stellaria, 357
stemmata, 37, 54
Stenelmis, 178
Stenopelmatidae, 184–5
Stenopelmatus fuscus, 185
Stephan's riffle beetle, 421
Sterictophora cellularis, 196
sterile insect technique, 323, 380
Sterna antillarum, 398
stick insects, 14, 18
stickleback, 152, 155
    ten-spined, 141
    threespine, 153
sticktight flea, 235, 332
Stilbometopa impressa, 330
stilt sandpiper, 138, 150
stilts, 259
sting of Hymenoptera, 45, 162, 193
stink bugs, 188
stomach bots, 326
stomach contents
    of birds, 126–139, 142–152
    of fish, 152–156
    of mammals, 107, 114–126
    of amphibians and reptiles, 105–113
stomach poisons, 345
Stomatitis, Vesicular, 313
Stomoxys calcitrans, 233, 318–320
stoneflies, 14, 16, 172–173, 199, 432
stork, 126
    wood, 334
Streblidae, 285–286
Strepsiptera, 15
Stricticimex, 304
stridulatory apparatus, 60
Strigiformes, 126, 151
striped field mouse, 297
striped mullet, 407
striped skunk, 118, 123, 126–127, 165,
    251, 272, 296–297, 353
Strix aluco, 151
Strix woodfordii, 151
Strobiloestrus, 328
Struthio camelus, 317
Struthioniformes, 142

sturgeon, 152
    Alabama, 141, 153, 155
Sturnella magna, 132, 143–144
Sturnella neglecta, 98, 352, 393
Sturnidae, 142, 150
Sturnus vulgaris, 136, 150, 250–251,
    322, 352, 396
Stygoparnus comalensis, 419
stylopids, 15
subcutaneous mites, 287
sublethal effects of insecticides, 354–355
subsocial behavior, 61
subspecies, 5
substances, secondary plant, 75
sucker, 152, 155, 407
    bigmouth buffalo, 141, 153
    shorthead redhorse, 141, 154
sucking lice, 14, 20, 298–302
sucking mouthparts, 37, 40–42
Suidae, 114
Sula variegate, 287
sun spiders, 6–8
sunbird
    amethyst, 401
    scarlet-chested, 401
    southern double-collared, 401
sunfish, green, 138, 153, 155
sunning, 301
surprising cave beetle, 420
Sus scrofa, 272, 355
sutures, 34
Swainson's hawk, 359
Swainson's thrush, 130
swallow bugs, 285, 288, 303–304
swallow-tailed kite, 151
swallow, 142, 149
    bank, 134, 332
    barn, 134, 249, 286
    cliff, 134, 304, 332
    tree, 134, 322
swallowtail, 192–193
    Homerus, 423
swan, 126, 259, 301, 310
swift, 126, 406
    common, 249
swift fox, 119, 258
Sylvatic Plague, 255–258
Sylvia atricapilla, 250
Sylvicapra grimmia, 268
Sylvilagus, 253, 313, 355
Sylvilagus audubonii, 251
Sylvilagus bachmani, 248
Sylvilagus brasiliensis, 248
Sylvilagus floridanus, 238, 248,
    253–254
Sylvilagus palustris, 238
symbionts, 241
symbiosis, 241, 400–402

symbiotic relationship, ant-acacia,
    401–402
symbiotic relationships between insects
    and wildlife, 400–402
sympatric speciation
Symphimus mayae, 112
symptom of disease, 241
synapse, 52
Syncerus caffer, 252, 254
Syntermes wheeleri, 211
Synxenoderus, 304
Syrphidae, 204
systemic pesticides, 345
systems, life
    alimentary, 47, 49
    central nervous, 52
    circulatory, 50–51
    digestive, 47–50
    excretory, 69
    female reproductive, 65–66
    glandular, 55–56
    male reproductive, 65
    muscular, 45, 47–48
    nervous, 52–54
    reproductive, 65
    ventilatory (respiratory), 51–52
    vision, 54–55

Tabanidae, 195, 285, 314–316, 403
Tabanus nigrovittatus, 315
Tabanus, 314, 315
Tachycineta bicolor, 134, 156, 322
Tadarida brasiliensis, 115, 116, 120, 125
Taeniosikya, 287
tagma, 35
tagmata, 35
tagmosis, 35
taiga tick, 296–297
Tamandua tetradactyla, 250
Tamandua, 164
tamandua, 164
tamarin, 124
Tamias striatus, 234, 254
tanager, 142
Tantilla gracilis, 107, 112
Tapeworm, Dog, 275
tapeworm
    dog, 279, 281
    double-pored (dog), 302
Tapeworm, Fowl, 275
tapir, Brazilian, 396
Tapiridae, 114
tapir, 114
Tapirus terrestris, 396
tarantula, 396
tarpon, 152
tarsier, 114, 124
    spectral, 120

Tarsius, 124
Tarsius spectrum, 121, 124
tarsus, 36, 41, 46
tasting, 37
Tatum cave beetle, 420
tawny owl, 151
taxa, 3
Taxidea taxus, 119, 123
taxon, 3
Tayassuidae, 114
teal
    blue-winged, 143, 147
    green-winged, 249
    grey, 136, 150
techniques, arthropod sampling, 86–91
tegmina, 181
temperature, high, and pesticides, 350
temporal shifts in wildlife feeding, 100
temporary parasites, 242
ten-spined stickleback, 141
tenaculum, 12
Tenebrio, 65
Tenebrio molitor, 99–100, 416
Tenebrionidae, 190
teneral, 35
Tennessee warbler, 392
tenrec, 114
tent caterpillar, 192
Tenthredinidea
Tephritidae, 192
terminal arborization, 52
termitaria, 209–212, 401
termite mound, 209–212
termite, 14, 18, 63, 179–180, 183,
    401
tern
    black, 136, 150
    least, 397
Terrapene, 398
terrapin egg fly, 325
terrestrial frog, 396
terrestrial insects as wildlife food, 179
Tertiary period, 29
testes, 65
Testudines, 106
Tettigoniidae, 181, 185
Texas horned toad, 398
Thamnophilidae, 142
thanatosis, 162
theory, optimal foraging, 93
therapy, maggot, 416
thermal biology, 69
thermal constant, 70
Thompson's gazelle, 292
thorax, 41, 44
thrasher, 142
    brown, 134
    pearly-eyed, 318

Thraupidae, 142
threatened species, 419–422
three-host tick, 293
three-toed sloth, 251
threespine stickleback, 153
threshold
    action, 381
    developmental, 70
    economic, 381
thrips, 14, 22
throughfall, 207
thrush, 142, 143
    hermit, 132
    Swainson's, 130
    wood, 130, 226
Thryothorus ludovicianus, 136, 143, 145,
    150
Thymallus arcticus, 152
Thysanoptera, 14, 22
tibia, 36, 41, 46
Tibicen, 187
tick, 6–8, 287, 293–298
    American dog, 295, 297
    blacklegged, 297–298
    cave, 295
    deer, 297–298, 434
    fowl, 252, 295
    lone star, 434
    taiga, 296–297
    wood, 295, 297
    hard, 285, 293
    one-host, 293
    soft, 285, 293
    three-host, 293
tick paralysis, 291, 296
tiger beetle
    Coral Pink Sand Dunes, 421
    Highlands, 421, 425
    northeastern beach, 421
    Ohlone, 421, 425
    Puritan, 421, 425
    Salt Creek, 421, 425
tiger
    Bengal, 271
    Siberian, 289
tilapia, 407
time budgets, 97
Tineidae, 205
Tingidae, 188
Tipula paludosa, 194
Tipulidae, 177, 192, 194
tissue, adipose, 47
titmice, 142
    tufted, 134, 150, 249
tit
    blue, 332
    great, 332
toad flesh fly, 325

toad, 105
    American, 359
    cane, 108, 396
    Japanese common, 108
    marine, 396
    Mexican marbled, 108
tolerance, 372
Tommotian epoch, 29
Tooth Cave ground beetle, 419
top-down regulation of populations, 390
topi, 210
topminnow, dispar, 396
tortoise, 106
    gopher, 398
Tortricidae, 215
towhee, eastern, 398
Townsend's mole, 115–116, 122
Townsend's solitaire, 130
toxicant, of pesticides, 343
toxicity, acute, of disease, 241
toxicity, acute, of pesticides, 344
toxicity categories of pesticides, 343–344
toxicity, chronic, of disease, 241
toxicity, chronic, of pesticides, 344
toxicity of common insecticides, 348
toxin, 162
Toxoplasma gondii, 264, 271–274
Toxoplasmosis, 264, 271–274
Toxostoma rufum, 134, 149
trachea, 51
Trachemys, 328
Trachemys scripta, 102
tracheole, 51
Tracheomyia, 328
Tragelaphus scriptus, 210, 268
Tragelaphus strepsiceros, 252, 317
transfer, horizontal, of pesticides, 345
translaminar pesticides, 345
transmission of disease agents to wildlife,
    233
traumatic insemination, 303
trawling long-fingered bat, 120
tree shrew, 114, 134
tree swallow, 322
treehopper, 188
trehalose, 51
Trematoda, 274
trematode, 274
Trialeurodes vaporariorum, 68
triangle, disease, 238
Triassic period, 27, 29
Triatoma, 234, 265–266, 302
Triatoma infestans, 265
Triatoma sanguisuga, 302
Triatominae, 302–303
Trichodectes canis, 279
Trichogramma brassicae, 380
Trichogramma minutum, 377

Tricholipeurus parallelus, 301
trichome, 371
Trichomycterus itacarambiensis, 141
Trichoplusia ni, 394
Trichoptera, 15, 22, 177, 179, 400, 432
Trilobita, 6
trilobites, 6
Trimerotropis infantilis, 421
Trinoton aserinum, 301
Trinoton querquedulae, 287
Trissolcus basalis, 380
tritocerebrum, 53
trochanter, 36, 41, 46
Trogidae, 205
Troglodytes aedon, 136, 150, 198, 226,
    249, 254, 322
Troglodytidae, 142, 150
trogon, 210, 406
Trogon melanurus, 210
Troides alexandrae, 420
Trombicula alfreddugesi, 290
trophic cascade, 198
trophic levels, 198
trophic pyramids, 198
Tropical Eyeworm, 275
tropical fire ant, 398
tropical fowl mite, 291
tropical sand lizard, 110
tropical screech owl, 151
Tropidolophus formosus, 184
trout, 152, 155
    brown, 138, 154
    cutthroat, 138
    lake, 407
    rainbow, 138, 152, 407
trumpeter, palewinged, 396
trumpet, respiratory, 307
Trypanosoma, 303, 330
Trypanosoma brucei, 266–268
Trypanosoma cervi, 264
Trypanosoma congolense, 268
Trypanosoma cruzi, 263–266, 302
Trypanosoma evansi, 318
Trypanosoma gambiense, 266, 268
Trypanosoma rhodesiense, 264, 267–268
Trypanosoma simiae, 268
Trypanosoma suis, 268
Trypanosoma theileri, 315
Trypanosoma uniforme, 268
Trypanosoma vivax, 268, 315
Trypanosomiasis, 264
    African, 266–270, 316–318, 403–404
    American, 263–267, 302–303
    East African, 264, 268
    West African, 268
tsetse fly, 285, 234–235, 316–318,
    403–404
tuatara, 106

tubenose, 406
tubular proboscis, 37, 42
tubule, Malpighian, 69
Tubulidentata, 114
tufted titmouse, 134, 150, 249
Tularemia, 238, 252–254, 291
tumbler, 307
tuna, 152
Turdidae, 142, 143
Turdus migratorius, 130, 149, 249, 251, 291, 352
turkey, 126, 269, 330, 332
    wild, 216, 248, 254, 272
Tursiops truncates, 272
turtle flesh fly, 325
turtle, 106, 315
    box, 398
    false map, 112
    Florida red-bellied, 398
    green, 398
    Krefft's river, 112
    loggerhead, 398
    Murray, 112
    Ouachita map, 112
    red-eared, 103
tymbal organ, 60
tympana, 191
Tympanuchus cupido, 146
tympanum, 36, 43, 60
Typhus
    Epidemic, 301
    Murine, 332
Typophorus nigritus, 191
tyrannid royal flycatcher, 401
Tyrannidae, 142, 143
Tyrannus forficatus, 130
Tyranus tyrannus, 130, 432
Tyranus verticalis, 130, 156
Tyrophagus, 239
Tyto alba, 272

Uma exsul, 110
Uncompahgre fritillary butterfly, 420
univoltine, 71
unnatural host, for disease, 239
Urocyon cinereoargenteus, 119, 121, 123, 238, 257, 272, 276, 292, 296–297
Urophora affinis, 430
Urophora quadrifasciata, 430
Uropygi, 6–8
Ursidae, 114
Ursus, 200, 276
Ursus americanus, 119, 121, 123, 125, 216, 238, 250, 272, 276, 292, 296
Ursus arctos, 272
Ursus arctos horribilis, 125
urticating hairs, 162

Vagrans egestina, 420
Vaillant's frog, 108
valley elderberry longhorn beetle, 419
value, LD$_{50}$, for pesticides, 343
Vampyrops lineatus, 115, 116
vane mite, 292
Varanus, 210, 317
varying hare, 253
vector competence, 240
vector-borne disease, 240
vector, seed, 218
veery, 130, 143
Veliidae, 175
venom, 162
ventilatory (respiratory) system, 51
Vermivora peregrine, 392
vertebrates for biological suppression of insects, 396–397
vertical transmission of disease, 240
Vesicular Stomatitis, 313
vesper mouse, 251
vesper sparrow, 393–394
Vespidae, 195–196
Vespula squamosa, 196, 333
vessel, dorsal, 50
Vipera, 106
Vipera kaznakovi, 106
Vipera lotievi, 106
Vipera orlovi, 106
Vipera renardi, 106
vireo, 142
Vireo atricapillus, 399
vireo, black-capped, 399
Vireonidae, 142
Virginia opossum, 118, 120, 123, 126, 238, 251, 265, 272, 291, 297, 332
virion, 245
virulence of disease agents 236
virulence, beneficial, 237
virulence, coincidental, 236
Virus, West Nile, 249
viruses, 245–252
Visceral Leishmaniasis, 313–314
visceral muscles, 47
vision, 54
visual communication, 60
Vitacea polistiformis, 375
viviparity, 66
vole, 115, 253, 301, 355
    bank, 254, 297
    California, 257
    common, 253
    field, 355
    grey-sided, 297
    red-backed, 253
    rock, 392
    water, 253
voltinism, 71

volume of diets, 93
Vombatus ursinus, 291
Vulpes fulva, 292
Vulpes velox, 119, 119, 121, 123, 238, 251, 257–258, 272, 276, 297, 355
vultures, 126, 406

wagtail, yellow, 396
walleye, 407
warble fly, 285, 325–329
warbler, 142
    bay-breasted, 392
    black-throated green, 392
    blackburnian, 392
    New World, 142
    northern parula, 392
    Tennessee, 392
    yellow-rumped, 134, 150
Warm Springs Zaitzevian riffle beetle, 419
warthog, 210, 268, 317
Wasmannia auropunctata, 398
Wasmannian mimicry, 160
wasp, 15, 24, 192–193, 196, 333–334, 401
    paper, 333
    social, 62
water boatman, 176
water bodies, pesticide contamination of, 353
water buffalo, Asian, 254
water bug, creeping, 400
water bug, giant, 400
water measurers, 175
water quality and aquatic insects, 433
water scavenger beetles, 177–178
water scorpions, 176, 400
water snake, 158
water strider, 175–176
water treader, 175
water vole, 253
water-penny beetle, 177
waterbuck, 114, 210, 291
wax gland, 55
waxwing, 142
    cedar, 143, 145
waxworm, 416
weasels, 114
webs, food, 198–199
webspinner, 14
webworms or snout moths, 192
weevil, 190
    alfalfa, 366
    palm, 416
Wekiu bug, 419
West African Trypanosomiasis, 268
West Nile Virus, 249
western bluebird, 130
western fence lizard, 354

western gorilla, 126
western harvest mouse, 296
western kingbird, 130
western meadowlark, 352, 393
western pine beetle, 392
western screech owl, 151
western spruce budworm, 392
western yellow-bellied flycatcher, 130
western yellow-bellied racer, 110
wheatear, northern, 301
whip scorpion, 8
whippoorwill, 126
whirligig beetle, 177
white crappie, 153
white crowned sparrow, 134
white grub, 371
white ibis, 334
white mullet, 407
white perch, 407
white rhino, 403–404
white-footed mouse, 117, 120, 254, 296, 393
white-lined bat, 115–116
white-tailed deer, 216, 236, 238–239, 249, 252, 277, 291, 301, 311–313, 323, 329, 330
white-tailed hawk, 151
white-throated sparrow, 392
whitefish
    common, 154–156
    lake, 407
whitefly, 371
whitesucker, 141, 154
Whulge checkerspot butterfly, 420
wigeon, Eurasian, 249
wiggler, 307
wild boar, 355
wild dog, 252
wild turkey, 248, 254, 272
wildebeest, 114, 254, 268, 292, 315
wildlife diets, determination, 91–97
wildlife feeding behavior
    folivorous, 98
    frugivorous, 98
    fungivorous, 98
    geographic shifts, 102
    granivorous, 97
    nectarivorous, 98
    ontogenetic shifts, 102
    temporal shifts, 100

wildlife food, 105–165
    insects as, 171–197, 200
wildlife-based recreation, 406–407
wildlife, destruction by ants, 397–400
wildlife, disease transmission to, 233
wildlife, symbiotic relationships with
    insects, 400–402
Wilson's snipe, 138, 143, 147, 150
wind scorpion, 6–8
wing veins, 41
wing, 36, 41
wireworm, 371
Wohlfahrtia, 324, 329
Wohlfahrtia magnifica, 324, 327
Wohlfahrtia opaca, 324
Wohlfahrtia vigil, 287, 324, 327
wolf, 114, 291–292
    Mexican, 119, 123
    red, 296, 301
wolf spider, 396
wombat, 115, 291–292
wood ant, 202
wood boring beetles, 425
wood duck, 136, 150
wood mouse, 254, 297
wood rat, 251, 254, 329
wood storks, 334
wood thrush, 130, 226
wood tick, 295, 297
wood-peewee, eastern, 432
woodchuck, 313
woodcock, American, 138,143, 147, 151
woodland jumping mouse, 117, 120
woodlice, 9, 206
woodpecker, 126, 392, 406
    American three-toed, 128
    California, 128
    downy, 128, 322
    hairy, 128, 143–144
    Lewis's, 128
    pileated, 128, 143
    red-bellied, 128
    red-cockaded, 128, 143
    red-headed, 128, 143–144
woodrat, 313
woodwasp, 216
woollybear, 192
worm, giant thorny-headed, 279–283
    gizzard, 275

wren, 142
    Carolina, 136, 143, 145, 145, 150
    house, 136, 150, 249, 254, 322
    rufus napped, 401
wrigglers, 307

Xanthocephalus xanthocephalus, 132
Xenopsylla cheopis, 257–258, 332
Xenosaurus grandis, 110
Xenosaurus newmanorum, 110
Xylosandrus, 202

yellow crazy ant, 226, 398
yellow fever, 250
yellow flies, 314, 403, 413
yellow jacket, 333
yellow perch, 138, 153, 155, 407
yellow pine chipmunk, 117, 120
yellow wagtail, 396
yellow-bellied bulbul, 396
yellow-bellied marmot, 251, 313
yellow-bellied sapsucker, 128
yellow-billed cuckoo, 132, 149
yellow-billed oxpecker, 395
yellow-headed blackbird, 132
yellow-leg frog, 359
yellow-legged gull, 249
yellow-necked mouse, 254
yellow-rumped warbler, 134, 150
yellowthroat, 254
    common, 432
Yersinia pestis, 253, 255–258
Yucátan cricket-eating snake, 112

Zaitzevia thermae, 419
Zalophus californianus, 272
Zamia pumila, 424
Zayante band-winged grasshopper, 421
zebra, 114, 210, 268, 328
    mountain, 396
zebrafish, 154–155
Zenaida macroura, 352
Zenaidura macroura, 251
Zonotrichia albicollis, 392
Zonotrichia leucophrys, 134
zoogeographic realms, 32
Zoraptera, 14
Zygentoma, 14
Zygoptera, 173